Modular Systems for Energy Usage Management

Sustainable Energy Strategies
Series Editor
Yatish T. Shah

Modular Systems for Energy Usage Management
Modular Systems for Energy and Fuel Recovery and Conversion
Thermal Energy: Sources, Recovery, and Applications
Chemical Energy from Natural and Synthetic Gas

Other related books by Yatish T. Shah
Energy and Fuel Systems Integration
Water for Energy and Fuel Production
Biofuels and Bioenergy: Processes and Technologies

Modular Systems for Energy Usage Management

Yatish T. Shah

CRC Press
Taylor & Francis Group
Boca Raton London New York

CRC Press is an imprint of the
Taylor & Francis Group, an **informa** business

CRC Press
Taylor & Francis Group
6000 Broken Sound Parkway NW, Suite 300
Boca Raton, FL 33487-2742

© 2020 by Taylor & Francis Group, LLC
CRC Press is an imprint of Taylor & Francis Group, an Informa business

No claim to original U.S. Government works

Printed on acid-free paper

International Standard Book Number-13: 978-0-367-40796-4 (Hardback)

This book contains information obtained from authentic and highly regarded sources. Reasonable efforts have been made to publish reliable data and information, but the author and publisher cannot assume responsibility for the validity of all materials or the consequences of their use. The authors and publishers have attempted to trace the copyright holders of all material reproduced in this publication and apologize to copyright holders if permission to publish in this form has not been obtained. If any copyright material has not been acknowledged please write and let us know so we may rectify in any future reprint.

Except as permitted under U.S. Copyright Law, no part of this book may be reprinted, reproduced, transmitted, or utilized in any form by any electronic, mechanical, or other means, now known or hereafter invented, including photocopying, microfilming, and recording, or in any information storage or retrieval system, without written permission from the publishers.

For permission to photocopy or use material electronically from this work, please access www.copyright.com (http://www.copyright.com/) or contact the Copyright Clearance Center, Inc. (CCC), 222 Rosewood Drive, Danvers, MA 01923, 978-750-8400. CCC is a not-for-profit organization that provides licenses and registration for a variety of users. For organizations that have been granted a photocopy license by the CCC, a separate system of payment has been arranged.

Trademark Notice: Product or corporate names may be trademarks or registered trademarks, and are used only for identification and explanation without intent to infringe.

Library of Congress Cataloging-in-Publication Data

Names: Shah, Yatish T., author.
Title: Modular systems for energy usage management / Yatish T. Shah.
Description: Boca Raton : CRC Press, [2020] | Series: Sustainable energy strategies | Includes bibliographical references and index. |
Summary: "This book examines the role of the modular approach for the back end of the energy industry- energy usage management. It outlines the use of modular approaches for the processes used to improve energy conservation and efficiency, which are the preludes to the prudent use of energy. Since energy consumption is conventionally broken down into four sectors, residential, transportation, industrial, and commercial, the discussions on energy usage management are also broken down in these four sectors in the book. The book examines the use of modular systems for five areas that cover all sectors described above: buildings, vehicles, computer and electrical/electronic products, district heating, and waste water treatment and desalination. The book also discusses the use of modular approach for energy storage and transportation. Finally, the book describes how modular approach facilitates bottom up, top down and hybrid simulation and modeling, of the energy systems from various scientific and socio/economic perspectives. Aimed at industry professionals and researchers involved in the energy industry, the book illustrates in detail how modular approach can facilitate the management of energy usage with the help of concrete industrial examples"—Provided by publisher.
Identifiers: LCCN 2019042093 | ISBN 9780367407964 (hardback ; acid-free paper) | ISBN 9780367822392 (ebook)
Subjects: LCSH: Energy conservation. | Energy consumption. | Modularity (Engineering)
Classification: LCC TJ163.3 .S4535 2020 | DDC 621.31/7—dc23
LC record available at https://lccn.loc.gov/2019042093

Visit the Taylor & Francis Web site at
http://www.taylorandfrancis.com

and the CRC Press Web site at
http://www.crcpress.com

This book is dedicated to my parents.

Contents

Series Preface xxi
Preface xxv
Author xxvii

Chapter 1 Modular Systems for Energy Conservation and Efficiency: Residential and Transportation Sectors 1

    1.1 Introduction 1
    1.2 Energy Conservation and Efficiency 3
    1.3 Energy Conservation and Efficiency in Residential Sector 5
    1.4 Modular Approaches for Energy Conservation and Efficiency Improvement in Buildings 7
        1.4.1 Modular Passive Solar Heating 7
        1.4.2 Energy-Saving Modules (ESM) 9
        1.4.3 Phase Change Material Based Modular Thermal Storage 10
            1.4.3.1 Phase Change Material Applications in Buildings 11
            1.4.3.2 A Modular Solar Envelope House 13
        1.4.4 Modular Hybrid Wall with Surface Heating and Cooling 15
        1.4.5 Energy Savings for a Wood-Based Modular Prefabricated Façade Refurbishment System 15
        1.4.6 Energy Star Modular Homes 17
        1.4.7 Modular Zero-Energy Building 18
            1.4.7.1 Design and Construction 22
            1.4.7.2 Energy Harvest 23
            1.4.7.3 Advantages and Disadvantages of ZEB 24
            1.4.7.4 Role of Geothermal Energy in Net-Zero-Energy Buildings 25
            1.4.7.5 Geothermal and Solar Net-Zero-Energy Building 28
    1.5 Energy Conservation and Efficiency in Transportation Sector 28
    1.6 Energy Efficiency of Modular Electric and Hybrid Vehicles 29
        1.6.1 Electrical Losses Reduction in EV 31
            1.6.1.1 Energy Efficiency of the Converters 31
            1.6.1.2 Energy Efficiency of the Electric Motor 32
            1.6.1.3 Super Capacitors as Energy Storage for Increasing Energy Efficiency 33
            1.6.1.4 Super Capacitors in Regulated Electrical Drives 34

		1.6.1.5	Fuel Cells 34

- 1.6.1.5 Fuel Cells 34
- 1.6.1.6 New Systems 35
- 1.6.1.7 Reduction of Losses in the Conductors and Connectors 35
- 1.6.1.8 Lighting and Heating of EV 36
- 1.6.2 Mechanical Loss Reduction in EV 37
  - 1.6.2.1 Tires Role in EV 37
  - 1.6.2.2 Vehicle Body 38
- 1.6.3 Additional Sources of Energy in EV and Hybrid Vehicles 38
  - 1.6.3.1 Solar Cells 38
  - 1.6.3.2 Improved Kinetic Energy Recovery Systems 39
  - 1.6.3.3 Additional Waste Heat Energy Recovery 40
  - 1.6.3.4 Reduction of Energy Loss from Adverse Airflow 41
  - 1.6.3.5 Hybrid Electric Vehicle 41
- 1.7 Efficiency Improvements of Modular Fuel Cell 42
  - 1.7.1 Design and Development of Modular Fuel Cell Stacks for Various Applications 42
  - 1.7.2 Novel Reversible Fuel Cell System for Very High Efficiency 43
  - 1.7.3 Efficiency Improvements by Pulsed Hydrogen Supply in PEM Fuel Cell Systems 44
- 1.8 Role of Nanotechnology for Energy Efficiency in Modular Vehicle Operations 45
  - 1.8.1 Lubricating Oils for Cars Using Nanoparticle Additives 45
  - 1.8.2 Nanotechnology for Improved Energy Storage in Cars 45
    - 1.8.2.1 Batteries 46
    - 1.8.2.2 Super Capacitors 47
    - 1.8.2.3 Fuel Cells 48
  - 1.8.3 Nanotechnology for Solar Cell in Automobile 50
  - 1.8.4 Nanotechnology-Based Catalyst for Reduction of Exhaust Emission 51
- 1.9 Integrated Modular Avionics 51
- 1.10 Advanced Modular Shipboard Energy Storage System for Energy Efficiency Improvement 54
  - 1.10.1 System and Technology Overview 55
- References 56

Chapter 2 Modular Systems for Energy Conservation and Efficiency: Industrial and Commercial Sectors 65

- 2.1 Energy Conservation and Efficiency in Industrial Sector 65

2.2	Modular Efficiency Improvement Measures for Water and Wastewater Facilities		66
	2.2.1	Managing Energy Costs in Wastewater Treatment Plants	69
	2.2.2	Benefits of Improving Energy Efficiency through Modular CHP in Water and Wastewater Facilities	70
	2.2.3	Modular Designs Recover Energy during Desalination	72
2.3	Modular Cogeneration in Biomass/Waste Industry		74
	2.3.1	BioMax Modular CHP Technology	74
	2.3.2	High-Efficiency Modular Biomass Gasification Systems	77
	2.3.3	AgriPower's Modular, Mobile, and Transportable CHP System	77
	2.3.4	Modular Compact CHP Using Local Heterogeneous Biomass Wastes	80
	2.3.5	Modular Power and Cogeneration Systems Using Landfill Gas	81
2.4	Cogeneration with Small Modular Nuclear Reactors for Industrial Applications		83
	2.4.1	Load Following Cogeneration Strategy for Typical Small Modular Reactors for Industrial Applications	83
	2.4.2	Examples of Cogeneration by Small Modular Reactors for Various Industrial Applications	86
		2.4.2.1 Nuclear Process Heat Systems	86
		2.4.2.2 Oil Recovery Applications	87
		2.4.2.3 Oil Refinery Applications	89
		2.4.2.4 Desalination	90
		2.4.2.5 Hydrogen Production	92
2.5	CHP Projects in Manufacturing Industries		95
2.6	Energy Conservation and Efficiency in the Commercial Sector		97
2.7	Modular Systems for Improving Energy Efficiency in Commercial Facilities		98
	2.7.1	Five Office Buildings Using Modular Passive Heating and Cooling Design	98
	2.7.2	Modular CHP for Energy Efficiency-Benefiting from a Modular Approach to Energy Efficient Heating	102
	2.7.3	Energy Efficiency for Data Centers	103
	2.7.4	Location-Independent Cogeneration Power Plants	107
	2.7.5	Ultra-Clean CHP Modules for Microgrid Demonstration Project	108
2.8	Modular CHP for Efficient Distributed Power and Heat in Commercial Sector		108

		2.8.1	Optimal Modular Design for Cogeneration for Hospital Facilities 113
		2.8.2	Large Scale Modular Geothermal District Heating.... 114
	2.9	Efficient Energy Use by Modular Micro-CHP.................... 115	
		2.9.1	Global Market Status for Micro-CHP..................... 117
	2.10	Modular Energy Efficiency Improved in Energy Industry by Nanotechnology... 118	
	References 123		
Chapter 3	Modular Systems for Energy Usage in Buildings 129		
	3.1	Introduction 129	
		3.1.1	Choice of Energy Source... 132
	3.2	Modular Solar Thermal HVAC for Buildings 133	
		3.2.1	The Design and Development of an Adaptable Modular Sustainable Commercial Building (Co_2nserve) for Multiple Applications 140
		3.2.2	Six Innovative Modular Rooftop Solar Technologies... 140
		3.2.3	Modular Passive Solar Heating System for Residential Homes... 142
		3.2.4	Modular Zero Carbon Emission School Building with Hybrid Solar Heating and Passivhaus Construction 144
		3.2.5	Modular Bright Built Home Brand 146
		3.2.6	Modular Building Integrated Solar Thermal Technologies and Their Applications....................... 147
			3.2.6.1 Working Principle of Typical BIST System.. 148
	3.3	Modular Geothermal HVAC Systems 151	
		3.3.1	Modular HydroLogic's Hydron Module for Geothermal System 153
		3.3.2	Modular Chiller System Using Geothermal Resources 153
		3.3.3	Modern versus Traditional Modular Chillers........... 154
		3.3.4	Modular Geothermal Measurement System 156
		3.3.5	Modular HISEER Geothermal Ground Source Heat Pumps 158
		3.3.6	Modular Geothermal Heat-Pump System Installation.. 159
		3.3.7	Modular Tube Bundle Heat Exchanger and Geothermal Heat Pump System 162
	3.4	Modular Design of Building Security System 166	
	References 167		

Contents   xi

Chapter 4   Modular Systems for Energy Usage in Vehicles 171
- 4.1   Introduction 171
- 4.2   Advanced Modular Power System Management for Affordable, Supportable Multi-Vehicles Space Mission 171
  - 4.2.1   Modular Power Management and Distribution in Space Mission 173
- 4.3   Modular Power Generation and Storage Technologies for Space Vehicles 174
  - 4.3.1   Solar Array Technology 174
  - 4.3.2   Modular Fuel Cell System 175
  - 4.3.3   Modular Battery Technology 178
- 4.4   Modular Proton Exchange Membrane (PEM) Fuel Cell for Stationary or Mobile Power System 179
- 4.5   Energy Technology Changes in Modular Automobiles 181
  - 4.5.1   Wheel Hub Motors 181
  - 4.5.2   Batteries 181
  - 4.5.3   Solid Oxide Fuel Cell 182
  - 4.5.4   Autonomy 182
- 4.6   Modular Energy Management in Electric Vehicles 183
  - 4.6.1   Novel Modular Power Management for Ground Electric Vehicles 183
    - 4.6.1.1   Plug-In Hybrid Electric Vehicles 185
    - 4.6.1.2   Very Large Batteries 186
    - 4.6.1.3   Fuel Cells 186
    - 4.6.1.4   Synthetic Fuel 186
  - 4.6.2   Modular Approach for Energy Consumption Estimation in Electric Vehicles (EVs) 186
  - 4.6.3   Modular Fuel Cells for Multiple Vehicle Applications 187
  - 4.6.4   Vehicle Systems Controller with Modular Architecture 188
- 4.7   Modular Integrated Energy Systems 189
  - 4.7.1   Modular Verd2GO—EV Connection 189
  - 4.7.2   Additive Manufacturing Integrated Energy 191
- 4.8   Modular Fuel Cell Stack System 194
- 4.9   Modular Solar Vehicles 197
  - 4.9.1   Solar Electric Car 197
    - 4.9.1.1   Plug-In Hybrid and Solar Car 199
    - 4.9.1.2   Electric Car with Solar Assist/Solar Taxi 199
    - 4.9.1.3   Solar Buses 200
  - 4.9.2   Personal Rapid Transit Vehicles 202
    - 4.9.2.1   Rail 202

		4.9.2.2	Modular Solar Photovoltaic Canopy System for the Development of Rail Vehicle Traction Power..............................202

        4.9.3   Solar Electric Aircraft..204
                4.9.3.1   Manned Solar Aircraft...............................204
                4.9.3.2   Solar Impulse 2 (HB-SIB).......................205
                4.9.3.3   Unmanned Aerial Solar Vehicles 205
        4.9.4   Solar-Powered Spacecraft 206
        4.9.5   Solar-Powered Boats 207
        4.9.6   Modular and Mobile Photovoltaics on Two Wheelers..208
   4.10 Modular Energy Storage and Usage in Ships........................209
        4.10.1 GE Modular Naval Vessel Electrification 210
References 211

Chapter 5   Modular Systems for Energy Usage in Computer and Electrical/Electronic Applications 217

   5.1   Introduction 217
   5.2   Modular Systems for Energy Usage in Data Centers............ 217
        5.2.1   Maintenance Costs 220
        5.2.2   Energy Costs 220
        5.2.3   Modular Data Center from Delta 227
        5.2.4   Efficiency Improvement by Modular Data Center at Sun's Computer Operation of Oracle 228
   5.3   Modular UPS Systems..230
        5.3.1   Intelligent Cooling Technology................................ 231
        5.3.2   Higher Power Density in the Rack.......................... 231
        5.3.3   ABB's Modular UPS Systems................................. 231
   5.4   Modular Fuel Cell for Data Centers 236
   5.5   Modular Supercomputer..237
        5.5.1   NASAs Module-Based Supercomputer....................237
        5.5.2   Optimizing Energy Efficiency for HPC...................237
        5.5.3   Deploying Energy-Optimized Solutions 238
   5.6   Modular Power Plants 239
        5.6.1   Power Plant Design Taking Full Advantage of Modularization...240
                5.6.1.1   Modular Plant Design...............................240
                5.6.1.2   Cost 243
        5.6.2   Modular Chiller Plant for a 540-MW Gas-Fired Power Plant in Mexico City......................................244
   5.7   Compact Modular™ Technology Cooling Properties for Lighting System..244
   5.8   Jenoptik Modular Energy Systems for Military Platforms 245
   5.9   Modular Micro- and Mini-Scale Energy Systems for Electronic Applications 246

Contents   xiii

	5.9.1	Power Consumption and Mode of Operation of MNSs..247
	5.9.2	Micro/Nanotechnology-Enabled Technologies for Energy Harvesting...248
	5.9.3	PV Technologies for Solar Energy Harvesting........249
	5.9.4	Self-Powered MNSs...249
	5.9.5	Comparison of Cylindrical and Modular Microcombustor Radiators for Micro-TPV System Application 251
	5.9.6	Micropower Management of Solar Energy Harvesting Using a Novel Modular Platform...........251
	5.9.7	Nanotechnology for Portable Energy Systems: Modular PV, Energy Storage, and Electronics........252
5.10	Modular, Integrated Power Conversion and Energy Management System..252
	5.10.1	High-Power Application Modular Multilevel Converter 254
References 254

Chapter 6 Modular Systems for Energy Usage in District Heating...................259

6.1	Introduction 259
	6.1.1	DH Advantages and Drawbacks..............................261
6.2	Small Modular Renewable Heating and Cooling Grids........264
6.3	Modular DH by Biomass...266
	6.3.1	DH with Modular Biomass CHP.............................269
	6.3.2	Examples of Modular Biomass Based DH Schemes 271
	6.3.3	Modular Biomass Based DH in Europe 273
6.4	Modular Geothermal DH 276
6.5	Decarbonizing DH with Modular Solar Thermal Energy.....279
6.6	Modular DH with Wind Energy..284
6.7	DH with Small Modular Nuclear Reactors 285
	6.7.1	Global Assessment of Modular Nuclear Heat Based DH 288
6.8	Modular DH by Industrial Waste Heat................................293
6.9	DH by Modular Hybrid Sources..299
	6.9.1	Hybrid Modular Geothermal Heat Pump for DH 299
	6.9.2	The Multisource Hybrid Concept............................299
		6.9.2.1 The Ground Heat Exchanger (GHX)........300
		6.9.2.2 Geothermal Heat Pumps 301
		6.9.2.3 Peaking Boiler 301
		6.9.2.4 Solar Thermal Recharge and Sewer Heat Recovery..301
	6.9.3	Hybrid Modeling and Cosimulation of DH Networks 302

		6.9.4	Hybrid Renewable Energy Systems for Buildings......302
		6.9.5	Small-Scale Hybrid Plant Integrated with Municipal Energy Supply System...........................304
		6.9.6	Modular Hybrid Solar and Biomass System for DH 309
	6.10	Role of Thermal Energy Storage in DH 310	
		6.10.1	Advantages and Disadvantages of Using TES in DE 311
		6.10.2	Energy Central 313
		6.10.3	Use of TES in LTDH.. 315
		6.10.4	TES-Solar Energy-DH Partnership......................... 316
	References 317		

Chapter 7 Modular Systems for Energy Usage for Desalination and Wastewater Treatment 325

	7.1	Introduction 325		
		7.1.1	Process Alternatives 326	
			7.1.1.1	Hybrid Thermal-Membrane Desalination.. 327
	7.2	Modular Desalination Plants 331		
		7.2.1	IDE Chemical-Free River Osmosis Desalination Plant... 331	
			7.2.1.1	IDE PROGREEN Plant 332
		7.2.2	MIT Study on Modular Desalination Plants............ 333	
			7.2.2.1	Melbourne Case Study.............................. 333
		7.2.3	NexGen Desal Modular Desalination Plant............. 334	
		7.2.4	KSB Modular Desalination Plant............................ 335	
	7.3	Use of Modular Renewable Energy Systems for Desalination.. 336		
		7.3.1	Solar-Assisted Modular Desalination Systems 337	
			7.3.1.1	Solar Thermal-Assisted Modular Systems 337
			7.3.1.2	Modular Solar Thermal Applications....... 342
		7.3.2	Modular PV-Based Desalination.............................. 343	
		7.3.3	Desalination Systems Driven by Wind 345	
		7.3.4	Desalination by Biomass and Geothermal Energy.....347	
		7.3.5	Advantages and Disadvantages of Technologies...... 348	
		7.3.6	Economic Analysis for Renewable Energy Desalination Processes...348	
		7.3.7	Modular Solar Desalination—Grid-Friendly Water Technology... 351	
		7.3.8	Modular Wave Energy System for Desalination 352	
	7.4	Desalination with Small Modular Nuclear Reactors............. 355		
		7.4.1	Global Landscape of Modular Nuclear Reactors with Desalination 357	

Contents xv

	7.5	Modular Systems for Wastewater Treatment 363
		7.5.1 Benefits of Reducing Water Treatment Plant Energy Consumption 365
		7.5.2 Modular Systems for Energy Usage-Water Treatment-AdEdge Technology 366
		7.5.3 Newterra Advantages for Wastewater Treatment 367
		7.5.4 Envirogen Modular Wastewater Treatment 367
		7.5.4.1 Membrane Biological Reactors 368
		7.5.5 Modular Tubular Microbial Fuel Cells for Energy Recovery during Sucrose Wastewater Treatment at Low Organic Loading Rate 369
		7.5.6 Modular Package, Field-Erected Wastewater Treatment Plants 369
		7.5.7 Evoqua Modular Water Technologies 369
		7.5.7.1 Evoqua Innovative Modular Desalination Technology 370
		7.5.8 Package Systems from ClearBlu 371
		7.5.9 PCS Package Wastewater Treatment Plants 372
		7.5.9.1 Advantages 373
	References 374

Chapter 8 Modular Systems for Energy and Fuel Storage 385

	8.1	Methods for Energy and Fuel Storage 385
		8.1.1 Energy Storage Technologies 385
		8.1.1.1 Electrochemical Storage (Batteries) 386
		8.1.1.2 Electrochemical Capacitors and Super Capacitors 388
		8.1.1.3 Mechanical Storage 390
		8.1.1.4 Thermal Storage 390
		8.1.1.5 Bulk Gravitational Storage 391
		8.1.1.6 Fuel Storage 392
	8.2	Modular Battery Storage 392
		8.2.1 Methods for Integrating and Controlling Modular Battery-Based Energy Storage Systems 392
		8.2.2 Modular Battery System for Community Energy Storage 398
		8.2.3 Containerized Energy Storage Power Stations 401
		8.2.4 Modular Flow Cell Batteries 405
		8.2.5 Other Notable Modular Battery Storage Systems 407
	8.3	Modular Electrical and Electromagnetic ESSs 410
		8.3.1 Modular Ultracapacitors or Supercapacitors 410
		8.3.2 Modular Double-Layer Capacitors 412
		8.3.3 Modular Integrated Energy Storage Device and Power Converter Based on Supercapacitors 412
		8.3.4 Supercapacitor Module SAM for Hybrid Buses 414

- 8.4 Modular Mechanical Storage Systems 416
  - 8.4.1 Modular Compressed Air ESSs 416
  - 8.4.2 Low-Pressure, Modular CAES System for Wind Energy Storage Applications 418
  - 8.4.3 Design of a Modular Solid-Based Thermal Energy Storage for a Hybrid Compressed Air ESS 419
  - 8.4.4 Modular Flywheel Energy Storage 419
- 8.5 Modular Thermal Storage 424
  - 8.5.1 TES Module Using Phase Change Composite Material 424
    - 8.5.1.1 PCM Modular System 426
  - 8.5.2 Modular IceBrick™ and Chillers Thermal Energy Storage Cell 427
  - 8.5.3 Modular TES Tanks Using Modular Heat Batteries 428
  - 8.5.4 Modular Molten Salt TES Plants 429
  - 8.5.5 Modular Thermochemical Heat Storage 432
  - 8.5.6 Modular "Thermal Capacity on Demand" in Rapid Deployment Building Solutions 433
  - 8.5.7 Scalable Modular Geothermal Heat Storage System 435
- 8.6 Economic Viability of Modular Hydropower Storage System 436
  - 8.6.1 Use of Wind-Powered Modular Pumped Hydropower in Canary Islands 442
- 8.7 Modular Hydrogen Storage 442
  - 8.7.1 Modular Metal Hydride Hydrogen Storage System 442
    - 8.7.1.1 Modular Heat Exchanger for Metal Hydride Hydrogen Storage 443
  - 8.7.2 Hydrogen Storage Module Based on Hydrides 444
  - 8.7.3 Calvera Modular Hydrogen Storage System 445
  - 8.7.4 Demonstration of a Microfabricated Hydrogen Storage Module for Micropower Systems 446
  - 8.7.5 Conception of Modular Hydrogen Storage Systems for Portable Applications 448
- 8.8 Modular Liquid Fuel Storage System 448
- 8.9 Modular Compressed Natural Gas Storage 449
  - 8.9.1 Modular Compressed Natural Gas (CNG) Station and Method for Avoiding Fire in Such Station 449
  - 8.9.2 Modular CNG Tank 450
- 8.10 The Future of Modular LNG Tanks 451
  - 8.10.1 Membrane Modular LNG Tank 455
  - 8.10.2 Vacuum-Insulated LNG Modular Storage Systems 457
- References 459

Contents　　　　　　　　　　　　　　　　　　　　　　　　　　　　　　　　xvii

Chapter 9 Modular Systems for Energy and Fuel Transport 471

9.1 Introduction 471
9.2 Microenergy Systems and Modular Microgrid 471
　　9.2.1　Modular Microgrids 474
　　9.2.2　Expandable, Modular Microgrid Unit 477
　　9.2.3　Modular Microgrids for Renewable Resources 478
　　9.2.4　Modular Microenergy Grids for Oil and
　　　　　 Gas Facilities 484
9.3 Concept of Modular Microenergy Internet 485
9.4 Modular Nanogrid 488
9.5 Modular Open Energy System 491
　　9.5.1　OES as Multi-Level DC Grid System 493
9.6 Modular Transport of Biomass 495
　　9.6.1　Modular System for Low-Cost Transport and
　　　　　 Storage of Herbaceous Biomass 495
　　9.6.2　Novel Modular Timber Transport Projects in
　　　　　 Sweden 496
　　　　　 9.6.2.1　ETT Project 496
　　　　　 9.6.2.2　Results from Studies of the ETT
　　　　　　　　　 Project 497
　　　　　 9.6.2.3　ST Project 497
　　　　　 9.6.2.4　Results from Studies of the ST Rigs 498
9.7 Modular Fuel System (MFS) 498
　　9.7.1　DRS MFS 499
　　9.7.2　ISO Tank Rack Design 500
　　9.7.3　Modular Canisters for Spent Nuclear Fuel 500
9.8 Modular Systems for Natural Gas Transport 501
　　9.8.1　Gas Compressor Modules 501
9.9 Modular Transport of Compressed Liquid and Solid
　　 Natural Gas 502
　　9.9.1　The Transport of CNG 502
　　9.9.2　Carriers for the Transport of CNG 503
　　　　　 9.9.2.1　Coselle Technology 503
　　　　　 9.9.2.2　VOTRANS Technology 504
　　　　　 9.9.2.3　CRPV Technology 505
　　　　　 9.9.2.4　PNG Technology 505
　　9.9.3　The Modular Transport of LNG 506
　　　　　 9.9.3.1　Tankers for the Transport of LNG 506
　　　　　 9.9.3.2　Tankers with Integrated Tanks 506
　　　　　 9.9.3.3　Tankers with Self-Supporting Tanks 507
　　　　　 9.9.3.4　The LNG Transport Fleet 507
　　　　　 9.9.3.5　LNG ISO Containers 508
　　　　　 9.9.3.6　LNG Transport Trailers 508
　　9.9.4　The Transport of NGH with GTS Technology 508

		9.9.4.1	Properties of Methane Hydrates and Options for Hydrate Storage and Transport 508

9.10 Transport Modules for Hydrogen (H_2) 510
    9.10.1 Tube Trailers, Cryogenic Liquid Trucks, Rail, Barges, and Ships 512
        9.10.1.1 Tube Trailers 512
        9.10.1.2 Liquid Hydrogen Trailers 513
    9.10.2 Modular Station with Hydrogen Produced On-Site 514
References 515

Chapter 10 Modular Approach for Simulation, Modeling, and Design of Energy Systems 523

10.1 Introduction 523
10.2 Scientific Modular Simulation and Modeling (Bottom-Up Approach) 525
    10.2.1 Module Types 526
    10.2.2 Standard Connections 527
    10.2.3 Connection Types 527
10.3 Standardized Modular Design—Aspen Platform and Others 528
    10.3.1 Customer Success: Sadara and DSM 532
10.4 Modular Global (Techno-Econometric/Social) Energy Models (Top-Down Approach) 533
    10.4.1 Top-Down versus Bottom-Up Approaches 534
10.5 Hybrid Global Energy System Models 535
10.6 Hybrid Models with Soft Links 536
    10.6.1 MARKAL–MSG 536
    10.6.2 MARKAL–EPPA 537
    10.6.3 TIMES–EMEC 537
    10.6.4 MESSAGE–MACRO 537
10.7 Hybrid Models with Hard Links 538
    10.7.1 The Decomposition Method in the Complementarity Framework 538
    10.7.2 MARKAL–MACRO 538
    10.7.3 MESSAGE–MACRO 539
10.8 Challenges and Further Perspectives on Hybrid Models 539
    10.8.1 Integrated Hybrid Models 542
10.9 Large-Scale Modeling Systems Combining Multiple Modules 542
10.10 Brief Review of Literature for Modular Approach for Simulation and Optimization 544
    10.10.1 Modular Simulations of Energy System Equipment and Process 544

Contents

      10.10.2 Modular Simulation of Buildings Construction and Energy Consumption 546
      10.10.3 Examples of Modular Approach for Renewable Power Supply 548
      10.10.4 Modular Simulation of Energy Consumption and Efficiency 550
      10.10.5 Novel Modular Approaches for Simulation and Optimization 552
  References 554

Index 563

Series Preface

SUSTAINABLE ENERGY STRATEGIES

While fossil fuels (coal, oil, and gas) were the dominant sources of energy during the last century, since the beginning of the twenty-first century, an exclusive dependence on fossil fuels has been believed to be a non-sustainable strategy due to (a) their environmental impacts, (b) their non-renewable nature, and (c) the dependence of their availability on the local politics of the major providers. The world has also recognized that there are in fact ten sources of energy: coal, oil, gas, biomass, waste, nuclear, solar, geothermal, wind, and water. These can generate chemical/biological, mechanical, electrical and thermal energies required to satisfy our needs. A new paradigm has been to explore greater roles of renewable and nuclear energy in the energy mix to make energy supply more sustainable and environmentally friendly. The adopted strategy has been to replace fossil energy by renewable and nuclear energy as rapidly as possible. While fossil energy still remains dominant in the energy mix, by itself, it cannot be a sustainable source of energy for the long future.

Along with exploring all ten sources of energy, sustainable energy strategies must consider five parameters: (a) availability of raw materials and accessibility of product market, (b) safety and environmental protection associated with the energy system, (c) technical viability of the energy system on the commercial scale, (d) affordable economics, and (e) market potential of a given energy option in the changing global environment. There are numerous examples substantiating the importance of each of these parameters for energy sustainability. For example, biomass or waste may not be easily available for a large-scale power system making a very large-scale biomass/waste power system (like a coal or natural gas power plant) unsustainable. Similarly, transferring power through an electrical grid to a remote area or onshore needs from a remote offshore operation may not be possible. Concerns of safety and environmental protection (due to emissions of carbon dioxide) limit the use of nuclear and coal-driven power plants. Many energy systems can be successful at laboratory or pilot scales, but they may not be workable at commercial scales. Hydrogen production using a thermochemical cycle is one example. Many energy systems are as yet economically prohibitive. The devices to generate electricity from heat such as thermoelectric and thermophotovoltaic systems are still very expensive for commercial use. Large-scale solar and wind energy systems require huge upfront capital investments which may not be possible in some parts of the world. Finally, energy systems cannot be viable without market potential for the product. Gasoline production systems were not viable until the internal combustion engine for the automobile was invented. Power generation from wind or solar energy requires guaranteed markets for electricity. Thus, these five parameters collectively form a framework for sustainable energy strategies.

It should also be noted that the sustainability of a given energy system can change with time. For example, coal-fueled power plants became unsustainable due to their impact on the environment. These power plants are now being replaced by gas-driven power plants. New technology and new market forces can also change sustainability of the energy system. For example, successful commercial developments of fuel cells and electric cars can make the use of internal combustion engines redundant in the vehicle industry. While an energy system can become unsustainable due to changes in parameters, outlined above, over time, it can regain sustainability by adopting strategies to address the changes in these five parameters. New energy systems must consider long-term sustainability with changing world dynamics and possibilities of new energy options.

Sustainable energy strategies must also consider the location of the energy system. On the one hand, fossil and nuclear energies are high-density energies and best suited for centralized operations in urban areas; on the other hand, renewable energies are of low density and well suited for distributed operations in rural and remote areas. Solar energy may be less affordable in locations far away from the equator. Offshore wind energy may not be sustainable if the distance from shore is too great for energy transport. Sustainable strategies for one country may be quite different from another depending on their resource (raw material) availability and local market potential. The current transformation from fossil energy to green energy is often curtailed by required infrastructure and the total cost of transformation. Local politics and social acceptance also play an important role. Nuclear energy is more acceptable in France than in any other country.

Sustainable energy strategies can also be size dependent. Biomass and waste can serve local communities well at a smaller scale. As mentioned before, the large-scale plants can be unsustainable because of limitations on the availability of raw materials. New energy devices that operate well at micro- and nanoscales may not be possible at a large scale. In recent years, nanotechnology has significantly affected the energy industry. New developments in nanotechnology should also be a part of sustainable energy strategies. While larger nuclear plants are considered to be the most cost effective for power generation in an urban environment, smaller modular nuclear reactors can be the more sustainable choice for distributed cogeneration processes. Recent advances in thermoelectric generators due to advances in nanomaterials are an example of a size-dependent sustainable energy strategy. A modular approach for energy systems is more sustainable at a smaller scale than at a very large scale. Generally, a modular approach is not considered as a sustainable strategy for a very large, centralized energy system.

Finally, choosing a sustainable energy system is a game of options. New options are created by either improving the existing system or creating an innovative option through new ideas and their commercial development. For example, a coal-driven power plant can be made more sustainable by using very cost-effective carbon capture technologies. Since sustainability is time, location, and size dependent, sustainable strategies should follow local needs and markets. In short, sustainable energy strategies must consider all ten sources and a framework of five stated parameters under which they can be made workable for local conditions. A revolution in technology (like nuclear fusion) can, however, have global and local impacts on sustainable energy strategies.

The CRC Press Series on Sustainable Energy Strategies will focus on novel ideas that will promote different energy sources sustainable for long term within the framework of the five parameters outlined above. Strategies can include both improvement in existing technologies and the development of new technologies.

Series Editor,
Yatish T. Shah

Preface

In the previous volume on the subject of modular systems, we demonstrated a wide use of modular approach in coal, oil, gas, biomass, waste, nuclear, solar, geothermal, wind, and hydro industries for energy fuel recovery and conversion processes. The book showed the growing industrial interests in implementing modular systems not only for distributed small-scale energy systems but also for portions of large-scale systems. To this extent, energy industry is following the same paradigm that building, vehicle, and computer industries have successfully followed for many years. With many industrial examples in all sectors of energy industry, the previous book illustrated that modular approach is more cost effective, produces better quality products, delivers products on time with very little waste, gives more flexibility, allows easy future modifications to the process, allows easy scale-up and scale-down, and allows easy integration of new innovations compared to conventional on-site construction. In short, off-site modular construction carries lot more advantages over stick-shift approaches. The book, however, did point out some disadvantages of modular approach and caution that the use of modular approach should be thoughtfully carried out.

Energy industry is a complex mix of service, products, and rapidly changing technologies. The previous book points that energy industry is going through rapid changes due to growing need for demand, increasing use of new resources, significant environment concerns, and rapidly changing technologies. Unlike in the past, new energy industry will require management to be agile, flexible, and aware of rapid changes in technology and competition. The industry will go through unprecedented changes due to acceleration in technology innovations, mobility revolution, and energy system fractionation. All of these will require future energy systems to be more modular so that they can adopt to these rapid changes. In the previous book, we examined the use of modular systems for the front end of the energy industry, i.e., energy and fuel recovery and conversion. In this book, we examine the role of modular approach for the back end of the energy industry, i.e., the energy usage management.

It is to the world's interest that energy is consumed thoughtfully and efficiently so that more resources are preserved, less harmful gases are emitted, and the needs for entire society are met. Energy usage management covered in this book contains several components. First, energy conservation and efficiency are the preludes to sensible management of energy usage. Both high conservation and efficiency of usage reduce energy consumption. While both achieve the same end results, conservation requires human actions and choices. Energy efficiency, on the other hand, depends on the equipment and processes chosen for energy usage. This book outlines the use of modular approaches for the processes used to improve energy conservation and efficiency. Since energy consumption is conventionally broken down into four sectors—residential, transportation, industrial, and commercial—the discussions on energy conservation and efficiency are also broken down into these four sectors. Chapter 1 examines the use of modular systems for energy conservation and efficiency for residential and transportation sectors, while Chapter 2 examines the same for industrial and commercial sectors.

Energy is used in almost all walks of life. It is difficult to cover all applications of energy usage. Instead in this book, we examine five areas that cover all sectors described above. These five areas namely, buildings, vehicles, computer and electrical/electronic products, district heating, and wastewater treatment and desalination cover (a) the lion's share of total energy consumption, (b) topics of significant societal interests, and (c) topics of significant growth and new innovations. Chapter 3–7 cover the use of modular systems for these five application areas, respectively.

The prudent use of energy requires efficient storage and transportation infrastructure for both energy and fuel. Energy storage has become more important with rapid infusion of renewable energy in the total energy mix. Both solar and wind energy are intermittent, and their use needs to be backed up by stored energy to handle uninterrupted and peak power supply. This book shows that both storage and transportation industries for energy and fuel are highly modular. They will play an increasingly important role as the contribution of distributed energy will grow to satisfy the needs of rural, remote, and isolated areas. Both storage and transportation industries are also very important for mobile use of energy. The increasing use of hydrogen and fuel cell for both stationary and mobile purposes will require modular storage and transportation facilities. The topics of storage and transportation are covered in Chapters 8 and 9, respectively.

Finally, the book points out the use of modular approach for bottom-up, top-down, and hybrid simulations of energy industry and its components. It shows how modular approach has facilitated scientific evaluation of energy systems by simulation, modeling, and optimization through modular approach. It also shows the complex socioeconomic assessment of energy industry, which incorporates technological progress through various modular-based simulation and modeling algorithms. This final topic is covered in Chapter 10.

As the first volume, this book will be a very good reference book for all the people involved in the energy industry as researchers, plant developers, or managers. It will also be a good reference for graduate students and researchers in government energy laboratories.

Author

Yatish T. Shah received his BSc in chemical engineering from the University of Michigan, Ann Arbor, USA, and MS and ScD in chemical engineering from the Massachusetts Institute of Technology, Cambridge, USA. He has more than 40 years of academic and industrial experience in energy-related areas. He was chairman of the Department of Chemical and Petroleum Engineering at the University of Pittsburgh, Pennsylvania, USA; dean of the College of Engineering at the University of Tulsa, Oklahoma, USA, and Drexel University, Philadelphia, Pennsylvania, USA; chief research officer at Clemson University, South Carolina, USA; and provost at Missouri University of Science and Technology, Rolla, USA, the University of Central Missouri, Warrensburg, USA, and Norfolk State University, Virginia, USA. He was also a visiting scholar at the University of Cambridge, UK, and a visiting professor at the University of California, Berkley, USA, and Institut für Technische Chemie I der Universität Erlangen, Nürnberg, Germany. He has previously written nine books related to energy, six of which are under the "Sustainable Energy Strategies" book series (by Taylor and Francis) of which he is the editor. The present book is another addition to this book series. He has also published more than 250 refereed reviews, book chapters, and research technical publications in the areas of energy, environment, and reaction engineering. He is an active consultant to numerous industries and government organizations in energy areas.

1 Modular Systems for Energy Conservation and Efficiency: Residential and Transportation Sectors

1.1 INTRODUCTION

In the previous volume, we examined a modular approach for energy systems related to energy and fuel recovery and conversion [1]. This is the front end of the energy industry. In this volume, we evaluate the importance of modular systems for the back end of the energy industry, namely, energy usage management. Once the energy or fuel are recovered and converted to its end use (such as electricity or other useful fuels), they are often stored and transported for their end uses. Energy usage management strategies depend on the sectors where energy is used, the source of energy, and the technology adopted for its use. Energy conservation and high usage efficiency are essential in prudent energy usage management. The United States is the second-largest single consumer of energy in the world. The U.S. Department of Energy [2,3] categorizes national energy use in four broad sectors: transportation, residential, commercial, and industrial. Energy usage in transportation and residential sectors (about half of U.S. energy consumption) is largely controlled by individual domestic consumers. Commercial and industrial energy consumption is determined by businesses entities and other facility managers. National energy policy has a significant effect on energy usage across all four sectors (see Figure 1.1).

This book evaluates all the steps involved in energy usage management and the modular systems attached to each step. For all types of usages, energy conservation and efficiency are the preludes to responsible energy consumption. The efforts made for energy conservation and for improving energy conversion and its usage efficiency would save fuels, help environments, and allow responsible and rational use of energy for generations to come. We, therefore, first address modular systems used to improve energy conservation and efficiency. It is important to note that though energy conservation and efficiency both result in saving energy, energy conservation is more human-initiated while energy efficiency is often related to equipment and processes used in energy conversion and usages. Often both conservation tactics and efficiency improvements simultaneously affect energy consumption. They may overlap in efforts to build zero-energy houses using both active and passive solar building designs. In this case, passive solar heating is initiated by human beings

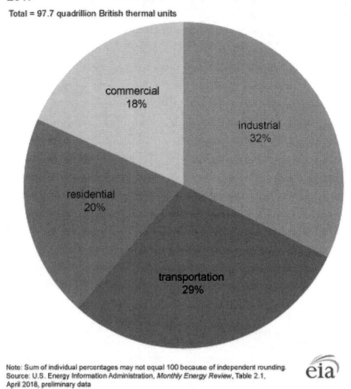

FIGURE 1.1 Energy usage by sector [3].

but the efficiency of active solar heating is more process-oriented. Both active and passive solar heating can be modular in nature, and if properly designed, they can reduce energy consumption. In other words, both active and passive solar heating can complement each other. It is important that both conservation and efficiency measures are fully implemented before the full usage of energy.

Furthermore, both conservation and efficiency are the best methods to reduce overall energy consumption and thereby reduce greenhouse gas (GHG) emissions if fossil fuels are used. With the finite level of energy generation at a given time, the increase in population growth, and the human desire to improve quality of life through better economics and convenient lifestyle will limit the overall global energy use per capita. The best method for achieving energy use neutrality is to reduce waste through means of conservation and efficiency. Thus, energy conservation and efficiency are the underpinnings of the strategies for energy usage management. In this chapter, we examine the use of a modular approach for energy conservation and efficiency for residential and transportation sectors. In the following chapter, we will examine the same subject for the industrial and commercial sectors. In line with the theme of the book, major emphasis is placed on the use of modular systems.

Further, to illustrate the role of modular operations in energy usages, we pick five application areas, namely, buildings, vehicles, computer and electrical/electronic applications, district heating and wastewater treatment, and water desalination. Collectively, these areas represent all four sectors of energy usages. These five applications are chosen because of their energy consumption levels and their interests in society. Building and vehicle industry consumes about half of the total energy usage. Computer and electrical/electronic industry is the fastest growing and most personalized industry. For electronic applications, although micro- and nano-level devices use small amount of energy per unit, the number of units is very large and rapidly growing. Besides electricity, heating and cooling is the largest segment of energy consumption. District heating is becoming more popular and it is an efficient method to provide heating and cooling to the communities in colder countries. In both Europe and North America, its use is rapidly expanding. Finally, the need for cleaner and potable water is increasing faster than ever. In addition, clean and potable water shortage has been predicted to be one of the important problems facing the global community in the 21st century. Furthermore, water desalination is one of the most energy-consuming industries. These five applications give a broad base picture of energy usage. As shown in this book, modular systems are becoming very popular in all of these applications.

Along with these applications, energy and fuel storage and transport are very important parts of the energy usage management process. Energy storage is particularly important for promoting renewable sources of energy like solar and wind, which are intermittent by their basic nature. Both storage and transport allow the steady use of energy from a wide variety of sources. The book will show that the modular approach is heavily infused in building meaningful storage and transport systems. The book ends with a brief demonstration of the usefulness of modular approach for simulation and modeling of the energy industry as whole and its parts. There are various types of modeling efforts being carried out in the energy industry. The book will show that the modular method of simulation, both at global and individual system levels, profoundly facilitates theoretical efforts.

1.2 ENERGY CONSERVATION AND EFFICIENCY

The first step in energy consumption management is energy conservation and efficiency. Energy conservation is an effort made to reduce the consumption of energy by using less of an energy service [1–10]. It is always desirable to reduce consumption by adopting various energy-saving mechanisms to make less use of fuel, less emission of GHGs, and achieve the same end results. Energy conservation, to the extent possible, is therefore very essential prerequirement to the overall process of energy consumption management.

While saving in energy consumption can be achieved either by using energy more efficiently (using less energy for a constant service) or by reducing the amount of service used (for example, by driving less) without expense of human comfort. While increasing efficiency of an energy system is tantamount to conserving energy, improving thermal energy efficiency (like in cogeneration) also means more usages for a given input of energy source. As pointed out above, while energy conservation efforts are initiated by humans, improving energy efficiency largely depends on the equipment or process. Energy conservation is a part of the concept of eco-sufficiency. It reduces the need for energy services and can result in increased environmental quality, national security, personal financial security, and higher savings. It is at the top of the sustainable energy hierarchy. It also lowers energy costs by preventing future resource depletion.

While improving energy efficiency results in similar outcomes, the means of achieving it are different. Energy use can be reduced by minimizing wastage and losses, and improving efficiency through technological upgrades and by improved operation and maintenance. On a global level, energy use can also be reduced by the stabilization of population growth. Energy can only be transformed from one form to other, such as heat energy to motive power in cars, or kinetic energy of water flow to electricity in hydroelectric power plants. These transformations involve energy efficiency which can be altered with innovative equipment or process designs. Machines and processes are required to transform energy from one form to other. The wear and friction of the components of these machines or ill-conceived processes can incur significant waste of energy. Machine running causes loss of quadrillions of British thermal unit (BTU) and revenue loss of $500 billion in industries only in the United States. It is possible to minimize these losses by adopting green engineering practices to improve the lifecycle of the components. The use of nonrenewable sources of energy can also be reduced by replacing them with more renewable, environment-friendly sources or clean nuclear energy. As shown later, this type of conservation of nonrenewable sources of energy is an integral part of the greening energy usages.

Some countries employ energy or carbon taxes to motivate energy users to reduce their consumption. Carbon taxes can force energy users to shift to nuclear power and other energy sources that carry different sets of environmental side effects and limitations. On the other hand, taxes on all energy consumption can reduce energy use across the board while reducing a broader array of environmental consequences arising from energy production. This method also pushes more energy conservation or efforts to improve the energy efficiency of equipment and processes. California employs a tiered energy tax whereby every consumer receives a baseline energy allowance that carries a low tax. As usage increases above that baseline, the tax increases drastically. Such programs aim to protect poorer households while creating a larger tax burden for high energy consumers [4].

While the efforts to conserve energy are made at global level, out of four sectors of energy consumption, building and transportation sectors (which are more human choice dependent) appear to have made significant progress in implementing various techniques to reduce or replace fossil energy consumption. Other sectors, particularly industrial, are, however, also making significant efforts to comply with new government regulations and improve their financial productivity through improved energy efficiency. In both industrial and commercial sectors, energy conservation and efficiency play very important roles to the company bottom-line profits.

Another aspect of energy conservation and efficiency is using Leadership in Energy and Environmental Design (LEED) [10]. This program is not mandatory but voluntary and has many categories in which energy and atmosphere prerequisite applies to energy conservation. This focuses on energy performance, renewable energy, and many more. This program is also designed to promote energy efficiency and a green building concept. The Energy Policy Act of 2005 included incentives which provided a tax credit of 30% of the cost of the new item with a $500 aggregate limit; the program extended to 2010. By Executive Order 13514, U.S. President Barack Obama mandated that by 2015, 15% of the existing federal buildings conform to new energy efficiency standards and 100% of all new federal buildings will be zero net energy (ZNE) by 2030.

Consumers are often poorly informed of the savings of energy-efficient products. A prominent example of this is the energy savings that can be made by replacing an incandescent light bulb with a more modern alternative. When purchasing light bulbs, many consumers opt for cheap incandescent bulbs, failing to take into account their higher energy costs and lower lifespans when compared to modern compact fluorescent and light-emitting diode (LED) bulbs. LED lamps use at least 75% less energy, and last 25 times longer, than traditional incandescent light bulbs. The price of LED bulbs has also been steadily decreasing in the past 5 years due to improvements in semiconductor technology. Estimates by the U.S. Department of Energy state that widespread adoption of LED lighting over the next 20 years could result in about $265 billion worth of savings in the United States' energy costs. All light bulb industries are modular and can be constantly improved through infusion of new innovation [1,10].

1.3 ENERGY CONSERVATION AND EFFICIENCY IN RESIDENTIAL SECTOR

The residential sector is all private residences, including single-family homes, apartments, manufactured homes, and dormitories. Energy use in this sector varies significantly across the country, due to regional climate differences and different regulations. On average, about half of the energy used in U.S. homes is expended on space conditioning (i.e. heating and cooling). Despite technological improvements on furnaces and air conditioners, many American lifestyle changes have put higher demands on heating and cooling resources. The average size of homes built in the United States has increased from 1,500 sq ft (140 m^2) in 1970 to 2,300 sq ft (210 m^2) in 2005. The single-person household has become more common, as has central air conditioning: 23% of households had central air conditioning in 1978, and that figure rose to 55% by 2001. Table 1.1 outlines average energy consumption by various parts of the home.

Energy usage in some homes may vary widely from these averages. For example, milder regions such as the Southern United States and Pacific Coast of the United States need far less energy for space conditioning than New York City or Chicago. On the other hand, air conditioning energy use can be quite high in hot-arid regions (Southwest) and hot-humid zones (Southeast). In milder climates such as San Diego, lighting energy may easily consume up to 40% of total energy. Certain appliances such as a waterbed, hot tub, or pre-1990 refrigerator use significant amounts of electricity. However, recent trends in home entertainment equipment can make a large

TABLE 1.1
Home Energy Consumption Averages

Sector	Percentage
Home heating systems	28.9
Home cooling systems	14.0
Water heating	12.9
Lighting	9.0
Home electronics	7.1
Refrigerators and freezers	5.9
Clothing and dishwashers	4.5
Cooking	3.7
Computers	2.2
Others	4.4
Non-end user energy expenditure	5.4

Source: Adapted and modified from Ref. [6].

difference in household energy use. For instance, a 50-inch LCD television (average on-time of 6 h a day) may draw 300 W less than a similarly sized plasma system. In most residences no single appliance dominates, and any conservation efforts must be directed to numerous areas to achieve substantial energy savings. However, ground, air and water source heat pump systems, solar heating systems, and evaporative coolers are among the more energy-efficient, environmentally clean and cost-effective, space conditioning, and domestic hot water systems currently available (Environmental Protection Agency) can achieve reductions in energy consumptions of up to 69%. All of these indicate that while all energy-consuming equipment used in residential sectors are modular, nature of modular equipment and the way they are used make significant difference in energy consumption.

Residential sector contributes heavily to energy consumption and GHG emission, if fossil energy is used for power and heating and cooling. Several measures can be taken to reduce fossil fuel-dependent energy consumption. The first step is to design a passive solar heating home that can make the maximum use of natural solar energy and its distribution within the home without any additional mechanical or electrical means of providing heating and cooling in the home. This method does not provide power within the home. The heating and cooling needs of the home can also be conserved and efficiently managed with various active changes in home such as using phase change materials (PCMs) for energy storage, insulating home in various ways and so on. These changes can all be modular in nature. Finally, power, heating, and cooling needs of a home can be satisfied with the use of solar, geothermal, and wind energy (which are free) instead of fossil energy which can lead to net-zero-energy building. These types of buildings are also called green or carbon-free buildings because they do not emit any GHG to the environment. Zero-energy buildings can also be net producers of energy. Modular approach is implicit in these discussions. These steps are described in more detail in the next section.

1.4 MODULAR APPROACHES FOR ENERGY CONSERVATION AND EFFICIENCY IMPROVEMENT IN BUILDINGS

The first step for energy conservation and efficiency improvements in buildings is designing the building for passive solar heating. Passive building design can help energy conservation and reduce heating and cooling requirements of the homes. Such design is modular and can be standardized to cut costs. As shown in this section, additional modular changes can also be made to improve energy conservation and efficiency in home energy consumption.

1.4.1 Modular Passive Solar Heating

Of all the steps outlined above for energy conservation and efficiency in homes, the passive modular solar building design does not require additional mechanical or electrical means for reduction in energy consumption. In modular passive solar building design, windows, walls, and floors are made to collect, store, and distribute solar energy in the form of heat in the winter and reject solar heat in the summer. This is called passive solar design or climatic design because, unlike active solar heating systems, it does not involve the use of mechanical and electrical devices. They are however modular in nature. In designing an active solar heating building or zero-energy building, passive solar building design forms a prelude or basis for building architecture. It reduces the energy load required for active solar energy and zero-energy building [8,9].

The key to designing a passive solar building is to best take advantage of the local climate. Elements to be considered include window placement and glazing type, thermal insulation, thermal mass, and shading. Passive solar design techniques can be applied most easily to new buildings but existing buildings can be retrofitted (see Figure 1.2).

Typically, passive solar heating involves:

- The collection of solar energy through proper-oriented, south-facing windows.
- The storage of this energy in "thermal mass," comprised of building materials with high heat capacity such as concrete slabs, brick walls, or tile floors.

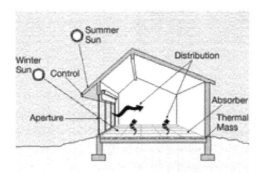

FIGURE 1.2 Elements of passive solar design, shown in a direct gain application [8].

- The natural distribution of the stored solar energy back to the living space, when required, through the mechanisms of natural convection and radiation.
- Window specifications to allow higher solar heat gain coefficient in south glazing.

Passive solar heating systems make use of the building components to collect, store, and distribute solar heat gains to reduce the demand for space heating (see Figure 1.2). A passive solar system does not require the use of mechanical equipment because the heat flow is by natural means, such as radiation, convection, and conductance, and the thermal storage is in the structure itself. A passive solar heating system is made up of the following key components, all of which must work together for the design to be successful (see Figure 1.2):

- Aperture (collector)
- Absorber
- Thermal mass
- Distribution
- Control

In a passive solar heating system, the aperture (collector) is a large glass (window) area through which sunlight enters the building. Typically, the aperture(s) should face within 30° of true south and should not be shaded by other buildings or trees from 9 a.m. to 3 p.m. each day during the heating season. The hard, darkened surface of the storage element is known as the absorber. This surface—which could consist of a masonry wall, floor, or partition (PCM), or a water container—sits in the direct path of sunlight. Sunlight then hits the surface and is absorbed as heat. The thermal mass is made up of materials that retain or store the heat produced by sunlight. The difference between the absorber and thermal mass, although they often form the same wall or floor, is that the absorber is an exposed surface whereas thermal mass is the material below or behind that surface. In some applications, fans, ducts, and blowers may help with the distribution of heat through the building. Elements to help control under- and overheating of a passive solar heating system include roof overhangs, which can be used to shade the aperture area during summer months, electronic sensing devices, such as a differential thermostat that signals a fan to turn on, operable vents and dampers that allow or restrict heat flow, low-emissivity blinds, and awnings [8,9].

Passive solar buildings are designed to let the heat into the building during the winter months, and block out the sun during hot summer days. Besides using all passive solar building design concepts mentioned above, incorporating shading concepts into the landscape design can help reduce the solar heat gain in the summer and reduce cooling costs. The leaves of deciduous trees or bushes located to the south of the building can help block out sunshine and unneeded heat in the summer. These trees defoliate in the winter and allow an increase in the solar heat gain during the colder days. Incorporating overhangs, awnings, shutters, and trellises into the building design can also provide shade. Effective thermal mass materials, like concrete, or stone floor slabs, have high specific heat capacities, as well as high density. It is ideally placed within the building where it is exposed to winter sunlight but insulated from heat loss. The material is warmed passively by the sun and releases thermal energy into the interior during the night.

The most important characteristic of passive solar design is that it is holistic and relies on the integration of a building's architecture, materials selection, and mechanical systems to reduce heating and cooling loads. It is also important to consider local climate conditions, such as temperature, solar radiation, and wind, when creating climate-responsive, energy-conserving structures that can be powered with renewable energy sources. The passive systems assisted by mechanical devices are referred to as hybrid heating systems. Some of these are described in the next section. To achieve a high percentage of passive solar heating, it is necessary to incorporate adequate thermal mass in buildings. Specific guidelines for this include the following:

- Confirm that the area of thermal mass is six times the area of the accompanying glazing (when possible). For climates with foggy or rainy winters, somewhat less thermal mass is needed.
- Place the mass effectively by ensuring that it is directly heated by the sun or is spread in thin layers throughout rooms in which there is a large quantity of solar collection.
- Disregard the color of the mass surface. However, natural colors (e.g., colors in the 0.5–0.7 absorption range) are quite effective.
- Incorporate thermal storage in floors or walls that consist of concrete, masonry, or tile. To reflect light and enhance the space, walls should generally remain light colored.

1.4.2 Energy-Saving Modules (ESM)

ESM reduce the *electricity consumption* (kWh) and maximum demand (kW) of *air conditioning* and *refrigeration* compressors. The concept was developed in Australia in 1983 by Abbotly Technologies and is now distributed by Smart Cool Systems, Inc. The system works in conjunction with existing heating, ventilation and air conditioning (*HVAC*) controls ensuring that compressors work at maximum efficiency while maintaining desired temperature levels. By preventing over-cycling, known as "compressor optimization" consumption of electricity is cut by 15%–25% [11–13].

Conventional controls, including building and energy management systems and state-of-the-art refrigeration controls, often operate only on reaching preprogramed static values to switch compressors off and on or adjust capacity. When the measured medium is within the dead band, the system and controllers remain idle. The "energy-saving module" is a computer that records the switching values of the primary controller and also measures the "rate of change" of both the rise and fall of temperatures during the operating cycle. With this data, the "energy-saving module" computes a reference heat load to match the cooling capacity and then calculates operating parameters. This calculation is used to minimize compressor operation within the switching values, with a resultant reduction in refrigeration and air conditioning compressor run time and reduced electricity consumption. By dynamically measuring the heat load and adjusting the control differential in proportion to the cooling demand, it is possible to dynamically control the cycle rate of the compressors. This is achieved while maintaining the desired *operating temperature* [11–13].

1.4.3 Phase Change Material Based Modular Thermal Storage

Velraj and Pasupathy [14] gave an excellent account of how PCM can be used as a modular storage device in a number of different ways to reduce energy consumption for heating and cooling of buildings. Here, we briefly summarize the findings of this study. Thermal energy storage can be used (either for passive or active solar buildings) to conserve energy. The most effective storage device has been found to be PCMs. This storage device is important for modular implementation of renewable sources of energy in buildings. Thermal energy storage systems provide the potential to attain energy savings, which, in turn, reduce the environmental impact related to energy use. In fact, these systems provide a valuable solution for correcting the mismatch that is often found between the supply and demand of energy. Modular latent heat storage is increasingly being considered for waste heat recovery, load leveling for power generation, building energy conservation, and air conditioning applications.

Electrical energy consumption varies significantly during the day and night according to the demand by residential activities. In hot and cold climate countries, the major part of the load variation is due to air conditioning and domestic space heating, respectively. This variation leads to a differential pricing system for peak and off-peak periods of energy use. Better power generation/distribution management and significant economic benefit can be achieved if some of the peak load could be shifted to the off-peak load period that can be achieved by thermal energy storage for heating and cooling in residential building establishments. The most common material used for this purpose is the PCM. As pointed out by Velraj and Pasupathy [14], the application of PCMs in buildings can be carried out either by using natural heat and cold sources or using man-made heat or cold sources. The storage of heat or cold is necessary to match availability and demand with respect to time and also with respect to power. The use of PCM for thermal energy storage can be done by inserting PCM in building walls, in other building components or in separate heat or cold stores. The first two are passive systems, where the heat- or cold-stored is automatically released when indoor or outdoor temperatures rise or fall beyond the melting point. The third ones are active systems, where the stored heat or cold is in containment thermally separated from the building by insulation. Therefore, the heat or cold is used only on demand and not automatically.

In building applications, only PCMs that have a phase transition close to human comfort temperature (20°C–28°C) can be used. Along with paraffin, fatty acids and polyethylene glycol, commercial phase change materials that have been developed by some of the manufacturers are generally suitable for building applications. Work is also continuing on integrating PCM into solar photovoltaic modules to reduce the operating temperature and thus improving their conversion efficiency. Recent tests have demonstrated temperature reductions of more than 10°C by incorporating a 29°C PCM as a backing of solar modules [14–23]. This has been achieved by using square metal tubes filled with the PCM. A significant requirement in this approach was to develop designs, which maintain good thermal contact between the PCM and the modules to facilitate timely heat storage by the PCM during the day and heat loss to the environment during nighttime.

The PCM incorporation can be done by (a) direct incorporation, (b) immersion, and (c) encapsulation. The third one can be defined as the containment of PCM within a capsule of various materials, forms, and sizes prior to incorporation so that it may be introduced to the mix in a convenient manner. According to Velraj and Pasupathy [14], there are two principal means of encapsulation. The first is microencapsulation, whereby small, spherical, or rod-shaped particles are enclosed in a thin and high-molecular-weight polymeric film. The coated particles can then be incorporated into any matrix that is compatible with the encapsulating film. It follows that the film must be compatible with both the PCM and the matrix. The second containment method is macroencapsulation, which comprises the inclusion of PCM in some form of package such as tubes, pouches, spheres, panels, or other receptacle. These containers can serve directly as heat exchangers or they can be incorporated in building products. As pointed out by Velraj and Pasupathy [14], for many applications, encapsulation method is preferred.

1.4.3.1 Phase Change Material Applications in Buildings

In general, PCM can be used in several different ways to capture solar heat. A PCM wall can capture a large proportion of the solar radiation incident on the walls or roof of a building. Because of the high thermal mass of PCM walls, they can minimize the effect of large fluctuations in the ambient temperature on the inside temperature of the building. They can be very effective in shifting the cooling load to off-peak electricity period. Arkar and Medved [22], however, designed and tested a latent heat storage system used to provide ventilation of a building. The spherical encapsulated polyethylene spheres were placed in a duct of a building ventilation system and acted as porous absorbing and storing media. The heat absorbed was used to preheat ambient air flowing into the living space of a building.

The "solar wall" is another application of PCM for thermal storage. In this case, the solar radiation that reaches the wall is absorbed by the PCM "buried" in the wall. In addition, Stritih and Novak [18] designed an "experimental wall" which contained black paraffin wax as the PCM heat storage agent. The stored heat was used for heating and ventilation of a house. The results of this work were very promising. In this system, short wave radiation passes through glass with TIM (transparent insulation material) which prevents convective and thermal radiation heat transfer. PCM in a transparent plastic casing made of polycarbonate absorbs and stores energy mostly as latent heat. The air for the house ventilation is heated in the air channel and it is led into the room. Insulation and plaster are standard elements used for this purpose.

PCM can also be integrated into wood lightweight concrete [16]. Wood lightweight concrete is a mixture of cement, wood chips, or sawdust, which should not exceed 15% by weight, water, and additives. This mixture can be applied for building interior and outer wall construction. For integration in wood lightweight concrete, two PCM materials Rubitherm GR 40 and GR 50 were investigated by Mehling et al. [16]. It was shown that PCMs can be combined with wood lightweight concrete and that the mechanical properties do not seem to change significantly. The authors reported that the incorporation of PCM has two additional reasons (a) to increase the thermal storage capacity (b) to get lighter and thinner wall elements with improved thermal performance.

In another study, Ismail [23] proposed a different concept for thermally effective windows using a PCM moving curtain. The window is double sheeted with a gap between the sheets and an air vent at the top corner. The sides and bottom are sealed except for two holes at the bottom which are connected by a plastic tube to a pump and the PCM tank. The pump is connected in turn to the tank containing the PCM, which is in liquid phase. The pump operation is controlled by a temperature sensor. When the temperature difference reaches a preset value the pump is operated and the liquid PCM is pumped out of the tank to fill the gap between the glass panes. Because of the lower temperature at the outer surface, the PCM starts to freeze, forming a solid layer that increases in thickness with time and hence prevents the temperature of the internal ambient from decreasing. This process continues until the PCM changes to solid. A well-designed window system will ensure that the external temperature will start to increase before the complete solidification of the enclosed PCM which has thermal conductivity λ between 0.15 and 0.75 W/m K. Thus, the proposed concept of the PCM filled window system is viable and thermally effective. It is also confirmed by Velraj and Pasupathy [14] that the PCM filling leads to filtering out the thermal radiation and reduces the heat gain or loss because most of the energy transferred is absorbed during the phase change of the PCM. The double-glass window filled with PCM is more thermally effective than the same window filled with air. The colored PCM is more effective in reducing radiated heat gains and the green color is the most effective of all.

UniSA (University of South Australia) [17] has developed a roof-integrated solar air heating/storage system, which uses existing corrugated iron roof sheets as a solar collector for heating air. A PCM thermal storage unit is used to store heat during the day so that heat can be supplied at night or when there is no sunshine. The system operates in three modes. During times of sunshine and when heating is required, air is passed through the collector and subsequently into the home. When heating is not required air is pumped into the thermal storage facility, melting the PCM and charging it for future use. When sunshine is not available, room air is passed through the storage facility, heated and then forced into the house. When the storage facility is frozen, an auxiliary gas heater is used to heat the home. Adequate amounts of fresh air are introduced when the solar heating system is delivering heat into the home. Kunping Lin et al. [15] put forward a new kind of under-floor electric heating system with shape-stabilized PCM plates. Different from conventional PCM, shape-stabilized PCM can keep the shape unchanged during the phase change process. Therefore, the PCM leakage danger can be avoided. This system can charge heat by using cheap nighttime electricity and discharge the heat stored at daytime. Furthermore, to investigate the thermal performance of the under-floor electric heating system with the shape-stabilized PCM plates, an experimental house with this system was set up in Tsinghua University, Beijing, China. The experimental house was equipped with the under-floor electric heating system including shape-stabilized PCM plates. The dimensions of the experimental house were 3 m (depth) × 2 m (width) × 2 m (height).

It had a 1.6 m × 1.5 m double-glazed window facing south, covered by a black curtain. The roof and walls were made of 100 mm thick polystyrene wrapped by metal board. The under-floor heating system included 120 mm thick polystyrene insulation, electric heaters, 15 mm thick PCM, some wooden supporters, 10 mm thick air layer, and 8 mm thick wood floor.

Presumably, free cooling investigated at the University of Nottingham [19] is a replacement of a full air conditioning system by the new system that is a nighttime cooling system, which is also easy to retrofit. It is ceiling-mounted with a fan to throw air over the exposed ends of heat pipes. The other end of the heat pipes is in a PCM storage module. During the day, the warm air generated in the room is cooled by the PCM i.e. heat is transferred to the PCM. During the night, the fan is reversed and the shutters are opened such that cool air from the outside passes over the heat pipes and extracts heat from the PCM. The cycle is then repeated the next day. The melt and freeze temperatures of the PCM are approximately 22°C and 20°C, respectively. Complete melting occurs over a period of about 8 h when the temperature difference between the PCM and the air is 2°C and over a period of about 3 h, the temperature difference is 3.5°C. The heat transfer rates are 80 and 200 W per unit, respectively, or 800 and 2,000 W for a room with 10 units.

While as shown above, PCM can be used as energy storage in a number of different ways to reduce energy consumption for heating and cooling of buildings, the optimization of various operational parameters is fundamental to demonstrate the possibilities and the extent of success of PCMs in building materials. Such optimization exercise also delineates the limits of this approach. The use of this approach can also be applied to commercial and industrial buildings and this is briefly described in the next chapter under the section of energy conservation and efficiency for the commercial sector.

1.4.3.2 A Modular Solar Envelope House

Old Dominion University and Hampton University joined forces to build a Solar Decathlon house called Unit 6 for the 2011 competition. Their goal was to create a solar net-zero-energy house that would contribute to the formation of walkable city neighborhoods [24].

The Tidewater team created a modular home (see Figure 1.3) made from four main sections and featured passive and active solar designs. The roof had modular solar photovoltaic (PV) panels to generate electricity, which were arranged in a square pattern. Located in the middle of the PV, a solar thermal panel collected solar heat to generate hot water for the home. The panels were mounted on racks to allow airflow behind them to help keep them cool as solar PV operates more efficiently within a lower temperature. The panels were integrated into the roof.

The home was designed to be passive solar so that the summer sun would not heat up the house. Its overhang and window placement block the summer sun's heat, while they allow the winter sun to enter the house. Its design and materials enable the house to collect solar heat and later release it to warm the home through the cooler evening and night. This design can be achieved by following a few passive solar guidelines. The home featured a southern facing porch space that could be utilized throughout the year. The porch was lined with a dark-colored tile floor. Tile was a

FIGURE 1.3 The Virginia team designed a home called Unit 6 for the 2011 Solar Decathlon. (Photo credit: Jim Tetro/U.S. DOE—Solar Decathlon [24].)

material that has thermal mass, but the house had additional thermal mass with the utilization of a PCM installed below the tile. A PCM is one that changes from a liquid to solid matter at specific temperatures having a high heat of fusion. With these capabilities, it can store and release large amounts of heat. The PCM was contained within plastic "dimples" located directly under the tile floor. It liquefied when it was warm and formed into a gelatinous material after cooling. The material was similar to water in having superior ability to store and release heat. During the winter when the sun shines on the tile, it warms it, allowing the solar heat to be stored both within the tile and the phase change material. Because these materials can store large amounts of heat, after the sun goes down, the heat is slowly released from the material, is transferred through the tile, and helps to keep the home warm through the cooler evening and night.

The Unit 6 home could be categorized as an *envelope house* because it uses air currents to cool the home. The motorized window in the front porch area seen above allows air to be pulled through the porch and into the kitchen. The home is naturally ventilated as the warmer air is pulled out through the windows located at the top of the home, as warm air naturally rises. The airflow takes place from the left to right side and exits the home from an open window on the upper right-hand side of the kitchen. To save energy, the HVAC installed can sense when any windows are open and will not operate under these conditions. This is a function that can be overridden, but over time, it is also one that saves enormous amounts of energy. The kitchen and bathroom were located in close proximity to keep the water lines short and energy-efficient.

Windows, sometimes referred to as "glazing", are another important aspect of an efficient passive solar home. The windows that Unit 6 utilized in the south-facing living room had the ability to swing open on the side, or from the bottom. The team used double-paned windows on the southern side of the building and triple-paned windows throughout the rest of home. Outside the windows, wooden overhangs block the summer sun's heat, while they allow the winter sun to warm the home.

Windows are often the culprit of energy loss; however, these windows have mechanical gaskets that can be seen in the gallery below. Instead of a door, the home had a moving bookshelf to close the bedroom off from the living room. That section of the shelf can slide to the right to meet the other bookshelf and provide a partition between the separate areas. Other facets of this Solar Decathlon home include a water cistern located in the back of the home. Water could be collected from the roof, stored in the cistern, and reused as irrigation water in the yard.

The utility room was located outside the home. It was accessible under a southwest-facing awning that was also helpful in shading the area from the afternoon sun. The Tidewater team integrated local materials such as the Cyprus wood from neighboring North Carolina in the home [24].

1.4.4 MODULAR HYBRID WALL WITH SURFACE HEATING AND COOLING

Erikci Çelik et al. [25] examined modular design of hybrid wall with surface heating and cooling system. In this study, it was proposed that wall elements, which are vertical building elements and constitute a broad area within the structure, are regulated with a different system concerning the reduction of building energy consumption ratio. Within the scope of this study, integration of modular wall elements with surface heating and cooling system, which are convenient for using hybrid energy, into the buildings was evaluated.

One of the aims of the study by Erikci Çelik et al. [25] was to determine the direct impact of the product on architectural design process and identify the issues that will affect the process and need to be resolved. In design, implementation, and usage phases, integration of technical combination and montage details of modular wall elements, together with issues regarding energy saving, heat-saving, and other environmental aspects will be discussed in detail. As a result, the ready-wall product with surface heating and cooling modules was created and defined as hybrid wall and was compared with the conventional system in terms of thermal comfort. The study evaluated performance of this novel concept. The results indicated viability of the concept.

1.4.5 ENERGY SAVINGS FOR A WOOD-BASED MODULAR PREFABRICATED FAÇADE REFURBISHMENT SYSTEM

There is a need to focus on energy efficiency for the existing buildings, especially buildings erected before the energy crises in the 1970s, as these buildings provide a massive potential for improvements in energy performance. The facades of these buildings are also facing an increased need for a face-lift. One problem is that these buildings must be refurbished at low costs and with limited disturbance to the users/tenants. In the study by Ruud et al. [26] the energy efficiency of a modular prefabricated façade renovation system based on wood and its comparison with two existing systems for on-site mounting of additional insulation was investigated. These existing alternatives require a lot of on-site works. Hence, prefabricated and modular solutions would, of course, offer advantages.

The prefab element only consists of an inner and an outer laminated board with insulation in between, interconnected with thin wooden bars. The amount of insulation can be increased, but the analysis of Rudd et al. [26] was limited to 50–150 mm of added insulation, as a complement to an estimate of 100 mm in poorly insulated buildings from the 1960s and 1970s. Thus, it was estimated that one can reach current levels of U-values which was found feasible in recent research regarding larger housing blocks [26–29]. The most common façade refurbishment solution today, in the Nordic countries, is based on plastered insulation, which is not suited for wooden substructures as it poses a well-known risk of moisture damages. There are a large number of such facades in all these countries. The alternative would be on-site works, but prefabricated solutions would, of course, offer advantages. There have been several attempts to develop such solutions in recent years [26–29] but so far none of them, based on their market share, seem to have provided an acceptable solution. The approach by Rudd et al. [26] was based on an alternative with small scale prefab elements with a simple assembly process. The energy analyses concentrated on thermal bridges and U-values, relating it to the total energy performance and comparing it to the current situation and comparable refurbishment methods with wooden structures. The study mainly focused on the wall structures but as a comparison also other energy-saving measures were also investigated. The authors have also conducted a study of energy efficiency regulations in different Nordic countries to see the effect of geographical location. The outcome was a verification of the energy efficiency of the chosen prefab structure as a solution for refurbishment. The advantages are to be found in the dry production process combined with a simple and fast assembly on site. Similarly, there are advantages in the reduced number of thermal bridges and in the potential for the use of recyclable materials.

The study by Rudd et al. [26] was part of a larger research project aiming at cost-efficient and sustainable solutions for the refurbishment of facades in a Nordic context. In the larger research project, the authors also conducted a cost-benefit analysis and studied moisture conditions to see how far this solution might be feasible. The ultimate aim was to test a product and compare the findings from demo projects in the three countries Norway, Sweden, and Finland. While the purpose of this study was to evaluate the energy savings of a wooden prefab facade element compared to two different conventional on-site façade refurbishment solutions, it also compared the energy savings with other types of refurbishment options, i.e. new windows, additional insulation of attic floor, ventilation heat recovery, and so on. Another objective was to study the influence of thermal bridges around the window fastenings. A typical low-rise multifamily building from the period from 1965 to 1975 was chosen as a reference building. It is a two-story-high building with 12 apartments. It has a concrete frame with infill walls (wooden studs and insulation) on the long sides. The windowless short sides consist of concrete, insulation, and bricks. The windows, the wooden paneling on the sides, and the bricks on the short sides were in poor condition and in need of replacement. The building had exhaust ventilation without heat recovery and the airtightness of the building envelope was poor. There are more than a million apartments in the Nordic countries built in a similar manner and in need of refurbishment [26–29]. The reference building was modeled in an energy calculation program and the specific energy use calculated. The influence on the specific energy use for three different façade refurbishment systems was calculated.

The study concluded that the most effective energy-saving measures for this typical multifamily building in the period from 1965 to 1975 were changing to new energy-efficient windows and installation of ventilation heat recovery. But the third most effective measure was the façade refurbishment. Insulation of the roof/attic floor was surprisingly only the fourth most effective measure. Air tightening and changing to energy-efficient door was the least effective single measures. However, if combined with balanced ventilation and heat recovery the air tightening was twice as effective. If one wants to reach really good energy performance, i.e. comparable to new buildings, then the walls also need additional insulation. Sometimes 50 mm additional insulation is enough to reach the requirements on energy and thermal comfort. The advantage of much more insulation is however that the required airtight layer then can be placed on the outside of the existing wall. The studied prefabricated wood-based refurbishment system has an integrated airtight layer for that purpose. Even 50 mm of additional insulation may provide better airtightness of the building envelope to avoid condensation and moisture in the outer parts of the wall. But only 50 mm additional insulation requires the airtight layer to be placed on the inside of the existing wall which is much more difficult and would mean major disruptions for the residents. Good airtightness of the building envelope is also required if one wants to reach the full potential of installing a ventilation heat recovery system. The length of the thermal bridge around the windows was by far the longest of all the thermal bridges in this type of building, and as the insulation of the wall increased, the U-value of this thermal bridge also tend to increase. If one is not attentive and careful, the thermal bridges may become more than double, leading to a significant and unnecessary heat loss. But as shown in this study, it is also possible to mount the windows in a way that keeps the thermal bridges at a reasonably low level [26–29].

1.4.6 Energy Star Modular Homes

One of the best parts about building a modular house is that you get superior home energy efficiency as part of the standard package. Evidence of this modular advantage is readily available from the Energy Star initiative, a federal program administered by the Department of Energy and the Environmental Protection Agency that promotes greater energy efficiency standards in new home construction. An important component of the Energy Star construction program is the tightness of the home. Tests performed for the energy efficiency program show that when a typical modular home is finished correctly by the set crew and general contractor, it does considerably better than a typical stick-built home. One can make one's modular home even more energy efficient than it already is by having it built to the full Energy Star construction specifications. Conventional builders need to take many extra steps at greater cost to their customers to meet these energy efficiency standards. For modular customers, the Energy Star construction steps require very little effort or expense [30–32].

A typical Energy Star modular home lowers utility bills by hundreds of dollars per year. It also improves a home's comfort and safety. The higher levels of insulation and lower levels of air infiltration make the interior surfaces of the modular home quieter and warmer. By closing off the air gaps in the walls, the tighter construction retards the spread of fire, which allows occupants more time to exit their homes safely and call the fire department. The reduced air infiltration also decreases unwanted moisture in the house, and it blocks bugs and rodents from gaining entry.

The mechanical ventilation included in an Energy Star modular home, which greatly improves indoor air quality, is another key feature. This is important because the reduced air infiltration, while enhancing energy efficiency in the home, can allow indoor air pollutants to build up. Household cleaners, carpet adhesives, carbon monoxide from heating systems, dust, and pet dander can accumulate in a well-insulated home. In addition, showering, cooking, and breathing produce humidity that can build up if not vented to the outside [30].

In fact, even new site-built homes that are not Energy Star rated tend to build up too many indoor air pollutants. Today's building codes create tighter homes even when the builder takes no extra steps to increase their energy efficiency practices. Tighter green homes are not only more energy efficient, which saves money; they are also much less drafty, which makes them more comfortable. The solution is to build tight energy-efficient modular homes but to let them breathe with controlled continuous ventilation. The Home Store can accomplish this by installing an appropriately rated bathroom fan connected to a variable-speed timer. This very affordable system will continually refresh indoor air for pennies a day. In fact, the cost to run the fan is only one-tenth the cost of heating the air in a leaky house. The most important extra step that some modular manufacturers need to take to build an energy-efficient modular house is to apply extra air-sealing techniques to the exterior shell. Although all modular homes are tight, not all of them are tight enough to comply with the Energy Star construction requirements. The modular manufacturer also needs to use recessed lights rated for low air infiltration. The manufacturer might need to insulate the basement stairwell, increase the ceiling insulation, upgrade to low-e windows, switch to a properly rated bathroom fan, and use a vented range-hood for a gas range. These are either standard features or routine options for The Home Store and its modular manufacturers. To complete the Energy Star modular home, the general contractor needs to install an approved Energy Star heating system and water heater. The General Contractor needs to seal the joints between ducts on a forced-air system with mastic and then insulate the ducts. Registers must also be sealed along with the basement plumbing and electrical penetrations, the bulkhead, knee walls, and sill plates. In addition, a door sweep on the bottom of the door should be inserted into the basement and insulate and seal the attic, including the hatch or pull-down stairs. If the basement ceiling rather than the walls are insulated, it should be done with R-30. Construction of any living spaces, such as a site-built bonus room, must be in compliance with Energy Star specifications. Energy-efficient windows or sealing air leaks should also be done when the home is constructed [30–32].

1.4.7 Modular Zero-Energy Building

A zero-energy building (see Figure 1.4), also known as a ZNE building, net-zero-energy building (NZEB), net-zero building, or zero-carbon building (ZCB), is a building with ZNE consumption which means that the total amount of energy used by the building on an annual basis is roughly equal to the amount of renewable energy created on the site, or in other definitions by renewable energy sources

Residential and Transportation Sectors 19

FIGURE 1.4 Zero-energy test building in *Tallinn, Estonia. Tallinn University of Technology.*

elsewhere. These buildings consequently contribute less overall GHG to the atmosphere than similar non-ZNE buildings. They do at times consume nonrenewable energy and produce GHGs, but at other times reduce energy consumption and GHG production elsewhere by the same amount. A similar concept approved and implemented by the European Union and other agreeing countries is nearly zero-energy building (nZEB), with the goal of having all buildings in the region under nZEB standards by 2020. Renewable energy use in these buildings is generally modular in nature. Moreover, these buildings often use modular passive solar energy measures to conserve energy and reduce energy consumption [4,5,33].

Most ZNE buildings get half or more of their energy from the grid and return the same amount at other times. Buildings that produce a surplus of energy over the year may be called "energy-plus buildings" [33] and buildings that consume slightly more energy than they produce are called "near-zero-energy buildings" or "ultra-low energy houses". Traditional buildings consume 40% of the total fossil fuel energy in the United States and European Union and are significant contributors of GHGs. The ZNE consumption principle is viewed as a means to reduce carbon emissions and reduce dependence on fossil fuels and although zero-energy buildings remain uncommon even in developed countries, they are gaining importance and popularity.

Most zero-energy buildings use the electrical grid for energy storage, but some are independent of the grid. Energy is usually harvested on-site through energy-producing technologies like solar, wind, or geothermal (like geothermal heat pump) while reducing the overall use of energy with highly efficient HVAC and lighting technologies. The zero-energy goal is becoming more practical as the costs of alternative energy technologies decrease, and the costs of traditional fossil fuels increase. As shown in the previous book, since the building industry is highly modular in nature, energy generation mechanisms for power (largely by solar and wind) and heating and cooling (by solar or geothermal) are generally modular in nature. The design of building for passive solar heating through building architecture, insulations in walls, floors, and other places also tend to be modular in nature. The development of modern zero-energy buildings became possible largely through the progress made in new energy and construction technologies and techniques. These include highly insulating spray-foam insulation, high-efficiency solar panels, high-efficiency heat pumps, and highly insulating low-E triple-glazed windows. These are all modular in nature and these innovations have also been significantly improved by academic research, which collects precise energy performance data on traditional and experimental buildings and provides performance parameters for advanced computer models to predict the efficacy of engineering designs. Recent developments of nanotechnology have also helped the efficient insertion of renewable technologies in the building industry [4,5,8,9,33].

Zero-energy buildings can be part of a smart grid. Some advantages of these buildings are:

- Integration of renewable energy resources
- Integration of *plug-in electric vehicles* called *vehicle-to-grid*
- Implementation of zero-energy concepts

Although the net-zero concept is applicable to a wide range of resources such as energy, water, and waste, energy is usually the first resource to be targeted because:

1. Energy, particularly electricity and heating fuel like natural gas or heating oil, is expensive. Hence reducing energy use can save the building owner money. In contrast, water and waste are inexpensive.
2. Energy, particularly electricity and heating fuel, has a high carbon footprint. Hence reducing energy use is a major way to reduce the buildings carbon footprint
3. There are well-established means to significantly reduce the energy use and carbon footprint of buildings. These include adding insulation, using heat pumps instead of furnaces, using low-e double or triple-glazed windows and adding solar panels to the roof.
4. There are government-sponsored subsidies and tax breaks for installing heat pumps, solar panels, triple-glazed windows and insulation that greatly reduce the cost of getting to a net-zero-energy building for the building owner. For instance, in the United States, there are federal tax credits for solar panels, state incentives (which vary by state) for solar panels, heat pumps, and highly insulating triple-glazed windows. Some states, such as Massachusetts, also offer zero-interest or low-interest loans to allow building owners to purchase heat pumps, solar panels and triple-glazed windows that they otherwise could not afford. The cost of getting an existing house to net-zero-energy has been reported as being 5%–10% of the value of the house. A 15% return on investment has been reported.

Despite sharing the name "zero net energy", there are several definitions of what the term means in practice, with a particular difference in usage between North America and Europe.

Zero net site energy use
In this type of ZNE, the amount of energy provided by on-site renewable energy sources is equal to the amount of energy used by the building.

Zero net source energy use

This ZNE generates the same amount of energy as is used, including the energy used to transport the energy to the building. This type accounts for energy losses during electricity generation and transmission. These ZNEs must generate more electricity than zero net site energy buildings.

Net-zero-energy emissions

Outside the United States and Canada, a ZEB is generally defined as one with ZNE emissions, also known as a ZCB or *zero-emissions building* Under this definition, the carbon emissions generated from on-site or off-site fossil fuel use are balanced by the amount of on-site renewable energy production. Other definitions include not only the carbon emissions generated by the building in use but also those generated in the construction of the building and the embodied energy of the structure.

Net-zero cost

In this type of building, the cost of purchasing energy is balanced by income from the sale of electricity to the grid of electricity generated on-site.

Net off-site zero energy use

A building may be considered a ZEB if 100% of the energy it purchases comes from renewable energy sources, even if the energy is generated off the site.

Off-the-grid

Off-the-grid buildings are stand-alone ZEBs that are not connected to an off-site energy utility facility. They require distributed renewable energy generation and energy storage capability.

Net-Zero-Energy Building

The overall conceptual understanding of net ZEB is an energy-efficient, grid-connected building enabled to generate energy from renewable sources to compensate for its own energy demand. The wording "net" emphasizes the energy exchange between the building and the energy infrastructure The net ZEB concept requires two actions: (a) reduce energy demand by means of energy efficiency measures and passive energy use; (b) generate energy from renewable sources.

Besides various definitions outlined above, the U.S. National Renewable Energy Laboratory (NREL) published a groundbreaking report titled Net-Zero Energy Buildings: A Classification System Based on Renewable Energy Supply Options. This classification system identifies the following four main categories of Net-Zero-Energy Buildings/Sites/Campuses [5]:

- NZEB: A — A footprint renewables Net-Zero-Energy Building
- NZEB: B — A site renewables Net-Zero-Energy Building
- NZEB: C — An imported renewables Net-Zero-Energy Building
- NZEB: D — An off-site purchased renewables Net-Zero-Energy Building

Applying this U.S. Government Net Zero classification system means that every building "can" become Net Zero with the right combination of the key Net Zero Technologies—PV (solar), GHP (geothermal heating and cooling, thermal batteries),

EE (energy efficiency), sometimes wind, and electric batteries. Many well-known universities have professed to want to completely convert their energy systems off of fossil fuels. An example of this is in the Net Zero Foundation's proposal at MIT to take that campus completely off fossil fuel use. This proposal shows the coming application of Net-Zero-Energy Buildings technologies at the District Energy scale.

1.4.7.1 Design and Construction

The most cost-effective steps towards a reduction in a building's energy consumption usually occur during the design process. To achieve efficient energy use, zero-energy design departs significantly from conventional construction practice. Successful zero-energy building designers typically combine time tested passive solar, or artificial/fake conditioning, principles that work with the on-site assets. Sunlight and solar heat, prevailing breezes, and the cool of the earth below a building can provide daylighting and stable indoor temperatures with minimum mechanical means. ZEBs are normally optimized to use passive solar heat gain and shading, combined with thermal mass to stabilize diurnal temperature variations throughout the day, and in most climates. All the technologies needed to create zero-energy buildings are available off-the-shelf today.

Sophisticated 3-D building energy simulation tools are available to model how a building will perform with a range of design variables such as building orientation, (relative to the daily and seasonal position of the sun), window and door type and placement, overhang depth, insulation type, and values of the building elements, airtightness (weatherization), the efficiency of heating, cooling, lighting and other equipment, as well as local climate. These simulations help the designers predict how the building will perform before it is built, and enable them to model the economic and financial implications on building cost-benefit analysis, or even more appropriate—life cycle assessment.

ZEBs are built with significant energy-saving features. The heating and cooling loads are lowered by using high-efficiency equipment (such as heat pumps rather than furnaces. Heat pumps are about four times as efficient as furnaces) added insulation (especially in the attic and in the basement of houses), high-efficiency windows (such as low-E triple-glazed windows), draft-proofing, high-efficiency appliances (particularly modern high-efficiency refrigerators), high-efficiency LED lighting, passive solar gain in winter and passive shading in the summer, natural ventilation, and other techniques. These features vary depending on climate zones in which the construction occurs. Water heating loads can be lowered by using water conservation fixtures, heat recovery units on wastewater, and by using solar water heating, and high-efficiency water heating equipment. In addition, daylighting with skylights or solar tubes can provide 100% of daytime illumination within the home. Nighttime illumination is typically done with fluorescent and LED lighting that use 1/3 or less power than incandescent lights, without adding unwanted heat. Miscellaneous electric loads can be lessened by choosing efficient appliances and minimizing phantom loads or standby power. Other techniques to reach net zero (dependent on climate) are Earth sheltered building principles, super insulation walls using straw-bale construction, Vitruvian built prefabricated building panels and roof elements plus exterior landscaping for seasonal shading.

Once the energy use of the building has been minimized it can be possible to generate all that energy on-site using roof-mounted solar panels. ZEBs are often designed to make dual use of energy including that from white goods. For example, using refrigerator exhaust to heat domestic water, ventilation air, and shower drain heat exchangers, office machines and computer servers, and body heat to heat the building. These buildings make use of heat energy that conventional buildings may exhaust outside. They may use heat recovery ventilation, hot water heat recycling, combined heat and power, and absorption chiller units [4,5,8,33].

1.4.7.2 Energy Harvest

ZEBs harvest available energy to meet their electricity and heating or cooling needs. By far the most common way to harvest energy is to use roof-mounted modular solar photovoltaic panels that turn the sun's light into electricity. Energy can also be harvested with modular solar thermal collectors (which use the sun's heat to heat water for the building). Modular heat pumps either ground source (otherwise known as geothermal) or air-source can also harvest heat and cool from the air or ground near the building. Technically, heat pumps move heat rather than harvest it but the overall effect in terms of reduced energy use and reduced carbon footprint is similar. In the case of individual houses, various microgeneration technologies may be used to provide heat and electricity to the building, using solar cells or wind turbines for electricity, and biofuels or solar thermal collectors linked to a modular seasonal thermal energy storage (STES) for space heating. An STES can also be used for summer cooling by storing the cold of winter underground. To cope with fluctuations in demand, ZEBs are frequently connected to the electricity grid, export electricity to the grid when there is a surplus, and drawing electricity when not enough electricity is being produced [7]. Other buildings may be fully autonomous.

Energy harvesting is most often more effective (in cost and resource utilization) when done on a local but combined scale, for example, a group of houses, cohousing, local districts, villages, and so on rather than on an individual basis. An energy benefit of such localized energy harvesting is the virtual elimination of electrical transmission and electricity distribution losses. On-site energy harvesting such as with rooftop-mounted solar panels eliminates these transmission losses entirely. Energy harvesting in commercial and industrial applications should benefit from the topography of each location. However, a site that is free of shade can generate large amounts of solar-powered electricity from the building's roof and almost any site can use geothermal or air-sourced heat pumps. The production of goods under net-zero fossil energy consumption requires locations of geothermal, micro hydro, solar, and wind resources to sustain the concept.

Zero-energy neighborhoods, such as the BedZED development in the United Kingdom, and those that are spreading rapidly in California and China, may use distributed generation schemes. This may in some cases include district heating, community chilled water, shared wind turbines, and so on. There are current plans to use ZEB technologies to build entire off-the-grid or net-zero-energy use cities. One of the key areas of debate in zero-energy building design is over the balance between energy conservation and the distributed point-of-use harvesting of renewable energy (solar energy, wind energy, and geothermal energy). Most zero-energy homes use a combination of these strategies. As a result of significant government subsidies for photovoltaic solar electric systems, wind turbines, and so on, some suggest that a ZEB is a conventional house with distributed modular renewable energy harvesting technologies. Entire additions of such homes have appeared in locations where PV subsidies are significant [20] but many so-called "Zero-Energy Homes" still have utility bills. This type of energy harvesting without added energy conservation may not be cost-effective with the current price of electricity generated with photovoltaic equipment (depending on the local price of power company electricity) [21]. The cost, energy, and carbon-footprint savings from conservation (e.g., added insulation, triple-glazed windows, and heat pumps) compared to those from on-site energy generation (e.g., solar panels) have been published in the literature for an upgrade to an existing house [4,5,8,33]. Since the 1980s, passive solar building design and passive house have demonstrated heating energy consumption reductions of 70%–90% in many locations, without active energy harvesting. The energy used in a building can vary greatly depending on the behavior of its occupants. An average widely accepted ratio of highest to lowest energy consumer in identical homes is about 3, with some identical homes using up to 20 times as much heating energy as the others.

Wide acceptance of zero-energy building technology may require more government incentives or building code regulations, the development of recognized standards, or significant increases in the cost of conventional energy. The Google photovoltaic campus and the Microsoft 480-kW photovoltaic campus relied on U.S. Federal, especially California, subsidies and financial incentives. California is now providing US$3.2 billion in subsidies for residential-and-commercial near-zero-energy buildings [4,5,8,33].

1.4.7.3 Advantages and Disadvantages of ZEB
ZEB carries certain advantages as well as disadvantages [4,5,8].

1.4.7.3.1 Advantages
- Isolation for building owners from future energy price increases
- Increased comfort due to more-uniform interior temperatures (this can be demonstrated with comparative isotherm maps)
- The reduced requirement for energy austerity
- The reduced total cost of ownership due to improved energy efficiency
- The reduced total net monthly cost of living
- Reduced risk of loss from grid blackouts
- Improved reliability—photovoltaic systems have 25-year warranties and seldom fail during weather problems—the 1982 photovoltaic systems on the Walt Disney World EPCOT (Experimental Prototype Community of Tomorrow) Energy Pavilion are still working fine today, after going through three recent hurricanes
- An extra cost is minimized for new construction compared to an afterthought retrofit
- Higher resale value as potential owners demand more ZEBs than available supply
- The value of a ZEB building relative to a similar conventional building should increase every time energy costs increase

- Future legislative restrictions and carbon emission taxes/penalties may force expensive retrofits to inefficient buildings
- Contribute to the greater benefits of the society, e.g. providing sustainable renewable energy to the grid, reducing the need for grid expansion

1.4.7.3.2 Disadvantages
- Initial costs can be higher—effort required to understand, apply and qualify for ZEB subsidies, if they exist.
- Very few designers or builders have the necessary skills or experience to build ZEBs [22]
- Possible declines in future utility company renewable energy costs may lessen the value of capital invested in energy efficiency
- New photovoltaic solar cells equipment technology price has been falling at roughly 17% per year—It will lessen the value of capital invested in a solar electric generating system—current subsidies will be phased out as photovoltaic mass production lowers future price
- Challenge to recover higher initial costs on the resale of building, but new energy rating systems are being introduced gradually [33].
- While the individual house may use an average of net zero energy over a year, it may demand energy at the time when peak demand for the grid occurs. In such a case, the capacity of the grid must still provide electricity to all loads. Therefore, a ZEB may not reduce the required power plant capacity.
- Without an optimized thermal envelope, the embodied energy, heating and cooling energy and resource usage are higher than needed. ZEB by definition does not mandate a minimum heating and cooling performance level, thus allowing oversized renewable energy systems to fill the energy gap.
- Solar energy capture using the house envelope only works in locations unobstructed from the sun. The solar energy capture cannot be optimized in the north (for northern hemisphere, or south for southern Hemisphere) facing shade, or wooded surroundings.

Because of the design challenges and sensitivity to a site that are required to efficiently meet the energy needs of a building and occupants with renewable energy (solar, wind, geothermal, etc.), designers must apply holistic design principles, and take advantage of the free naturally occurring assets available, such as passive solar orientation, natural ventilation, daylighting, thermal mass, and nighttime cooling. In recent years as shown in Figures 1.5 and 1.6, the ZEB concept has been implemented to office buildings and laboratories.

1.4.7.4 Role of Geothermal Energy in Net-Zero-Energy Buildings

Along with solar energy, geothermal energy systems are a promising alternative to conventional fossil fuel systems for homes. In the case of geothermal energy, the installed system can provide a residential or commercial space with efficient, cost-effective heating or cooling with low emissions [35]. The earth's resources are abundant and heat energy that is stored just below the earth's crust provides

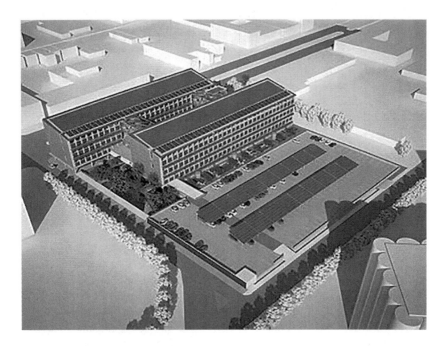

FIGURE 1.5 Net zero emission office building prototype in St. Louis, Missouri [4].

FIGURE 1.6 Zero-Energy Lab construction on UNT campus in Denton, Texas [4].

untapped potential for energy conversion with limited ground disturbance focusing on geothermal in indirect heating and cooling applications for residential settings. Indirect heating and cooling systems gather energy from low-temperature geothermal resources. Low-temperature resources are of interest because it is the most accessible resource around the world. In addition, it needs minimal area and depth to extract heat energy [34,36]. Geothermal energy can be used in a wide range of applications from electricity, to direct heating, to indirect heating, and cooling [36]. These systems operate by using the low temperature from the ground and raising it through a series of compressors and pumps to the desired temperature for heating or cooling uses. Geothermal energy is most useful for home HVAC system through geothermal heat pumps. Unlike wind energy (not often used for homes) which can generate power, geothermal energy can generate both power and heating/cooling for homes.

For indirect ground source heating and cooling systems, there are two main types of setups, namely, horizontal and vertical loop systems. Each of these approaches has drawbacks. The vertical loop system often ranges from 45 to 75 m deep for residential cases and exceeds 150 m deep for commercial applications [36]. This depth is not practical for easy installation or maintenance of the system, leading us to try to minimize this depth to efficiently be able to capture the heat and have the installation depth as shallow as possible. The other typical type of geothermal system arrangement is the horizontal loop. This setup can look differently depending on the space available in the location, but typically, these systems are shallow in the ground and very vast in surface area [36,37]. There is a potential to improve on this system as well, in the sense that the expansive surface area could be minimized, which would dispel space limitations of where these systems could be located. The new design [34–37] can encompass these spacing optimizations as well as potential improvements to maintain the systems and add additional adjacent systems to increase the energy production if needed. Currently, some geothermal heat pump systems consist of pipes placed between 300 and 900 ft below ground. These pipes contain a heat transfer fluid and are designed in either closed-loop or open-loop configurations. During winter, the fluid absorbs the underground heat, as the temperature profile under the surface of the Earth is a relatively consistent 50°F. The fluid that stores the absorbed energy is then compressed through a series of pumps using electricity to provide heat at a temperature suitable for space heating and a hot water source.

The heating process explained above can be reversed to provide cooling in summer. Heat is extracted from within the building and is transferred back to the earth through the working fluid within the pipes [34–37]. During the summer months, the residual heat energy can still be used to provide hot water for little to no additional cost. These features increase the ground source heat pump (GSHP) marketability as an "all-in-one" residential temperature control solution. Existing geothermal heat pump technology is marketed as a fossil fuel alternative for heating and cooling. As found through the Massachusetts Department of Energy Resources (Massachusetts DOER), Geothermal heat pumps are 3.5–5 times as efficient as the most efficient fossil fuel furnace. Instead of burning a combustible fuel to make heat, they simply transport heat that already exists. By doing so, they provide 3.5–5 units of energy for every unit used to power the heat-pump system [34–37]. The systems also have a large lifespan, with a secondary component lifespan of approximately 25 years and an underground component lifespan of more than 50 years [34–37].

1.4.7.5 Geothermal and Solar Net-Zero-Energy Building

White Oak Modular is one of the leading builders of energy-efficient homes in the United States. This company is proficient in the installations of solar electric panels to power home and geothermal heating and cooling systems. Combining both solar and geothermal systems allows to have free heating and cooling [38]. The solar panels generate all of the electricity needed to run the geothermal system. During sunny days, one generates more electricity than one uses, and this results in net surplus of energy. Geothermal heating and cooling systems use the 56° temperature of the earth to heat or cool your home, using a fraction of the energy used by a conventional system. While conventional furnaces and boilers burn fuel to generate heat, geothermal systems use electricity to simply move heat from the earth into your home during winter and transferring it out of your home and back into the ground during summer. The most efficient fuel-burning heater can reach efficiencies around 95% but a geothermal heating system can move up to four units of heat for every unit of electricity needed to power the system, resulting in a practical equivalence of over 400% efficiency [38].

1.5 ENERGY CONSERVATION AND EFFICIENCY IN TRANSPORTATION SECTOR

The transportation sector includes all vehicles used for personal or freight transportation. Of the energy used in this sector, approximately 65% is consumed by gasoline-powered vehicles, primarily personally owned. Diesel-powered transport (trains, merchant ships, heavy trucks, etc.) consumes about 20% and air traffic consumes most of the remaining 15%. In 1975, the two oil supply crises of the 1970s spurred the creation of the federal Corporate Average Fuel Economy (CAFE) program [2,39], which required auto manufacturers to meet progressively higher fleet fuel economy targets. In addition to the CAFE program, the U.S. government has tried to encourage better vehicle efficiency through tax policy. Since 2002, taxpayers have been eligible for income tax credits for gas/electric hybrid vehicles. A "gas-guzzler" tax has been assessed on manufacturers since 1978 for cars with exceptionally poor fuel economy. While this tax remains in effect, it generates very little revenue as overall fuel economy has improved. Another focus in gasoline conservation is reducing the number of miles driven. An estimated 40% of American automobile use is associated with daily commuting. Many urban areas offer subsidized public transportation to reduce commuting traffic and encourage carpooling by providing designated high-occupancy vehicle lanes and lower tolls for cars with multiple riders. A vehicle's gas mileage decreases rapidly with increasing highway speeds, normally above 55 mph (though the exact number varies by vehicle), because aerodynamic forces are proportionally related to the square of an object's speed (when the speed is doubled, drag quadruples). According to the U.S. Department of Energy (DOE), as a rule of thumb,

Residential and Transportation Sectors

each 5 mph (8.0 km/h) one drives over 60 mph (97 km/h) is similar to paying an additional $0.30 per gallon for gas. The exact speed at which a vehicle achieves its highest efficiency varies based on the vehicle's drag coefficient, frontal area, surrounding airspeed, and the efficiency and gearing of a vehicle's drive train and transmission [2,39].

Energy conservation and efficiency improvements in the vehicle industry have taken a number of different paths. Besides all the improvements mentioned above, significant other efforts are made to improve energy efficiency in automobile. The waste energy emitted from a car contributes to more than 60% of the total energy consumed by fuel. Efforts are being made to use this waste energy in a variety of ways such as for energy required for heating and cooling of car cabin and engine and the energy required for various technologies used within the car. Efforts are also made to use solar energy (with highly efficient solar panels) to generate power for automobiles. Major changes made to reduce the use of fossil fuel in the car industry are in the basic use of the internal combustion engine. New hybrid cars, FCV (fuel cell-driven car), or electric cars are becoming more popular to make the car industry greener and less oil dependent. Hybrid cars are extremely fuel efficient. FCV can use hydrogen as fuel which makes car operation carbon-free. Finally, electric cars are completely emission-free and most environment-friendly [2]. The use of modular fuel cell in the car industry is described in detail in Chapter 4. Modular approach for electric vehicles is also described in detail in Chapter 4. Here, we briefly examine the efficiency aspects related to electric vehicles.

1.6 ENERGY EFFICIENCY OF MODULAR ELECTRIC AND HYBRID VEHICLES

As mentioned in the previous book, all vehicles are modular in nature. Because of that, all energy systems that are embedded in the vehicles tend to be modular in nature as well. Due to environmental concerns, automobile industry is going through significant changes. Internal combustion engine-based automobiles which are generally operated by oil, are slowly replaced by more cleaner hydrogen-based fuel cell cars, hybrid cars, or all-electric cars (see Figure 1.7).

It appears that hydrogen (based fuel cell) and electricity are the future of the automobile industry. In this section, we examine the efficiency issues related to electric and hybrid cars. The measures described in this section, if implemented, will become standardized and modularized for all-electric and hybrid vehicles. The most important possibilities for improving energy conservation and efficiency of electric and hybrid vehicles include [41,42]:

1. Using energy under braking
2. Using waste heat energy
3. Additional power supply by solar cells
4. Improved mechanical energy transmission system
5. Improved cars shell design
6. Increasing of efficiency of power convertors
7. Special design of electric engines

8. Using supercapacitors, fuel cells and new-generation batteries which are modular in nature
9. Route selection on the criterion of minimum consumption in real time
10. Parameter monitoring inside and outside of the vehicle and computerized system control with optimization of energy consumption

Some of the measures mentioned above are human-initiated such as route selection, while others are purely equipment or process-oriented such as using waste heat energy, improved mechanical energy transmission system, and so on. Some also involve both human initiatives and process design. Today, the problem of energy becomes so important that an entire industry is turning towards clean and renewable energy (solar energy, wind energy, etc.). Prototypes of hybrid vehicles with the announcement of mass production scheduled for the near future have become an everyday occurrence. In addition, many cars are designed to use only electricity as motive power, which reduces emissions to zero level [41,42].

Modular photocells in a glass roof generate electricity, even at a lower intensity of solar radiation can operate a fan in a vehicle. In this way, the vehicle interior has a constant supply of fresh air and pleasant temperatures (up to 50% lower), even when the motor vehicle is turned off so that fuel economy is evident. The solar roof is only the beginning; city cars are going towards the development of full solar energy-based vehicles prototype. A modular solar vehicle is an electric vehicle powered completely or significantly by direct solar energy. Usually, photovoltaic (PV) cells contained in solar panels convert the sun's energy directly into electric energy. The term "solar vehicle" usually implies that solar energy is used to power all or part of a vehicle's propulsion. Solar power may be also used to provide power for communications or controls or other auxiliary functions.

FIGURE 1.7 Front view, from the driver side of Tesla electric car [40].

Another concept that has been developing over the years is a kinetic energy recovery system, often known simply as KERS. KERS is an automotive system for recovering a moving vehicle's kinetic energy under braking. The recovered energy is stored in a reservoir (for example a flywheel or a battery or super-capacitor which are all modular in nature) for later use under acceleration. Electrical systems use a motor-generator incorporated in the car's transmission which converts mechanical energy into electrical energy and vice versa. Once the energy has been harnessed, it is stored in a battery and released when required. The mechanical KERS system utilizes flywheel technology to recover and store a moving vehicle's kinetic energy which is otherwise wasted when the vehicle is decelerated. Compared to the alternative of electrical-battery systems, the mechanical KERS system provides a significantly more compact, efficient, lighter, and environmental-friendly solution. There is one other option available, i.e. hydraulic KERS, where braking energy is used to accumulate hydraulic pressure which is then sent to the wheels when required.

Development of new components, improved connections, and electric engine control algorithms allow an increase of efficiency of power convertors, therefore, the electric engine itself, to the maximum theoretical limits. New-generation improvements of electric engine system increases the initial price of the car; however, investment quickly pays off during the operation phase. Major efforts are invested in the development of high-energy-density batteries with minimum ESR. Moreover, current research shows that fuel cells have reached required performances for commercial use in electric vehicles. Supercapacitors that provide high power density increases the acceleration of vehicles as well as collects all the energy from instant braking; therefore, improvements in the characteristics of power supply are made. Modern electric vehicles have a full information system that has constant modifications and does monitoring of inside and outside parameters to achieve maximum energy savings. Except for smart sensors, it is highly important to process GPS signals and route selection on the criterion of minimum energy consumption. By combining these technologies, concepts and their improvements, we are slowly going towards energy-efficient vehicles, which will not only simplify our lives to a great extent but save energy as well. In considering energy conservation and efficiency of electric and hybrid cars, three parameters are important; electric loss reduction, mechanical loss reduction, and alternate methods of adding energy within the vehicle systems. These parameters are briefly described here [41,42].

1.6.1 Electrical Losses Reduction in EV

1.6.1.1 Energy Efficiency of the Converters

The energy efficiency of the converters can be increased by optimizing their configuration and control, as well as, by choosing the adequate components. Converter configuration depends on the type of the electric motor (DC or AC), possible braking energy recovery, drive dynamics and so on. For DC motor supply, chopper voltage reducers are mostly used. The converter part of the AC drive of the vehicle consists of the inverters, regulators, and control set. The inverter is part of the drive inverter that inverts DC voltage to AC voltage, necessary waveform to ensure the required control of electric motors. The three-phase inverter consists of three inverter bridges with two switching elements in each bridge, therefore, a total of six switches. By controlling the moments of switching off the particular switches, and by controlling the length of their involvement, the appropriate waveforms at the output of inverter are achieved.

There may be a large energy saving by selecting the suitable power switching elements, whose development is in high demand. Also, switch elements in the inverters and choppers, high-power bipolar transistors, MOS (Metal Oxide Semiconductor) transistors, or IGBTs (Insulated Gate Bipolar Transistor) are used. High-power bipolar transistors have very low collector-emitter resistance in the conducting state, while their control must provide sufficient supply base it requires for a relatively high power for control. On the other hand, MOS transistors have very high input resistance, but to control them only the appropriate value of the voltage between the gate and source is required. Therefore, the MOS transistor control current is almost zero and there is no power dissipation in the control circuit. Lack of MOS transistors requires relatively high resistance in ON state. IGBT belongs to the family BiMOS transistors and combines these fine qualities of high-power bipolar and MOS transistors [43].

In recent years, multi-axis distributed control systems are developed where sensors, actuators, and controllers are distributed across networks. System features, system synchronized control, and high-speed serial communications using fiber optic channels for noise immunity. In addition, communication protocols have been developed that monitor data integrity and can sustain operation in the event of a temporary loss of communication channel. Engineers can design a system to meet exact customer requirements [43]. In this way, the optimization of the drive can be achieved by the criteria of the dynamics and energy efficiency, while following the user's request. For supply of certain components, particularly in hybrid vehicles, high power supplies of constant current or current impulses are needed. Precise management and optimization of such sources are exclusively microprocessor-controlled [44].

1.6.1.2 Energy Efficiency of the Electric Motor

The electric motor is the most important part of the electrical drive and the last link in the chain of energy conversion. DC motors, because of their good qualities, control of the rotation speed and torque, have been irreplaceable part of the controlled electric motor drives. In recent years, due to the advanced control techniques, asynchronous motors have replaced DC motors in regulated drives because of their good properties. Electric motor drive is designed and optimized with the help of the known parameters of the engine. The latest methods for minimizing the power loss in real time by reducing the level of flux does not require the knowledge of all engine parameters and can be applied to asynchronous motor drives with scalar and vector control. Optimization of efficiency of asynchronous motors is based on adaptive adjustment of flux levels to determine the optimum operating point by minimizing losses [43].

For EV, switched reluctance motor (SRM) is gaining much interest as a candidate for electric vehicle (EV) and hybrid electric vehicle, (HEV) electric propulsion due to its simple and rugged construction, ability of extremely high-speed operation, and insensitivity to high temperatures. However, due to SRM construction with doubly salient poles and its nonlinear magnetic characteristics, the problems of acoustic noise and torque ripple are more severe than those for other traditional motors. Power electronics technology has made the SRM an attractive choice for many applications. The SRM is a doubly salient, singly excited synchronous motor. The rotor and stator are comprised of stacked iron laminations with copper windings on the stator. The motor is excited with a power electronic inverter that energizes appropriate phases based on shaft position. The excitation of a phase creates a magnetic field that attracts the nearest rotor pole to the excited stator pole in an attempt to minimize the reluctance path through the rotor. The excitation is performed in a sequence that steps the rotor around.

1.6.1.3 Super Capacitors as Energy Storage for Increasing Energy Efficiency

Super-capacitors are a relatively new type of capacitors distinguished by the phenomenon of electrochemical double-layer, diffusion, and large effective area, which leads to extremely large capacitance per unit of geometrical area (in order of multiple times compared to conventional capacitors). They are placed in the area in-between lead batteries and conventional capacitors. In terms of specific energy (accumulated energy per mass unity or volume) and in terms of specific power (power per mass unity or volume) they take place in the area that covers the order of several magnitudes. Super-capacitors fulfill a very wide area between accumulator batteries and conventional capacitors taking into account specific energy and specific power [41,42]. Batteries and fuel cells are typical devices of small specific power, while conventional capacitors can have specific power higher than 1 MW/dm^3 but at a very low specific energy. Electrochemical capacitors improve batteries' characteristics considering specific power or improve capacitors' characteristics considering specific energy in combination with them [45–47]. In relation to other capacitor types, super-capacitors offer much higher capacitance and specific energies [45–47].

The principal supercapacitor characteristic that makes it suitable for using in energy storage systems (ESS), is the possibility of fast charge and discharge without loss of efficiency, for thousands of cycles. This is because they store electrical energy directly. Supercapacitors can recharge in a very short time having a great facility to supply high and frequent power demand peaks [42]. Most strict requirements are related to supercapacitors applying in electric haulage, i.e. for vehicles of the future. Nowadays, batteries of several hundred-farad capacitance working voltages of several hundred volts have been produced. Beside great capacitance and relatively high working voltage, these capacitors must have great specific energy and power (because of limited space in vehicle). Considering their specific power, they have a great advantage in relation to accumulator batteries, but, on the other side, they are incomparably weaker considering specific energy. Hence, the ideal combination is a parallel connection of accumulator and condenser batteries. In an established regime (normal drawing) vehicle engine is supplied from Accu-battery, and in the case of rapidly speeding, from the supercapacitor. In the case of abrupt braking, complete mechanical energy could be taken back to system by converting into electrical energy only in presence of supercapacitor with great specific power [41,42].

1.6.1.4 Super Capacitors in Regulated Electrical Drives

Regulated electrical drives are more than 30% of all electric drives. They are developing quickly and present to constructors stricter and stricter speed regulation (and position) and torque. From energy point of view, it is desirable to have their more participation, since optimal speed setting or required torque can lead to reduction of used energy [41,42,45–49]. DC source voltage is performed by means of a DC-DC converter (chopper). Nowadays, there are a great number of standard batteries that can be used for EV, however, every single type has disadvantages that affect the performance of the vehicle. Therefore, compromises are often made between cost and quality, at the expense of energy efficiency almost all the time. Batteries in combination with supercapacitors are significant improvement and for now, this is the system that has the best perspective for future EV. When the cost per Wh and Specific Energy (Wh per kg) for various types of batteries is presented, it is not surprising that the well-known lead-acid storage batteries head the list. Alkaline cells may be recharged literally dozens of times using the new technology. Recharging alkaline, nickel-cadmium, and nickel-metal hydride cells side-by-side in one automatic charger opens up new possibilities for battery selection economy [48]. Costs of lithium-ion batteries are falling rapidly in the race to develop new electric vehicles. The $0.47 price per Wh above is for the Nissan Leaf automobile, and they predict a target cost of $0.37 per Wh. Tesla Automobiles uses a smaller battery pack, and they are optimistic about reaching a price of $0.20 per Wh soon [48]. Lithium thionyl chloride is limited in the relative amount of current it can deliver. However, it has even higher energy storage per kg, and its temperature range is extreme, from $-55°C$ to $+150°C$. It is used in extremely hazardous or critical applications. The specifications for Lithium thionyl chloride are $1.16 per Wh, 700 Wh/kg [48].

Several parameters can be considered for selecting the more adequate battery typology: specific energy, specific power, cost, life, reliability, and so on. In addition, it is to be considered that batteries for hybrid electric vehicles require higher powers and lower energies than batteries for pure electric vehicles. Among the previously listed typologies, lead-acid and nickel-cadmium and sodium-nickel chloride batteries are normally used onboard electric vehicles due to their low specific powers [49].

1.6.1.5 Fuel Cells

While several types of fuel cells are available in the market, for vehicle propulsion, polymer electrolyte fuel cell (PEFC) systems, fed by air and pure hydrogen stored aboard, seem to be highly preferable over other types, mainly because their reduced operating temperature (65°–80° depending on the cell design), very fast start-up times and eases the thermal management. A polymer electrolyte fuel cell is an electrochemical device that converts chemical energy directly into electrical energy without the need of intermediate thermal cycles. It normally consumes H_2 and O (typically from the air) as reactants and produces water, electricity, and heat. Since cell voltage is so low (less than 1 V), several cells are normally connected in series in a modular form to build a fuel cell stack with a voltage and power suitable for practical applications.

A fuel cell electric vehicle (FCEV) has higher efficiency and lower emissions compared with the internal combustion engine vehicles. But, the fuel cell has a slow dynamic response. Therefore, a secondary power source is needed during startup and transient conditions. Ultracapacitor can be used as a secondary power source. By using ultracapacitor as the secondary power source of the FCEV, the performance and efficiency of the overall system can be improved. In this system, there is a boost converter, which steps up the fuel cell voltage, and a bidirectional DC-DC converter, that couples the ultracapacitor to the DC bus [49,50].

1.6.1.6 New Systems

The priority in the EV future development and its commercial success is the optimization of its electric power supply. Besides the usual combinations (batteries and supercapacitors, and supercapacitors), researches are going towards new systems that integrate favorable characteristics of the previously used systems. Typically, standard ultracapacitors can store only about 5% as much energy as lithium-ion batteries. New hybrid system can store about twice as much as standard ultracapacitors, although this is still much less than standard lithium-ion batteries. However, the advantage of ultracapacitors is that they can capture and release energy in seconds, providing a much faster recharge time compared with lithium-ion batteries. In addition, traditional lithium-ion batteries can be recharged only a few hundred times, which is much less than the 20,000 cycles provided by the hybrid system. In other words, the hybrid lithium-ion ultracapacitors have more power than lithium-ion batteries but less energy storage. In the future, the hybrid lithium-ion ultracapacitor could also be used for regenerative braking in vehicles, especially if it could be scaled up to provide greater energy storage. Since vehicle braking systems need to be recharged hundreds of thousands of times, the hybrid system's cycle life will also need to be improved [41,42].

Using new processes central to nanotechnology, researchers create millions of identical nanostructures with shapes tailored to transport energy as electrons rapidly to and from very large surface areas where they are stored. Materials behave according to physical laws of nature. The Maryland researchers exploit unusual combinations of these behaviors (called self-assembly, self-limiting reaction, and self-alignment) to construct millions—and ultimately billions—of tiny, virtually identical nanostructures to receive, store, and deliver electrical energy [41,42].

1.6.1.7 Reduction of Losses in the Conductors and Connectors

From the viewpoint of energy efficiency, choice of supply voltage, as well as quality contacts in the connectors and cable section is very important. The designer is limited by other factors such as the security problem (for battery overvoltage), limited space, and cost. Therefore, it is necessary to optimize the supply voltage and the conductor section with given constraints. It is similar to the choice of connectors. Hybrid and electric vehicles have a high-voltage battery pack that consists of individual modules and cells organized in series and parallel. A cell is the smallest, packaged form a battery can take and is generally on the order of 1–6 V. A module consists of several cells generally connected in either series or parallel. A battery pack is then assembled by connecting modules together, again either in series or parallel [41,42]. The pack operates at a nominal 375 V, stores about 56 kWh of electric energy and delivers up to 200 kW of electric power. These power and energy capabilities of the pack make it essential that safety be considered a primary criterion in the pack's design and architecture [41,42]. Recent battery fires in electric vehicles have prompted automakers to recommend discharging lithium-ion batteries following serious crashes. However, completely discharging a vehicle's battery to ensure safety will permanently damage the battery and render it worthless. Self-discharge effects and the parasitic load of battery management system electronics can also irreversibly drain a battery.

Zero-volt technology [51] relies on manipulating individual electrode potentials within a lithium-ion cell to allow deep discharge without inflicting damage to the cell. Quallion [51] has identified three key potentials affecting the zero-volt performance of lithium-ion batteries. First, the zero–crossing potential (ZCP) is the potential of the negative electrode when the battery voltage is zero. Second, the substrate dissolution potential (SDP) is the potential at which the negative electrode substrate begins to corrode. Finally, the film dissolution potential (FDP) is the potential at which the SEI begins to decompose. The crucial design parameter is to configure the negative electrode potential to reach the ZCP before reaching either the SDP or the FDP at the end of discharge. This design prevents damage to the negative electrode which would result in permanent capacity loss.

Connector contacts are very important, both in terms of energy efficiency (when it comes to high power), and in terms of reliability and security. In recent years, the copper alloy with silver and/or gold is used but other combinations of metals are to be explored [41,42]. So, the compromise between good electrical and mechanical properties, on the one hand, and reasonable prices, on the other, is required. More recently Cu-Ag alloys with an annealing temperature range of 140°C–400°C and hardness being increased with the degree of performance [45,46] are being explored.

1.6.1.8 Lighting and Heating of EV

With the rapid development of high-intensive LED technology, it enables large savings in energy consumption. That fact is crucial for EV. LED and power consumption of exterior vehicle lighting indicated that an all-LED system employing the current generation of LEDs would result in general power savings of about 50% (nighttime) to about 75% (daytime) over a traditional system. This means that while the long-term fuel cost savings (money) were higher for the gasoline-powered vehicle, long-term distance savings (range) favored the electric vehicle. Now, automotive lighting producer Osram comes to strengthen the idea mentioned above, stating that "microhybrids" or mild hybrids, which feature engine stop/start mechanisms to boost the efficiency of conventional vehicles, will benefit greatly from LED lighting by reducing power draw and battery drain, as well as increasing light output during low power mode and startups [41,42]. Today's roads have very little actual technology incorporated into their design and function. There are many types of technologies that could be incorporated in actual roads. Since EVs are becoming increasingly popular, while their batteries are still too weak to assure an anxiety-free drive on the highway, the induction charging (wireless) will begin to be incorporated into one of the lanes, so that these all-electric cars will be able to drive on the highway without using their onboard batteries at all, as they will get their power straight from underneath the road surface. The idea of inductive charging is simple, and various companies and universities are testing the system now, in view of future mass implementation [41,42].

Electric vehicles generate very little waste heat and resistance electric heat may have to be used to heat the interior of the vehicle if the heat generated from battery charging/discharging cannot be used to heat the interior. While heating can be simply provided with an electric resistance heater, higher efficiency and integral cooling can be obtained with a reversible heat pump (this is currently implemented in the hybrid Toyota Prius). Positive temperature coefficient (PTC) junction cooling [52] is also attractive for its simplicity—this kind of system is used for example in the Tesla Roadster. Some electric cars, for example, the Citroën Berlingo Electrique, use an auxiliary heating system (for example gasoline-fueled units manufactured by Webasto or Eberspächer) but sacrifice "green" and "zero emissions" credentials. Cabin cooling can be augmented with solar power, most simply and effectively by inducting outside air to avoid extreme heat buildup when the vehicle is closed and parked in the sunlight (such cooling mechanisms are available as aftermarket kits for conventional vehicles). Two models of the 2010 Toyota Prius include this feature as an option [41,42].

1.6.2 Mechanical Loss Reduction in EV

1.6.2.1 Tires Role in EV

Large impact on the fuel consumption of the cars, in general, has tires on its wheels. A wide range of research and development is carried out for tire optimization with respect to energy efficiency criteria, acceptable stability, comfort, and durability. The tires are a race car's biggest single performance variable. The racing tire itself is constructed from very soft rubber compounds which offer the best possible grip against the texture of the racetrack but wear very quickly in the process. All racing tires work best at relatively high temperatures. EV can benefit from the years of research and usage of this kind of tires. The Goodyear Efficient Grip prototype tire for electric vehicles delivers a range of benefits, including top-rated energy efficiency and excellent noise and wet braking performance levels in combination with Goodyear's latest generation of Run On Flat Technology for continued mobility after a puncture or complete loss of tire pressure [53]. The design of this tire is uniquely suited to complement the performance requirements of electric vehicles. The tire's narrow shape in combination with its large diameter leads to reduced rolling resistance levels and to a reduced aerodynamic drag and thus reduced energy consumption.

Rolling resistance is mainly caused by energy loss due to the deformation of the tire during driving. Less deformation means less energy loss and hence, less rolling resistance. Goodyear research indicates that the larger rim diameter reduces the overall amount of rubber needed, which leads to less rubber deformation during driving. The large tire diameter requires fewer tire rotations for a certain distance, which in turn results in less heat buildup and tire deformation, which again leads to lower rolling resistance levels and less energy consumption [53]. Electric engines often provide a relatively constant torque, even at very low speeds, which increases the acceleration performance of an electric vehicle in comparison to a vehicle with a similar internal combustion engine. This requires the development of a modified tread design in combination with a new tread compound to ensure excellent grip especially on dry, and to provide high levels of mileage [53]. This Efficient Grip concept tire provides extremely low rolling resistance and noise levels in combination with a very high level of wet performance. Fitted on a standard car this tire would give 30% less rolling resistance which leads to about 6% less fuel consumption compared to an average standard summer tire [53].

The effect of tire pressure on either fuel consumption with regular cars or EV consumption should be emphasized. To this end, researches were done in the United States in the last few years. For the control test, the pressure was set at the factory recommended 33 psi in each tire. The subsequent test was done with the pressure set at 45 psi. For each test, the vehicle has driven a total of 550 miles over the course of 1 week traveling back and forth between the same two cities via the same route. The fuel tank was filled twice per week. Measurement of the quantity of fuel used was taken from the readout on a gas pump at each fill-up. The number of miles traveled was taken from the vehicle's trip odometer. Results showed that during the control period, the number of miles traveled per gallon of gasoline consumed was 27. With the tire pressure at 45 psi, the vehicle traveled 30 miles per gallon of gasoline consumed; a difference of 11% [54].

1.6.2.2 Vehicle Body

The convergence of more and more electronics with controls and mechanics makes the design process for EV very complex [55]. Besides all electrical issues described here, the energy efficiency of EV also depends on the vehicle body. In recent years simulation programs allow the optimization of vehicle body shapes from the standpoint of energy efficiency. On the other hand, simulations and experiments in the wind tunnel achieve significant energy savings by introducing an air turbine, which provides airflow into electricity. These simulation programs of complex systems are simplified with the use of modular simulations. Drag reduction system used in race cars can also be useful for EV [56]. The materials used in these systems require great precision and generally made of titanium, aluminum, or stainless steel. The design of diffuser is also important for energy savings.

1.6.3 Additional Sources of Energy in EV and Hybrid Vehicles

1.6.3.1 Solar Cells

Today, the world recognizes the synergy between solar panels and electric cars. As a matter a fact there are several car companies that plan to install solar panels in their newer hybrid vehicles. Solar panels at the moment don't have that much of an impact on a hybrid and electric car's efficiency. Solar panels are also made out of silicon, which is too expensive for automakers to use as a viable source [57]. However, there are companies such as Toyota, one of the pioneers in this field, which uses the solar roof panel. Nowadays, the roof panel will power at least part of the hybrid Toyota Prius' air-conditioning unit. Smaller, less power-hungry systems seem to work better with solar power [57]. The most common type of solar panel uses single- or multi-crystalline silicon wafers. Creating the silicon crystal is by far the most energy-intensive part of the process, followed by various and sundry manufacturing steps, such as cutting the silicon into wafers, turning the wafers into cells, and assembling the cells into modules [58].

Today's electric vehicles consume about 150 Wh/km. If the average distance per day is 50 km, then it would be 18,250 km/year. For this calculated consumption, an electric vehicle would need to generate 2.75 MWh/year. By this math, monocrystalline solar panels generate about 263 kWh/m^2 per year in the United States. Therefore, about 10.5 m^2 of solar panels are required to completely offset the energy consumed by today's electric vehicles [58]. The only practical place to put panels on the Roadster is the roof (about 1 m^2). Ideally, this would then generate 263 kWh/year. However, the Roadster won't always be in the sun, and it won't be at its ideal angle. A 60% d-rating would be generous to account for shade and suboptimal angles, so the panel would generate about 150 kWh/year—driving the car an additional 3 km/day [58]. However, there is a possibility to put solar cells on the other part of the vehicle's surface. The surface from the vehicle's nose, across the hoods, and all the way to the roof. Also, technology development will without a doubt make progress in increasing solar energy efficiency. New research in thermoelectric-photovoltaic hybrid energy systems is also proposed and implemented for automobiles. The key is to newly develop the power conditioning circuit using maximum power point tracking so that the output power of the proposed hybrid energy system can be maximized. This experimental concept can be easily implemented in electric vehicles [59]. The use of a thermoelectric system for waste heat energy recovery is further discussed in Section 1.6.3.3.

1.6.3.2 Improved Kinetic Energy Recovery Systems

A kinetic energy recovery system (KERS) is an automotive system for recovering a moving vehicle's kinetic energy under braking. As mentioned earlier, the recovered energy is stored in a reservoir (flywheel or a battery or/and supercapacitor) for later use under acceleration. The device recovers the kinetic energy that is present in the waste heat created by the car's braking process. It stores that energy and converts it into power that can be called upon to boost acceleration [35]. The concept of transferring the vehicle's kinetic energy using flywheel energy storage was postulated by physicist Richard Feynman in the 1950s. It is exemplified in complex high-end systems such as the Zytek, Flybrid, Torotrak, and Xtrac used in F1 and simple, easily manufactured, and integrated differential based systems such as the Cambridge Passenger/Commercial Vehicle Kinetic Energy Recovery System (CPC-KERS) [35]. Xtrac and Flybrid are both licensees of Torotrak's technologies, which employ a small and sophisticated ancillary gearbox incorporating a continuously variable transmission (CVT). The CPC-KERS is similar as it also forms part of the driveline assembly. However, the whole mechanism including the flywheel sits entirely in the vehicle's hub (looking like a drum brake). In the CPC-KERS, a differential replaces the CVT and transfers torque between the flywheel, drive wheel, and road wheel [41,42]. KERS technology is based on a completely new design capable of accumulating power and keeping it in store for the right moment. As mentioned earlier, there are three types of systems available for KERS: battery (electrical), flywheel (mechanical), and hydraulic pressure based.

1.6.3.3 Additional Waste Heat Energy Recovery

In recent years, there has been active research on exhaust gas waste heat energy recovery for automobiles. According to the recent studies, General Motors is using shape memory alloys that require little as a 10°C temperature difference to convert low-grade waste heat into mechanical energy. When a stretched wire made of shape memory alloy is heated, it shrinks back to its prestretched length. When the wire cools back down, it becomes more pliable and can revert to its original stretched shape. This expansion and contraction can be used directly as mechanical energy output or used to drive an electric generator. Shape memory alloy heat engines have been around for decades, but the few devices that engineers have built were too complex, required fluid baths, and had insufficient cycle life for practical use. Around 60% of all energy in the United States is lost as waste heat; 90% of this waste heat is at temperatures less than 200°C and termed low grade because of the inability of most heat-recovery technologies to operate effectively in this range. The capture of low-grade waste heat, which turns excess thermal energy into useable energy, has the potential to provide consumers with enormous energy savings [60]. For practical use, parts of the automotive industry nowadays are working to create a prototype that is practical for commercial applications and capable of operating with either air or fluid-based heat sources.

Thermal energy recovery systems for better fuel efficiency proposes solutions for fuel economy and lower CO_2-emissions on combustion engines by making use of their exhaust waste heat. This fuel economy is accessible for engines running on gasoline, diesel, biofuels, hydrogen, or any other type of fuel. This solution proposes high power density for mobile applications and rugged solutions for power generation and marine applications, also being recognized by the motorsport world as an important technology for the future in racing and finally a technology that will contribute to the development of electric vehicle [61]. Plug-in hybrid electric vehicles are already noted for their environmental advantages and fuel savings—but now a new breakthrough technology could mean their fuel economy is boosted by a further 7% [62]. Most vehicle waste heat recovery systems that are currently being developed utilize a thermoelectric converter to create electricity. An effective waste recovery system requires three elements:

1. a thermoelectric material package
2. an electric power management system, which directs the electricity injected into the vehicle's electrical system to the place where it will do the most good at any given time
3. a thermal management system, which is essentially a sophisticated heat exchanger [63]

Some other systems in hybrid electric vehicles reduce fuel consumption by replacing a significant portion of the required electric power normally produced by the alternator with electric power produced from exhaust gas waste heat conversion to electricity in a thermo-electric generator module [64].

1.6.3.4 Reduction of Energy Loss from Adverse Airflow

A vehicle body can be designed to reduce down force and otherwise adverse airflow. During forward motion of an electrically powered vehicle, the air is captured at the front of the vehicle and channeled to one or more turbines. The air from the turbines is discharged at low-pressure regions on the sides and/or rear of the vehicle. The motive power of the air rotates the turbines which are rotatable engaged with a generator to produce electrical energy that is used to recharge batteries that power the vehicle. The generator is rotatable, engaged with a flywheel for storing mechanical energy while the vehicle is in forward motion. When the vehicle slows or stops, the flywheel releases its stored energy to the generators, thereby enabling the generator to continue recharging the batteries. The flywheel enables the generators to provide a more stable and continuous current flow for recharging the batteries [65].

1.6.3.5 Hybrid Electric Vehicle

Generally, hybrid vehicles could be described as vehicles using a combination of technologies for energy production and storage. Two types of vehicles are in consideration—so-called parallel and linear hybrids. Parallel type possesses a mechanical connection between the power generators and drives wheels, while in linear one such connection does not exist. Serial hybrids have significant advantages in relation to parallel ones because of their mechanical simplicity, design flexibility, and the possibility for simple incorporation of new technologies [41,42]. Hybrid electric vehicles (HEVs) combine the internal combustion engine of a conventional vehicle with the high-voltage battery and electric motor of an electric vehicle. As a result, HEVs can achieve twice the fuel economy of conventional vehicles. In combination, these attributes offer consumers the extended range and rapid refueling they expect from a conventional vehicle, as well as much of the energy and environmental benefits of an electric vehicle. HEVs are inherently flexible, so they can be used in a wide range of applications—from personal transportation to commercial hauling. Hybrid electric vehicles have several advantages over conventional vehicles:

- Greater operating efficiency because HEVs use regenerative braking which helps to minimize energy loss and recover the energy used to slow down or stop a vehicle;
- Lighter engines because HEV engines can be sized to accommodate average load, not peak load, which reduces the engine's weight;
- Greater fuel efficiency because hybrids consume significantly less fuel than vehicles powered by gasoline alone;
- Cleaner operation because HEVs can run on alternative fuels (which have lower emissions), thereby decreasing our dependency on fossil fuels (which helps ensure our national security); and
- Lighter vehicle weight overall because special lightweight materials are used in their manufacture.

Hybrid electric vehicles are becoming cost-competitive with similar conventional vehicles and most of the cost premium can be offset by overall fuel savings and tax incentives. Some states even offer incentives to consumers buying HEVs [66].

1.6.3.5.1 Optimized Modular Design for Hybrid Vehicle Energy Efficiency

The modular design has made an important contribution to the industrial revolution, increase of quality of products and goods, and economic development. It has produced an important evolution in design (technical modularity), in the organization of production and of companies. It allowed going beyond vertical integration, by fostering vertical specialization in both manufacturing and innovation. Several authors have, however, indicated that the enthusiasm for modularity has gone too far. The study by Trancossi and Pascoa [67] agrees with some arguments in cautioning against errors that can be produced by pervasive modularity. It aims to present a possible way for coupling modular design with energy optimization in the case of an electric vehicle. The initial inspiration can be of this case study is Bejan's preliminary modular definition of constructal optimization, which can fit perfectly with an industrial modular design. Even if this modular optimization does not have the ambition of defining the best possible solution to a complex design problem, such as multidisciplinary design optimization has, it allows defining configuration that can simply evolve over time by means of a step-by-step optimization of the critical components that influence the behavior of a complex industrial system. It reveals then to be applicable to the concept of a vehicle platform that is widely in use today. The specific test case is the design of an electric city vehicle, which has been optimized by a step applying this modular optimization approach.

1.7 EFFICIENCY IMPROVEMENTS OF MODULAR FUEL CELL

While efficiency of fuel cell is being constantly improved, in this section we briefly describe three examples of efficiency improvements of modular fuel cell. It should be noted that there are more examples of this type are constantly reported.

1.7.1 Design and Development of Modular Fuel Cell Stacks for Various Applications

Polymer electrolyte membrane fuel cell (PEMFC) stacks are conventionally made by assembling a large number of cells together connected electrically in series. The number of cells and the area of the electrodes determine the capacity of such a stack. One of the reasons for nonproliferation of the fuel cell systems in many applications is the high cost involved in making these units. Cost reduction of the fuel cell components, fuel cell stacks, and fuel cell systems are being extensively pursued. One of the options for cost reduction is through fabricating the fuel cells in smaller-scale industrial units as their overheads are low. To attract small scale entrepreneurs (SSE) the capital investment has to be minimum. If compact fuel cell stacks of size ranging from 200 to 300 W can be produced comfortably, the SSE's can be encouraged to set up manufacturing units and they can become original equipment manufacturer

(OEM) supplier to system integrators who could develop fuel cell systems of various capacities by simply connecting these compact stacks electrically in series or parallel depending on the end use [68].

Such modular architecture would reduce the cost and improve the reliability of manufacture and increase the range of applications. This modular construction not only allows the materials and manufacturing technologies for components and stacks for uniformity but also suitable for homogenous large volume production. The geometry of the stacks allows easy installation even in crowded or compact areas. Centre for Fuel Cell Technology (CFCT) has demonstrated PEMFC stacks of 1–3 kW capacities. These stacks are single units with fixed voltage and current capabilities. CFCT is now embarked on a program to explore the possibility of introducing modular architecture for various applications of fuel cell systems. In this context, CFCT has recently developed a 250 W module which can be cooled either by air or water. The study by Rajalakshmi et al. [68] discusses the concept and design of a modular architecture. This subject is further discussed in Chapter 4.

1.7.2 Novel Reversible Fuel Cell System for Very High Efficiency

Researchers at Forschungszentrum Jülich have commissioned a highly efficient fuel cell system that achieves an electrical efficiency in hydrogen operation of over 60%. The newly developed reversible high-temperature fuel cells can not only generate electricity but can also be used for the production of hydrogen by electrolysis [69]. Reversible fuel cells, or "reversible solid oxide cells" or rSOC for short, combine virtually two devices into one. The cell type is therefore particularly suitable for the construction of plants that can store electricity in the form of hydrogen, and these can back-flow at a later date again. Such storage technology could play an important role in the energy transition. It is needed to compensate for fluctuations in renewable energies and to counteract the divergence between supply and demand. In addition, it can be used for remote stations on islands and mountains, to ensure a self-sufficient energy supply [69].

The exceptional property of reversibility is exhibited only by high-temperature fuel cells, SOFC for short, which operate at around 800°C. Due to the high temperature, less noble and less expensive materials than for low-temperature fuel cells may be used for this type of fuel cell. At the same time, high-temperature fuel cells work extremely efficiently. Unlike low-temperature systems, whose efficiency with hydrogen is limited to around 50%, high-temperature fuel cells can also achieve significantly higher efficiency. Researchers at Forschungszentrum Jülich have now succeeded further in increasing their efficiency and achieving a value of more than 60% for the first time. For their system, the researchers determined in the test operation an electrical efficiency of 62%. This was made possible by an improved stack design in conjunction with an optimized and highly integrated system technology, which electrochemically converts more than 97% of the supplied hydrogen. One of these improvements lies in the dimensioning of the converter unit ("stack"). Stack in this study comes with a power of 5 kW, which could be about the power consumption of two households. In normal modular operation, one has to combine several units in kW to achieve comparable performance. The researcher hopes that this larger stack in modular units can also reduce the manufacturing costs since fewer units are needed for the construction of high-performance equipment [69].

In electrolysis mode, when the system produces hydrogen, the Jülich system can even drive at a much higher output. With a current consumption of the stack of 14.9 kW, it then generates 4.75 cubic meters (Nm^3/h) of hydrogen per h, which corresponds to a system efficiency of 70%. As a result, the pilot plant is already operating more efficiently than alkaline and polymer electrolyte electrolyzers, which account for 60%–65% and are now standard. According to Julich scientists, while the electrolysis works quite well for the beginning, there is still a potential for improvement. High-temperature systems from other developers, which have been specially optimized for electrolysis, today achieve efficiencies of over 80%. In fuel cell mode, however, they may not work as efficiently as the new Jülich system. The Jülich researchers have already considered further optimizations with which they want to further increase the so-called "round-trip" efficiency. The figure describes the efficiency that remains in the recuperation, i.e. after the production of hydrogen and reconversion. The scientists want to improve the value from the current 43% to more than 50%. For a hydrogen storage, this value would be sensational, even if the technology cannot keep up with battery storage in this regard, which sometimes comes to over 90%. The fuel cell systems offer other benefits. Since the energy converter, the fuel cell, and the energy carrier hydrogen are clearly separated from each other, the system can repeatedly supply or derive hydrogen. This helps since the size of the storable amount of energy is limited [69].

1.7.3 Efficiency Improvements by Pulsed Hydrogen Supply in PEM Fuel Cell Systems

The membrane electrode assembly of a PEM fuel cell has to be wet for the fuel cell to work efficiently. However, too much water impedes the delivery of reactants and impairs the electrochemical reaction. In the study by Rodatz et al. [70], an arrangement is presented which uses pressure waves to remove water droplets and inert gas blankets from the fuel cell. Experiments showed good short- and long-term performance of the proposed configuration. In contrast to static fuel cell systems, fuel cells in mobile application are operated under transient conditions. Therefore, it is very difficult to obtain satisfactory humidification of the membrane. Under these circumstances, a device by which excess water is removed from the fuel cell is highly desirable. Two pressure waves are used to clear the hydrogen side of the fuel cell from water droplets. Compared to conventional dead-end systems a substantial increase in system efficiency is achieved with the described hydrogen supply arrangement. In addition, experiments showed good long-term stability of the arrangement. Water droplets that are formed on the airside are more easily removed from the fuel cell because of the much larger mass flow. Therefore, the need for additional devices is less urgent. However, it is assumed that pulses will assist the removal of water droplets and/or nitrogen blankets. Hence further research will focus on the expected gains of the scavenging effect on the airside of the fuel cell.

1.8 ROLE OF NANOTECHNOLOGY FOR ENERGY EFFICIENCY IN MODULAR VEHICLE OPERATIONS

As mentioned in the previous volume [1], one advantage of modular approach is the ability to insert new innovations in the existing design and construction. In recent years nanotechnology has introduced many advances in materials used in automobile industry; such as lubricating oils, batteries, supercapacitors, fuel cells, solar cells, and automobile catalysts for exhaust treatments. All of these changes create fuel savings for the car industry. These advances are continuing. Some of these advances are outlined below. Modular approach can easily accommodate these changes [71–135].

1.8.1 Lubricating Oils for Cars Using Nanoparticle Additives

Lubricants like mineral oil are used to reduce friction and wear in automobile engine. The pistons' movement in the cylinder of an engine produces frictions as a result of metals wear. This may lead to reduced engine energy efficiency as well as lowered engine life. Oils are used as lubricant to reduce friction. The conventional oils need to be exchanged after a special engine working time. In fact, the oil lubricant properties will be gradually reduced. In recent years, researchers are engaged to produce better oils with longer life. Nanotechnology is one of the most effective ways of fulfilling this target [79,123]. It has been shown that nanoparticles could improve lubricant behavior of conventional oils. Particles' shape, size, and concentration are influential parameters affecting wear and friction reduction. It has been shown that gold particles having particle size of 20 nm has the best lubricating effects [79,123]. Dialkyldithiophosphate modified copper nanoparticles are also shown as an effective nanoparticle with high ability of improving anti-wear ability of metal surface by producing an anti-friction film. Diamond and inorganic fullerene-like (IF) particles are other examples of anti-wear nanoparticles being used as additives for lubrication. The most important mechanisms, which result in friction reduction are colloidal effects, rolling effects, protective film, and third body.

In recent studies, Diamond nanoparticles were added to the oil to improve its anti-wear ability. This nanoparticle has found to improve oil lubricant behavior via various mechanisms including (a) ball bearing effects of the spherical particles existed between rubbing surfaces, (b) the surface polishing and (c) increasing surface hardness. Adding CuO nanoparticles to oil could also significantly reduce friction coefficient. Ball bearing at high temperature and viscous effect at low temperature are the reasons CuO nanoparticles can improve anti-wear behavior of oil. The nanoparticles depositions at worn surfaces would be responsible for shear stress reduction leading to tribological properties improvement of the surface [77,79,112,123,129].

1.8.2 Nanotechnology for Improved Energy Storage in Cars

To replace combustion engines, different strategies and methods have been developed. Among them, electrochemical energy production/storage is the most important option owing to sustainability and being environmentally friendly [113].

The so-called electrochemical energy storage and conversion systems include fuel cells, batteries, and supercapacitors. Nanomaterials are improving storage capabilities of these systems [72,114].

1.8.2.1 Batteries

Although fuel cell-powered cars have been driven for more than 45 years, the current efficiency for replacing the internal combustion engines hinders widespread use of FCV (additionally lack of hydrogen storage infrastructures and cost are real barriers) [134]. In addition, from practical point of view, because of low power density of FCs, FC-powered automobiles require an energy storage device to deliver required energy in power-peak demands. As a result, currently automakers have come to this decision to launch new versions of hybrid cars before FC cars, which has resulted to introduction of new generation of automobiles with considerably low gasoline combustion reaching to record less than 1 L/100 km [83]. Therefore, at the heart of the upcoming automobiles, energy storage devices play a key role. Batteries are blooming in different markets and automobile industry is not an exemption. Among different classes of batteries, lithium-ion batteries have higher potential for employing in the next generation of cars due to higher energy density. Similar to other batteries, Li-ion batteries also consist of a cathode, an electrolyte, and an anode. Due to principles governing the electrochemistry of Li-ion batteries wide ranges of materials can be used in this class of devices, which significantly affect the cell potential, energy density, and safety of batteries. Therefore, huge amount of attention has been attracted towards developing high-performance Li-ion batteries from both academic communities and industrial firms. The efforts are focused on improving the capacity, safety, and the charging rate. Nanoscopic materials are presumed to have great contribution in the world's $56 billion battery market in near future [114].

The cathode and anode materials must be able to be intercalated with Li-ions having high Li hosting capacity and also high electron conduction [119]. Among the various materials employed as cathode for Li-ion batteries, iron (or other metal) phosphate is a promising and safe replacing candidate for conventional cathode material, cobalt oxide [108]. This case is interesting as nanoengineering helped a lot to see marketable version of this material. In fact, due to the low electrical conductivity of $FePO_4$, the energy charging rate is very low. Unless it is found that doping with other transition metals can enhance the material's conductivity, $FePO_4$-based materials found their way in marketplace when Chiang's group at MIT uncovered that nano-sized $FePO_4$ particles can store and deliver energy much faster than usual size (~10 μm) due to higher surface area which facilitates intercalation of Li ions [81,91].

In fact, conventional Li-ion batteries equipped with cobalt oxide which are prone to catch fire due to thermal runaway, phosphates can be used to fabricate larger Li-ion batteries much more suitable for automobiles. The basic structure of a Li-ion battery is such that in this battery lithium-ion intercalation into anode and cathode during charge and discharge process, respectively, is employed to store electrochemical energy [119]. Nano sizing the cathode and anode materials are now tremendously followed in different battery materials. Silicon, one of the most promising anode materials, may find somewhere in market if researchers could overcome instability of these materials during charge-discharge process through the nanostructuring of this element [73]. Different nanostructures of Si such as nanoparticles [96,97], nanowire [76], nanotube [125,135], hierarchical nanoporous structures [103] and their composites with nanocarbons [82,96,97] have shown to have exceptionally high capacity and stability raising hopes to have commercial batteries with Si-based anode. In the realm of cathode materials for Li-based batteries, sulfur boosts the capacity of Li-ion batteries with one order of magnitude higher theoretical capacitance [109].

Therefore, the Li-S batteries may succeed Li-ion batteries as their energy density is extremely high plus low cost and density of sulfur [91]. However, its capacity fades away during charge-discharge of the cell due to polysulfide ions (the reaction intermediates) dissolution in electrolyte causes irreversible loss of active materials diminishing the capacitance a few times lower than theoretical value. In 2009, by employing an innovative technique through the nanostructuring the sulfur inside the mesoporous carbon, capacitances near the theoretical limits were attained [90]. After that, different carbon nanostructures such as hollow carbon nanofibers [128], graphene oxide [89] and pyrolyzed PAN/graphene [126] were used to immobilize sulfur. This class of batteries (Li-S) would find market in automobile industries as also claimed by Daimler in its concept vehicles, Mercedes-Benz F125 using this type of battery is having exceptional high range of 1,000 km.

1.8.2.2 Super Capacitors

Despite the advantages of using batteries as energy storage devices in hybrid cars including high energy density, challenges associated with employing batteries especially timely recharging, safety and lifetime bring another electrochemical storage device as candidate for the same purpose, i.e. supercapacitors. Supercapacitors store energy by forming a double layer of electrolyte ions on the surface of conductive electrodes, called EDLCs [105]. The widespread applications of supercapacitors are limited by their low energy density (1–5 Wh/kg) comparing to batteries (10–500 Wh/kg) and as a result high cost of energy storage. But the fact that supercapacitors can be charged and discharged in less than a minute over a million cycles motivates scientific communities to enhance energy density of supercapacitors [80,115]. It is envisioned at energy density of 40 Wh/kg, the supercapacitors would be an improvement over the batteries used in some hybrid vehicles. In addition, the concept of supercapacitor powered urban bus which recharges at each bus stop in a minute is another intriguing idea. At the heart of the current EDLCs, nanoporous carbon acts as electrode. Emergence of graphene has revolutionized this field, as this material is the thinnest imaginable carbon allotrope [117]. The graphene-based pseudo-capacitors are still in infancy stage, but initial results confirm high capacity of graphene-based EDLCs having improved energy density. In May 2011, Ruoff's group [133] at university of Texas-Austin claimed they have developed graphene-based EDLCs through an industrially viable method having energy density of 75 Wh/kg which is more than one order magnitude higher than conventional supercapacitors. It is foreseen by 2020, half of graphene's market (~$675 million) belongs to supercapacitors which clearly illustrate the impact of graphene-based materials on this field (www.bccresearch.com/report/AVM075A.html). It should be noted that hybrid systems of batteries and supercapacitors are identified as the most effective and reliable solution for applications where lifecycle and reliability are vital including cars.

In June 2010, researchers at MIT introduced a new concept of energy storage device, having both high density (comparable to Li-ion batteries) and power density (even higher than supercapacitors) [96,97]. Again nanocarbons performed excellently as electrode. The whole idea was to exchange Li-ions between the surfaces of two nanostructured carbon electrodes having functional groups [88]. In fact, charging of Li-ion batteries is timely because Li-ions must intercalate into cathode and anode which takes time. This strategy to store energy may be implemented in next generation of automobiles but further investigation is required to clarify the exact chemistry governing the device and also confirm feasibility and other issues.

1.8.2.3 Fuel Cells

High efficient energy conversion, safety, high energy density, and nonpolluting are the advantages of employing fuel cells as energy source for driving a car [87]. However, high cost, low volumetric power density, low durability, and cell life plus high sensitivity to purity gas stream and complex operation are disadvantage of using fuel cells. Thus, hybrid configuration of fuel cells with batteries or supercapacitors is being developed in order to supply power for peak-power demands such as acceleration and start-up and also recovering braking energy [95]. In fact, despite comparable energy density of fuel cells (100–1,000 Wh/kg) relative to combustion engines, they have few orders of magnitude lower power densities, rendering them as steady energy source. On the other hand, supercapacitors possess high power density comparable to combustion engines (Figure 1.8). Therefore, hybrid systems of fuel cells and supercapacitors or batteries are an efficient configuration for replacing the combustion engines. Figure 1.8 describes Ragone plot of the energy storage domains for the various electrochemical energy conversion systems [122,136].

Although one can drive a car powered by some classes (PEFCs) of fuel cells, there are still some challenges associated with employing them which are mostly high cost, fuel supply/storage and lifetime. Generally the fuel choice is hydrogen and oxygen which finally exhaust water. Hydrogen supplication, refueling infrastructure, and storage of hydrogen are still ongoing challenges [75]. Although hydrogen has

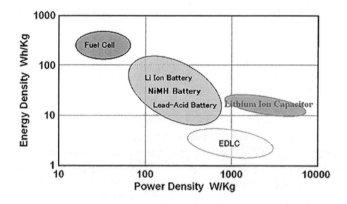

FIGURE 1.8 Ragone plot of the energy storage domains for the various electrochemical energy conversion systems [136].

very high gravimetric energy density, its' application hindered due to low volumetric energy density. Storing hydrogen in liquid state requires employing highly expensive cryogenic tanks. Compressing hydrogen gas also urges using costly storage facilities. There has been considerable research to develop new materials enabling storage of hydrogen at high enough concentration at not too high pressure and too low temperature.

More recently, the efforts are focused on the physical entrapping of hydrogen in porous materials. Physisorption of H_2 allows fast loading and unloading. Nanostructured materials are the sole candidates for this purpose. Novel nanoporous materials, MOFs, have been the center of attention for gas adsorption. These materials are product of reaction of metals ions with rigid organic molecules. Due to exceptionally high surface area and tunable chemical structure of MOFs, high potential for high enough H_2 uptake is envisioned. In September 2011, Daimler introduced a concept vehicle, Mercedes-Benz F125!, which was pictured to be derived by 2025 (http://media.daimler.com). The most interesting technology of this conceptual car is its source of energy which is hybrid of Li-S battery (See below for more details) and hydrogen fuel cell. Despite current technology used in available fuel cell vehicles, liquid or compressed hydrogen, storage facility of hydrogen is based on MOFs materials in this concept vehicle. The manufacturer claims one will be able to derive up to 1,000 km with maximum speed of 220 km/h before it is needed to be refilled.

Although interest to fabricate solid-phase hydrogen reservoir using CNTs is now quenched, researches on developing high-surface-area, graphene-based materials are ongoing and more time requires to confirm whether or not graphene-based nanoporous materials are able to solve the mystery of the hydrogen storage [118]. Another major issue that hinders widespread application of fuel cell vehicles is high cost associated with their production. The expensive constituents of fuel cells including the catalyst and electrolyte membrane are the origin of high cost of fuel cells [116]. In addition, performance and lifetime of these classes of fuel cells (PEFC, DMFC) can be remarkably improved by nano-engineering of the catalyst and electrolyte membrane. In fact, the commercialization of fuel cells will very much rely on the ability to reduce the cost and improve the performance of catalyst, membrane and other expensive parts to launch a fuel cell-powered vehicle at a competitive cost and driving capabilities [72]. To reach this goal, nanotechnology plays a dominant role.

The challenges associated with implementing hydrogen as the next generation green fuel including hydrogen source and production, storage infrastructure, fuel tank, and highly efficient fuel cells [120]. The use of platinum catalysts for electrodes increases the cost of fuel cells. In order to reduce cost, new efforts have dominantly followed three major strategies including (a) improving the Pt-based catalyst (b) developing new class of nonprecious catalysts using other transition metals and finally (c) metal-free catalyst materials. Today Pt-carbon catalysts which are widely used in PEFCs and DMFCs, are nanoparticles of Pt decorated on carbon support (e.g. carbon black) [85]. Pt nanoparticles' activity increases as the particle size decrease reaching a minimum of ~3 nm [72]. Rational nano-engineering of Pt alone or with other metal atoms into specific arrangement of nanostructured alloys such as core-shell nanoparticles or nanowire [93,94,101,134] is a highly effective tool to synthesis new generation of catalyst if fundamentals of governing the performance of catalyst are understood. Core-shell nanoparticles of Pt-Cu, for example, are more active than Pt which consists of cores made of a Cu-Pt alloy and Pt-rich outer shell [104]. Such structure is obtained through a controlled dealloying the hybrid nanoparticles. However, many challenges associated with using precious Pt-based catalysts still remain untouched through these strategies.

Despite quite ineffective early catalysts, recent developments through nanoscale engineering of nanostructured catalysts revive the hopes to have a nonprecious replacement of Pt catalysts. Catalysts based on thermally annealed precursors comprising nitrogen, carbon and transition metals, especially Fe and Co (Fe (or Co)/N/C), have attracted more attention due to high activity and performance [74,98,110]. This class of catalysts consists of metal nanoparticles embedding in nanostructured nitrogen-doped graphitic carbon [124]. Metal-free catalysts have been also synthesized and evaluated as catalyst. Among them, carbon-based nanostructures are the most studied systems. In 2009, Dai's group [84] at Case Western University showed gas-phase N-doping of CNTs would result in a metal-free electrocatalyst. After that, many studies revealed promising performance of N-doped carbon nanomaterials such as SWNTs [127], graphene [111], mesoporous graphitic array [102] and carbon quantum dots [100], for replacing Pt-based catalysts [78,84,121]. Findings in Fe (or Co)/N/C systems and N-doped carbon nanomaterials may be combined possibly through using CNTs or graphene as support in Fe (or Co)/N/C catalysts instead of carbon black. However, it seems that carbon nanomaterials would have much more contribution in the next generation of catalyst in fuel cells than what they have in current Pt/C commercial catalysts. Rational design of nanostructure of the so-called upcoming catalyst would be the key issue.

1.8.3 Nanotechnology for Solar Cell in Automobile

Solar cells are used to produce electricity from sunlight. This system has been gradually developed in different industries as it is an environmentally benign method of producing electricity and helps industries to reduce fuel consumption. In recent years, research to find new sources of energy in automobiles are being carried out [99,107]. It has been found that solar cells can be used as an additional source of energy supporting some of the electronic devices used in an automobile. Silica-based solar system is a conventional system for producing electricity from sunlight. However, production and large-scale use of this system is complicated and costly. Therefore, attempts have been carried out to produce a new solar system usable for automobile economically. In silicon-based solar cells, the electron needed will be supplied by silicon after exposure to sunlight. The produced electron can be transferred to semiconductors from electrodes. In solar cells, an organic material like chlorophyll can be used as substrate. A large surface area layer based on nanoporous titanium oxide is used for the transmission to the electrodes. This system consists of two glass plates, each of which has a transparent electrode [99,107].

In this system, one of the plates covered with a layer of dye-sensitive titanium oxide and another one is coated with platinum as catalyst. However, in conventional silicon-based solar cells, the efficiency is low due to light reflection at the solar glass pan. The reduction is approximately 10% for even high-efficiency solar cells.

This problem has been solved in new generation of solar cells using sol-gel method. Using this technique, a coating layer is applied over the glass panes. This coating could reduce light reflection from glass pans resulting in an increase in solar cell efficiency up to 6%. The sol used for this purpose includes a mixture of silica balls at two different sizes. To obtain the best anti-reflective properties from coating, a mixture of particles with diameter of 10 and 30 nm should be used. In order to apply coating layer over the cells, the glass pans should be immersed in the tank containing nano SiO_2 sol. The optimum antireflective properties can be obtained at 120 nm thickness. The sol becomes dry using gel and nanoporous layer can be obtained after hardening glass-coated pans at 600°C–700°C. This novel coating layer has very low refractive index (of only 1.25) resulting in light reflection at 400–2,500 nm. In this way, solar transmission will be increased from 90% for conventional one to 95% for this solar cell [99,107].

1.8.4 Nanotechnology-Based Catalyst for Reduction of Exhaust Emission

The catalytic materials are able to convert exhaust pollutants to nitrogen, steam, and carbon dioxide. The three most important polluting elements that exhaust include carbon monoxide, hydrocarbons, and nitric oxides. To eliminate or reduce these pollutants emission from exhaust gas, three kinds of catalysts are needed. Nanotechnology can play an important role in converting toxic pollutants into nontoxic gases. It is well known that increasing surface area of catalyst enhances its catalytic activity. Designing catalytic materials to absorb nitric oxides from exhausted gas has become a big challenge for car manufacturers. To solve this problem, new generation of catalyst with high capability of NOx-absorbing are developed. The mechanism by which this system works is described in the literature [86,92,106,130]. Three-way nano-structured catalyst for cleaning exhaust from pollutants [86]. Recent researches revealed that Au nanoparticles having sizes lower than 5 nm are very effective catalysts. There are many different mechanisms indicating catalytic activity of nano-sized Au particles. The most important of them are the quantum size effects, charge transfer to or from the support or support induced strain, oxygen spill-over to or from the support, oxidation state of Au and the role of very low-coordinated Au atoms in nanoparticles. The metal nobility depends on the metal surface ability to oxidize or chemisorption of oxygen. From periodic table of elements, Au is the only metal with endothermic chemisorption energy. Because of this behavior, Au is a metal that could not bind with oxygen at all. Au is a very good catalyst for oxidation of carbon monoxide (CO) presented in exhausted gas in automobile. The activity of Au depends on particle size as the best activity can be seen at particle sizes <5 nm [86].

1.9 INTEGRATED MODULAR AVIONICS

The Integrated Modular Avionics (IMA) concept, which replaces numerous separate processors and line replaceable units (LRU) with fewer, more centralized processing units, promises a significant weight reduction and maintenance savings in the new generation of commercial airliners. Boeing indicated that by using the IMA approach, it was able to save 2,000 pounds off the avionics suite of the new 787 Dreamliner as compared to previous similar aircraft. Airbus indicated its IMA approach cuts in half the part numbers of processor units for the new A380 avionics suite [137–143].

It's not just the IMA modules themselves, and reducing the number of LRUs, IMA brings a more efficient network for the aircraft. From an airline standpoint, fewer types and varieties of spares should drive higher reliability, and therefore less maintenance. Some believe the IMA concept originated in the United States with the new F-22 and F-35 fighters and then migrated to the commercial jetliner arena. Others say the modular avionics concept, with less integration, has been used in business jets and regional airliners since the late 1980s or early 1990s. IMA is the trend of the future due to the economies in fuel savings derived from less weight and lower maintenance costs. It also offers an open architecture allowing for the use of common software, which makes upgrades and changes both cheaper and easier to accomplish [137–143].

An IMA operator can upgrade software without having to upgrade the hardware and vice versa. Using elements common to different computer modules makes maintenance of the computer less expensive. Since the same part (or card) can be used in any of the IMA computers, inventory in the shop is smaller. The advantage is less expensive maintenance. While adapting the general concept of "shared resources," the Boeing 787 and the Airbus A380 approaches to IMA differ. Both aircraft have applications for specific LRUs that are on the plane and individual computers for certain systems. Key to the B787 avionics suite, which Boeing developed with partners Smiths Aerospace, Rockwell Collins and Honeywell, is a central computing system Boeing calls the Common Core System (CCS), which eliminated more than 100 different LRUs. The A380 Super Jumbo, which touts 15%–20% lower operating costs than previous airliners, applies the IMA concept with computers capable of hosting different functions and integrated modular avionics connected by a network. This approach differs from Boeing's 787 central computing system in that it does not rely on a single (or dual) central processor to run most of the aircraft systems. Instead, the A380's IMA approach relies on eight processing modules, some tailored for specific applications, but all tied together by a common Avionics Full-Duplex Switched Ethernet (AFDX), ARINC 664 standard network. Seven of the three MCU computers are core processing input/output modules (CPIOM); the eighth is an input/output module (IOM). Even if all these computers are connected to the AFDX, and the majority of the communications is done through the network, there is some specificity on the module depending on which function they are in. There is a slight difference in the input/output of the computers, which is why they have different part numbers [137–143].

Although the Airbus IMA computers have eight different part numbers, memory and power supply cards are common to all the computers. It is only the input/output card that is different, depending on what type of system the computer interfaces with. The use of a central processing system is not exactly the way A380 utilizes IMA. There are seven CPIOMs doing different types of functions. A380 has preferred to develop what one calls an open IMA—some computing resources on which one can have different functions hosted. Rockwell Collins is providing the AFDX network for the A380 along with overall network integration support to Airbus and to third-party suppliers that have network "end systems" embedded into products they are delivering for the A380. The end system, a chipset, is required in a manufacturer's box to have a network connection [137–143]. Three functional domains of IMA are cockpit (electrical flight control, communications, and warning); cabin (air conditioning and pneumatics); and utilities; including energy, fuel functions, and landing gear functions. There are 30 line replaceable modules, all 3-MCU boxes, associated with the IMA platform, and 22 software functions hosted in the CPIOMs. France's Thales and Airbus Avionique each are providing CPIOMs. Some 11 suppliers provide software functions hosted within the IMA, ranging from communications to landing gear extension and retraction. They include Fairchild Controls (pneumatic), Parker Aerospace (fuel applications), and Hamilton Sundstrand (air conditioning).

As a supplier of the B787 pilot controls, Rockwell Collins delivered the first fully operational system to Boeing to develop an integrated test vehicle. The pilot controls include interfaces to the aircraft's fly-by-wire flight control system. Rockwell Collins completed initial deliveries of precertification hardware to Boeing for its integration facilities and began deliveries for the first B787 flight test aircraft. A large amount of avionics integration work has been accomplished at the Rockwell Collins lab in Cedar Rapids, Iowa, including integrating functions from other suppliers. Honeywell provides the flight management system software that resides on the B787's IMA system as part of its navigation component, which also includes its air data inertial reference unit and multimode GPS receiver. The Honeywell crew information/maintenance system evolved from the central maintenance computing and aircraft condition monitoring functions the company developed for the Boeing 777. Boeing 777's AIMS (Aircraft Information Management System) was the first actual IMA implementation, in commercial aviation. The IMA concept is all about weight and power savings. With new aircraft getting more software-based functionalities, and computers becoming more powerful, it doesn't make sense to add another box with its own computer every time you want to add a new feature or function. It makes a lot of sense to share computing resources. Honeywell also provides the flight control package for the B787's fly-by-wire system.

With a takeoff weight of 1.2 million pounds, the A380 is the largest and heaviest commercial airplane ever built [137–143].

The study by Schneider [144] presents an innovative approach for modular and flexible positioning systems for large aircraft assembly, for instance, the manufacturing of the fuselage sections from shell panels and floor grids, the alignment of the sections to build the fuselage, and the joining of wings and tail units to the fuselage. The positioning system features a modular, reconfigurable, and versatile solution for various aircraft dimensions and different applications. This includes the positioning units, the controls, the measurement interface, and the product supports. It provides the customer with a holistic solution that considers the specific positioning task taking into account high absolute positioning accuracy, repeatability and synchronization of the motion for all manipulators that constitute the positioning system. Various tools and method which were used during the development process are introduced and the developed standardized Positioning Technology is briefly explained.

1.10 ADVANCED MODULAR SHIPBOARD ENERGY STORAGE SYSTEM FOR ENERGY EFFICIENCY IMPROVEMENT

DOD has adopted the practice of trail shaft operations to reduce propulsion fuel usage when feasible. However, in the case of the ship's service electrical plant, the need for assured electrical power requires the operation of redundant generators which has limited the ability of Commanding Officers to operate the electrical plant at optimal efficiency. In the case of the Arleigh Burke Class Destroyers, like the CG 47 Ticonderoga Class Cruisers and earlier Spruance and Kidd classes, ship service electrical power is provided by three installed gas turbine generator sets (GTG). While the DDG 51 Class peace time ship electrical load is less than the generator rating for a significant fraction of operating time, standard practice is to have two GTGs on line at all times to ensure continuity of service should there be a system fault or casualty to one of the GTGs. While saving fuel is an important consideration, loss of power due to a system fault, or casualty to that single GTG is unacceptable and would mean going dead-in-the water (DIW) with the consequent loss of all ship systems, including weapons, sensors, communications, navigation, as well as propulsion and auxiliary systems, except those which have battery backup to prevent system damage and simplify restart. Because the restart time for critical sensitive electronics can be considerable, many systems have dedicated uninterruptible power supplies (UPS) that are found throughout the ship, adding battery systems that take up space, weight, and require on-going maintenance. The potential for loss of power is the primary reason that standard procedures require two engine operations.

Gas turbine engines are the most fuel-efficient (lowest specific fuel consumption [SFC] in pounds of fuel burned per horsepower-hour) when operating at or near full power. When a ship operates with two partly loaded generators providing power they are less efficient, with higher SFC (i.e. burning more fuel). Therefore, operating one GTG that is highly loaded can result in significant fuel savings. For the purposes of the BAA, the assumed average ship's electrical load was 2,525 kW and 4,000 h/year operation. The calculated annual fuel savings at this operating point is nearly 8,000 BBL of fuel or $1.28M at $160/BBL at the Energy rate charged the Navy in January 2012. Actual operating hours could be lower perhaps 2,500–3,000 h/year. However, as the delivered cost of fuel continues to increase the dollar savings will also increase. These figures represent >25% improvement in overall electric plant fuel efficiency.

Ships force can calculate fuel savings in gallons/h (GPH) using a nomograph for the 501-K34 fuel usage found in the NAVSEA Shipboard Energy Conservation Guide. The resulting savings from calculation carried out by Mahoney et al. [145] is ~100 GPH at the load conditions specified above. This is consistent with but slightly higher than the results obtained using the 501-K34 SFC curves and reinforces the concept that enabling single generator operations, while providing backup power to support the full ships' electrical load in case of a GTG casualty would save a significant amount of fuel. The fleet average DDG 51 underway fuel usage is on the order of 25.2 Bbls/h (average of 2010 PACFLT and LANTFLT DDG-51 Class fuel usage from Navy ENCON program). The electrical fuel usage for two generator operations at the load specified above is ~7.5 Bbls/h, or 30% of the total fuel usage and roughly

5.5 Bbls/h for the single generator case. The resulting "electrical fuel savings" for this Single Genset with ESM mode of operation of ~2 Bbls/h translates to almost 30% savings of "electrical" fuel, and ~8% of overall ships underway fuel usage.

1.10.1 SYSTEM AND TECHNOLOGY OVERVIEW

The ESM modules developed under the ONR contract consist of a modular bidirectional AC/DC power conversion system. This is based on significant improvements to the RCT Systems PCM-2 ship service inverter modules (SSIMs) that were successfully tested as part of the NAVSEA DDG(X) integrated fight through power (IFTP) program to develop the power conversion components (AC/DC, DC/DC, and DC/AC) that were the prototypes for the DDG-1000 low-voltage power distribution system. The PCM-1 and PCM-2 ship service converter modules (SSCM) were thoroughly tested at NSWC Philadelphia land based test site (LBTS). The ESM also includes the necessary controls and system interface devices, along with modular energy storage (ES). The system interfaces with the Ships 450V AC distribution and includes a high-voltage DC link for connection with the battery or other ES. While the module size is notionally 600 kW (five would support a 3 MW load), the system can be scaled to meet shipboard power and space needs concepts for DDG 51 or other ship classes where fuel efficiency and energy storage are requirements [145–151].

The overall system objectives were derived from the estimated requirements to provide continuity of power for up to 10 min given the loss of a single generator. A distributed, modular, and redundant system was developed based on the desire for a 3 MW system (current RR 501-K34 GTG). Each module is rated at 600 kW. At the rating of 600 kW. Once again, five modules would be required for a full 3 MW system (at the 150 kW LRU rating or four modules at the full LRU rating of 200 kW/LRU). At present the module weight is estimated to be 9,000 lbs, assuming a Li-ion battery system weight of 4,000 lbs (based on the 10 min storage requirement). Given the maximum weight for the purposes of ship structural considerations, that leaves a margin of 3,800 lbs for growth for shock mounts, and other structural, or containment systems [145–151].

The development of multiple line replaceable units (LRUs) for the bidirectional AC/DC inverters led to a tested 200 kW/LRU design. This equates to power densities of 1.78 kW/L, respectively (up from 0.90 kW/L for IFTP converters). For the purpose of the demo, the LRU's were rated at 150 kW, so the resulting "module" now consists of 4–150 kW LRUs, plus batteries for a notional module power. In the system development, all appropriate shipboard shock, electrical (MIL-STD-1399 Section 300 Type I power quality), EMI/EMC (MIL-STD-461), and related standards were incorporated into the design but since this was a prototype development program, compliance testing was not done. A MIL-STD 882 risk analysis was conducted, and all potential mishaps were addressed. System simulations were conducted to look at the response to step load changes (e.g., loss or startup of a generator), short circuits, and autonomous paralleling of ESM modules with the generator. While the system is agnostic to the energy storage technology (battery, fuel cell, flywheel, etc.), the cabinets and system have been designed around advanced lithium-ion battery technology [145–151]. In addition to the direct "electric plant" fuel savings and reduced operating hours on the Gensets and consequently reduced maintenance burden, the addition of an energy storage system such as the one proposed here has other synergistic benefits, including it enables the adoption of advanced fuel-efficient gas turbine generators.

While the development of this modular Energy Storage System (ESM) was targeted to the DDG 51 Class as a fuel efficiency improvement, the technology has wider applicability to other ship classes for energy storage (UPS) needs, energy efficiency improvements, and as an enabler for the NGIPS architecture and future weapons and sensor systems. The distributed, modular, movable system can go anywhere on the ship where space is available since it needs no special support other than saltwater cooling. The completion of testing to TRL 4+ of the prototype DDG 51 ESM in December 2010 provided the initial demonstration of an advanced shipboard energy storage system that enabled significant fuel savings for the DDG 51 Class, as well as other current and future ship classes. Testing leading to TRL 5+ at NSWC-CD Philadelphia at the DDG 51 Land-Based Engineering Site (LBES), and sea testing in a DDG 51 Class ship were completed in 2012 [145–151].

REFERENCES

1. Shah, Y.T. (2019). *Modular Systems for Energy and Fuel Recovery and Conversion*. CRC Press, New York.
2. Energy Conservation, Wikipedia, The free encyclopedia, last visited 16 June 2019 (2019).
3. Use of Energy in the United States—Energy Explained, Your Guide to, Independent Statistics and Analysis, U.S. Energy Information Administration, DOE, Washington, www.eia.gov›EnergyExplained (29 May 2018).
4. Zero Energy Buildings, Wikipedia, The free encyclopedia, last visited 9 June 2019 (2019).
5. Gillies, B. (September 2015). A Common Definition of Zero Energy Buildings. A report by National Institute for Building Sciences (NREL) for Department of Energy, Energy efficiency and renewable energy, Washington, DC.
6. Why Energy Efficiency Upgrades | Department of Energy—Energy.gov. A website DOE report, Washington, DC www.energy.gov/eere/why-energy-efficiency-upgrades (2011).
7. DSIRE. Residential Energy Efficiency Tax Credit. A website report by Department of Energy, Washington, DC (2015).
8. Passive Solar Building Design, Wikipedia, The free encyclopedia, last visited 26 June 2019 (2019).
9. Passive House, Wikipedia, The free encyclopedia, last visited 1 July 2019 (2019).
10. U.S. Green Building Council. (2013). LEED Reference Guide for Building Design and Construction (v4 ed.). U.S. Green Building Council, pp. 318–466, ISBN 1932444181.
11. Energy Saving Module, Wikipedia, The free encyclopedia, last visited 3 March 2018 (2018).
12. ESM™ Energy Saving Module | Incubation | Pas Reform Hatchery. A website report www.pasreform.com/en/solutions/2/incubation/.../esmtm-energy-saving-modulz... (2015).
13. Energy Saving Module - PowerCalc. A website report www.powercalc.co/more-on-energy-savings (2017).

14. Velraj, R. and Pasupathy, A. (June 2006). Phase Change Material Based Modular Thermal Storage for Energy Conservation in Building Architecture, Institute of Energy Studies, Chennai, India. A website report; also Pasupathy, A., Velraj, R., and Seeniraj, R.V. (2008). Phase change material-based building architecture for thermal management in residential and commercial establishments. *Renewable and Sustainable Energy Reviews*, Elsevier, 12(1), 39–64; Janu Ismail, K.A.R. and Henriquez, J.R. (2001). Thermally effective windows with moving phase change material curtains. *Applied Thermal Engineering*, 21, 1909–1923.
15. Lin, K., Zhang, Y., Xu, X., Di, H., Yang, R., and Qin, P. (2005). Experimental study of under-floor electric heating system with shape-stabilized PCM plates. *Energy and Buildings*, 37, 215–220.
16. Mehling, H., Krippner, R., and Hauer, A. (2002). Research Project on PCM in Wood-Lightweight-Concrete. *IEA, ECES IA Annex 17, Advanced Thermal Energy Storage through Phase Change Materials and Chemical Reactions—Feasibility Studies and Demonstration Projects. 2nd Workshop*, 3–5 April, Ljubljana, Slovenia.
17. Saman, W.Y. and Belusko, M. (1997). Roof Integrated Unglazed Transpired Solar Air Heater. *Proceedings of the 1997 Australian and New Zealand Solar Energy Society*, Lee T. (Ed). Paper 66, Canberra, Australia.
18. Stritih, U. and Novak, P. (2002). Thermal Storage of Solar Energy in the Wall for Building Ventilation. *Second Workshop: IEA, ECES IA Annex 17, Advanced Thermal Energy Storage Techniques—Feasibility Studies and Demonstration Projects*, 3–5 April, Ljubljana, Slovenia.
19. University of Nottingham. www.nottingham.ac.uk/sbe/research/ventcool/objectives.htm (2002).
20. Vakilaltojjar, S. and Saman, W. (2000). Domestic Heating and Cooling with Thermal Storage. *8th International Conference on Thermal Energy Storage*, 28 August–1 September 2000, Stuttgart, Germany, pp. 381–386.
21. Velraj, R., Anbudurai, K., Nallusamy, N., and Cheralathan, M. (2002). PCM Based Thermal Storage System for Building Air Conditioning at Tidel Park, Chennai. *Proceedings of WREC*, Cologne.
22. Arkar, C. and Medved, S. (2002). Enhanced Solar Assisted Building Ventilation System Using Sphere Encapsulated PCM Thermal Heat Storage. *IEA, ECES IA Annex 17, Advanced Thermal Energy Storage Technique—Feasibility Studies and Demonstration Projects 2 Workshop*, 3–5 April, Ljuubljana, Slovenia.
23. Ismail, K. (December 2001). Thermally effective windows with moving phase change material curtains. *Applied Thermal Engineering*, 21(18), 1909–1923. doi: 10.1016/S1359-4311(01)00058-8.
24. A Modular Solar Envelope House | Green Passive Solar Magazine. A website report from Old Dominion University, Norfolk, VA https://greenpassivesolar.com/2012/02/modular-solar-envelope-house/ (20 February 2012).
25. Erikci Çelik, S.N., Gedik, G.Z., Parlakyıldız, B., Çetin, M.G., Koca, A., and Gemici, Z. (2017). The Performance Evaluation of the Modular Design of Hybrid Wall with Surface Heating and Cooling System. *9th International Conference on Sustainability in Energy and Buildings, SEB-17*, 5–7 July 2017, Chania, Crete, Greece.
26. Ruud, S., Ostmann, L., and Oradd, P. (2016). Energy savings for a wood based modular pre-fabricated façade refurbishment system compared to other measures. *Energy Procedia*, 96, 768–778. doi: 10.1016/j.egypro.2016.09.139.
27. Lattke, F. (2014). smartTES-Innovation in Timber Construction for the Modernization of the Building Envelope. Final project report. ebook: www.woodwisdom.net, Germany.
28. Jokisalo, J. (2012). Rakennuksen energiankulutus muuttuvassa ilmastossa (Building Energy Consumption in a Changing Climate). Power point presentation (in Finnish), Aalto University, Finland.

29. Heikkinen, P., Kaufmann, H., Winter, S., and Larsen, K.E. (2011). TES EnergyFaçade-Prefabricated Timber Based Building System for Improving the Energy Efficiency of the Building Envelope. Final project report. ebook: www.tesenergyfacade.com, Finland.
30. A website report by Energy Star Home, The Home Store, Inc., Whately, MA (2015).
31. Fosdick, J. (11 November 2016). Tierra Concrete Homes. Updated by U.S. Department of Energy Federal Energy Management Program (FEMP).
32. ENERGY STAR® | Department of Energy—Energy.gov. A website report by Department of Energy, Washington, DC www.energy.gov/eere/buildings/energy-star (2015).
33. Energy Plus House, Wikipedia, The free encyclopedia, last visited 27 May 2019 (2019).
34. Wu, R. (2009). Energy efficiency technologies—Air source heat pump vs. ground source heat pump. *Journal of Sustainable Development*, 2, 14–23.
35. Chua, K., Chou, K., and Yang, W. (December 2010). Advances in heat pump systems: A review. *Applied Energy*, 87(12), 3611–3624. doi: 10.1016/j.apenergy.2010.06.014.
36. Self, S., Reddy, B., and Rosen, M. (January 2013). Geothermal heat pump systems: Status review and comparison with other heating options. *Applied Energy*, 101, 341–348. doi: 10.1016/j.apenergy.2012.01.048.
37. Omer, A. (February 2008). Ground-source heat pump systems and applications. *Renewable and Sustainable Energy Reviews*, 12(2), 344–371. doi: 10.1016/j.rser.2006.10.003.
38. White Oak Farm Geothermal | EnergySage. A website report by White Oak www.energysage.com/project/6288/white-oak-farm-geothermal/ (2013).
39. Gellings, C.W. and Parmenter, K.E. Efficient Use and Conservation of Energy—Vol. II - Efficient Use and Conservation of Energy in the Transportation Sector—Encyclopedia of Life Support Systems. A website report on Efficient Use and Conservation of Energy in the Transportation Sector. www.eolss.net/Sample-Chapters/C08/E3-18-03.pdf (2003).
40. Electric Car, Wikipedia, The free encyclopedia, last visited 13 August 2019 (2019).
41. Stevic, Z. and Radovanovic, I. (2012). Energy Efficiency of Modular Electric and Hybrid Vehicles, licensee InTech. http://creativecommons.org/licenses/by/3.0, doi: 10.5772/55237.
42. Stević, Z. and Vujasinović, M.R. (2011). Supercapacitors as a power source in electrical vehicles, In: Soylu, S. (ed) *Electric Vehicles—The Benefits and Barriers*. Intech, Rijeka, pp. 119–134.
43. Ellabban, O. (2001). Review of Torque Control Strategies for Switched Reluctance Motor in Automotive Applications. ieeeexplore.net/search/.
44. Stević, Z. and Rajčić-Vujasinović, M. (2006). Chalcocite as a potential material for supercapacitors. *Journal of Power Sources*, 160, 1511–1517.
45. Stević, Z., Rajčić-Vujasinović, M., Bugarinović, S., and Dekanski, A. (2010). Construction and characterisation of double layer capacitors. *Acta Physica Polonica A*, 117(1), 228–233.
46. Stević, Z. (2001). Supercapacitors Based on Copper Sulfides, Ph.D. Thesis, University of Belgrade.
47. Arbizzani, C., Mastragostino, M., and Soavi, F. (November 2001). New trends in electrochemical supercapacitors. *Journal of Power Sources*, 100(No 1–2), 164–170. Energy Efficiency of Electric Vehicles 131. doi: 10.5772/55237.
48. www.allaboutbatteries.com/Battery-Energy.html.
49. Miulli, C. (2012). Testing Modeling and Simulation of Electrochemical Systems, Doctoral Thesis, University of Pisa.
50. Samosir, A.S. and Yatim, A. (2008). Dynamic Evolution Control of Bidirectional DC-DC Converter for Interfacing Ultracapacitor Energy Storage to Fuel Cell Electric Vehicle System. *Power Engineering Conference, AUPEC '08*, 14–17 December 2008, Australasian Universities, Sydney, NSW, Australia, pp. 1–6.

51. Fay, A. and Nagata, M. (11–15 May 2014). (Quallion LLC), Zero-Volt and Long Life Chemistry Enhancements to Rechargeable Batteries for Medical Devices. A paper presented at 225th ECS Meeting, Orlando, FL.
52. Electrical PTC Heating Device. (1999). US Patent, US 5889260, Golan, Gad & Yuly Galperin.
53. Goodyear Tire and Rubber Company, Wikipedia. The free encyclopedia, last visited 11th Nov., 2019, https://en.wikipedia.org › wiki › Goodyear_Tire_and_Rubber_Company.
54. http://thesmartdrive.com/2009/07/the-effect-of-tire-pressure-on-fuel-economy/.
55. Knorr, U. and Juchem, R. (2003). A Complete Co-Simulation-Based Design Environment for Electric and Hybrid-Electric Vehicles, Fuel-Cell Systems, and Drive Trains, Ansoft Corporation, Pittsburgh, PA.
56. www.racecar-engineering.com/articles/f1/drs-the-drag-reduction-system/.
57. http://auto.howstuffworks.com/fuel-efficiency/hybrid-technology/hybrid-cars-utilize-solar-power1.htm.
58. www.teslamotors.com/blog/electric-cars-and-photovoltaic-solar-cells.
59. Zhang X., Chau, K.T., Chan, C.C., and Gao, S. (2010). An Automotive Thermoelectric-Photovoltaic Hybrid Energy System. *Vehicle Power and Propulsion Conference (VPPC)*, 2010 IEEE, 1–3 September 2010, Lille, France, pp. 1–5.
60. www.heat2power.net.
61. www.f1technical.net/forum/viewtopic.php?f=4&t=5751.
62. www.thegreencarwebsite.co.uk/blog/index.php/2012/10/12/huge-break-through-for-plug-in-hybrid-electric-vehicles/.
63. www.triplepundit.com/2010/07/vecarious-will-turn-you-car%E2%80%99s-waste-heat-into-energy/.
64. LaGrandeur, J., Crane, D., Hung, S., Mazar, B., and Eder, A. (2006). Automotive Waste Heat Conversion to Electric Power Using Skutterudite, TAGS, PbTe and BiTe. Thermoelectrics. *25th International Conference on ICT '06*, 6–10 August 2006, Vienna, Austria, pp. 343–348.
65. Energy Efficiency of Electric Vehicles 133. (2011). US Patent 5,680,032. doi: 10.5772/55237.
66. www.supercars.net/cars/5835.html. Story by Daimler AG. (2013).
67. Trancossi, M. and Pacoa, J. (2016). Optimized Modular Design for Energy Efficiency: The Case of an Innovative Electric Hybrid Vehicle Design. Paper No. IMECE2016-65430, p. V012T16A023; 16 pages doi: 10.1115/IMECE2016-65430, *ASME 2016 International Mechanical Engineering Congress and Exposition*, Volume 12: Transportation Systems, Phoenix, Arizona, USA, 11–17 November 2016, ISBN 978-0-7918-5066-4.
68. Rajalakshmi, N., Pandiyan, S., and Dhathathreyan, K.S. (January 2008). Design and development of modular fuel cell stacks for various applications. *International Journal of Hydrogen Energy*, 33(1), 449–454. doi: 10.1016/j.ijhydene.2007.07.069.
69. FuelCellsWorks. (31 December 2018). Fuel Cell System Breaks Efficiency Record.
70. Rodatz, P., Tsukada, A., Mladek, M., and Guzzella, L. Efficiency Improvements by Pulsed Hydrogen Supply in PEM Fuel Cell Systems. Swiss Federal Institute of Technology (ETH), Zurich, Switzerland; Paul Scherrer Institut (PSI), Villigen, Switzerland, Copyright © IFAC Terms & Conditions (2002).
71. Mohseni, M., Ramezanzadeh, B., Yari, H., and Gudarzi, M.M. (2012). The Role of Nanotechnology in Automotive Industries. Additional information is available at the end of the chapter. doi: 10.5772/49939.
72. Arico, A.S., Bruce, P., Scrosati, B., Tarascon, J.-M., and van Schalkwijk, W. (2005). Nanostructured materials for advanced energy conversion and storage devices. *Nature Materials*, 4(5), 366–377.

73. Armand, M. and Tarascon, J.M. (2008). Building better batteries. *Nature*, 451(7179), 652–657; Banerjee, A.N. (2011). The design, fabrication, and photocatalytic, utility of nanostructured semiconductors: Focus on TiO_2-based nanostructures. *Nanotechnology, Science and Applications*, 4, 35–65.
74. Bashyam, R. and Zelenay, P. (2006). A class of non-precious metal composite catalysts for fuel cells. *Nature*, 443(7107), 63–66.
75. Chan, C.C. (2007). The state of the art of electric, hybrid, and fuel cell vehicles. *Proceedings of the IEEE*, 95(4), 704–718.
76. Chan, C.K., Peng, H., Liu, G., McIlwrath, K., Zhang, X.F., Huggins, R.A., and Cui, Y. (2008). High-performance lithium battery anodes using silicon nanowires. *Nature Nanotechnology*, 3(1), 31–35.
77. Chen, S., Liu, W., and Yu, L. (1998). Preparation of DDP-coated PbS nanoparticles and investigation of the antiwear ability of the prepared nanoparticles as additive in liquid paraffin. *Wear*, 218, 153–158.
78. Chen, P., Xiao, T.-Y., Li, H.-H., Yang, J.-J., Wang, Z., Yao, H.-B., and Yu, S.-H. (2012). Nitrogen-doped graphene/ZnSe nanocomposites: Hydrothermal synthesis and their enhanced electrochemical and photocatalytic activities. *ACS Nano*, 6(1), 712–719.
79. Chinas-Castillo, F. and Spikes, H.A. (2003). Mechanism of action of colloidal solid dispersions. *Transactions of the ASME*, 125, 552–557.
80. Chmiola, J., Yushin, G., Gogotsi, Y., Portet, C., Simon, P., and Taberna, P.L. (2006). Anomalous increase in carbon capacitance at pore size less than 1 nanometer. *Science*, 312(5794), 1760–1763.
81. Chung, S.-Y., Bloking, J.T., and Chiang, Y.-M. (2002). Electronically conductive phospho-olivines as lithium storage electrodes. *Nature Materials*, 1(2), 123–128.
82. Cui, L.-F., Yang, Y., Hsu, C.-M., and Cui, Y. (2009). Carbon-silicon core-shell nanowires as high capacity electrode for lithium ion batteries. *Nano Letters*, 9(9), 3370–3374.
83. Demirdeven, N. and Deutch, J. (2004). Hybrid cars now, fuel cell cars later. *Science*, 305(5686), 974–976.
84. Gong, K., Du, F., Xia, Z., Durstock, M., and Dai, L. (2009). Nitrogen-doped carbon nanotube arrays with high electrocatalytic activity for oxygen reduction. *Science*, 323(5915), 760–764.
85. Greeley, J., Stephens, I.E.L., Bondarenko, A.S., Johansson, T.P., Hansen, H.A., Jaramillo, T.F., Rossmeisl, J., Chorkendorff, I., and Nerskov, J.K. (2009). Alloys of platinum and early transition metals as oxygen reduction electrocatalysts. *Nature Chemistry*, 1(7), 552–556.
86. Hvolbk, B., Janssens, T.V.W., Clausen, B.S., Falsig, H., Christensen, C.H., and Nørskov, J.K. (2007). Catalytic activity of Au nanoparticles. *Nano Today*, 2(4), 14.
87. Jacobson, M.Z., Colella, W.G., and Golden, D.M. (2005). Cleaning the air and improving health with hydrogen fuel-cell vehicles. *Science*, 308(5730), 1901–1905.
88. Jang, B.Z., Liu, C., Neff, D., Yu, Z., Wang, M.C., Xiong, W., and Zhamu, A. (2011). Graphene surface-enabled lithium ion-exchanging cells: Next-generation high-power energy storage devices. *Nano Letters*, 11, 3785–3791.
89. Ji, L., Rao, M., Zheng, H., Zhang, L., Li, Y., Duan, W., Guo, J., Cairns, E.J., and Zhang, Y. (2011). Graphene oxide as a sulfur immobilizer in high performance lithium/sulfur cells. *Journal of the American Chemical Society*, 133(46), 18522–18525.
90. Ji, X., Lee, K.T., and Nazar, L.F. (2009). A highly ordered nanostructured carbon-sulphur cathode for lithium-sulphur batteries. *Nature Materials*, 8(6), 500–506.
91. Kang, K., Meng, Y.S., Breger, J., Grey, C.P., and Ceder, G. (2006). Electrodes with high power and high capacity for rechargeable lithium batteries. *Science*, 311(5763), 977–980.

92. Kim, J., Samano, E., and Koel, B.E. (2006). Oxygen adsorption and oxidation reactions on Au(2 1 1) surfaces: Exposures using O_2 at high pressures and ozone (O_3) in UHV. *Surface Science*, 600(2006), 4622.
93. Koenigsmann, C., Santulli, A.C., Gong, K., Vukmirovic, M.B., Zhou, W.-P., Sutter, E., Wong, S.S., and Adzic, R.R. (2011). Enhanced electrocatalytic performance of processed, ultrathin, supported Pd-Pt core-shell nanowire catalysts for the oxygen reduction reaction. *Journal of the American Chemical Society*, 133(25), 9783–9795.
94. Koenigsmann, C., Sutter, E., Chiesa, T.A., Adzic, R.R., and Wong, S.S. (2012). Highly enhanced electrocatalytic oxygen reduction performance observed in bimetallic palladium-based nanowires prepared under ambient, surfactant less conditions. *Nano Letters*, 12(4), 2013–2020.
95. Kötz, R., Müller, S., Bärtschi, M., Schnyder, B., Dietrich, P., Büchi, F.N., Tsukada, A., Scherer, G.G., Rodatz, P., Garcia, O., Barrade, P., Hermann, V., and Gallay, R. (2001). Supercapacitors for peak-power demand in fuel-cell-driven cars. *Electrochemical Society Proceeding*, 21, 564–575.
96. Lee, J.K., Smith, K.B., Hayner, C.M., and Kung, H.H. (2010). Silicon nanoparticles-graphene paper composites for Li ion battery anodes. *Chemical Communications*, 46(12), 2025–2027.
97. Lee, S.W., Yabuuchi, N., Gallant, B.M., Chen, S., Kim, B.-S., Hammond, P.T., and Shao-Horn, Y. (2010). High-power lithium batteries from functionalized carbon-nanotube electrodes. *Nature Nanotechnology*, 5(7), 531–537.
98. Lefevre, M., Proietti, E., Jaouen, F.D.R., and Dodelet, J.P. (2009). Iron-based catalysts with improved oxygen reduction activity in polymer electrolyte fuel cells. *Science*, 324(5923), 71–74.
99. Levitsky, I.A. (2010). Hybrid solar cells based on carbon nanotubes and nanoporous silicon. *IEEE Nanotechnology Magazine*. doi: 10.1109/MNANO.2010.938654.
100. Li, Y., Zhao, Y., Cheng, H., Hu, Y., Shi, G., Dai, L., and Qu, L. (2011). Nitrogen-doped graphene quantum dots with oxygen-rich functional groups. *Journal of the American Chemical Society*, 134(1), 15–18.
101. Lim, B., Jiang, M., Camargo, P.H.C., Cho, E.C., Tao, J., Lu, X., Zhu, Y., and Xia, Y. (2009). Pd-Pt bimetallic nanodendrites with high activity for oxygen reduction. *Science*, 324(5932), 1302–1305.
102. Liu, R., Wu, D., Feng, X., and Müllen, K. (2010). Nitrogen-doped ordered mesoporous graphitic arrays with high electrocatalytic activity for oxygen reduction. *Angewandte Chemie International Edition*, 49(14), 2565–2569.
103. Magasinski, A., Dixon, P., Hertzberg, B., Kvit, A., Ayala, J., and Yushin, G. (2010). High- performance lithium-ion anodes using a hierarchical bottom-up approach. *Nature Materials*, 9(4), 353–358.
104. Mayrhofer, K.J.J., Juhart, V., Hartl, K., Hanzlik, M., and Arenz, M. (2009). Adsorbate-induced surface segregation for core–shell nanocatalysts. *Angewandte Chemie International Edition*, 48(19), 3529–3531.
105. Miller, J.R. and Simon, P. (2008). Electrochemical capacitors for energy management. *Science*, 321(5889), 651–652.
106. Nilsson, A., Pettersson, L.G.M., Hammer, B., Bligaard, T., Christensen, C.H., and Nørskov, J.K. (2005). The electronic structure effect in heterogeneous catalysis. *Catalysis Letters*, 100(2005), 111.
107. Ong, P.-L.; Euler, W.B., and Levitsky, I.A. (2010). Hybrid solar cells based on single-walled carbon nanotubes/Si heterojunction. *Nanotechnology*, 21(10), 105203.
108. Padhi, A.K., Nanjundaswamy, K.S., and Goodenough, J.B. (1997). Phospho-olivines as positive-electrode materials for rechargeable lithium batteries. *Journal of the Electrochemical Society*, 144(4), 1188–1194.

109. Peramunage, D. and Licht, S. (1993). A solid sulfur cathode for aqueous batteries. *Science*, 261(5124), 1029–1032.
110. Proietti, E., Jaouen, F.D.R., Lefevre, M., Larouche, N., Tian, J., Herranz, J., and Dodelet, J.P. (2011). Iron-based cathode catalyst with enhanced power density in polymer electrolyte membrane fuel cells. *Nature Communications*, 2, 416.
111. Qu, L., Liu, Y., Baek, J.-B., and Dai, L. (2010). Nitrogen-doped graphene as efficient metal-free electrocatalyst for oxygen reduction in fuel cells. *ACS Nano*, 4(3), 1321–1326.
112. Rapoport, L., Feldman, Y., Homyonfer, M., Cohen, H., Sloan, J., and Hutchison, J.L. (1999). Inorganic fullerene-like material as additives to lubricants: Structure–function relationship. *Wear*, 225–229, 975–982.
113. Schlapbach, L. and Zuttel, A. (2001). Hydrogen-storage materials for mobile applications. *Nature*, 414(6861), 353–358.
114. Serrano, E., Rus, G., and Garcea-Martenez, J. (2009). Nanotechnology for sustainable energy. *Renewable and Sustainable Energy Reviews*, 13(9), 2373–2384.
115. Simon, P. and Gogotsi, Y. (2008). Materials for electrochemical capacitors. *Nature Materials*, 7(11), 845–854.
116. Steele, B.C.H. and Heinzel, A. (2001). Materials for fuel-cell technologies. *Nature*, 414(6861), 345–352.
117. Stoller, M.D., Park, S., Yanwu, Z., An, J., and Ruoff, R.S. (2008). Graphene-based ultracapacitors. *Nano Letters*, 8(10), 3498–3502.
118. Subrahmanyam, K.S., Kumar, P., Maitra, U., Govindaraj, A., Hembram, K.P.S.S., Waghmare, U.V., and Rao, C.N.R. (2011). Chemical storage of hydrogen in few-layer graphene. *Proceedings of the National Academy of Sciences*, 108(7), 2674–2677.
119. Tarascon, J.M. and Armand, M. (2001). Issues and challenges facing rechargeable lithium batteries. *Nature*, 414(6861), 359–367.
120. Tollefson, J. (2010). Hydrogen: Fuel of the future? *Nature*, 464, 1262–1264.
121. Wang, X., Ding, B., Yu, J., and Wang, M. (2011). Engineering biomimetic superhydrophobic surfaces of electrospun nanomaterials. *Nano Today*, 6, 510.
122. Winter, M. and Brodd, R.J. (2004). What are batteries, fuel cells, and supercapacitors? *Chemical Reviews*, 104(10), 4245–4270.
123. Wu, Y.Y., Tsui, W.C., and Liu, T.C. (2007). Experimental analysis of tribological properties of lubricating oils with nanoparticle additives. *Wear*, 262, 819–825.
124. Wu, G., More, K.L., Johnston, C.M., and Zelenay, P. (2011). High-performance electrocatalysts for oxygen reduction derived from polyaniline, iron, and cobalt. *Science*, 332(6028), 443–447.
125. Wu, H., Chan, G., Choi, J.W., Ryu, I., Yao, Y., McDowell, M.T., Lee, S.W., Jackson, A., Yang, Y., Hu, L.B., and Cui, Y. (2012). Stable cycling of double-walled silicon nanotube battery anodes through solid-electrolyte interphase control. *Nature Nanotechnology*, 7, 309–314.
126. Yin, L., Wang, J., Lin, F., Yang, J., and Nuli, Y. (2012). Polyacrylonitrile/graphene composite as a precursor to a sulfur-based cathode material for high-rate rechargeable Li-S batteries. *Energy & Environmental Science*, 5. doi: 10.1039/c2ee03495f.
127. Zhang, M. and Dai, L. (2012). Carbon nanomaterials as metal-free catalysts in next generation fuel cells. *Nano Energy*, 1, 514–517, advance online publication.
128. Zheng, G., Yang, Y., Cha, J.J., Hong, S.S., and Cui, Y. (2011). Hollow carbon nanofiber-encapsulated sulfur cathode for high specific capacity rechargable lithium batteries. *Nano Letters*, 11(10), 4462–4467.
129. Zhou, J., Yang, J., Zhang, Z., Liu, W., and Xue, Q. (1999). Study on the structure and tribological properties of surface-modified Cu nanoparticles. *Materials Research Bulletin*, 34(9), 1361–1367.

130. Zhou, W., Wu, H., Hartman, M.R., and Yildirim, T. (2007). Hydrogen and methane adsorption in metal organic frameworks: A high-pressure volumetric study. *The Journal of Physical Chemistry C*, 111(44), 16131–16137.
131. Zhou, X., Zhu, Y., Liang, J., and Yu, S. (2010). New fabrication and mechanical properties of styrene-butadiene rubber/carbon nanotubes nanocomposite. *Journal of Materials Science and Technology*, 26(12), 1127–1132.
132. Zhu, Y., Murali, S., Stoller, M.D., Ganesh, K.J., Cai, W., Ferreira, P.J., Pirkle, A., Wallace, R.M., Cychosz, K.A., Thommes, M., Su, D., Stach, E.A., and Ruoff, R.S. (2011). Carbon-based supercapacitors produced by activation of graphene. *Science*, 332(6037), 1537–1541.
133. Turner, J.A. (2004) Sustainable hydrogen production, *Science*, 305(5686), 972–974.
134. Hong, J.W., Kang, S.W., Choi, B.S., Kim, D., Lee, S.B., and Han, S.W., (2012). Controlled synthesis of Pd-Pt alloy hollow nanostructure with enhanced catalytic activities for oxygen reduction, *ACS Nano*, 6(3), 2410–2419.
135. Wu, H., Zheng, G.Y., Liu, N.A., Carney, T.J., Yang, Y., and Cui, Y. (2012). Engineering empty space between Si nanoparticles for lithium-ion battery anodes. *Nano Letters*, 12, 904–909.
136. Ragone Plot, Wikipedia, The free encyclopedia, last visited 11 April 2019 (2019).
137. Ramsey, J.W. Integrated Modular Avionics, Integrated Modular Avionics: Less Is More—Avionics. A website report www.aviationtoday.com/2007/02/01/integrated-modular-avionics-less-is-more/ (1 February 2007).
138. Aleksa, B.D. and Carter, J.P. (1997). Boeing 777 Airplane Information Management System Operational Experience, IEEE.
139. Alena, R.L., Goforth, A., Figueroa, F., Ossenfort, J., and Laws, K.I. (2007). Communication for Integrate Modular Avionics. *IEEE Aerospace Conference*, 2007, Big Sky, MT.
140. Black, R. and Fletcher, M. (2006). Simplified Robotics Avionics System: An Integrated Modular Architecture Applied across a Group of Robotic Elements, IEEE.
141. Gangkofer, M., Kader, H., Klockner, W., and White, C.G. (2000). Transitioning to Integrated Modular Avionics with a Mission Management System. *RTO SCI Symposium*, Budapest, Hungary.
142. Watkins, C.B. and Walter, R. (2007). Transitioning from Federated Avionics Architectures to Integrated Modular Avionics, IEEE.
143. Di Vito, B.L. (2002). A Model of Cooperative Noninterference for Integrated Modular Avionics, IEEE.
144. Schneider, T. Innovative Approach for Modular and Flexible Positioning Systems for Large Aircraft Assembly. SAE Technical Paper 2015-01-2503. doi: 10.4271/2015-01-2503, p. 5, SAE 2015 AeroTech Congress & Exhibition. doi: 10.4271/2015-01-2503 (2015).
145. Capt. Dennis Mahoney, USN (ret) (RCT Systems), Don Longo and John Heinzel (NSWCCD SSES), and Jeff McGlothin (PEO SHIPS PMS 320) (23–24 May 2012). Advanced Shipboard Energy Storage System, *ASNE EMTS Symposium*, Philadelphia, PA.
146. Defense Science Board (DSB). (2001). More Capable Warfighting through Reduced Fuel Burden, task force report.
147. Defense Science Board (DSB). (2008). More Fight—Less Fuel, task force report.
148. ONR BAA 07-029. (2007). Fuel Efficient and Power Dense Demonstrator for the USS Arleigh. Burke (DDG 51) Flight IIA Class Ship.
149. Petersen, L.J., Ziv, M., Burns, D.P., Dinh, T.Q., and Malek, P.E. (2011). U.S. Navy Efforts Towards Development of Future Naval Weapons and Integration into an All Electric Warship (AEW). Paper Presented at the *2011 Engine as a Weapon IV Conference*, September 2011, Plymouth, England.

150. Petersen, L.J., Hoffman, D.J., Borraccini, J.P., and Swindler, S. B. (2010). Next-Generation Power and Energy: Maybe Not so Next Generation. Paper and presentation from ASNE Engineering the Total Ship and as published in the *Naval Engineers Journal*, December 2010, 122(4), 59–74.
151. Shipboard Energy Conservation Guide. (2005). SL101-AA-GYD-010 [0910-LP-103-3613 Rev 3 of 1 February 2005] Naval Sea Systems Command.

2 Modular Systems for Energy Conservation and Efficiency: Industrial and Commercial Sectors

2.1 ENERGY CONSERVATION AND EFFICIENCY IN INDUSTRIAL SECTOR

The industrial sector represents all production and processing of goods, including manufacturing, construction, farming, water management, and mining. Increasing costs have forced energy-intensive industries to make substantial efficiency improvements in the last 30 years. For example, the energy used to produce steel and paper products has been cut 40% in that time frame, while petroleum/aluminum refining and cement production have reduced their usage by about 25%. These reductions are largely the result of recycling waste material and the use of cogeneration equipment for electricity and heating [1,2]. Figure 2.1 describes the breakdown of energy consumption by various industry sectors [3].

Another example of efficiency improvements is the use of products made of high-temperature insulation wool (HTIW) which enables predominantly industrial users to operate thermal treatment plants at temperatures between 800°C and 1,400°C. In these high-temperature applications, the consumption of primary energy and the associated carbon dioxide (CO_2) emissions can be reduced by up to 50% compared with old-fashioned industrial installations. The application of products made of HTIW is becoming increasingly important against the background of the dramatic rising cost of energy. U.S. agriculture has doubled farm energy efficiency in the last 25 years.

The energy required for delivery and treatment of fresh water often constitutes a significant percentage of a region's electricity and natural gas usage (an estimated 20% of California's total energy use is water-related). In light of this, some local governments have worked toward a more integrated approach to energy and water conservation efforts. To conserve energy, some industries have begun using solar panels to heat their water. Unlike the other sectors, the total energy use in the industrial sector has declined in the last decade. While this is partly due to conservation efforts, it is also a reflection of the growing trend for U.S. companies to move manufacturing operations overseas [1,2].

There are numerous examples of the use of modular systems to conserve and improve energy efficiency in industrial sector. Major efforts are directed toward

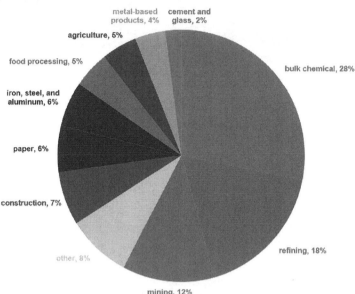

FIGURE 2.1 United States industrial sector energy consumption by type of industry in 2018 [3].

efficiency improvement using modular cogeneration (combined heat and power (CHP)) approach. Here we examine some of the successful case studies in this regard. CHP production is becoming more acceptable in the energy industry due to its favorable effect on energy efficiency. This approach allows good use of waste heat coming from fossil fuel, biomass/waste, solar, nuclear and geothermal power plants. The approach is easier to implement at a smaller scale than at a larger scale because heat cannot be transported to a long distance. While in previous book on modular systems [4], some reference to the modular approach to cogeneration processes has been made, due to its importance, in this chapter we examine in some details perspectives and examples of modular cogeneration processes.

2.2 MODULAR EFFICIENCY IMPROVEMENT MEASURES FOR WATER AND WASTEWATER FACILITIES

Specific power consumption of state-of-the-art wastewater treatment plants should be between 20 and 45 kilowatt hour (kWh)/(PE·a) [PE = Population equivalent or unit per capita loading]. The lower figure applies for large plants serving >100,000

PE, while the higher figure applies for around 10,000 PE. The smaller a plant, the higher is its specific power consumption. The figures refer to power consumption and do not take on-site power generation into account [5–29]. Power consumption depends not only on a plant's size but also on its design. The above figures apply for plants with N and P removal and anaerobic sludge digestion. Plants without nutrient removal consume less energy, and plants without anaerobic digesters consume more. Of course, efforts to save energy must not jeopardize wastewater and sludge treatment quality, but improved energy efficiency usually concurs with more effective treatment and operation. A typical industrial wastewater plant is illustrated in Figure 2.2.

Thorough investigations of 23 treatment plants in Germany and Switzerland showed that their power consumption could be reduced by between 20% and 80% and on average by about 67%. Such savings could be achieved in spite of the fact that most of these plants had been built or modernized recently, had already modest power consumption, and were provided with power-heat-cogeneration (PHC) systems. In all cases, the achievable power cost savings far exceeded the required investment and operation costs—on average by about 50%. Energy efficiency improvements are almost always economical [5–27].

Energy use can account for as much as 10% of a local government's annual operating budget [17]. A significant amount of this municipal energy use occurs at water and wastewater treatment facilities. With pumps, motors, and other equipment operating 24 h a day, 7 days a week, water and wastewater facilities can be among the largest

FIGURE 2.2 Overview of the wastewater treatment plant of Antwerpen-Zuid, located in the south of the agglomeration of Antwerp (Belgium) [30].

consumers of energy in a community—and thus among the largest contributors to the community's total greenhouse gas (GHG) emissions. Nationally, the energy used by water and wastewater utilities accounts for 35% of typical U.S. municipal energy budgets [14]. Electricity use accounts for 25%–40% of the operating budgets for wastewater utilities and approximately 80% of drinking water processing and distribution costs [14]. Drinking water and wastewater systems account for approximately 3%–4% of energy use in the United States, resulting in the emissions of more than 45 million tons of GHGs annually [25].

Increase economic and environmental costs can be reduced by improving the energy efficiency of water and wastewater facilities' equipment and operations, promoting the efficient use of water, and capturing the energy in wastewater to generate electricity and heat. Improvements in energy efficiency allow the same work to be done with less energy; improvements in water use efficiency reduce demand for water, which in turn reduces the amount of energy required to treat and distribute water. Capturing the energy in wastewater by burning biogas from anaerobic digesters in a CHP system allows wastewater facilities to produce some or all of their own electricity and space heating, turning them into "net zero" consumers of energy. Local governments can also reduce energy use at water and wastewater facilities through measures such as water conservation, water-loss prevention, stormwater reduction, and sewer system repairs to prevent groundwater infiltration. Measures to reduce water consumption, water loss, and wastewater lead to reductions in energy use and result in savings associated with recovering and treating lower quantities of wastewater and treating and delivering lower quantities of water.

Energy E kinky is guide focuses primarily on strategies for improving energy efficiency in water and wastewater facilities. Opportunities for improving energy efficiency in these facilities fall into three basic categories: (a) equipment upgrades, (b) operational modifications, and (c) modifications to facility buildings. Equipment upgrades focus on replacing items such as pumps and blowers with more efficient models. Operational modifications involve reducing the amount of energy required to perform specific functions, such as wastewater treatment. Operational modifications typically result in greater savings than equipment upgrades and may not require capital investments [18]. Modifications to buildings, such as installing energy-efficient lighting, windows, and heating and cooling equipment, reduce the amount of energy consumed by the facility buildings themselves. Processes of pumping and treatment are the largest consumers of energy in the water-use cycle. In most cases, pumping is the largest source of energy use before, during, and anaerobic treatment of water. For wastewater, where energy-intensive technologies, such as mechanical aerators, blowers, and diffusers, are used to keep solids suspended and to provide oxygen for biological decomposition, treatment accounts for the largest share of energy use [29]. Facility managers can perform energy audits or install monitoring devices that feed into their supervisory control and data acquisition opportunities for energy efficiency improvements.

The most effective way for communities to improve energy efficiency in their water and wastewater facilities is to use a systematic, portfolio-wide modular approach that considers all of the facilities within their jurisdiction. This approach allows communities to prioritize resources, benchmark and track performance across all facilities, and establish cross-facility energy management strategies. A portfolio-wide modular approach not only results in larger total reductions in energy costs and GHG emissions but enables communities to offset the upfront costs of more substantial energy efficiency projects with the savings from other projects. Adopting a portfolio-wide approach can also help local governments generate greater momentum for energy efficiency programs, which can lead to sustained implementation and continued savings. Before developing a portfolio-wide modular approach, local governments first need to understand the steps involved in identifying and implementing energy efficiency improvements at individual facilities. This guide is designed to help local governments understand how to work with municipal or privately owned water and wastewater utilities to identify energy efficiency opportunities. It provides information on how water and wastewater utilities have planned and implemented programs to improve energy efficiency in existing facilities and operations, as well as in the siting and design of new facilities.

2.2.1 Managing Energy Costs in Wastewater Treatment Plants

There are approximately 17,000 municipal wastewater treatment facilities in the United States. Wastewater and municipal drinking water treatment systems account for about 3% of the total U.S. energy consumption and approximately 35% of the total energy consumed by municipalities. A breakdown of electricity consumption at a typical municipal activated-sludge wastewater treatment facility is shown in Table 2.1 [7]. Electricity accounts for 90%–95% of the total energy consumed by these facilities; the remainder is fuel (fuel oil or natural gas) used for backup electricity generation or natural gas used to heat buildings [5–7].

There are quick fixes like turning things off like computers and monitors, other office equipment and plug loads, lights, chilled-water drinking fountains, vending machines, and motors. All of these could be achieved by walk-through audits. Some of the things could be turned down such as heating, ventilation, and air conditioning (HVAC) temperature setbacks, particularly in peripheral and backrooms, window shades and blinds are put down, and so on. The energy consumption can also be reduced by monitoring the operation and maintenance of the plant. This involves

TABLE 2.1
Electricity Consumption by End-Use

Aeration	54%
Wastewater pumping	15%
Anaerobic digestion	14%
Lighting and buildings	8%
Belt press	4%
Clarifiers	3%
Others	2%

Source: Data obtained from the U.S. Department of Energy report [7].

checking motors, inspect fans, bearings and belts and conducting repeated audits. The long-term solution involves monitoring energy usage; auditing aeration and pumping systems, aeration system controls, and adjustable-speed drive (ASD) on pumps; upgrading to light-emitting diode (LED) lighting; using turbo blowers; taking advantage of CHP systems; and adding solar photovoltaic (PV) for power and heat.

2.2.2 Benefits of Improving Energy Efficiency through Modular CHP in Water and Wastewater Facilities

Wastewater facilities can use biogas generated within the anaerobic digester to produce heat and, in many cases, electricity as well. As a rule of thumb, each million gallons per day of wastewater flow can generate enough biogas in an anaerobic digester to produce 26 kilowatts (kW) of electric capacity and 2.4 million British thermal unit (BTU)/day of thermal energy in a CHP system [24]. Not all wastewater facilities use anaerobic digesters, so CHP is not an option for all wastewater plants. Furthermore, some facilities with anaerobic digesters must rely on supplemental sources to provide enough energy for electricity generation in their CHP system. Improving energy efficiency through CHP in water and wastewater facilities can produce a range of environmental, economic, and other benefits, including

1. **Reduce air pollution and GHG emissions:** Improving energy efficiency in water and wastewater facilities can help reduce GHG emissions and criteria air pollutants by decreasing consumption of fossil fuel-based energy. Fossil fuel combustion for electricity generation accounts for approximately 40% of the nation's emissions of CO_2, a principal GHG. It also accounts for 67% and 23% of the nation's sulfur dioxide and nitrogen oxide (NO) emissions, respectively. These pollutants can lead to smog, acid rain and airborne particulate matters that can cause respiratory problems for many people [20,21].

    Example:
    Green Bay, Wisconsin Metropolitan Sewerage District has two treatment plants that together serve more than 217,000 residents. One of the treatment plants installed new energy-efficient blowers in its first-stage aeration system, reducing electricity consumption by 50% and saving 2,144,000 kWh/year, enough energy to power 126 homes and avoiding 1,480 metric tons of CO_2 equivalent, roughly the amount emitted annually by 290 cars [19,22,23].

    With more than two-thirds of the up-front installation and maintenance costs covered by the State of Minnesota and a local utility, the Albert Lea Waste Water Treatment Plant in Albert Lea, Minnesota developed a 120 kW microturbine CHP system, which saves the plant about $100,000 in annual energy costs. About 70% of the savings resulted from reduced electricity and fuel purchases, and the remainder from reduced maintenance costs. The installation of the CHP system raised awareness at the plant about energy use in general and led to a number of other energy efficiency improvements and additional cost savings [23].

Industrial and Commercial Sectors 71

2. **Support economic growth through job creation and market development:** Investing in energy efficiency can stimulate the local economy and spur the development of energy efficiency service markets. The energy efficiency services sector accounted for an estimated 830,000 jobs in 2010, and the number of jobs was growing by 3% annually [8]. Most of these jobs are performed locally by workers from relatively small local companies because they typically involve installation or maintenance of equipment [8,13]. Furthermore, facilities that reduce their energy costs through efficiency upgrades can spend those savings elsewhere, thereby contributing to the local economy [13].
3. **Demonstrate leadership:** Investing in energy efficiency epitomizes responsible government stewardship of tax dollars and sets an example for others to follow.
4. **Reduce energy costs:** Local government can achieve significant cost savings by increasing the efficiency of the pumps and aeration equipment at a water or wastewater treatment plant. A 10% reduction in the energy use of U.S. drinking water and wastewater systems would collectively save approximately $400 million and 5 billion kWh annually [24]. Facilities can also use other approaches to reduce energy costs, such as shielding energy use away from peak demand times to times when electricity is cheaper or (for wastewater plants) using CHP systems to generate their own electricity and heat from biogas.

   Example:
   In an initiative led by the city's mayor, a group of residents and city staff led an initiative in 2008 to develop a plan to make the City of Franklin, Tennessee, more environmentally sustainable. This group created the city's 2009 Sustainability Community Action Plan, which called for reductions in energy use and GHG emissions and directed Franklin's utilities to become more involved in energy efficiency audits. As part of its efforts to meet the action plan's energy goals, Franklin participated in the Tennessee Water and Wastewater Utilities Partnership, cosponsored by EPA Region 4. The partnership helped Franklin's water department identify and implement opportunities to reduce energy costs by more than $194,000 per year—a 13% reduction—through changes in operations and installing energy-efficient lighting. The improvements have avoided more than 1,280 metric tons of GHG emissions, equivalent to the annual emissions from powering 125 homes [9,10].
5. **Improve energy and water security:** Improving energy efficiency at a water or wastewater treatment facility reduces electricity demand, avoiding the risk of brownouts or blackouts during high energy demand periods and helping avoid the need to build new power plants. Water efficiency strategies reduce the risk of water shortages, helping ensure a reliable and continuous water supply.

   Example:
   East Bay Municipal Utility District (EBMUD), which provides drinking water to 1.3 million customers and handles wastewater for 650,000

customers in the San Francisco Bay Area, transformed itself from an energy consumer to a net energy producer. By 2008 the district had brought its GHG emissions back to their 2000 level and then reduced them by an additional 24% the following year, all while insulating itself from energy price fluctuations and supply uncertainties [11]. EBMUD started its energy transformation by cutting its energy use requirements to the point where its facilities now use 82% less energy than the California average for delivering 1 million gallons of drinking water from source to tap. It accomplished these improvements through design features, such as delivering drinking water via downhill pipes rather than using electric pumps, and through energy efficiency upgrades, such as installing microturbine CHP units. EBMUD's remaining energy needs are met by renewable energy systems, including hydropower, solar, and biogas. Excess power produced by the renewables provides a source of income through sales of electricity into the grid [12].

6. **Extend the life of infrastructure/equipment:** Energy-efficient equipment often has a longer service life and requires less maintenance than older, less efficient technologies [26]. Efforts to improve water efficiency or promote water conservation can also extend the life of existing infrastructure due to lower demand and can avoid the need for costly future expansions.
7. **Protect public health:** Improvements in energy efficiency at water and wastewater facilities can reduce air and water pollution from the power plants that supply electricity to those facilities, resulting in cleaner air and human health benefits [27]. Equipment upgrades may also allow facilities to increase their capacity for treating water or wastewater or improve the performance of treatment processes, reducing the potential impacts of sea-level rise, treatment failures, and risk of waterborne illness.

Example:
Millbrae, California implemented a program to divert inedible kitchen grease from the city's wastewater system, where it could clog sewer lines and cause releases of raw sewage into the environment, posing risks to public health. Waste haulers collect the grease daily from area restaurants and deliver it to the wastewater treatment facility, where it is processed in digester tanks to create biogas. Before the program was implemented, the grease ended up in area landfills where its decomposition produced methane emissions. The treatment plant's digester system produces enough biogas to generate about 1.7 million kWh of electricity annually, meeting roughly 80% of the plant's power needs [16].

2.2.3 MODULAR DESIGNS RECOVER ENERGY DURING DESALINATION

More than 1 billion people live in arid regions with little to no access to fresh water. Two-thirds of the Earth is covered in water, but 97% of this water is seawater. Land-based freshwater sources, from the developing world to more industrialized countries, have been nearly depleted as populations grow and demand increases. Seawater desalination can address the world's demand for consumable water.

For many countries, desalination is the only source of potable water. However, the equipment in the desalination process—from the pump to the motor—consumes vast amounts of energy [28]. Equipment efficiency is the key consideration in desalination plant design and operation because it ultimately determines the final cost of the water. The core hydraulic system of a reverse osmosis (RO) plant is the high-pressure pump. This pump pushes pretreated seawater through a membrane at a pressure that exceeds osmotic pressure. Because of the higher power requirements, the pump accounts for most of the facility energy consumption and determines the system's overall efficiency. A new technology combining a high-pressure pump, an electric drive, a booster pump, and an energy-recovery device can save up to 75% of energy costs in RO desalination plants.

The combined pumping unit includes a lubricated free axial piston pump and an axial piston motor as an energy recovery device. The piston pump produces the water flow to the RO membrane. This flow depends on the engine speed and the pump's displacement volume. The axial piston pump and the axial piston motor are connected through an axle driving shaft. When the redirected brine concentrate flows through the piston motor, it supports the electric motor in powering the piston pump. The relation between the displacement volume of the piston pump and the piston motor is defined by the recovery rate. The operating pressure is adjusted automatically by the membranes. The combined pumping unit receives pretreated seawater with a concentration of about 40,000 parts per million (ppm) from a feed pump. The pressure center raises the seawater pressure from 36 to 1,015 psipound force per square inch (psi) and transfers the water into the RO membrane. The membrane separates the brine from the water to produce permeate.

Modular design and energy efficiency are key characteristics for high-pressure pumps in non-industrial desalination applications such as seaside resort in Turkey. The modular desalination pumping unit will produce up to 70,000 gallons per day of fresh water for about 400 guests at the hotel. The resulting high-pressure brine returns to the pressure center where its mechanical energy is recovered. The recovered energy contributes to raising the pressure of the incoming pretreated seawater. The brine energy recovery system contributes to the high rates of energy efficiency during the process, with an energy consumption of less than 2 kWh/m^3.

The international desalination industry has experienced growing demand for modular design desalination plants. Modular design allows hotels, resorts, and even manufacturing sites to be independent of regional water supply companies. Modular RO desalination systems quickly and efficiently produce low-cost drinking water while saving space in facilities. A modular RO pumping unit responds to international demand by combining several components into one. The unit's single axial piston motor serves as an integrated energy recovery device. The motor creates high pressure, compensates pressure losses, and recovers energy. Because the system runs on a single electric motor and frequency inverter, a separate booster pump is not required [28]. By removing the need for a separate booster pump and motor, the number of components associated with other systems is reduced. Fewer components save on capital and installation costs, such as pipes between individual components, as well as maintenance and lifecycle costs. When using this integrated system, the only other mechanical elements required are a seawater feed pump and a membrane.

In addition, the combined pumping unit functions without lubricants, instead of using seawater. There is no fluid exchange between the feed water and brine, avoiding energy-wasting mixing. By removing the need for conventional lubrication, the unit saves up to 150 tons of CO_2 emissions per year.

In early 2013, a German company was awarded the order to supply and deliver a modular RO system with low operating costs to provide fresh water to feed cooling towers for an oil and gas company in North Africa. The high concentration of salt in the Red Sea required a high-pressure pump to desalinate the seawater and provide fresh water. The company needed a technology with low-energy consumption and low cost of operation to produce consistent high-quality water. The company built a cost-effective modular RO system equipped with combined pumping units with energy consumption of 2 kWh/m^3. The energy recovery device reduced operational cost by 25%–40% compared with conventional RO systems [28].

2.3 MODULAR COGENERATION IN BIOMASS/WASTE INDUSTRY

2.3.1 BioMax Modular CHP Technology

There is a need to develop small, modular, distributed-power generation systems that can utilize biomass feedstock from a relatively small harvesting radius, where the electricity and recovered waste heat can be efficiently utilized within that same radius. Community Power Corporation (CPC) develops and commercializes small, modular, distributed-power generation systems that can use waste biomass and convert it into electricity and thermal energy for such tasks as drying agricultural crops or space heating. To be economically feasible, these small systems must be fully automated to minimize labor costs and operate continuously at their full-rated power levels [31].

The ability to fully connect to the electrical grid to export excess electricity is critical to the economic viability of this technology due to the wide variety of electrical loads at farms or small businesses that could benefit from this technology. Utilities are required to connect alternative electrical energy from solar and dairy biogas to the grid, but no such requirement exists for similar power from biomass. The goal of the project by CPC was to demonstrate the feasibility of small-scale, modular gasification of biomass materials such as wood chips or agricultural residues to fuel an engine generator to produce CHP. The widespread application of the BioMax technology could produce a significant amount of distributed power that could be locally used, reducing electrical transmission loads and losses in the grid. The recovery of waste heat from the process could provide economic benefits by further reducing the use of premium fossil fuels, such as propane and natural gas. Use of fossil fuels and their GHG emissions could also be reduced if this technology becomes more widely used [4].

The modular BioMax technology contains three modules: feed preparation module, gasification module, and power production module. The feed preparation module consisted of a feed hopper that was loaded by a dump-bin mounted on a manually operated fork-lift. The feed hopper must be refilled several times a day. The feed hopper automatically delivered as-received, wet wood chips to a

SWECO separator to remove excessively large wood chips and excessively small sawdust, dirt, and pebbles from the desired wood chips. The SWECO separator used oscillatory vibration and gravity to move the wood chips across two screens, one above the other. The oversized chips were retained on the coarse upper screen and were automatically removed to an "overs" drum. The correctly sized chips were retained on the lower screen and were removed to a conveyor that takes them to drier. The fines fell through the lower screen and were removed to a "fines" drum. The middle fraction of the wet chips exited the SWECO separator and fell past an air knife to separate the low-density wood chips from high-density rocks, nuts, bolts, and other tramp materials. The desired middle fraction of the screened wood chips was then conveyed to the dryer. The hot air recovered from the producer-gas heat exchanger in the gasification and gas cleanup module was used to dry the wood chips to a desired moisture level between 8% and 15% (wet basis). The chips slowly moved through the dryer on a conveying system. The chips then fell into a surge bin, which provided warm, dry chips to a Flexicon feed auger, which automatically delivers wood chips to the gasifier as needed to maintain the level of wood chips in the gasifier. The feed preparation module used level sensors and two thermocouples acting through a computer to turn motors on and off and to adjust the speed of the tempering air blower. This module delivered wood chips on demand to the gasifier.

A gasifier vibrator was operated intermittently to settle the char bed, collapsing bridges, and rat holes. A grate mechanism was also operated intermittently to remove excess char/ash and friable clinkers. Three thermocouples were used to monitor the heat exchanger and automatically control the speed of the cooling air blower. The hot air produced in the heat exchanger was ducted to the dryer. This hot air could be used for space heating as part of the CHP load, if dry wood chips (<15% moisture content, wet basis) were available. Caps with sanitary fittings were removed for inspection and cleaning of the heat exchanger tubes, as needed during scheduled maintenance. During startup, electric heaters in the air duct preheated the heat exchanger prior to gasifier light off to avoid water condensation in the filter. These heaters were automatically shut off after ignition of the gasifier [4]. The filter consisted of five filter elements using a propriety filter media to remove the char and ash particles. The five filter elements were individually and sequentially valved off from the main gas flow to automatically remove the accumulated filter cake from the filter media. The char and ash fell to the bottom of the filter housing, where an auger continuously moved it out of the filter into a char-receiving drum lined with a plastic bag for easy disposal. The char-receiving drum was valved off temporarily from the filter auger and the drum switched out during continuous operation, as needed. Additional electrical resistance heaters were present in the filter enclosure to preheat the filter and avoid water condensation during startup. These heaters were automatically shut off after feeding had begun.

The flow rate of producer gas was determined with a Venturi meter mounted in the filter enclosure, through the use of pressure transducers and a thermocouple. The computer used these measurements to calculate the flow rate of producer gas and display it in real time. An oxygen sensor was mounted at the exit of the filter and monitored by the computer to warn of combustible mixtures in the gas cleanup system. A roots blower moved the gas through the system and sent it to a vortex flare. The vortex action in the flare mixed the producer gas with air to form a combustible mixture. The flare was equipped with an electrically powered glow plug to ensure ignition of the producer gas, as soon as it was combustible. In BioMax technology, the generator used was rated to produce up to 60 kilowatt-electricity (kWe) of three-phase 240 volts alternating current (VAC) 60 hertz (Hz) electricity. This commercial system was modified to control the flow of producer gas and combustion air. A commercial oxygen sensor was used to control the air/fuel ratio with CPC proprietary software and controllers but with commercial control valves. The produced electricity powered a variable load bank that was monitored with an Ohio Semitronics power meter. The waste heat was recovered from the engine coolant and the exhaust gases for space heating (and potentially cooling) purposes [4].

The tests of the BioMax 50 prototype were designed to demonstrate the successful integration of the three-component modules: feed preparation module, gas production module, and power production module. This was to show that this integration included sufficient automation so that one operator would be able to operate the system 24/7. CPC would measure and report the system's CHP values and conversion efficiencies to compare with the previous BioMax 12.5 kWe system operated on the Hoopa Indian Reservation. The energy commission previously supported the successful demonstration of a small BioMax 12 1/2 kW system at Hoopa Valley in northern California that converted forestry waste to electrical and thermal power. That technology evolved to the BioMax 15 system. The approach taken in this project was to scale-up that small system to a BioMax 50 system to deliver 50 kWe and over 80 kilowatts thermal (kWth). This project documented the performance and reliability of that scaled-up system with a 2-year field demonstration. CPC accumulated over 75 stop–start cycles and over 360 h of operation prior to shipping the BioMax 50 for field testing. The purpose of these repeated tests was to highlight any issues relating to thermal stresses on the system during heat-up and cool-down. The tests simulated approximately 18 months of operation since thermal cycling was the primary limiting factor to component life [4].

Most system performance measures were excellent. Stable gas flow rates were established and gains in electrical (gas conversion) efficiency revealed that system capacity could be increased significantly simply by boosting the pressure of the air/producer-gas mixture into the engine. Gasifier temperatures and flame front remained extremely stable. System pressure drops were consistently low, the gas quality was excellent with extremely low residual tars, and the heat exchanger and filter modules required very little cleaning or maintenance. Char/ash yields were on the order of 1.2% at CPC, which meant that the biomass fuel was very efficiently converted to producer gas. Moreover, these solid char/ash wastes were certified as nonhazardous. Low volume and low toxicity of the char/ash implied reduced handling efforts and disposal costs and also reduced hazards to personnel throughout the disposal process. The BioMax 50 system proved to be very reliable with a demonstrated high availability over the 2-year period of the field test. The electricity produced was grid quality. The gas produced was also used directly in modified propane burners to dry walnuts. Waste heat from the process was recovered and was also used to dry walnuts.

2.3.2 HIGH-EFFICIENCY MODULAR BIOMASS GASIFICATION SYSTEMS

Thompson Spaven [32] provides a series of partially or fully automatic modular gasifier systems for off-grid and grid-connected power generation systems. The standard modules are 50 and 100 kWe, but the company produces gasifiers in the range of 10–400 kWe. There are also modules for gas conditioning using sealed tank water scrubbing, water clean-up, and CHP applications. The company has a fully modular range of spark ignition gensets fully adapted to producer gas, with electronic air/gas mix control. Unit sizes are available from 10 up to 500 kWe. The company also supplies a series of nonproprietary add-on modules to deal with feedstock preparation and drying and for waste heat conversion to electricity.

2.3.3 AGRIPOWER'S MODULAR, MOBILE, AND TRANSPORTABLE CHP SYSTEM

The complete AG-375 CHP system consists of (a) the fuel feed hopper module, (b) the combustion chamber module, (c) the fans and filters module, and (d) the automated control module, all set on a tractor-trailer chassis. A second trailer chassis contains (e) AgriPower's proprietary T3 Technology Module (it stands for thermal treatment and transfer), (f) the 250 kW turbine generator module, and (g) the 125 kW waste heat generator. The systems provide flexibility as they can be operated on tractor-trailers or "permanently" installed (they can always be easily dismantled and relocated). The installation, set-up, and tear-down time are only 1–2 days. Modularity, ease of transport, and rapid set-up and tear-down times are significant benefits of the system [4,31].

AgriPower has developed a series of "biomass-fueled" (as described below), low-operating cost, modular, mobile and transportable, environmentally friendly, safe, and reliable electric generation systems (the "Systems"). The Systems provide a great deal of flexibility regarding the materials they can use for fuel and can utilize wood, sawdust, cardboard, paper, and most agricultural by-products and animal wastes ("Biomass") and many types of plastic as fuel to generate electricity and clean hot air suitable for cogeneration (as described below). The Systems are highly efficient, each 5 BTUs of fuel yields approximately 1 BTU of electricity and more than 3 BTUs of valuable cogeneration that can be used for heating buildings, making hot water, air conditioning and refrigeration, and for drying applications (a total of more than 4 BTUs or more than 80% efficiency). The Systems' ability to use virtually free or low-cost biomass and plastic waste as fuel makes them an attractive method of power and heat production for enterprises that produce or have access to this waste material (and thereby avoid paying expensive tipping fees and disposal taxes to have their waste brought to a landfill) to reduce the need to purchase natural gas or propane to produce heat and hot water or to replace expensive-to-operate diesel generators. The Systems are available with power outputs from 200 to 750 kW and from 2 up to 65 metric million British thermal unit (MMBTU)/hour (h) of heat output.

The Systems use a unique and proprietary clean air, high-temperature technology, space-age metals and ceramics, and a robust and proprietary personal computer (PC)-based software operating system. The Systems use low air pressure and are noncondensing (i.e., they do not use water or steam in their operation). Other biomass systems in the market use steam (which need ultra-clean water to operate) or gasifiers (that use dirty air that coats the turbine blades with contaminants) and thereby increase maintenance costs and down-time to drive their turbines. By contrast, AgriPower's "clean, hot air" technology contains no contaminants that can coat its turbine blades; hence, it is a highly reliable and relatively maintenance-free power generation system. The System's unique modular and mobile design enables it to be brought to where the fuel source is located rather than having to transport the fuel to a centrally located furnace, thereby saving vehicle transport and manpower expenses and eliminating the carbon footprint.

Biomass and plastic fuel is fed into the combustion chamber by a series of automated screw augers from a large fuel hopper. Its PC-based software system (using various embedded monitoring devices that take multiple readings every second) controls it and regulates its fuel intake, furnace temperature, and inlet air temperature to the gas turbine and waste heat generator. The System's high-temperature technology and extended fuel dwell time design provide a clean burn and minimize ash, which is automatically and safely removed from the System [4,31].

The Systems provide numerous significant benefits as follows:

1. **Virtually Free/Low-cost Fuel:** The Systems combust most types of biomass and many types of plastic waste. One of its best fuels is the leftover residue of whatever crops are being grown locally. This provides an abundant, renewable, carbon-neutral, and virtually free or low-cost fuel source.
2. **Significant Fuel Cost Savings:** Using biomass and plastic waste as fuel represents a dramatic fuel cost saving. Fuel savings or reduced electric and gas purchases from the local power company from using the System usually provide a financial payback period of about 3 years (before tax benefits). Fuel savings are estimated to be $10–$15 million or more over its useful life of 20+ years compared with the fuel costs of a comparably sized diesel generator set or from avoided tipping fees and disposal taxes by using virtually free or low-cost biomass and plastic waste as fuel in the Systems.
3. **Environmentally Friendly:** The Systems are virtually pollution-free. Certain types of contaminated materials (such as wood waste containing paint and creosote) can be used for fuel by adding an inexpensive mini scrubber that captures the contaminants during the combustion process and enables their proper disposal. Because of their design and size, the Systems are frequently exempt from local permitting requirements [4,31].
4. **Mobility, Reduced Fuel Transport Costs, Rapid Assembly, Size, and Weight:** The Systems are prefabricated, modular, mounted on trailers and shipped preassembled and pretested in or as standard 20′ or 40′ shipping containers, making them highly mobile and easily and inexpensively transportable to even the most remote areas (including disaster sites) or to where the fuel is located (thereby reducing or eliminating fuel transport costs). They are skid mounted and contain bolted fittings for rapid assembly and

can be installed on-site and producing electricity only 1–2 days after delivery. When installed, each System measures about 8′ wide by 40′ long by 12′ high and weighs about 80,000 pounds (40 tons).

5. **Cogeneration or Additional Power:** The Systems produce a considerable amount of clean hot air perfectly suited for cogeneration and that provides an additional and free energy source. Cogeneration can be used to operate distillation, desalinization, and water purification equipment; operate ice machines, refrigeration, and air conditioning units; generate heat for buildings; produce hot water for laundries, kitchens, room showers and baths, and heated swimming pools; and for industrial heating, bonding, and drying processes, such as drying paint and wood prior to shipment or drying extremely wet biomass prior to being used as fuel. Alternatively, the hot air can be used to generate an additional 125–250 kW of electricity.

6. **Ease of Use; Remote Monitoring; and Digital Proof:** The Systems have been designed to be easily and safely used by unskilled and inexpensive labor and are extremely user-friendly. The proprietary, fully-automated control system, with its fully-developed PC-based software program, and embedded sensors installed throughout the System, constantly monitors key control parameters (i.e., fuel feed and speed, load requirements, fan speeds, oxygen levels, etc.) resulting in a fully automated power plant, thus, eliminating the need for an expensive on-site engineer or technician. The System can be remotely monitored via telephone, the Internet, or satellite by the customer or AgriPower to confirm it is being properly operated, and digital proof can be obtained and used for carbon credit billing and payment purposes.

7. **Proven and Reliable Technologies; Low Operating Costs:** The Systems use proven combustion and gas turbine technologies that have been widely used for power and aircraft propulsion and for combusting a wide variety of fuels for more than 40 years. The reliability of these proven technologies translates into extremely low operating and maintenance costs and reduced downtime.

8. **Proprietary and Unique Technology Provides Considerable Benefits:** The Systems' proprietary clean, hot air technology separates the fuel combustion products from the gas turbine cycle, thereby reducing turbine wear and downtime and maintenance and operating costs. AgriPower's unique and proprietary air-to-air technology accommodates a wide variety of fuels and provides lower emissions than comparable furnace technologies.

9. **High Efficiency:** The System's use of proprietary technology and space-age metals and ceramics enables it to be highly efficient. Its high operating temperatures and long fuel dwell time enable it to completely combust the fuel and produce high output for its size and weight; for each 5 BTUs of fuel it consumes, it produces approximately 1 BTU of electricity and more than 3 BTUs of cogeneration (a total of more than 4 BTUs, or more than 80%

efficiency). Ash residue is minimal: between 1% and 3% depending on the type of fuel used [4,31].
10. **Distribution and Service Channels:** The System's proprietary technology has few moving parts and uses standard off-the-shelf components (several readily available fans and blowers) that can be inventoried on-site and easily repaired or replaced by AgriPower's extensive network of authorized sales and service representatives, distributors, dealers, strategic partners, and repair technicians.

2.3.4 Modular Compact CHP Using Local Heterogeneous Biomass Wastes

Enerkem from Montreal, Canada developed a proprietary thermochemical technology platform that powers the production of renewable biofuels and chemicals from waste rich in biogenic carbon. Over the last 10 years, the company has been validating its gasification technology by sourcing municipal solid waste (MSW) from numerous municipalities and other types of biomass feedstock and using it to produce a synthetic gas (syngas) that can then be further processed into chemicals. Enerkem has developed processes and business models to profitably produce cellulosic ethanol from MSW. Enerkem operates a commercial demonstration facility in Westbury, its first full-scale commercial waste-to-biofuels plant opened in June 2014, in Edmonton, Alberta. The company plans on developing similar facilities in Varennes (Quebec), Rotterdam (The Netherlands), Barcelona (Spain), and several other locations in the United States. These plants will produce methanol and ethanol and enable Enerkem to enter the fuels and commodities market [33,34]. Enerkem also proposed the project "Modular Compact Combined Heat and Power (CHP) Using Heterogeneous Biomass Wastes" for ecoEII funding. The project was awarded $2.913 million to determine whether the use of syngas derived from a mixed feedstock that is typical of that available in remote communities, could fuel an internal combustion engine (ICE) and produce CHP—a low GHG alternative to existing diesel-fueled systems [33,34].

A simulated mixed feedstock was prepared using available biomass and waste representative of that generated by a typical remote community. Source materials were procured in the Eastern Townships (Quebec), at close proximity to Enerkem's Westbury gasification demo plant. A processing or "finishing step" was then implemented and the result was a rather uniform shredded fluff. The fluff was introduced into Enerkem's Westbury plant via an innovative feeding system, which eliminated the need for the feed to be pelletized. Gasification of the prepared fluff was performed by Westbury's bubbling fluidized bed gasification reactor, with air as the partial oxidation agent. A low calorific (<6 megajoule (MJ) per normal m^3) clean bio-syngas was obtained. This low calorific syngas resulted in a 30%–40% lowering of the ICE's nominal power output as compared with that from hydrocarbon fuels, such as natural gas. Small-scale power generation was not effective, and large scale (1 megawatt (MW) system) system will have problem with available required biomass in the isolated northern region. It was, therefore, determined that the production of biofuels would be more cost-effective if done on a large scale at central conversion hubs (much like Enerkem's plant in Edmonton). Distribution to remote communities could then be achieved by way of existing fossil fuel transportation networks. The use of biofuels in place of diesel fuel to power ICEs in remote communities has the potential to reduce GHG emissions—by 1.72 tons carbon dioxide equivalent (CO_2e) per hour for each megawatt electrical (MWe) of power produced or 13,760 tons/year (over 8,000 operating hours). Enerkem intends to further develop and optimize their gasification platform and pursue the installation of a methanol-powered ICE demonstration system at a remote community [33,34].

2.3.5 Modular Power and Cogeneration Systems Using Landfill Gas

The landfill in the Three Rivers County generates a significant amount of landfill gas (LFG) that can last for many years. Initially, the Tree Rivers Waste Management Authority planned to construct a power generation building to house generator sets. During the public bid process, the waste management authority quickly learned that a modular system without constructing a power generation building with the engine room is much more cost-effective. The project owner decided to purchase a 100% premanufactured and modular LFG power generation plant. This "all-in-one" module was placed next to an office and warehouse building, with direct access to the power generation module [35,36].

At Tarboro, North Carolina, Edgecombe County analyzed their LFG availability for the next 20 years. Based on those numbers, the landfill owner decided to apply two generators, sized between 370 and 400 kWh each. Initially, the county planned to construct a power generation building with an engine room. During the public bid process, it was determined that it would save the county more than $850,000 by selecting modular premanufactured LFG power generation systems. The originally designed building plans were discarded. The county not only saved a significant amount of money, but the landfill owner was also able to reduce the technical risk and project lead times quite significantly. Also, a 100% premanufactured, "all-in-one," and "plug and play" LFG power generation plant was installed in approximately 3 days.

The Business Technology Center of Rockingham County in North Carolina planned to construct a power generation building with an engine room for their local landfill. After carefully analyzing all available options and based on the results from the public bid process, Rockingham County changed their plans. Instead of the traditional approach to constructing such a building, the decision was made to purchase a modular LFG power generation and CHP system. Using this approach, a substantial cost savings was achieved. Once again, this was a 100% premanufactured, "all-in-one," and "plug and play" LFG power generation plant.

In the 1980s, the European CHP cogeneration and on-site power generation industry initiated a transition from custom-built to modular premanufactured "all-in-one" and "plug and play" cogeneration systems. Modular CHP and on-site power generation systems were completely engineered and designed for most efficient utilization, applying best practice, standardized, consistently production-line manufactured, and did not require any additional engineering. A professional CHP cogeneration and landfill (LFG) energy conversion system contains the entire technology required to function most efficiently and effectively. An advanced gas engine is just one component. The complete design includes the "all-in-one" and "plug and play" package. This includes a closed-loop heat extraction technology if needed, "Best Available Control Technology" for exhaust emissions; sophisticated electronics; and many other vital high-tech components. Larger systems up to 2,000 kW single or multiple, for example, 3 MW, 4 MW, and so on, are possible. A containerized solution with container size of 9.8 ft^3 is possible. Enlarged 10 ft wide and 10 ft high with various lengths are also possible.

The heart of a modular on-site power generation system is a comprehensive and very advanced digital control technology that is monitored on-line around the clock. The electrical grid interconnection is also pre-engineered and genuine "all-in-one" modules are adequately prepared for synchronization and paralleling. If thermal energy is required, all modules are combined into one central thermal heat distribution assembly. Genuine "all-in-one" and "plug and play" modules can be connected very easily to the grid. The entire grid paralleling technology is an integral part of the systems control. The advanced grid interconnection protection relay technology is optionally 100% fully integrated into the CHP and power generation module controls. The interconnection is also monitored 24/7 around the clock. This example shows that the CHP market is currently in a transition phase from "customary design built" to "modular premanufactured" [35,36].

Kramer et al. [37] evaluated methods and designs for a modular waste processing system that utilized an anaerobic process to produce hydrogen from food, animal, or human waste. This hydrogen was used to produce electricity in a reciprocating engine or fuel cell. A solar energy system was designed and tested to provide heat for pre and postprocessing of waste and production of potable water. Potentially harmful pathogens from the waste were isolated from the environment and are drastically reduced by a thermal process. It was anticipated that this combined waste processing and renewable energy unit would be constructed in a standard shipping container for use in undeveloped and/or remote locations or at disaster sites.

The use of microbial organisms to produce hydrogen has many advantages over conventional techniques. Remote locations place a premium on the availability of electricity, heat, and potable water. Methane production by biological means is often used for producing electricity. Using microorganisms that produce hydrogen rather than methane significantly reduces GHG emissions for the overall process. By using this hydrogen in a reciprocating engine or fuel cell, the major end products would be electricity, water, and heat. To produce hydrogen rather than methane anaerobically it is necessary to first thermally pretreat the feed material. The developed solar energy system had consistently produced temperatures above 115°C. Typical hydrogen concentrations produced in the fermentation using food waste were 22% after 48 h. These efforts included the use of a statistical experimental design to determine optimal operating parameters and a preliminary modular energy system design. The details of this study are described in the published conference paper by Kramer et al. [37].

2.4 COGENERATION WITH SMALL MODULAR NUCLEAR REACTORS FOR INDUSTRIAL APPLICATIONS

In this section, we briefly illustrate methods to use cogeneration strategy for efficiency improvement with small modular nuclear reactors for various industrial applications. High, low, and mid-temperature operations are considered. The section also examines both load following (LF) and no-LF strategies. The global scenario is briefly examined. This subject is also treated in detail in my previous book [4].

2.4.1 Load Following Cogeneration Strategy for Typical Small Modular Reactors for Industrial Applications

Locatelli et al. [38] examined LF cogeneration strategy using two model small modular reactors (SMR) representing pressurized water reactor (PWR) and the very high-temperature gas-cooled reactors (VHTGRs) categories, namely, the International Reactor Innovative and Secure (IRIS) [39] and GTHTR300 (gas turbine high-temperature reactor) [40,41]. The VHTGR was resized to make the economic assessment over the same power output so that the two reactors can provide the same electric energy to the grid over 1 year or the same thermal power to the auxiliary plant.

As explained by Locatelli et al. [38], a key advantage of adopting multiple SMR instead of a single large reactor (LR) is the intrinsic modularity of an SMR site. In particular, it is possible to operate all the primary circuits of the SMR fleet at full capacity and switch the whole thermal power of some of them for the cogeneration of suitable by-products. Therefore, the LF strategy is realized at the site level by diverting 100% of the electricity produced or 100% of the thermal power generated by some SMR units to different cogeneration purposes and letting the remaining units produce power for the electricity market. Either in the case of full electricity conversion or in the case of full cogeneration operation mode, the efficiency would be maximized by design: SMR could run at full nominal power and maximum conversion efficiency and cogeneration plant size could be optimized for the thermal power rate. Considering four IRIS units, the power rates at site level could be approximately 0%, 25%, 50%, 75%, and 100%, and these steps are suitable for the general LF requirement by a baseload plant. Gas plants could provide the fine matching with the electricity market demand, as usual. By using SMR smaller than 335 megawatt electric (MWe) size, the possible power rate steps of the nuclear power station could be smoother. Even if the IRIS plants do not house multireactors in the same reactor building, the concept still applies.

For the sake of the reasoning, the study compares a site with four "independent IRIS SMR of 335 MWe" vs a site of same total power (1,340 MWe (megawatt electricity)) represented by a single LR. If during the night, the power needs to be reduced by about 50%, two IRIS can be disconnected from the grid and used for the cogeneration of other products while the remaining two IRIS will continue to produce electricity at full power rate and maximum efficiency. In the case of a 1,340 MWe, the 50% power reduction will cause some components (including pumps and turbine) to work outside the most efficient operating conditions, with lower efficiency of the cogeneration process. Therefore, when operating in LF mode, the four IRIS would be more efficient than a single stand-alone LR at plant level. The detailed analysis considering the coupling with a desalination plant is presented by Locatelli et al. [38].

Criteria for selecting cogeneration

The challenge for the "LF by cogeneration" strategy is to find an external system whose characteristics allow the coupling with a nuclear power plant (NPP). In particular, there are both economic and technical criteria for selecting the cogeneration system. From the economical perspective, the main criterion is that the investment profitability of the NPP-cogeneration combined facility, i.e. has a "Net Present Value" is above zero. Along with this essential criterion, there are a number of other criteria that need to satisfy the scrutiny of investors, such as the "payback time," the "value at risk," the "uncertainty of costs and revenues," and so on.

All the economic performances include the capital cost of the facility, the operation cost (including the opportunity cost related to the electricity), and the revenue(s). These parameters are "market specific": for instance, in case of a desalination plant, the value of the fresh water (and therefore the overall investment) would be different if the plant is located in the United Kingdom or Sweden (two countries with abundant low-cost fresh water) or a country with a desert climate and very limited fresh water. In the study by Locatelli et al. [38], the investment appraisal has the character of a "feasibility study," highlighting scenarios that might be relevant for a future detailed analysis and ruling out scenarios that are not worthy of future investigations.

Essentially, every system requiring electricity could be coupled with NPPs, if:

1. its power demand is large enough (670 MWe, i.e., the half of 1,340 MWe, which is the nominal power of 2 PWR SMR modules);
2. it is flexible enough to work at full power during the night and be switched off (or operated at a much lower load, consuming less electricity) during the day.

Systems using thermal energy are more demanding. The technical criteria for these systems are as follows:

1. requiring a large thermal power supply (1,000–2,000 megawatt thermal (MWth), i.e. approximately three times the excess electric power, due to the characteristic conversion efficiency of a light-water reactor (LWR) or an equivalent combination of electrical and thermal);
2. requiring relatively low-temperature heat (except for the coupling with a VHTGR, but currently the large majority of NPPs worldwide are PWRs and boiling water reactors (BWRs));
3. do not having a relevant thermal inertia;
4. allowing daily load variations, with rather fast dynamics.

Cogeneration, therefore, should not be a continuous process, but a "batch" type production and the last two characteristics are essential for the flexibility required by the LF operation. The actual daily power output profile for an NPP varies case by case and strongly depends on the local power supply and demand structure. The analysis was based on the following hypothesis:

1. the electric power required by the grid is equal to the nominal power during the day (8.00 a.m.to 12.00 p.m.) and to the 50% of the nominal power during the night (0.00 a.m. to 8.00 a.m.). This means that the available power for the auxiliary plant will be 670 MWe (or, in case of thermal application: 2,000 MWth with the PWR, 1,196 MWth with the VHTGR) for 8 h;
2. all 365 days of the year are considered identical in terms of energy required by the grid;
3. NPP availability is 95% (5% is lost for refueling and maintenance).

Although a few commercial NPPs worldwide provide energy to nonelectrical applications, nuclear energy is primarily used only for base-load electricity production. Of the nominally 440 commercial nuclear plants operational worldwide, 59 units in 9 different countries (Bulgaria, Czech Republic, Hungary, India, Romania, Russia, Slovakia, Switzerland, and Ukraine) are being used for district heating/process heat and 12 units in 3 countries (India, Japan, and Pakistan) are being used or considered for water desalination [42].

Locatelli et al. [38] examined the following eight systems for cogeneration with nuclear power:

1. Seawater desalination plant
2. Gasoline production plant
3. Oil sand extraction facility
4. Algae to biofuel production plant
5. District heating
6. Diesel-like fuel production from waste plastic pyrolysis
7. Waste wood palletization plant
8. Hydrogen production from water splitting plant

The results of the analysis by Locatelli et al. [38] offered the following assessments:

1. PWR has a number of options, such as seawater desalination and district heating, while VHTGR might access to further interesting options.
2. Among these, the production of hydrogen without using electricity in the process is the most interesting.
3. The thermal applications are preferable since they use steam before conversion into electricity, avoiding a loss of efficiency.

In the study by Locatelli et al. [38], a preliminary analysis was conducted on different possible systems to verify the feasibility of coupling them with an NPP operating in the LF mode. Some processes seem suitable for coupling with an NPP operated in the LF mode. Seawater desalination process is flexible enough to be coupled with a nuclear power source, and the largest existing desalination plants have a size compatible with cogeneration purpose by NPP. This is extremely relevant since many countries in the Middle East have plans for the construction of NPPs and they need fresh water.

The gasoline production can use nuclear energy, but this would be not economically competitive, given the current gas price and CO_2 emissions fee. It might be suitable just for "stable cogeneration," but not for LF. The preliminary calculation shows that for the plastic pyrolysis and wood pelletization the feedstock procurement is limited. Hydrogen has a theoretically infinite feedstock availability, and the current production is 1,000 times the output obtainable with the energy supply of a single NPP. Thus, hydrogen is a by-product that deserves a deeper feasibility analysis.

The study concludes that the technologies that might be most relevant for the LF, especially with SMR, are:

1. district heating, particularly in countries where there are several months with low temperature;
2. desalination, particularly for countries where the electricity price between day and night is different (at least by 100%) and water is high-priced;
3. hydrogen, in particular, if the electricity price during the night decreases getting close to zero or new high temperatures.

2.4.2 Examples of Cogeneration by Small Modular Reactors for Various Industrial Applications

While the analysis presented in the previous section indicates the suitability of three cogeneration systems with LF strategy, globally, numerous efforts are made for modular cogeneration of multiple systems including these three. These efforts are briefly summarized here and described in detail in my previous book [4] and by others [43,44].

2.4.2.1 Nuclear Process Heat Systems

Experience with nuclear process heat systems was gained in Canada, Germany, and Switzerland. In Canada, steam from the Bruce Nuclear Power Development (BNPD) is supplied to heavy water production plants and an adjacent industrial park at the Bruce Energy Center (BEC). The BNPD has successfully operated the nuclear steam and electricity-generating complex for over 20 years. It includes eight Canada Deuterium Uranium (CANDU) nuclear reactors, the world's largest heavy water plant, and the Bruce Bulk Steam System (BBSS). The BBSS can produce 5,350 MWth of medium pressure process steam. The cost of steam is reported to be significantly lower than the cost of heat from burning natural gas, which is the closest competitor. The six private industries established at the Bruce industrial park are (a) a plastic film manufacturer, (b) a 30,000 m² (7.5 acres) greenhouse, (c) a 12 million L/year ethanol plant, (d) a 200,000 ton/year alfalfa dehydration, cubing and pelletizing plant, (e) an apple juice concentration plant, and (f) an agricultural research facility.

In Germany, since December 1983, the Stade NPP PWR, 1,892 MWth, 640 MWe supplied steam for a salt refinery, which is located at a distance of 1.5 km. The salt refinery requires 45 ton/h process steam with 190°C at 1.05 megapascal (MPa), i.e.

30 MWth or 1.6% of the thermal output of the NPP. Since 1983, the steam supply by Stade NPP has a very high time availability, and the operating experience with process steam extraction is very good. In Switzerland, the 970 MWe PWR of Gosgen provides process steam for a nearby cardboard factory since 1979. The process steam (1.37 MPa, 220°C) is generated in a tertiary steam cycle by live steam from the PWR. It is then piped over a distance of 1,750 m to the cardboard factory. A maximum process steam extraction of 22.2 kg/s is possible, which corresponds to about 54 MWth or about 2% of the total thermal power of the PWR [4,43,44].

2.4.2.2 Oil Recovery Applications

For both enhanced oil recovery (EOR) as well as recovery of heavy oil, bitumen and oil shale (contained in solid kerogen), a process called "steam-assisted gravity drain" (SAGD) is heavily used. In developing nuclear-assisted heavy oil recovery, new technology challenges will have to be confronted in the adaptation and optimization processes. Massive steam production and its delivery at relatively high temperatures and pressures, inexpensive treatment of large quantities of raw water, ground stability problems, together with operational optimization of a multipurpose nuclear plant, are some of the main tasks that are confronted in the implementation of SAGD process. In addition, oil field characteristics and petroleum properties change from place to place, resulting in varying demand for steam volume, as well as varying steam conditions. A probably shorter oil field production time than reactor life also represents a new challenge. Many of these aspects point in the opposite direction from the standardization of nuclear plant design and construction with the intended purpose of reducing costs. The nuclear steam supply system and other plant components may, however, still be capable of some standardization.

A Canadian study on the use of CANDU reactors covered the application of nuclear power in heavy oil extraction from oil sands by steam injection in reasonable detail. At the time of the study (1980), cost savings from 25% to 50% with respect to burning coal for steam production were estimated. An organic-cooled CANDU reactor was also proposed for oil sand deposits deeper than 650 m, which required higher steam pressures. However, as already noted and despite the several advantages of LWRs and HWRs, these reactors are still limited in temperature and pressure capability for oil recovery from deep deposits. In 1981, the General Electric Company (GE) in the United Kingdom proposed the use of Magnox reactors for heavy oil recovery. The use of natural uranium as fuel and unsophisticated materials in the reactor fabrication was very attractive features of this reactor concept, especially for developing countries. However, the reactor's low uranium resource utilization and its relatively large plutonium production were the main drawbacks. Companies, such as General Atomics in the United States; the European ASEA Brown Boveri (ABB) and more recently, Siemens in the Federal Republic of Germany, have performed extensive studies on the design and application of high-temperature gas-cooled reactors (HTGRs) for EOR, including heavy oil recovery. Other countries, such as the Union of Soviet Socialist Republics (USSR), the People's Republic of China, and Japan, have also carried out design studies for HTGRs to be used in a process heat generation, the first two countries with specific interests in heavy oil recovery [4,43,44].

The HTGR-type reactors are capable of producing heat and steam at temperatures and pressures even higher than required for heavy oil recovery. They are, thus, capable of simultaneously producing high-quality steam for both processing and electricity generation, together with the injection steam. Such a cogeneration scheme adds versatility to the operation of a plant as oil field steam demand variations could be accommodated by diverting steam to electricity production, with a part of the electricity satisfying the plant demand and the excess for export. In a heavy oil project requiring large amounts of injection steam, steam diversion directly from the secondary cycle would not be feasible as conditions in the secondary cycle require water of much higher purity and quality, and hence, a more expensive water treatment than demanded injection steam. However, the use of reboilers solves this problem with, of course, an added capital cost and a minor penalty in injection steam conditions.

There is an expanded scheme for using nuclear power in both the recovery and upgrading operations, corresponding to the hydride–dehydride (HDH) process for Venezuelan extra-heavy crudes. In this scheme, an HTGR could provide to most of the process heat needs with temperatures in the range of 500°C–700°C, besides high-pressure injection steam. The higher temperature attained with more advanced versions of the developing HTGR (as proceeding in Japan and other countries), with temperatures around 900°C or more, would permit the capability of using process heat from the reactor to also power a steam methane plant for hydrogen production in addition to providing steam and electricity production. With the recent emphasis on the smaller output, the modular versions of the HTGR also provide additional flexibility in terms of oil field demand requirements, the ability to dedicate different modules to different types of service, if necessary, and more benign behavior in terms of operation and safety. Other reactor designs have also been considered for process heat applications. Liquid metal cooled reactors, able to produce 500°C heat, and particularly the small modular versions under development in the United States could certainly have possibilities for application in future heavy oil exploitation. Reactor cores that could sustain the most adverse circumstances with extremely low risks of even reaching fuel-melting temperatures (so there is no significant fission product release in case of an accident) are now the design goal of the advanced reactor concepts under development for commercial application [4].

A cogeneration option for the heat source is generally preferred due to the need for a modest amount of electricity because the reliability of the heat source is important, but it is not as critical as required for refinery applications, as discussed in the next section. Long disruptions in the heat-up process would become expensive if the rock formations are allowed to cool down significantly. Temperature requirements for EOR processes generally range from 250°C to 350°C, which is achievable with an LWR, such as NuScale, if heat recuperation or temperature boosting is used. The small module size and scalability of the NuScale plant provide unique opportunities for advanced oil recovery processes. In this case, however, challenges may be dominated by economic and logistical considerations. First, unlike the refinery application in which a centralized multimodule plant could provide the necessary electricity and steam for the entire refinery, the oil recovery application requires a more geographically distributed array of smaller energy sources. This might dictate that the modules be deployed as single or few module clusters, which could dramatically increase the cost of construction, operations, security, and so on.

The other logistical consideration for in-situ oil recovery is the longevity of the field operations in relation to the anticipated NuScale plant lifetime. Using enhanced recovery processes to extract heavy oil or oil from tar sands is likely to fully deplete a given field in 10–15 years. This depletion time may be somewhat lengthened with the widespread use of horizontal drilling that can significantly extend the reach of a well. Even so, the demand for a heat source at a fixed location is likely to be significantly shorter than the 60-year lifetime of a nuclear facility. Currently, oil companies use mobile natural gas units to provide energy. Comparable nuclear options might include the development of a mobile nuclear plant or a less enduring plant with a 10–20 year design lifetime. These options have their own set of challenges and are long-term solutions at best. In the case of oil shale, there are indications that the deposits are sufficiently massive, and the heating process sufficiently protracted that harvesting the oil from these formations may require many tens of years; hence, it is a better match for a nuclear plant with a traditional design life.

Although these economic and logistical considerations cause a nuclear option for in-situ oil recovery to be less obvious, ex-situ recovery, that is the shale oil is mined and processed elsewhere, is a potential application that overcomes the distributed and migratory issues of in-situ recovery. Furthermore, oil recovered from tar sands is of such low quality, that it requires processing in upgrader facilities located near the oil fields. Upgraders are basically in-field refineries that can service a large oil recovery area. As the local recovery operations migrate to new areas of the larger field, the oil is piped over progressively longer distance to the upgrader. The upgrader has a much longer lifetime and energy demand characteristics similar to finishing refineries. Therefore, the arguments for NuScale's suitability for refineries that are discussed in the next section apply equally well to upgraders [4,43,44].

2.4.2.3 Oil Refinery Applications

The energy requirements of a refinery represent a more practical and potential application of a NuScale plant. Refineries are large, energy-intensive industrial complexes with extended lifetimes similar to NPPs. Although the initial design lifetime of a refinery maybe 20 years, they are frequently upgraded as technology improves or product markets evolve and, typically, operate for several decades. One of the longest-running refineries in the United States is the Casper Refinery near Rawlins, Wyoming, and it has been operating for 90 years. Also, many refineries are in less populous areas and have industrial exclusion zones. In 2007, there were 145 U.S. refineries with the average refinery using roughly 650 MWth, which is distributed as 8% steam, and 17% using in excess of 2,000 MWth.

The small module size and modular nature of the NuScale plant are well-suited for this application. A multimodule, multioutput plant can be easily customized to the needs of a specific refinery while maintaining a highly standardized nuclear power module design. A single NuScale power module produces roughly 245,000 kg/h steam with an outlet temperature of approximately 300°C. As superheated steam has limited use for process heating, a secondary heat transport medium would be used like high-pressure water or a specially designed heat transfer fluid, such as

DOWTHERMTM. An intermediate heat exchanger (IHX) transfers heat to the secondary fluid stream (for use in preheating refinery process inputs) and provides additional isolation between the reactor and refinery. The end-use heated fluid characteristics can be adjusted as needed to match the requirements of a specific process. An initial estimate is that a single 160 MWth module can provide for the preheating of several refinery process input streams to 288°C (550°F).

Another attractive feature of the NuScale plant design for this application is the staggered refueling of modules. The plant is designed with a high level of independence between modules, including the power conversion systems so that the other modules can continue to produce power (or steam) while one of the modules is off-line for refueling. Many refinery processes become very inefficient if disrupted; therefore, they have a high-reliability requirement. A multimodule NuScale plant uniquely provides for redundancy and availability of energy supply. The output of a NuScale module is in the lower range of process temperature requirements. A variety of hybrid cycles have been suggested in the literature that could be used to boost the end-use steam temperature. The suggested approaches should use electrical heaters powered by an electricity-generating module or a natural gas-fired heater. In the latter case, although some feedstock is consumed for energy generation, feedstock usage is much less than if used to achieve the full steam temperature. Although advanced high-temperature reactors appear capable of reaching the required temperatures directly, an LWR-based nuclear system can be available in the near-term and provide a high level of confidence in deployment and operational reliability [4].

In addition to the liquid fuels produced in a refinery, there are many oil-derived petrochemical products that can be processed at temperatures within the range of a NuScale plant. Depending on the rise and fall of demand for liquid fuels, many refineries can shift to petrochemical production to maintain refinery capacity. Challenges for the viability of using a NuScale plant to provide process energy at a refinery are primarily regulatory—both for the nuclear plant and the refinery. The potential impact of an accident at either plant on the other plant will need to be carefully analyzed. The low-risk factor and high level of robustness in the NuScale design, which results from many best-in-class plant design features, will help reduce the regulatory and social–political hurdles for placing a nuclear facility near a refinery [4,43,44].

2.4.2.4 Desalination

Several SMR designs are being developed worldwide, including in the United States. One design, which is being developed by NuScale Power, LLC with the financial backing of Fluor Corporation, is the most modular of the U.S. designs with the smallest power unit size and the largest number of reactor modules in a single plant (up to 12 modules). The flexibilities afforded by the high level of modularization of the NuScale plant, coupled with a significantly enhanced level of plant safety and robustness, make it uniquely suitable for desalination applications in a wide variety of locations and coupling with multiple desalination technologies. This SMR has a wide range of industrial applications, which require a low- and medium-range of heat. There are several key features of the NuScale plant that collectively distinguish it from many other SMRs being developed today and make it especially well-suited for application to water desalination and other processes [4,43,44], they are as follows:

1. **Output reliability:** Using a small integral design for each reactor module enables significant simplification of the entire power train, thus eliminating many potential failure modes and reducing plant maintenance issues. For example, the natural circulation of the primary coolant eliminates several pumps, pipes, and valves. Additionally, other modules continue to produce power while one module is undergoing refueling or maintenance, which provides a high plant capacity factor. This ensures that power is always available to support the coupled desalination plant and other applications.
2. **Light water reactor technology:** The NuScale plant can be licensed within the existing LWR regulatory framework, thus, drawing on a vast body of operational data, proven codes and methods, and existing regulatory standards. This will facilitate expeditious licensing of the plant for near-term deployment to support the rapidly growing desalination market.
3. **Nuclear modularity:** While most new nuclear builds utilize modular construction practices, the NuScale design extends this approach to the nuclear steam supply system. Each power module is contained within a compact, factory-manufactured containment vessel and provides output steam to a dedicated and independent power conversion system. Also, the scalability of the plant from 1–12 modules further enhances plant economics and deployment flexibility for coupling to desalination plants of varying sizes and other applications.
4. **Dedicated power trains:** As each power module, including the power conversion system, is independent of other modules, it is possible to operate the plant in such a manner that some modules produce only electricity while other modules produce only steam for thermal heat applications. This allows the plant to cogenerate at the plant level, without the additional complexities of steam extraction from one or more turbine stages to support multiple desalination technologies and other applications.

The synergy created by these unique features of output reliability, reliance on existing light water technology, and the plant-level flexibilities afforded by the multimodule configuration, all combine to position the NuScale plant for early and successful application to water desalination. Of additional importance is the high level of plant resilience afforded by the small unit size, which improves the system response to upset conditions. As stated earlier, the majority of existing desalination plants use seawater as the water source; hence, they are located on coastlines and can be hit by a tsunami. The terrible earthquake-induced tsunami that struck Japan in March 2011 destroyed four of the six nuclear reactors that comprised the Fukushima Daiichi nuclear power station on Japan's eastern coast. As a result of this accident, a higher level of scrutiny on new nuclear plants located on coastlines can be expected, along with a higher standard for plant resilience to such extreme events. Although not discussed in detail here, the NuScale design offers an unparalleled level of plant resilience to the type of events that happened in Japan [7]. A more detailed review of global activities on nuclear heat desalination is given in Chapter 7.

2.4.2.5 Hydrogen Production

The transition to a hydrogen economy will require a significant expansion in the production and use of hydrogen. The use of hydrogen for all our transportation energy needs would require a factor of 18 more hydrogens than currently used. The use of hydrogen for all our nonelectric energy needs would require a factor of 40 increase. Clearly, new sources of hydrogen will be needed. Hydrogen produced from water using nuclear energy can be one of the sources and would avoid the use of fossil fuels and GHG emissions.

Hydrogen could be produced from nuclear energy by several means [4,40,45,46]. It can be produced by steam reforming of methane or coal gasification. Both of these methods, however, generate undesirable CO_2. Electricity from nuclear power can separate water into hydrogen and oxygen by electrolysis. The net efficiency is the product of the efficiency of the reactor in producing electricity, times the efficiency of the electrolysis cell, which, at the high pressure needed for distribution and utilization, is about 75%–80%. For LWRs with 32% electrical efficiency, the net efficiency is about 24%–26%. If an advanced high-temperature reactor with 48% electrical efficiency is used, the net efficiency could be about 36%–38%. Thermochemical water-splitting processes offer the promise of heat-to-hydrogen efficiencies of ~50%. While hydrogen has received much attention recently, there are few industrial facilities for large-scale hydrogen production [10]. SMR nuclear power offers a unique solution to hydrogen production processing, with several distinct advantages [4,40,45,46], including:

1. The new hydrogen production technologies powered with nuclear energy offer increases in efficiency and dramatic reductions in pollution when compared with traditional hydrogen production.
2. Using nuclear energy offers advantages in each phase of the lifecycle of an energy medium, including collection, production, transmission and distribution, and end-use. Production via nuclear is reliable and safe.
3. It appears to be economically viable.

While conventional LWR can be used readily to deliver electricity for the electrolysis process (however, at a very low total efficiency), HTGR with their helium coolant outlet temperature of up to 950°C would allow the direct utilization of the hot gas, which transfers its heat to the chemical process. The nuclear reactor and hydrogen plant can be separated from each other by using an IHX between the primary helium circuit of the reactor and hydrogen production system. The intermediate circuit serves the safety-related purpose of preventing primary coolant to flow through the (conventionally designed) hydrogen production plant and, on the other hand, product gas to access the nuclear reactor building.

The steam-methane reforming process as the most widely applied hydrogen production method was subjected to a long-term research and development (R&D) program in Germany with the goal to utilize HTGR process heat required as an energy input for methane splitting. The necessary heat exchanger components (IHX, reformer, and steam generator), with respect to their dimensions of the 125 MWth power class, were successfully tested in terms of reliability and availability in a 10 MW test loop over 18,400 h. The steam reforming of methane was investigated in the ethyl vinyl acetate (EVA) test facilities under nuclear conditions with dimensions typical for industrial plants. Also, EVA's counterpart, ADAM, a facility for the remethanation of the synthesis gas generated in EVA, was constructed and operated, demonstrating successfully the closed-cycle energy transportation system based on hydrogen as the energy vector. A corresponding experimental program on nuclear steam reforming was conducted and recently completed by the Japan Atomic Energy Research Institute (JAERI), Japan. In the United States, the hydrogen industry using steam reforming produces 11 million tons of hydrogen a year with a thermal energy equivalent of 48 gigawatts thermal (GWt). In so doing, it consumes 5% of the U.S. natural gas usage and releases 74 million tons of CO_2.

Nuclear coal gasification processes were investigated in the German long-term project prototype nuclear process heat (PNP), which has eventually resulted in the construction and operation of pilot plants for the gasification of brown coal (lignite) and stone coal, respectively, under nuclear conditions. Catalytic and noncatalytic steam-coal gasification of hard coal was verified in a 1.2 MW facility operated for about 23,000 h with a maximum throughput of 230 kg/h. The hydrogasification process was realized in a 1.5 MW plant operated for about 27,000 h with a throughput of 320 kg/h of lignite. These approaches are, however, in general, believed to be nonpractical.

One of the most promising concepts is the VHTR with its characteristic features of direct cycle gas turbine plant for high efficiency and a coolant outlet temperatures of 1,000°C. The top candidate production method is the sulfur–iodine (S–I) thermochemical cycle, considered presently as a reference method by various countries. Most advanced in this respect is the Japan Atomic Energy Agency (JAEA), which is planning to connect the S–I process to their high-temperature engineering test reactor (HTTR) and demonstrate for the first time nuclear hydrogen production foreseen for 2010. The United States is currently designing a "Next Generation Nuclear Plant" (NGNP). This government-sponsored demo program is based on a 400–600 MWth full-scale prototype gas-cooled reactor to provide electricity and process heat at 900°C–1,000°C. The 100 MW is planned to be used for hydrogen production using the I-S process as a reference method or alternatively high-temperature electrolysis (HTE). In addition, in China and Korea, ambitious programs have been started with the goal to bring nuclear hydrogen production to the energy market.

Schultz [46] of General Atomics outlined a modular helium reactor (MHR) to produce hydrogen. Selection of the helium gas-cooled reactor (GCR) for coupling with the S–I hydrogen production process allows us to propose a design concept and do preliminary cost estimates for a system for nuclear production of hydrogen. The latest design for the helium GCR is the gas turbine-modular helium reactor (GT-MHR) [9]. This reactor consists of 600 MWth modules that are located in underground silos. The direct-cycle gas turbine power conversion system is located in an adjacent silo. This new generation of the reactor has the potential to avoid the difficulties of earlier generation reactors that now have stalled nuclear power in the United States. The GT-MHR has high-temperature ceramic fuel and a core design that provides passive safety. A catastrophic accident is not possible. Under all conceivable accident conditions, the reactor fuel stays well below failure conditions with no actions required by the plant operators or equipment. By avoiding the need for massive active safety backup systems, the capital cost of the GT-MHR is reduced. The high-temperature fuel also allows high-efficiency power conversion. The gas turbine cycle is projected to give 48% efficiency. The high helium outlet temperature also makes possible the use of the MHR for production of hydrogen using the S–I cycle. By replacing the gas turbine system with a primary helium circulator, an IHX, an intermediate helium loop circulator, and the intermediate loop piping to connect to the hydrogen production plant, the GT-MHR can be changed into the "H_2-MHR."

Schultz [46] has estimated the economics of hydrogen production from nuclear energy using the S–I thermochemical cycle. He estimated the total capital and operating costs of the integrated hydrogen plant to calculate the cost of the hydrogen produced. The costs include all direct and indirect costs, plus interest during construction. The hydrogen plant's operating costs included normal operation and maintenance costs plus the cost of high purity water. Based on his analysis, he concluded that the nuclear production of hydrogen using the modular helium reactor could thus be competitive. As the price of natural gas rises, nuclear production of hydrogen would become more and more cost-effective. In summary, nuclear energy is an attractive potential source of hydrogen for the hydrogen economy. Nuclear production of hydrogen by the S–I thermochemical water-splitting cycle coupled to the MHR (the H_2-MHR) is an attractive candidate system for hydrogen production.

Richards et al. [45] also examined the use of MHR for hydrogen production. For electricity and hydrogen production, an advanced reactor technology receiving considerable international interest is a modular, passively safe version of the HTGR, known in the United States as the MHR, which operates at a power level of 600 MWth. For hydrogen production, the concept is referred to as the H_2-MHR. Two concepts that make direct use of the MHR high-temperature process heat are being investigated to improve the efficiency and economics of hydrogen production. The first concept involves coupling the MHR to the S–I thermochemical water splitting process and is referred to as the S–I-based H_2-MHR [45]. The second concept involves coupling the MHR to HTE and is referred to as the HTE-based H_2-MHR [45].

Yan et al. [40] presented a small modular HTR50S reactor design for multiple energy applications. The HTR50S is an SMR system based on HTGR. It is designed for a triad of applications to be implemented in successive stages. In the first stage, a base plant for heat and power is constructed of the fuel proven in JAEA's 950°C, 30 MWth test reactor HTTR and a conventional steam turbine to minimize development risk. While the outlet temperature is lowered to 750°C for the steam turbine, thermal power is raised to 50 MWth by enabling 40% greater power density in 20% taller core than the HTTR. However, the fuel temperature limit and reactor pressure vessel diameter are kept the same. In the second stage, a new fuel that is currently under development at the JAEA will allow the core outlet temperature to be raised to 900°C for the purpose of demonstrating more efficient gas turbine power generation and high-temperature heat supply. The third stage adds a demonstration of nuclear-heated hydrogen production by a thermochemical process. The low initial risk and the high long-term potential for performance expansion attract the development of the HTR50S as a multipurpose industrial or distributed energy source.

2.5 CHP PROJECTS IN MANUFACTURING INDUSTRIES

The U.S. Department of Energy (DOE) planned to allot up to $10 million to seven CHP projects (see Figure 2.3), some of them were CHP microgrids, to explore the technology's use by small to midsize manufacturers [48].

CHP is considered a highly efficient form of distributed energy because it uses the waste heat created during power production. In conventional generation systems, the waste heat is discarded. But, in CHP systems, it is used to heat or cool buildings or water or to create steam. CHP systems, therefore, use one fuel (usually natural gas) to produce two forms of energy—power and heat. Figure 2.3 illustrates finished regenerative oxidizer in a manufacturing plant.

In announcing the award, the DOE noted CHP's role in supplying supplemental power during natural disasters, lowering energy costs, easing the strain on the grid, and balancing the intermittency of renewable energy. The technology is widely used by large industrial facilities and college campuses that have operators to run the plants, but less so by small to midsize facilities without expert staff. So the DOE is

FIGURE 2.3 Finished regenerative thermal oxidizer at the manufacturing plant [47].

focusing the funds on the smaller facilities. The $10 million will be used for research on CHP technologies, including CHP microgrids, in terms of power electronics and control systems and electricity generation components.

The DOE listed the award winners as:

1. **Clemson University, Clemson, SC:** To develop a power conditioning system converter and a corresponding control system for flexible CHP systems. It will enable high-speed gas turbines to more effectively provide grid support functions and could be readily applied in new CHP installations or potentially retrofit some applications.
2. **ElectraTherm, Flowery Branch, GA:** To develop a high-temperature organic Rankine cycle (ORC) generation unit to provide additional power when needed by the grid, while also maintaining useful thermal energy for use in CHP applications. The ORC developed under this project will overcome the current limitation of useful thermal energy after the bottoming cycle.
3. **GE Global Research, Niskayuna, NY:** To develop a set of full-size grid-interface converter system and control solutions to interconnect small to midsize CHP engines to the low-voltage and the MV utility grid. The enhanced microgrid controller would enable engagement of a CHP system operator with the electric power grid operator through generator or microgrid controls.
4. **Siemens, Princeton, NY:** To develop an improved CHP system by demonstrating key novel components with computer simulations of standard technologies. The project will use a supercritical CO_2 bottoming cycle to increase electrical output to respond to grid requests. The approach will optimize the design of power systems, develop some key components (advanced heat exchangers), and demonstrate their performance in actual rig tests to prove the feasibility of the complete system.
5. **Southwest Research Institute, San Antonio, TX:** To expand the operational window of gas turbines for greater turndown, allowing more flexibility in the power/heat ratios, and enable grid support by CHP systems. This will be accomplished by developing a low-emission combustion system capable of sustaining combustion during high turndown operation. The project focuses on the Solar Titan 130 combustor and will expand the operating window to allow for turndown to 30%–40% load.
6. **University of Tennessee, Knoxville, TN:** To develop a power conditioning system converter and corresponding control system for flexible CHP systems. The power conditioning system (converter and controller will support different kinds of CHP sources and be scalable to form at needed power to serve as the interface connector between CHPs and a medium voltage (MV) grid. This project provides foundational work that could allow various CHPs/distributed energy resources to work together and interface directly with the utility MV grid. The technology could support multiple device CHP microgrids in the future.
7. **Virginia Polytechnic Institute, Blacksburg, VA:** To modular, scalable MV power converter featuring stability-enhanced grid-support functions for

future flexible CHP systems operating in small to midsize U.S. manufacturing plants. The project provides foundational work on power electronics and control systems enabled by advanced wide band gap (WBG) technology. It is capable of being implemented into a variety of existing and future CHP systems using a wide range of prime mover technologies [48].

2.6 ENERGY CONSERVATION AND EFFICIENCY IN THE COMMERCIAL SECTOR

The commercial sector consists of retail stores, offices (business and government), restaurants, schools, and other workplaces. The energy in this sector has the same basic end uses as the residential sector, in slightly different proportions. Space conditioning is again the single biggest consumption area, but it represents only about 30% of the energy use of commercial buildings. Lighting, at 25%, plays a much larger role than it does in the residential sector. Lighting is also generally the most wasteful component of commercial use. A number of case studies indicate that more efficient lighting and elimination of over-illumination can reduce lighting energy by approximately 50% in many commercial buildings [1,2]. As shown in Figure 2.4, the Empire State Building in New York was once the tallest LEED-certified building in the world [49].

Commercial buildings can greatly increase energy efficiency by thoughtful design, with today's building stock being very poor examples of the potential of systematic (not expensive) energy-efficient design. Commercial buildings often have professional management, allowing centralized control, and coordination of energy conservation efforts. As a result, fluorescent lighting (about four times as efficient as incandescent lighting) is the standard for most commercial space, although it may

FIGURE 2.4 Receiving a Gold rating for energy and environmental design in September 2011, the Empire State Building once was the tallest and largest LEED-certified building in the United States and Western Hemisphere [49].

produce certain adverse health effects. Potential health concerns can be mitigated by using newer fixtures with electronic ballasts rather than older magnetic ballasts. As most buildings have consistent hours of operation, programmed thermostats and lighting controls are common. Many corporations and governments also require the Energy Star rating for any new equipment purchased for their buildings. Solar heat loading through standard window designs usually leads to high demand for air conditioning in summer months. An example of building design overcoming this excessive heat loading is the Dakin Building in Brisbane, California, where fenestration was designed to achieve an angle with respect to sun incidence to allow maximum reflection of solar heat; this design also assisted in reducing interior over-illumination to enhance worker efficiency and comfort.

Advances include the use of occupancy sensors to turn off lights when spaces are unoccupied, and photosensors to dim or turn off electric lighting when natural light is available. In air conditioning systems, the overall equipment efficiencies have increased as energy codes and consumer information have begun to emphasize year-round performance rather than just efficiency ratings at maximum output. Controllers that automatically vary the speeds of fans, pumps, and compressors have radically improved the part-load performance of those devices. For space or water heating, electric heat pumps consume roughly half the energy required by electric resistance heaters. Natural gas heating efficiencies have improved through the use of condensing furnaces and boilers, in which the water vapor in the flue gas is cooled to the liquid form before it is discharged, allowing the heat of condensation to be used. In buildings where high levels of outside air are required, heat exchangers can capture heat from the exhaust air to preheat incoming supply air [1,2].

A company in Florida tackled the issue of both energy-conservation and enhancing its workplace environment by implementing a conveyor system that is 40%–60% quieter than traditional systems, emitting a noise level of only 55–50 decibels (dB), equivalent to a soft-rock radio station. Lighting was addressed by not only programming the lighting console so that isolated lights could be switched on and off in designated areas of the warehouse, but also by enhancing natural lighting through the use of skylights and a high-gloss floor [1,2].

2.7 MODULAR SYSTEMS FOR IMPROVING ENERGY EFFICIENCY IN COMMERCIAL FACILITIES

2.7.1 Five Office Buildings Using Modular Passive Heating and Cooling Design

Passive heating and cooling refer to techniques to manage the internal temperature and air quality of a building without using power. Here are three examples of new buildings where such techniques have been used. In each case, modeling of the effects of heat gains throughout the year is first undertaken. For example, the solar gain experienced by the building is a function of the total daily irradiation on the building surface, glazing area, angle of incidence at which the sun hits the window, transmittance value (g) of the glazing, and area of floor or wall reached by the sunlight, as well as its airtightness, U-value, and thermal mass [50–52].

1. **The Energon passive office building in Ulm, Germany:** This is a triangular, compact building with five storeys, and it has a physically curved facade enclosing a glass-covered atrium at the center. This provides ventilation and daylight. The building is a reinforced concrete skeleton construction with facades made of prefabricated wooden elements of largely equal dimensions. Insulation is 20 cm-thick under the foundation slab, 35 cm in the facade, up to 50 cm in the roof. The windows are thermally insulated triple glazing. The heat pumps and thermal stores help moderate the temperature [50].
2. **The office building at the Building Research Establishment, England:** "The solar chimneys" on an office building at the Building Research Establishment, England are automatically opened when required to release unwanted hot air. Their height and metal composition allow them to be heated by the sun, which heats the air internally. This rises through the chimney, drawing up air from within the building [50].
3. **The Solar XXI building in Lisbon, Portugal:** This building is the National Energy and Geology Laboratory's (LNEG) combined office and laboratory. This 1,500 m^2 (16,146 ft^2) multipurpose building in Lisbon, Portugal, is naturally ventilated and functions as a near-zero energy building. Its cost is said to be little more than a conventional building of the same size. The office space is on the south side of the building to take advantage of daylighting and solar heating. Spaces with intermittent use, such as laboratories and meeting rooms, are on the other side of the building. Office spaces are in use from 9 a.m. to 6 p.m. weekdays, and the ventilation pattern was arranged to suit this. The heat output of the PV modules is ingeniously used to supplement ventilation. There is an air gap behind each panel with openings to indoor and outdoor air at both high and low levels, where heat from the rear of the panel causes a convective flow [50].

    In winter, the upper opening takes air indoors, either from outside or from the room, through the lower opening to be heated. In summer, the upper opening lets the warmed air outdoors. The lower opening can either be open to the room to provide ventilation or to outside to provide cooling for the PV panels only. The building has a high thermal capacity and external installation on the walls and roof. The south facade supports 100 m^2 of solar PV modules and the majority of the glazing. Additional space heating is provided by 16 m^2 (172 ft^2) of roof-mounted solar thermal that also supplies hot water, which can be supplemented by a gas boiler.

    The 18 kilowatt peak (kWp; the rated power output under standard test conditions) grid-connected PV arrays supply electricity, and further panels are located in a car park, where they also provide shade. The entire system satisfies heating requirements of 6.6 kWh/m^2 and cooling requirements of 25 kWh/m^2. The annual electricity use for the building is about 17 kWh/m^2, of which 12 kWh/m^2 is from the PV arrays, leaving 30% to be drawn from the national grid. Natural lighting is encouraged. In the center of the building, a skylight provides light for corridors and north-facing rooms on all three storeys. The installed artificial lighting load is e8 W/m². There is no need for an active (powered) cooling system. Venetian blinds are outside the glazing to limit direct solar gain. Natural ventilation is promoted through the use of openings in the facade and between internal spaces, together with clerestory windows at the roof level, which help create a crosswind and stack effect [50].

Assisted ventilation is provided by convection from the PV module heat losses. To supplement this, in the cooling season, incoming air can be precooled by being drawn by small fans through an array of underground pipes. The openings are adjustable, and the air is allowed to rise through the central light well. The vents are manually operable, and staff needed to be educated in their use. In other buildings, such vents can operate automatically, governed by sensors. The building's occupants were surveyed, and they expressed 70%–95% satisfaction in terms of air quality and temperature. These buildings show that with a little thought, buildings can drastically reduce their heating and cooling costs and increase the comfort of occupants.

4. **Phase change material (PCM) integrated modular design by SEC, Australia:** The Sustainable Energy Centre (SEC) at the University of South Australia (2000) started work with PCMs in the mid-1990s with the development of a storage unit that can be used for both space heating and cooling. The nighttime charging and daytime utilization process during both heating and cooling seasons for a storage system comprising of two different PCMs are integrated into a reverse cycle refrigerative heat pump system utilizing off-peak power. As the air is forced through the system, it undergoes a two-stage heating or cooling process. It first goes through one PCM and then the second.

The melting/freezing point of the first material is below comfort temperature, while the second material has a melting/freezing point above comfort temperature. During the winter, the airflow is adjusted so that the system stores heat at night (by both materials melting) and releases heat at a temperature above comfort conditions (by freezing) at daytime. During summer, the airflow direction is reversed, and the system stores cold energy at night and it releases the cool air below comfort temperature at daytime [52].

The amount of reduction in the required capacity for the air conditioner and the amounts of the heating and cooling loads transferred to off-peak hours were reported by them using the computer model for the storage system. The annual energy cost savings were also provided by them. Using a thermal storage system containing two different PCMs can reduce the required capacity and the initial cost of an air conditioner for a residential house. It also can shift a portion of the heating and cooling loads to off-peak hours, when electricity cost is lower. The calculations for a typical house in Adelaide showed that a storage system consisting of 100 kg of 29°C PCM and 80 kg of 18°C PCM reduced the nominal rate of the air conditioner required by 50% of the total load. Also, the annual electricity cost was reduced by 32% due to shifting the load to off-peak time. If the proposed storage system is used on a large scale, the utility company could benefit by the shift of 52% and 41% of the air conditioning loads during the cold and warm seasons by reduced generation and transmission capacities [52].

5. **PCM-based cool thermal energy storage in building air conditioning system in Tidal Park, Chennai, India:** Velraj et al. [51] presented a detailed study on PCM-based cool thermal energy storage (CTES) integrated with building air conditioning systems in Tidel Park, Chennai, India. Tidel Park is a software office complex with 12 storeys and a building carpet area of about 92,900 m^2. The storage system in Tidal Park is the largest in the south Asian region and third largest in the world. Their study has been made on the existing large PCM-based cool storage, which is 24,000 TRH (303,840 MJ) integrated with a 3,000 TR (10,550 kW) chillers system. TR and TRH are units used in chiller design. The total capacity is split into four parallel paths by chiller banks A, B, C, and D, each comprising 750 TR. Each of the 750 TR-capacity chiller banks is provided by three 250 TR units. All the chiller banks of the air conditioning unit are connected to three plate heat exchangers (PHEs) of each 2,000 TR capacity. The installed capacity of the CTES system is 24,000 TRH. This is provided by four cool energy storage tanks, each of 6,000 TR capacity. Of these, one tank is kept as a standby and all the tanks are connected in parallel to the three PHEs. The PHE receives cold heat transfer fluid (Brine solution) from the chiller/CTES system and transfers the energy to the chilled water, which in turn transfers the energy to the air in the air handling unit (AHU). The modes of operation of such a system for load management have been discussed in detail in their study [51].The study illustrated following five features:

1. The peak cooling load demand can be reduced. In the present case, the CTES capacity of 24,000 TRH reduced the installation requirement of a centralized air-cooled vapor-compression air conditioning system from 6,000 to 3,000 TR. This reduces the electricity demand by approximately 4,000 kilovolt-ampere (kVA). The power distributor (Tamil Nadu Electricity Board, India) charges INR 300/kVA/month; hence, there is a saving in demand charge of INR 14.4 million/year (4,000 × 300 × 12 months) achieved due to this cool thermal storage.
2. The tariff difference during peak hours and off-peak hours can be exploited.
3. The performance of the chiller plant is high if the system is operated during the night hours when the surrounding temperature is low. The CTES system can be charged during the night hours and the stored energy can be retrieved during the daytime.
4. The chiller plant can be operated always under full load conditions; hence, the efficiency of the system is high.
5. The diesel generator set operation can be avoided for airconditioning load during a power failure.

Velraj et al. [51] suggested that the CTES system can be introduced economically for air conditioning in residential/commercial establishments.

2.7.2 MODULAR CHP FOR ENERGY EFFICIENCY-BENEFITING FROM A MODULAR APPROACH TO ENERGY EFFICIENT HEATING

An efficient commercial heating design requires matching both peak and low loads without oversizing the plant and wasting energy. This is where a modular boiler system comes into play. Modular boilers are designed as an alternative to large single boilers. Each module can be a separate boiler installed alongside another in a horizontal arrangement or as a vertical stack of boiler modules one above another [53,54]. A combined high turndown ratio of these modules—the operational range of a boiler, defined as the ratio of maximum to minimum capacity—offers a very efficient approach to commercial heating. As an example, each boiler module could have a turndown ratio of 5 to 1, meaning a stack of three modules will have a 15 to 1 turndown. If each module has an output of 250 kW, a vertical stack of three would have 750 kW output, giving a substantial range of outputs from 50 up to 750 kW to match demand. This ensures the load is matched to warm the building up, and in low load conditions, the boilers are not constantly cycling and wasting energy.

This accurate load matching has become increasingly important when designing heat networks, which typically have diverse load requirements and where boilers are used with renewables and CHP. The load requirements must meet minimum summer loads when mainly hot water requirements are being met with morning and evening spikes and demand in winter loads with additional heating needs. When installing CHP, designers need to look at the minimum load very closely as the unit will be running continuously to meet the baseload heating requirements. Boilers will provide a top-up to meet peaks in demand at different times, with their minimum output matching the baseload covered by CHP. However, they also need to cover the full load to ensure the security of supply to the buildings in the heat network.

Sufficient turndown of the boilers is essential to accurately match the baseload without wasteful cycling. However, the boilers also need to be running at or near the minimum modulation to ensure very high system operating efficiency and cost and carbon savings for the owner of the network. As system loads increase, the return water temperature decreases as more energy is consumed from the circuit. A modular approach means the increasing load can be met by multiple boilers operating at part-load conditions. Coinciding with favorable system temperatures, this allows high part-load condensing performance.

Boilers capable of operating with large differential temperatures (the difference between flow and return temperature from and to the boiler, e.g. 80°C/50°C) can closely match system dynamics throughout the year. The chartered institute of building service engineers (CIBSE) Guidance AM12—CHP for Buildings, makes reference to designing district heating schemes with a minimum of 30°C differential temperature (delta T) for efficient operation. Originally, larger differential temperatures were only possible from large water content boilers, without the benefits of condensing capability, higher turndown ratios, and accurate load matching compared with modular condensing boilers. But now, some modern modular condensing boilers

Industrial and Commercial Sectors

with low water content are able to operate up to 40° delta T, maximizing condensing operation and increasing energy and cost savings, while being compact enough to fit through a single doorway. This gives a considerable advantage compared with large water content single boilers, which take up more room and require installation before the building is built. The supporting water pipes and pumps can also be smaller, helping to reduce installation costs, and gaining further energy savings.

In the future, this system is expandable. A lot of projects apply the phased approach for purchase and installation. In particular, when a building needs to keep running during a refurbishment, such as a hospital, a modular boiler system can provide this flexibility. All equipment does not need to be purchased and installed right away, but it can be split over budget years to ease the financial burden, which means an expansion of the system at a later point can be easily carried out. Another possibility is to fit everything at the start but only use the boilers that are needed, switching them on incrementally when energy demand increases, helping control costs and energy use. Modular boilers tick all the boxes for heat networks, energy centers, as well as basement plant rooms, rooftop plant rooms, schools, hospitals, and city center developments. Buildings, such as the St Paul's Cathedral, can face great challenges when a boiler replacement is required due to strict regulations regarding the modification of the building structure. Compact modular boilers can provide a space- and energy-saving solution with flexibility and accessibility in terms of installation.

For several years, Tecnoimpianti Srl has been producing generators that are distinguished by their extreme robustness, low noise, and reliability for their innovative design. These generators are available in either fixed positions or on wheels. The bogies are approved for towing fast roads suitable for operation under severe environmental and prohibitive temperatures between −20° and +55°. The absence of welds makes the means extremely qualitative. The soundproofing of Technoimpianti equipment is achieved using galvanized steel painted materials which are extremely resistant to rust and corrosion. The noise level, independent of the engine used, is in full compliance with the regulations in force [54]. The coating material is a high-quality class 1 fire reaction. The control panels and control of generators can be manual, automatic, remote-controlled, or suitable for parallel operation. Tecnoimpianti Srl also realizes modular cogeneration systems for the production of electricity and thermal energy recovery from the cooling system and the exhaust gases of the engine first [53,54].

Cogeneration systems, while providing heat and electricity, allow not only to increase energy independence of the user, but also to be combined with water heating systems, regeneration for air conditioning systems and water purification systems. Among other features, this CHP system has a centralized air suction system, air compressed distribution, air conditioning, and humidification system. The system is very suitable for dyeing and textile industry.

2.7.3 Energy Efficiency for Data Centers

Data centers use nearly 2% of the world's supply of electricity at any given time, and 37% of that amount is used to keep computing equipment cool.

Data centers are a lynchpin of our modern economy. Keeping up with the explosive growth of digital content, big data, e-commerce, and Internet traffic is making these facilities one of the fastest-growing consumers of electricity in developed countries. Data-center power consumption is on the rise, increasing 56% worldwide and 36% in the United States from 2005 to 2013. Not only is this a drain on the power grid, but it also taxes water supplies. A 15 MW data center can use as much as 360,000 gallons of water/day—that's more than half the water in an Olympic-sized swimming pool [55,56]. A large data center for Google is illustrated in Figure 2.5 [57].

Due to the rising power costs, energy efficiency is growing as a top concern of data center managers. DellTM PowerEdgeTM M1000e infrastructure directly improves the energy efficiency of current and future data centers through a combination of hardware and software improvements. Through higher power density, load balancing, and improved energy efficiency, hardware directly affects a data center's operating cost. The Dell Chassis Management Controller (CMC), which is a critical part of all Dell blade servers, provides software features that help the 2,700 watts (W) power supply perform at maximum efficiency. Features dynamic power supply engagement (DPSE), maximum power conservation mode, and power monitoring help data center managers push the limits of the M1000e modular blade server enclosure. With data centers increasing in density and requiring more performance per watt, more power at higher efficiency is required. The Dell PowerEdge M1000e infrastructure meets the demand of the modern data center and pushes power supply technology to the cutting edge. Higher alternating current-direct current (AC-DC) conversion efficiency, extreme capacity, and innovative algorithms have ushered the 2,700 W power supply to the next level of modular infrastructure. The AC-DC conversion has reached a new high of 94% efficiency, with an increase of up to 13% in

FIGURE 2.5 Google Data Center, The Dalles, Oregon.

power output over prior consumption. When these modular hardware advantages are coupled with the power features in the Dell ChCMC, the amount in power efficiency is very high.

The 2,700 W power supply is designed to improve performance per watt for Dell's modular solutions. One key to improving the ratio is to maximize the amount of energy applied to the modular servers when converting AC power to DC power. The AC-DC efficiency conversion is engineered to meet certification levels that are maintained by an industry consortium run by the Environmental Protection Agency (EPA). This consortium, called 80 PLUS, is an initiative by the technology industry driven by the government to promote advancements in energy efficiency in power supplies. With over 200 power supplies certified by the 80 PLUS program, this initiative has helped drive efficiency levels from 89% to 96% over the years. Dell's 2,700 W power supply meets the 80 PLUS Platinum level, which is a peak efficiency of 94%.

One of the main advantages of modular systems is shared resources and infrastructure. This infrastructure includes fans, power supplies, and switches. As the blades, switches, and fans create a power draw, each power supply contributes more power for better asset utilization. The higher power density enables higher density in the newest generation of blade servers. For example, Dell's current blade servers offer 24 dual in-line memory modules (DIMMs) in half-height blades compared with 16 DIMMs in just the previous generation. Full-height blades have 48 DIMMs with four processors. Not only is the DIMM and processor density higher, but both items are consuming more power, faster than the base operating frequency. A higher capacity allows information technology (IT) managers to take advantage of dynamic loading and turboboost created by processors. Dynamic loading can happen at any time. When workload for a high performance compute cluster starts—for example, the public is looking at breaking news—this surge on blade servers and blade switches creates a sharp rise in the load on power supplies. The M1000e with the 2,700 W power supply can handle the load with efficiency and elegance. Load balancing is one method the power supplies use to handle power surges neatly. All power supplies in the M1000e provide the same amount of power, which adds to the longevity of the power supplies and helps maintain power redundancy.

Another advantage of high-capacity supplies is that the 2,700 W power supply has been built with additional ride-through capability for situations where power is momentarily lost. It can act as a buffer while the data center backup power comes online. This feature complements data center infrastructure like uninterruptible power supplies (UPSs) by providing power quickly in the event of a brownout or power grid glitch. The Dell CMC, which manages the M1000e infrastructure, implements various power features to ensure the 2,700 W power supply is used most efficiently. These features include [55,56]:

- The DPSE mode
- Maximum power conservation mode
- Power monitoring

For DPSE, the CMC monitors total enclosure power allocation and moves the power supplies that are not required into a standby state, causing the total power allocation of the chassis to be delivered through fewer power supplies. This improves asset utilization of the online power supplies, and at the same time, improves the longevity of the standby power supplies.

With the maximum power conservation mode feature of CMC, one can place the M1000e into the lowest power consumption mode possible. This mode is used in combination with DPSE and the 2,700 W power supplies. In this mode, the system can place all but one power supply into standby during periods of low utilization, saving power by turning other power supplies off. Power monitoring provides information on system input power, peak system power, minimum system power, and many more valuable power statistics. Since the 2,700 W power supply increased the power to the chassis by 13% over the 2,360 W, the CMC increased the power cap to match. The increased power cap enables the chassis to accept any power load.

One of the advantages of the M1000e is its thermal design. The thermal architecture within the system ranges from the layout of each blade to optimized fans, and it also includes the CMC that manages the infrastructure cooling. The architecture focuses on air management that provides parallel air paths for ambient air to cool blades, switches, power supplies, and other infrastructure hardware. Because ambient air is delivered to the infrastructure with minimal preheating, the amount of power required by the M1000e's nine fans to cool the system is minimized. As energy efficiency improvements are a primary goal of many data centers, Dell's latest modular fan improves on the initial M1000e fan design by reducing the maximum DC (per unit) [55,56].

Aligned Energy, Danbury, CT, an integrated technology platform, has also developed a solution that eliminates infrastructure complexity and waste, heightens visibility and control, and improves reliability in data centers. One of Aligned Energy's subsidiary companies, Inertech, set out to address the key drivers of cost in data centers as (a) over-building a data center, (b) underusing an existing data center, and (c) using cooling technology inefficiently. With 80% of a data center's costs going toward the electrical and mechanical systems, Inertech determined that the only way to effect real change was to drive down the cost of the cooling system and electrical blocks. Using the Danfoss, Baltimore, portfolio of products and application expertise, Inertech personnel were able to develop a solution that scales mechanical and energy infrastructure directly to servers and storage use, which has yielded significant savings in water and electricity costs. The majority of a data center's upfront costs are in building chiller infrastructure. The average data center is constructed to a "perceived build," based on the anticipated IT capacity. Companies try to predetermine the size of chiller plants needed to support IT. However, these calculations are highly complex and difficult to accurately predict. Often, companies significantly overbuild data centers, unnecessarily inflating their capital costs [58,59].

Inertech's patented model has been able to reduce 80%–85% of the cost of starting a data center. On the operational side, because Inertech's cooling systems are 90% more efficient than a traditional chiller plant, it is able to drastically cut the electrical infrastructure that supports that data center for its customers. Inertech built a platform of small modular cooling blocks that can be scaled to actual IT use. It worked with Danfoss to identify critical components that would enable it to

Industrial and Commercial Sectors   107

maximize efficiencies for energy and water use. The system design supports data center needs in a much more cost-effective delivery model than a traditional chiller plant, as the smaller platforms can be installed exactly when they are needed, or "just-in-time," without interrupting IT online operations.

Inertech's cooling cycles were designed modularly, in the equipment as well as in the physical infrastructure, enabling data centers to scale over time. Working in a supply chain of 4–6 weeks, Danfoss and Inertech deploy and hook up preassembled modular units to data centers based on their actual IT use, which has resulted in the energy efficiency of 80%–90%, versus a normal chiller plant [58,59].

A traditional 10 MW data center would typically require 20 MW to have sufficient power to get the chiller plant back online in the event of a power outage. Modularizing the system with small Turbocor® blocks resulted in very low in a rush, only using the compressors as needed, which reduced the electrical infrastructure required from 100% overhead to 15%. Inertech's CACTUS® units use about 80%–85% less water than a traditional chiller plant. This particular unit also affords data centers the ability to run dry, providing added flexibility to compensate for the atmosphere and surrounding conditions. Modular chiller plant or cooler solution in 350–500 kW modular blocks create efficiencies in both energy and space. Inertech worked closely with Danfoss to reduce amperage on the Turbocor® compressor in the way Inertech applies it in its patented cycle, and by doing so, Inertech has created efficiencies beyond what the original product intended [58,59].

2.7.4 Location-Independent Cogeneration Power Plants

Due to the modular design of the power plant elements, the MWM Modular Power Plant can be set up and installed within 12 days per unit. The MWM Modular Power Plant TCG 2032 gas engines run reliably in all-natural gas and biogas applications. Among others, the main advantages of MWM cogeneration power plants are that it has advantages of credit financing due to the mobile system, short payback time of the cogeneration plant and economical power generation and reduced emissions. The MWM Modular Power Plant is a location-independent complete system with perfectly matched components. All the modules are preconfigured for quick installation and delivered to the installation site prefabricated. This facilitates the installation of the modular cogeneration power plant on-site and shortens set-up times, allowing precision assembly even in locations with limited infrastructure [60,61].

The gas engine of the MWM TCG 2032 V16 series with an output of 4.5 MW_e makes up the heart of each module and can be reliably deployed in all-natural gas and biogas applications. The power generators are designed for the highest electrical and thermal efficiency, low operating and service costs, as well as high reliability and availability. There are already more than 600 gas engines with an overall output of some 2,200,000 kW_e installed in various applications worldwide. Each MWM Modular Power Plant can be expanded step by step to include up to six serially switched units, delivering a rated output of up to 27 MW_e. Wherever you as an investor are planning your location to be, one thing is for sure, you can rely on a proven CHP plant for highly efficient, decentralized, and mobile power generation with an overall efficiency of over 86% in natural gas operation [60,61].

2.7.5 ULTRA-CLEAN CHP MODULES FOR MICROGRID DEMONSTRATION PROJECT

In 2011, Tecogen Inc., the leading manufacturer of advanced modular CHP systems, supplied to the Sacramento Municipal Utility District (SMUD) three advanced CHP modules each rated at 100 kW electrical output, to be utilized year-round at the headquarter facility for supplying a significant portion of its electrical, cooling, and hot water requirements. Most significantly, these modules are equipped with two important, advanced features that will be demonstrated for the first time in a commercial application [62–64]. Specifically, the modules are equipped with sophisticated control algorithms that enable the units to be operated as a microgrid in the electric circuit configuration being prepared at the site. Per the DOE definition of a microgrid, the units will have the ability to operate normally in conjunction with the utility's grid, but then in the event of a utility outage, the microgrids must be able to seamlessly continue to power a selected portion of the utility site. The reverse process of reconnecting to the utility grid is likewise seamless. The second groundbreaking feature incorporated into these modules is a newly developed emissions reduction system that will enable the units to operate at extremely low levels of "criteria" (regulated) pollutants. While not required to do so in the Sacramento location, these units will be used to validate long-term emissions levels that meet or exceed those required under the California Air Resources Board's (CARB) distributed generation emissions standards for 2007. The capabilities of the INV-100 modules are exceptional, and they are cost-competitive, modular CHP modules, with UPS/outage capability and emissions on par with the best technology available to much larger, multi-megawatt power plants [62–64]. A more detailed discussion on microgrid is given in Chapter 9.

2.8 MODULAR CHP FOR EFFICIENT DISTRIBUTED POWER AND HEAT IN COMMERCIAL SECTOR

In general, plug and play units expand the U.S. market for small-scale CHP. The emergence of factory pre-engineered and packaged CHP systems is improving the return on investment (ROI) for customers and accelerating the take-up of small-scale CHP in the United States. Duffy and Purani [65], in their excellent article on use of CHP for efficiency improvement, show that cogeneration and CHP project developers and general contractors, working either with energy supply companies (ESCOs) or independently, can now provide smaller CHP system installations that offer a good ROI for their end-users, with either natural gas, digester gas, LFG, or syngas as the fuel source.

While the development of modular CHP has faced many obstacles in the past, what has changed is the development and full-scale deployment of reliable and efficient smaller systems that are not only pre-engineered by their manufacturer but are also prepackaged and pretested (although only a few manufacturers actually build specifically designed units at their factories). These units come complete with a prime mover, a generator mounted on a base rail with pumps, heat exchangers, and emission control devices, along with robust digital, on-board controllers capable of many functions and protective features. Thus, both engineering and installation costs are greatly reduced, while total system cost is far less than the energy-saving provided, especially when applications feature favorable conditions. Moreover, the gains in reliability from avoiding site-building—combined with new opportunities for quality control—further enhance the ROI by allowing for lengthier hours of operation and long-term service intervals, with consequent further reduction in grid dependence.

While much of the manufacturing, installation, and operation for smaller systems have been standardized to cut costs, each application remains custom-engineered, with the know-how and experience from application selection, through start-up and continued operation. While realizing a ROI from smaller CHP systems hinges on proper sizing, design, manufacturing, and operation, good application selection is still essential. For example, where electricity rates are high relative to fuel costs, good "spark spread" results. Meanwhile, the larger the size of the facility the unit is serving, the lower will be the first cost per kilowatt. Moreover, the longer the length of its operating hours, the higher is the use made of the energy savings the investment is providing.

When a facility has a mission requirement for additional power capacity and/or more reliable supply, system value is further enhanced. Central heating and/or cooling plant provides for thermal load, and a good coincidence between electric and thermal loads allows for the use of all available energy products. The opportunity to realize proper sizing for a small-scale CHP has been enhanced by the development and availability of a range of modular, standard-sizes, and their capability of being "stacked" to extend the standard range. The basic standard range is 55–400 kWe for natural gas systems and 60–350 kWe for biogas types. Very reliable and robust engines, the prime movers for CHP modules, are now available to provide better fuel efficiency, as well as extended hours of operation between overhauls. They can also feature low emissions of NOx and CO, the technology for which continues to advance [65,66].

Pre-engineering allows for an interface carefully customized to the end user's facility and application, including automated full digital control for the CHP unit(s), as well as for other plant equipment, such as pumps and heat exchangers. This customized interface, which may also cover proper utility interconnection and the integration of related infrastructure improvements, can provide multiple benefits, such as more efficient serviceability. Pre-engineering may also help maximize heat recovery, which can significantly mitigate the impact of natural gas prices on project costs. Similarly, maximizing savings on electricity costs can often easily cancel out any increases in fuel cost. In this area, careful consideration of the load levels the CHP unit will be connected to, both for electricity and heat recovery, can pay great dividends. Prepackaging modules eliminate the need to challenge general contractors with the task of procuring and site assembling 150–200 loose pieces. Greater availability of prepackaged and assembled systems also means end-users no longer need to install systems that are, essentially, just prime movers that were modified elsewhere before delivery.

Cost-effective noise-reducing enclosures have also been standardized, helping to extend application diversity, while modules without enclosures remain available for indoor installation. Increased concern for noise control in a variety of equipment sectors has made such control for CHP systems more easily attainable. Higher levels of uptime, which bring direct savings through lower grid dependence, can now be enhanced by the availability of long-term maintenance agreements. Such contracts may provide a fund for scheduled overhauls and/or replacement of major components. Pre-engineering can also help further increase uptime, by allowing for more efficient serviceability. A fully digital and automatic design can provide for full remote access and monitoring. The premanufacturing of small-scale CHP systems allows selection from a variety of stock engines and generators, as well as a manufacturing process that includes, engine generator base frame; engine generator assembly; exhaust heat exchanger; mechanical system piping and flanges; assembly and wiring of control and breaker cabinets; inhouse controller programming and enclosure assembly; and complete preshipping system testing [65,66].

At a multiple-building, public institution facility, a 250 kWe unit, stacked with a 150 kWe unit, has been operating at full load, 24/7, since 2008. The prepackaged unit includes supplying hot water as a vehicle for BTU recovery. The integrated controls included in this premanufacturing capability include a paralleling breaker that allows for single unit baseload to a utility or multiunit parallel with load add, load share, and load shed features. These controls also offer end users an opportunity for engine start/stop sequencing and engine monitoring, among general control functions, such as those to help determine the degree of reliance on the grid at any moment. Moreover, utility intertie controls have now been standardized, and switchgear and genset can be preintegrated at the factory. These fully designed and tested, plug and play CHP systems can include automated, preprogrammed preventive maintenance notification, including spark plug change, oil filter change, and belt/filters change.

According to Duffy and Purani [65], at a state office building, a 500 kWe unit, with a synchronous generator, has been operating since early 2010. Factory-trained and certified service technicians, when available 365 days/year, have also helped maximize run time, especially when they are fully equipped to perform field testing and can facilitate testing, repairs, and/or overhauls inhouse or in-frame. Similarly, when manufacturers have large stocking facilities, staffed by factory-trained and certified parts specialists, run time can be extended further. Manufacturers with well-trained service technicians and parts specialists may also provide good end user training, toward further maximizing system run time.

Examples of cost-effective installations of small CHP systems installed by Kraft Power include [65,66]:

1. **A public institution, multiple buildings:** A 250 kWe unit, stacked with a 150 kWe unit, has been operating at full load, 24/7, since 2008. Full utilization of combined electricity and thermal contribution potential is being realized, as the CHP system is also providing domestic hot water service in

addition to grid relief. The prepackaged unit includes supplying hot water as a vehicle for BTU recovery.

A second prepackaged CHP system was installed at a similar facility nearby. The system consists of two 150 kWe units with hot water recovery.

2. **The municipal wastewater treatment plant, Northeast:** Two 180 kWe synchronous units, equipped with external heat dump radiation add-ons, have been operating since early 2011. Biogas from the plant's anaerobic digester is collected, cleaned, and fed to the engines, with heat recovered from the engine used for heating the digester and also providing for the plant's domestic hot water needs.
3. **State office building:** A 500 kWe unit, with a synchronous generator, has been operating since early 2010. The system includes the supply of hot water as a vehicle for use of by-product heat.
4. **Hospital, New York City:** A 250 kWe CHP unit in a soundproof container was delivered to a private hospital to displace its high-cost electricity. Due to stringent interconnect requirements by the local utility company, the unit was manufactured utilizing an induction generator. The module also included a hot water heat recovery system to supplement the hospital's hot water requirements. As this was a prepackaged, fully automated, plug and play system, it was installed on the roof of the hospital, where the installing contractor only had to make gas, water, and electrical connections.
5. **Hospital, Massachusetts:** A 250 kWe CHP unit was installed for a Massachusetts hospital. The indoor installation has a state-of-the-art digital control system for hot water heat recovery. The module and its components are housed in a soundproof enclosure.

A study by Broccardo et al. [67] proposes the analysis of two case studies about the optimized design of modular small CHP plants. The analysis was carried out using DiOGene, an algorithm designed to simulate the operation of such plants and also considering the integral constraints imposed by Italian regulations. Two hospitals were the subject of case studies; in fact, such kind of public building is usually an ideal site for a CHP plant due to its peculiar energetic loads. Results show that by carefully choosing the plant configuration, it is possible to achieve significant economic and fuel savings, consequently reducing CO_2 emissions. The analysis also points out how the modular approach, due to its intrinsic flexibility that increases the number the degrees of freedom, is the most suitable one for an effective matching between the site characteristics and machines.

6. **Wastewater treatment plants, Wisconsin and Illinois:** In spite of low electricity costs, incentives had led to projects in the state of Wisconsin and Illinois, where 180 kWe-sized CHP modules were supplied. A few more biogas CHP projects, in the range of 180–350 kWe, were also installed.
7. **Airport energy center:** Here, a larger system was available from a manufacturer that specialized primarily in smaller systems to gain the benefits of standardized design and pre-engineering and preassembly. It was not

supplied as a separately housed unit, as the airport already had a building to house it. Four units provided a total of 5.66 MW of electricity, in addition to 96°C hot water.

The installation began operation in 2002 and has provided complete outsourcing of total energy requirements. Airport heating and cooling are now assisted by reclaimed heat from the units' engine jacket coolants. Meanwhile, the project has benefited from incentives from the local electric power utility that encourage on-site generation.

Market potential
The industrial applications for smaller CHP systems that have been identified so far include food processing, textiles, wood, pulp and paper, petroleum, and chemicals. According to data acquired from the Mid-Atlantic CHP Application Center, U.S. application potential also includes [65,66]:

- Hospitals: 8,000 sites
- Convention hotels: 7,000 sites
- High-rise residential complexes: 12,000 MW
- Office buildings: 25,000 MW
- Restaurants: 1,000 MW and 20,000 sites
- Shopping centers or malls: 12,000 MW and 8,000 sites
- Schools and colleges: 2,600 MW and 13,000 sites
- Supermarkets: 8,400 MW and 28,000 sites
- Casinos

Currently, about 95% of CHP projects are electric-driven. The thermal load profile is also matched to facility thermal demand, thereby optimizing system performance. Applications include space heat, dryer operations, boiler preheating, absorption air conditioning, process heat, sludge digestion, domestic hot water, laundry operations, and washing operations. However, in some cases, projects are driven by thermal loads, and not by electricity demand.

General outlook
A few years ago, there was great uncertainty about the outlook for CHP in the United States. Today, many industry analysts are envisioning some strong trends in the direction of CHP/DG. Environmentally sensitive and energy-intensive customers in the Northeast and Midwest, as well as on the West Coast, are recognizing the advantages and benefits of premanufactured, plug and play-type, efficient, modular CHP plants. Experience has shown that smaller CHP systems, when acknowledged by their end users as reliable and beneficial, will continue to operate for a long time. Use for 20 years and longer is increasingly common. This longevity, coupled with continuing advances by smaller CHP system manufacturers in preengineering, prepackaging, manufacturing technology, system integration, and ongoing service and maintenance capability—and combined with incentives from grid operators that encourage on-site power generation—strongly suggest a good opportunity for electricity generation project developers and general contractors, working either with ESCOs or independently.

As an example of modular CHP, 2G cogeneration systems are a modular, all-in-one solution for any facility. Their "connection ready" design allows fast and cost-effective installation with no need for upfront engineering. The 2G engine technology offers industry-leading efficiency that is continuously improving through partnerships with prestigious universities, research institutes, and 2G's own research and development company, 2G Drives. This system can operate with biogas/natural gas/dual fuel or fuel blending operation, it is custom designed and manufactured, has an adjustable part-load operation, island mode capability, highest efficiency, and lowest emission, and it requires no upfront engineering. As a solution provider, 2G delivers a complete package and integrates additional components, such as absorption chillers, gas treatment, production of steam and hot water, and emission control systems, into the system. The 2G CHP systems can regulate their output within a very short time frame. They are infinitely adjustable in the power range between 50% and 100% load. With the help of modern control technology, they adjust to actual energy demand. The 2G power systems can be installed in various ways—depending on local conditions and sound attenuation and esthetics requirements. They can be incorporated in existing buildings, heating systems, or can be set up in a container or an engine room. With the appropriate sound attenuation package, noise emissions can be as low as 35 dB (A) at a distance of 33 ft. 2G provides a container for its CHP system [35,55]:

2G CENERGY installed a CHP cogeneration plant at Simonds Intl. (the oldest cutting tool manufacturer in North America) in Fitchburg, Massachusetts. The cogeneration process results in overall electrical and thermal efficiencies close to 90%, compared with most utility power plants operating in the 33% efficiency range. The plant is an integrated package, fully containerized, and will be supplied as a unique "all-in-one" and "connection-ready" module. Benefits over conventional gas engine gensets include much higher overall efficiency, reliability, durability, extended life, fast installation, and less maintenance cost. The modular 2G® natural gas-based cogeneration system to be installed was rated 1,600 kWh (1.6 MWh) consisting of three 600 kWh 2G® avus® series CHP modules with MWM engines fully integrated into the unique and modular 2G® cogeneration technology package. The plant was expected to produce enough electricity to operate the entire manufacturing plant. The annual output was 13,280 MWh and 62.5 billion BTU's of thermal energy. 2G®'s CHP modules also include an advanced combustion management system. The automation and control technology-enabled Simonds to monitor their energy efficiency and lower the environmental load, reducing CO_2 emissions. 2G®'s output-optimized cogeneration CHP modules have been installed at more than 1,500 locations around the world [35,55].

2.8.1 Optimal Modular Design for Cogeneration for Hospital Facilities

The study by Gimelli et al. [68] evaluated the potential of cogeneration based on reciprocating gas engines for some Italian hospital buildings. Comparative analyses were conducted based on the load profiles of two specific hospital facilities and through the study of the cogeneration system-user interaction. To this end, a specific methodology was set up by coupling a specifically developed calculation algorithm to a genetic optimization algorithm, and a multiobjective approach was adopted. The results from the optimization problem highlight a clear trade-off between total primary energy savings (TPESs) and a simple payback period (SPB). Optimized plant configurations and management strategies showed TPES exceeding 18% for the reference hospital facilities and multigas engine solutions along with a minimum SPB of approximately 3 years, thereby justifying the European regulation promoting cogeneration.

The proposed methodology was enhanced to focus on some innovative aspects. In particular, this study proposed an uncommon and effective approach to identify the most stable plant solutions through multiobjective robust design optimization. In particular, the sensitivity of the expected results to possible difficulties in finding commercially available CHP gas engines with sizes reasonably close to the optimal numerical solutions was estimated. The results indicated that the economic sensitivity was often higher than the energetic sensitivity for most of the optimal solutions, with standard deviation (SD) accounting up to 7% of its mean value for the SPB, whereas that percentage is always under 3% for the TPES [68].

Furthermore, the study highlighted how the expected results obtained through a deterministic definition of the input decision variables could be overestimated compared to the robust design approach. The proposed research also highlighted how optimized CHP plants can be characterized by reasonable levels of energetic and economic sensitivity to changes in the variable quantities including selling price of electricity, reference efficiency of the Italian thermoelectric generation, and selling price of the energy efficiency certificates recognized by the Italian legislation. Indeed, compared with optimal solutions indicated that the SD for the SPB was always less than 3.5% of its mean value, while this percentage was always under 7% for the TPES.

2.8.2 LARGE SCALE MODULAR GEOTHERMAL DISTRICT HEATING

Large scale modular district heating using groundwater geothermal energy is developed in Milan (Italy) by AEM, the local public company, which meets a large part of the city energy needs, including gas, electricity and district heating. AEM's experience in cogeneration and heat pumps and the groundwater availability in Milan has led to its deciding to opt for the design of a standard plant that combines the advantages of both heat pump and district heating technologies; the new plant design is being implemented in its new district heating systems, which will perform better both in terms of environmental protection and energy-saving. The engineering solution allows for the rationalization of primary conventional energy sources used, such as electricity and natural gas [69] and to the use of a renewable geothermal resource. AEM has therefore launched a development plan for a new line of heat pump plants that will be constructed on five of its own sites (Canavese, Gonin, Ricevitrice Nord, Ricevitrice Sud, and Bovisa), currently used for technological purposes. This modular, standard type of plants will also be proposed for other existing district heating initiatives in the built-up areas of Garibaldi-Repubblica, Quartiere Santa Giulia, and Bocconi.

The unified project uses a modular, standard size and shapes for its plant, which serves the purpose of simplifying its construction and management. It is made up of a cogeneration section, a heat pump system, a boiler section, and a heat storage tank. The heat pump section, which represents the most innovative part of the project, guarantees a significant heat production, integrated due to the contribution of the cogeneration and heat boilers sections. Heat pumps are powered by electrical energy; connection to the electricity network combined with heat storage tanks ensures that the plant's functioning can be modulated. It should be noted that the extensive use of heating systems powered by electricity, especially heat pump systems with heat storage tanks, guarantees an important contribution to equilibrating the national electric demand increasing the load at night-time [69].

The feasibility study was set up for the purpose of identifying the best configuration for the plant in terms of main machines and accumulators size, taking account of potential consumers that present a thermal load profile with the typical characteristics found in Milan. The thermal profile of users has been defined, on the basis of AEM's experience in district heating. The basic data on the standard users are taken into consideration were 90 MWth thermal power absorbed at plant outlet, 150 GWhth/h thermal energy distributed, and 1,700 equivalent hours per year. To identify the optimal configuration for the AEM plant, the method adopted was mathematical programming and specifically mixed integer linear models. This model was developed to guide the design toward a plant configuration, which, in addition to complying with binding environmental norms, also has to be economically sustainable and profitable and offer state-of-the-art technical solutions. By developing the mathematical model in a specific software environment, it has been possible, using optimization algorithms, to make comparisons between different plant configurations, starting from process parameters and operative conditions. The input data for the model was the thermal load of users, known for every day of the year and calculated on the basis of day degrees of the area in which the plant will be constructed; the objective function to optimize is the profit of the investment, while complying with binding supply obligations for energy carriers and environmental norms. The new modular geothermal district heating is the largest groundwater heat pump powered system and the second largest (in terms of connected clients) geothermal system in the world (after Paris) [69].

2.9 EFFICIENT ENERGY USE BY MODULAR MICRO-CHP

Micro combined heat and power or micro-CHP or mCHP is an extension of the idea of cogeneration to the single/multifamily home or small office building in the range of up to 50 kW. This is generally modular. The local generation has the potential for higher efficiency than traditional grid-level generators as it lacks the 8%–10% energy losses from transporting electricity over long distances. It also lacks the 10%–15% energy losses from heat transfer in district heating networks due to the difference between the thermal energy carrier (hot water) and the colder external environment. The most common systems use natural gas as their primary energy source and emit CO_2 [66].

CHP systems for homes or small commercial buildings are often fueled by natural gas to produce electricity and heat. A micro-CHP system usually contains a small fuel cell or a heat engine as a prime mover used to rotate a generator that provides electric power, while simultaneously utilizing the waste heat from the prime mover for an individual building's heating, ventilation, and air conditioning. A micro-CHP generator may primarily follow heat demand, delivering electricity as the by-product, or may follow electrical demand to generate electricity and use heat as the by-product. When used primarily for heating, micro-CHP systems may generate more electricity than is instantaneously being demanded in circumstances of fluctuating electrical demand. As electricity can be transported practically, it is more efficient to generate electricity near where the waste heat can be used. So in a "micro-combined heat and power system" (micro-CHP), small power plants are instead located where the secondary heat can be used, in individual buildings. After the year 2000, micro-CHP has become cost-effective in many markets around the world, due to rising energy costs. The development of micro-CHP systems has also been facilitated by recent technological developments of small heat engines. This includes improved performance and cost-effectiveness of fuel cells, Stirling engines, steam engines, gas turbines, diesel engines, and Otto engines. A 2013 UK report from Ecuity Consulting stated that MICRO-CHP is the most cost-effective method of utilizing gas to generate energy at the domestic level. Delta-ee consultants stated that with 64% of global sales the fuel cell micro-CHP passed the conventional engine-based micro-CHP systems in sales in 2012. For proton-exchange membrane cell fuel cell (PEMFC) units with a lifetime of 60,000h, which shut down at night, this equates to an estimated lifetime of between 10 and 15 years [66].

As both MiniCHP and CHP have been shown to reduce emissions [26] they could play a large role in the field of CO_2 reduction from buildings, where more than 14% of emissions can be saved using CHP in buildings [27]. The University of Cambridge reported a cost-effective steam engine micro-CHP prototype in 2017, which has the potential to be commercially competitive in the following decades [2]. There are many types of fuels and sources of heat that may be considered for micro-CHP. The properties of these sources vary in terms of system cost, heat cost, environmental effects, convenience, ease of transportation and storage, system maintenance, and system life. Some of the heat sources and fuels that are being considered for use with micro-CHP include natural gas, LPG, biomass, vegetable oil (such as rapeseed oil), woodgas, solar thermal, and lately also hydrogen, as well as multifuel systems. The energy sources with the lowest emissions of particulates and net- CO_2 include solar power, hydrogen, biomass (with two-stage gasification into biogas), and natural gas. The majority of cogeneration systems use natural gas for fuel, because natural gas burns easily and cleanly, it can be inexpensive, it is available in most areas and is easily transported through pipelines, which already exist for over 60 million homes [66].

Reciprocating ICEs are the most popular type of engine used in micro-CHP systems [2]. Reciprocating ICE based systems can be sized such that the engine operates at a single fixed speed, usually resulting in a higher electrical or total efficiency. However, as reciprocating ICEs have the ability to modulate their power output by changing their operating speed and fuel input, micro-CHP systems based on these engines can have varying electrical and thermal output designed to meet changing demand [23]. Natural gas is suitable for ICEs, such as Otto engine and gas turbine systems. Gas turbines are used in many small systems due to their high efficiency, small size, clean combustion, durability, and low maintenance requirements. Gas turbines designed with foil bearings and air-cooling operate without lubricating oil or coolants. The waste heat of gas turbines is mostly in the exhaust, whereas the waste heat of reciprocating ICEs is split between the exhaust and cooling systems. External combustion engines can run on any high-temperature heat source. These engines include the Stirling engine, hot "gas" turbocharger, and the steam engine. Both range from 10%–20% efficiency, and as of 2014, small quantities are in production for micro-CHP products. Other possibilities include the ORC, which operates at lower temperatures and pressures using low-grade heat sources. The primary advantage to this is that the equipment is essentially an air conditioning or refrigeration unit operating as an engine, whereby the piping and other components need not be designed for extreme temperatures and pressures, reducing cost and complexity. Electrical efficiency suffers, but it is presumed that such a system would be utilizing waste heat or a heat source such as a wood stove or gas boiler that would exist anyway for purposes of space heating [66].

Fuel cells generate electricity and heat as a byproduct. The advantages of a stationary fuel cell application over stirling CHP are no moving parts, less maintenance, and quieter operation. The surplus electricity can be delivered back to the grid. PEMFCs fueled by natural gas or propane use a steam reformer to convert methane in the gas supply into CO_2 and hydrogen; the hydrogen then reacts with oxygen in the fuel cell to produce electricity. A PEMFC-based micro-CHP has an electrical efficiency of 37% lower heating value (LHV) and 33% higher heating value (HHV) and a heat recovery efficiency of 52% LHV and 47% HHV with a service life of 40,000 h or 4,000 starts/stop cycles which is equal to 10-year use. An estimated 138,000 Fuel cell CHP systems below 1 kW had been installed in Japan by the end of 2014. Most of these CHP systems are PEMFC based (85%) and the remaining are solid oxide fuel cell (SOFC) systems [66].

2.9.1 GLOBAL MARKET STATUS FOR MICRO-CHP

The largest deployment of micro-CHP is in Japan wherein 2009 were over 90,000 units in place, with the vast majority being of Honda's "ECO-WILL" type. Six Japanese energy companies launched the 300 W–1 kW PEMFC/SOFC ENE FARM product in 2009, with 3,000 installed units in 2008, a production target of 150,000 units for 2009–2010 and a target of 2,500,000 units in 2030. 20,000 units were sold in 2012 overall within the Ene Farm project making an estimated total of 50,000 PEMFC and up to 5,000 SOFC installations. ECO-WILL is sold by various gas companies and as of 2013, installed in a total of 131,000 homes. Manufactured by Honda using their single-cylinder EXlink engine capable of burning natural gas or propane. Each unit produces 1 kW of electricity and 2.8 kW of hot water [66].

In Europe, the project ene.field aimed to deploy by 2017 up 1,000 residential fuel cell CHP (micro-CHP) installations in 12 EU member states. Powercell Sweden is a fuel cell company that develops environmentally friendly electric generators with a unique fuel cell and reformer technology that is suitable for both existing and future fuel. In Germany, ca 50 MW of micro-CHP up to 50 kW units have been installed in 2015 [71]. The German government is offering large CHP incentives, including a market premium on electricity generated by CHP and an investment bonus for micro-CHP units. The German testing project Callux has 500 micro-CHP installations per November 2014. North Rhine-Westphalia launched a 250 million subsidy program for up to 50 kW lasting until 2017 [66,70,71].

It is estimated that about 1,000 micro-CHP systems were in operation in the United Kingdom as of 2002. These are primarily Whispergen using Stirling engines, and Senertec Dachs reciprocating engines. Of the 24 million households in the United Kingdom, as many as 14–18 million are thought to be suitable for micro-CHP units. Two fuel cell varieties of micro-CHP cogeneration units are almost ready for mainstream production and are planned for release to commercial markets in early in 2014. With the UK Government's Feed-In-Tariff available for a 10-year period, a wide uptake of the technology is anticipated. The Danish micro-CHP project 2007–2014 with 30 units is on the island of Lolland and in the western town Varde. Denmark is currently part of the Ene.field project. In Netherland, the micro-CHP subsidy was ended in 2012. To test the effects of micro-CHP on a smart grid, 45 natural gas SOFC units (each 1.5 kWh) from Republiq Power (Ceramic Fuel Cells) will be placed on Ameland in 2013 to function as a virtual power plant [66,70,71].

In the United States, the federal government is offering a 10% tax credit for smaller CHP and micro-CHP commercial applications. Marathon Engine Systems, a Wisconsin company, produces a variable electrical and thermal output micro-CHP system called the ecopower with an electrical output of 2.2–4.7 kWe. The ecopower was independently measured to operate at 24.4% and 70.1% electrical and waste heat recovery efficiency, respectively [66,70,71].

2.10 MODULAR ENERGY EFFICIENCY IMPROVED IN ENERGY INDUSTRY BY NANOTECHNOLOGY

Nanotechnology is finding applications in traditional energy sources and is greatly enhancing alternative energy approaches to help meet the world's increasing energy demands. Many scientists are looking into ways to develop clean, affordable, and renewable energy sources, along with means to reduce energy consumption and lessen toxicity burdens on the environment. Some of the notable advances made by nanotechnology can be listed as follows [66,72–76]:

1. Nanotechnology is improving the efficiency of fuel production from raw petroleum materials through better catalysis. It is also enabling reduced fuel consumption in vehicles and power plants through higher-efficiency combustion and decreased friction.
2. Nanotechnology is also being applied to oil and gas extraction through, for example, the use of nanotechnology-enabled gas lift valves in offshore operations or the use of nanoparticles to detect microscopic down-well oil pipeline fractures.
3. Researchers are investigating carbon nanotube "scrubbers" and membranes to separate CO_2 from power plant exhaust.

Industrial and Commercial Sectors

4. Researchers are developing wires containing carbon nanotubes that will have much lower resistance than the high-tension wires currently used in the electric grid, thus reducing transmission power loss.
5. Nanotechnology can be incorporated into solar panels to convert sunlight to electricity more efficiently, promising inexpensive solar power in the future. Nanostructured solar cells could be cheaper to manufacture and easier to install, since they can use print-like manufacturing processes and can be made in flexible rolls rather than discrete panels. Newer research suggests that future solar converters might even be "paintable."
6. Nanotechnology is already being used to develop many new kinds of batteries that are quicker-charging, more efficient, lighter weight, have a higher power density and hold electrical charge longer.
7. An epoxy containing carbon nanotubes is being used to make windmill blades that are longer, stronger, and lighter-weight than other blades to increase the amount of electricity that windmills can generate.
8. In the area of energy harvesting, researchers are developing thin-film solar electric panels that can be fitted onto computer cases and flexible piezo-electric nanowires woven into clothing to generate usable energy on the go from light, friction, and/or body heat to power mobile electronic devices. Similarly, various nanoscience-based options are being pursued to convert waste heat in computers, automobiles, homes, power plants, and so on, to usable electrical power.
9. Energy efficiency and energy-saving products are increasing in number and types of applications. In addition to those noted above, nanotechnology is enabling more efficient lighting systems; lighter and stronger vehicle chassis materials for the transportation sector; lower energy consumption in advanced electronics; and light-responsive smart coatings for glass.
10. Engineers have developed a thin-film membrane with nanopores for energy-efficient desalination. This molybdenum disulfide (MoS_2) membrane filtered two to five times more water than current conventional filters.
11. Nanoparticles are being developed to clean industrial water pollutants in groundwater through chemical reactions that render the pollutants harmless. This process would consume less energy and cost less than methods that require pumping the water out of the ground for treatment.
12. Researchers have developed a nanofabric "paper towel" woven from tiny wires of potassium manganese oxide that can absorb 20 times its weight in oil for cleanup applications. Researchers have also placed magnetic water-repellent nanoparticles in oil spills and used magnets to mechanically remove the oil from the water.
13. Many airplane cabins and other types of air filters are nanotechnology-based filters that allow "mechanical filtration," in which the fiber material creates nanoscale pores that trap particles larger than the size of the pores. The filters also may contain charcoal layers that remove odors. Better filters are more energy efficient.
14. Nanotechnology offers the promise of developing multifunctional materials that will contribute to building and maintaining lighter, safer, smarter, and

more efficient vehicles, aircraft, spacecraft, and ships. Lighter vehicles will be more energy efficient. In addition, nanotechnology offers various means to improve the transportation infrastructure.

Nano-engineered materials in automotive products include polymer nanocomposites structural parts, high-power rechargeable battery systems, thermoelectric materials for temperature control, lower rolling-resistance tires, high-efficiency/low-cost sensors and electronics, thin-film smart solar panels, and fuel additives and improved catalytic converters for cleaner exhaust and extended range. Nano-engineering of aluminum, steel, asphalt, concrete and other cementitious materials, and their recycled forms offers great promise in terms of improving the performance, resiliency, and longevity of highway and transportation infrastructure components while reducing their life cycle cost. New systems may incorporate innovative capabilities into traditional infrastructure materials, such as self-repairing structures or the ability to generate or transmit energy [66,72–76].

Nanoscale sensors and devices may provide cost-effective continuous monitoring of the structural integrity and performance of bridges, tunnels, rails, parking structures, and pavements over time. Nanoscale sensors, communications devices, and other innovations enabled by nanoelectronics can also support an enhanced transportation infrastructure that can communicate with vehicle-based systems to help drivers maintain lane position, avoid collisions, adjust travel routes to avoid congestion, and improve drivers' interfaces to onboard electronics [66,72–76].

"Game-changing" benefits from the use of nanotechnology-enabled lightweight, high-strength materials would apply to almost any transportation vehicle. For example, it has been estimated that reducing the weight of a commercial jet aircraft by 20% could reduce its fuel consumption by as much as 15%. A preliminary analysis performed for NASA has indicated that the development and use of advanced nanomaterials with twice the strength of conventional composites would reduce the gross weight of a launch vehicle by as much as 63%. Not only could this save a significant amount of energy needed to launch spacecraft into orbit, but it would also enable the development of a single stage to orbit launch vehicles, further reducing launch costs, increasing mission reliability, and opening the door to alternative propulsion concepts [66,72–76].

An important subfield of nanotechnology related to energy is nanofabrication. Nanofabrication is the process of designing and creating devices on the nanoscale. Creating devices smaller than 100 nm opens many doors for the development of new ways to capture, store, and transfer energy. The inherent level of control that nanofabrication could give scientists and engineers would be critical in providing the capability of solving many of the problems that the world is facing today related to the current generation of energy technologies [1]. Researchers have already begun developing ways of utilizing nanotechnology for the development of consumer products. Benefits already observed from the design of these products are an increased efficiency of lighting and heating, increased electrical storage capacity, and a decrease in the amount of pollution from the use of energy. Recently, previously established and entirely new companies such as BetaBatt, Inc. and Oxane Materials are focusing on nanomaterials as a way to develop and improve upon older methods

Industrial and Commercial Sectors 121

for the capture, transfer, and storage of energy for the development of consumer products [66,72–76].

ConsERV, a product developed by the Dais Analytic Corporation, uses nanoscale polymer membranes to increase the efficiency of heating and cooling systems and has already proven to be a lucrative design. The polymer membrane was specifically configured for this application by selectively engineering the size of the pores in the membrane to prevent air from passing while allowing moisture to pass through the membrane. ConsERV's value is demonstrated in the form of an energy recovery, a device which pretreats the incoming fresh air to a building using the energy found in the exhaust air steam using no moving parts to lower the energy and carbon footprint of existing forms of heating and cooling equipment. Polymer membranes can be designed to selectively allow particles of one size and shape to pass through while preventing other. This makes for a powerful tool that can be used in all markets—consumer, commercial, industrial, and government products from biological weapons protection to industrial chemical separations. Dais's near term uses of this "family" of selectively engineered nanotechnology materials, aside from ConsERV, include (a) a completely new cooling cycle capable of replacing the refrigerant-based cooling cycle the world has known for the last 100+ years. This product, under development, is named NanoAir. NanoAir uses only water and this selectively engineered membrane material to cool (or heat) and dehumidify (or humidify) air. There are no fluorocarbon producing gasses used, and the energy required to cool a space drops as thermodynamics does the actual cooling [66,72–76].

It should also be noted Dais received a U.S. Patent (Patent Number 7,990,679) in October 2011 titled "Nanoparticle Ultracapacitor." This patented item again uses the selectively engineered material to create an energy storage mechanism projected to have performance and cost advantages over existing storage technologies. The company has used this patent's concepts to create a functional energy storage prototype device named NanoCap. NanoCap is a form of ultra-capacitor potentially useful to power a broad range of applications including most forms of transportation, energy storage (especially useful as a storage media for renewable energy technologies), telecommunication infrastructure, transistor gate dielectrics, and consumer battery applications (cell phones, computers, etc.) [66,72–76].

A New York-based company called Applied NanoWorks, Inc. has been developing a consumer product that utilizes LED technology to generate light. LEDs use only about 10% of the energy that a typical incandescent or fluorescent light bulb uses and typically last much longer, which makes them a viable alternative to traditional light bulbs. While LEDs have been around for decades, this company and others like it have been developing a special variant of LED called the white LED. White LEDs consist of semi-conducting organic layers that are only about 100 nm in distance from each other and are placed between two electrodes, which create an anode, and a cathode. When voltage is applied to the system, light is generated when electricity passes through the two organic layers. This is called electroluminescence. The semiconductor properties of the organic layers are what allow for the minimal amount of energy necessary to generate light. In traditional light bulbs, a metal filament is used to generate light when electricity is run through the filament. Using metal generates a great deal of heat and therefore lowers efficiency [66,72–76].

Research for longer-lasting batteries has been an ongoing process for years. Researchers have now begun to utilize nanotechnology for battery technology. mPhase Technologies in conglomeration with Rutgers University and Bell Laboratories have utilized nanomaterials to alter the wetting behavior of the surface where the liquid in the battery lies to spread the liquid droplets over a greater area on the surface and therefore have greater control over the movement of the droplets. This gives more control to the designer of the battery. This control prevents reactions in the battery by separating the electrolytic liquid from the anode and the cathode when the battery is not in use and joining them when the battery is in need of use [66,72–76].

Thermal applications also are future applications of nanotechnology creating a low-cost system of heating, ventilation, and air conditioning, changing molecular structure for better management of temperature. The relatively recent shift toward using nanotechnology with respect to the capture, transfer, and storage of energy has and will continue to have many positive economic impacts on society. The control of materials that nanotechnology offers to scientists and engineers of consumer products is one of the most important aspects of nanotechnology. This allows for an improved efficiency of products across the board. A major issue with the current energy generation is the loss of efficiency from the generation of heat as a by-product of the process. A common example of this is the heat generated by the ICE. The ICE loses about 64% of the energy from gasoline as heat and an improvement of this alone could have a significant economic impact. However, improving the ICE in this respect has proven to be extremely difficult without sacrificing performance. Improving the efficiency of fuel cells through the use of nanotechnology appears to be more plausible by using molecularly tailored catalysts, polymer membranes, and improved fuel storage [66,72–74].

In order for a fuel cell to operate, particularly of the hydrogen variant, a noble-metal catalyst (usually platinum, which is very expensive) is needed to separate the electrons from the protons of the hydrogen atoms. However, catalysts of this type are extremely sensitive to carbon monoxide reactions. To combat this, alcohols or hydrocarbons compounds are used to lower the carbon monoxide concentration in the system. This adds an additional cost to the device. Using nanotechnology, catalysts can be designed through nanofabrication that are much more resistant to carbon monoxide reactions, which improves the efficiency of the process and may be designed with cheaper materials to additionally lower costs. Fuel cells that are currently designed for transportation need rapid start-up periods for the practicality of consumer use. This process puts a lot of strain on the traditional polymer electrolyte membranes, which decreases the life of the membrane requiring frequent replacement. Using nanotechnology, engineers have the ability to create a much more durable polymer membrane, which addresses this problem. Nanoscale polymer membranes are also much more efficient in ionic conductivity. This improves the efficiency of the system and decreases the time between replacements, which lowers costs.

Another problem with contemporary fuel cells is the storage of the fuel. In the case of hydrogen fuel cells, storing the hydrogen in gaseous rather than liquid form improves the efficiency by 5%. However, the materials that we currently have available to us significantly limit fuel storage due to low-stress tolerance and costs.

Scientists have come up with an answer to this by using a nanoporous styrene material (which is a relatively inexpensive material) that when super-cooled to around −196°C, naturally holds on to hydrogen atoms and when heated again releases the hydrogen for use [66,72–76].

For portable energy systems, nanotechnology-enabled high-power and high-energy-density microelectronics, power electronics, energy sources, and energy storage solutions were examined by Smith [74]. These energy systems are used in various applications including aerospace, biomedical, communication, electronics, micro-electromechanical systems (MEMS), robotics, and so on. Proof-of-concept autonomous power systems were designed, and, practical solutions were substantiated and proposed. The study performed applied research and technology developments in the design of efficient energy harvesting, energy management, and energy storage systems. To guarantee modularity and compatibility, the following components and modules were examined: (a) low-power microelectronics; (b) high-power-density power electronics with high-current, high-voltage, and high-frequency power MOSFETs; (c) high-efficiency energy harvesting sources such as solar cells and electromagnetic generators; (d) high specific energy and power density electrochemical energy storage devices; and (e) optimal energy management systems. The most advanced technology-proven nanoscale microelectronics, MEMS sensors, and energy conversion solutions were also studied. The modular design, enabling topologies and optimization schemes resulted in high-performance micro- and mini-scale energy systems [66,72–76].

REFERENCES

1. Energy Conservation, Wikipedia, The free encyclopedia, last visited 16 June 2019 (2019).
2. Clark Gellings, K.E. Parmenter and Patricia Hurtado, Efficient Use and Conservation of Energy – Vol. I - Efficient Use and Conservation of Energy in the Industrial Sector, Encyclopedia of Life Support Systems. A website report www.eolss.net/Sample-Chapters/C08/E3-18-01.pdf (2013).
3. Energy Use in Industry—Energy Explained, Your Guide to … EIA. A website report www.eia.gov›EnergyExplained›UseofEnergy (2018).
4. Shah, Y.T. (2019). Modular Systems for Energy and Fuel Recovery and Conversion. CRC Press, New York.
5. Daw, J., Hallett, K., DeWolfe, J., and Venner, I. (January 2012). Energy Efficiency Strategies for Municipal Wastewater Treatment Facilities. Technical Report NREL/TP-7A30-53341 under Contract No. DE-AC36-08GO28308, National Renewable Energy, Golden, Colorado.
6. Energy Efficiency in Water and Wastewater Facilities, Local Government Climate And Energy Strategy Guides, A Guide to Developing and Implementing Greenhouse Gas Reduction Programs, U.S. Environment Protection Agency, Washington (2013).
7. Energy Data Management Manual for the Wastewater Treatment Sector, a report by better building initiative, Department of Energy, Washington, DC. DOE/EE-1700 (December 2017).
8. American Council for an Energy-Efficient Economy (ACEEE). (2012). Energy Efficiency and Economic Opportunity. Available: www.aceee.org/les/pdf/fact-sheet/ee-economic-opportunity.pdf. Accessed 9/20/12.

9. City of Franklin. (2009). Sustainable Community Action Plan. Available: www.franklin-gov.com/Modules/ShowDocument.aspx?documentid=5877. Accessed 10/25/12.
10. City of Franklin. (2012). Franklin Water Management Department Projected to Save Over $190,000. Available: www.franklin-gov.com/index.aspx?page=25&recordid=1667&returnURL=%2Findex.aspx. Accessed 10/25/12.
11. East Bay Municipal Utility District (EBMUD). (2010). Climate Change Update. Available www.ebmud.com/sites/default/les/102610_energy_sta_reports_1.pdf. Accessed 2/7/13.
12. East Bay Municipal Utility District (EBMUD). (2012). A Commitments to the Environment. Available: www.ebmud.com/sites/default/les/pdfs/energy-factsheet-03-12.pdf. Accessed 10/24/12.
13. Lawrence Berkeley Laboratory. (2010). Energy Efficiency Services Sector: Workforce Size and Expectations for Growth. Available: http://eetd.lbl.gov/ea/emp/reports/lbnl-3987e.pdf. Accessed 9/20/12.
14. NYSERDA. (2008). Statewide Assessment of Energy Use by the Municipal Water and Wastewater Sector. New York State Energy Research and Development Authority. Available: www.nyserda.ny.gov/~/media/Files/EERP/Commercial/Sector/Municipal%20Water%20and%20Wastewater%20Facilities/nys-assess-energy-use.ashx?sc_database=web. Accessed 5/9/11.
15. NYSERDA. (2010). Existing Facilities Program. New York State Energy Research and Development Authority. Available: www.nyserda.org/Programs/Exist-ing_Facilities/pdfs/1219ponbrochure.pdf. Accessed 6/1/11.
16. Renewable Energy World. (2006). Turning Kitchen Grease into Biogas. Available: www.renewableen-ergyworld.com/rea/news/article/2006/11/turning-kitchen-grease-into-biogas-46585. Accessed 11/15/12.
17. U.S. DOE. (2005). A State Energy Program: Projects by Topic—What Are State and Local Government Facility Projects in the States? Available: www.eere.energy.gov/wip/sep.html. Accessed 5/22/11.
18. U.S. EPA. (2002). e Clean Water and Drinking Water Infrastructure Gap Analysis. Available: http://water.epa.gov/aboutow/ogwdw/upload/2005_02_03_gapre-port.pdf. http://water.epa.gov/aboutow/ogwdw/upload/2005_02_03_gapreport.pdf. Accessed 5/9/11.
19. U.S. EPA. (2010). Evaluation of Energy Conservation Measures for Wastewater Treatment Facilities. EPA 832-R-10-00. Available: http://water.epa.gov/scitech/wastetech/upload/Evaluation-of-Energy-Conservation-Measures-for-Wastewater-Treatment-Facilities.pdf. Accessed 6/1/11.
20. U.S. EPA. (2011a). U.S. Greenhouse Gas Inventory. Available: www.epa.gov/climatechange/emissions/usinventoryreport.html. Accessed 11/7/08.
21. U.S. EPA. (2011b). Air Emissions. Available: www.epa.gov/cleanenergy/energy-and-you/aect/air-emissions.html. Accessed 9/15/11.
22. U.S. EPA. (2011c). Greenhouse Gas Equivalencies Calculator. Available: www.epa.gov/cleanenergy/energy-resources/calculator.html#results. Accessed 6/1/11.
23. U.S. EPA. (2011d). Opportunities for Combined Heat and Power at Wastewater Treatment Facilities: Market Analysis and Lessons from the Field. Available: http://epa.gov/chp/documents/wwtf_opportunities.pdf. Accessed 11/14/12.
24. U.S. EPA. (2011e). ENERGY STAR for Wastewater Plants and Drinking Water Systems. Available: www.energystar.gov/index.cfm?c=water.wastewater_drinking_water. Accessed 5/31/11.
25. U.S. EPA. (2012). State and Local Climate and Energy Program: Water/Wastewater. Available: www.epa.gov/statelocalclimate/local/topics/water.html. Accessed 10/12/12.
26. U.S. EPA Region 9. (2012a). Water and Energy Efficiency in Water and Wastewater Facilities: Energy Efficient Equipment, Technology, and Operating Strategies. Available: www.epa.gov/region9/waterinfrastructure/technology.html. Accessed 7/10/12.

27. U.S. EPA Region 9. (2012b). Water and Energy Efficiency in Water and Wastewater Facilities: Environmental Benefits. Available: www.epa.gov/region9/water-infrastructure/enviro-benet.html. Accessed 7/10/12.
28. Golembiewski, W. and Orchard, B. Modular Designs Recover Energy While Supplying Consistent Water Service. A website report www.pumpsandsystems.com/.../mayjune-2015-modular-designs-recover-energy-... (13 May 2015).
29. California Energy Commission. (2005). California's Water-Energy Relationship. Available: www.energy.ca.gov/2005publications/CEC-700-2005-011/CEC-700-2005-011-SF.PDF. Accessed 9/18/12.
30. Wastewater Treatment, Wikipedia, The free encyclopedia, last visited 10 August 2019 (2019).
31. Modular and Transportable System by Agripower Incorporated. A website report www.environmental-expert.com/.../modular-and-transportable-system-432189 (2015).
32. Miles, T. Thompson Spaven High Efficiency Modular Biomass Gasification Systems. A website report (February 2010).
33. Modular Compact Combined Heat and Power (CHP) Using Local. A website report www.nrcan.gc.ca/.../modular-compact...chp-using-local-heterogeneous-biomas... (August 2018).
34. Enerkem: From Waste to Cellulosic Ethanol, Biomethanol | Disruptive. A website report https://enerkem.com/ (August 2018).
35. Nielsen, U. Innovative LFG Energy Project Case Studies The Benefits of Modular Power Generation Systems. A website report by 2G CENERGY (33) Innovative LFG Energy Project Case Studies the Benefits of Modular www.epa.gov/sites/production/files/2016-05/documents/29_nielen.pdf (2015).
36. Faucette, R. Three Rivers Regional Solid Waste Management Authority Landfill Gas to Energy Project. A website report www.mdeq.ms.gov/wp-content/uploads/2017/05/3Rivers_LFGTE_Profile.pdf also www.tva.com/greenpowerswitch/partners/index.htm (2016).
37. Kramer, R., Pelter, L., Kmiotek, K., Branch, R., Colta, A., Popa, B., Ting, E., and Patterson, J. (30 October–1 November 2011). Modular Waste/Renewable Energy System for Production of Electricity, Heat, and Potable Water in Remote Locations. 2011 IEEE Global Humanitarian Technology Conference, 15 December 2011 INSPEC Accession Number: 12439854. doi: 10.1109/GHTC.2011.13, Seattle, WA, USA.
38. Locatelli, G., Boarin, S., Pellegrino, F., and Ricotti, M.E. (2015). Load following with Small Modular Reactors (SMR): A real options analysis. *Energy*, 80. doi: 10.1016/j.energy.2014.11.040.
39. Carelli, M. (2009). The exciting journey of designing an advanced reactor. *Nuclear Engineering and Design*, 239, 880–887. doi: 10.1016/j.nucengdes.2008.10.032.
40. Yan, X., Yan, X., Tachibana, Y., Ohashi, H., Sato, H., Tazawa, Y., and Kunitomi, K. (2013). A small modular reactor design for multiple energy applications: HTR50S. *Nuclear Engineering and Technology*. doi: 10.5516/NET.10.2012.070.
41. Yan, X., Kunitomi, K., Nakata, T., and Shiozawa, S. (2003). GTHTR300 design and development. *Nuclear Engineering and Design*, 222, 247–262. doi: 10.1016/S0029-5493(03)00030-X.
42. Nuclear Desalination | IAEA. A website report www.iaea.org/topics/non-electric-applications/nuclear-desalination (2018).
43. Kupitz, J. and Podest, M. (1998). Nuclear Heat Applications: World Overview. IAEA Bulletin, 26(4), 18–21. www.iaea.org/sites/default/files/publications/magazines/.../26404781821.pdf.
44. Verfondern, K. Overview of Nuclear Cogeneration in High Temperature Industrial Process Heat Applications. A paper presented at OECD-IAEA workshop, Paris (4–5 April 2013).

45. Richards, M., Shenoy, A., Schultz, K., Brown, L., Fukuie, M., and Harvego, E. (2006). The Modular Helium Reactor for Hydrogen Production. 15th Pacific Basin Nuclear Conference (15PBNC), October 2006, INL/CON-06-11320 PREPRINT INL, Idao.
46. Schultz, K.R. (2003). Use of the Modular Helium Reactor for Hydrogen Production. World Nuclear Association Annual Symposium, 3–5 September 2003, London, pp. 1–11.
47. Manufacturing, Wikipedia, The free encyclopedia, last visited 9 August 2019 (2019).
48. Elisa Wood US DOE. Allots $10M to Combined Heat & Power Projects for Manufacturers, Combined Heat & Power Projects for Manufacturing Industries. A website report in microgrid knowledge https://microgridknowledge.com/chp-microgrids-manufacturers-doe/ (11 September 2018).
49. Efficient Energy Use, Wikipedia, The free encyclopedia, last visited 4 August 2019 (2019).
50. Thorpe, D. Three Office Buildings Using Modular Passive Heating and Cooling Design. A website report www.smartcitiesdive.com/.../three-office-buildings-using-passive-heating-and-c… (2015).
51. Velraj, R., Anbudurai, K., Nallusamy, N., and Cheralathan, M. (2002). PCM Based Thermal Storage System for Building Air Conditioning at Tidel Park, Chennai. Proceedings of WREC, Cologne.
52. Bruno, F. Using Phase Change Materials (PCMs) for Space Heating and Cooling in Buildings. A paper presented at the 2004 AIRAH Performance Enhanced Buildings Environmentally Sustainable Design Conference, Sydney, Australia. For copies of the conference proceedings, contact the AIRAH office on 03 8623 3000, ECOlibrium. Also Using Phase Change Materials (PCMs) for Space Heating … - CiteSeerX , citeseerx.ist.psu.edu/viewdoc/download?doi=10.1.1.466.6725&rep=rep1… (March 2005).
53. Modular CHP for Energy Efficiency-Benefiting from a Modular Approach to Energy Efficient Heating. A website report in EM Magazine www.energymanagermagazine.co.uk›Boilers/Burners (1 June 2017).
54. Generators/Modular Cogeneration - Tecno Impianti. A website report www.tecnoimpianti.link/en/generators-modular-cogeneration.html (2015).
55. 2G Energy | Modular CHP Cogeneration Systems for Biogas. A website report www.2g-energy.com/ (2015).
56. Michael Hobbs Modular System Team. Increasing Energy Efficiency through Modular Infrastructure of UPS by Dell for Data Center Energy Conversion and Management, Volume 134, pp. 20–31 (15 February 2017).
57. Data Center, Wikipedia, The free encyclopedia, last visited 8 August 2019 (2019).
58. Modular Cooling System Enables On-Demand Data Center Capacity. A website report www.datacenterknowledge.com/.../modular-cooling-system-enables-demand-d… (19 October 2015).
59. Modular Approach Curbs Data Center Energy Use - Commercial. A website report www.commercialarchitecturemagazine.com/modular-approach-curbs-data-cent… (5 June 2017).
60. Location-Independent, Flexible Power+Heat Generation … - MWM. A website report www.mwm.net›Press›PressReleases (11 July 2019).
61. Modular Power Plant—MWM. A website report www.mwm.net/mwm-chp…cogeneration/…power-plants/modular-power-plant/ (2019).
62. Tecogen Supplies Three Ultra-Clean CHP Modules to Sacramento. A website report www.tecogen.com/news…/tecogen-supplies-three-ultra-clean-chp-modules-to (1 March 2011).
63. CHP/Microgrids—L & S Energy Services. A website report https://ls-energy.com/chpmicrogrids/ (2015).

64. Katmale, H., Clark, S., Bialek, T., and Abcede, L. Energy Research and Development Division. Final project report, Borrego Springs: California's First Renewable Energy-Based Community Microgrid | CEC-500-2019-013, California Energy Commission, CA www.energy.ca.gov/2019publications/CEC-500-2019.../CEC-500-2019-013.pdf (February 2019).
65. Duffy, O. and Purani, U. Plug and Play Units Expand US Market for Small-Scale CHP. www.cospp.com, co-generation of on-site power generation, January–February, 2012. A website report www.kraftpower.com/wp-content/uploads/2012/.../1201_COSPP_P35-41_KRAFT.pd (2012).
66. Micro Combined Heat and Power, Wikipedia, The free encyclopedia, last visited 23 May 2019 https://en.wikipedia.org/wiki/Micro_combined_heat_and_power (2019).
67. Broccardo, M., Girdinio, P., Moccia, S., and Molfino, P. (November 2010). Quasi static optimized management of a multinode CHP plant. *Energy Conversion and Management*, 51(11), 2367–2373. doi: 10.1016/j.enconman.2010.04.011.
68. Gimelli, A., Muccillo, M., and Sannino, R. (May 2017). Optimal design for cogeneration for hospital facilities. *Progress in Nuclear Energy*, 97, 153–161.
69. Sparacino, M. and Camussi, M. Mauro Colombo - AEM SpA Roberto Carella, CIaudio Sommaruga - U.G.I. The World's Largest Geothermal District Heating Using Ground Water Under Construction in Milan (Italy): AEM Unified Heat Pump Project. A website report www.geothermal-energy.org/pdf/IGAstandard/EGC/2007/143.pdf (2007).
70. Worldwide Micro CHP Market Study for 2018 to 2024 Providing. A website report www.reuters.com/brandfeatures/venture-capital/article?id=69825 (20 December 2018).
71. Global Micro Combined Heat and Power (micro-CHP). Market 2019–2023. A website report www.businesswire.com/.../Global-Micro-Combined-Heat-Power-micro-CHP-M... (22 April 2019).
72. Shaik, S., Rama Krishna, K.S., and Vaddi, R. (2017). Nano-scale transistors with circuit interaction for designing energy-efficient and reliable adder cells at low VDD. *IETE Technical Review*, 35(5), 456–466. doi: 10.1080/02564602.2017.1327826.
73. Li, Y., Yang, J., and Song, J. (2017). Nano energy system model and nanoscale effect of graphene battery in renewable energy electric vehicle. *Renewable and Sustainable Energy Reviews*, 69, 652–663. doi: 10.1016/j.rser.2016.11.118.
74. Smith, T.C. (2016). Sergey Edward Lyshevski, Nanotechnology for Portable Energy Systems: Modular Photovoltaics, Energy Storage and Electronics, IEEE 36th International Conference on Electronics and Nanotechnology (ELNANO), 19–21 April 2016, IEEE, Kiev, Ukraine. Electronic ISBN: 978-1-5090-1431-6. doi: 10.1109/ELNANO.2016.7493077.
75. The Growing Developments of Nanotechnology in Consumer Products. A website report www.nanotechetc.com/growing-developments-nanotechnology-consumer-prod... (10 September 2017).
76. Growing Application of Commercial Nanotechnology-based Products. A website report https://statnano.com/news/50098 (24 July 2015).

3 Modular Systems for Energy Usage in Buildings

3.1 INTRODUCTION

As shown in my earlier book [1], the building industry has extensively used the modular approach to cut cost, improve efficiency and quality, and reduce the time for construction. Buildings also account for approximately 40% of the worldwide annual energy consumption [2]. The total global energy consumption in 2007 was 495 quadrillion British thermal units (Btu), out of which the buildings sector consumed about 198 quadrillion Btu. According to the Energy Information Agency, worldwide energy consumption is expected to increase 1.4% per year through 2035, implying that buildings will consume 296 quadrillion Btu by 2035 [3].

Fossil fuels meet a majority of world energy needs, and because buildings are a large energy consumer, they are also a major contributor to global carbon emissions and greenhouse gas (GHG) production. It is now largely recognized that addressing energy use in buildings can reduce total fossil fuel consumption and the associated GHG emissions. Benefits such as decreased building operational energy costs have prompted growing interest among policy makers, the technical community, and the general public in addressing building energy issues and investigating solutions for reducing energy consumption of buildings.

While energy efficiency is being incorporated into new construction, existing buildings account for a majority of the building stock that will be in place in the foreseeable future. In his 2009 presidential address, American Society of Heating, Refrigerating, and Air-Conditioning Engineers (ASHRAE) presidential member Gordon Holness stated that 75%–80% of the buildings that will exist in 2030 already exist today [4]. This statistic suggests that there is an opportunity for reducing the building sector's contribution toward global energy consumption through reduction of energy use in existing buildings. Significant efforts are needed to at least reduce fossil energy consumption in building industry.

Reducing existing building energy consumption (by fossil energy) consists of two synergistic approaches: (a) to reduce the need for energy through implementation of energy efficiency or conservation measures, and (b) to offset the remaining building energy needs through use of renewable energy systems. It is important to note that building energy efficiency measures should be considered first as the cost to invest in efficiency (and conservation) measures is approximately half the cost of installing renewable energy generating capacity equal to what the efficiency measures offset [5]. It is advised that all energy efficiency opportunities are explored and as many are implemented as is feasible before or in conjunction with renewable energy projects for existing buildings. The subject of energy efficiency and conservation was discussed in detail in Chapter 1. Numerous efforts have been made in the literature [6] to examine how renewable energy (largely solar and geothermal) can replace the use of fossil energy to provide the energy needs of the buildings. If possible, wind and water can also be used to provide power for buildings. Often microgrids (or nanogrids) are used to supply the energy needs of buildings or building complexes. These efforts are largely modular in nature and are the subject of a detailed examination in this chapter.

As shown in my previous book [1], there are two types of building structures: permanent and temporary. Problems of the temporary structures are generally dealt with by the use of modular buildings. These actually meet the terms of low costs, as opposed to the terms of convenience of use, or energy efficiency in operation. Using the latest technologies in the production of modular buildings has improved the operation sufficiently; it is now possible to use them entirely for purposes associated with the use of the buildings. Office buildings, warehouses, and conference rooms have become the common standard. In Eastern Europe, particularly in a country like Slovakia, this has become a normal part of cities and municipalities, social housing, schools, and kindergartens, which are all built using this technology. During the assessment phase of these buildings, energy efficiency is always the priority. The study by Tauš et al. [6,7] was aimed at establishing the economic potential of modular buildings using renewable energy sources in Slovakia. This study outlined the parameters that need to be considered in evaluating the usefulness of renewable energy sources for energy in modular buildings.

The modular construction of temporary buildings have progressed far enough that now highly efficient construction of residential modules, alternatively called "containers," are pursued [8]. These buildings are often wrongly called building units. Construction of container modules in Europe is not new; it has been used for more than 50 years, and their success is due to the speed of construction, minimal environmental disruption, and high variability. Today, when it is necessary to respond to situations flexibly and rapidly, the modular construction directly offers an appropriate solution [9]. The most prominent representative of container architecture in Europe is the German architect Han Slawik, who has built some successful container modules. The use of renewable energy sources like solar and geothermal has only been sporadically used to curtail the use of fossil energy. The aim of the study by Tauš et al. [6,7] was to assess the parameters that will lead to the real possibilities of using renewable energy in these buildings [10].

Modular architecture responds to needs specific to the production of economically affordable housing, the lack of space in cities, and an eco-friendly way of life. The modular design is smart but is also a comfortable solution for today's lifestyle [11]. The modules are a means of standardization of constituent component of the building. Modules used to build today's modular objects are referred to as prefabricated space modules. The most detailed element of the module is the Le Corbusier module that expresses the specified system, which had merged a measure of human scale with the conventional scale. In a modular design, the module is composed of multiple materials in a so-called sandwich system construction. The whole building

Systems for Energy Usage in Buildings 131

is assembled from these modules of the same size. Modular constructions were born in the United States, where a relatively high percentage of houses and residential buildings are built using this method. The most developed country in this particular field in Europe is the United Kingdom, where there are thus constructed primarily schools, kindergartens, nursing homes, and other similar buildings. Another appropriate use of modular buildings is in the business sector, during the construction of hotels, restaurants, shops, car repair shops, and so on.

Modern age demands modern approaches. The speed and flexibility of this system often determine the reasons why investors opt for modular construction projects. Rapid return on investment is their reward for progressive thinking. Modular constructions appeal to investors by their mobility with the possibility of disassembly and reassembly at the new location. This method of construction is also environmentally friendly because the construction site is quiet and clean. Air pollutants, which occur in the conventional constructions, are minimized. Use of modular construction is friendly for the surrounding area, with no large claims to the territory for setting up the construction site. However, the operation of modular buildings requires high costs. The largest share of operating costs is in ensuring thermal comfort as well as power supply. Heating and cooling of modular buildings cost just as much as the rental price of such buildings. At present, the standard method of heating in temporary modular buildings is a direct electrical heating unit. Cooling is provided by a separate air conditioning unit.

The energy demands represent the amount of energy required to operate the building. Here, we count every energy input entering the building for its purposeful use [12]. This includes the energy for heating, heat loss, cooling - reducing heat gain, hot water, ventilation, lights and power necessary for the operation of electrical appliances. Energy demands for buildings involve two components. The investment energy demands are difficult to determine and include energy consumed in the production of materials for the construction of the building and its liquidation. In contrast, the operating energy demands are becoming a major indicator of the economic operation of the building. Energy performance of the building represents the amount of energy consumed for its operation. Operating energy values may be affected already in the project phase by factors that cannot be altered, such as placement in the field, shape, methods of deployment of a glass surface, and the composition of envelope structures. Factors that can be modified in built objects are internal heating systems, hot water systems, air treatment and ventilation, air conditioning or heat gain utilization, and electrical and power supply. Energy consumption provided by energy performance of buildings may differ from the actual values because they are considered as a standardized use of the building. In determining the energy performance of buildings, we need to address the areas that can correctly assess the current situation and recommend improvements [13]. These areas are as follows:

1. **Description and evaluation of the initial state:** Except basic identification data of the building—the location, size, shape, age—it is necessary to identify the ways of energy consumption and supply. It also requires characterization of the physical and structural conditions of the object, as to the composition and characteristics of the envelope structure's condition, roof,

floor, and apertures. Another area is the technical condition of the building, which addresses an energy supply system, the manner of their use, and their technical specifications. It deals with the technical condition of the heating system, its parameters, hot water supply, and lighting method. All these factors are at the heart of energy conservation evaluation.
2. **Energy balance:** For determining the energy intensity of the building, it is necessary to balance the type, amount, and purpose of energy, including losses in the supply and distribution networks due to the influence of efficiency.
3. **Selection of the energy source:** The optimal solution is choosing an energy source for each building separately. It should be considered in addition to the technical characteristics of the building, the site of a building location, the purpose of its use, the time of its operation, and so on [14]. It is not appropriate to standardize the energy source. Renewable energy sources are usually financially more demanding than traditional energy sources; it is, therefore, appropriate to consider their use in some aspects [15].
4. **Economic assessment:** The considered options should also be evaluated in terms of their economic efficiency of the use of investment funds and the saving effect achieved. This is based on the value of the building before and after the implementation of austerity measures and the determination of the internal rate of return, which is the ratio of the increase in value to the funds spent which helps determine the payback period for investments. It is clear that the optimal variant is the one that will bring the highest internal rate of return and the shortest payback period.

The course of electricity consumption is similar in most modular buildings for administrative and educational uses. Therefore, it is intended to work primarily with renewable resources to enable comparison with the model used so far to respond flexibly to the consumer requirements and enable the production of heat as well as cold. While energy efficiency and conservation measures apply to all four areas mentioned above, it is the choice of energy source (fossil or renewable) that has most environmental effects. The subject of energy efficiency and conservation is discussed in detail in Chapter 1.

3.1.1 Choice of Energy Source

The categorical imperative for the choice of energy source is in the nature of temporary modular buildings. This means that the priority is focused on an easily accessible link to the local energy and local renewable energy sources. Because of a need for labor resources for operating, cleaning, and regulatory regimes, they have primarily excluded "primitive" renewable energy sources such as biomass and its derivatives [16,17]. Similarly, the use of heat pump water/water is excluded because it is particularly expensive to install because of the need for two water wells, boreholes, or meanders. The use of biomass, small modular nuclear reactor, wind, and water sources for energy supply for buildings makes most sense with the use of microgrid or nanogrid for the buildings or the community. This topic is discussed in detail in Chapter 9.

Systems for Energy Usage in Buildings 133

Major energy needs of buildings are electricity and heating, ventilation, and air conditioning (HVAC). As shown in this chapter, the best choices of renewable resources that fulfill the requirements of modularity, simplicity of installation, and operation and are structurally acceptable are photovoltaic (PV) technologies that use solar energy and geothermal technologies; however, the latter is mostly used for the HVAC needs of the building [18]. The use of wind and water for energy supply is very site-specific, and they are best implemented through microgrids. In this chapter, therefore, we focus on modular systems for PV and geothermal technologies for energy consumption of buildings.

3.2 MODULAR SOLAR THERMAL HVAC FOR BUILDINGS

Businesses are changing every day, whether that means a new client occupying a given piece of office space or an existing business adding more employees or equipment to accommodate new needs. These changes can affect the output needs of the HVAC system. Figure 3.1 illustrates a North Carolina home where passive solar design and solar thermal system supplies domestic hot water and a secondary radiant floor heating system. Modular units allow one to meet these requirements simply by adding or subtracting individual units, rather than replacing the entire system. Many business buildings-whether an office space or simply a commercial space-host multiple residents, who may have vastly different heating and air conditioning needs. Modular units allow office managers to meet these needs simply by routing the output for different units to different locations. A breakdown in HVAC system can be catastrophic to a business. Modular units mean that the vast majority of breakdowns occur in one unit, allowing the rest to shoulder the load until it can be fixed instead of shutting the

FIGURE 3.1 This North Carolina home gets most of its space heating from the passive solar design, but the solar thermal system supplies both domestic hot water and a secondary radiant floor heating system. (Photo courtesy of Jim Schmid Photography, NREL.)

whole system down and turning your space into an oven or a freezer in the process. There are three advantages of a modular commercial HVAC system that one should consider when installing commercial air conditioning in the office.

- A. **Adaptable:** Modular systems contain a series of separate units, which can be added or removed from the system very easily. That means if your business grows and you need more air conditioning power, you don't need to replace the whole system. You just need to add another modular unit. The versatility allows you maximum flexibility while ensuring that you're never paying for more of a system than you need.
- B. **Space saving:** Because commercial units are placed on the roof, you don't need to use any valuable square footage on it. That leaves more room for storage, extra personnel, or whatever your business needs to remain competitive.
- C. **Efficient:** Commercial HVAC systems are sturdy and reliable, which means they can handle the hottest days as well as the coldest nights. They're built of solid materials and can handle much larger loads to boot. While repairs are sometimes necessary, a regular schedule of maintenance and timely repairs will allow them to run for a good long line. There is a difference between commercial and residential HVAC system. A commercial HVAC needs to heat or cool the air in a much larger space, as well as tailoring the specifics to match multiple occupants in some cases and adjusting the temperature based on the needs of the specific business. A modular HVAC system that utilizes multiple units harnessed in common purpose instead of one big unit handling the whole building provides a number of benefits.

In the first place, modular units allow one to increase or decrease the power of the entire system without necessitating a wholesale removal and installation. Heating and cooling units need to be balanced to match the power needs of the business in question—neither too powerful nor insufficiently powerful—and given the ever-changing needs of business, that number can fluctuate quite a bit. Modular units allow you to keep pace with those changes quickly and easily. If there is a single unit and it suffers a breakdown, it will knock out heating and cooling for the entire space, which can be disastrous. With modular units, however, the issue becomes much less of a concern. The chances of multiple units breaking down at the same time is minimal, and when one unit fails, the others can pick up the slack long enough for you to schedule repairs.

Most commercial HVAC systems are mounted on the roof to save space for other aspects of the business. A centralized unit would need to be placed in a specific point, reinforced to reduce the stress of the weight. Modular units, on the other hand, distribute the weight evenly and reduce stress on the building's infrastructure. Solar thermal technologies can be used anywhere in the United States. However, some regions naturally receive more intense and more reliable solar energy than others, depending on latitude, typical weather patterns, and other factors. Solar thermal technologies absorb the heat of the sun and transfer it to useful applications such as heating buildings or water. There are several major types of solar thermal technologies in use [20,21,22–36]:

1. Unglazed solar collectors
2. Transpired solar air collectors
3. Flat-plate solar collectors
4. Evacuated tube solar collectors
5. Concentrating solar systems

In addition to the solar thermal technologies mentioned above, technologies such as solar PV modules can produce electricity, and buildings can be designed to capture passive solar heat. We briefly describe these options below.

1. **Unglazed Solar Collectors:** An unglazed solar collector is one of the simplest forms of solar thermal technology. A heat-conducting material, usually a dark metal or plastic, absorbs sunlight and transfers energy to a fluid passing through or behind the heat-conducting surface. The process is similar to how a garden hose, laying out in the open, will absorb the sun's energy and heat the water inside the hose (see Figure 3.2). These collectors are described as "unglazed" because they do not have a glass covering or "glazing" on the collector box to trap heat. The lack of glazing creates a trade-off. Unglazed solar collectors are simple and inexpensive, but without a way to trap heat, they lose heat back to the environment and they operate at relatively low temperatures. Thus, unglazed collectors typically work best with small-to-moderate heating applications or as a complement

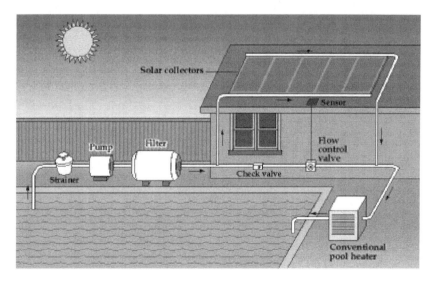

FIGURE 3.2 Unglazed solar collector for swimming pool heating [19].

to traditional heating systems where they can reduce fuel burdens by preheating water or air.

Solar pool heating collectors are the most commonly used unglazed solar technology in the United States. These devices often use black plastic tubular panels mounted on a roof or other support structure. A water pump circulates pool water directly through the tubular panels, and then returns the water to the pool at a higher temperature. Although used primarily for pool heating, these collectors can also preheat large volumes of water for other commercial and industrial applications. In this type of heating, sunlight hits the dark material in the collector, which heats up. Cool fluid (water) or air circulates through the collector, absorbing heat. The warmer fluid is used for applications such as pool heating. The potential applications include pool heating and space heating. Its key end use sectors are single-family homes, multiunit housing, lodging, schools, and municipal government buildings.

2. **Transpired Solar Air Collectors:** Transpired solar air collectors (see Figure 3.3) typically consist of a dark-colored, perforated metal cladding material mounted on an existing wall on the south side of a building. A fan pulls external air into the space behind the metal cladding through the perforations, where the air heats to as much as 30°F–100°F above the ambient air temperature. The fan then pulls the air into the building, where it is distributed through the building's ventilation system.

   The transpired solar collector (see Figure 3.3) is a proven but still emerging solar heating technology. This type of technology is best for heating air and ventilating indoor spaces. It can also be applied to several manufacturing and agricultural applications such as crop drying.

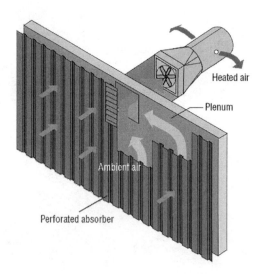

FIGURE 3.3 Transpired solar collector [21].

In this type of heating, sunlight hits the dark perforated metal cladding, which heats up. A circulation fan pulls air through the perforations behind the metal cladding, heating the air, which is then pulled into the building for distribution. Its potential applications include space heating. Its key end use sectors are multiunit housing, lodging, schools, and municipal government buildings.

3. **Flat-Plate Solar Collectors:** Most flat-plate collectors (see Figure 3.4) consist of a copper tubing and other heat-absorbing materials inside an insulated frame or housing covered with clear glazing (glass). The heat-absorbing materials may have a special coating that absorbs heat more effectively than an uncoated surface. Glazed flat-plate collectors can operate efficiently at a wider temperature range than unglazed collectors. Flat-plate collectors are often used to complement traditional water boilers, preheating water to reduce fuel demand. They can also be effective for space heating. Using a heat exchange system, they can reliably produce hot air for large buildings during daylight hours. In this type of heating, sunlight travels through the glass and hits the dark material inside the collector, which heats up. A clear glass or plastic casing traps heat that would otherwise radiate out. This is similar to the way a greenhouse traps heat inside. Cold water or another fluid circulates through the collector, absorbing heat. Its potential applications include space heating and hot water heating. Its key end use sectors are single-family homes, multiunit housing, lodging, restaurants, schools, and municipal government buildings.

4. **Evacuated Tube Solar Collectors:** Evacuated tube collectors (see Figures 3.5–3.7) feature thin, copper tubes filled with a fluid, such as water, housed inside larger vacuum-sealed clear glass or plastic tubes. Evacuated tubes use the sun's energy more efficiently and can produce higher temperatures than flat-plate collectors for a few reasons. First, the tube design increases the surface area available to the sun, efficiently absorbing direct

FIGURE 3.4 Two flat-plate solar collectors side-by-side [21].

FIGURE 3.5 Evacuated tube solar [21].

FIGURE 3.6 Two types of evacuated tube [21]. (a) Direct flow evacuated tube. (b) Heat pipe evacuated tube.

FIGURE 3.7 An array of evacuated tube collectors on a roof [21].

sunlight from many different angles. Second, the tubes also have a partial vacuum within the clear glass enclosure, which significantly reduces heat loss to the external environment. Figure 3.6 illustrates two types of evacuated tubes.

In this type of heating, sunlight hits a dark cylinder, efficiently heating it from any angle. A clear glass or plastic casing traps heat that would otherwise radiate out. This is similar to the way a greenhouse traps heat inside. A copper tube running through each cylinder absorbs the cylinder's stored heat, causing fluid inside the tube to heat up and rise to the top of the cylinder. Cold water circulates through the top of the cylinders, absorbing heat. Though evacuated tube systems are typically more expensive than flat-plate collectors, they are more efficient and can produce higher temperatures. Evacuated tubes can reliably produce very hot water for batch or on-demand water heating and for many industrial processes. Moreover, they can produce enough heat to handle almost any space heating or space cooling application. Its potential applications include space heating, hot water heating, space cooling, and industrial process heat. Its key end use sectors are single-family homes, multiunit housing, lodging, breweries, restaurants, industrial processes, schools, and municipal government buildings.

5. **Concentrating solar system:** Concentrated solar power (also called concentrating solar power, concentrated solar thermal, and CSP) systems generate solar power by using mirrors or lenses to concentrate a large area of sunlight, or solar thermal energy, onto a small area. Electricity is generated when the concentrated light is converted to heat, which drives a heat engine (usually a steam turbine) connected to an electrical power generator or powers a thermochemical reaction (experimental as of 2013). CSP had a global total installed capacity of 4,815 MW in 2016, up from 354 MW in 2005. As of 2017, Spain accounted for almost half of the world's capacity at 2,300 MW, making this country the world leader in CSP. The United States follows with 1,740 MW. Interest is also notable in North Africa and the Middle East, as well as India and China. The global market has been dominated by parabolic-trough plants, which accounted for 90% of CSP plants at one point. The largest CSP projects in the world are the Ivanpah Solar Power Facility (392 MW) (which uses solar power tower technology) and the Mojave Solar Project (354 MW) (which uses parabolic troughs) in the United States.

   In most cases, CSP technologies currently cannot compete on price with PV solar panels, which have experienced huge growth in recent years due to falling prices and significantly smaller operating costs. CSP generally needs large amount of direct solar radiation, and its energy generation falls dramatically with cloud cover. This is in contrast with PVs, which can produce electricity also from diffuse radiation. However, the advantage of CSP over PV is that as a thermal technology, running a conventional thermal power block, a CSP plant can store the heat of solar energy in molten salts, which enables these plants to continue to generate electricity whenever it is needed, whether day or night. This makes CSP a dispatchable form of solar energy. This is particularly valuable in places where there is already a high penetration of PV, such as California, because an evening peak is being exacerbated as PV ramps down at sunset.

CSP has other uses than electricity. Researchers are increasingly investigating solar thermal reactors for the production of solar fuels, making solar a fully transportable form of energy in the future. These researchers use the solar heat of CSP as a catalyst for thermochemistry to break apart molecules of H_2O, to create hydrogen (H_2) from solar energy with no carbon emissions. By splitting both H_2O and CO_2, other much-used hydrocarbons—for example, the jet fuel used to fly commercial airplanes—could also be created with solar energy rather than from fossil fuels. In 2017, CSP represented less than 2% of worldwide installed capacity of solar electricity plants. However, in recent years, falling prices of CSP plants are making this technology competitive with other base-load power plants using fossil and nuclear fuel even in high moisture and dusty atmosphere at sea level, such as the United Arab Emirates. Base-load CSP tariff in the extremely dry Atacama region of Chile reached below ¢5.0/kWh in 2017 auctions.

3.2.1 THE DESIGN AND DEVELOPMENT OF AN ADAPTABLE MODULAR SUSTAINABLE COMMERCIAL BUILDING (CO_2NSERVE) FOR MULTIPLE APPLICATIONS

Co_2nserve is a commercially sponsored project aimed at developing a cost-effective modular building that is able to meet sustainability, energy and carbon efficiency needs, as well as adaptable to meet a wide range of occupancy applications. Building applications include light industrial production, warehousing, office accommodation, educational needs, and health sector requirements. The study by Giulliani et al. [37] describes the building development and design process. Maximum use is made of passive measures including natural ventilation and natural cooling. Structural and envelope elements have been designed for factory assembly using sustainable and/or recyclable components. A key part of the design process was to establish a multidisciplinary design group at the start of the project. The design has achieved a Class A Energy Performance Certificate under the requirements of the European Energy Performance of Buildings Directive.

3.2.2 SIX INNOVATIVE MODULAR ROOFTOP SOLAR TECHNOLOGIES

Falling manufacturing costs and increasing demand have led to a number of fascinating new solar products in recent years, including roof shingles with integrated solar cells, modular solar arrays, and even efficient thermal tiles made from glass. As the technology improves and more people get turned on to the benefits of renewable energy, more innovative products are expected to hit the market in the coming months and years, pushing forward the envelope for solar power production and use of solar energy for HVAC. Here, we briefly describe six innovative products. All of these are modular in nature [38].

1. **Dow Chemical Powerhouse solar shingles:** In 2010, Dow Chemical unveiled a line of solar-integrated rooftop shingles that were a marked improvement over the existing technologies. The sleek plastic-coated Powerhouse shingles were capable of converting 13% of the sun's energy into electricity—a full 2% increase over other solar shingles on the market at that time. The shingles were expensive when they debuted ($10,000 for 250 shingles spread over 1,000 sq ft), and an array that size would only make a small dent in energy usage for a typical household. However, Dow promised they would pay for themselves within a decade, and the product was an important step forward for integrated solar power systems.
2. **Solar Power Tiles:** SRS Energy launched a product that promised to make installation of solar panels very easy. Its curved Solé Power Tiles were designed to mimic the shape of interlocking mission-style clay or cement shingles. In this case, the solar shingles had the same barrel design as their traditional counterparts, so they could be easily integrated into existing mission roofs. This adaptability would enable homeowners to replace as little or as much of their roof with the unique solar shingles as they liked, without having to rip off the entire roof.
3. **SolTech Energy glass mission tiles:** Solar shingles continue to evolve not only in efficiency but also in design. Sweden-based SolTech Energy created a stunning example of the best of both worlds with its translucent glass mission tiles, which, when installed across an entire building, give the illusion of a roof tiled with ice. The shimmering SolTech roof tiles capture solar heat and use it to warm air beneath the tiles, which is then used to heat water and warm the home during the winter. The company claims the gorgeous roof tiles can produce about 350 kWh of heat per square meter (10.7 sq ft), depending on weather conditions and the angle of the roof.
4. **Solar shingles roof:** Once a unique way to add solar power production to your rooftop may actually become a thing of the past. This is due to the emergence of new roof technologies that integrate solar cells so fully that they're actually part of the roof rather than just installed atop it. Elon Musk promised that SolarCity, which is being acquired by Tesla Motors through a $2.6 billion merger deal, will create such a roof. In a similar effort, SunTegra's integrated solar systems have been installed on homes in the northeastern United States and in California, two prime spots for making the most of the sun's energy. SunTegra's solar roof (available in tiles or shingles) currently costs 15% more than typical rooftop solar panels, but the company claims it's just as durable and weather-resistant as traditional roof coverings.
5. **SolarPod's Grid Tied solar array:** This system by SolarPod can be mounted to nearly any type of roof and requires no drilling of holes. Since holes are the last thing you want in your roof, this is a fairly clever solution to a common installation challenge. SolarPod's Grid Tied solar array is an integrated and modular plug-and-play solar power system that includes a prefabricated frame made from corrosion-resistant steel that holds the solar

panels. Because the frame floats above the existing roof, it is also easy to adjust the angle of the solar array to capture the maximum amount of sunlight for that particular location, thereby increasing solar energy production.

6. **SoloPower flexible solar panel:** In a completely different approach to easing the woes of installation, SoloPower developed a flexible solar panel that can be unfurled as easily as a carpet. The thin-film solar panels, linked together in long strips, boast an 11% energy conversion rate and a smooth installation process, due to their lightweight and flexible composition. In theory, the flexible solar panels could be unrolled right over the top of an existing roof, in any quantity desired, without the sort of expensive glass and aluminum frames required by most rooftop solar arrays. Although the desire for integrated solar roofs may drive innovation faster, it would be nice to see more flexible—and potentially portable—options hit the market as well.

3.2.3 Modular Passive Solar Heating System for Residential Homes

A modular passive solar energy storage system comprises a plurality of heat tubes arranged to form a flat-plate solar collector and are releasably connected to a water reservoir by, and are part of, double-walled heat exchangers that penetrate to the water reservoir and enhance the heat transfer characteristics between the collector and the reservoir. The flat-plate collector-heat exchanger disassembly, the collector housing, and the reservoir are integrated into a relatively lightweight, unitary structural system in which the reservoir is a primary structural element. In addition to lightweight, the system features high efficiency and ease of assembly and maintenance. The invention by Hunter Billy [39] relates to solar heating systems and, in particular, to a modular passive solar heating system which uses a heat pipe or heat tube absorber. Passive solar water heating systems are well established in the art and are desirable in many applications because of their characteristic lack of need (or minimum requirements) for electric power, controls, and pumps. The so-called water thermosyphon may well be the prevalent passive solar water heating system in terms of worldwide usage. In a typical and simple arrangement, the water thermosyphon uses a non-tracking water absorber for absorbing energy in the form of heat from incident solar radiation, and then transfers this thermal energy via an inlet manifold to a water reservoir or tank mounted on the absorber.

U.S. Pat. No. 4,217,882, issued August 19, 1980, discloses a two-phase thermosyphon which is an improvement over the relatively heavy freeze-susceptible, conventional water thermosyphons. The '882 patent adapts the well-known heat pipe or heat tube concept, originally developed approximately 20 years ago, to a tracking parabolic trough concentrator, as well as to fixed, cylindrical, and compound parabolic concentrators. As applied to such embodiments, the heat pipe comprises an elongated, small diameter tube which at one end partially surrounds or penetrates a water tank. The tube is partially filled with a vaporizable working fluid and is supported at a slight incline relative to the horizontal. Incident solar radiation focused on the tube by the surrounding concentrator structure heats and evaporates the fluid. The resulting vapor rises to the upper tank end of the heat tube where, due to thermal coupling, the solar heat is transferred from the tube fluid to the tank fluid, which typically is water. Upon giving up its latent heat of evaporation, the condensed working fluid is returned by gravity to the lower concentrator end of the tube.

The above-described heat tube imparts several advantages to passive solar water heating systems. The heat pipe has excellent thermal conductance in that it has very high heat transfer capability over even a relatively small temperature gradient. In addition, the evaporation–condensation cycle provides highly anisotropic, essentially one-way heat transfer along the tube. It may help to consider the situation at night or during other periods of low incident solar radiation for better understanding. At such times, there is little or no evaporation and condensation of the working fluid; the fluid in its liquid state pools at the low incident solar radiation. At such times, there is little or no evaporation and condensation of the working fluid; the fluid in its liquid state pools at the low incident solar radiation. The resulting discontinuity in the conduction path between the absorber/concentrator and the tank essentially eliminates heat loss via the working fluid. The combined result of the excellent thermal conductance characteristics and the one-way heat transfer characteristics is very efficient heat transfer along the heat pipe into the reservoir with little outward heat loss. Other advantages of adapting heat tubes to passive solar collectors, not exhaustive, include relatively lightweight; adaptability to freeze protection, since only the reservoir tank contains water; and a high percentage net usable system energy, since little or no parasitic power consumption is required to operate the system.

One known alternative to the '882 system (US Patent No. 4,217,882) comprises a flat-plate collector array of finned heat tubes which wrap around an integral storage tank. The heat tubes and the tank wall in effect comprise a double-walled heat exchanger so that heat is transferred from the working fluid in each heat tube through the heat tube and tank walls and into the storage medium. Despite the above-mentioned substantial advantages over prior art passive solar water heating systems, it has not been possible to fully utilize the potential of the heat tube concept. In particular, the unitary heat collector system exemplified in the '882 patent utilizes a relatively complex collector/concentrator, which, like the other prior art systems, is integrated with the tank. It is difficult to disassemble all or part of the '882 system or the alternative system for inspection or repair. It is also difficult to use such systems in primary applications of passive systems, namely, roof top applications on residential dwellings. In short, while incorporation of the heat tube technology into solar heat collector systems represents an advance in passive solar energy technology, the potential of heat tube systems for combined simplicity, durability, and ease of maintenance has not been fully realized.

The objects of simplicity, durability, and ease of maintenance are realized in present modular passive solar energy storage system. The system described by Hunter Billy [39] comprises a reservoir for storing heat in a fluid medium such as water, one or more finned heat tube absorbers assembled into a flat-plate collector system for absorbing incident radiation and heating an evaporizable medium contained therein, and a double-walled heat exchanger releasably mounted to and projects into the reservoir and in turn releasably mounts the heat tube absorber for effectively transferring heat stored in the heat tube medium into storage in the reservoir medium. In a preferred embodiment, the heat tube forms the inner wall of the double-walled heat exchanger. A preferred embodiment of my solar energy storage system also includes a support housing for the flat-plate collector integrally mounted to the reservoir. The housing is adapted to mount and support the heat tube absorbers in a desired orientation and permits sliding insertion of the heat tubes there through for engagement with the heat exchanger. This cooperative, modular structural arrangement is unique in that it incorporates the system reservoir as a primary structural support element in a structure of relatively low weight and excellent structural integrity, in which the heat tubes and/or collector housing are easily removed and replaced individually or collectively for inspection, maintenance, or repair. In a preferred structural embodiment, the collector housing comprises a lower base plate having a peripheral wall structure extending upwardly therefrom. Apertures or slots are formed in the wall on opposite ends of the housing so that the finned heat tube absorbers can be inserted into and removed from the heat exchanger and reservoir. A support plate extends between the side walls of the housing substantially transverse to the longitudinal axes of the heat tube absorbers. The support plate has a slot formed therein which has the shape of the transverse cross-sectional profile of the heat tubes. As a result, the plate supports and orients the tubes in an essentially flat array.

3.2.4 Modular Zero Carbon Emission School Building with Hybrid Solar Heating and Passivhaus Construction

The UK's 2008 Budget announced the government's ambition that all new non-domestic buildings should be zero carbon from 2016. Montgomery Primary School, a 420-pupil facility, was the site of one of these projects; the new school was designed to replace the existing facilities which had become outdated and no longer in use. In theory, the simplest way to produce a zero-carbon design would be to build a typical school, replace the gas boiler with biomass for heating, and buy electricity via a green tariff for power and lighting. The design team that worked on Montgomery School considered such an approach unsustainable and unrealistic as it relies on continued use of precious resources. The team also believed that passive design is not about applying ashy green technology to a standard product: it is an integrated design process based on a holistic approach that focuses on reducing energy demands. Montgomery therefore designed to use the minimum amount of resources, including fossil fuel. Its conception was based on a modular design approach, utilizing off-site prefabrication techniques with multiple repeatable units where appropriate, and all the energy required for heating, lighting, and power are generated on site [40].

As the school design was based on the Passivhaus standard, the heating system was not be able to make up for any wastefulness on behalf of the occupants: if they fail to maintain sensible practices such as shutting windows and doors and cease to control the plant they will notice a drop in temperature. It is hoped that by understanding how the building works and its positive impact on costs, users will be encouraged to change their behavior in relation to energy and sustainability. Zero carbon signifies that all emissions from the building and the activities that take place within it must have a net energy balance of zero over a year.

As Passivhaus is a performance-based "energy" assessment, the following targets standard needed to be met to qualify for certification. The standard requires that the primary energy demand target is met in all cases. This figure must include space heating, domestic hot water, lighting, fans, and pumps, as well as all the projected appliance consumption. In addition to the primary energy demand, the standard permits that either the specific heating demand (SHD) or the specific heating load (SHL) must be met. Thermal comfort is also a very important issue: a certified Passivhaus should not fall below 16°C even without heating during the coldest winter months [40]. In designing school, the philosophy was to use on-site renewable energy sources to meet the "zero carbon" in use target, and the first step was to minimize the building's overall energy consumption by preconstruction modeling. Passivhaus has a proven capability of achieving low energy in buildings. After consideration, PV panels were chosen as the most appropriate solution since the site did not lend itself to wind or water-based power generation.

A precast concrete panel solution was not critical to achieve a Passivhaus building as it is recognized that a wide variety of construction products and techniques can be used. Concrete becomes important when it provides high thermal mass in the step-up from Passivhaus to zero carbon, "future proofed" to 2080 as the design needed to ensure that the school would not overheat under the most aggressive climate change criteria. The thermal inertia of concrete allows it to absorb and store surplus heat or cold and then release these back into the air as part of a designed thermal strategy. Concrete also has excellent sound suppression and vibration dampening properties as it absorbs both low- and high-frequency sounds. Moreover, it displays air tightness, safety, and security benefits related to its massiveness and density, provided that wet joints are constructed with cast-in-place concrete poured between the precast panels.

Structural junctions and the interfaces of different building elements can be problematic because of the requirement to provide a continuous insulation wrap around and under the building. Air movement through or around the insulation bypasses its effectiveness and reduces its performance. All insulation has been tightly sealed with tapes and joined with low-expanding-rate foam. Junctions at the top and bottom of insulation panels are also sealed on a thin bead of foam, thereby eliminating thermal bypass. In addition, it was important to ensure that the insulation was continuous and unbroken as even modern products can produce a cold bridge, reducing thermal performance and introducing possible condensation risks. Where it became necessary to break the insulation layer, specialist thermally efficient structural fixings were installed [40]. The school's Passivhaus design and solar-generating electrical power plant enable its electricity bill to be zero each year, with excess electricity providing a small income to maintain the power cells and equipment. The Montgomery School was awarded a Quality Approved Passivhaus certificate in February 2012, essentially for "the comfort and quality of the internal environment with extremely low energy consumption." It was the first Passivhaus school in the United Kingdom.

Passive use of solar energy may be a significant factor in Passivhaus design. Windows (glazing and frames) should have U-values not exceeding 0.80 W/m^2/K, with solar heat-gain coefficients around 50%. Major rooms orientated north to avoid overheating/glare. South-facing mono-pitch roof for PV array placement was used. Energy-efficient, triple-glazed, thermally broken window was installed with a conductivity of 0.80 W/m²/K. The air permeability values for completed buildings frequently exceed these design limits. Air leakage through unsealed joints measured 0.28 times the building volume per hour. The majority of new buildings did not achieve sufficient air permeability values to warrant the incorporation of a whole building ventilation system—thus, trickle vents, extract fans, or passive stack ventilation is commonly used. Fresh air may be brought into the building through underground ducts that extract heat from the soil. This preheats fresh air to a temperature above 5°C (41°F) even on cold winter days. Low energy refrigerators, stoves, freezers, lamps, washers, dryers, etc. are indispensable in a Passivhaus design. No passive heating of fresh air occurs. Ventilation was accomplished via controlled mechanically ventilated heat recovery (MVHR). In winter the cold intake air is warmed by the regenerator units. In the summer natural ventilation is via open-able windows. Highly efficient heat recovery from exhaust air was accomplished using an air-to-air heat exchanger. Most of the perceptible heat in the exhaust air is transferred to the incoming fresh air (heat recovery rate over 80%). Using mechanically ventilated ultra-high efficient heat recovery air handing unit, heat recovery rate over 82% can be achieved. Dedicated low-energy lights are provided in a number of rooms in a new building. If appliances are supplied, they will be generally C-rated or perhaps "Energy Saving Recommended" in some instances. Low-energy, A-rated appliances are specified throughout. Similar results were obtained in the lifecycle analysis of hybrid active and passive solar heating of another school building by Paya-Marin et al. [41].

3.2.5 Modular Bright Built Home Brand

In mid-2013, Kaplan Thompson Architects in Portland, Maine launched the Bright Built Home (BBH) brand—a line of high-performance, modular homes designed to attain zero net energy consumption with 5–10 kW PV systems. Though Kaplan Thompson Architects had designed several custom, zero net energy homes for clients in the Northeast, they were eager to make zero energy homes more affordable and available to a wider range of clients.

Although modular homes certainly offer more savings than site-built homes, there are some additional challenges associated with modular construction—especially with high-performance homes. Optimizing the BBH wall system, for example, involved not only achieving desired R-values (at least 35 ft²h °F/Btu) and durability at reasonable costs but also the wall systems needed to be resilient during shipment. In addition, wall construction at the factory could not be allowed to slow the modular production line [42].

With funding from the Building America Program, part of the U.S. Department of Energy Building Technologies Office, the Consortium for Advanced Residential Buildings (CARB) worked with BBH to evaluate and optimize building systems. CARB's work focused on a home built by Black Bros. Builders in Lincolnville, Maine (International Energy Conservation Code Climate Zone 6). As with most BBH projects to date, modular boxes were built by Keiser Homes in Oxford, Maine. Modeling showed that the home in Lincolnville achieved 41% source energy savings without

PV and 77% with PV. Modeling also showed $2,400 in annual energy savings compared to the Building America benchmark. In brief, BBH systems include [42]:

- R-19 minimum foundation insulation—details may be determined by the builder; the Lincolnville home used R-20 insulated concrete forms (ICFs).
- Double, 2×4 walls with staggered studs built on 2×8 plates with dense cellulose; 2 inch of rigid foam is installed outside of the oriented strand board sheathing.
- R-60 ceiling/roofs, which are challenging to achieve with insulated roofs. The Lincolnville home uses 4 inch of extruded polystyrene (XPS) above the deck and 2×6 rafters packed with cellulose (R-42).
- Triple-pane windows (U-values from 0.18 to 0.25 Btu/ft^{2}h °F)
- Heating and cooling provided by inverter-driven, air-source heat pumps; different homes use different combinations of single-zone, multi-zone, ductless, and ducted heat pumps.
- A heat recovery ventilator that provides whole-building ventilation. Local ventilation strategies vary; the Lincolnville home used the heat recovery ventilator for bathroom exhaust but used a dedicated kitchen exhaust hood.

Beginning in January 2015, BBH started meeting Home Energy Rating System (HERS) ratings and ENERGY STAR certification standards for their homes. The demand for these types of homes have been increasing every year.

3.2.6 MODULAR BUILDING INTEGRATED SOLAR THERMAL TECHNOLOGIES AND THEIR APPLICATIONS

Solar energy has enormous potential to meet the majority of present world energy demand by effective integration with local building components. One of the most promising technologies is building integrated solar thermal (BIST) technology. The study by Zhang et al. [43] presents a review of the available literature covering various types of BIST technologies and their applications in terms of structural design and architectural integration.

The review covers detailed description of BIST systems using air, hydraulic (water/heat pipe/refrigerant), and phase-changing materials (PCM) as the working medium. The fundamental structure of BIST and the various specific structures of available BIST in the literature are described. Design criteria and practical operation conditions of BIST systems are illustrated. The state of pilot projects is also fully depicted. Current barriers and future development opportunities are therefore concluded. Based on the thorough review, it is clear that BIST is very promising with considerable energy saving prospective and building integration feasibility. This review facilitates the development of solar-driven service for buildings and helps the corresponding saving in fossil fuel consumption and the reduction in carbon emission. Solar energy is one of the most important renewable sources locally available for use in building heating, cooling, hot water supply, and power production. Truly BIST systems can be a potential solution toward the enhanced energy efficiency and reduced operational cost in contemporary built environment. All BIST systems can be modular.

According to the vision plan issued by European Solar Thermal Technology Platform (ESTTP), by 2030 up to 50% of the low and medium-temperature heat will be delivered through solar thermal [44]. However, currently, the solar thermal systems are mostly applied to generate hot water in small-scale plants. When it comes to applications in space heating, large-scale plants in urban heating networks, the insufficient suitable-and-oriented roof of most buildings may dictate implementation of solar thermal technology. For a wide market penetration, it is therefore necessary to develop new solar collectors with feasibility to be integrated with building components. Such requirement opens up a large and new market segment for the BIST system, especially for district or city-level energy supply in the future.

BIST is defined as the "multifunctional energy facade" that differs from conventional solar panels in that it offers a wide range of solutions in architectural design features (i.e., color, texture, and shape), exceptional applicability and safety in construction, as well as additional energy production. It has flexible functions of buildings' heating/cooling, hot water supply, power generation, and simultaneous improvement of the insulation and overall appearance of buildings. This façade-based BIST technology would boost the energy efficiency of buildings and turn the envelope into an independent energy plant, creating the possibility of solar thermal deployment in high-rise buildings.

3.2.6.1 Working Principle of Typical BIST System

The system normally comprises a group of modular BIST collectors that receive solar irradiation and convert it into heat energy, whereas the heating/cooling circuits could be further based on the integration of a heat pump cycle, a package of absorption chiller, a modular thermal storage, and a system controller. In case of some unsatisfied weather conditions, a backup/auxiliary heating system (e.g., boiler) is also integrated to guarantee the normal operation of system. In the typical BIST system, the overall energy source is derived from solar heat, which is completely absorbed by modular BIST collectors. This part of heat is then transferred into the circulated working medium and transported to the preliminary heat storage unit, within which heat transfer between the heat pump refrigerant and the circulating working medium occurs. This interaction decreases the temperature of the circulating medium, which enables the circulating medium absorbing heat in the facades for next circumstance.

Meanwhile in the heat pump cycle (compressor-condenser expansion valve-evaporator), the liquid refrigerant vaporizes in the heat exchanger, which, driven by the compressor, subsequently converts into higher temperature and pressure, supersaturated vapor, and further releases heat energy into the tank water via the coil exchanger (condenser of the heat pump cycle), increasing the temperature of the tank water. Further, the heat transfer process within the coil exchanger results in the condensation of the supersaturated vapor, which gets downgraded into lower temperature and pressure liquid refrigerant after passing through the expansion valve. This refrigerant then undergoes the evaporation process within the heat exchanger in the initial heat storage again, thus completing the heat pump operation. When the water temperature in the tank accumulates to a certain level, i.e. 45°C, water can be directly supplied for utilization or under-floor heating system. For the cooling purpose, the system is coupled with an additional appliance of absorption chillers.

BISTs can be classified into air, hydraulic (water/heat pipe/refrigerant) and PCM-based types according to the heat transfer medium. The air-based type is characterized by lower cost, but lower efficiency due to the air's relatively lower thermal mass. This system usually uses the collected solar heat to preheat the intake air for the purpose of building ventilation and space heating. Hydraulic-based BISTs are most commonly used BIST devices that enable the effective collection of the striking solar radiation and its conversion into heat for hot water production and space heating. The PCM-based type is usually operated in combination with air, water, or other hydraulic measures that enable storing parts of the collected heat during the solar-radiation-rich period, and releasing them to the passing fluids (air, water, or others) during the poor solar radiation period to achieve a longer period of BIST operation. The details of these technologies and their comparative performances along with BIST structural design in terms of architectural element are given by Zhang et al. [43]. The architectural element includes wall-based, window-based, balcony-based, roof-based, and sunshield-based BIST technology. The design criteria and standards for BIST technology as well as operating conditions for different types of BIST technology are also outlined by Zhang et al. [43].

Based on the advantages and disadvantages outlined by Zhang et al. [43], some conclusions can be made [22–36,43]. First, the opaque facade is usually composed of multiple layers with functions of external protection and insulation. Such features exactly offset the limitations of flat-plate BIST, which is less flexible in translucency and module thickness. Therefore, the wall-based BIST application is especially common in a renovation project as a cladding element with further insulation and protections of weather and mechanical stress. Second, the transparent and translucent window-based application concerns more on daylight transmission, outdoor visual relation, and partial sun shading. In such cases, the lightweight glazed/unglazed collector or evacuated tubes are recommended to integrate with glazing in an alternating or interlaced pattern to have partial sun shading or create a dummy effect. Third, the evacuated tube shows more promising applications in the balcony-based integration for its lightweight, higher efficiency, and convenience in assembly and pipe connections. Finally, roof has the great popularity for the installation of BIST systems. It provides superiority in dissimulating solar collector, higher solar thermal yield, and convenient installation methods. Zhang et al. [43] have described 21 different successful practical applications of BIST technology across Europe. Five representative applications from the study of Zhang et al. [43] are described below.

1. **"Home for life" concept house, Aarhus, Denmark:** This project dissimulates BIST within the roof to form a kind of active house with a living area of $190\,m^2$. The total solar collecting area of $6.7\,m^2$ is integrated in the lowest part of the roof surface with an auxiliary heat pump system, which directly supplies 50%–60% of the yearly household hot water heating demand, as well as a supplement for downward room heating [43].
2. **Retrofitted Office Building in Ljubljana, Slovenia:** This is a detail designed retrofitting project with BIST application. The air heating vacuum tube collectors replace the balustrade on the fifth floor, while the transparent solar thermal collectors are attached to the stairwell. Both collector

areas face almost south, the solar collectors 15° toward east and the air-heating tubes 15° toward west. Both components are developed to be a substitute for the building skin as well as the thermally activated building system of 100 m² office space with fluid at temperatures above 35°C during the heating season [43].
3. **Granby Hospital, Granby (Quebec)/Canada:** This application is one of the few air-based BIST systems. The transpired solar fresh air heating system in this hospital is specially designed to satisfy the demand of large amount of fresh air. The solar system is aesthetically presented with 82 m²-curved facades to provide 8,160 m³/h to the space below. The working principle is that the perforated metal absorber draws in heated fresh air off the surface of south-facing walls, where it is then distributed throughout the building as preheated ventilation air. Overall, the system shows operating efficiencies up to 70% with total paybacks within 5 years with energy savings about 149.1 GJ [43].
4. **Social housing in Paris, France:** This is the first social house building with BIST application. The new multifunctional semi-transparent collector encapsulated into a double-skin facade, which weighs 45 kg, offers complete privacy from the passengers commuting nearby, ensures light penetrating into the back of room, and restricts noise flow. Besides, the solar panel captures solar energy to produce enough power to meet 40% of the domestic hot water needs, providing 44% of domestic hot water needs [43].
5. **School building in Geis, Switzerland:** This is another successful prefabricated BIST application. Because of early design phase intervention, designers paid considerable attention to the facade design, layout, size, and fixed modular dimensions of the solar collectors. The total 63 m² collector field fully respects the rhythm of window openings and the color of both window frame and concrete bricks, giving a convincing result [43].

The BIST devices mentioned here has created a considerable energy saving perspective, and endorses an innovative solar thermal approach for building integration with efficient and durable devices, variable choices of color and texture, and mature processing techniques. Diverse directions with miscellaneous motives have led to a lot of different BIST building integration variations. The core element of a BIST system in the above buildings has exhibited a high energy capacity, and the end users therefore benefit of an attractive payback time on their investment. The shortfalls in the current BIST systems are primarily: (a) most concepts of absorber parts in BIST are directly inherited from the conventional solar thermal collectors, which exhibit instability under long-term weather exposure, difficulty in both on-site assembly and practical application, and complexity in fluid channel structure, resulting in bulk volume and fragility for BIST application; (b) most BIST designs only function as a structural cladding element of glass curtain-wall, rooftop, or traditional wall surface; (c) limited considerations are given to the irregular geometry, color, and texture leading to boring building appealing; (d) lack of building-related studies in terms of lighting pollution, acoustic effect, structural load, and thermal performance, etc. Typical strategies can be assigned to dissimulation into building envelope, special placement, and modular building component design. Furthermore, majority of new building projects are solar houses or passive houses, while renovation projects provide a new direction of multifunctional transformation instead of sole repairing. More focus has been given on threatening resource shortage, comfort living environment, as well as position architectures themselves in the former niche. It is worth mentioning that as an innovative choice of multifunctional building envelope, BIST has superiority in good insulation, ability to capture solar heat, and high compactness, which fully satisfies the current boom in green buildings and zero energy buildings. Some additional useful information for BIST technology is outlined in Refs. [22–36,43–48].

To alleviate above-mentioned barriers, Zhang et al. [43] suggest that future work should focus on: (a) integration of structural and finishing materials to work as true building material; (b) compulsory structural/rigidity test for the BIST serving as a load bearing structural element; (c) development of lightweight and long-life polymer materials to replace the current promising materials, such as metal, glass, and ceramic, to minimize loads on existing architectural structure; and (d) integration of BIST design into architectural or lifecycle design tools (such as building information modeling, BIM) to quickly assess the appropriate structure and integration method of a BIST system for buildings [22–36,43–48].

3.3 MODULAR GEOTHERMAL HVAC SYSTEMS

Geothermal systems, which use the relatively stable and moderate temperature of the ground as an energy source, are piquing the interest of green-minded builders as an efficient and clean alternative to conventional heating and cooling systems. Geothermal heat pumps (GHPs) accounted for about 50,000 residential and (mostly) commercial installations nationwide in 2006. This is less than 1% of the overall heating and cooling equipment market, according to a recent report issued by the Freedonia Group, a Cleveland-based research firm. GHPs work typically by exchanging or transferring heat via liquid-filled tubing loops that run between the house and the ground or a nearby body of water. However, that same report forecasts 6.5% annual growth for the technology through 2011, setting a new bar of 70,000 installations that year. By 2016, the report predicts nearly 100,000 GHPs will be put in place each year. In addition, a GHP is a heating and cooling machine-in-one, eliminating the outdoor air conditioning or air-to-air heat pump compressor from the spec sheet. Fitted with a standard blower and filter, it leverages the same distribution network of ducts and supply/return registers as any other air-forced system. GHPs also mitigate seasonal fluctuations in performance (unlike air-source heat pumps), run on about half the amount of electricity of a conventional system, deliver effective humidity control, and can be specified within the same unit footprint to heat the home's water supply, in-floor radiant heating system, and swimming pool [49].

While builders will need an excavation or drilling crew and likely a certified installer to trench and hook up the underground loop of circulation tubes that feed the system, any HVAC contractor can connect the rest of the equipment. The systems are generally applicable in almost any climate, due to the consistent temperature of the ground of around 70°F, at about 8 ft below the surface. There are caveats, of course, chief among them an installed cost that's about two to three times that of a conventional system, primarily to excavate for and install the underground loop of high-density polyethylene tubing. For a new home, the straight return on an upfront investment of about $2,500 per ton of capacity plus perhaps another $5,000-plus to install a closed-loop, ground-source system is at least 4 years and probably more, even in the most expensive utility markets. Some of that premium might be recouped from local utility rebates, state-sponsored grants, and federal tax credits, though it may require proof of a system that meets the minimum performance standards and certified installation to qualify. The effectiveness of a modular GHP also relies on a well-built and insulated shell. If the (building) envelope is not designed for high-efficiency equipment, a ground-source heat pump will not allow significant energy and cost savings [49].

GHP efficiency is defined in two different ways: COP (coefficient of performance, for heating efficiency) and EER (energy efficiency ratio, for cooling performance). In both cases, the higher the rating, the better the energy performance. These two ratios should be used as a relative gauge among geothermal systems—and within that, similar types of GHPs such as open- versus closed-loop—and reflect a "steady-state," or factory-tested performance rather than what's likely to be found in the field. The ratings are similar, but not directly comparable, to efficiency ratios calculated for conventional HVAC equipment. That being said, GHPs are typically three to four times as efficient in heating mode and at least 50% better in cooling mode than a furnace and air conditioner, respectively. To ensure consistency, the ratings were standardized for the industry in 2000. The Air-Conditioning and Refrigeration Institute regulates them, mitigating discrepancies in what manufacturers include in their energy use calculations to arrive at the ratios. Further, relative to conventional HVAC equipment, which also is tested and rated in a steady state, GHPs are clearly—and in some cases dramatically—more energy efficient. There are, however, variations on that basic model. Among closed-loop systems, the network of tubing can be installed horizontally or vertically, usually depending on site conditions and available land around or even under the building footprint, as well as the comparative costs of trenching (for horizontal) or drilling (for vertical). Closed-loop systems also can tap the heat stored in a water well or nearby pond, which is called a water-to-water system.

Lastly, an open-loop (or groundwater) GHP eliminates the antifreeze carrier, instead using the solar-heated water of a pond or well. The heat is then exchanged at the pump inside the house, as with a closed-loop system, or used directly for hydronic space (and perhaps water) heating. For various reasons, some municipalities regulate the drilling for and discharge water of open-loop systems, especially large-scale installations far exceeding those of an individual home. For open-loop systems, especially, consult the local building department and other appropriate regulatory agencies about what is allowed and how to best handle the discharge of water that's been through the heat exchanger. Once the heat from the ground or water source makes it to the heat pump inside the house, the system works the same as any forced-air system. In cooling modes, the process is reversed, with the heat pump taking hot air out of the house and exchanging it with the ground temperature—which, in the summer, is far cooler than the outside air—and using an ozone-safe refrigerant to cool it further and distribute it throughout the house [49].

3.3.1 MODULAR HYDROLOGIC'S HYDRON MODULE FOR GEOTHERMAL SYSTEM

Water Furnace's HydroLogic for radiant heating integrates seamlessly into a geothermal system. The main component is a prepiped, prewired modular mechanical panel that supports cooling, dehumidification, and multiple zones of radiant heating while optimizing heat pump performance by automatically adjusting water temperatures based on indoor feedback and outdoor temperature via the included sensor. HydroLogic's intelligent heat/cool switchover reduces energy waste and maximizes the overall system performance. Features and benefits include easy equipment selection; powerful communicating controls that integrate both forced-air and radiant heating; simple wiring; professional appearance; optional insulated models for chilled water; and zones that are easy to add on. Hydron Module geothermal systems are manufactured by Enertech Global [50,51].

A Hydron Module geothermal system is an excellent heating and cooling solution for almost any application. Geothermal systems perform well in cool northern climates as well as warm southern conditions. They are ideal for both large and small properties, and in new construction or existing homes. Hydron Module solution meets any heating and cooling needs. In addition to being unique in the way that they heat and cool, geothermal systems also provide distinctive benefits that set them apart from conventional systems [50,51] such as no use of fossil fuels, efficiency ratings of 400%–550%, simple, quite operation, reduce water heating cost by up to 50%, single-unit, multi-family and commercial applications, and industry leading warranty options.

3.3.2 MODULAR CHILLER SYSTEM USING GEOTHERMAL RESOURCES

The ClimaCool SHC onDEMAND modular chiller units were the heart of the 150-ton geothermal HVAC system at The Charles Machine Works' product development center in Perry, Oklahoma. CMW has recently developed horizontal directional drilling (HDD) equipment, which is being used in the growing geothermal industry [52–54]. CMW also designed its latest geothermal project, a retrofit system for its product development center, with innovative new simultaneous heating and cooling heat pump units from neighboring Oklahoma City-based ClimaCool Corp. The three 50-ton ClimaCool SHC onDEMAND modular chiller units were specified to replace an antiquated reciprocating chiller-driven HVAC system installed during the building's construction in 1978. The project goal was to save money. Each of the three, six-pipe SHC onDEMAND units also featured a unique modular design with built-in redundancy with separate module electrical feeds and dual independent refrigeration circuits, allowing for the unit to maintain operation while individual modules are being serviced. The new SHC onDEMAND modular chiller unit offered dramatic energy saving benefits—potentially more than 50% when compared to the traditional boiler/chiller systems. It also featured a patent-pending six-header design that eliminated the required space between and external to the modules, creating the smallest system operating footprint when compared to a typical simultaneous system [52–54].

This ClimaCool configuration incorporated several notable features that maintain precise chilled and hot water temperatures, building loads and compressor run time equalization for ultimate operational efficiency. The SHC onDEMAND heat pump unit is an exceptional piece of equipment when it comes to reducing energy consumption. Features such as the CoolLogic Control System and integral motorized valves for variable pumping can result in cooling efficiencies up to 25 EER and heating efficiencies up to 5 COP. This opens a lot of doors for us in supplying to projects that demand a high level of energy efficiency, including sustainable building projects, and the flexible nature of the system means it can be integrated with cooling towers, geothermal loops, or hybrid systems [52–54].

The new 150-ton system also incorporates two existing air handlers with replacement Wilo brushless DC motorized pumps for load demand sensing. The controls system and fresh air vents with an economizer, considered state-of-the art when originally installed in 1978, would also be replaced with a web-based building automation system (BAS). The new BAS provides much more control than the previous system, and really brought mechanical operations to a whole new level in the building. Initial geothermal field drilling for the project began in September 2011. Overall, the field includes 168,400-ft deep boreholes with HDPE double U-bend pipe installed throughout. The drilling was conducted in tandem with the construction of a separate foam-insulated 20-ft-by-24-ft mechanical building that would house the new ClimaCool units and ancillary HVAC system equipment. Upon completion of the loop field in March 2012, the ClimaCool unit was installed and initial flow testing was conducted.

Thousands of ClimaCool units are currently in operation throughout North America, with a dominant and growing concentration of successful simultaneous heating and cooling unit applications. All ClimaCool modular chillers are designed to minimize installation time and costs, with individual modules that can fit through standard doorways and have low centers of gravity for easy transport via pallet jacks and forklifts. Modular chillers from ClimaCool are additionally engineered to streamline maintenance, with single-point electrical connections and waterside isolation valves that allow for the servicing of an individual module while the remaining modular chillers in the bank continue to operate. Their non-proprietary designs also afford contractors the ability to service the units without proprietary parts or factory technicians [52,53]. With a focus on system efficiency, ClimaCool has engineered its units with a holistic view on cooling, heat recovery, heat pump operation, geothermal capabilities, and simultaneous heating and cooling applications in mind [52–54].

3.3.3 MODERN VERSUS TRADITIONAL MODULAR CHILLERS

Modular chillers have come a long way in the last 6–8 years. What once was a viable solution only for retrofits, cooling, and tight spaces has become one of the industry's most flexible and versatile HVAC systems [55]. Whether a mission critical, commercial building, or a hospitality application any downtime is costly and stressful. It is inevitable, however, that at some point the chiller will require maintenance or one will experience an unexpected point of failure. With backup generators that aren't always sized properly, this elevates the need to look for a system that offers true redundancy. Traditional modular chillers, that retain their design from the 1990s, have a single power source, so if one unit needs to be serviced the whole bank would go offline. New modular chillers are designed with separate electrical feeds providing a built-in redundancy. If one module fails or needs to be shut down for maintenance, the other modules can continue to operate to keep your building running at full capacity.

Chillers, as the name implies, have long been seen as a staple for their cooling capabilities. Unlike first-generation modular chillers, modern chillers have the ability to provide heating, cooling, heat recovery, and heat pump technology in each module. Some even harness both air and water to achieve simultaneous heating and cooling, without geothermal. This means there's no need to throw away the heat these chillers generate as today's modular chiller technologies recover and harness it. Not only does this provide tremendous flexibility and multifunctional configurations but it also offers energy efficiencies and cost savings unattainable by any other system. This ability, through individual modules, enables the system to recover heat from one of the cooling modules, allowing users to harness the heat already produced in another part of the building. This could mean cooling one room in your building while supplying recovered heat in another, customizing low-temperature chilled water requirements for an operating room suite without compromising efficiency in the rest of the hospital for comfort cooling, or simply redistributing heat to make domestic hot water [55].

When selecting a modular chiller, consider the expected total cost of ownership throughout the lifespan of the equipment. This can save you money in the long run. When it comes to controlling costs and boosting energy efficiency, modern modular chillers have the ability to outperform the rest because they use only the energy needed, and nothing more. While they function as one piece of equipment, the modules are enabled individually based upon leaving chilled water and hot water control temperature set points. This improves turndown capacity when a building is partially occupied, during off-peak hours, or during the spring–summer start up period and fall–winter changeover days. They are powerful enough to cool or heat an entire building on peak days, but can also be turned down to power only one or two units, significantly reducing the total cost of ownership. Moreover, when you consider modular chillers can eliminate the costs and maintenance of boilers or cooling towers, the cost savings really start to add up.

Whether a renovation or a new construction, modern modular chillers make expanding the capacity easy with friendlier component configuration and streamlined water piping. First-generation designs make modular chillers top heavy, but modern ones have moved the compressors from the top of the unit to the front and placed them lower for a low center of gravity to ease handling during installation. In addition, modern modular chillers, especially those that have the ability to recover heat for simultaneous heating and cooling, are designed so that all water piping connections can be made on one end. Expansion onto the chiller bank occurs at the other end without the need to disconnect and weld new pipes together as is the case with traditional modular chillers. New modular chillers also make expandability easier as they can complement conventional systems, allowing building owners to expand capacity and obtain all the cooling and heating benefits of truly tailored turndown and redundancy associated with them, while leveraging their existing cooling and heating systems [55].

Not only do modern modular chiller technologies offer the ability to service modules individually, eliminating downtime, but their design makes them inherently easier to maintain. With the compressors placed in front and lower, it is easier to access when servicing is required. In addition, headers no longer block the heat exchangers as they've been relocated from the front and back of the unit to the top. Once connected, they don't have to be taken apart for easy serviceability. Off-the-shelf components and simplicity of design eliminates the need for factory technicians for servicing. However, equipment startup should always be performed by factory-certified technicians. Modular chillers utilize brazed plate heat exchangers which need to be back flushed to clean them. Modern units are designed with individual flush ports for each heat exchanger, which make it easy to clean the individual module heat exchanger while the rest of the modular bank continues to operate. This not only maintains energy efficiency of the building but it eliminates the need to shut down the full bank of chillers, reduces the risk of damaging the heat exchanger, and reduces costs from needing to hire outside service support that is frequently required with traditional modular chillers.

Inherent to all modular chillers—traditional and modern—but frequently overlooked is the ease of compliance with ASHRAE Standard 15 mechanical ventilation standards. While refrigerant leaks are potentially dangerous, modern modular chiller systems are compliant with today's standards because of their small charge in comparison to volume of the mechanical room. Considering that each module contains two refrigerant circuits, the largest circuit from a 250-ton modular chiller can contain only 25 tons of refrigerant, which is typically well within the acceptable range of concentration in the mechanical room space. This advantage eliminates the need for costly design and installation of ventilation, monitoring, and safety equipment required with conventional chillers, further lowering the total cost of ownership [55].

3.3.4 Modular Geothermal Measurement System

Conventional air-source reverse-cycle heat pump systems are commonly used for providing heating and/or cooling to building environmental spaces, manufacturing processes, and a variety of other uses. Properly used, such systems can be quite effective in environments where the ambient temperature is not extreme. Although generally acceptable performance is obtained in such moderate ambient temperature conditions, such systems leave a lot to be desired during extreme fluctuations in ambient temperatures, wherein substantial reductions in heating and cooling capabilities and in operating efficiencies are seen. In recent years, heat pump systems have been developed which use ground-source heat exchangers whereby the earth is utilized as a heat source and/or sink, as appropriate. Heat pump systems utilizing the more moderate temperature range of the earth provide efficiencies which are substantially improved over those obtained from air-source heat pump systems. Such earth exchange systems are based on the concept that useful thermal energy could be transferred to and from the earth by the use of subterranean tubes in flow communication with various above-ground components. In a direct exchange ground loop system, refrigerant coolant pumped through such tubes by a compressor serves as a carrier to convey thermal energy absorbed from the earth, as a heat source, to the above-ground components for further distribution as desired for heating purposes. Similarly, the coolant carries thermal energy from the above-ground components

Systems for Energy Usage in Buildings 157

through the subterranean tubes for dissipation of heat energy into the earth, as a heat sink, for cooling purposes.

Unfortunately, a number of major complications may arise when refrigerant is pumped through the subterranean tubes. First, lubricant oil which characteristically escapes from the compressor while the system is operating is carried along with the refrigerant throughout the system and tends to accumulate in the tubes, substantially reducing the ability of the subterranean tubes to perform their originally intended function. Second, when an energy demand cycle is completed, the system would shut down while waiting for a subsequent demand for energy transfer. As a result, a certain amount of liquid refrigerant then passing through the subterranean tubes would lose its momentum and revert to liquid, leading to low-pressure fault conditions. A third problem, which was generally observed for prior art heat pumps, was the absence of a mechanism for achieving refrigerant pressure equalization subsequent to system shutdown for reducing start-up loads. Because of the absence of such pressure equalization, the service life of the compressor was reduced. Previous attempts to circumvent some of the aforesaid problems generally followed either of the two approaches: (a) using a vertically disposed, single-closed loop, subterranean exchanger, or (b) using a plurality of closed loop systems working in combination, with one of such loops horizontally or vertically disposed subterraneously.

The vertical single-closed-loop approach generally utilized downwardly or vertically inclined subterranean tubes. Such a system could generally be designed to operate in either a heating mode or a cooling mode. Unfortunately, however, the same system generally would not properly function when operated in a reverse mode due to the difference in specific density of gaseous refrigerant relative to that of liquid refrigerant. Specifically, while the transition from liquid to gas could be designed to occur while the refrigerant was passing downwardly in one mode of operation, such transition could occur while the refrigerant was passing upwardly in the reverse mode of operation. Another shortcoming of a prior art vertical loop system was the entrapment of oil at the lower extremities of the vertically oriented tube, thereby depriving the compressor of essential lubricant oil.

The plural loop approach generally utilized indirect heat exchange rather than direct heat exchange. That approach basically employed two or more distinct and separate, cooperating, closed-loop systems. A first one of such closed loop systems was sequentially routed through the earth and through an interim heat exchanger transferring thermal energy there between. The other one of such closed-loop systems, which was sequentially routed through the interim heat exchanger and through a dynamic load heat exchanger for further distribution as desired via techniques commonly known in the heating and cooling industry, operably interacted with the first such closed-loop system in the interim exchanger. Thus, through the cooperative effort of the two separate closed-loop systems, thermal energy was indirectly transferred between the earth and the environmental or process load.

By using a plural-loop approach, the oil deprivation problem was partially resolved by eliminating transmission of the refrigerant and oil through the subterranean tubes, thereby minimizing the quantity of oil which could be drained away from the compressor or by using a non-phase-change heat transfer fluid in the subterranean portion of such a plural-loop system. The liquid-based earth-source systems provided improvements over air-source systems by using the earth's thermal mass as a heat source/sink and by eliminating them to ambient air as a heat source/sink. An attempted solution to the stressing problem included the augmentation of a liquid-source heat pump with a liquid-heat exchanger loop which integrated both a liquid-based subterranean heat exchanger and a liquid-based fan coil in an attempt to boost the performance of the liquid-source heat pump. In another approach, a third fan coil was integrated with a refrigerant-based subterranean heat pump design. These and many other efforts were made to improve the performance of heat pumps.

The invention by Baller [56] and the various embodiments described and envisioned in the patent provides a modular geothermal measurement system that provides for the pumping of a heat transfer fluid and that simplifies on-site installation time, allows for growth of the system over time, increases ground loop pumping power while providing energy transfer data specific to each thermal load, and allows the beneficiary of a geothermal investment to be billed for their benefit, enabling the investor to capture the economic benefit of the investment. It is, therefore, an object of the present invention to provide new, useful, unique, efficient, non-obvious systems and methods for providing energy to an end user from a ground energy transfer system, and, in one aspect, from an energy transfer loop system. Such systems and methods include metering and quantifying energy delivery for use in various later calculations and transactions. Such systems and methods further include measuring energy transfer for each of a variety of heating or cooling loads for use in later calculations and transactions. It is another object of the present invention to provide new, useful, unique, efficient, non-obvious systems and methods for combining heating and cooling sources to improve overall system performance. It is another object of the present invention to provide new systems and methods for measuring carbon dioxide reduction to communicate the social or economic benefits of such reduction.

A modular geothermal measurement system that provides for the pumping of a heat transfer fluid was devised in this patent. The modular unit simplifies on-site installation time by reducing the number of distinct components to be installed and allows for the optional incorporating of additional heat sources or sinks, whereby the length of the ground loop can be reduced, further reducing installation costs. The modular measurement system further allows for the growth of the system over time by adding modules, increasing the ground loop pumping power while providing energy transfer data specific to each thermal load. A controller having an energy control module provides energy control points. Such a system allows the beneficiary of a geothermal investment to be billed for their benefit, enabling the investor to capture the economic benefit of the investment.

3.3.5 Modular HISEER Geothermal Ground Source Heat Pumps

To heat (or cool) home with geothermal energy one needs a heat pump unit (In the same way that fridge uses refrigerant to extract heat from the inside, keeping a ground source heat pump extracts heat from the ground, and uses it to heat the home or business.), a loop of refrigerant filled piping buried outside, ductwork inside, a circulating pump, and other mechanical and electrical items which complete the system. Though initially expensive, ground source heat pumps are very cheap to run and maintain, and they are also very efficient and environmentally friendly. HISEER commercial water to water heat pump, GSWW60-DNNNS, geothermal ground source heat pump is a electrically powered system that takes advantage of the earth's constant ground temperature to provide heating, cooling, and hot water for home or business [57,58].

HISEER commercial ground source heat pump can be easily connected together for any size heating and cooling system at the project spot. A maximum of 64 units can be worked together for the largest 3,840 KW output capacity. HISEER geothermal ground source heat pump, designed by European standard EN 14511, approved by TUV with its high efficiency [57,58].

3.3.6 MODULAR GEOTHERMAL HEAT-PUMP SYSTEM INSTALLATION

The invention by Lambert [59] relates to an economical and easy to install geothermal heathuge pump system which utilizes horizontal ground loops. The system is environmentally friendly and can be installed in most already developed locations, such as existing apartment complexes, as well as to-be-developed areas. The invention is directed to novel heat exchange loops, the method of installing the loops and devices for use in the installation process, and in association with the loops.

The invention involves the use of the earth as a constant source of heat to be extracted by a heat pump. Geothermal or ground-source heat pumps, although costly to install, have been found to more efficiently heat and air condition building spaces than other heat pumps. It is much more efficient to extract heat from a substance such as earth, which has a near-constant temperature, than from air which can be subject to severe temperature variations. Prior art geothermal systems have utilized ground loops that have been installed horizontally using open trenches. Horizontal installation, however, causes significant damage to the environment. Nature has suffered from root-system damage and removal of vegetation caused by the huge displacement of earth required by horizontal ground loop installation. Landscaping has often been destroyed by the large displacement of earth, removal of trees, shrubs, structures, and grass. Parking lots, driveways, sidewalks, and curbs have been removed, damaged, or their installation delayed for long periods of time to allow for the settling that must occur after massive displacements of earth. Moreover, polluted runoff from the large excavations has disturbed the environmental in areas beyond the job site. Furthermore, the huge excavating equipment is destructive in its weight, size, and polluting use of fossil fuels.

The installation of prior horizontal ground loops requires the subcontracting of big, expensive equipment and specialized personnel to perform very time-consuming drilling, excavation, and installation. The equipment used for these excavations is extremely expensive and is not owned by many HVAC installers. The man-hours required to install prior art horizontal loops is extensive and costly. Deep, dangerous ditches are dug and painstakingly prepared. Workers then spend many hours installing specialized pipes and fittings. Finally, the ditches are carefully filled and left for settling. Land that has had a prior art ground loop system installed must remain untouched for as much as a year and a quarter to allow for settling. This is an unacceptable delay to the installation of landscaping, parking lots, sidewalks, curbs, driveways, etc. The untouched ground is not only unsightly but provides dust which is carried by the wind to undesirable places (i.e., indoor surfaces, wet paint and caulk, lungs, eyes, etc.). Many owners of modern homes and commercial buildings, as well as town houses, condominiums, apartments, etc., have land areas that are too restrictive for prior art horizontal ground loop installation. Many homeowners wishing to change to geothermal heating systems forego the conversion due to the destruction or existing landscaping and wooded areas as well as other improvements. The ditch excavation required for prior art ground loops is simply not feasible for homes located on rocky land.

The obvious next step is to install vertical ground loops. Vertical ground loop installation requires the use of large cumbersome 6 inch vertical boring machinery mounted on large trucks weighing in at 15 tons or more. Few people want these monstrous machines in their yards to destroy their driveways and landscaping. These machines are noisy and leave large piles of cuttings and muddy streams of run-off water. The vibrations caused by the machinery can crack foundations and basement walls when drilling near buildings. The depth of vertical bore-holes can penetrate subterranean caverns and the water aquifer. State water control boards have expressed a preference for horizontal instead of vertical ground loops because of the greater threat to drinking water contamination posed by the vertical loop installation. Furthermore, in the case of cavern penetration, well inspectors will require cement trucks to fill a large cavern. Cement is much too expensive to waste on cavern filling. The earth's crust is full of caverns and underground rivers, creating money pits for vertical ground-loop installers.

As with horizontal ground loops, the cost of vertical ground loops is prohibitive. Drilling or trenching equipment is not typically owned by HVAC professionals because it is unique and costs thousands to hundreds of thousands of dollars for one machine. The cost of casing, pipe, fittings, cement, bits, and drill stems required for vertical ground loops can be high. Substantial expense is further incurred in the man-hours required to install the vertical loop system in the bore-holes before they cave-in. During rainy seasons, a sea of mud can fill bore-holes the minute the drill bit is pulled, rendering the bore-holes useless. The large, 6 inch bore-holes must also be filled with some substance to facilitate the conduction of heat between bore-hole walls and the heat transfer medium-carrying pipe. This substance is a costly one not needed in the instant invention. Although vertical ground loops have been put in places where prior art horizontal loops have not been feasible, the small yards of many homes have still been off limits to huge drilling equipment. Thus, because of destruction to landscaping and size and weight of water well construction drill rigs, in some rocky soil, vertical ground loops have not been feasible or desired in many cases. In addition, vertical ground loops have suffered from design problems, i.e. poor flow distribution, velocity problems are liquid or oil accumulating in the bottom of the vertical ground loops. Moreover, a simple, inexpensive way of preventing flash gas from occurring when supply and return conduits are in the same bore-hole has heretofore been unobtainable.

Also of paramount importance is the superiority of direct-exchange (DX) geothermal heat pump systems over indirect-exchange systems. In indirect-exchange systems (water-source), additional pumps to circulate a liquid other than the refrigerant in an additional indoor heat exchanger results in greater pump horsepower being

Systems for Energy Usage in Buildings 161

required. An additional heat exchanger is required because the transfer of heat goes from ground to ground-loop liquid (water) to refrigerant to air. In DX systems, however, the heat goes from ground to refrigerant to air, thus eliminating not only a heat exchanger and various pumps but also the bothersome water and antifreeze mixture in the ground-loop. Furthermore, the plastic pipe used in prior art water-source ground loops has been large, cumbersome, crinkled easily, and provided too much resistance when being inserted into bore-holes. The negative aspects inherent in the prior art are the dominant factors, which economically and environmentally rule out ground source heat pumps as the installation of choice in new structures or to retrofit existing structures. The instant invention overcomes the negative aspects of the prior art by providing a low cost, environmentally friendly geothermal system. Installation of the ground loops of the instant system is microsurgery compared to prior art installations.

The geothermal heat transfer system comprises a plurality of heat exchange loops placed in the ground at less than a 20° angle with the surface of the ground but greater than a 5° angle. Each loop has an outflow line and a return line, and a U-turn juncture at the juncture of the outflow line and the return line. The juncture has an inlet in fluid tight communication with the outflow line and an outlet in fluid tight communication with the return line. The inlet means and outlet means are on the same side of the juncture, providing a flow chamber between the inlet and outlet means. The juncture has a tapered leading edge opposite the inlet and outlet. A distributor member has an inlet at a lower end and a plurality of outlet members at the upper end. The inlet is in fluid tight contact with a return line, and each of the outlet members are, through the distributor member, in fluid tight engagement with the return line. A heat exchange device which has heat exchange conduits in a floor, the floor conduits being in heat exchange with heat exchange medium in the return line. The heat exchange medium in the floor conduits flowing through the heat exchange loop.

The installation of the geothermal heat transfer system utilizes a plurality of heat exchange loops offset from a line parallel to the surface of the earth above the loops. A trench hole is dug and a plurality of bore holes drilled, commencing on one side of the trench hole, passing through the trench hole at a first level, and continuing on the opposite side of the trench hole at a lower level. The bore hole is at an angle ranging from about 5° to less than 25° with a line parallel to the surface of the earth above the bore hole. A pair of conduits are inserted into each drilled bore hole, each of the pair being joined, in fluid tight communication by a juncture member, having a tapered leading edge. The pair of conduits are inserted into the drilled bore hole by forcing the tapered leading edge into the hole. The tapered edge clears the path for the pair of conduits being inserted. Within the trench hole, the inlet lines are connected to a first flow distributor and the outlet conduits are connected to a second distributor.

A liquid–oil–gas separator having a housing with a top and bottom. An inlet member is mounted in the housing top for delivering a liquid–oil–gas mixture to the separator. An outlet member mounted in the housing top removes accumulated oil from the housing. A deflector plate is mounted parallel to, and spaced from, the top thereby providing a flow path from the interior of the housing, around the deflector plate, and to the outlet member. The inlet member is positioned to deliver liquid–oil–gas mixture to a cup-shaped member which is spaced from, and below, the deflector. A fluid flow path is provided from the inlet member toward the cup between the deflector member and the cup to the outlet member and to the housing interior. Fluid is thereby delivered toward the cup, deflected toward the deflector plate, deflected off of the deflector plate and sprayed in a rain-like pattern toward the bottom of the housing. The liquid–oil–gas separator has at least one oil delivery conduit with an inlet and proximate to, but spaced from, the housing bottom and an outlet end positioned interiorly of the oil delivery conduit. The oil delivery conduit is at least a pair of L-shaped elements extending parallel to the housing bottom and having a plurality of oil inlet holes. The oil delivery conduit outlet end forms a Venturi fluid inlet, and the deflector and cup form a Venturi-member. Whereby a suction force is developed at the oil delivery conduit outlet, siphoning accumulated oil from the housing interior and delivering the oil to the fluid flow stream from the liquid–oil–gas mixture inlet.

The oil is separated from the liquid–oil–gas mixture by delivering a downward stream of a liquid–oil–gas mixture to the interior region of a cup-shaped member. This causes the liquid–oil–gas mixture to have its flow directed radially outward and upward off of the cup. The radially outward flow is deflected off of a deflector plate, which deflects the flow radially outward and downward off of the deflector plate to the bottom region of the housing. A Venturi suction effect is produced between the cup and the oil tubes. Accumulated oil is siphoned from the housing through oil tubes which have multiple inlets at the bottom portion of the housing. The oil is delivered to the cup-shaped member by developing a Venturi effect vacuum in the region of the cup. The heavier oil is pulled off deflector plate by suction line pressure in the return pipe.

A bore hole drilling device comprises an elongate hollow drilling bit with a cutting member affixed at one end, a central fluid passage, and connecting means at the other end. The connecting means is preferably a spiral thread. The cutting member is a substantially planar member having a leading cutting tip and presenting a triangular cutting region. A fluid inlet to the central fluid passage is located at the other end from the fluid outlet proximate the first end. The cutting bit has a radial dimension greater than the radial dimension of the elongated hollow drilling bit. The bore-hole drilling device further comprises a swivel member with a body portion and a housing element. The housing element has a central bore and a inlet, water line which connects to a water line to the inlet. The body portion is positioned within the central bore and mounted for rotational motion within the housing element. The body portion has an inlet and an outlet. The inlet is a hole through the body portion and is positioned proximate the housing element inlet. The body portion has a central bore hole, a first seal for providing a water tight rotation seal between the body portion and the housing. A second seal means provides a second water tight rotation seal between the body portion and the housing. The housing inlet and body portion inlet are positioned between the first and the second seal.

3.3.7 Modular Tube Bundle Heat Exchanger and Geothermal Heat Pump System

It is an object of the invention by Galiyano et al. [60] to provide a ground source heat exchanger which reduces the drilling and backfilling costs and problems associated with vertically oriented and slanted exchangers. It is also an object of the invention to provide an in-ground heat exchanger for use with a heat pump-based heating/cooling system, which eliminates large land area requirements and excavation/backfill steps associated with horizontally arrayed, or substantially horizontal heat exchangers. The invention also provides a compressor mechanism for a ground source heat pump system, which eliminates refrigerant phase and compressor oil problems, and substantially eliminates the need for a refrigerant accumulator and/or an oversized refrigerant accumulator. The invention provides a compressor and in-ground heat exchanger arrangement, which is insensitive to equalization problems encountered with prior art reciprocating piston compressors. It also provides oversized air handler coils for more efficient thermal transfer of the extra heat generated via the subject in-ground heat exchange coil design. A cathodic protection to these unique in-ground metal ground source exchange coils is provided so as to prevent metal decay and so as to extend system life.

To accomplish above mentioned objectives, arrays of modular, in-ground, heat exchangers are provided, each comprising tubes which are readily placed in trenches formed simply using a backhoe or the like. Trenches formed could be U-shaped or, to avoid danger of trench collapse when installing coils, V-shaped. A reciprocating, or preferably a scroll type compressor drives flow of refrigerant through the heat pump circuit. The heat exchanger arrays include grouped, connected tubes which are vertically placed in trench excavations along the sides of the trenches. The coils have relatively small inner and outer diameters whereby they may be rolled-up prior to installation. The tubes are rolled up prior to shipping and unrolled at the time of installation and placed along the trench walls. Stakes having preformed coil attachment parts are evenly disposed intermittently along the trench wall and hold the coils against the trench walls. The individual coils are held the requisite distance apart by the attachment parts of the stakes. Connections between particular tubes are provided generally made at each tube end, for example, using a header structure that defines the successive paths of the refrigerant in alternating directions through the tubes in the array, the fluid traversing each tube in the array. This provides substantial fluid-tube and tube-ground surface contact for maximum thermal interaction between the refrigerant within the tube and the ground. The tubes are generally aligned in a vertical plane relative to each other, with one set of tubes on each respective side of its' subject trench. The tubes are held on the sides of the trench via their affixation to stakes, which also serve to keep the tubes spaced equidistant from one another to avoid tube-to-tube contact, crossovers, and resulting inefficiencies.

The modular bundle of tubes is particularly adapted to be placed in a simple earth excavation such as the trough formed by a backhoe. Therefore, the tube bundles are preferably placed along each side of a 2 ft wide trench, corresponding to a standard width of a backhoe bucket. The relatively short length and width of the trenches for the subject copper coils allows this design's installation in small land areas and without major disruption to alternate land usage.

Alternately, the tube configuration can be installed in a V-shaped trench formed in the earth by appropriate machinery, such as a backhoe digging a wider width at the top and a narrower width at the bottom. A V-shaped is advantageous as less prone to collapse than a U-shaped trench. It is appropriate to provide standard length trenches, e.g. 10–50 ft, etc.

The subject tube bundles are preferably placed in the earth to dispose the top layer of tubes below the frost line. This provides for efficient interaction between refrigerant in the tubes and the relatively constant temperature existing below ground, and prevents damage to the tubes and connections due to frost heaving, i.e., periodic ground contraction and expansion as the ground alternately freezes and thaws. In warmer climates wherein the heat pump is primarily used to cool the load, the bundle can be placed closer to the surface. Typically, the coils are placed so that the top layer of tubes is about 4 ft beneath the ground surface where the ground remains relatively cool and freezing rarely or never occurs. However, in extremely hot, desert areas, the coils need to be buried deeper to avoid the extreme usual ground surface heat. When installing in locations having a rock substrate, the coils can be longer and buried shallower.

A plurality of individual tubes of each of one or more heat exchanger units are coupled to a distributer for interfacing the tubes to the system conduits and compressor. The distributer has a number of plates and fittings adapted for interfacing to a plurality of tube ends, whereby the refrigerant in the system is evenly distributed into the interfaced tubes and substantially equal quantities of refrigerant are passed through each tube of the bundle during system operation. If a large capacity is required, a plurality of interconnected, physically separate bundles are linked serially or in parallel. In a parallel arrangement, a secondary distributor having plates and apertures can be used to distribute equal flows of refrigerant to each of the plurality of individual bundles in the same manner that the distributor for a given bundle distributes flow to the individual tubes. A heat exchanging cartridge preferably comprises a plurality of substantially parallel spaced tube members for carrying refrigerant, defining a cartridge unit. The cartridge is placed in the trench whereby the tubes are disposed adjacent to the trench walls. It is advantageous for improved heat transfer to provide the tubes with either or both of rifled inner surfaces and finned outer surfaces to increase the surface area of refrigerant to ground interaction.

To address operating problems of reciprocating piston compressors of known heat pump systems, a scroll-type compressor is preferably employed. Reciprocating-type compressors, as noted above, are generally designed to pump gas as opposed to liquid, and can be overloaded or overheated in the event of an uneven refrigerant distribution throughout the system, as typically occurs during the reversing process or shortly thereafter. When the system seeks to achieve an even distribution of refrigerant in the required direction of pumping, a temporary low pressure condition exists at the input to the compressor. The low-pressure loading conditions are not readily distinguishable from loading conditions due to critical system problems as might be caused by blockage or loss of refrigerant. Control systems that sense pressure conditions, compressor loading and the like in a piston compressor system, and time out to shut down the compressor, may be necessary to avoid compressor damage, but too often shut down during reversing operation of the system.

Piston compressors also suffer from accumulation of oil from the compressor in the heat exchangers and conduits of the system, often resulting in insufficient quantities of oil returning to the compressor and eventual compressor failure. According to the invention, a scroll compressor can be provided. The scroll compressor has an involute spiral impeller member which, when matched with a mating fixed scroll member, defines a series of crescent-shaped gas pockets between the two members. During compression, one of the members (the fixed scroll member) remains stationary, while the other member (the impeller member) is allowed to orbit, but not rotate, relative to the fixed scroll member. As this motion occurs, the pockets defined between the two members are pushed to the axial center of the spirals between the two scroll members. The pockets simultaneously are reduced in volume. When a pocket reaches the center of the spirals, the gas in the pocket, now at high pressure, is discharged from a port located at the center. During compression, several pockets are compressed simultaneously, resulting in a very smooth process. Both the suction action (at the radially outermost point between the scroll members) and the discharge action (at the center) are continuous. Compressors of this type are available from the Copeland Corporation.

By-pass and direction reversing valves are used to provide alternative opposite flow directions of the refrigerant as a means to alter the heat pump system function from, for instance, heating to cooling, and vice versa. Thermal energy is carried from the ground source coils in the earth via a pipe network into the area to be heated in the winter, and vice versa in the summer. In the conditioned area, a fan is used to move air over a second coil system. The second coil system and fan combination is used to deliver the thermal energy into the area to be heated in the winter, and is used to remove thermal energy in the summer. The second coil system and fan combination is known as an air handler. Conventional ground source heat pump systems utilize air handlers of a size comparable to that used in an air-to-air heat pump of a similar power rating. In general, ground source heat pump systems utilizing conventional air handlers provide coefficients of performance (COP) in the range of 2's–5's.

The present invention uses oversized coils in the air handler, i.e., having an unconventionally large surface area as compared to compressor capacity. This is a unique feature of the invention. It has been found, via extensive testing, that larger than expected COPs result from the use of an oversized coil in the air handler. The measured COP of the invention is in the range of 4's–7's, representing a significant improvement over known ground source systems. The trenched ground source heat exchanger of the invention provides more energy to the refrigerant, or extracts more energy, than other ground source techniques, enabling use of the larger air handler heat exchange coil on the indoor heat exchanger. Copper coils, or those of other metal, when used with ground source heat pumps provide superior thermal conductivity between the carried refrigerant and the ground. Unfortunately, when buried in the ground, copper and other highly conductive metals are prone to corrode via chemical reactions involving a loss of electrons, especially in acidic soils having a pH below 5.0. The corrosive process is known as oxidation and is similar to the reaction in a battery, resulting in a net electron flow from the copper metal to the acid in the soil. As the reaction continues over time, the copper becomes excessively oxidized, whereby it becomes brittle and susceptible to holes or breakage.

The invention uses cathodic protection to protect the copper heat transfer coils and to prevent oxidation from providing a thermal barrier between the heat conductive copper and the surrounding soil. Cathodic protection involves coupling a quantity of a metal dissimilar to copper to the copper coils with a conductive wire. The dissimilar metal, which is buried in the ground near the copper coils, is generally zinc or magnesium. The dissimilar metal forms a sacrificial anode, and as connected to the copper tubing reacts preferentially with the acidic soil. The difference in electron valence between the dissimilar metals is such that the two metals comprise a type of dry cell battery. The copper acts as the cathode of the battery. Whereas charge needed for oxidation is more available at the sacrificial anode than at the copper, corrosion occurs at the anode rather than at the copper. The copper is continuously supplied with electrons by the sacrificial anode, and does not oxidize. The sacrificial anode oxidizes readily due to its net loss of electrons, hence the descriptive adjective "sacrificial." The anode is sacrificed to save the copper. No other ground source heating/cooling system has shown or taught the necessity of utilizing cathodic protection with such a unique heating/cooling system.

The invention includes a heat-resistant soaker hose. The soaker hose can be substantially buried in the soil with the ground coils and preferably is disposed in the soil just above the ground coils. The soaker hose has an end coupleable to water sources, such as a water spigot, or a drain for condensation from the air handler coils. The soaker hose has a plurality of apertures disposed along its length. Water from the water source flows out of the apertures to moisten the soil surrounding the ground coils. The wet soil compacts around the ground coils to increase the efficiency of heat transfer. The soaker hose is preferably constructed of a heat-resistant material and is unaffected by heat radiated from the ground coils.

The soaker hose affords immediate ground settlement and compaction, thus enhancing immediate high system efficiency and performance. In addition, moisture removed from interior cooled air via the air handler can be usefully drained into the soaker hose to keep the ground adjacent to the buried coils moist, enhancing heat dissipation. System operation, such as compressor cycling, compressor shut-down and thermostat monitoring are supervised by a system controlling microprocessor. The microprocessor provides an efficient method for matching compressor operation to operating conditions. This is a unique control method heretofore unknown in a direct exchange ground source heating/cooling system. As a further method of matching compressor operation to operating conditions, a variable frequency drive modulates the compressor speed. Thus, the compressor can be operated at a rate consistent with the amount of heating or cooling required by the load.

3.4 MODULAR DESIGN OF BUILDING SECURITY SYSTEM

EnOcean's energy harvesting wireless technology for self-powered wireless switches, sensors, and controls provides home automation and energy saving security system. This flexible smart home solution can be new built and retrofit from a simple switch "all on/all off" to a gateway-connected system (TCP/IP) controlled via smartphone. Due to interoperable devices, users can easily combine EnOcean-based products from different vendors according to their individual needs—for increased security and comfort, reduced energy consumption, or technical assistance in old age [61,62].

EnOcean offers a modular security concept to meet the power requirements of energy harvesting systems. Device manufacturers can combine rolling code and other encryption mechanisms to suit individual needs and implement different security levels according to the requirements of the respective solutions. Energy harvesting wireless technology is ideal for retrofitting smart homes. If one wants to renovate or enlarge an attic, or need to add switches later on, EnOcean technology is the best suited and value-for-money alternative to the routing of cables and all the costly and time-consuming chores like caulking that go along with it. The EnOcean smart home is the good way to integrate into future smart grids and to create assisted living concepts. EnOcean offers self-powered wireless sensors and switches in the smart home [61,62]. The user benefits of the system are increased comfort, convenience and security, reduced energy consumption, maximum flexibility due to battery-free, wireless devices, plug and play operation and easy system expansion, extensive cost savings, and modular enhanced security with rolling code and encryption [61,62].

REFERENCES

1. Shah, Y.T. (2019). *Modular Systems for Energy and Fuel Recovery and Conversion.* CRC Press, New York.
2. World Business Council for Sustainable Development (WBCSD). A website report www.wbcsd.org/ (2009).
3. U.S. Energy Information Administration (EIA)—Sector. A website report, Department of Energy, Washington, DC www.eia.gov/outlooks/aeo/archive.php (2010).
4. Goel, S. ANSI/ASHRAE/IES Standard 90.1-2010 Performance Rating Method Reference Manual, PNNL-25130. Prepared for the U.S. Department of Energy under Contract DE-AC05-76RL01830, Pacific Northwest National Laboratory Richland, Washington 99352, www.pnnl.gov/main/publications/external/technical_reports/PNNL-25130.pdf (May 2016).
5. International Energy Agency. A website report, Paris, France www.iea.org/newsroom/news/2006/ (2006).
6. Tauš, P., Taušová, M., Šlosár, D., Jeňo, M., and Koščo, J. (2015). Optimization of energy consumption and cost effectiveness of modular buildings by using renewable energy sources. *Acta Montanistica Slovaca*, 20(3), 200–208.
7. Tauš, P. and Taušová, M. (2009). Economical analysis of FV power plants according installed performance. *Acta Montanistica Slovaca*, 14(1), 92–97, ISSN 1335-1788.
8. Rafayová, V. and Buc, D. (2012). Kontajnerové moduly aj energeticky efektívne. In: *Komunálna energetika*. Roč. 4, č. 4, s. 55, ISSN 1337-9887.
9. Tomčejová, J. (2012). Energeticky nezávislý dom—Endom. In: *SolarTechnika*. Č. 1, s. 18–19, ISSN 1338-0524.
10. Tkáč, J. and Hvizdoš, M. (2012). Netradičné zdroje energie, 1. vyd. - Košice : TU - 127 s. [CD-ROM]. ISBN 978-80-553-0924-8.
11. Buc, D. (2012). Modulárna výstavba na vzostupe. In: *Komunálna energetika*. Roč. 4, č. 3, s. 62–65, ISSN 1337-9887.
12. Braunmiller, G., Horbaj, P., and Jasminska, N. (2009). Geothermal energy and power generation in Germany. *Komunikacie*, 11(1), 64–67.
13. Durdán, M. and Stehlíková, B. (2015). Modelling of the Heating Process Massive Batch. In: *ICCC 2015*. Danvers: IEEE, pp. 112–116. ISBN 978-1-4799-7369-9.
14. Papučík, Š., Lenhard, R., Kaduchová, K., Jandačka, J., Koloničný, J., and Horák, J. (2014). Dependence the amount of combustion air and its redistribution to primary and secondary combustion air and his depending on the boiler. *AIP Conference Proceedings*, 1608, pp. 98–102.
15. Rybar, R., Kudelas, D., Beer, M., and Horodnikova, J. (2015). Elimination of thermal bridges in the construction of a flat low-pressure solar collector by means of a vacuum thermal insulation bushing. *Journal of Solar Energy Engineering, Transactions of the ASME*, 137(5), 54501.

16. Cehlár, M., Jurkasová, Z., Kudelas, D., Tutko, R., and Mendel, J. (2014). Using wood waste in Slovakia and its real energy potential. *Advanced Materials Research*, 1001, 131–140.
17. Šebo, J., Treburňa, P., and Šebo, D. (2007). Biomass boilers as alternative energy source and its utilization in Slovakia. *Intercathedra*, 23, 123–126. ISSN 1640-3622 207.
18. Tošer, P., Bača, P., and Neoral, J. (2014). The ways how to measure the characteristics of a solar panel. *ECS Transactions*, 48(1), 297–302.
19. Solar Swimming Pool Heaters | Department of Energy, A website report www.energy.gov/energysaver/solar-swimming-pool-heaters (2015).
20. Solar Energy, Wikipedia, The free encyclopedia, last visited 8 July 2019 (2019).
21. Solar Thermal Collectors, Wikipedia, The free encyclopedia, last visited 29 June 2019 (2019).
22. Kalogirou, S.A. (2004). Solar thermal collectors and applications. *Progress in Energy and Combustion Science*, 30, 231–295.
23. Tripanagnostopoulos, Y. (2007). Aspects and improvements of hybrid photovoltaic/thermal solar energy systems. *Solar Energy*, 81, 1117–1131.
24. Zhang, X., Zhao, X., Smith, S., Xu, J., Yu, X. (2012) Review of R&D progress and practical application of the solar photovoltaic/thermal (PV/T) technologies. *Renewable and Sustainable Energy Reviews*, 16, 599–617.
25. Xi, C., Hongxing, Y., Lu, L., Jinggangb, W., and Weic, L. (2011) Experimental studies on a ground coupled heat pump with solar thermal collectors for space heating. *Energy*, 36, 5292–5300.
26. Zhao, X., Zhang, X., and Riffat, S. (2011) Theoretical investigation of a novel PV/e roof module for heat pump operation. *Energy Conversion and Management*, 52, 603–614.
27. Deng, Y., Zhao, Y., Wang, W., Quan, Z., Wang, L., and Yu, D. (2013) Experimental investigation of performance for the novel flat plate solar collector with micro-channel heat pipe array (MHPA-FPC). *Applied Thermal Engineering*, 54, 440–449.
28. Leon, M.A. and Kumar, S. (2007). Mathematical modeling and thermal performance analysis of unglazed transpired solar collectors. *Solar Energy*, 81, 62–75.
29. Gunnewiek, L.H., Hollands, K.G.T., and Brundrett, E. (2002) Effect of wind on flow distribution in unglazed transpired-plate collectors. *Solar Energy*, 72, 317–325.
30. Fleck, B.A., Meier, R.M., and Matovic, M.D. (2002). A field study of the wind effects on the performance of an unglazed transpired solar collector. *Solar Energy*, 73, 209–216.
31. Kumar, R. and Rosen, M.A. (2011). A critical review of photovoltaic–thermal solar collectors for air heating. *Applied Energy*, 88, 3603–3614.
32. Shukla, A., Nkwetta, D.N., Cho, Y.J., Stevenson, V., and Jones, P. (2012). A state of art review on the performance of transpired solar collector. *Renewable and Sustainable Energy Reviews*, 16, 3975–3985.
33. Arkar, C. and Medved, S. (2015). Optimization of latent heat storage in solar air heating system with vacuum tube air solar collector. *Solar Energy*, 19, 10–20.
34. MunariProbst, M.C. and Roecker, C. (2007). Towards an improved architectural quality of building integrated solar thermal systems (BIST). *Solar Energy*, 81, 1104–1116.
35. Xu, P., Zhang, X, Shen, J., Zhao, X., He, W., and Li, D. (2015) Parallel experimental study of a novel super-thin thermal absorber based photovoltaic/thermal (PV/T) system against conventional photovoltaic (PV) system. *Energy Reports*, 1, 30–35.
36. Musall, I., Weiss, T., Voss, K., Lenoir, A., Donn, M., Cory, S., and Garde, F. (2010). Net Zero Energy Solar Buildings: An Overview and Analysis on Worldwide Building Projects. IEA Solar Heating & Cooling Programme.
37. Giuliani, I., Aston, W., and Stewart, A. (29 March 2016). The Design and Development of an Adaptable Modular Sustainable Commercial Building (Co_2nserve) for Multiple Applications, pp. 123–133.

38. Cat DiStasio Six Innovative Modular Rooftop Solar Technologies, A website report Six innovative rooftop solar technologies—Engadget www.engadget.com/2016/08/27/six-innovative-rooftop-solar-technologies/ (27 August 2016).
39. Hunter Billy, D. (19 March 1985). Modular Passive Solar Heating System. US 4505261 A.
40. Tatchell, A. (2012). Modular Zero Carbon Emission School Building with Hybrid Solar Heating and Passivhaus Construction, ISSN 2072-7925, CELE Exchange 2012/1 © OECD.
41. Paya-Marin, M., Lim, J., and Sengupta, B. (2013). Life-cycle energy analysis of a modular/off-site building school. *American Journal of Civil Engineering and Architecture*, 1, 59–63. doi: 10.12691/ajcea-1-3-2.
42. Aldrich, R. and Butterfield, K. (2016). Modular Zero Energy Bright Built Home, A report by NREL, Office of energy efficiency and renewable energy, Department of Energy, Washington (March 2016). Available electronically at SciTech Connect www.osti.gov/scitech, NREL Technical Monitor: Stacey Rothgeb Prepared under Subcontract No. KNDJ-0-40342-05 www.nrel.gov/docs/fy16osti/65299.pdf.
43. Zhang, X., Shen, J., Hong, Z., Wang, L., and Yang, T. (2015). A review of building integrated solar thermal (Bist) technologies and their applications. *Journal of Fundamentals of Renewable Energy and Applications*, 5, 182. doi: 10.4172/2090-4541.1000182.
44. ESTTP European Solar Thermal Technology Platform. (2009). Solar Heating and Cooling for a Sustainable Energy Future in Europe. European Solar Thermal Technology Platform.
45. Zhang, X., Shen, J., Lu, Y., He, W., Xu, P., Zhao, X., Qiu, Z., Zhu, Z., Zhou, J., and Dong, X. (2015). Active solar thermal facades (ASTFs): From concept, application to research questions. *Renewable and Sustainable Energy Reviews*, 50, 32–63.
46. Probst, M.C.M. (2008). Architectural Integration and Design of Solar Thermal Systems. Ph.D. Thesis of écolepolytechniquefédérale de lausanne.
47. Giovanardi, A. (2012). Integrated Solar Thermal Facade Component for Building Energy Retrofit. Ph.D. Thesis of Doctoral School in Environmental Engineering, UniversitàDegliStudi Di Trento in collaboration with Eurac Research.
48. Payakaruk, T., Terdtoon, P., and Ritthidech, S. (2000). Correlations to predict heat transfer characteristics of an inclined closed two-phase thermosyphon at normal operating conditions. *Applied Thermal Engineering*, 20, 781–790.
49. Binsacca, R. Geothermal Heating and Cooling Systems Are a Viable Alternative to Traditional HVAC Gaining Ground, A website report in Architect www.architectmagazine.com/.../geothermal-heating-and-cooling-systems-are-a-... (May 2008).
50. Hydron Module Geothermal Heat Pump Systems, A website report commercial. hydronmodule.com/ (2015).
51. Hydron Module Heating, Cooling, & Hot Water Systems, A website report https://hydronmodule.com/ (2015).
52. SHC onDEMAND® Modular Chillers from…—ClimaCool Corp, A website report www.climacoolcorp.com/sites/climacool/uploads/documents/Sales…/Ditch_Witch.pdf (2013).
53. All ClimaCool Chiller Products—ClimaCool Corp, A website report www.climacoolcorp.com/sites/climacool/uploads/documents/All_Chiller_Products_40… (2013).
54. Linder, J. Modular Chiller System Using Geothermal Resources, A website report www.nationaldriller.com/authors/2078-judy-linder (1 February 2014).
55. McDermott, T. (26 July 2017). Redefining Modularity: Modern vs. Traditional Modular Chillers, A website report www.climacoolcorp.com/.../redefining-modularity-modern-vs-traditional-modular-chi….
56. Baller, E.H. (1 January 2013). Modular Geothermal Measurement System. US20100223171A1.
57. Hiseer Geothermal Heat Pump Wholesale, Pump Suppliers—Alibaba, A website report www.alibaba.com›HomeAppliances›pump (2017).

58. Soil-Water Geothermal Ground Source Heat Pumps Boreholes for Sale, A website report chineseheatpump.sell.everychina.com/p-90276598-soil-water-geothermal-ground-sou… (2017).
59. Lambert, K. (3 June 1997). Modular Geothermal Heat-Pump System Installation. US 5634515 A.
60. Galiyano, M.P., Galiyano, M.J., Ryland Wiggs, B., and Jeffrey, T. (6 July 1993). Aspacher Modular Tube Bundle Heat Exchanger and Geothermal Heat Pump System. US 5224357 A.
61. Modular Design of Building Security System EnOcean Makes the Smart Home Smarter, EnOcean's Perpetuum Magazin—The World of Self-Powered Wireless, A website report www.enocean.com/fileadmin/redaktion/pdf/…/perpetuum_02_2018_EN.pdf (2018).
62. Smart Home Applications for Home Automation Using…—EnOcean, A website report www.enocean.com/en/internet-of-things…/smart-home-and-home-automation/ (2018).

4 Modular Systems for Energy Usage in Vehicles

4.1 INTRODUCTION

In this chapter, we examine energy usage management in vehicle systems. While the major emphasis is placed on vehicles for space missions and automobiles, modular energy storage for electrification of ships is also considered. The energy management in a space mission is a very critical and modular approach and has been used to not only conserve energy but also control the required power in an uninterrupted way. The energy usage in automobiles is similarly complex and important to manage. There are several diverse issues for energy management facing the car industry. First, new technology changes are offering more challenges and opportunities for automobiles to be more energy efficient. Second, new technology such as fuel cells and the use of solar energy for the car are becoming more prominent. Finally, electric cars are becoming more and more reality in the future. These changes in the car industry and resulting management of energy usage are facilitated by the modular approach. Lastly, energy savings and electrification of ships are significantly enhanced with the use of a modular approach to energy storage.

4.2 ADVANCED MODULAR POWER SYSTEM MANAGEMENT FOR AFFORDABLE, SUPPORTABLE MULTI-VEHICLES SPACE MISSION

An excellent review by Oeftering et al. [1] describes the use of advanced modular power system (AMPS) to manage power needs of a multi-vehicles space endeavor in an affordable and supportable way. This type of modular approach for power consumption and management allows the management of space operations with multiple objectives. The review by Oeftering et al. [1] describes various aspects of power management for space mission. We briefly summarize important concepts outlined by Oeftering et al. [1].

In order to address the need for designing vehicles for various space missions such as explorations of Moon, Mars, and Near-Earth Asteroids, an effort was made by NASA to design these vehicles such that they are elements of a larger segmented spacecraft rather than separate spacecraft flying in formation [2]. It was understood that the future need for multi-vehicle exploration architecture required for exploration of different places with different objectives creates the need to establish a global power architecture that is common across all vehicles that reduces the overall energy consumption. In order to satisfy this need, NASA Glen Research Center (GRC) created and managed the AMPS project [1,2]. The objective of this project was to create and manage modular power system architecture that allows power systems needed for these multi-vehicle operations to be built from a common set of modular building blocks.

The goal of AMPS project was multifold. Besides reducing the overall energy consumption, the project aimed at developing and analyzing by demonstration of key modular power technologies that are expected to minimize non-recurring development costs, and reduce recurring integration costs, as well as mission operational and support costs. Also, modular power was expected to enhance mission flexibility, vehicle reliability, scalability, and overall mission supportability. The project was thus designed to support and enhance multi-mission capability by field demonstration of vehicles. These demonstrations not only evaluated effectiveness of modular technology and approach but also assured the progress towards truly flexible and operationally supportable modular power architecture. The modular power architecture was composed of technologies for power generation, energy storage, power distribution, and health management (HM) that will reduce the cost of future space systems. Since power needs are different in different vehicles, the modular approach was designed such that while investment is preserved in basic modular building blocks the approach provides adaptability to serve an array of future vehicles with varying power needs. This was accomplished by designing standardized modules that encapsulate a common set of reusable software, embedded diagnostics, and HM functions and features that assure interoperability. Since NASA mission involved travel to various distances with multiple crew and varying vehicle design depending on length of the mission, in order to reduce overall cost of various missions the AMPS strategy included the ability to salvage key elements to be reconfigured and repurposed for secondary applications [1–3,27].

AMPS common design approach extended to all spacecraft within the space exploration architecture such that power hardware can be interchangeable among a group of spacecraft. Commonality allows a program to pursue a single development effort that is then used by multiple spacecraft. This approach eliminated duplication and assured compatibility of different spacecraft to common modular power architecture. AMPS approach was to design modular power architecture with flexibility, scalability, interoperability, and operational supportability for wide variety of vehicles with varying missions [1]. Future operations may be complex with multiple objectives and multiple durations. In these varying missions, the modular flexibility enables the crew to select the best power generation, energy storage, and power distribution options [5]. Modular operation also provided system availability to minimize spare mass, maintenance equipment mass, operational time dedicated to maintenance, and crew skip and training. It also allowed crew to effectively and independently respond to problems. The supportability features that support these goals were essential for effective modular commonality, scalability, and flexibility. The modular operation also provided interoperability which is often referred as "plug-and-play." AMPS also anticipated a broad impact on program affordability by addressing non-recurring, recurring, and long-term operational costs. AMPS worked

Systems for Energy Usage in Vehicles 173

to modularize lower levels of assembly to reduce spares mass. The embedded HM diagnostics and prognostics capability also reduced the crew time required to isolate a fault [3].

Besides vehicles for getting in and out of space and their AMPS, solar electric propulsion vehicle for unmanned space mission was unique since it had a power level exceeding 300 kW. As a low-thrust high-specific impulse vehicle, it was primarily used well beyond low Earth orbit. For human missions, lunar landers were actually composed of two modules. A descent module (DM) for landing and an ascent module (AM) that returns the crew to orbit. For unmanned cargo missions, the vehicle had only a DM. The power configuration [10–26] for two types of lunar landers depended on their rendezvous site with multi-purpose crew vehicle. For lunar orbit rendezvous (LOR), Lunar Lander was equipped with large solar arrays to power the cryo-coolers that preserve propellants during a long loiter period in lunar orbit. On the other hand, for Earth orbit rendezvous (EOR), Lunar Lander had a very short loiter time and it used fuel cell-generated power.

For all these different space vehicles, AMPS carried out power management and distribution using advanced solar array, battery, and fuel cell technologies. Modularity in power management and distribution assured flexibility, supportability, and reusability [11–26].

4.2.1 MODULAR POWER MANAGEMENT AND DISTRIBUTION IN SPACE MISSION

Space power systems are composed of power generation, energy storage, and power distribution and management (PMAD). Power generation includes solar arrays, fuel cells, and thermodynamic engines. Energy storage includes batteries but may include the evolving flywheel technology. PMAD connects the power generation and energy storage to the user loads. It regulates the power and handles the delivery of power. It also provides the primary system fault detection, fault isolation, and rerouting of power. These are typically designed as separate subsystems. The AMPS approach to modularity integrated certain PMAD regulator and HM functions into power generation and energy storage. Embedding these functions combined with plug-and-play features enabled them to act as independent self-contained modular subsystems. This made them more portable so they can be moved to different parts of the vehicle or another vehicle entirely [1].

There are a number of factors that affect the effectiveness of the modular PMAD design. For example, international space station (ISS) requires high degree of modularity driven by scalability and supportability needs. The ISS design incorporates modularity using assembly level of hardware commonly known as orbital replacement units (ORU) [9], which is the preferred level of replacement for ISS [10–12]. In general, AMPS defined a component-level module as an encapsulated unit that cannot be further disassembled with common tools. This may be a single discrete part or a combination of multiple parts that are encapsulated in a manner that makes them a monolithic module. The component level serves as the common fundamental power building block. A number of different functions can be provided by the electrical arrangement of these blocks. These independent encapsulated blocks incorporate sensors and other devices that support diagnostics, which in turn are connected to higher level modules that carry out the diagnostics and HM functions. Both low- and high-level modules are portable and have embedded functions and related software.

Using many small blocks to satisfy the need of a spacecraft means many interconnections, which drives up the mass of harnessing while reducing overall reliability. Using large blocks means a tendency for overcapacity and excessive mass. An assortment of modular elements with varied capacity enables designers to mix and match modules to arrive at a system that meets mission needs. For the AMPS project, the generic term "scaling increment" was defined as the incremental capacity of a modular building block element. An assortment of several "scaling increments" was needed to provide the flexibility to group elements together to meet the mission needs. Power generation, energy storage, and power management each had distinct scaling increments (10, 15, 22, 33, 47, 68).

Current practice in Power Management and Distribution design is to develop mission-specific solutions. Each mission has unique loads requirements, power quality specifications, operational requirements, etc. PMAD elements (e.g., load and bus regulators, energy storage subsystem interfaces, and protection) are designed to optimally meet these unique system and subsystem requirements, and cannot be used in any other vehicles. PMAD system is composed of two types of electronic devices: power electronics and control electronics. The power electronics (includes electromechanical devices) conduct and direct the current, and tend to scale in relation to the power loads. Control electronics manages the power electronics with low-power mixed signal (digital and analog) devices that scale somewhat independently of power loads. A modular building block approach must account for differing scaling factors. While different control techniques are required for different applications, regulation functions can all be addressed with common power converter designs. That is, solar array regulation, bus regulation, battery charge/discharge control, and load power regulation can all be done with common DC-DC converters (referred to as flexible power modules). These modules are used in sufficient numbers to meet the operational requirements [1,15–20]. An advantage of the modular approach is the ability to continue operation but at a reduced capacity after a failure. Oeftering et al. [1] points out that this modular approach for PMAD has worked very effectively for NASA.

4.3 MODULAR POWER GENERATION AND STORAGE TECHNOLOGIES FOR SPACE VEHICLES

The power in space mission is basically managed by solar array, fuel cell, and battery technologies. The first two are power generation technologies, while the last one is the energy storage technology. All three are advanced modular technologies.

4.3.1 SOLAR ARRAY TECHNOLOGY

The key parameters that characterize solar array performance are specific power (W/kg) and a real power (W/m2). Various mission-related environmental effects and degradations affect the effectiveness of solar array technology. A key performance characteristic for solar array technology is the acceleration capability. The size of solar array depends on the required acceleration capability. Normally, large solar arrays are designed for low accelerations because they are heavy and require added structure to maintain their integrity. The large structure limits vehicles acceleration capability needed for initial launch phase.

AMPS adopted two approaches for the solar cells. (1) Due to its mass and size limitations, the SEP vehicle used IMM (inverted metamorphic multi-junction) cells with an assumed beginning of life (BOL) efficiency of 34%. These cells are one-tenth the thickness of traditional cells. These cells are assumed to be available in quantity for the large SEP vehicle. (2) Thinned gallium arsenide triple-junction solar cells with a BOL efficiency of 30% were adopted for use in the solar arrays of the remaining vehicles primarily because they were available in large quantities at lower cost than IMM cells. A wing composed of multiple, modular panels could be utilized, although these were typically used for low acceleration (<<1 g) and/or low required spacecraft power levels. For the MMSEV (multi-mission space exploration vehicle) rover, a modular approach was able to replace the rigid panels with an adequately designed roll-out solar array (ROSA) [1]. This type of solar array is best seen in examples such as Mars Phoenix [13] and the CEV Orion [14–16] spacecraft.

In modular approach, solar arrays are composed of numerous mass-producible cell module building blocks. Cell modules are typically a series connection of solar cells (i.e., a string) to provide the voltage required by the power system. The cell module is rectangular and includes the cover glasses for radiation shielding, coatings, substrate structure (if any), and mechanical/electrical interfaces. These cell modules are assembled into wings to obtain the required power levels with each vehicle's wing sharing the maximum number of parts across the range of vehicles. For ROSA-like wings, the strategy was to design the roll-out tubes to have common diameters, but obtain more power by increasing the length (i.e., adding more solar cell modules). More details on ROSA like wings are given by Oeftering et al. [1].

The cell modules are designed such that they can be used for different vehicles. Standardizing the number of ROSA module widths with their associated roll-out tubes (diameter, material, and thickness) has the potential of reducing development and testing costs for the wing structures. Selection of optimal cell module sizes enabled automated testing and inspection of modules. Modules that fail can be swapped out with ease. The SEP vehicle due to its size and number of wings [17] can be considered modular on that standpoint alone. SEP as part of a multi-vehicle mission can serve as a source for spares for other vehicles. Once SEP's mission phase is complete, the SEP arrays are suitable for salvage and reuse. The multiple deployment capability of the ROSA array simplifies the salvage process, and its compact roll-up form is suitable for stowing and redeployment [15–20]. In short, modular solar array technology has worked very effectively for various NASA missions.

4.3.2 Modular Fuel Cell System

AMPS project used fuel cell as another source for power generation. The factors that drive the design of a fuel cell power plant are primarily mission driven. The peak and nominal power required defines the size of the fuel cell stack and balance of plant (BOP). This requirement defines not only the number of cells and/or stacks to be included, but also the size of the cells or stacks. Typically, a power plant is characterized by the specific power (W/kg) and power density (W/l) as a whole. The voltage to be delivered to the vehicle bus by the fuel cell power plant also impacts the design of the fuel cell stack. Higher voltages require a larger number of cells and/or stacks and how those cells or stacks are arranged, i.e., series/parallel arrangements [1,26]. Total system energy, in Watt-hours (Wh), is driven by mission power level and mission duration. In long-duration missions, the system mass is dominated by the mass of the reactants and storage tanks. The fuel cell hardware may represent a relatively small fraction of the mass. Specific power (kW/kg), system efficiency, power density, and desired peak to nominal power delivery ratio are typical parameters imposed by the mission. Fuel cells, however, impose their own requirements on the vehicle including reactant pressure and flow rate, reactant purity, and heat loads handled by the vehicle thermal system. Future missions beyond Earth orbit will have additional requirements such as fuel flexibility [26].

The flexibility of a fuel cell technology depends on its ability to utilize available reactants. This includes reactants scavenged from propulsion systems or reactants extracted from in-situ sources. Where some designs require pure hydrogen/oxygen, others provide flexibility to operate on relatively impure reactants or hydrocarbon fuels. Reliability and redundancy requirements also affect the design of the power plant as a whole. Higher required reliability not only impacts how many redundant components are included within the power plant design but also how those redundant components are handled, i.e., actively operating at all times within the power plant or unpowered but available on a standby basis to take over a key function within the power plant.

Vehicles and their design reference mission requirements influence the selection of fuel cell technology for the modular fuel cell [1,26]. The voltage and current produced by a fuel cell is determined by the design's cell stack series and/or parallel arrangement and the size of the individual cells. Unlike batteries, the fuel cell continues to provide power as long as fuel and oxidant continue to be fed into the fuel cell stack. Solar arrays and fuel cells are often seen as alternative power sources. Solar energy is seen as renewable source, while fuel cells are limited by the supply of the reactants. A hybrid system composed of both fuel cells and solar arrays can exploit the benefits of both. In a regenerative fuel cell system, the reactants and water are part of a closed-loop system where the water is reconverted back to reactants by a solar-powered electrolyzer. This is similar to rechargeable batteries except that the regeneration can occur concurrently while the fuel cell continues to produce power without diminishing as long as reactant production stays ahead of power consumption [1].

As pointed out by Oeftering et al. [1], only three types of fuel cell chemistries have actively been investigated for space applications, specifically, alkaline, proton exchange membrane (PEM), and solid oxide fuel cells (SOFC). Alkaline fuel cells were the workhorse power source from Apollo to Shuttle Orbiter. Concerns with usable life, cost, and sensitivity to contaminants drove the investigation into other options. PEM fuel cells operate at relatively low temperatures (80°C) and can bootstrap themselves to operational status as needed. PEM fuel cells typically operate using hydrogen as a fuel (although they have been shown to operate on methanol/air) and are generally intolerant of most reactant impurities. Three basic types of supporting BOP designs have been under investigation for PEM fuel cell systems: active, passive, and non-flow through (NFT). SOFC operate at higher temperatures (>600°C) and need to be pre-heated to that temperature before the fuel cell reaction can begin. Like PEM and alkaline fuel cells, SOFCs can operate directly on hydrogen and air or oxygen, but also operate with impure hydrogen or reformate from hydrocarbon fuels and methane. SOFCs are also very tolerant of most impurities, while the waste heat from SOFCs is high enough to support cogeneration (via Stirling, turbines, etc.) for additional power and potentially increasing the overall system efficiency to greater than 70%.

Fuel cell power plants have traditionally been designed incorporating one fuel cell stack with one BOP. The BOP delivers preconditioned reactants to the stack, and removes water and waste heat. The BOP provides the interfaces with the vehicle reactant stores, thermal control, vent lines, and external water processing. The total power plant is normally scaled to meet the entire power requirement of the vehicle. System redundancy is provided by additional fuel cell power plants operating in parallel. In order to meet the application power requirements, the fuel cell stack scales linearly by adding cells and/or rescaling the basic cell size. An alternative method is to expand the power plant by adding additional stacks. The power plant has a variable output, and thus, BOP fluid lines and control components are sized to cover a range of flow rates [26]. Components are sized for maximum flow with margin. Beyond that the lines and controls jump to the next commercially available size step. As a result, the BOP follows a continuous increase in power with a stepwise increase in component size. Therefore, a given BOP design can typically accommodate a range of power levels about some nominal value before stepping up to the next size increment. The increment in stack has to always be balanced against increase in vehicle weight. Generally, increment is made in smaller quantities (like 4 kW). The present practice for many missions involves the use of three 4 kW fuel cells. In case of failure of one or more fuel cells, additional power is obtained from remaining ones or use of power is curtailed.

A spacecraft fuel cell is literally an electrochemical power plant that often runs on the fuel/oxidizer reactants that it taps from the propulsion system. Their innate complexity makes them a challenge to integrate into a modular package. Modularization of a fuel cell power plant begins with packaging the modular components to simplify the interfaces. In the case of the NFT fuel cell technology, the process of separating the water from the oxygen is done within the fuel cell stack rather than within the BOP, potentially simplifying the interface between stack and BOP. NASA Glenn Research Center has been working with Infinity Fuel Cell and Hydrogen, Inc. to develop NFT PEM fuel cell power systems. The goal of the NFT fuel cell technology development is to develop and produce units of progressively higher capability. The next steps are 1 and 3 kW units. The 3 kW NFT fuel cell developed by Infinity Fuel Cell and Hydrogen, Inc. is composed of 144 cells with a nominal stack voltage of 120 Vdc to produce a nominal stack power level of 3 kW and a peak level of 6 kW [1]. Along with solar array technology, modular fuel cell will remain an important device for power generation of future space crafts.

4.3.3 Modular Battery Technology

The battery is an ideal example of a system composed of simple modular blocks. Battery capabilities are defined by the characteristics of the fundamental cell unit. The selection of the cell chemistry and geometry translates into battery characteristics and performance [1,21–25]. Major battery design drivers include the Amp-hour (Ah) capacity, peak and average currents, and voltage that the battery is required to deliver, as well as the cycle life, operational life, and redundancy requirements on the battery. Ah capacity, current, and voltage define the size of the battery. Fixed cell size establishes the Ah, current capacities, and voltage increments. Ah capacity and the current scale up with the number of cells in parallel. Voltage scales by the number of cells in series. Life requirements define the extra margin required to be able to perform the required functions at the end of life [21]. Redundancy requirements determine the number of extra strings or batteries required to meet reliability, loss-of-crew, or loss-of-mission requirements. Overall, the primary attribute used to choose between design options for a given set of requirements is usually mass [1].

Current spacecraft batteries are mainly designed using lithium-ion battery technology for any mission that requires recharging. Lithium-ion technology has a variety of choices available in "space-qualified" cell designs. Lithium-ion cells are available in prismatic, cylindrical, and pouch formats. Pouch and prismatic cells require compression to achieve their optimum performance, while cylindrical cells do not. Pouch cells also have more specific handling requirements to avoid damage to the internal components [21–25]. Lithium-ion cells are available commercially in a large range of Ah capacities. The Ah capacity and current requirement of the battery can be met by using a large number of small Ah capacity cells or by using a small number of large Ah capacity cells. For safety reasons, overheating of any cell in a lithium-ion battery should be avoided.

AMPS modular batteries were closely coupled with charge/discharge functions provided by PMAD. Modular batteries also incorporate HM largely through extensions of the charge/discharge controls. Not only HM monitors battery health, it keeps the system apprised of cell level changes and in some instances provides the capability to isolate a faulty cell while keeping the remaining cells safe and in operations. The battery beginning-of-life Ah capacity requirements (including redundancy and the extra margin required to meet end-of-life requirements) depend on the nature of the mission. All missions had a requirement to support a 120 V bus. Generally, the need of the most missions was satisfied by designing two module types of 27 Ah and 150 Ah using lithium-ion cells, both of which operated at approximately 120 V [21–25]. A common modular battery also reduced recurring logistics cost. If every spacecraft in a given mission uses the same types of modular battery, then they could share a common source of flight spares. This reduces the overall spares inventory and improves supportability. Modular batteries would be primary targets for hardware-salvaging operations. Once again, in short, battery-based modular energy storage worked very well for various NASA missions.

Power generation and energy storage technologies described above are constantly improved to increase their efficiencies, reduce costs, and increase durability.

In recent years, the use of nanotechnology has accelerated these improvements. More details on power generation and energy storage technologies are given by Oeftering et al. [1].

4.4 MODULAR PROTON EXCHANGE MEMBRANE (PEM) FUEL CELL FOR STATIONARY OR MOBILE POWER SYSTEM

The invention by Richards [31] is directed to a modular fuel cell system that can be configured in an assembly having module components capable of being easily removed and replaced. As a preferred embodiment, the module components are capable of being replaced and removed while the fuel cell stack is under electrical load. This provides continuous duty/uninterruptible operation of a series and/or parallel fuel cell stack configuration comprising the modules of the present invention.

PEM fuel cells are used for power generation and each of the fuel cells has fuel and air requirements for operation. When a number of individual fuel cells are assembled together to constitute a module, problems can develop with the modules, and in particular, with the supply of fuel (H_2) and air, as well as with heat dissipation and power collection. In U.S. Patent No. 6,030,718, a PEM fuel cell power system is disclosed which enables individual fuel cell modules to be connected to racks within a housing. The modules have a hydrogen distribution rack with a terminal end that engages a valve on the rack that supplies hydrogen gas to the module. The rack or housing has many slots and each slot accepts a module. Accordingly, there are valves for supplying hydrogen gas and a return for each slot. According to the present invention, a "modular," "building-block" array of nominal 1 kW output capacity fuel cell assemblies is provided. The modular fuel cell blocks may be individually removed and/or replaced in an array or assembly that enables continuous operation without interruption or significant disruption of the power supplied by the assembly.

It was an objective of the patent to provide a lightweight and easily held fuel cell assembly of nominal 1 kW output capacity, which may be gripped with ease and minimal force applied, to accomplish the safe removal and/or replacement of "modular" fuel cell assembly from a similar "modular" manifold block assembly. It is a further object to provide the "modular" manifold block assembly with such features that both fuel and reactant gas feed and return lines provide positive, leak-tight shutoff, and/or isolation of all porting connections to/from the "modular" fuel cell assembly, by the employment of Failsafe (spring-loaded to close or "off" position) "On-Off" control action cartridge-type plug valve (or similar), and as effected by operator access to the exposed front face (accessible portion) of the 1 kW fuel cell assembly.

It is a further object to provide the "modular" manifold block assembly with nominal 1 kW output capacity with such features to assure that associated electrical loads may be safely and reliably disconnected and subsequently reconnected, by the consideration of "make-break" current levels being constrained to values of approximately 20 A (25), up to a maximum of 50 A per electrical plug connection. It is a further object to provide the "modular" manifold block assembly with features to allow porting and distribution exhaust reactant gas flows uniformly over the exposed periphery of the fuel cell assembly, such that convective air cooling measures may be effected, and in conjunction with water management/cooling processes within the fuel cell stack assembly envelope, shall limit the maximum external surface temperatures, and thereby allow for the safe removal of said fuel cell stack assembly by "hand" without discomfort.

It is a further object to provide the fuel cell stack assembly with internal fuel and reactant gas distribution features such that both the fuel cell stack electrical output capability is maximized, and the internal water management/cooling processes are facilitated. This object may be accomplished, in a preferred embodiment of the invention, by the use of slotted versus circular internal gas distribution supply, return passages, and tapered distribution headers, and thereby allowing for the lowest possible velocity head losses, and the highest possible uniformity (laminar flow) in gas flow volumes to/from all active unit area increments of the conductive (electro-chemically active) region of the fuel cell stack assembly.

It is a further object to provide a uniform compressive clamping force within the conductive (electrochemically active) region of the fuel cell stack assembly, such that associated contact voltage drop effects are minimized, and which results in the achievement of optimal resistivity performance of the gas diffusion media, and which therefore yields the greatest possible output power versus internal heat generated. It is a further object to provide all of the above features in a freestanding array consisting of symmetric 1kW "modular" fuel cell stacks capable of being installed in a series manifold of up to 20 "plug-in" elements, thereby allowing capability to operate at 480 VDC, and via symmetric configuration, yielding capability to provide up to 40-kW per array. The envelope of this freestanding array of 1kW "modular" fuel cell assemblies would be less than 9.75" W × 80.0" H × 12.0" B. Larger-sized output capacity arrays would be accomplished by mounting of these freestanding arrays in parallel to realize units with 240 kW (and larger) output capacities and providing a very small footprint.

It is yet a further object to provide such large-scale arrays with preventative maintenance/fault-location features to facilitate the rapid location of defective fuel cell assemblies, such that they may be rapidly removed and replaced. These features would preferably consist of the threshold detection of increased temperature levels within the individual 1 kW module, such that illumination of LEDs would indicate the operating status. By way of example, green would indicate operation within room (or ambient) temperature up to 150°F, yellow would indicate incipient failure at greater than 150°F for 10 min or longer, and red for temperatures of greater than 180°. Both local and remote alarms would be triggered based on performance/operational necessity and safety considerations. It is further objective of the invention to include a bipolar plate (BPP) configuration, in accordance with embodiments of the present invention having tapered-width microchannel grooving, which provides a significantly improved level of fuel and reactant gas distribution uniformity over the active area of an individual cell itself, between a plurality of cells within a fuel cell module.

The modular fuel cell of the present invention that includes the BPPs having tapered-width microchannel grooving facilitates the achievement of substantially reduced concentration gradient variations over any subject unit area of the active region of a cell (also minimizing gas flow volume/gas velocity variations per unit area), thereby maximizes the vaporization capability of the reactant gas stream, and which subsequently minimizes development of "hot-spots."

4.5 ENERGY TECHNOLOGY CHANGES IN MODULAR AUTOMOBILES

As shown in the previous book [4], automobile production is heavily empowered by modular strategy. In fact, the design and manufacturing processes for all automobiles are highly modular. In the development of modular strategies for automobiles, insertion of new technologies in an energy-efficient way is very important. Vehicle technology is constantly changing making it necessary for vehicle design to be flexible to accommodate insertion of new technology. Since the insertions of all technologies within vehicle will involve some level of energy consumption, it is important to note how modularity can accommodate these technological changes. Some of the technological changes outlined by Rue [32] are briefly summarized below.

4.5.1 Wheel Hub Motors

The wheel hub motor's performance is a testament to the viability of using an electric wheel hub motor in the design of future ground vehicles. The motors evaluated in the study by Rue [32] were 63 kW (85 HP) each and for the modular vehicle (MODV) motors of 7.5–10 kW (10–12 HP) are required. The power of the hub motors is not in question, rather the availability in the 10 kW (12 HP) range. As the quality and availability of WHM improve, they will adequately meet the performance demands of the MODV concept.

4.5.2 Batteries

For the past decade both energy and power density for batteries have been increasing by about 6% per year. The total weight is calculated based on the assumption that there will be batteries specializing in power delivery and energy delivery. Intelligent use of the batteries based on the operating conditions helps optimize the weight of the vehicle. The study by Rue [32] indicates that the weight budget could be reduced by 25% if the energy and power density increase by 6% per year. The basic rule of battery design is that given a certain battery chemistry it can be optimized more toward energy or power. Therefore, the likely solution will be a combination of batteries specializing in certain load requirements. The CPU of the MODV can, either by user input or autonomously, optimize the power use. During silent operation where low power is required, the power can be drawn mainly from the high capacity batteries. When the MODV operates in a hybrid electric vehicle (HEV) mode or needs high acceleration then power can be drawn from the high-power batteries [32]. Engineering the optimal battery setup will require in-depth analysis. Different loading conditions must be considered such as towing up an incline, high accelerations on differing terrain, and acceptable accelerations during silent mode. It is possible, however, with the MODV's modular nature that the battery setup will be modular as well. It is easily conceivable that a battery setup can be outlined for certain mission specifications. All this would require is for the batteries to be physically secured, and power and communications cables hooked up. Then, the CPU would have the information on the number of batteries and their characteristics. If a logistics MODV were needed then the vehicle would see significant weight savings. There would no longer be a need for all of the high capacity batteries required for silent mode which contribute to the bulk of the weight. Only 10 kg is needed for the power requirements while 110 kg is needed for the silent mode requirements. Less weight for batteries means more weight for suppliers. Nanotechnology is already being used to develop many new kinds of batteries that are quicker-charging, more efficient, lighter weight, have a higher power density and hold electrical charge longer [32].

4.5.3 Solid Oxide Fuel Cell

A SOFC-driven vehicle can offer the extended range that so many EVs fall short of. Many consumers choose not to have a purely EV because the range is limited. Even though conventional vehicles only have 30% efficiency, a tank of gas has so much energy in it that a vehicle running off of gasoline will have a greater range. Technically a vehicle powered by SOFCs isn't purely electrical because the energy is from a liquid fuel, but the electrical energy in the batteries of current EVs comes from an outside source as well. In any case, a vehicle supplemented or completely powered by SOFCs would be very desirable for the average American consumer because the cost per mile of operation would be vastly superior to cars using internal combustion engines (ICEs).

The confidence that can be found in the advancement of SOFCs stems from the fact that this fuel cell system is not theoretical, it actually works and is in use today. This has not been lost on major companies who are rushing to take advantage of the system's excellent performance. The main hindrances for its use in ground vehicles are its energy and power density and its high operating temperatures. These issues are not holding back major companies form investing in SOFCs for industrial power application. Companies such as GE are already investing in applying SOFC systems, so they will be invested in research to advance the technology in the future. With this large investment it follows that these systems will see increases in power and energy density, and see a reduction in cost. A ground vehicle that could be powered by a SOFC and battery system would have many advantages over conventional vehicles, especially if the MODV concept were powered by such an efficient and reliable power plant. The SOFC system has no moving parts, and this coupled with wheel hub motors would mean a very low maintenance vehicle. All in all, it is in the MODV concept's best interest to integrate a SOFC into the system as soon as the technology can be reasonably used on a ground vehicle [32].

4.5.4 Autonomy

Cisco's technology trend watchers stated that in 5–7 years it could cost people more to drive their own cars than to let the cars drive themselves [33]. This is a best-case but possible estimate. Autonomous vehicles will be able to avoid accidents easier than human drivers which should drive the cost of owning a car down. In the United States, vehicular accidents inflict $450 billion in annual costs and it is estimated that self-driving cars could reduce human-caused crashes by 80% [33]. This should lead to lower insurance rates and car ownership rates in general. As far as the number of self-driving cars on US road goes, IHS estimates 230,000 by 2025 and 12 million by 2035. With such a large savings to be had and the ample number of vehicles predicted, it is a certainty that vehicle autonomy will see heavy investments and advances in the commercial market. The future for autonomous vehicles is looking bright. Research and development of autonomous ground vehicles is here to stay with massive demands from the consumer and defense industries which provide a baseline for off-road autonomous applications. Autonomous capabilities will improve and be a valuable asset to the MODV concept [32].

Besides the ones mentioned above, Rue [32] examined structural health monitoring, energy harvesting shock absorbers, non-pneumatic tires, protection from electromagnetic pulses, and drive-by-wire for MODV. His conclusions were that while it is possible and reasonable to use SHM on the MODV, but the unknown is how that solution will be realized. For EHSA, 5 years down the road we should yield proven EHSA systems that can properly harvest energy and aid in the control of the vehicle. For non-pneumatic tires, Rue [32] concluded that it is worth looking into a tire that can fit the mission needs of the MODV. Shielding from EM (electromagnetic) pulses should be given special attention in the MODV because of the breadth of electronics used in the MODV that are not used in many vehicles at this time. Finally, drive-by-wire is a proven technology that must be implemented in the MODV because it allows quick attachment of an autonomous appliqué kits. The fault tolerance and reliability of drive-by-wire systems should also increase as the technology matures, making it even more attractive for use [32].

4.6 MODULAR ENERGY MANAGEMENT IN ELECTRIC VEHICLES

4.6.1 Novel Modular Power Management for Ground Electric Vehicles

EP Tender and Nomadic Power are proposing paradigm-shifting innovation to design EV for typical use and provide a network of energy modules (Tenders, Nomads) mounted on small trailers available to rent at the point of use (e.g., motorway service stations and commercial centers) for longer trips (ca. 600 km). By providing a modular approach the consumer gains affordability, convenience, and peak range whilst not sacrificing environmental credentials [28,29,33–51]. It is noteworthy to point out a key difference between ICE vehicles and EVs: for the former, the fuel tank is one of the cheapest components (just a little more plastic); for the latter the batteries are the most expensive [28]. In consumer electronics, cost has been optimized with batteries that provide power only for typical usage (a typical smart phone's battery lasts a single day). For extra power needs, external battery modules or chargers are used. Significantly, the consumer only pays for typical power use. In this way, cost, weight, and volume are optimized for common use.

Energy storage technology is progressing at a faster rate than electric motors or other major vehicle components (cost/performance halves every 10 years [39]). As a result, a key advantage to this approach is that the Tender/Nomad can take advantage of technological progress and offer it to the consumer without the need to replace their vehicles. EP Tender is based on an ICE, while Nomadic Power is based on a battery. They may evolve to be based on fuel cells or metal-air batteries as technologies mature. A full EV thus gets Omni-hybrid capability when having access to Tenders and Nomads. Revenue from EV charging services is expected to reach $2.9 billion annually by 2023 [40]. This is a major opportunity where EP Tender and Nomadic Power, by complementing for long distances the static charging services with more convenient on-road charging, can play a meaningful role and develop a profitable industrial and commercial activity. EVs can be run for many years and benefit from upgraded Tenders and Nomads carrying newer technology for energy storage (e.g., fuel cells), without having to replace the whole car. The benefits of this approach are many fold:

1. A car remains optimal for daily usage (five passengers, trunk >300 L, 150–200 km effective range, weight <1,400 kg).
2. Car is affordable and will become a high margin with volume.
3. Range can be increased on demand.
4. The marginal cost of range increase is affordable as the energy modules can be shared and rented only as needed, and Nomads can be used for home storage of energy.
5. Life cycle footprint is minimized as the add-on energy module used for long-distance trips (ICE Rex, fuel cell, or large battery) is shared by a number of cars, instead of being onboard each car: these are the benefits of the functional economy concept, of which we are part.
6. Full EVs may become Omni-hybrid when they have a hitch.
7. The burden on the electricity grid remains mostly off peak on residential slow charging and office normal charging and generates far less peak fast/superfast charging demand [33,40–42].
8. Due to a wider and accelerated public adoption, EVs (and Nomads) can be used to smooth grid demand enhancing the sourcing of renewable but intermittent energy (Nissan-Endesa initiative [42], San-Diego [33]).

Milestones already achieved to date demonstrate the credibility of the mobile energy approach and the ability of the EP tender and Nomadic Power to attain objectives. Both EP Tender and Nomadic power have chosen to power the EV with DC and are connected in parallel to the EV's battery. They are not using the onboard charger of the EV. Through the controller area network (CAN) communication one receives data from the battery management system (BMS), which at all times indicates the maximum power acceptable by the battery and the actual power it receives. The power produced by the modules is continuously adjusted accordingly. As a result, they can provide energy while driving, without interfering in any way with the vehicle's systems: the vehicles are 100% standard and require no modifications, apart from adding a tow bar and connecting to the 400 V line and to the CAN bus.

Vehicle diversity is managed by adding a "CAN Merger Box" along with the high voltage junction box in order to select and reformat the vehicle's CAN messages into a vehicle independent format. When the energy module carries a battery or a fuel cell, a DC-DC power electronics module will provide the necessary voltage adjustments. When the energy module carries an ICE generator, the generator's output is equipotential with the vehicle's battery. The energy module can provide a continuous power which equals the average energy consumption of the vehicle. Peak power can be provided by its battery, which acts as a buffer. As a result, the energy module is optimized on the energy capacity parameter, with a moderate power requirement.

As mentioned earlier, EP Tender and Nomadic have taken slightly different routes for the initial implementation of modular energy for EVs. For example, Nomadic Power has chosen to store energy in a battery, whereas EP Tender has chosen to use a classic combustion engine and a fuel tank. The energy is provided by gasoline. Electricity is generated by a small ICE and an alternator. Peak power is 20 kW, with a fuel tank of 35 L providing 85 kWh, and the total weight is 250 kg. EP Tender allows the public to switch, with current technology, from 100% petrol to 98% electric, at a total cost of ownership (TCO) and convenience which are equivalent to an ICE vehicle! Attaching the Tender to an EV is achieved in one go, by punching the car with the hitch, and backing maneuvers are made easy due to the self-steering feature embedded into EP Tender's platform. The QR codes are linking to videos demonstrating both features [28,29,33–53].

Future versions will be self-driving. Self-driving Nomads can be ordered automatically while driving on the highway. They will track the vehicle autonomously and transfer energy by induction. Auto-Nomad solutions will be developed for electric passenger vehicles and for autonomous driving electric trucks. AC propulsion has developed an early demonstrator of the future Tesla Roadster and its powertrain, as well as a very interesting self- steering mobile range extender, from 1992 to 2001 [47]. www.tzev.com/1992_1995_2001_acp_rxt-g_html. Future versions of EP Tender will include fuel cell or batteries. Hitching to the EV will become fully automatic within a few years. Nomadic Power (www.nomadic- power.com) has focused on a 100% ZE approach by mounting a high power density battery on its Nomads, with a capacity of 85 kWh. The Nomads will be either rented or owned. They can be integrated into the grid and provide storage as well as backup power. Their battery is designed to provide both DC (electric mobility) and AC (home power). Following are the alternatives [28,29,33–53]:

4.6.1.1 Plug-In Hybrid Electric Vehicles

Quite a few new products (A3, Golf GTE, BMW i3, New Volt) and many high-tech range extender prototypes with equipment makers (AVL, KSPG, Lotus-Fagor, Polaris-SwissAuto, Mahle, Obrist, and so on) have come to the market. The difficulty is to keep an acceptable level of cost, given the complexity involved. This is acceptable for large or premium cars, but not reasonable for smaller and affordable cars. The other significant downsides are that the full-electric range tends to be lower than a full EV (Golf GTE = 40 km effective electric range [29], or the gasoline range is less than a classic ICE (BMW i3 = 9 L fuel tank), and the trunk gets smaller in order to accommodate for all the technology (only 270 L in the Golf GTE vs 370 L in the e-Golf and 225 L in the BMW i3).

4.6.1.2 Very Large Batteries

Tesla Model S is probably the best electric car in the world and is loved by affluent enthusiasts. But such a large, heavy, and expensive car is not ideal in towns, and is not a good cruiser, requiring to stop 35 min every 225 km (assuming driving at 130 km/h and 80% battery charge on supercharger). Recent examples of cross-continental trips in the United States show that total trip duration is just above 30% higher than the same trip with an ICE. Recent academic work from Oak Ridge National Laboratory [50] confirms that most users are better off with 100 miles range. Model S and the new Model X are anyhow a great symbol helping to enhance the credibility of EVs.

4.6.1.3 Fuel Cells

This technology remains very expensive in itself, as well as requiring H_2 distribution infrastructure. There is also the debate on how to manufacture clean H_2 at an acceptable efficiency. It will take some years to emerge as a viable stand-alone business, as expensive and scarce infrastructure has to be used by expensive and scarce vehicles. … Toyota Mirai. Mobile energy modules, equipped with a fuel cell and rented on demand, might change that picture by allowing a concentrated and progressive deployment of the infrastructure used by numerous and affordable vehicles.

4.6.1.4 Synthetic Fuel

Possibly the best of all solutions could be the production of synthetic fuel from solar energy, hydrogen and carbon capture. Energy density and ease of storage would no longer be an issue but this still remains highly hypothetical. EV as main car + ICE car rental: this solution is perfect in the context of multimodal trips. But costs are higher, and door to door trips are often more convenient than having to fetch a vehicle at the car rental (unless as a complement to plane or train). EP Tender's research [51] shows that only 11.5% of EV owners prefer this solution.

The fundamental benefit of designing a car for optimal daily usage and adding on-demand an energy module for occasional long distance will remain as long as the marginal cost and weight of a larger battery remain favorable to a larger fuel tank [52], and the time for filling it. The energy modules are also part of the grid when not in use and are providing energy storage or microgeneration. The cars are optimized for theirs daily usage and are supplemented by the most suitable energy module given the infrastructure available on a long-distance trip (grid charging, H_2, inductive road [34], biofuel).

4.6.2 MODULAR APPROACH FOR ENERGY CONSUMPTION ESTIMATION IN ELECTRIC VEHICLES (EVs)

EV energy consumption is variable and dependent on a number of external factors such as road topology, traffic, driving style, ambient temperature, and so on. The goal of the study by Cauwer et al. [54] was to detect and quantify correlations between the kinematic parameters of the vehicle and its energy consumption. Real-world data of EV energy consumption were used to construct the energy consumption calculation models. Based on the vehicle dynamics equation as an underlying physical model, multiple linear regression was used to construct three models. Each model used a different level of aggregation of the input parameters, allowing predictions using different types of available input parameters. One model used aggregated values of the kinematic parameters of trips. This model allowed prediction with basic, easily available input parameters such as travel distance, travel time, and temperature. The second model extended this by including detailed acceleration data. The third model used the raw data of the kinematic parameters as input parameters to predict the energy consumption. Using detailed values of kinematic parameters for the prediction in theory increased the link between the statistical model and its underlying physical principles, but required these parameters to be available as input in order to make predictions. The first two models showed similar results. The third model showed a worse fit than the first two but had a similar accuracy. The authors indicated that this model has great potential for future improvement.

Rosario et al. [55,56] presented a modular approach for estimating energy consumption in EV having dual energy sources. The primary design challenges in EVs having multiple energy storage systems lie in controlling the net energy expenditure, determining the proportional power split, and establishing methods to interface between the energy systems so as to meet the demands of the vehicle propulsion and auxiliary load requirements. The study by Rosario et al. [55,56] presents a modular approach to the design and implementation of a power and energy management system (PEMS) for EVs. The model EV is powered by dual-energy sources consisting of batteries and ultracapacitors. Operation of the PEMS has been structured into modular hierarchical process shells. The energy management shell (EMS) handles the longer-term decisions of energy usage in relation to the longitudinal dynamics of the vehicle. The process within the power management shell (PMS), however, handles the fast decisions to generate power split ratios between the batteries and ultracapacitors. Finally, the power electronics shell (PES) handles the ultra-fast switching functions that facilitate the active power sharing between the two sources. Within the EMS, a fuzzy inference system (FIS) is employed as an intelligent decision engine. Simulations are presented to exemplify the functions of the PMS and EMS. The modular structure approach is design-implementation oriented, with the objective of contributing toward a more unified description of EV power and energy management [55,56].

4.6.3 Modular Fuel Cells for Multiple Vehicle Applications

Most new technologies with the potential to disrupt aerospace get their start in other sectors. For hydrogen fuel cells, that will certainly be in personal, commercial, and military vehicles. But General Motors believes its fuel cells will find a place in aviation, in auxiliary power units (APUs), distributed power generation and ground-support vehicles. GM has launched SURUS (for Silent Utility Rover Universal Platform), a flexible and autonomous vehicle platform [57–59]. Like competing automakers, GM is adding EVs and other alternative fuel vehicles to their fleet. They're also investing heavily in autonomous driving tech, which can shift freight around ports and warehouses, moving supplies around construction sites, and even shuttling humans to the hospital when they require medical attention. GM's newest EV concept is built to handle all of these tasks [57–59].

Because the SURUS is powered by fuel cells it can move around stealthily, producing minimal noise and very little heat. That would make it an excellent choice for running resupply missions and shuttling troops (both healthy and wounded). Fuel cells give SURUS an impressive range of about 400 miles, and it can cover the whole distance without a human driver. Its computerized brain also allows multiple SURUS units to form a convoy for tackling bigger jobs, and it's designed to handle rough terrain with ease. GM has equipped SURUS with external plugs so you can trade some of that 400-mile range in for juice. Plug-in tools, lights, pumps, or anything else that you need to get the job done [57–59]. SURUS is a modular platform, and GM envisions offering several different add-ons for specific applications. It can be a rolling medical facility, move around shipping containers, and haul workers and materials around the job site. The SURUS is intended as a "blank slate" platform. The GM has devised a few of the possibilities, but the future customers could have plenty of other usages. The army will probably come up with dozens of different ways to utilize SURUS [57–59].

4.6.4 Vehicle Systems Controller with Modular Architecture

Vehicle systems controllers (VSCs) are devices used within automotive vehicles, such as an HEV, in order to control various vehicle systems, processes, and functions and are often part of the powertrain controller. One type of HEV, commonly referred to as a "power-split" type HEV, includes three powertrain subsystems which cooperatively provide the torque necessary to power the vehicle, and a vehicle system controller which controls the three subsystems. A "parallel-series" type HEV includes an engine subsystem (e.g., an ICE and controller), a generator subsystem (e.g., a motor/generator and controller), and a motor subsystem or an "electric drive subsystem" (e.g., an electric motor and controller) [60].

This hybrid configuration provides improved fuel economy and reduced emissions since the ICE can be operated at its most efficient/preferred operating points by use of the various subsystems. Additionally, this configuration can achieve better drivability and may extend vehicle performance relative to a comparative conventional vehicle. In order to achieve the goal, appropriate coordination and control between subsystems in the HEV are essential. This goal is achieved by use of the VSC and hierarchical control architecture. The VSC is typically used to interpret driver inputs (e.g., gear selection, accelerator position, and braking effort) to coordinate each of the vehicle subsystems and to determine the vehicle system-operating state. The VSC generates commands to the appropriate subsystems based on driver inputs and control strategies and sends the generated commands to the respective subsystems. The generated commands sent to the respective subsystems are effective to cause the subsystems to take appropriate actions to meet the driver's demands [60].

Due to the numerous types of vehicle subsystems and processes which may vary from vehicle to vehicle, conventional VSCs are relatively complex and are designed to serve and/or function within a specific type of vehicle. Due to this complexity and design, it is relatively difficult to modify a conventional VSC to operate with a new vehicle system or functionality. For example, and without limitation, if one were to replace the braking system or functionality within an HEV having a conventional VSC with a different type of system of functionality (e.g., series versus parallel regenerative braking), many control features within the powertrain controller would have to be modified or reprogrammed. This increases the cost and time required to make such a modification. Moreover, each different type of HEV typically requires a VSC with a somewhat different functionality, thereby reducing the uniformity among HEVs and increasing the overall cost of the HEVs. There is, therefore, a need for a modular VSC which is partitioned into portions that corresponds to and/or provide a logical grouping of vehicle functions, thereby allowing the VSC to be easily modified to conform to new vehicle functions or features.

The first non-limiting advantage of the patent by Ramaswamy and McGee [60] is that it provides a VSC for an HEV, which overcomes at least some of the previously delineated drawbacks of prior VSCs or powertrain controllers. The second non-limiting advantage of this invention is that the present invention provides a modular VSC which includes various portions that correspond to a logical grouping of vehicle functions thereby allowing the vehicle functionality to be relatively easily modified. The third non-limiting advantage of the invention is that the present invention provides a VSC that is partitioned to take into account a logical grouping of vehicle functions while maintaining a hierarchy of control within the VSC. According to the first aspect of the present invention, a MODV system controller is provided for use with an HEV. The MODV system controller includes a plurality of portions, wherein each of the plurality of portions corresponds to a certain vehicle functionality. According to the second aspect of the present invention, a method of organizing a vehicle system controller for use with an HEV is provided. The method includes a step of partitioning the controller into a plurality of control portions, each of the plurality of control portions corresponding to a particular vehicle functionality.

4.7 MODULAR INTEGRATED ENERGY SYSTEMS

4.7.1 MODULAR VERD2GO—EV CONNECTION

This distributed energy system, which bundles EV range extension with personal energy storage, has some intriguing elements to it. One of the big missing links in home renewable energy production is battery storage, and the current state of home- and EV-scale energy storage systems on the market is still mostly a matter of too little capacity for too high a price. With *EV-to-building* and vehicle-to-grid (V2G) systems still serving as fringe applications and not for mass adoption quite yet, there doesn't seem to be any practical and affordable pathways to truly portable clean energy that can be used for any electrical demand (the energy in EV batteries isn't readily accessible for anything other than powering the car itself) [61].

One potential system, called Verd2GO, appears to be yet another battery-swapping scheme on the surface, but upon closer look, is actually akin to a swappable EV "range extender" that doubles as a portable power solution. At its heart, the Verd2GO system could effectively "modularize" clean energy (which is another way of saying it's a system of power packs/batteries) in a standard format that can be swapped out as needed in an EV, or pulled from the EV into a home to use for power. The Indiegogo campaign page calls this system of portable energy packs a "standardized currency" that can provide "an easy and shareable way to store clean, renewable energy for when it's needed." With the Verd2GO system, power packs made up of multiple battery "cartridges" could be charged with solar energy and then be dispensed through a rental or subscription scheme at kiosks, giving EV drivers an instant extra range of 40 or so extra miles. These power packs could also be used in solar (or nonsolar) homes as a portable power supply that can reduce peak demand charges for home electric bills by being charged or rented at a standard rate, and then used instead of grid electricity during peak hours.

Verd2GO facilitates the ability to take renewable energy from and to one's home, to power one's car while on your way to a picnic, where you can charge all your devices at once, without the necessity of dirty grid power. The way we use electricity right now is an ever-on, on-demand system that's as easy and close as finding the nearest electrical outlet (and to which our monthly utility budget is vulnerable, considering the way our electricity rates tend to be tiered). With this "modular" system from Verd2GO, electricity would be more manageable/customizable (although probably also a lot more hands-on, in the sense that the power packs are meant to be physically moved and replaced), and would be a distributed energy system that could actually be "distributed." Verd2GO removes the barriers to entry for average folks to actively participate in the business of harnessing, distributing, and selling clean power. Once there is a convenient, simple, and intuitive way to sell energy into the grid, we can address our important energy issues and clean power plans [61].

Accessing the energy stored in EVs, when not in use, could play a vital role in advancing further the energy connectivity landscape under the V2G concept, which refers to the reciprocal flow of power between any electric or hybrid vehicle and the grid. In addition to the low carbon future that V2G supports, it also introduces the prospect of financial incentives for the consumer, through offering frequency regulation and energy storage facilities to the grid. The energy stored in the vehicle can be used to avoid peak tariffs at times of high demand and optimize the value of electricity generated from domestic renewables, such as solar panels, to reduce home electricity bills. According to research produced by the global engineering consultancy Ricardo and the National Grid on grid balancing measures, V2G could provide $1,000–$14,000 of income each year for vehicle owners. This would be a significant incentive for additional EV uptake, not to mention that EVs are also road tax free. The government's Plug-In Car Grant which provides a 35% grant toward the cost of new EVs for new buyers provides further encouragement.

Innovations such as home battery energy storage products offer the opportunity to scale-up dispersed energy storage capacity and maximize local energy use in a fully connected home energy network. A 4 kWh battery, for example, would be able to provide a third of the energy needs of a typical home in the UK. In light of the fact that the RAC Foundation stated that privately owned vehicles in the UK are parked 96% of the time, adding to grid storage capacity by utilizing the batteries in electric cars plugged in when not in use makes perfect sense. Clearly, the EV must still be available with sufficient charge when needed for transport, and there are a number of early-stage smart charger and V2G trials ongoing looking at how this all might work in practice. The existence of the trials show that the technology exists on and off the car but it isn't always suited to every application. There are challenges yet to be overcome, for example, in UK-based domestic charging applications. UK-based *Cenex*, a low-carbon and fuel cell technologies consultancy, is leading a project to integrate V2G technology into existing energy infrastructure at the district and city scale in Birmingham, Berlin, and Valencia. The project aims to deliver a reliable assessment of network impacts and business cases for V2G [61].

The dramatic increase in EV sales over the past 2 years, according to recent figures released by the *Society of Motor Manufacturers and Traders (SMMT)*, combined with rapidly declining technology costs mean that EVs can move away from being another challenge for our aging grid infrastructure to becoming managed loads and ultimately dispersed storage capacity as part of a revolutionary decentralized energy system. Energy is stored close to where it will end up being used and local resources are utilized in the most effective way possible. The automotive sector is taking seriously the idea that owners can be compensated for using their vehicles as a way of achieving cost parity between EVs and gasoline or diesel vehicles with regulatory pressure to deliver zero-emissions vehicles. The battery in an EV is the most expensive component—typically 227/kWh in 2016, according to a recent report by McKinsey—and designed to the stringent standards of the automotive industry. It can comfortably handle the demands from the grid and, in some cases, this functionality has already been enabled [61].

EVs currently manufactured by Nissan and Mitsubishi already offer onboard V2G compatibility to help balance supply and demand when used alongside smart chargers. A recent collaborative project between Nissan, energy company Enel, and California-based company *Nuvve* is the first of its kind to commercially integrate and host V2G units at its headquarters in Copenhagen. A number of EVs and charging stations are now providing commercial V2G services in the city. Other automakers such as BMW and Honda are following suit through building relationships with commercial participants and research and development hubs.

4.7.2 Additive Manufacturing Integrated Energy

As such, much of the urban planning and architecture industries are focused on ways to radically subvert this inherited infrastructural wisdom. A unique way to reform the residential-transit complex is not to discard existing models in favor of mass transit and dense high-rises, but instead simply to change the relationship between homes and cars using integrated energy systems. That's the supposition made by Additive Manufacturing Integrated Energy (AMIE) demonstration project, a collaboration among the Department of Energy's (DOE) Oak Ridge National Laboratory (ORNL) in Tennessee, the Chicago office of Skidmore, Owings & Merrill LLP (SOM), and 19 other partners. AMIE 1.0 pairs up ORNL's hybrid natural gas/electric car with a 210-sq ft modular structure that both generate and share power with each other [62–64] (see Figures 4.1–4.4).

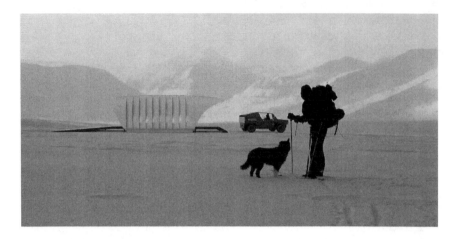

FIGURE 4.1 The Additive Manufacturing Integrated Energy (AMIE) car and building in winter [62–64].

FIGURE 4.2 Assembly of the AMIE structure. (Courtesy: ORNL [62–64].)

Both 3D-printed, the structure and vehicle use a wireless coupling station to move electricity, with 85% efficiency. The structure's roof is coated in thin photovoltaic (PV) panels, which it can use to power the building and recharge the car's battery. The car's engine is electric, but its battery can be charged with an onboard natural gas generator. Excess power from the car can be sent to the structure's battery to power it when sunlight isn't sufficient. Even the most seasoned mega commuter's car is active for only a few hours at a time, and AMIE looks to make vehicles and buildings work together all day. The structure (which contains a kitchenette and a fold-out Murphy bed) is made up of a series of brilliant white vaguely biomorphic ribs, tied

Systems for Energy Usage in Vehicles

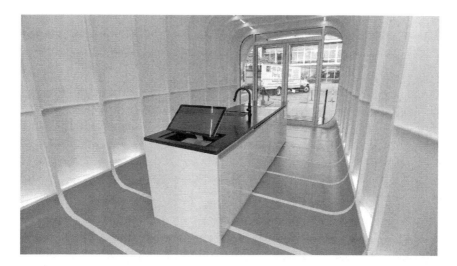

FIGURE 4.3 Interior of the AMIE structure. (Courtesy: ORNL [62–64].)

FIGURE 4.4 The entire integrated structure [62–64].

together with post-tensioned steel rods. Each rib features gill-like windows that are hidden from view as you approach from the rear of the structure but visible from the entrance—transparent or opaque depending on which direction you face [62]. The structure's insulation system is a modular vacuum-sealed foam board called modified atmospheric insulation. It's extremely efficient, only an inch thick, and flexible enough to slot into the structure's curved geometry. The entire wall system integrates all its structural, façade, and insulation systems into one component: sections of 3D-printed resin—only 2 inches thick—reinforced with carbon fiber. The chassis of the car is made from the same material. Oak Ridge National Laboratory says that the use of 3D printing allowed his team to integrate material around these groundbreaking energy-sharing systems at hyper speed. From conception to execution, the entire project took only 1 year [62–64].

But in many ways, AMIE is a conservative, incrementalist way to reform transit and housing infrastructure. It doesn't question the supremacy of detached homes or of private automobiles. And because users are now dependent on fossil-fuel-aided vehicles to power their homes, it entrenches these two infrastructure models together into a new (albeit more efficient) economic relationship. AMIE is not meant as a definitive solution to mitigating climate change and carbon emissions. The whole idea is to demonstrate issues of energy and how you can move (it) around. It's intended as exploratory research with future iterations that could be wildly different. It would be interesting to see this technology deployed in a more urban model. If a small structure and car can power each other, can a subway train team up with a high-rise?

The most important innovation from AMIE is intermodal power sharing across transit and residential infrastructure, not the specific forms that transit and building take. Using a car and using a structure was one convenient way to describe it [62–64].

4.8 MODULAR FUEL CELL STACK SYSTEM

The polymer electrolyte membrane fuel cell (PEMFC) is still not an economically attractive prime-mover due to a number of key challenges that have yet to be fully resolved; such as degradation to cell components resulting in inadequate lifetimes, specialized manufacturing processes, and poor gravimetric/volumetric energy densities. The study by Scott et al. [65] presents a stack concept that replaces the conventional BPP, a component that is responsible for a large proportion of stack cost and volume in traditional fuel cell stack designs. The stack architecture comprises active and passive components, which are suited to mass manufacture and maintain functionality that the BPP fulfilled. Furthermore, the design allows the implementation of a fault-tolerant system (FTS) which can bypass faulty cells while still ensuring electrical output. The study presents and characterizes the stack architecture over a number of operating scenarios. The experimental studies suggest that the performance of the new design is similar to that of traditional stacks over a number of operating conditions. As mentioned in Chapter 1, Rajalakshmi et al. [66] also discuss the modular architecture of PEMFC stacks in detail.

Horizon [67,68] modular fuel cell system has also followed the similar concept for the self-sustaining energy supply of small consumers in the range up to 300 W. The main goal of the implementation is to design and provide effective and low-cost PEM fuel cell stacks, which are hydrogen-supplied and its balance-of-plant (BoP) components. The system is scalable, built in 30 W increments, in order to cover all the power categories up to 300 W. Great importance is attached to the construction of an insensitive and robust system because the portable system should be able to operate under a variety of environmental conditions. The modular fuel cell system consists of a modular fuel cell stack, a modular designed electronic circuit board, and a customizable BoP according to the total power requirement. Similarly, the study of Ebaid and Mustafa [69] deals with the design of a modular fuel cell that can be mass-produced and used to set up a larger fuel cell stack for stationary applications (6 kW) which is capable of powering a medium-sized household. The design for 100 W fuel cell module includes the calculations for the main dimensions of the fuel cell components, mass flow rate of reactants, water production, heat output, heat transfer, and the cooling system. The study is intended to facilitate material and process selection prior to manufacturing wide-scale production.

In principle, a modular fuel cell system consists of a modular fuel cell stack, a modular designed electronic circuit board, and a customizable BoP according to the total overall power. The modular fuel cell stack distinguishes itself from a standard fuel cell stack in two important features:

1. the design of the fuel cells based on metal pole plates
2. the hydraulic compression of the fuel cell.

The advantages of point 1 are obvious, metal pole plates are more robust than graphitic pole plates. In addition, the pole plates can be quickly molded by a molding press, which leads to a low-cost fabrication [30,66–73]. The smallest unit of the modular fuel cell stacks, the mono cell (the fuel cell itself) consisting of two metal pole plates with integrated membrane-electrode assembly. This stack structure with parallel gas supply route and hydraulic compression has the advantage of being able to remove a damaged fuel cell quickly from the stack with minimal effort and without damaging the other units. In conventional fuel cell stacks, an exchange of the fuel cell is often not possible because the plates are pressed or bonded with each other which leads to damage while disassembling. In the case of the modular fuel cell system, the replacement of one cell is quite simple. In order to perform a cell switch, the housing has to be relieved, which is easily done via a simple screw mechanism. Subsequently, the cover can be removed and the cell can be easily replaced. Evan a nonspecialist is able to replace a faulty fuel cell within minutes. The information regarding which cell has to be changed is displayed by the electronic circuit board [30,66–73].

The second feature of hydraulic compression is exercised in commercial MOD-FC [30]. Due to the complete flushing of separate individual cells with the hydraulic medium, a homogeneous pressing of the inner cell components is ensured. In addition, the hydraulic medium can be directly used as a cooling medium. This ensures a homogeneous electric current generation among the surface and also that no life-shortening hot spots appear [30]. By the nearly ideal operating conditions MOD-FC are capable to examine in-situ membrane electrode units. First functional models are already in use. A plurality of simultaneously reproducible samples can be examined under the same operating conditions, which improve the quality of test results and reduces significantly the number of sampling tests. Moreover, individual elements can be replaced by separate individual cells without great effort, due to the modular design of the stacks. In contrast to the conventional fuel cell stacks, MOD-FC can thus be maintained on a cellular basis. MOD-FC can be used in all areas of application [30].

There are a variety of components such as valves, pumps, and so on offered at the market, suitable for operation with hydrogen. However, the selection and integration of corresponding components are still a challenge. A great deal of importance was attached to the high availability of the BoP, which is applied in modular fuel cell systems and the system efficiency optimization. Since the fuel cell stack is scalable, BoP of the system has to be adapted depending on the system output power. The appropriate platform for an individual adjustment is realized by standardized housing and mounting fixtures. In order to supply a load with electrical energy efficiently, an electronic circuit board is required in addition to the fuel cell BoP, which has to provide a stable output voltage by an optimized efficiency. Fuel cell stacks produce a DC output with a 2:1 variation in output voltage from no load to full load. The output voltage of each fuel cell is about 0.4 V at full load, and several of them are connected in series to construct a stack. An example 100 V fuel cell stack consists of 250 cells in series and to produce 300 V at full load requires 750 cells stacked in series. Since fuel cells actively convert the supplied fuel to electricity, each cell requires proper distribution of fuel, humidification, coupled with water/thermal management needs. With this added complexity, stacking more cells in series decreases the reliability of the system. For example, in the presence of bad or malperforming cell/cells in a stack, uneven heating coupled with variations in cell voltages may occur. Continuous operation under these conditions may not be possible or the overall stack output power is severely limited.

Marx et al. [72] examine the current state of the art of multi-stack fuel cell systems. Different electrical and fluidic architectures are designed for MFC systems. The effects of these architectures on the performances of the system are presented. Depending on the chosen architecture, a more reliable system is obtained through the enabling of degraded mode operation. Different electrical, fluidic, and thermal configurations for ancillaries are depicted. Impacts of multi-stack fuel cell architectures on the performances are detailed. DC-DC converter architecture for multi-stack systems is compared. For a multi-stack fuel cell system, often a two-stage DC-DC converter is designed. It is possible to operate a second fuel cell stack in addition to the main stack at the entrance of the DC-DC converter. The second stage of the switched power amplifier offers an efficient increase of the DC voltage to output voltage level, as well as electrical galvanic separation. The output voltage of the DC-DC converter is adjustable to the needs of the load. In addition, the galvanic separation allows switching the output of a further modular fuel cell system in series, thus doubling the output voltage is possible. It is also apparent that great value was attached to a high degree of modularity and flexibility while designing the system. Another essential element is the control board, which constitutes the controlling "heart" of the whole system. Especially for this purpose a microcontroller board was developed. This board has high power outputs for the BoP, such as the pumps and the magnetic valves. All the control and automation interventions are initiated by the microcontroller [57,67,68,70,71]. In addition, the control unit supervises the fuel cell stack, for example, individual cell voltages are continuously read. Thus, the real functions of the fuel cell are monitored and if necessary, a message will be displayed. Finally, on this board, the monitoring, as well as the communication with the external monitoring point, are possible. In this way, online monitoring is possible for a given network connection or GSM. The study of Palma and Enjeti [73] takes this concept one step forward. In this study, a modular fuel cell powered by a modular DC-DC converter is proposed. The proposed concept electrically divides the fuel cell stack into various sections, each powered by a DC-DC converter. The proposed modular fuel cell powered by a modular DC-DC converter eliminates many of these disadvantages, resulting in an FTS. A design example is presented for a 150 W, three-section fuel cell stack and DC-DC converter topology. Experimental results obtained on a 150 W, three-section PEM fuel cell stack powered by a modular DC-DC converter are discussed in this study [73].

4.9 MODULAR SOLAR VEHICLES

4.9.1 Solar Electric Car

A solar vehicle is an EV powered completely or significantly by direct solar energy. Usually, PV cells contained in solar panels convert the Sun's energy directly into electric energy. The term "solar vehicle" usually implies that solar energy is used to power all or part of a vehicle's propulsion. Solar power may be also used to provide power for communications or controls or other auxiliary functions. All solar powered (fully or partially) are modular in their design. Solar vehicles are not sold as practical day-to-day transportation devices at present but are primarily demonstration vehicles and engineering exercises, often sponsored by government agencies. However, indirectly solar-charged vehicles are widespread and solar boats are available commercially. Solar cars depend on PV cells to convert sunlight into electricity to drive electric motors. Unlike solar thermal energy which converts solar energy to heat, PV cells directly convert sunlight into electricity [74–76].

The design of a solar car is severely limited by the amount of energy input into the car. Solar cars are built for solar car races and also for public use. Even the best solar cells can only collect limited power and energy over the area of a car's surface. This limits solar cars to ultra-light composite bodies to save weight. Solar cars lack the safety and convenience features of conventional vehicles. The first solar family car was built in 2013 by students in the Netherlands. This vehicle was capable of 550 miles on one charge during sunlight. It weighed 850 pounds and had a 1.5 kW solar array. Solar vehicles must be light and efficient. 3,000-pound or even 2,000-pound vehicles are less practical. Stella Lux, the predecessor to Stella, broke a record with a 932-mile single-charge range. The Dutch are trying to commercialize this technology.

During racing, Stella Lux is capable of 700 miles during daylight. At 45 mph Stella Lux has infinite range. This is again due to high efficiency including a coefficient of drag of .16. The average family who never drives more than 200 miles a day would never need to charge from the mains. They would only plug-in if they wanted to return energy to the grid. Solar cars are often fitted with gauges and/or wireless telemetry to carefully monitor the car's energy consumption, solar energy capture, and other parameters. Wireless telemetry is typically preferred as it frees the driver to concentrate on driving, which can be dangerous in such a small, lightweight car. The solar EV system was designed and engineered as an easy to install (2–3 h) integrated accessory system with a custom-molded low-profile solar module, supplemental battery pack, and a proven charge controlling system. As an alternative, a battery-powered EV may use a solar array to recharge; the array may be connected to the general electrical distribution grid. The Venturi Astrolab in 2006 was the world's first commercial electro-solar hybrid car and was originally due to be released in January 2008 [10]. In May 2007, a partnership of Canadian companies led by Hymotion altered a Toyota Prius to use solar cells to generate up to 240 W of electrical power in full sunshine. This is reported as permitting up to 15 km extra range on a sunny summer day while using only the electric motors. PV modules are used commercially as APUs on passenger cars in order to ventilate the car, reducing the temperature of the passenger compartment while it is parked in the sun. Vehicles such as the 2010 Prius, Aptera 2, Audi A8, and Mazda 929 have had solar sunroof options for ventilation purposes.

The area of PV modules required to power a car with a conventional design is too large to be carried onboard. A prototype car and trailer "solar taxi" has been built. According to the website, it is capable of 100 km/day using 6 m^2 of standard crystalline silicon cells. Electricity is stored using a nickel/salt battery. A stationary system such as a rooftop solar panel, however, can be used to charge conventional EVs. It is also possible to use solar panels to extend the range of a hybrid or electric car, as incorporated in the Fisker Karma, available as an option on the Chevy Volt, on the hood and roof of "Destiny 2000" modifications of Pontiac Fieros, Italdesign Quaranta, Free Drive EV Solar Bug, and numerous other EVs. In May 2007 a partnership of Canadian companies led by Hymotion added PV cells to a Toyota Prius to extend the range. SEV claims 20 miles per day from their combined 215 W module mounted on the car roof and an additional 3 kWh battery.

It is also technically possible to use PV technology (specifically thermophotovoltaic (TPV) technology) to provide motive power for a car. Fuel is used to heat an emitter. The infrared radiation generated is converted to electricity by a low bandgap PV cell (e.g., GaSb). A prototype TPV hybrid car was even built. The "Viking 29" [17] was the world's first TPV-powered automobile, designed and built by the Vehicle Research Institute (VRI) at Western Washington University. Efficiency would need to be increased and cost decreased to make TPV competitive with fuel cells or ICEs. Solar panels are solar systems that produce electricity directly from the sunlight. Solar panels produce clean, reliable electricity without consuming fossil fuels. Solar panel systems are an excellent way to generate energy in remote locations (the oceans) that are not connected to the electric grid. The benefits of solar panels are:

1. Solar panels are highly reliable and easy to maintain. Solar panels have no moving parts, so visual checks and servicing are enough to keep systems up and running. Solar panels are built to withstand hail impact, high wind, and freeze-thaw cycles. Solar panel systems can produce power in all types of weather. On partly cloudy days, they produce as much as 80% of their potential energy. Even on extremely cloudy days, they can still produce about 25% of their maximum output.
2. Virtually no environmental impact. Solar panel systems burn no fuel and have no moving parts. They are clean and silent, producing no atmospheric emissions of greenhouse gases (GHGs) that are harmful to the Earth.

Systems for Energy Usage in Vehicles

3. Modular and flexible in terms of size and applications. Solar panel systems can be built to any size in response to the energy needs at hand. They can be enlarged or moved easily.

4.9.1.1 Plug-In Hybrid and Solar Car

An interesting variant of the EV is the triple hybrid vehicle—the PHEV that has solar panels as well to assist. The 2010 Toyota Prius model has an option to mount solar panels on the roof. They power a ventilation system while parked to help provide cooling. There are many applications of *PVs in transport* either for motive power or as APUs, particularly where fuel, maintenance, emissions or noise requirements preclude ICEs or fuel cells. Due to the limited area available on each vehicle either speed or range or both are limited when used for motive power. There are limits to using PV cells for cars. Power from a solar array is limited by the size of the vehicle and area that can be exposed to sunlight. This can also be overcome by adding a flatbed and connecting it to the car and this gives more area for panels for powering the car. While energy can be accumulated in batteries to lower the peak demand on the array and provide operation in sunless conditions, the battery adds weight and cost to the vehicle. The power limit can be mitigated by use of conventional electric cars supplied by solar (or other) power, recharging from the electrical grid [77–79].

While sunlight is free, the creation of PV cells to capture that sunlight is expensive. Costs for solar panels are steadily declining (22% cost reduction per doubling of production volume). Even though sunlight has no lifespan, PV cells do. The lifetime of a solar module is approximately 30 years [52]. Standard PVs often come with a warranty of 90% (from nominal power) after 10 years and 80% after 25 years. Mobile applications are unlikely to require lifetimes as long as building integrated PV and solar parks. Current PV panels are mostly designed for stationary installations. However, to be successful in mobile applications, PV panels need to be designed to withstand vibrations. Also, solar panels, especially those incorporating glass, have significant weight. In order for its addition to be of value, a solar panel must provide energy equivalent to or greater than the energy consumed to propel its weight [77–79]. A typical solar car is illustrated in Figure 4.5.

4.9.1.2 Electric Car with Solar Assist/Solar Taxi

A Swiss project, called "Solartaxi," has circumnavigated the world. This is the first time in history an EV (not self-sufficient solar vehicle) has gone around the world, covering 50,000 km in 18 months and crossing 40 countries. It is a road-worthy EV hauling a trailer with solar panels, carrying a 6 m^2 sized solar array. The Solartaxi has Zebra batteries, which permit a range of 400 km without recharging. The car can also run for 200 km without the trailer. Its maximum speed is 90 km/h. The car weighs 500 kg and the trailer weighs 200 kg. According to initiator and tour director Louis Palmer, the car in mass production could be produced for 16,000 Euro. Solartaxi has toured the World from July 2007 till December 2008 to show that solutions to stop global warming are available and to encourage people in pursuing alternatives to fossil fuel. Palmer suggests the most economical location for solar panels for an electric car is on building rooftops

FIGURE 4.5 A typical solar car [74].

though, likening it to putting money into a bank in one location and withdrawing it in another. Solar Electrical Vehicles is adding convex solar cells to the roof of HEVs [74,75,80].

4.9.1.3 Solar Buses

Solar buses (see Figure 4.6) are propulsed by solar energy, all or part of which is collected from stationary solar panel installations. The Tindo bus is a 100% solar bus that operates as a free public transport service in Adelaide City as an initiative of the City Council [8]. Bus services which use electric buses that are partially powered by solar panels installed on the bus roof, intended to reduce energy consumption and to prolong the life cycle of the rechargeable battery of the electric bus, have been put in place in China [75]. Solar buses are to be distinguished from conventional buses in which electric functions of the bus such as lighting, heating or air-conditioning, but not the propulsion itself, are fed by solar energy. Such systems are more widespread as they allow bus companies to meet specific regulations, for example the anti-idling laws that are in force in several of the US states and can be retrofitted to existing vehicle batteries without changing the conventional engine. The Venturi Astrolab in 2006 was the world's first commercial electro-solar hybrid car and was originally due to be released in January 2008. In May 2007 a partnership of Canadian companies led by Hymotion altered a Toyota Prius to use solar cells to generate up to 240 W of electrical power in full sunshine. This is reported as permitting up to 15 km extra range on a sunny summer day while using only the electric motors. An inventor from Michigan, USA built a street legal, licensed, insured, solar-charged electric scooter in 2005.

FIGURE 4.6   Solar-powered minibus in Berlin, Germany [76].

It had a top speed controlled at a bit over 30 mph and used fold-out solar panels to charge the batteries while parked [75–77,82].

PV modules are used commercially as AUPs on passenger cars in order to ventilate the car, reducing the temperature of the passenger compartment while it is parked in the sun. Vehicles such as the 2010 Prius, Aptera 2, Audi A8, and Mazda 929 have had solar sunroof options for ventilation purposes. The area of PV modules required to power a car with a conventional design is too large to be carried onboard. A prototype car and trailer has been built Solar Taxi. According to the website, it is capable of 100 km/day using 6 m^2 of standard crystalline silicon cells. Electricity is stored using a nickel/salt battery. A stationary system such as a rooftop solar panel, however, can be used to charge conventional EVs. It is also possible to use solar panels to extend the range of a hybrid or electric car, as incorporated in the Fisker Karma, available as an option on the Chevy Volt, on the hood and roof of "Destiny 2000" modifications of Pontiac Fieros, Italdesign Quaranta, Free Drive EV Solar Bug, and numerous other EVs, both concept and production. In May 2007 a partnership of Canadian companies led by Hymotion added PV cells to a Toyota Prius to extend the range. SEV claims 20 miles per day from their combined 215 W module mounted on the car roof and an additional 3 kWh battery [75–77,82].

It is also technically possible to use PV technology, (specifically TPV technology) to provide motive power for a car. Fuel is used to heat an emitter. The infrared radiation generated is converted to electricity by a low bandgap PV cell (e.g. GaSb). A prototype TPV hybrid car was even built. The "Viking 29" was the World's first TPV-powered automobile, designed and built by the VRI at Western Washington University. Efficiency would need to be increased and cost decreased to make TPV competitive with fuel cells or ICEs.

4.9.2 Personal Rapid Transit Vehicles

Several personal rapid transit (PRT) concepts incorporate PV panels.

4.9.2.1 Rail

Railway presents a low rolling resistance option that would be beneficial for planned journeys and stops. PV panels were tested as APUs on Italian rolling stock under EU project PVTRAIN. Direct feed to DC grids avoids losses through DC to AC conversion. DC grids are only to be found in electric-powered transport: railways, trams, and trolleybuses. Conversion of DC from PV panels to grid AC was estimated to cause around 3% of the electricity being wasted, PVTRAIN concluded that the most interest for PV in rail transport was on freight cars where onboard electrical power would allow new functionality:

- GPS or other positioning devices, so as to improve its use in fleet management and efficiency.
- Electric locks, a video monitor, and a remote control system for cars with sliding doors, so as to reduce the risk of robbery for valuable goods.
- Anti-lock braking system (ABS) brakes, which would raise the maximum velocity of freight cars to 160 km/h, improving productivity.

The Kismaros–Királyrét narrow-gauge line near Budapest has built a solar-powered railcar called "Vili." With a maximum speed of 25 km/h, "Vili" is driven by two 7 kW motors capable of regenerative braking and powered by 9.9 m^2 of PV panels. Electricity is stored in onboard batteries [21]. In addition to onboard solar panels, there is the possibility to use stationary (off-board) panels to generate electricity specifically for use in transport [75,77,81,82].

A few pilot projects have also been built in the framework of the "Heliotram" project, such as the tram depots in Hannover Leinhausen and Geneva (Bachet-de-Pesay). The 150 kW$_p$ Geneva site injected 600 V DC directly into the tram/trolleybus electricity network provided about 1% of the electricity used by the Geneva transport network at its opening in 1999. On December 16, 2017, a fully solar-powered train was launched in New South Wales, Australia. The train is powered using onboard solar panels and onboard rechargeable batteries. It holds a capacity for 100 seated passengers for a 3 km journey. Recently, Imperial College London and the environmental charity 10:10 have announced the renewable traction power project to investigate using track-side solar panels to power trains. Meanwhile, Indian railways announced their intention to use onboard PV to run air conditioning systems in railway coaches. Also, Indian Railways announced it is to conduct a trial run by the end of May 2016. It hopes that an average of 90,800 L of diesel per train will be saved on an annual basis, which in turn results in reduction of 239 tones of CO_2 [75,77,81,82].

4.9.2.2 Modular Solar Photovoltaic Canopy System for the Development of Rail Vehicle Traction Power

The invention by Dearborn [81] provides a means by which rail (railroad) transportation operators can generate megawatts of carbon-free electrical power from the space over rail tracks and right-of-way without buying or leasing significant amounts of sun accessible space or power transmission right of way. Public rail transportation is becoming increasingly important in the effort to lower carbon emissions from automobile usage. Public rail transportation is migrating from diesel-electric locomotive to electric module unit (EMU) vehicles as they operate more efficiently, create less noise, and are less dependent on carbon-based energy. Electrified rail systems depend on large quantities of electric power, much of which is derived from coal, natural gas, or carbon-based energy sources generating tens of thousands of tons of carbon dioxide and other GHGs.

Periods of hot weather or peak electrical demand strain electrical producing capacities or cause reductions in service. Population growth and increasing urban densification will require more power generation plants, typically dependent on carbon-based sources of energy. U.S. DOE projections state that carbon-free renewable energy generating plants such as wind, solar, nuclear, or other will not come on line fast enough to meet this growing demand.

The U.S. DOE regulates the generation and distribution of electrical power. The U.S. Department of Transportation (DOT) and the Federal Railroad Administration (FRA) regulate rail and public transportation. The current body of regulations and assumptions of these and state-level regulations do not address or expressly permit the concept embodied in the preferred invention. Global warming, the rising cost of carbon-based energy, rapid advances in solar PV technology, and large-scale automated manufacturing are rapidly changing the way we view energy, transportation, and the way we live. Research, analysis, and design work embodied herein relate directly to these recent developments.

A system of solar PV canopy modules positioned over rail track scan provide a method by which rail system operators can become carbon neutral or make some portions of those systems carbon neutral; and in the process, offer rail operators an increased level of control over their carbon footprint and cost of traction power. Just as no one or two of the developments noted above (see background) make the case for favoring this invention; no single or pair of benefits make the case favoring modular solar PV canopy over rail power generation systems. The combination of changing environmental conditions and social values together with various benefits of this invention provide the foundation for its considerations and deployment. Example of one benefit: energy production from an urban or regional solar PV canopy over rail right of way during periods of peak demand coincides with the canopy's periods of peak power output. During periods of peak demand, a carbon-neutral modular solar PV canopy system over railroad tracks or right of way can produce 40–50% more electrical power than required for traction power of said rail vehicles. This surplus energy going into the local grid will reduce the need for carbon-based peak generating plants and the probability of reduced electrical supply service levels.

This invention and the long-term fixed cost of energy (levelized cost of energy (LCOE)) offered by same would allow an electrified rail system to:

   a. become carbon neutral faster than the local power grid,
   b. assist the local power grid supplier in reducing its carbon footprint,
   c. fix its average annual cost of rail traction power for 25–30 years,

d. insulate itself from uncontrolled market-driven pricing variations of carbon-based energy,
e. produce local jobs in design, construction, installation, and maintenance in clean and sustainable technology,
f. and make a greater contribution to a nation's energy independence.

4.9.3 Solar Electric Aircraft

Solar ships can refer to solar-powered airships or hybrid airships. There is considerable military interest in unmanned aerial vehicles (UAVs); solar power would enable these to stay aloft for months, becoming a much cheaper means of doing some tasks done today by satellites. In September 2007, the first successful flight for 48 h under constant power of a UAV was reported. This is likely to be the first commercial use for PVs in flight. Many demonstration solar aircraft have been built, some of the best known by AeroVironment [75,78,83,84].

4.9.3.1 Manned Solar Aircraft

A number of manned solar aircraft have been developed. These are Gossamer Penguin, Solar Challenger, Sunseeker, Solar Impulse, and Solar Stratos [75,78,83,84]. Here we briefly describe the development of solar impulse.

Solar Impulse is a Swiss long-range experimental solar-powered aircraft project, and also the name of the project's two operational aircraft. Bertrand Piccard initiated the Solar Impulse project in November 2003 after undertaking a feasibility study in partnership with the École Polytechnique Fédérale de Lausanne [75,78,83]. The aircraft is a single-seated monoplane powered by PV cells; they are capable of taking off under their own power. The prototype, often referred to as *Solar Impulse 1*, was designed to remain airborne up to 36 h. It conducted its first test flight in December 2009. In July 2010, it flew an entire diurnal solar cycle, including nearly 9 h of night flying, in a 26 h flight. Piccard and Borschberg completed successful solar-powered flights from Switzerland to Spain and then Morocco in 2012, and conducted a multistage flight across the United States in 2013.

A second aircraft, completed in 2014 and named *Solar Impulse 2*, carries more solar cells and more powerful motors, among other improvements. On 9 March 2015, Piccard and Borschberg began to circumnavigate the globe with *Solar Impulse 2*, departing from Abu Dhabi in the United Arab Emirates. The aircraft was scheduled to return to Abu Dhabi in August 2015 after a multi-stage journey around the world. By June 2015, the plane had traversed Asia, and in July 2015, it completed the longest leg of its journey, from Japan to Hawaii. During that leg, the aircraft's batteries sustained thermal damage that took months to repair. *Solar Impulse 2* resumed the circumnavigation in April 2016, when it flew to California. It continued across the United States until it reached New York City in June 2016. Later that month, the aircraft crossed the Atlantic Ocean to reach Spain. It stopped in Egypt before returning to Abu Dhabi on 26 July 2016, more than 16 months after it had left, completing the approximately 42,000 km (26,000-mile) by a piloted fixed-wing aircraft using only solar power. The aircraft's major design constraint is the capacity of the lithium polymer batteries. Over an optimum 24 h cycle, the motors can deliver a combined average of about 8 HP (6 kW), roughly the power used by the Wright brothers' Flyer, the first successful powered aircraft, in 1903. In addition to the charge stored in its batteries, the aircraft uses the potential energy of height gained during the day to power its night flights.

4.9.3.2 Solar Impulse 2 (HB-SIB)

The wingspan of *Solar Impulse 2* is 71.9 m (236 ft), slightly less than that of an Airbus A380, the world's largest passenger airliner, but compared with the 500-ton A380, the carbon-fiber Solar Impulse weighs only about 2.3 tons (5,100 pounds), little more than an average SUV. It features a nonpressurized cockpit 3.8 m^3 (130 ft^3) in size and advanced avionics, including an autopilot to allow for multi-day transcontinental and transoceanic flights (see Figure 4.7). Supplemental oxygen and various other environmental support systems allow the pilot to cruise up to an altitude of 12,000 m (39,000 ft).

4.9.3.3 Unmanned Aerial Solar Vehicles

Besides solar impulse, a number of unmanned solar vehicles were built. Some of these are Pathfinder and Pathfinder-plus, Helios, and Zephyr. The solar cell- and fuel cell-powered Helios set a world record for flight at 96,863 ft (29,524 m). Zephyr, built by Qinetiq, set the unofficial world record for the longest duration unmanned flight at over 82 h on 31 July 2008. Just 15 days after the Solar Impulse flight mentioned above, on 23 July 2010, the Zep claimed the endurance record for a UAV. It flew in the skies of Arizona for over 2 weeks (336 h). It has also soared to over 70,700 ft (21.5 km).

FIGURE 4.7 *Solar Impulse 2* at the Payerne Air Base in November 2014 [83].

4.9.4 Solar-Powered Spacecraft

Solar-powered spacecraft (see Figure 4.8) was described earlier in this chapter. Solar energy is often used to supply power for satellites and spacecraft operating in the inner solar system since it can supply energy for a long time without excess fuel mass. A communications satellite contains multiple radio transmitters which operate continually during its life. It would be uneconomic to operate such a vehicle (which may be on orbit for years) from primary batteries or fuel cells, and refueling in orbit is not practical. Solar power is not generally used to adjust the satellite's position, however, and the useful life of a communications satellite will be limited by the onboard station-keeping fuel supply.

A few spacecraft operating within the orbit of Mars have used solar power as an energy source for their propulsion system. All current solar-powered spacecraft use solar panels in conjunction with electric propulsion, typically ion drives as these give a very high exhaust velocity and reduce the propellant over that of a rocket by more than a factor of 10. Since propellant is usually the biggest mass on many spacecraft, this reduces launch costs. Other proposals for solar spacecraft include solar thermal heating of propellant, typically hydrogen or sometimes water is proposed. An electrodynamic tether can be used to change a satellite's orientation or adjust its orbit. Another concept for solar propulsion in space is the light sail; this doesn't require conversion of light to electrical energy, instead relying directly on the tiny but persistent radiation pressure of light. Perhaps the most successful solar-propelled vehicles have been the "rovers" used to explore surfaces of the Moon and Mars. The 1977 and the 1997 Mars Pathfinder used solar power to propel remote-controlled vehicles. The operating life of these rovers far exceeded the limits of endurance that would have been imposed, had they been operated on conventional fuels [75,78,83].

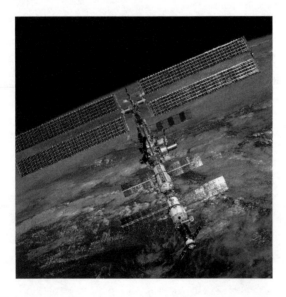

FIGURE 4.8 Photovoltaic (PV) on the International Space Station [75].

4.9.5 Solar-Powered Boats

PlanetSolar, the world's largest solar-powered boat and the first-ever solar EV to circumnavigate the globe (in 2012). Solar-powered boats have mainly been limited to rivers and canals, but in 2007 an experimental 14 m catamaran, the Sun21 sailed the Atlantic from Seville to Miami and from there to New York. It was the first crossing of the Atlantic powered only by solar. Japan's biggest shipping line Nippon Yusen KK and Nippon Oil Corporation said solar panels capable of generating 40 kW of electricity would be placed on top of a 60,213-ton car carrier ship to be used by Toyota Motor Corporation [75,85,86].

In 2010, the Tûranor PlanetSolar, a 30 m long, 15.2 m wide catamaran yacht powered by 470 m^2 of solar panels, was unveiled. It is, so far, the largest solar-powered boat ever built. In 2012, PlanetSolar became the first-ever solar EV to circumnavigate the globe. Various demonstration systems have been made. Curiously, none yet takes advantage of the huge power gain that water cooling would bring. The low power density of current solar panels limits the use of solar-propelled vessels, however, boats that use sails (which do not generate electricity unlike combustion engines) rely on battery power for electrical appliances (such as refrigeration, lighting, and communications). Here, solar panels have become popular for recharging batteries as they do not create noise, do not require fuel, and often can be seamlessly added to existing deck space.

Solar-powered boats (see Figure 4.9) get their energy from the sun. Using electric motors and storage batteries charged by *solar panels* and *PV cells*, solar-powered boats can significantly reduce or eliminate their use of *fossil fuels*. In 2007, five Swiss sailors piloted a solar-powered boat across the Atlantic Ocean. Using solar

FIGURE 4.9 *Solar Sailor* of Australia [86].

power only (via solar panels), the "sun21" made the first motorized crossing of the Atlantic Ocean in order to promote the great potential of renewable energy for ocean navigation and to combat climate change. The "sun21" arrived in New York City on May 8, 2007, having covered 7,000 sea miles.

The "sun21" is a 45.9 ft long specially built solar-powered boat known as a catamaran. On its canopy-like roof are 48 silicon PV cells, which collect energy from sunlight and transmit it to a device in one of the narrow cabins. This device transmits the energy to the 3,600 pounds of storage batteries below the deck. The 11-ton solar boat was powered on the energy needed to light 10,100 W light bulbs. The typical speed was 3.5 knots. The solar boat has two engines that can go up to 107 nautical miles a day in good weather. The "sun21" has sleeping accommodation for six people and has room for large groups for visits or short excursions. The kitchen is in one hull and the bathroom is in the other. A passenger ferry in Sydney Harbor, Australia, called the *Solar Sailor*, can run on wind, sun, battery, or diesel, or in any combination. It uses patented solar wing technology to simultaneously capture sunlight for electricity and for wind power. The wings move automatically via computer, tracking the sun for optimal solar collection and the wind for optimal sail power. If it gets really windy, the wings fold down against the solar boat. Much like a hybrid car, large batteries onboard the solar energy boat store electricity generated by the diesel generators and collected by the solar panels. The electricity then powers the electric motors. Both passengers and the environment benefit from this new design. The solar boat is very quiet; there are no fumes, low vibration, no GHG emissions, and no water pollution. Two of these solar-powered ferries (using the same design as the Sydney Harbor Ferry) are planned for passengers in the San Francisco Bay. They will accommodate 600 passengers.

4.9.6 Modular and Mobile Photovoltaics on Two Wheelers

An inventor from Michigan, USA built a street legal, licensed, insured, solar-charged electric scooter in 2005. It had a top speed controlled at a bit over 30 mph and used fold-out solar panels to charge the batteries while parked. The first solar "cars" were actually tricycles or quadracycles built with bicycle technology. These were called solar mobiles at the first solar race, the Tour de Sol in Switzerland in 1985. With 72 participants, half-used solar power exclusively while the other half used solar-human-powered hybrids. A few true solar bicycles were built, either with a large solar roof, a small rear panel, or a trailer with a solar panel. Later, more practical solar bicycles were built with foldable panels to be set up only during parking. Even later the panels were left at home, feeding into the electric mains, and the bicycles charged from the mains. Today, highly developed electric bicycles are available and these use so little power that it costs little to buy the equivalent amount of solar electricity. The "solar" has evolved from actual hardware to an indirect accounting system. The same system also works for electric motorcycles, which were also first developed for the Tour de Sol [75,77,78,82,87].

Dutch startup Solar Application Lab has designed a prototype solar e-bicycle that integrates multiple solar panels on its front wheel enough to recharge the e-bike's battery over the course of the day. Sporting a 250 W front-wheel-integrated electrical engine, a 400 W, 36 V, 11 A battery and 60 mini solar panels which provide the S-bike with an average of 35–40 Wh during daytime under the fairly clouded Dutch skies, the prototype was proving enough for the startup to secure strong partnerships with well-established bicycle manufacturers (see Figure 4.9). The first prototype was designed based on monocrystalline cell technology but the company is working in close cooperation with the Delft University of Technology to develop a second prototype based on amorphous silicon, hopefully ready within the next 6–12 months.

Although organic flexible PVs are on its radar, the startup sees them as a longer-term prospect, knowing that its energy harvesting architecture is cell-technology agnostic and could accommodate any new PV configuration. To make the most of each individual panel, the S-bike relies on a patented nanoinverter architecture which the company says makes the cluster of cells more efficient and ensures a stable power output, even under partial shading or at varying light angles. Having proven the concept, Solar Application Lab hopes to codevelop a solar e-bike with partners from the cycle industry, to whom it would then license the nanoinverter IP [75,77,78,82,87]. The nanoinverter can work with any type of solar cell in any cluster configuration, on a small scale. The Solar Cycle is a starter, aimed at booming an e-bike market. The lab wants to design an application-specific integrated circuit (ASIC) at the 130 or 90 nm nodes, or maybe two or three product versions to cater for 80%–90% of the market for small-scale modular PVs. For example, derivative products enabled by such an all-silicon nanoinverter could include unique PV tiles designed to be laid out in clusters of any size. In volume production, a chip solution could take the costs down dramatically compared to today's discrete board designs, estimates Peters, it would also help the company diversify its market, say to design readily connectable PV tiles [88,89].

4.10 MODULAR ENERGY STORAGE AND USAGE IN SHIPS

Energy storage is a major green investment for a shipowner. Returns are maximized when the system is correctly dimensioned for the specific ship and includes intelligent power control. Rolls-Royce has been delivering ship energy storage systems since 2010, however, the actual energy storage units (ESUs) were previously supplied by an external party. Rolls-Royce now offers SAVe Energy, a cost-competitive, highly efficient, and liquid-cooled battery system with a modular design that enables the product to scale according to energy and power requirements. SAVe Energy complies with international legislations for low- and zero-emission propulsion systems. The development work has been partly funded by the Norwegian Research Council of Norway's ENERGIX program. The three ship-owning companies Color Line, Norled, and the Norwegian Coastal Administration Shipping Company have been partners in the development, ensuring that the energy storage system covers a wide variety of marine applications, including ferries, cruise vessels, and multipurpose vessels [90–92]. SAVe Energy can be applied to several areas including peak shaving, spinning reserve, and battery-powered vessels. Combined with a liquefied natural gas- (LNG) or diesel-powered engine in a hybrid solution, it increases efficiency and reduces emissions and can be coupled with most types of propulsion units. In a hybrid set up, SAVe Energy handles the peak load, while the main power generators will relate to the average load and not reduce the propulsion units thrusting capabilities.

SAVe Energy is an ESU system and was recently class approved by DNV GL, confirming that SAVe Energy has been developed in compliance with the newest 2018 ruleset and is accepted for installation on all vessels classed by DNV GL [90–92].

The electrification of ships is building momentum. From 2010, Rolls-Royce has delivered battery systems representing about 15 MWh in total. However, due to SAVe Energy, in 2019 alone this number is 10–18 MWh. Battery systems have become a key component of ship power and propulsions systems, and SAVe Energy is being introduced on many of the projects Rolls-Royce is currently working on. This includes the upgrade program for Hurtigruten's cruise ferries, the advanced fishing vessel recently ordered by Prestfjord, and the ongoing retrofits of offshore support vessels [90–92].

4.10.1 GE Modular Naval Vessel Electrification

GE Energy Connections signed a contract with French electrical engineering company Cegelec in Brittany, France to supply a complete static power solution for the French Navy. The solution uses GE's advanced static frequency converters (SFCs) technology to provide high conversion efficiency and a safe and reliable power transfer from the electric power grid to the French naval vessels while in port. In order to keep vessels charged fast and on time, the French Navy requires a reliable and stable supply of electricity. GE's converter technology allows an efficient and secure transition of energy [93]. The public grid in France operates on a frequency of 50 Hz and it must be converted for use of electrical systems on board of French naval vessels, which can function on different frequencies according to the type of ship. GE's power solution has very low harmonics levels, ensuring high power quality. This clean electricity helps avoid disturbance on the transmission line, which could create network faults and energy losses. In fact, the conditions of coupling a vessel to the onshore grid are critical and it is necessary to ensure that during these operations, the power demand does not lead to the destabilization of the grid upstream. This power demand must be gradual, controlled, and synchronized through "intelligent" automation [93].

Modular design allows a smaller footprint and is therefore highly flexible during installation and also allows for future power upgrades. The compact design also means a higher power density of the equipment, ensuring the conversion of electricity in a more efficient, secure, and reliable manner. GE's SFC allows several loads, meaning several ships can be charged at the same time. Each converter can hold an overload of 150% for a short period, leaving enough time for the automatic coupling of a supplementary SFC. This means that when a new ship is charged, the immediately increased load will not disturb other ships that are already charging, allowing the smooth operation without load impact. This project has the possibility of seamless future expansion. One can adapt and optimize the harmonic filter upstream to the converter frequency according to the grid evolution, allowing easy implementation of extra power capacity if needed in the future, all while keeping the benefit of a simple installation. A special storage device and smart control are also being developed to provide reliable energy supply when microcuts occur, ensuring continuation of the power supply. GE's smart management during the events of microcuts is proven to be able to avoid tripping clients' installations several times per year [93].

REFERENCES

1. Oeftering, R.C., Kimnach, G.L., Fincannon, J., Mckissock, B.I., Loyselle, P., and Wong, E. Advanced Modular Power Approach to Affordable, Supportable Space Systems, NASA/TM-2013-217813, AIAA-2013-5253 https://ntrs.nasa.gov/search.jsp?R=201400003332019-07-20T17:10:08+00:00Z (June 2013).
2. Culbert, C. Human space flight architecture team overview. *GER Workshop, NASA Johnson Space Flight Center* http://www.nasa.gov/pdf/603232main_Culbert-HAT%20Overview%20for%20GER%20Workshop.pdf (November 2011).
3. Oeftering, R.C. (2010). The impact of flight hardware scavenging on space logistics. *AIAA Space 2010 Conference*, Anaheim, CA.
4. Shah, Y.T. (2019). *Modular Systems for Energy and Fuel Recovery and Conversion*. CRC Press, New York.
5. International Space Station Familiarization, NASA Mission Operations Directorate. Space Flight Training Division, ISS FAM C TM 21109 (31 July 1998).
6. Oeftering, R.C. and Struk, P.M. (2009). A lunar surface system supportability technology development roadmap. *AIAA-2009-6425, Space 2009 Conference*, Pasadena, CA.
7. NIST Framework and Roadmap for Smart Grid Interoperability Standards Release 1.0. National Institute of Standards and Technology (January 2010).
8. Lyke, J., Fronterhouse, D., Cannon, S., Lanza, D., and Byers W. Space Plug-and Play Avionics. Presented at AIAA 3rd Responsive Space (2005).
9. Wiczer, J. and Lee, K. A Unifying Standard for Interfacing Transducers to Networks—IEEE-1451.0. ISA Expo 2005 Automation+Control, Chicago, IL (2005).
10. Struk, P.M., Sefcik, R., and Anderson, E. Component Level Electronics Assembly Repair (CLEAR) Life Cycle Cost Impact Study. NASA Glenn Research Center (December 2007).
11. Kring, D.A. Lunar Electric Rover (LER) and Crew Activities, Black Point Lava Flow. NASA Directorate Integration Office (DIO) (December 2009).
12. Somervill, K., et al. Constellation Program (CxP) Common Avionics and Software (CAS) Operational Concept Study. NASA Langley Research Center (May 2010).
13. Button, R.M. (11–16 August 1996). An advanced photovoltaic array regulator module. *31st Intersociety Energy Conversion Engineering Conference cosponsored by IEEE*, AIChE, ANS, SAE, AIAA, and ASME Washington, DC.
14. Coyne, J., Jackson, W., and Lewicki, C. (2–5 August 2009). Phoenix electrical power subsystem—Power at the Martian pole. *AIAA-2009-4518, 7th International Energy Conversion Engineering Conference*, Denver, CO.
15. Paterson, J. (14–17 September 2009). Technology development for the Orion program. *AIAA-2009-6449, AIAA SPACE 2009 Conference and Exposition*, Pasadena, CA.
16. Murphy, D., Manson, J., Eskenazi, M., Wilder, N., and Steele, K. (22 September 2011). Advanced Solar Array Technology. *SPRAT XXII*, Cleveland, OH.
17. Eacret, D. and White, S. (4–7 January 2010). ST8 Validation Experiment: Ultraflex-175 Solar Array Technology Advance: Deployment Kinematics and Deployed Dynamics Ground Testing and Model Validation. *AIAA-2010-1497, 48th AIAA Aerospace Sciences Meeting Including the New Horizons Forum and Aerospace Exposition*, Orlando, FL.
18. Murphy, D. (23–26 April 2012). MegaFlex—The scaling potential of UltraFlex technology. *AIAA-2012-1581, 53rd AIAA/ASME/ASCE/AHS/ASC Structures, Structural Dynamics and Materials Conference, 20th AIAA/ASME/AHS Adaptive Structures Conference, 14th AIAA*, Honolulu, Hawaii.
19. Brophy, J., Gershman, R., Strange, N., Landau, D.F., Merrill, R., and Kerslake, T. 300-kW Solar Electric Propulsion System Configuration for Human Exploration of Near-Earth Asteroids. AIAA Paper 2011–5514 (2011).

20. Mercer, C.R., Oleson, S.R., Pencil, E.J., Piszczor, M.F., Mason, L.S., Bury, K.M., Manzella, D.H., Kerslake, T.W., Hojnicki, J.S., and Brophy, J.P. Benefits of Power and Propulsion Technology for a Piloted Electric Vehicle to an Asteroid. NASA/TM-2012-217274, AIAA-2011-7252 (February 2012).
21. Merrill, J.M. Air Force Space Power Technology Development. http://arlevents.com/energy/presentations/8th/Merrill.pdf (2010).
22. JSC-20793, Crewed Space Vehicle Battery Safety Requirements, Rev. B Linden, D. (April 2006). *Handbook of Batteries and Fuel Cells*, 3rd Edition, McGraw-Hill Book Company, New York.
23. Gallup, D. Trends in Lithium-ion Batteries for Aerospace Applications. *NASA Battery Workshop*, Marshall Space Flight Center (November 2007).
24. VES 100 Rechargeable Lithium Battery, (Product Literature) Doc. No. 33017-2-0608, Saft Specialty Battery Group. (June 2008). Available at: http://www.saftbatteries.com.
25. VES 180 Rechargeable Lithium Battery, (Product Literature) Doc. No. 33019-2-0608, Saft Specialty Battery Group. (June 2008). Available at: http://www.saftbatteries.com.
26. Lithium-Ion Batteries and Cells for Space. (Product Literature), Doc No. 1100.33011.2, Saft Specialty Battery Group. Available at: http://www.saftbatteries.com. (2014).
27. Scheidegger, B.T., Burke, K.A., and Jakupca, I.J. Non-Flow-Through Fuel Cell System Test Results and Demonstration on the SCARAB Rover. NASA/TM-2012-217693. (2013).
28. Battery Cost. http://www.greencarreports.com/news/1074183_how-much-and-how-fast-will-electric-car-battery-costs-fall (2016).
29. VW Golf GTE Tested by Auto Plus! http://news.autoplus.fr/news/1491103/Essai-vid%C3%A9o-hybride-rechargeable-Volkswagen-Golf-GTE (2018).
30. MOD-FC—Modular Fuel Cell Stack—PROvendis. A website report https://provendis.info/en/technologies/all/offer/mod-fc-modular-fuel-cell-stack/ (2016).
31. Richards, W.R. Modular Fuel Cell. WO 2004023588 A1 (18 March 2004).
32. Rue, T.J. (18 November 2014). Modular Vehicle Design Concept, M.S. Thesis, Virginia Polytechnic Institute and State University, Blacksburg, VA.
33. San Diego Utility Integrates EVs into California's Wholesale Energy Market. http://chargedevs.com/newswire/san-diego-utility-integrates-evs-into-californias-wholesale-energy-market/. (2015).
34. Powering Electric Vehicles on England's Major Roads. Transport Research Laboratory (TRL), p. 63. http://assets.highways.gov.uk/specialist-information/knowledge-compendium/2014–2015/Feasibility+study+Powering+electric+vehicles+on+Englands+major+roads.pdf. (2009).
35. EP Tender's H2020 Abstract. https://ec.europa.eu/easme/en/sme/5287/innovative-range-extending-service-electric-vehicles-v-based-modular-range-extender. (2016).
36. Nomadic Power's H2020 Abstract. https://ec.europa.eu/easme/en/sme/5519/mobile-energy-system-recharging-energy-buffering-and-long-distance-traveling. (2018).
37. Transport & Environment Letter to the EU Commission. http://www.transportenvironment.org/sites/te/files/publications/2015%2001%2027%20Electrification%20strategy_0.pdf. (2015).
38. Long Distance Trips in France, p. 2. http://www.statistiques.developpement-durable.gouv.fr/fileadmin/documents/Produ its_editoriaux/Publications/Chiffres_et_statistiques/2014/chiffres-stats590-mobilite-longue-distance-des-francais-en-2013-decembre2014.pdf. (2014).
39. Mercedes Interview. http://www.autoexpress.co.uk/mercedes/90698/mercedes-ev-boss-battery-breakthrough-is-close. (2018).
40. Navigant Research. http://www.reuters.com/article/2015/02/19/co-navigant-research-idUSnBw195136a+100+BSW20150219. (2015).

41. EPRI. http://www.epri.com/Press-Releases/Pages/EPRI,-Utilities,-Auto-Manufacturers-to-Create-an-Open-Grid-Integration-Platform.aspx. (2016).
42. Electric Vehicle Owners Could Soon Sell Unused Energy Back to the Grid. http://www.clickgreen.org.uk/news/international-news/125683-electric-vehicle-owners-could-soon-sell-unused-energy-back-to-the-grid.html. (2018).
43. Pearre, N.S., Kempton, W., Guensler, R.L., and Elango, V.V. (2011). Electric vehicles: How much range is required for a day's driving? *Transportation Research Part C*, 19, 1171–1184.
44. EP Tender Short Demo. https://www.youtube.com/watch?v=UnN4khtqa-U. (2017).
45. EP Tender Platform. https://www.youtube.com/watch?v=jnHX_L06Thk. (2014).
46. Nomadic Power Demos. http://nomadicpower.de/?page_id=102. (2019).
47. Gage, T.B. and Bogdanoff, M.A. (1997). Low-emission range extender for electric vehicles. *SAE Transactions*, 106(no. 2), 3319. http://www.tzev.com/files/rxt-g_acp_white_paper_range_extending_trailers.pdf.
48. Better Place's Failure. https://www.wsj.com/articles/SB10001424127887323855804578507263247107312. (2019).
49. Tesla Battery Swapping: Useful Service or Minimal Effort for Extra Income? https://www.greencarreports.com/news/1097214_tesla-battery-swapping-useful-service-or-minimal-effort-for-extra-income. (2019).
50. You're Better Off Buying 100-Mile Electric Cars, Report Says; Shoppers May Disagree. www.greencarreports.com/news/1093941_youre-better-off-buying-100-mile-electric-cars-report-says-shoppers-may-disagree. (2014).
51. EP Tender Market Research, p. 9. https://www.dropbox.com/s/eubmu5r0nfs4yh3/EP%20Tender%20market%20research.pdf?dl=0. (2016).
52. Rover, J.L. Hybrid Electric Vehicle Batteries: Getting the Most Out of Cell Chemistry through Engineering. http://futurepowertrains.co.uk/2015/assets/downloads/presentations/chris-lyness.pdf. (2015).
53. Segard, J.-B. (2–4 December 2015). Manfred Baumgärtner Novel Modular Power Management for Ground Electric Vehicles European Battery. *Hybrid and Fuel Cell Electric Vehicle Congress Brussels*, Belgium.
54. Cauwer, C.D, Mierlo, J.V., and Coosemans, T. (2015). Energy consumption prediction for electric vehicles based on real-world data. *Energies*, 8(8), 8573–8593. doi: 10.3390/en8088573.
55. Rosario, L., Luk, P., Economou, J., and White, B. (2006). A modular power and energy management structure for dual-energy source electric vehicles. *2006 IEEE Vehicle Power and Propulsion Conference, VPPC 2006*, pp. 1–6. doi: 10.1109/VPPC.2006.364291.
56. Rosario, L.C. (2007). Power and Energy Management of Multiple Energy Storage Systems in Electric Vehicles, Ph.D. Thesis, Department of Aerospace Power & Sensors Cranfield University, DCMT Shrivenham Swindon, Wiltshire, SN6 8LA, United Kingdom.
57. Warwick, G. Aviation Week & Space Technology Modular Fuel Cells for Multiple Vehicle Applications GM Eyes Aerospace Applications for Modular Fuel…— Aviation Week. A website report aviationweek.com/technology/gm-eyes-aerospace-applications-modular-fuel-cells (12 October 2017).
58. GM Outlines Possibilities for Flexible, Autonomous Fuel Cell Electric. A website report https://media.gm.com/media/us/en/gm/news.detail…/1006-fuel-cell-platform.html (6 October 2017).
59. Silent Utility Rover Universal Superstructure (SURUS)—Silver Winner. A website report https://drivenxdesign.com/NOW/project.asp?ID=16846 (2017).
60. Ramaswamy, D. and McGee, R. Vehicle Systems Controller with Modular Architecture. US20040044448A1 (4 March 2004).

61. Irish, S. V2G: The Role for EVs in Future Energy Supply and Demand. A website report www.renewableenergyfocus.com›Features (17 March 2017).
62. ZACH MORTICE Integrated Energy Systems: This Building and Car Create a Symbiotic Relationship to Leave the Electric Grid Behind. A website report, ORNL, Oak Ridge, TN (21 January 2016).
63. AMIE Demonstration Project—Oak Ridge National Laboratory. A website report by ORNL, OakRidge, TN https://web.ornl.gov/sci/eere/amie/ (2016).
64. AMIE. A website report by ORNL, Oak Ridge, TN https://web.ornl.gov/sci/eere/amie/media/AMIE-Demonstration-Project.pdf (2016).
65. Scott, P., Chen, Y., Calay, R., and Bhinder, F. (2015). Experimental investigation into a novel modular PEMFC fuel cell stack. *Fuel Cells*, 15(2), 306–321. doi: 10.1002/fuce.201200212.
66. Rajalakshmi, N., Pandiyan, S., and Dhathathreyan, K.S. (2008). Design and development of modular fuel cell stacks for various applications. *International Journal of Hydrogen Energy*, 33(1), 449–454. doi: 10.1016/j.ijhydene.2007.07.069.
67. Horizon 300W PEM Fuel Cell—Fuel Cell Store. A website report www.fuelcellstore.com/horizon-300watt-fuel-cell-h-300 (2016).
68. Horizon 300W PEM Fuel Cell | Fuel Cell Earth. A website report www.fuelcellearth.com›…›FuelCellStacks›Horizon300WPEMFuelCell (2016).
69. Ebaid, M.S.Y. and Mustafa, M.Y. (2011). Design methodology of a proton exchange membrane modular fuel cell of 100 W power output. *Journal of Fuel Cell Science and Technology*, 8(6), 061017–061017. doi: 10.1115/1.4004506.
70. Brodmann, M. and Greda, M. (2010). Modular fuel cell system. In Stolten, D. and Grube, T. (eds) *18th World Hydrogen Energy Conference 2010- WHEC 2010 Parallel Sessions. Fuel Cell Basics/Fuel Infrastructures Proceedings of the WHEC*, 16–21 May 2010, Essen Schriften des Forschungszentrums Jlich/Energy & Environment, Vol. 78-1 Institute of Energy Research - Fuel Cells (IEF-3) Forschungszentrum Jlich GmbH, Zentralbibliothek, Verlag, ISBN 978-3-89336-651-4.
71. Brodmann, M. and Edelmann, A. (12 October 2009). Modular Designed Robust Fuel Cell Stack. *Proceedings: 9th Annual Meeting of the Fuel Cell and Hydrogen Network NRW*, Düsseldorf.
72. Marx, N., Boulon, L., Gustin, F., Hissel, D., and Agbossou, K. (2014). A review of multi-stack and modular fuel cell systems: Interests, application areas and on-going research activities, *International Journal of Hydrogen Energy*, 39(23), 12101–12111. doi: 10.1016/j.ijhydene.2014.05.187.
73. Palma, L. and Enjeti, P. (2009). A modular fuel cell, modular DC-DC converter concept for high performance and enhanced reliability. *IEEE Transactions on Power Electronics*, 24(6), 1437–1443.
74. Solar Car, Wikipedia, The free encyclopedia, last visited 6 July 2019 (2019).
75. Solar Vehicle, Wikipedia, The free encyclopedia, last visited 17 July 2019 (2019).
76. Solar Bus, Wikipedia, The free encyclopedia, last visited 16 July 2019 (2019).
77. Modular and Mobile Photovoltaics on Wheels | SEMI.ORG. A website report prod7.semi.org/en/modular-and-mobile-photovoltaics-wheels (2016).
78. Hybrid Electric Vehicles, Wikipedia, The free encyclopedia, last visited 8 July 2019 (2019).
79. Plug in Hybrid, Wikipedia, The free encyclopedia, last visited 1 July 2019 (2019).
80. Solartaxi. A website report www.solartaxi.com/ (2018).
81. Dearborn, D.D. Modular Solar Photovoltaic Canopy System for Development of Rail Vehicle Traction Power. US20100200041A1 (12 August 2010, 6 February 2019).
82. Modular and Mobile Photovoltaics: On Wheels. A website report www.mwee.com/news/modular-and-mobile-photovoltaics-wheels-0/page/0/1 (21 March 2016).
83. Solar Impluse, Wikipedia, The free encyclopedia, last visited 24 June 2019 (2019).

84. Electric Aircraft, Wikipedia, The free encyclopedia, last visited 12 July 2019 (2019).
85. Solar Boat, Wikipedia, The free encyclopedia, last visited 5 January 2019 (2019).
86. List of Solar Powered Boats, Wikipedia, The free encyclopedia, last visited 10 July 2019 (2019).
87. Electrical Bicycle Wikipedia, The free encyclopedia, last visited 20 July 2019 (2019).
88. Nanoinverter, Wikipedia, The free encyclopedia, last visited 3 July 2018 (2018).
89. Solar Inverter, Wikipedia, The free encyclopedia, last visited 10 July 2019 (2019).
90. Rosario, L.C. and Luk, P.C.K. Implementation of a Modular Power and Energy Management Structure for Battery—Ultracapacitor Powered Electric Vehicles. *IET—The Institution of Engineering and Technology Hybrid Vehicle Conference 2006.* Date Added to IEEE Xplore: 29 January 2007. Print ISBN: 0-8634-17485. INSPEC Accession Number: 9309439. IET, Coventry, UK (12–13 December 2006).
91. Johnson, E. Growing Interest in Energy Storage for Maritime Applications. A website report www.pv-magazine.com/.../growing-interest-in-energy-storage-for-maritime-ap... (3 July 2019).
92. Research Partnership Aims to Enable Zero Emission Ships—Rolls-Royce. A website report www.rolls-royce.com/.../29-11-2017-research-partnership-aims-to-enable-zero-... (29 November 2017).
93. Seth, A. Rolls-Royce, EVP Electrical, Automation and Control—Commercial Marine, Said: GE Modular Naval Vessel Electrification in Brest, France (8 July 2015).

5 Modular Systems for Energy Usage in Computer and Electrical/Electronic Applications

5.1 INTRODUCTION

One of the major uses of energy is in the computer and power industries, both of which use large amounts of energy that is rapidly increasing as more electricity and technology permeates our society. In this chapter, we examine the use of modular systems for data centers, uninterrupted power supply (UPS), supercomputers, power plants, fuel cell for data centers, lighting applications, micro and nanoelectronic devices, and integrated energy systems used in power and computer industries. With the help of numerous concrete examples, the chapter demonstrates the increasing role of modular approach for energy usage in these two industries.

5.2 MODULAR SYSTEMS FOR ENERGY USAGE IN DATA CENTERS

Standardized, preassembled, and integrated data center modules, also referred to in the data center industry as containerized or modular data centers, allow data center designers to shift their thinking from a customized "construction" mentality to a standardized "site integration" mentality. Prefabricated modules are faster to deploy, more predictable, and can be deployed for a similar cost to traditional stick-built data centers. The study by Bouley et al. [1] compares both scenarios, presents the advantages and disadvantages of each, and identifies which environments can best leverage the prefabricated module approach. Data center stakeholders are often faced with the challenges of how to manage the growing demand and the need to expand their data center capacity. Typically, they are faced with the decision to invest in a new facility or to expand or retrofit the existing facility. This usually results in conducting a difficult financial and technical analysis to determine which is the best decision for the business. Today, an alternative option for data center construction exists that can be more cost-effective and less complex in the deployment of prefabricated data center modules for information technology (IT), power, and cooling infrastructure.

Prefabricated modules are pre-engineered, preassembled/integrated, and pretested data center physical infrastructure systems (i.e., racks, chillers, pumps, cooling units, UPS, power distribution units (PDUs), switchgear, transformers, etc.) that are delivered as standardized "plug-in" modules to a data center site. This contrasts with the traditional approach of provisioning physical infrastructure for a data center with unique one-time engineering, as well as all assembly, installation, and integration occurring at the construction site. The benefits of prefabricated modules include more predictable costs and efficiencies, time savings, space savings, simplified planning, improved reliability, improved agility, and a higher level of vendor accountability. Prefabricated modules may take the form of traditional ISO shipping containers or, more likely, the form of a purpose built enclosure. Modules can also be built on a skid. Deployment of prefabricated modules can be 40% faster but has approximately the same capital cost when compared to a traditional build out of the same infrastructure. The cost analysis demonstrates that, while the material cost comes at a premium for prefabricated systems, the installation and space savings offset the premium and overall costs are approximately equal.

When comparing a prefabricated solution to a traditional brick and mortar data center build, the physical infrastructure hardware (racks, PDUs, cable trays, switchgear, UPS, panel boards, heat exchanger, air-cooled chiller, pumps, filters, lighting, security, and fire suppression) are basically the exact same components. Generally speaking, the costs of these infrastructure components are the same regardless of where they are installed. Since the modules include the infrastructure, it is difficult to separate and compare the costs to traditional data center construction. The initial system costs are higher for prefabricated modules because of the cost of additional materials (such as the container shell) and the cost of preassembling/integrating the hardware, software, and controls together. Prefabricated modules are designed in a research and development area, are verified with the customer, and then are released to manufacturing. Once in manufacturing, the design is built in a factory and shipped to the end user. In the traditional approach, multiple parties play a role in developing the design. Numerous meetings are held as electrical contractors, mechanical contractors, designers, end users, facilities departments, IT departments, and executives are all involved. Design points are argued back and forth, politics plays a significant role, and decisions often have to be made serially.

"Design" costs include two types of costs: equipment selection and layout, and site plan design/engineering. For prefabricated modules, equipment selection and layout is already done in the factory (rolled into the system cost) and site plan design/engineering requirements are reduced. In the traditional building approach, site layout and planning is simpler and generally involve five trades—structural engineer, civil engineer, electrical engineer, mechanical engineer, and an architectural review. In traditional data center builds, site plan design/engineering can be 5% of the total project expense. "Installation" costs include all work performed in the field to assemble, integrate, and commission the system for operation. Specifically, this includes the following:

- **Systems project management:** The cost to oversee the project is significantly less for a prefabricated facility based on the decreased complexity of the project and having a single vendor to manage the entire physical infrastructure.
- **Site prep and site project management:** This includes steps like digging trenches for pipes and electrical conduits, grading and laying concrete pads, and other general site expenses. This type of work must happen regardless of the approach taken, and therefore, the cost is approximately the same.

Electrical/Electronic Applications 219

- **Site installation:** Typical infrastructure hardware installation includes the expense of unpacking components, taking inventory, laying out and assembling components, making interconnections between components, and starting the system up. For prefabricated modules, many of these tasks are eliminated (work consists only of placing the modules on cement slabs, wiring the modules up to the existing building switchboards, plumbing for the cooling, and starting up the systems). Another related expense savings is shipping. It is significantly less expensive to ship a preassembled module compared to shipping individual parts and pieces of a traditional field-assembled system. Simpler, consolidated shipping also results in less shipping damage, which can be an added expense and an unwelcome time delay.
- **Management/controls installation and programming:** In a traditional data center, installation and programming of the management software and controls system can be a significant expense ($0.30/W or more). This includes the cost to integrate the management system dashboard/interface with the power and cooling infrastructure and to tune the controls of the system to achieve desired performance (i.e., controlling cooling set points for optimal energy economizer mode hours and energy consumption). For many custom data centers, this is an end goal that is never achieved because of the complexities in controlling the system. For prefabricated facilities, this expense is brought primarily into the factory, where programming and optimization of software and controls are standardized, nearly eliminating on-site work and improving the operating performance of the data center.
- **Commissioning:** Commissioning involves documenting and validating the results of the data center's design/build process. The detailed steps of commissioning vary from data center to data center, but often includes steps like factory witness testing, quality assurance and quality control, start-up, functional testing, and integrated systems testing. For prefabricated facilities, steps like factory witness testing and quality assurance are generally brought into the factory, which reduces on-site time and expense.
- **Space cost:** Another key cost benefit of the prefabricated module approach is the reduction or elimination of on-site "brick and mortar" facility construction to house the physical infrastructure. This can be costly (on the order of $1,076 for normal facility operations).

With prefabricated modules, field installation is significantly reduced as the construction work is less invasive and less complex, and no core and shell needs to be built or interior space retrofitted. A prefabricated module is placed on-site on a cement pad with a crane. Once in place, electrical power is connected to the main switchgear, to the cooling facility module, and to the IT space, and chilled water piping is connected to the air handlers in the IT space. In a scenario where facility power and cooling modules support the data center, much of the traditional up-front design and construction management burden shifts from the data center owner/end user to the solution provider (note: a design-build firm could perform all of these tasks in the traditional approach). The manufacturer designs and then "stamps and repeats" data center power and cooling modules for multiple customers. The data center power and cooling physical infrastructure becomes part of the manufacturing supply chain instead of an on-site custom build. This has a significant impact on the installation expense.

In a traditional approach, the owner/end user is responsible for either developing the design, assembling the components of the solution, engaging the various vendors for equipment acquisition, or for hiring and managing contractors to perform this work. In contrast, since prefabricated modules are prebuilt in the factory, the owner/end user avoids time-consuming tasks (no need to chase down the individual pieces of equipment needed, one or few delivery schedules to manage, very few, if any, construction contractors to interface with). The above discussion focused on capital cost, but there are also operating expense considerations when comparing prefabricated modules to traditional power and cooling.

5.2.1 Maintenance Costs

The potential to reduce prefabricated module maintenance costs exists. Even though maintenance must now be done in a tighter space, the end user would save by contracting for "one stop shop" module maintenance. Rather than having to write up an assortment of terms and conditions with different vendors, only one contract could be drawn up to support one or two "big box" prefabricated modules. In such a scenario, one organization would be held accountable for the proper functioning of the module. This is a simplified approach as the data center owner no longer has to preoccupy himself with trying to track down which organization is responsible for resolving a mishap. In a traditional data center, many of the parts and pieces (plumbing, electrical, power system, cooling system, and racks) are supported by different suppliers and finger pointing is a common occurrence. Cost savings could also extend to software/management upgrades. Instead of custom written code for a large assortment or products, prefabricated data center modules could make available to the customer one set of standard firmware upgrades.

5.2.2 Energy Costs

Traditional mechanical and electrical rooms consume more energy than comparable power and cooling prefabricated modules. Energy savings exists primarily because the pre-engineered design of the modules allows for better integration of power and cooling system controls (this advantage is especially pronounced when it comes to the coordination of the cooling system controls).

Consider the example of controls for a chiller plant. The programming required to properly coordinate chillers, cooling towers, pumps, and valves, for example, is extensive. Adding economizer modes increases the complexity. In fact, often times, economizer modes are disabled in designs because of this complexity, which results in additional energy expense. The American Society of Heating, Refrigerating and Air-conditioning Engineers (ASHRAE) publishes standards for Coefficient of Performance (COP) for chiller plants. A higher COP indicates a better overall system performance. Although the individual parts that make up the chiller plant may achieve the published standards, most chiller plants achieve a much lower COP. This is a symptom of problems encountered when attempting to integrate the controls of the various components involved. The ineffectiveness of custom designed/integrated controls implemented in the field often means significantly less time operating in economizer mode and higher energy consumption overall. The complexity of the controls makes it difficult to predict power usage effectiveness (PUE) within a traditional setting. The PUE of a prefabricated data center is predictable, however, because the equipment has been extensively pretested using standard components and the controls have been coordinated ahead of time. Consider the PUE of a traditional 1 MW data center located in St. Louis, Missouri, USA at 50% load with an average density of 6 kW per rack, raised floor, chillers, variable frequency drives (VFD), water control, and economizers. In such a data center, a PUE of about 1.75 would be typical. Comparable prefabricated module configurations have been tested and analyzed and a measured PUE of 1.4 or better is expected. That difference translates into an electricity bill reduction of 20%.

Data center owners who choose to install prefabricated modules should check with local authorities because permitting processes vary greatly in different geographies. With high up-front capital cost and extensive field assembly, installation, and integration, a data center can be designed, delivered, installed, and operational within 8 months or less. A data center can be built out in large kilowatts building blocks of premanufactured power and cooling capacity. Regulatory approvals on an ad-hoc basis for the various steps of the infrastructure layout often results in delays that impact the initiation of downstream construction.

From a physical infrastructure perspective, a retrofit can be more complex and more invasive than building a new data center. Infrastructure components need to be installed and started up individually and then commissioned. Specialized equipment (such as a crane) is needed to maneuver prefabricated modules. Since much of the hardware installation is done in the factory, on-site work is dramatically simplified. The solution is assembled on site from various parts and pieces provided by multiple vendors. This increases the need for coordination and creates more chances for human error. Modular construction has more predictable performance because components are prewired and are factory acceptance tested before shipping. Smaller modules reduce the risks of human error, since the entire data center doesn't go down during a failure. Existing structures often limit the electrical efficiencies that can be achieved through optimized power and cooling distribution; complex custom configured controls often result in suboptimal cooling operation, reducing efficiency. Prefabricated modules can utilize standard modular internal components and can be specified to a target PUE. Constraints of a building's infrastructure can impact the PUE resulting in suboptimal energy consumption. Steel and aluminum produce about half the carbon emissions of concrete. In modular construction, concrete is only used to pour a support pad. Significantly less concrete is needed for prefabricated modules as opposed to a comparable building shell data center. Optimized and tested controls ensure a predictable PUE, thereby reducing energy consumption. Servicing is more limited because of space constraints. Traditional data centers have more room for service people to maneuver. Beyond the cost advantages of prefabricated modules, data center owners have additional reasons for pursuing the prefabricated approach:

- **Predictable efficiency:** The module approach allows the consumer to specify and the manufacturer to publish expected efficiencies based on real measurements of the design. This predictability is attractive for businesses with a focus on energy efficiency initiatives. A frequent challenge with new data center construction is managing unforeseen site issues and changes that have negative results in achieving the expected efficiency as designed. Prefabricated modules are much more likely to achieve efficiency targets because the solution is not subject to as many on-site construction issues.
- **Predictable cost:** The more complex a construction project, the higher the likelihood of change orders and cost overruns. Since the majority of the data center infrastructure is integrated within a controlled factory environment, there is significantly less chance for changes and delays on site.
- **Portability:** If portability represents a high value, then the prefabricated modules may make some sense. Consider the example of a business that needs to deploy data center power and cooling but whose lease runs out in 18 months. If their lease is not renewed, they can physically move their data center physical infrastructure (power and cooling) with them instead of leaving it behind.
- **Financial flexibility:** From an accounting standpoint, prefabricated modules could be classified as "equipment" as opposed to being designated as a "building." This would likely offer tax, insurance, and financing benefits. Obviously, tax law, insurance policies, and purchase/leasing contracts vary from place to place and from region to region. So, this potentially substantial benefit should first be verified before assuming it exists for your given circumstances.
- **Hedge against uncertainty:** Prefabricated modules are a viable option if a high degree of uncertainty exists regarding future growth. The flexibility of scaling and rightsizing helps minimize risk.
- **Reclamation of valuable space:** In many cases, the physical space that a data center occupies may be better suited to support an organization's core function. Universities, hospitals, call centers, and factories are primary examples where a data center expansion can threaten to consume value building space and create challenges to justify financially. Moving the data center outside the building or to a less important space can be a financial win-win for the organization's CIO and CFO
- **Speed of deployment:** Traditional data centers can take up to 2 years, from concept to commissioning, depending on the size and complexity of the data center. Speed of implementation is oftentimes critical to a business. Cost of time is important to organizations that place a high value on early delivery (e.g. companies who want to be first to market with new products). Data centers built with prefabricated modules can be deployed (concept to commissioning) up to 40% faster. Utilizing subsystems that have already been designed and integrated together can save 50% of the design/engineering time because the prefabricated module vendor absorbs the design process

for the client, using repeatable processes. In addition, with prefabrication, more tasks can be done in parallel saving significant time. For instance, the site prep work can be done at the same time the modules are being built in the factory. Testing, installation, and commissioning in the field is also completed faster because modules are pretested in the factory and delivered as a subsystem instead of individual components.

Simplified training: The prefabricated module approach allows the training of the staff to be greatly simplified since the modules are standardized with a system-level interface. This also means there is less risk to the data center operation when transitions in staff occur.

High standardization, higher quality: High standardization of components and processes enables mass production; this means lower costs, higher quality, easier repair, and shorter lead time. The high degree of standardization in modular data centers is one of their most important advantages: It is about 90%, making it the most important reason for short deployment times, cost efficiency, and scalability. When it comes to costs, there are three primary factors that lead to savings compared to conventional data centers. The first is the efficient use of space and the associated reduction in construction costs, the second is reliable forecasting of module costs, and the third is high energy efficiency.

Data centers built using prefabricated and standardized components already have significantly cheaper construction costs: The costs of design, management, and turnkey construction are 30% below those for conventional data centers. Pure construction time is usually just a few weeks. Previous estimates from construction to start-up were usually around 400 days; for modular data centers, this period is just a few months. For modular UPS (UPS for data centers) systems, one can use individual power modules to increase or reduce power capacity, if necessary, without affecting $N + 1$ redundancy or available backup time. If the cooling solution is part of the "modular design," this can also be adapted with a scalable approach. The main advantage is the ability to implement just as much infrastructure as is required, while maintaining the option to flexibly adapt the system in all areas—power, cooling, and management. There are now many options for implementing a modular data center concept. These range from individual rack module solutions, to containers as entire data center units, to energy cells. Procurement of modular UPS is often the first step as this facilitates and simplifies conversion of the data center. In all these approaches, the modular design ensures predictable construction, installation, service, and maintenance cost.

Mass production of the modules and standardized connection technology are the main reasons for shorter construction times and faster start-up: The modules are tested in advance, thereby drastically reducing the time for configuration and connection on site. Modular data centers also provide configuration solutions such as N, $N + 1$, and $2N$, and they support power reserves according to customer requirements. The way in which cables are laid and managed also has a direct impact on the setup and stability of the entire IT system and the costs. In the modular data center, all data and power cables are standardized and preconfigured so that they can be laid, managed, and repaired easily and cost-effectively. In addition, due to the modular, standardized, and highly integrated design, modular data centers are even more stable, especially if the modules come from the same provider and are configured and tested before start-up. All modules such as power supply, cooling, racks, cabling, and a monitoring system for energy and ventilation management are designed to interact perfectly. The modular data center also allows the following:

- **High expandability, greater flexibility:** The modular structure allows for easy expansion and adjustment of current IT requirements. The installation, updating, and reconfiguration of independent components and interfaces makes expansions much cheaper.
- **Shorter mean time to repair (MTTR):** Modularity, flexibility, and plug-in connection feat mean that many operations—e.g. precabling and distributor equipment before delivery or repairs of standard modules—can be performed by the manufacturer with optimum quality and cost-effectiveness. A module configured or repaired in the plant is very unlikely to be the origin of any faults.

While modular data center offers many benefits as mentioned above, prefabricated modules can also present the following challenges:

- **Existing investment in a building:** When evaluating prefabrication for the IT space, it is important to note the existence or planned investment in a traditional building, as it will have a strong impact on the cost analysis. Since a prefabricated module already has a weather-proof shell, the added cost of locating them in a "finished" space has minimal cost benefit.
- **Distance between the prefabricated modules and the internal data center:** In cases where outdoor facility power and cooling modules supply an indoor IT space, distance is an important factor. If the indoor IT space is located next to an outdoor perimeter wall or a roof, expense to connect the data center to the prefabricated modules is minimized. However, if the data center is located deep within the building, the cost of running cable and piping (breaking through multiple walls, floors, and/or ceilings) could quickly become prohibitive.
- **Physical risks:** Prefabricated modules can be exposed to outside elements such as severe weather, malicious intent, vehicle traffic (if placed in parking lot), and animal/insect infestation. Risks for a particular site should be assessed before choosing to deploy prefabricated modules. In these cases, similar considerations must also be made for other support systems such as generators, chillers, or condensers.
- **Arrangements for power provisioning and network connectivity:** When prefabricated modules are installed, arrangements for additional power distribution (additional breakers/switchgear) and fiber connections need to be established.
- **Restrictive form factor:** Prefabricated modules are large and heavy, and although mobile, they do present some challenges when it comes to

Electrical/Electronic Applications

placement and relocation. Modules may be too heavy to place on the roof of a building or may present logistical challenges in tight city streets.

Local code compliance: Since prefabricated modules present a new technology, local municipalities may not yet have established guidelines for restrictions on modules. Inconsistencies could exist regarding how different municipalities classify power, cooling, and IT modules. Local codes impact the level of module engineering and customization required to secure Authority Having Jurisdiction (AHJ) approvals.

Transportation: The Transportation Security Administration (TSA) stipulates width (11.6 ft, 3.5 m) and length limitations in the United States for truck and train loads to pass over curved roads, under bridges, and through tunnels. Outside of North America, roads can be even smaller, further restricting the mobility of containers. Non-standard wide loads require special permits and in some cases escorts, which increases the cost of transporting the prefabricated modules.

Schneider Electric [2,3] classifies prefabricated modules based on the following three attributes that together define the majority of prefabricated modular data centers:

- Functional block
- Form factor
- Configuration

Functional block: The functions of a data center can be broken down into three major categories—the power plant, the cooling plant, and the IT space. Prefabricated data center modules sometimes provide multiple functions (referred to as all-in-one configurations), but often they provide one function of the data center

Form factor: This refers to the type of structure, size, and shape. The form of a particular solution impacts its transportability, placement, and location (inside vs. outside, on the ground vs. on a rooftop). The three general forms of prefabricated data center modules are ISO container, enclosure, and skid-mounted.

Configuration: There are several ways the functional blocks can be implemented in a data center. These approaches fall under three main categories:
- **Semi-prefabricated:** A data center comprising a combination of prefabricated functional blocks and traditional "stick built" systems.
- **Fully prefabricated:** A data center comprising prefabricated IT, power, and cooling modules.
- **All-in-one:** A data center that is self-contained in a single enclosure, with IT, power, and cooling systems.

The prefabricated modular data center is most useful under the following conditions:

1. **Colocation facilities seeking faster, cheaper ways to "step and repeat" computer power and support systems for their customers:** Prefabricated modules provide clues with a solution to cost-effectively upsize and

downsize in large kilowatts modular building blocks when demand for their services fluctuates as a result of market conditions.
2. **Data centers that are out of power and cooling capacity or out of physical space:** Prefabricated modules can quickly add cooling and power capacity so that additional servers can be placed into existing racks, creating a higher density per rack, which can now be handled by the supplemental power and cooling.
3. **New facilities with tight time constraints:** Time is important to organizations that place a high value on early delivery (e.g., companies who want to be first to market with new products).
4. **Data center operators in leased facilities:** If a business has a lease, they may not want to pour money into a fixed asset that they would have to leave behind. If their lease is not renewed, modules can physically move with them.
5. **IT departments whose staff is willing to manage power and cooling:** Not relying on the stretched resources of corporate facility departments can leverage prefabricated facility modules to control their own chilled water supply.
6. **Data center facilities saddled with existing infrastructure characterized by poor PUE:** These facilities may only be marginally improved within the constraints of their existing physical plant. Adding prefabricated modules provides an alternative to help solve problems inherent to the inefficient data center design they may have inherited.
7. **An organization with vacant space:** An empty warehouse space can populate the space with a series of prepackaged modules. They leverage utilization of the space and avoid the delays and construction costs of building a new brick and mortar wing.
8. **An organization with limited space:** Trying to financially justify constructing a new building or expanding an existing space can be difficult for a data center operator. This is even more challenging in organizations where the building space can be used to generate revenue (hospital, university, or factory). For these organizations, a prefabricated data center offers an alternative to traditional construction. The ideal applications for prefabricated modules are:
   1. A new data center seeking faster ways to "step and repeat" computer power and support systems (especially when load growth is uncertain).
   2. An organization with vacant space (i.e., warehouse space) that can be leveraged for a more quickly deployed new data center without the expense of brick and mortar construction.
   3. Existing data centers that are constrained by space and power/cooling capacity.

In summary, modular data centers are faster, more flexible, and more energy-efficient. Cloud computing, big data, and the Internet of Things (IoT) require data centers with significantly higher performance and flexibility. The requirements range from simple and short planning phases, to rapid deployment of applications and seamless scaling during ongoing operations, to greater energy efficiency and cost effectiveness.

Electrical/Electronic Applications　　227

One approach for managing increasing complexity and greater requirements is the building of modular data centers, as shown by Dell [4–6]. In practice, this means cooling, power supply, and energy management systems are divided into individual components and modules. Any subarea of the data center is designed according to a uniform standard based on its size, workload, and configuration. This ensures fault-free, independent operation without shared resources. Some typical illustrations of modular data centers are illustrated in Figures 5.1 and 5.2.

5.2.3 Modular Data Center from Delta

Delta's modular data center [8,9] is a smart and flexible architecture suitable for small and medium-enterprise applications. In addition to the integration of main systems of power, cooling, racks, and monitoring, the cold or hot aisle containment also

FIGURE 5.1　A 40-ft portable modular data center [7].

FIGURE 5.2　A modular data center (Baselayer Edge) connected to the power grid at a utility substation [7].

provides superior air flow management to enhance cooling efficiency. A traditional data center takes 18–24 months to build, whereas a modular data center requires only a few weeks or months for deployment. Delta's modular data center, with its fully modular design concept, is flexible and offers pay-as-you-grow with short installation time to meet the rapidly growing needs of data storage and processing. The modular data center from Delta offers the following advantages [8,9]:

1. Modular data centers use an overhead wiring system, and the internal cabling rack is attached to the top of the modular racks. High- and low-voltage electricity are separated.
2. The power distribution cabinets provide the power supply for the network cabinets.
3. The battery/rechargeable power pack can be installed in the battery compartment.
4. Modular data centers are monitored with or without the color touch panel that is assigned to each module. An information upload is run via a touch panel for the former or via Data Acquisition Mainframe (DAM) for the latter.
5. Modular data centers are equipped with class B and C lightning protection systems.
6. The touch panel (optional) displays information about the power distribution cabinets, precision cooling, environment in the racks, water leaks, temperature, humidity, smoke detection, door status, and much more.
7. The UPS and the precision cooling have the same shape, size, and color. All cabinets have a highly integrated structure.
8. All UPS, precision cooling, and monitoring systems are developed by Delta. The core technologies and expertise are from Delta's research and development center.
9. The power supply systems, power distribution, and batteries for small and medium-sized modular data centers are integrated within a single cabinet (Module on DPH 75 kW series UPS) to reserve more valuable space for major IT equipment.

However, the growing computing power per rack requires special cooling and power supply systems. This requires a combination of cold/hot aisle concepts combined with water cooling and modular PDUs in the rack. This makes it possible to reduce the energy requirements of the systems by about one-third per rack. The new modular structures offer the greatest possible flexibility to meet future challenges and changes.

5.2.4 Efficiency Improvement by Modular Data Center at Sun's Computer Operation of Oracle

When Sun's Global Lab & Datacenter Design Services (GDS) organization was created by Oracle, the objective was to reduce Sun's data center space and power requirements. To achieve this objective, Sun designed a modular set of components that could be used and reused throughout Sun's technical infrastructure spread across 1,533 rooms worldwide. The path of least resistance would be to create a customized data center design for each location at the lowest cost according to its business needs. This would have involved making decisions on power, cooling, racks, and network connectivity for each location individually. While this approach may provide short-term solutions that would satisfy customer needs, in a world where the most important requirement is to enable agility with minimum economic and environmental impacts, such approach may limit the ability to capitalize on new business opportunities and innovations in technology [10–12]. In the end, GDS chose a modular, pod-based design that created energy-efficient building blocks that could be duplicated easily in any size or tier-level data center worldwide. A pod is typically a collection of up to 24 racks with a common hot or cold aisle along with a modular set of power, cooling, and cabling components. With a pod representing a small modular building block, Sun could build large data centers by implementing a number of pods. Likewise, a smaller data center can be implemented with a single pod, allowing data centers of all sizes to benefit from the modular design.

The pod design allows the different modules to be configured to best meet the client requirements as the data centers change over time. It also allows easy insertion of new technologies in the existing data centers. The standard infrastructure is sized so that it can accommodate rapid change, growth, and increase in server and storage densities. The Sun invested 10%–15% in piping and electrical infrastructure to "future proof" the data center. This allowed each data center to support today's requirements and adopt rapidly increasing server densities and changing business requirements without significant up-front costs. In some cases, this small investment can have a payback in as little as 18 months, even if the anticipated changes never occur [10–12]. Sun sees a role for energy-efficient designs in both buildings and container-based deployments. For example in Bay area, Sun used office space in Santa Clara as data center due to a strategic business deal. However, in many other locations, container-based deployment was used. Sun found that Project BlackBox provides provide the same benefits as building-based designs. Project BlackBox is a highly optimized, pod-based design within an enhanced shipping container, and thus can be deployed indoors or outdoors as an independent data center. For many of the data centers GDS designed, the building approach was the best choice given the "high touch" requirement of the engineering organization and the need to optimize Sun's real estate portfolio. For data centers that are running production workloads that require less human interactions and can be operated remotely, or are needed for temporary or transition missions that need to be deployed for months rather than years, Project BlackBox plays a big role. Sun's long-term strategy includes deploying data centers that use a mixture of both building and container-based components [10–12].

The modular nature of the pod approach allows the elimination of most of the custom design and re-engineering for each site, reducing costs and shrinking project times. This standardized approach permits the recognition and correction of mistakes as they are discovered. It also allows easy pod replacement as new technology emerges. This helps to avoid the pitfalls of "doing it wrong," which includes the loss of flexibility, scalability, and ultimately shortened lifespan for a facility.

The additional investment in future proofing accommodates today's requirements while making it easy to scale up or down as server densities increase and as business requirements change. This future-proof model allows Sun to easily add capacity without substantial capital outlay or major downtime. Modularity directly contributes to energy efficiency as each pod's power and cooling can be closely matched to the requirements of the equipment it hosts. It is no longer necessary to increase an entire room's cooling just to accommodate a small section of the equipment. Heat removal is now focused at the source and capacity is increased in smaller incremental units within the data center space. This requires only the energy needed to power and cool computer equipment today while enabling rapid and cost-effective expansion [10–12].

In the end, these energy-efficient, pod-based data centers may look different on the surface, but underneath they are all built of the same modular components. Even between building and container-based data centers, the underlying components that contribute to efficiency are surprisingly similar. With a uniformly flexible and scalable design, these data centers are ideally positioned to meet the needs of Sun's core business. This highly flexible and scalable pod architecture facilitates the quick and effective adaptation to ever-changing business demands. This is a competitive advantage for Sun, enabling products and services to be delivered faster and in a more energy-efficient way. GDS has built and used data centers very effectively using this modular approach all over the world, including their Santa Clara Campus, Guillemont Park Campus in United Kingdom, Prague Campus in Czech Republic, Bengaluru Campus in India, and Louisville in Colorado, project BlackBox [10–12].

5.3 MODULAR UPS SYSTEMS

Modular UPS systems are not only scalable but energy-efficient with lower power costs and reliable energy supply. The power costs in the data center make up the largest component of the total cost of ownership (TCO). The PUE (the total power consumption of a data center divided by the IT power consumption) is usually high in conventional data centers. It is often at around 2.0 or higher. This means: only half of the energy is spent on the IT workload. The other half is consumed by the physical and critical infrastructure such as power supply, cooling, and lighting [13–15]. As shown earlier, modular data centers are much more effective: they allow optimal coordination between the capacity and workload of the energy and cooling systems to improve efficiency and avoid excessive configuration. With modular UPS, for example, it is easy to simply increase the number of power supply modules when you need to expand the data center. This means: further growth makes a "seamless expansion" of the UPS easy to implement. To save energy, the efficiency of the components must be increased and the workload must be reduced. Modern UPS can be used to significantly reduce power costs compared to conventional UPS. Data centers are usually installed with $N + X$ redundant power supplies or even $2N$ configurations (dual BUS) to ensure reliable operation. This means that the percentage of the workload is usually only around 30%–40%, or even less. An **electrical bus** bar is defined as a conductor or a group of conductor used for collecting **electric power** from the incoming feeders and distributes them to the outgoing feeders.

Electrical/Electronic Applications 231

According to a report in 2013 from the Gartner market research company, in addition to UPS efficiency at full workload, it is particularly important to also consider the efficiency curve at a workload of between 20% and 100% to identify the ideal state of "high efficiency at low workload." If one takes into account the workload differences in the data center for day and night, then a highly efficient modular UPS—for example, with a capacity of 200 kVA—consumes about 5% less power than a conventional UPS.

5.3.1 Intelligent Cooling Technology

Modular data centers use separate hot and cold aisles and use cooling units in series to ensure the cooling of compact integrated racks and reduce cooling energy losses. The density of individual racks can be increased by more than 20 kW. If one attaches blanking plates at rack locations where no servers are installed, 95% of the cold air passes directly via the cold aisles in the server racks. The hot air emitted from all racks is then fed back through the cooling unit and does not flow to the front of the rack. A containment system for the cold or hot aisle in the modular data center increases utilization and reduces consumption. This "in-row" cooling method improves cooling performance because the cold air is guided directly to the hot points. Furthermore, the cold aisle containment is isolated to prevent the mixing of cold and hot air and to prevent the formation of hot islands. Cooling performance in modular data centers is 12% greater compared to conventional data centers. When chilled water-type cooling and free cooling are combined, the PUE can be reduced to <1.5. Finally, intelligent management systems help reduce energy costs. For example, big data analyses can be used to develop energy savings plans and continuously adapt them dynamically [13–15].

5.3.2 Higher Power Density in the Rack

As the central component of the data center's IT service, the rack consists of the physical structure, support structure, installation and compatibility of the IT equipment, thermal management (air inlet, air outlet, air flow management), power distribution (dual power supply, rack-mountable distribution cabinet), cable management (power and data cables), and much more.

Due to the standardization of these components, it is possible to increase the packing density of the racks in modular data centers—they need about 35% less energy, and they deliver the same performance over a smaller area. This means that, in a few years, a modular data center covering an area of around 800 m² and with a heat output of 20 kW per rack will achieve the same performance values as a 3,000 m² data center today. The savings—based on construction costs—are between 800,000 and 2.5 million dollars [13–15].

5.3.3 ABB's Modular UPS Systems

The effect of a power failure in a data center can be disastrous. So great care is taken to make sure that the very best back-up power scheme is in place. A reliable and efficient UPS is a mainstay of such a scheme. As shown in Figure 5.3, in principle,

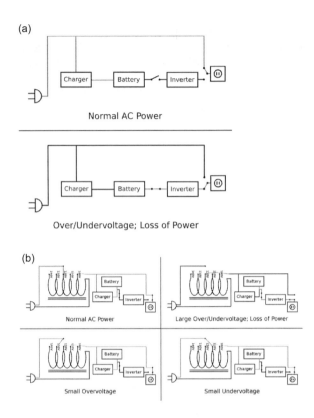

FIGURE 5.3 Two types of UPS systems [16]. (a) Offline/standby UPS: The dotted line illustrates the flow of electric power. Typical protection time: 0–20 min. Capacity expansion: Usually not available. (b) Line-interactive UPS: The dotted line illustrates the flow of electric power. Typical protection time: 5–30 min. Capacity expansion: several hours.

there are two types of UPS systems. Once the UPS is installed, however, it becomes a focus of reliability. ABB, as one of the leading suppliers of UPS, has invested much effort in developing and refining a UPS design that optimizes the availability and total cost of ownership. In ABB's range of modular UPS, each UPS module has all the hardware and software needed for autonomous operation—rectifier, inverter, battery converter, static bypass switch, back-feed protection, control logic, display, and mimic diagram for monitoring and control. With all the critical components distributed between individual units, potential single points of failure are eliminated.

If redundancy is provided for, there are more modules than needed to supply the critical load. In a redundant system, all modules are active and share the load equally. Should one module fail, the remaining modules take over the load smoothly. The system is fault-tolerant and there are no single points of failure. ABB calls this modular approach decentralized parallel architecture (DPA). DPA not only provides the best availability but also the best serviceability, scalability, and flexibility. Taken together, these features all deliver a low TCO. Modular UPS has no common system components. In centralized systems, failure of one point can bring down the entire system; in DPA, each UPS module has all the hardware and software it needs for autonomous operation—there are no shared critical elements and no single points of failure. For all applications, availability is the most crucial UPS parameter. It is a measure of how much time per year a system is up and available. UPS power availability is measured by mean time between failures (MTBF) and MTTR. The best ways to increase power availability are to increase the MTBF and decrease the MTTR of the power protection system. The nature of the modular design lends itself very well to achieving this objective [13–15]. The surest way to increase availability of power is to add redundancy to the UPS system and to minimize its maintenance and repair time. One major advantage of modularity is the ease with which redundancy can be accommodated. Usually, adding redundancy merely involves configuring one UPS module more than is necessary to cover the basic load. This is then switched in automatically when required [14,15]. The availability of modular UPS can be increased by:

- Adding redundancy
- Minimizing the chance for human error
- Selecting high-quality, reliable equipment and minimizing downtime
- Standardizing the service concept

Modular UPS is more reliable because of its ability to run without failures. Its mean time between failures (MTBF is very high. Modular UPS can be maintained and serviced faster and more easily. The best ways to maximize power availability are to increase MTBF and decrease MTTR of the power protection system. The surest way to increase power availability is to add redundancy to the UPS system and minimize its maintenance and repair time. One major advantage of modularity is the ease with which redundancy can be accommodated. Usually, adding redundancy merely involves configuring one UPS module more than is necessary to cover the basic load. This is then switched in automatically when required.

ABB's UPS modules can be online-swapped, i.e., removed or inserted, without risk to the critical load and without the need to power down or transfer to the mains. This procedure is simple and quick to perform and introduces no risk for system operation. Each module can be individually switched off before removing it from the system. This makes the service safe for the technician and ensures absolutely no disturbance to the system. As the same modular UPS can be used across different applications and load segments, the service technicians do not need to be educated on several different platforms, but can apply the same practices and procedures on all UPS equipment [14,15].

In ABB's DPA UPSs, the incoming AC is first converted to DC. The output AC is then synthesized from this DC, giving a clean sinusoid. These two conversion steps give the term "double conversion" and isolate the output voltage waveform from any disturbances on the input AC side. With over 20 years of experience in modular UPS, ABB's Swiss-made DPA delivers unrivaled UPS availability, as well as the serviceability, scalability, flexibility, and low energy usage made possible by the modular DPA approach deliver a very attractive TCO. There are no better UPS architectures available to those users whose critical electrical loads represent a valuable commercial asset that must be kept powered at all costs [14,15]. Because the UPS modules in a DPA are independent, they can be online-swapped without risk to the critical load and without the need to power down or transfer to raw mains supply. So engineers can work on the UPS without interrupting operations. Swap-out time is only 15–20 min and is very safe, and you never have to switch off your load. Online-swappable modules directly address availability requirements, significantly reduce MTTR, reduce inventory levels of specialist spare parts, and simplify system upgrades.

DPA modules are standardized. This keeps costs low: a straightforward, standardized modular concept simplifies and speeds up every step of the deployment process—from planning, through installation and commissioning, to final use. High-quality, standardized products significantly reduce intervention time during maintenance or in the event of failure—components can be changed quickly and easily and service is simplified. The better quality that results from the mass production and testing of standardized modules has a direct positive impact on reliability and, thus availability: modular systems with standardized connections can be prewired and field-configured at the factory, allowing for more thorough testing, and standardized connections and front access reduce the risk of bad connections in the field.

TCO is the sum of capital expenditure (CapEx) and operating expenditure (OpEx). The CapEx of a UPS comprises the UPS itself and the battery bank, the surrounding infrastructure, and the installation and commissioning costs. Energy consumption and maintenance are the two big contributors to OpEx. TCO can be optimized by optimizing battery capacity, saving valuable floor space, reduce installation and maintenance costs, and saving energy costs. UPS capacity can be changed with changing load, reducing the need to oversize. Modularity makes it easy to add modules and increase the power capabilities, if needed. Power consumption is a topic of great concern for most operators, and the energy savings made by the modular approach over the service life of the UPS are substantial. Run-time and battery sizing can be exactly fitted to what is required. A separate battery allows the system to be upgraded and autonomy preserved, without compromising availability. Full redundancy is only achieved with a redundant battery. If a common battery is required, ABB's modular UPS allows flexible blocks per string.

Modularity lends itself well to keeping UPS footprint small. A modular UPS rack has a small footprint, and when extra modules are added, no extra floor space is taken up. The modular approach makes installation and commissioning easy. Standardized modules reduce inventory levels of specialist spare parts and simplify system upgrades. This approach also pays off when it comes to serviceability and availability as service personnel do not need special skills and human error is reduced. Spares can be held on-site or at a nearby service center. Not only does this improve availability but also reduces cost as service engineers spend less time on site, and any risks of data or production loss are minimized. The only UPS elements common to all modules are contained in the mechanical frame that accommodates the UPS modules—I/O connection, customer interface signaling, maintenance bypass, and, in some models, a system display. These elements are standardized to minimize maintenance costs. The ABB design for module connectors makes swapping modules very easy and safe.

The modularity and scalability described have a major positive impact on achieving a low cost of ownership, but costs are held down too by designs that have best-in-class energy efficiency. ABB's Concept Power DPA 500, for example, operates with an efficiency of up to 96%. Its efficiency curve is very flat so there are significant savings in every working regime. Further, cooling costs can be substantial, and, because less power is consumed, high-efficiency modular UPSs require less cooling effort, creating further savings. The modular concept permits a range of different power protection solutions for all types of IT applications. The UPS function can be located centrally or beside each row of servers ("end-of-rack row"). A centralized power protection concept is appropriate, in most cases, for large data centers with servers using a single power supply. A distributed power protection concept may be applicable in small or large data centers with decentralized power protection demands.

In distributed power protection concept, the power demand grows from 40 kW ($N + 1$) to 120 kW ($N + 1$). The UPS can be easily adapted to meet the power demands of the growing infrastructure by adding four 20 kW modules. The power demand in the sample configuration increases from 40 kW ($N + 1$) to 120 kW ($N + 1$). The UPS is perfectly capable of adapting to this increase in power demand. The system flexibility allows power capacity to be upgraded to 3 MW. Critical applications are best served using modular UPS technology with an $N + 1$ configuration.

ABB's decentralized parallel architecture provides full redundancy and fault tolerance in a way that is unique among UPS vendors. UPS modules can be swapped online, which means fast and easy service and no downtime at all. DPA and modularity result in many knock-on advantages guaranteeing that ABB's modular UPS has the lowest TCO while providing the operator with a flexible, reliable, agile, and environmentally attractive infrastructure. The overwhelming benefits of modular UPS systems speak for themselves. ABB sees standardization and modularity as key elements in the set-up. Its unique UPS design is based on the concept of true modularity. With all the critical components distributed between individual units, potential single points of failure are eliminated [14,15]. If redundancy is provided for, there are more modules than are needed to supply the critical load. In a redundant system, all modules are active and share the load equally. Should one module fail, the remaining modules take over the load smoothly. The system is fault-tolerant and there are no single points of failure. ABB calls this modular approach DPA. DPA not only provides the best availability but also the best serviceability, scalability, and flexibility. Taken together, these features all deliver a low TCO.

ABB's UPS modules can be online-swapped, i.e., removed or inserted, without risk to the critical load and without the need to power down or transfer to the mains. This procedure is simple and quick to perform and introduces no risk for system operation. Each module can be individually switched off before removing it from the system. This makes the service safe for the technician and ensures absolutely no disturbance to the system. As the same modular UPS can be used across different applications and load segments, the service technicians do not need to be educated on several different platforms, but can apply the same practices and procedures on all UPS equipment. In ABB's DPA UPSs, the incoming AC is first converted to DC. The output AC is then synthesized from this DC, resulting in a clean sinusoid.

These two conversion steps give the term "double conversion" and isolate the output voltage waveform from any disturbances on the input AC side. ABB's Swiss-made modular DPA delivers UPS with high degree of serviceability, scalability, flexibility, and low energy usage and very attractive TCO. Because the UPS modules in a DPA are independent they can be online-swapped without risk to the critical load and without the need to power down or transfer to raw mains supply. So engineers can work on the UPS without interrupting operations. Swap-out time is only 15–20 min and is very safe, and load does not need to be switched off. Online-swappable modules directly address availability requirements, significantly reduce MTTR, reduce inventory levels of specialist spare parts, and simplify system upgrades [14,15].

5.4 MODULAR FUEL CELL FOR DATA CENTERS

Fuel cells have been around since the early days of NASA and have been a primary source of power for many missions. Nowadays, closer to home, they are becoming an increasingly attractive source of standby backup or primary power for modular data centers. One of the main concerns for large-scale data center operators over the last few years has been how to provide their IT facilities with clean energy at guaranteed future pricing. Investments in solar photovoltaic (PV) farms and wind turbines, hydro, and geothermal energy sources have to be done on a scale that can only really be financed by mega data center operators such as Google, Facebook, and Apple. However, the power horizon is changing with a rapidly advancing power-generation technology that could one day become the standard for modular data centers. It is one that can be used as a primary power source or as an emergency source of power for a UPS. The technology is that of the "fuel cell" [17–19].

In modular data centers, fuel cells, when combined with modular UPS systems, may also be the perfect choice for "Colo" data centers because they allow power protection to be right-sized to suit IT demands. That is, scaled vertically as the need comes to power more server racks to meet client demands. As described in great detail in the literature, the operating principle of fuel cells relies on an electromechanical process, forcing a gas through a catalyst membrane to generate electricity. There is also a by-product whether the fuel is pure hydrogen or natural gas. For hydrogen fuel cells, the "waste" product is water, and for natural gas fuel cells, it is carbon dioxide, but on a lower scale than coal-fueled power plants. The catalyst itself typically uses precious metals like Nickel and is one of the key cost components [17–19].

Several key factors are driving the fuel cell revolution within the data center market. The price of fuel cells is dropping. This is partly due to innovation and research into less costly catalyst materials and construction, as well as scale of economies as fuel cell technology becomes more widely used. Electricity costs are also rising in the United Kingdom with a less reliable electricity distribution grid as power-generation capacity falls. Some are forecasting parity with more carbon-intensive power generation in the years to come. From an infrastructure point of view, fuel cells can also benefit from local facilities and distribution. Hydrogen can be stored in specialist cylinders and delivery systems on site, with companies like BOC providing a well-known gas bottling service including delivery, collection, and refill. For natural gas-powered fuel cells, data centers can be constructed near to grid connection points or facilities that produce gas as a part of their operation such as biogas plants. Based on size, either approach may be less costly, more energy-efficient, and environmentally friendly than a traditional diesel-powered generator or sealed lead acid battery sets, which have short life expectancy and replacement cycles (3–5 or 7–8 years when used with a UPS system) [17–19]. What makes fuel cells appealing for modular data centers is the fact that they are modular designs themselves. To operate as cost effectively and efficiently as possible, every aspect within a modular data center has to be right-sized to meet the demands within a given time frame. Modular UPS systems allow this, and the modular build of fuel cells is complimentary to this approach. Both modular UPS and modular fuel cells tend to scale vertically.

In summary, fuel cell adoption by smaller and modular data centers is becoming increasingly attractive, especially when coupled with a modular UPS system. Their life expectancy of 15–20 years is also appealing when it comes to finance options, and they offer probably the most eco-friendly approach to power generation when fueled by hydrogen. While specialist skills are required to deploy either a hydrogen or natural gas-powered fuel cell within a data center environment, these skills are now easily available [17–19].

5.5 MODULAR SUPERCOMPUTER

5.5.1 NASAs Module-Based Supercomputer

As organizations tackle increasingly complex and data-heavy challenges, high-performance computing (HPC) systems are working overtime to quickly execute workloads and streamline data center operations. As a result, developers are striving to deploy a new breed of HPC systems that combine extreme speed and density with superior energy efficiency. Organizations with large-scale data centers are beginning to adopt a new approach to energy usage, leveraging powerful and eco-friendly solutions to turbocharge operational performance.

As the needs and densities of servers and data centers increase, organizations are turning to advanced cooling methods to bolster their HPC environments. Today, NASA is achieving new levels of water and power efficiency and performance of the Modular Supercomputing Facility (MSF) at NASA's Ames Research Center in Silicon Valley. Using technologies from Hewlett Packard Enterprise (HPE) and other partners, NASA's first module-based supercomputer, Electra, ranked 33rd on the November 2017 TOP500 list of the world's most powerful supercomputers. The MSF uses a combination of natural resources and adiabatic technology, consuming <10% of the energy used in traditional supercomputing facilities [20].

5.5.2 Optimizing Energy Efficiency for HPC

NASA has adopted a novel approach to cooling that not only enhances data center performance but also conserves electricity and water. Electra's new module employs a combination of outdoor air and adiabatic coolers on the roof to rapidly cool the

FIGURE 5.4 NASA's Electra, a high-performance computer [21].

system—more specifically, warm air is drawn through water-moistened pads, and as the water evaporates, the air is chilled and pushed out. In addition, the new module adds four HPE E-cells to deliver even greater efficiency. An E-cell is a sealed unit that uses a closed-loop cooling technology to release heated air outside the data center to ensure 100% heat removal. By transporting facility-supplied water into the system, E-cells help to rapidly and continuously cool the system, without compromising on performance or cost.

In its first year, Electra (see Figure 5.4) used 95% less water than a traditional data center environment, and it is expected to save 1,000,000 kWh and 130,000 gallons of water each year. This will provide users an additional 280 million hours of compute time each year, as well as an additional 685 million hours of compute time to augment its sister system, Pleiades. According to NASA, this is a different way to perform supercomputing in a cost-effective manner. It makes it possible for us to be flexible and add computing resources as needed, and we can save about $35 million dollars—about half the cost of building another big facility [20].

5.5.3 Deploying Energy-Optimized Solutions

Electra's new module is based on the HPE SGI 8600, a scalable, high-density clustered supercomputer that utilizes liquid cooling to achieve maximum efficiency and substantial savings in energy usage. This leading-edge system is based on E-cells, each containing two 42U-high E-racks separated by a cooling rack. The E-cells comprise 1,152 nodes with dual 20-core Intel® Xeon® Gold 6148 processors, increasing Electra's theoretical peak performance from 1.23 to 4.79 peta flops. With 24 racks (or 2,304 nodes), 78,336 total cores, and 368 terabytes of memory, Electra is engineered with the speed and robust compute capabilities to handle NASA's most challenging workloads. Based on the success of the new module, NASA is considering an expansion of up to 16 times the current capabilities of the

Electrical/Electronic Applications

modular environment. This effort would enable scientists and engineers nationwide to harness Electra for their research supporting NASA missions [20,22,23]. Along with NASA, Atos and Intel have also put out high-performance computing architectures for modular supercomputers [22,23]. More work on energy-efficient computing architecture is ongoing [24].

5.6 MODULAR POWER PLANTS

More than 10 years ago, power plants were traditionally stick-built, with each building custom designed and made for that particular plant. The major benefits of this approach were maintenance access and lowest equipment pricing, since a substantial portion of the work was being completed in the field. In a previous book [25], we examined modular coal, gas, and biomass/waste-driven power plants. Here, we briefly examine modular power plants from the perspectives of distributed power dissemination. We also evaluate more detailed layout of the generalized modular power plant.

In recent years, three factors started instigating a change in philosophy: centralized organizations, rising construction costs, and real estate issues. Companies started consolidating and sharing resources with a corporate office. This moved decision makers away from plant personnel so maintenance access was often an afterthought. Construction companies often ran unchecked, and owners were subjected to price escalations due to labor uncertainties inherent with field work, mainly productivity (driven by labor force availability/quality and site weather). This increased schedule risk and uncertainty which drive up cost.

The concept of distributed generation, as well as the increased challenge/cost of acquiring land has reduced power plant site acreage dramatically. The result was the rise of a new buzzword: modularity. This concept eliminated (or mitigated) all three of these factors, increasing acceptance and popularity of modular design. For the next several years, the focus shifted to these concepts [26–31]:

1. **Repeatability:** Design the system once and then use it at every site, reducing engineering costs.
2. **Minimize field time:** Deliver the system as building blocks that can be assembled both faster and easier, thus reducing site schedule risk. Labor productivity can be maximized in the controlled environment of a shop or a factory.
3. **Scalability:** Relegate equipment to catalog items that could be scaled up or down for a specific opportunity.

This philosophy continued to dominate the industry for the better part of a decade, and in some cases still does. However, as these highly modularized plants start approaching mid-life, maintenance requirements are increasing. As plant personnel attempt to perform these services, they are introduced harshly to one of the biggest weaknesses of modularization: access. There simply isn't enough space to perform the required work, which means intended "permanent" structures or equipment is being temporarily relocated to improve access. This inevitably increases the time and cost of maintenance. The result is that, during the next planning phase, there are often decisions made resulting in sacrifices between aspects of stick-built versus aspects of modularization.

There will inevitably be examples from the ends of the spectrum, both where traditional stick-built has its merits and where full-blown modularization is required. However, for the majority of projects falling in between these extremes, it is time to evolve the thinking of the either/or, mutually exclusive relationship to a hybrid design that offers the best of both stick-built and modular approaches. While the general philosophy should be identified, customization and maintenance access need not be sacrificed for project schedules and scalability. Modularization of power plants need to be done with caution [26–31].

5.6.1 Power Plant Design Taking Full Advantage of Modularization

In some parts of the world, coal-fired power plants are still a viable choice. Solid fuel plants must, however, overcome several obstacles, including higher capital costs, longer construction periods, and higher emissions. While emissions control technology is reasonably well advanced, longer construction periods and the associated higher capital costs still need to be addressed. Modular design represents a promising approach. The concept of modularizing pulverized coal-fired and fluidized-bed power plants has been around since the mid-1980s, but the use of modular or skid designs has not been fully exploited. By maximizing the use of modular design and construction, significant cost and schedule savings can be realized. The study by Gotlieb et al. [31] discusses modular plant design and how it differs from conventional stick-built design, including cost and project schedules. Modularization in this study refers to the use of shop assemblies, subassemblies, and full-scale modular packages.

For the sake of comparing modular and stick-built power facilities, a 300-MW net size installation was selected. The reason for this size selection was that the fluidized-bed boiler manufacturers are comfortable with this size for a single-unit design. Larger solid fuel plants are possible, of course, but a single-unit case is the objective of this analysis. The two plants are coal-fired installations, one a pulverized coal-fired facility and the other a circulating fluidized-bed boiler plant. Both plants are designed to meet all environmental requirements while burning a sub-bituminous coal. The pulverized coal plant employs a dry scrubber, whereas the fluidized-bed plant uses limestone in the combustor for sulfur capture. NO_x control for both cases are accomplished by a selective non-catalytic reduction (SNCR) system.

5.6.1.1 Modular Plant Design

Modular plant designs are very unique compared to stick-built facilities in that the amount of property required for a modular design is approximately one-third more than for a stick-built design. By removing verticality from a conventional plant, congestion during construction is significantly minimized. The complete modular design also relies extensively on modular fabrication at off-site facilities, where labor is generally two-thirds less costly than field labor and quality can be controlled more accurately. The most significant difference between backbone modular and conventional plant designs is in the full use of modular or skid-mounted mechanical and electrical equipment. The design involves locating nearly all equipment modules at grade and connecting them to a common "backbone" pipe rack that runs through the length of the plant. The modular components include major mechanical and electrical equipment as well as minor equipment of different systems grouped together, i.e., control rooms, HVAC, etc. [31].

All major equipment skids are subassembled in off-site fabrication facilities and shipped to the site for placement along the backbone. Even large items such as the steam generator can be broken down into subassemblies and shipped to the site in large modular pieces. In the case of a pulverized coal (PC) boiler, subassemblies could include entire low-NO_x burner fronts completely assembled with all piping, electrical, and instrumentation and routed to a single interface point. The selected turbine generator has a guaranteed nameplate rating at a generator outlet of 325 MW, with a net plant rating of approximately 300 MW. The turbine generator is lowered from the conventional 40 ft elevation to approximately 25 ft elevation. This allows the entire turbine bay structure height to be reduced. Further, instead of using a typical turbine pedestal approach, the modular design uses a steel pedestal much lower to grade and a side exhaust with complete condenser in two pieces. The turbine deck or operating floor and the mezzanine floor have been eliminated [31].

Another change is that the deaerator has been moved to a lower elevation to eliminate the typical enclosure on top of the turbine building. While this arrangement requires the addition of a booster pump to the feedwater system, it is considered to be more economical. The extra operating costs associated with the booster pump are less than the additional annualized structural costs associated with positioning the deaerator at a higher elevation. High- and low-pressure feedwater heaters are modularized and vertically stacked in a conventional cascading formation and will sit very close to the turbine generator and backbone pipe rack to interface with their connecting systems.

The circulating water is cooled in an evaporative eight-cell mechanical draft-cooling tower, which is modularized and constructed using aluminum and fiberglass along with a synthetic fill. The overall plant water balance is based on a conservative zero discharge station design. The flue gas exiting from the steam generator passes through a dry SO_2 removal system for the PC case, whereas SO_2 removal for the fluidized-bed case is accomplished via the limestone bed. The dry scrubber is modularized in several sections, including the inlet and outlet ducts, spray dryer assembly, and vessel assemblies for each parallel system. The flue gas finally passes through either one or two modular bag houses, and clean gas is extracted via two induced draft fans where it exits up the chimney.

The bag house modules are fabricated into modular compartments with pulse jet headers, controls, hoppers/hopper heaters, and ash discharge valves attached. Much of the materials handling systems and equipment is amenable to modularization because of its repetitive and consistent design. Fuel for the boiler is delivered to the site by rail car and is unloaded in a continuous car dumper. Coal is moved automatically by belt to concrete live storage silos with an alternate route to inactive or dead storage. Mobile equipment is used to reclaim coal from inactive storage. For the most part, all coal handling and conveying equipment is modularized into individual structural cages that include rollers, idlers, emergency trip switches, and electrical and instrumentation hardware. The conveying equipment is shipped in 60–80 ft lengths, depending on the site and local conditions.

The lime reagent or limestone for the SO_2 collection processes is delivered either by truck or rail cars, unloaded onto a transfer tank, and then conveyed to lime/limestone storage silos by a pneumatic transport system. All lime/limestone equipment—blowers, silos, feeders—is skid-mounted, while transfer lines or conveyors are modularized in shippable lengths. Bottom ash from the economizer section of the boiler and ash from fabric filters are collected, conveyed to holding silos, and transported to an off-site disposal area. The bottom ash conveyor and ash conveying systems are completely modularized and delivered on-site using preinstalled blowers/vacuum pumps, valves, motors, and instrumentation. The plant water/wastewater treatment systems, including fire water, are all modularized onto skids complete with piping, structural, valving, electrical, and instrumentation and controls. The water treatment system (demineralizer system) is fabricated into four modules (regeneration/degasifier, anion, cation, polisher) that plug directly into the backbone pipe rack, which in turn feeds the boiler make-up system. A major advantage of modularizing control and instrumentation for equipment comes from moving the instrument modules out of the field. The time and cost savings are achieved by shipping split assemblies of control panels; prewiring from control panel to direct digital controller, printers, relay and recorder board, logic cabinets, and data logger; installing in-line instruments and sensors on the modules; and prewiring to instruments from the direct digital controller (DDC). The electrical switchgear, motor control centers, and power distribution centers are skid-mounted onto modules to support the systems that each electrical system supplies. For example, the power distribution center for the ash handling system is located close to the ash handling equipment itself and is fed from a main switchgear center close to the main auxiliary transformer. Even though the main power distribution center is in two modules instead of one, this approach results in a net loss in wiring because of the close proximity between the equipment and the power distribution center [31].

Warehouse, administration building, machine shop, and electrical shop space has been removed from the power building block. Space for these functions can be more economically provided in a separate structure. The same building codes and design load criteria apply for the backbone design. The main difference is that the structure for the main power block and individual modules is much different as a result of the unique layout. Shop assembly of the critical piping module provides better access for the heat treatment and radiographic inspection of the main steam and hot and cold reheat piping welds. This permits better quality control for the assembly of critical piping since field welding and inspection will be reduced. Piping engineering and design involved with this type of plant consists of assembling spool pieces on appropriate modules at a central facility and then shipping the modules to the plant site for erection. Foundations are similar for both cases, i.e., spread footings, mats or pile, and caisson supports, if required. The building column foundations are smaller due to the smaller building profile and lighter loads. The floor slab may require additional thickening due to numerous floor-mounted pipe supports, movement during the installation of the skid-mounted equipment modules, and the profusion of equipment on the grade slab.

The plant design has not compromised the good engineering practices of generous equipment laydown and maintenance areas throughout the facility. This includes pull areas for heat exchanger tubes, pumps, compressors, and electrical equipment. Ease of equipment maintenance is also enhanced by this approach since significantly fewer pipes need be routed around equipment, thereby increasing available access area to any particular component [31].

A modularized power plant from the start of engineering and design through completion of plant start-up requires 34 months for completion. On the other hand, a standard stick-built power plant would require 43 months for completion, resulting in a saving of 9 months in the total project duration with the modular approach. The bulk of the time savings comes from three significant factors:

1. Use of prefabricated and pretested modules that only require installation and final piping and electrical hookup.
2. Reduced amount of field labor associated with assembly and construction.
3. Reduced time required for plant start-up as modules would have been shop-tested prior to shipment.

The modular assembly approach displaces 40% of the stick-built field man hours, which can be expended more efficiently in a shop environment than in the field. The engineering schedule is different for the backbone case. The main power block is fairly independent of the equipment details, and design may proceed after the initial size requirements are established. Module designs may proceed after initial size requirements are established. The pipe rack modules may be designed as general design, prior to final print layouts, thereby shortening the structural design schedule. In addition, the logistics effort for setting modules into the building is minimized. The backend facility design is similar to the conventional case [26–31]. The construction schedule for the main power block in the modular design approach is also different from that of the conventional case. Underground utilities are either in standard trenches and ducts or are run overhead on racks. The building foundations and buildings are smaller than in conventional designs and can be installed at an early date. The module mat foundations can be oversized and poured, enabling the concrete foundation contractor to leave the site early. Modules can then be supplied and installed on their own completion schedule.

5.6.1.2 Cost

Current cost estimates for a stick-built, solid fuel-fired facility are in the range of $1,250–$1,360/kW, whereas the modular plant costs are in the range of $1,080–$1,190/kW. Specific cost savings for a modularized power plant over a stick-built power plant come in a number of areas:

1. The total labor cost savings for a modularized plant are driven by the displaced stick-built field hours at an improved efficiency of 15% and a bare wage rate 20% less than the field labor rate. The resulting labor savings are in the range of $15–20 million.

2. Equipment rental, small tools, expendable supplies, and temporary facilities for the stick-built portion of the plant are significantly reduced due to the lower amount of field labor expended and the shorter overall duration in the field for both direct construction and plant start-up. These savings amount to $5–8 million.
3. The construction staff cost savings, due to the overall shorter duration in the field for construction and plant start-up, are in the range of $2.5–4 million.
4. The two offsets to the above savings are increased home office and engineering costs due to the modular design, as well as increased structural costs for the modules and modular logistics and transportation costs. The increased costs are in the $1.5–2.5 million range.
5. Other savings associated with modularization of the piping, electrical, buildings, and structures equate to $23–28 million.

5.6.2 Modular Chiller Plant for a 540-MW Gas-Fired Power Plant in Mexico City

Stellar Energy was chosen to supply a modular chiller plant, providing 1,800 tons of refrigeration (TR), to General Electric Co. (GE). The Stellar Energy-designed and fabricated module provided cooling for a turbine inlet air chilling (TIAC) system applied to the Valle de Mexico combined cycle gas turbine power plant in Mexico City, Mexico. The TIAC system was applied to the plant's GE LM6000 aeroderivative turbine. The 1800 TR modular chiller plant was consist of five [5] air-cooled chillers, one power distribution module (PDM), one pumping skid, and one control system [32,33].

Stellar Energy has supplied modular chiller plants for GE's LM6000 turbines at several power plants, including El Paso Merchant Energy, Kings River Conservation, Black Hills Power, Bryan Texas Utilities, Silicon Valley Power, Western Farmers Electric, Braunig Peaker Plant, Mariposa Energy, and Petrobras Regap. In September, Stellar Energy announced that it was selected to supply two modular chiller plants for a TIAC system that will be applied to the new 400-MW gas-fired power plant being developed by GE for Symbion Power Tanzania [32,33]. TIAC improves power producers' profitability by increasing a gas turbine's power output by up to 30%. By mechanically chilling the inlet air before it enters the compressor, TIAC gives the turbine a boost when it needs it the most, i.e., in hot weather. In addition to increasing power output, TIAC improves the turbine's heat rate, increasing efficiency and lowering emissions. Stellar Energy has designed, manufactured, and installed TIAC systems on more than 100 gas turbines worldwide, and has more than 1 million TRs of chilled-water plant experience [32,33].

5.7 COMPACT MODULAR™ TECHNOLOGY COOLING PROPERTIES FOR LIGHTING SYSTEM

Orion Energy Systems, Inc. was awarded a patent for the cooling properties incorporated in the power technology enterprise's Compact Modular™ high-intensity fluorescent (HIF) lighting platform. The conductive, radiant, and convective heat transfer properties of Orion's Compact Modular™ lighting system result in a cooler operating fixture, which enhances ballast longevity, increases the overall efficiency of the device, and eliminates cooling time that is typically required prior to performing routine maintenance [34,35]. Compared to the industry-standard, high-intensity discharge fixtures, or HIDs, that operate at ~1,000°, Orion's Compact Modular™ system operates at a cool 100°. Because of Orion's expertise in managing thermal energy, the Compact Modular™ platform has been deployed effectively in most extreme applications, including environments with temperatures up to 131° and in freezer applications as low as 40° below zero. By keeping the Compact Modular™ cool, companies can decrease their air-conditioning costs, which are often increased as they try to offset the heat emitted by HIDs [34,35].

The Compact Modular™ system is engineered to effectively manage heat by minimizing energy input and maximizing light output. Because of its superior thermal and optical efficiencies, Orion's technology consistently outperforms other HIF technology on the market when included in test sites at commercial and industrial facilities. Orion's technology reduces a facility's base load lighting costs by 38 cents/ft^2, on average, and reduces its energy use for lighting by 4.9 kWh/ft^2 per year. The result is a 6.5-lb/ft^2 reduction of carbon dioxide annually. When the Compact Modular™ platform is integrated with Orion's InteLite(R) wireless control system and direct renewable solar Apollo(R) light pipe, Orion's customers reduce energy consumption for lighting by up to 80%. The design of the Compact Modular™ fixture, its reflector, and its modularity are already patented [34,35]. Orion has deployed its energy management systems in 5,082 facilities across North America, including 120 Fortune 500 companies. Since 2001, Orion's technology has displaced more than 477 MW, saving customers more than $710 million and reducing indirect carbon dioxide emissions by 6.1 million tons [34,35].

5.8 JENOPTIK MODULAR ENERGY SYSTEMS FOR MILITARY PLATFORMS

Jenoptik plans, develops, and builds electrical energy systems for a wide range of applications in civilian and military vehicles, mobile platforms, and stationary systems. Jenoptik systems provide a powerful, reliable energy supply as well as lower fuel consumption. These systems secure the supply of power, either independently or in combination with the main engine. Even if the engine or main unit is switched off, electrical systems such as the on-board electronics, sensor systems, ventilation, and air conditioning continue to function to ensure that the overall system remains available and ready for operation. At the same time, factors such as fuel consumption, heat generation, and engine noise are all reduced to a minimum. The additional installation of an energy accumulator increases the range of the vehicle, providing the potential to extend the mission [36–38].

The increasing need to protect troops is just one of the factors driving up the energy requirements of military platforms. All energy systems supply the platform with electricity in exactly the form in which it is required. Power, capacity, voltage, current, and frequency are all optimally tuned to particular consumers in question, as well as innovative high-voltage consumers. To achieve this, Jenoptik offers products of the latest technological standard that have already proven successful in both national and international applications. When used in the particular vehicle in question, Jenoptik subsystems, such as generator sets, alternators, power electronics, or electrical components, are optimally tuned to one another and enable the energy for the platform to be managed in an intelligent and extremely efficient manner [36–38].

Jenoptik offers both customized complete systems/system solutions and standardized subsystems and components. The latter are designed to be flexible, enabling us to adapt them in line with customer requirements with very little effort. In this way, one can modernize the fleet at lower cost and enhance the level of performance and availability offered by one's vehicles. All newly developed systems are subjected to a rigorous process of testing and optimization in a "hardware-in-the-loop" (HiL) simulation, in which the real-world environment is replicated. By adopting this approach, the customer obtains an individual energy system that exactly meets their requirements within the shortest of time frames. The benefits of Jenoptik military platforms include the following [36–38]:

1. **Secure investment:** Low costs throughout the entire lifecycle of the product.
2. **Reliable:** Maximum availability and operability.
3. **Flexible:** Update existing systems in a quick and simple process.
4. **Efficient:** Reduce fuel consumption while achieving a high power density and a high level of efficiency.
5. **Proven:** Tried and tested for use in numerous national and international projects.
6. **Modular:** Adapt standard products or plan and implement specific systems.

These platforms are used for security and defense technology, as well as and power supply for military needs, such as ventilation and air conditioning in military vehicles.

5.9 MODULAR MICRO- AND MINI-SCALE ENERGY SYSTEMS FOR ELECTRONIC APPLICATIONS

A dramatic technological trend today is the rapid growth of personal and mobile electronics for applications in communication, health care, and environmental monitoring. Individually, the power consumption of these electronics is low; however, the number of such devices deployed can be huge. Currently, the powering of electronic devices still relies on rechargeable batteries. The number of batteries required increases in proportion with the increase in the number and density of mobile electronic devices used, and may result in challenges for recycling and replacement of batteries as well as concerns about potential environmental pollution. To effectively extend the lifetime of batteries and even completely replace them in some cases, a global effort has begun toward the development of technologies for harvesting energy from our living environment, such as solar energy, wind energy, thermoelectricity, mechanical vibration, and biofuels. A study by Wang and Wu [39] focused on recent advances in both the scientific understanding and the technological development of energy harvesting, specifically for powering future functional micro and nano sensors (MNSs).

There has been an increasing need for the design/development of MNSs for wireless applications, and as pointed out by Wang and Wu [39], it is predicted that the market for wireless sensor networks (WSNs) alone will be about 2 billion USD in 2021 Enormous efforts have been focused recently on integrating individual micro/nanodevices with diversified functionalities into multifunctional MNSs, and further into large-scale networks for applications in numerous chemical, environmental, and health-related sensing devices. To address these application needs, each device/system node within the network should consist of a low-power, microcontroller unit, high-performance data-processing/storage components, a wireless signal transceiver, ultrasensitive sensors based on a micro/nanoelectromechancial system (MEMS/NEMS), and most important imbedded power units. The integration of these conventionally discrete devices with dedicated functionality toward smart and self-powered systems is considered to be one of the major future roadmap for micro and nanotechnology in self-powered electronic devices. A shear number of these devices dictate that the standardization and modular approach suits well for this purpose to reduce cost and improve quality.

As technology advances, the miniaturized dimensions of nanomaterials and the ability to modulate their composition in a well-controlled manner to develop properties that are not possible in their bulk counterparts provide the potential to address some of the critical challenges faced by silicon-based microelectronics. Furthermore, these characteristics enable the incorporation of diversified functionalities into systems to complement digital signal/data processing with augmented functional capabilities, such as interactions between the machine and humans or the environment. Numerous nodes of such devices or systems can be spatially distributed and embedded virtually anywhere, from a remote field to civil structures and even the human body, to fulfill their respective purposes. It is therefore essential that these MNSs have an extended lifespan, especially for applications with limited human accessibility, such as monitoring/tracking in a remote or hazardous environment [39].

5.9.1 POWER CONSUMPTION AND MODE OF OPERATION OF MNSs

Micro/nanodevices exhibit operational advantages such as small electrical/thermal time constants, enhanced sensitivity/responsivity, and high integrated complexity with respect to their conventional counterparts. In addition to the above merits, another key characteristic of micro/nanodevices is low operation power, with typical values in the range of microwatts to milliwatts. Typical power consumption in MNSs also includes contributions from other operations of the system such as signal conditioning, information processing/storage, and data communication. For most application purposes, digitization of the sensed signal is critical during front-end signal conditioning and for analog-to-digital converters with sub-microwatt and microwatt power consumption. Logic processing and storage of the conditioned information are also indispensable for MNSs.

Nanomaterials based on logic operations with power consumption in the nanowatt to microwatt range have been demonstrated. Such memory devices are highly suitable for data storage in future wireless MNSs owing to their long-term data retention and low power consumption. In general, the transmission/communication of data consumes the majority of power in a wireless MNS. For example, as pointed out by Wang and Wu [39], modular products for low-power medical, industrial, and consumer applications, such as TI CC2560 from Texas Instruments and BCM4329 from Broadcom, operate at a rate of 3–5 Mbps. Therefore, long-term sustainable operation of MNSs requires either data transmission modules with even lower power consumption or a new MNS operation scheme. Although the power consumption of MNSs could be small, it is an indispensable component of the entire energy spectrum. More importantly, for small electronic devices, energy consumption is not solely measured in terms of cost, as in the case of a large-scale power requirement, but energy is required to enable the operation of the MNS. Pacemakers are a typical example.

Practically, in applications such as environmental and biomedical monitoring, a sampling rate of once every few minutes should be adequate, whereas in other scenarios, such as infrastructure and health monitoring, a measurement rate of once every few hours could be appropriate. The operation of a wireless MNS for such applications should therefore consist of both active and standby modes with low-duty cycles: most of the time, the MNS is dormant with minimum energy consumption, and active operation of the system is only required periodically according to the preset sampling rate. Accumulated power harvested by the system during the standby period between two active periods might be sufficient to drive the operation of the MNS in its active mode. The main requirement for energy harvesting may thus be minimization of the power consumption of the device in its standby mode and during the transitions from the standby mode to the active mode [39].

5.9.2 Micro/Nanotechnology-Enabled Technologies for Energy Harvesting

There are a variety of sources available for energy scavenging by MNSs from the ambient environment, including, but not limited to, energy in natural forms, such as wind, water flow, ocean waves, and solar power; mechanical energy, such as vibrations from machines, engines, and infrastructures; thermal energy, such as waste heat from heaters and joule heating of electronic devices; light energy from both domestic/city lighting and outdoor sunlight; and electromagnetic energy from inductor coils/transformers as well as from mobile electronic devices. Moreover, the human body itself provides a tremendous amount of energy that is available for harvesting and potential utilization in self-powered MNSs: mechanical energy resulting from vibration/motion from body movement, respiration, and even blood flow in vessels; thermal energy from body heat; and biochemical energy generated during physiological processes and metabolic reactions. Wang and Wu [39] have described several mechanisms for energy harvesting for MNSs systems such as PV technologies for solar energy harvesting, nanostructured thermoelectric materials for heat to electricity, piezoelectric generator, bio and microbial fuel cell using nanomaterials, triboelectric, and pyroelectric nanogenerators. Wang and Wu [39] have also described the present state-of-the-art of hybrid mechanisms such as (a) solar and mechanical energy, (b) biomechanical and biochemical energy, and (c) solar and thermal energy. The MNS can also be self-powered. Here, we briefly examine one example of energy harvesting and the perspectives on self-powered MNS and its future. More details on energy harvesting are reported by Wang and Wu [39].

5.9.3 PV Technologies for Solar Energy Harvesting

Solar energy is by far the most abundant exploitable renewable energy resource. More energy is provided to earth by sunlight irradiation within 1 h than is consumed by human society in 1 year. Semi-conductor materials that exhibit a PV effect can be used to convert solar radiation into electricity through a PV process. PV technology has been growing and expanding rapidly. The total global energy production by PV processes reached 64 GW by 2011. Despite this considerable capacity, however, PV technology only accounts for 0.1% of global electricity generation largely as a result of the inability of existing PV technologies to produce electricity with an efficiency that fulfils the grid parity set by conventional power-generation routes. Enormous efforts and resources have therefore been devoted to the development of new-generation PV technologies that operate with enhanced efficiency at a lower cost. Practically, all PV devices, or solar cells, incorporate a p–n junction. Such junctions occur in various possible configurations. Solar cells containing multiple p–n junctions have recently been investigated intensively for the more efficient absorption of light with different wavelengths with the aim of reducing the inherent sources of energy loss in conventional single-junction cells. A conversion efficiency of 42.3% was achieved by combining multi-junction cells and concentration technology. However, high efficiency of multi-junction cells is offset by their increased complexity and manufacturing cost, which limit their application mainly to aerospace exploration, for which a high power-to-weight ratio is desirable. The dominance of PV technology historically by inorganic solid-state junction devices is now being challenged by the emergence of a new generation of PV technologies built, for example, on nanostructured materials or conducting polymers. Such technologies offer the prospect of converting solar energy into electricity at low cost. Wang and Wu [39] described advances in different kinds of solar cells such as dye-sensitized solar cells, organic solar cells, quantum-dot and plasmonic solar cells, and solar cells based on low-dimensional nanostructures for efficient energy harvesting. These modular devices have made enormous progress on PV cell technology and their potential applications in MNSs.

5.9.4 Self-Powered MNSs

The current rapid advancement of micro/nanotechnology will gradually shift its focus from the development of discrete devices to the development of more complex integrated systems that are capable of performing multiple functions, such as sensing, actuating/responding, communicating, and controlling, by the integration of individual devices through state-of-the-art microfabrication technologies. Furthermore, it is highly desirable for these multifunctional MNSs to operate wirelessly and self-sufficiently without the use of a battery, especially in applications such as remote sensing and implanted electronics. This operation scheme will not only extend the lifespan and enhance the adaptability of these MNSs, while greatly reducing the footprint and cost of the entire system, but it will also increase the adaptability of these MNSs to the environment in which they are deployed.

As the dimensions of individual devices shrink, the power consumption decreases accordingly to a reasonably low level, so that the energy scavenged directly from the ambient environment is sufficient to drive the devices. The concept of self-powered nanotechnology was first proposed and developed by Wang and his co-workers with the aim of building a system that operates by harvesting energy from the ambient vicinity of the system and converting it into usable electrical power for wireless, self-sufficient, and independent operations. A typical self-powered MNS should consist of a low-power, microcontroller unit, high-performance data-processing/storage components, a wireless signal transceiver, ultrasensitive sensors based on MEMSs/NEMSs, and most importantly the embedded powering/energy-storage units. Wang and Wu [39] have given numerous examples of self-powered sensors and systems.

It can be anticipated that self-powered MNSs will play a critical role in the implementation of implantable electronics, remote and mobile environmental sensors, nanorobotics, intelligent MEMSs/NEMSs, and portable/wearable personal electronics. Self-powered MNSs are also key components of large-scale, fault-tolerant sensor networks. When traditional discrete sensors are replaced by a large number of sensor nodes distributed in a field, the statistical analysis of signals collected through the network of distributed sensors can provide precise and reliable information for tracking and monitoring purposes. An IoT that can correlate objects/products and devices with databases and networks is expected to revolutionize the future of health care, medical monitoring, infrastructure/environmental monitoring, logistics, and smart homes.

One decisive factor for the implementation of self-powered MNSs is the successful development of energy-harvesting technologies to provide appropriate power sources that operate over a broad range of conditions for extended time periods with high reliability. Despite the excellent progress that has been made in the emerging field of self-powered micro/nanotechnology, several issues still need to be addressed appropriately for the promised potential of self-powered micro/nanotechnology to be fully realized. The design/fabrication flow for the development of future self-powered MNSs should be amenable to scale up and, critically, be compatible with the microfabrication technology. Almost all prototypes demonstrated to date were fabricated at a level unsuitable for mass production. State-of-the-art microfabrication technologies were hardly used, which severely prohibits the broader implementation of current self-powered MNSs. Moreover, future self-powered MNSs should be implemented in such a way that multiple types of energy can be harvested synergistically by hybridized systems with sufficiently high outputs. Finally, to enable not only self-sufficient but also sustainable operation of the deployed devices/systems in applications, such as wireless biomedical sensing, self-powered MNSs should be developed using materials which are environmentally friendly, biocompatible, and biodegradable.

5.9.5 COMPARISON OF CYLINDRICAL AND MODULAR MICROCOMBUSTOR RADIATORS FOR MICRO-TPV SYSTEM APPLICATION

One example of harvesting energy at microlevel is to develop micro-thermo-photo voltaic (TPV) system. Yang et al. [40] introduced a modular micro-TPV system to enhance the efficiency of thermophotovoltaics. Compared to the cylindrical structure, the modular TPV system has the advantages of easier fabrication and assembly. A microcombustor is a key component of the micro-TPV system. To maximize power output, higher wall temperature and uniform distribution along the combustor wall is desirable. A microcylindrical combustor (inner diameter $d=3.56$ mm) and three kinds of micromodular combustors with widths of 1.0, 1.5, and 2.0 mm have been experimentally examined. The results indicated that the micromodular combustor with a width of 1 mm has a much higher radiation efficiency than the microcylindrical combustor. The wall temperature decreased with an increase in width due to the reduced heat transfer between the wall and hot gases. The performance of the micromodular TPV system was also predicted in this study.

5.9.6 MICROPOWER MANAGEMENT OF SOLAR ENERGY HARVESTING USING A NOVEL MODULAR PLATFORM

As mentioned above, a micropower management (μPM) is essential to supply power to WSNs from energy harvesting sources. There are various ways of realizing μPM, either with discrete electronic components or with commercial power management integrated circuits (PMICs). It is a challenging task to find the optimal solution; although the efficiency of individual components can be determined at certain operation points, it is not clear which operating point is the most dominant in interplay with other components and in a realistic scenario. Moreover, additional losses may originate from component interaction. Furthermore, the question arises of how to measure the energy balance and benchmark different systems under realistic and reproducible boundary conditions. Kokert et al. [41] examined this issue in details.

Several publications [42–47] document that μPMs are realized in various ways. In these publications, system evaluation was usually limited to a single test run of a few days in a non-repeatable, real-world scenario. Different approaches to address this issue and estimate system performance in an adequate way were presented. In battery-powered WSNs, the lifetime can be used as a performance indicator as reported in [42–47]. The authors show that using a dc–dc converter [48] prolongs the lifetime to 30%. Lee et al. [42] implement an energy-aware duty cycle and consider the number of active time slots as a figure of comparison. The boundary conditions are vague using a real solar panel and a fluorescent lamp. The authors of reference [45] present a energy harvesting platform for evaluation of a double-layer capacitor (EDLC, short: supercaps) and thin-film batteries in conjunction with power converters. A flaw is that the energy harvester (a solar cell) is oversimplified to a constant voltage power supply [42–47].

Energy extraction and voltage supply require power converters. The energy extraction block is required to extract as much energy from a harvester as possible. Hence, the output impedance of the harvester and the input impedance of the extraction block need to match. The voltage supply block is required to supply a constant voltage (e.g., 3.3 V) to a load. The solar thermal generator is sized for a footprint compatible with roof terraces in the developing world—or an American urban high-rise balcony. It uses a reflective trough (aluminum sheeting formed by a simple wooden truss, covered with 99% reflective Mylar, and coated with "Saranwrap" heat-resistant, non-polluting plastic sheeting) to focus sunlight onto copper tubes that form the heating element of a gaseous heat engine. Here, emphasis is again on minimizing cost and construction complexity, and then obtaining the best figure of merit consistent with those decisions. Thus, tracking motors and exotic collector shapes are avoided. Kokert et al. [41,43] gave numerous other examples demonstrating the importance of µPM platform.

5.9.7 Nanotechnology for Portable Energy Systems: Modular PV, Energy Storage, and Electronics

For portable energy systems, the study by Smith and Lyshevski [49] examined nanotechnology-enabled high-power and high-energy density microelectronics, power electronics, energy sources, and energy storage solutions. These energy systems are used in various applications including aerospace, biomedical, communication, electronics, MEMS, and robotics. Proof-of-concept autonomous power systems are designed, and practical solutions are substantiated and proposed. The study performed applied research and technology developments in the design of efficient energy harvesting, energy management, and energy storage systems. To guarantee modularity and compatibility, the following components and modules were examined: (a) low-power microelectronics; (b) high-power-density power electronics with high-current, high-voltage, and high-frequency power metal–oxide–semiconductor field-effect transistor (MOSFETs); (c) high-efficiency energy harvesting sources such as solar cells and electromagnetic generators; (d) high specific energy and power density electrochemical energy storage devices; (e) and optimal energy management systems. The most advanced technology-proven nanoscale microelectronics, MEMS sensors, and energy conversion solutions are studied. The modular design enabling topologies and optimization schemes result in high-performance micro and mini-scale energy systems [49].

5.10 MODULAR, INTEGRATED POWER CONVERSION AND ENERGY MANAGEMENT SYSTEM

The invention by Schienbein et al. [50] relates to power conversion and energy management systems for distributed energy resources (DER). This invention further relates to electrical power conditioning, controlling, and/or metering devices, including, but not limited to, UPS systems, remote power systems, backup power systems, harmonic filters, and voltage or frequency regulators or adapters. Broadly speaking, DER includes all power generators and energy storage systems other than medium and large-scale conventional power plants. Medium and large-scale power plants include, for instance, coal-fired steam turbine generators and hydroelectric generators that are rated at hundreds of megawatts up to 1,000 MW. The fundamental

Electrical/Electronic Applications 253

and distinguishing feature of DER power systems is that they can be largely or completely factory manufactured, assembled, and tested. They can also be easily commissioned, shipped, and installed as complete modules. Hence, they can go into service very quickly, where and when needed. Lead times are very short compared to those required to bring a large-scale power plant on-line. In other words, DER, or distributed generation, includes the use of small generators (typically ranging in capacity from 1 to 10,000 kW), scattered throughout a power system, to provide the electric power needed by electrical consumers. DG typically includes all uses of small electric power generators whether located on the utility system, at the site of a utility customer, or at an isolated site not connected to the power grid.

Dispersed generation is a subset of DG, which refers to generation that is located at customer facilities off the utility system. Dispersed generation is also typically understood to include only very small generation units, of the size needed to serve individual households or small businesses, in the capacity range of 10–250 kW. Most DGs utilize traditional power-generation paradigms—for example, diesel, combustion turbine, combined cycle turbine, low-head hydro, or other rotating machinery. DG also, however, includes the use of fuel cells and renewable power-generation methods such as wind, solar, or low-head hydrogeneration. These types of renewable generators are included in DG because their small size makes them very convenient to connect to the lower voltage (distribution) parts of the electric utility grid [48,50].

According to one embodiment of the invention, a modular, integrated power conversion and energy management system preferably includes a plurality of independent power module(s) integrated with communication module(s) and a configurable controller. In a preferred embodiment of the invention, for example, an integrated power conversion and energy management system includes an integrated controller, one or more standard modules, and a custom (or semi-custom) backplane. The integrated controller preferably accommodates one or more power or communication modules, and uses those modules to control power quality and/or flow to one or more input and/or output connections. The integrated controller preferably includes controller software, control circuits, power circuits, protection circuits, external electrical connections, an interface with one or more integrated or modular communications module, and an interface with one or more power modules [48,50]. The standard modules are preferably power modules. Each power module can include power circuits and a driver circuit, if necessary, to receive control signals from an external source. Each module also preferably includes a memory that can be polled by the backplane to identify the module and provide "plug-and-work" functionality. In other words, the memory can pass all of the module's design parameters to the controller on the backplane, thereby informing the controller how best to operate it.

For each application, a custom (or semi-custom) backplane can be developed to accommodate the standard modules. The backplane preferably includes everything but the power electronics. The backplane can, for example, include the external connectors, disconnects, fusing, and so forth, that are required for the application. The backplane can further include the various internal DC [48,50] and AC buses that are required to interconnect the modules. The backplane also preferably includes the communications module, a user interface, a supervisory controller, a socket for control firmware, and an appropriate number of slots for accommodating the standard modules. For semi-custom backplanes, the control firmware preferably provides full customization for each application.

5.10.1 High-Power Application Modular Multilevel Converter

High voltage levels reduce voltage stress and lower semiconductor losses. A new modular multilevel converter (MMC) can reach twice the voltage levels as conventional electrical energy transmission topologies. The modular structure of this MMC enables it to reach three voltage levels by adding a bidirectional switch with half the voltage stress, resulting in low switching frequency and near sinusoidal voltage waveforms. Benefits of these submodule topologies include reduced voltage stress, considerably lower semiconductor losses, and a compact size (and therefore a smaller footprint). These attributes translate into lower operation costs for high-voltage power applications such as high-voltage direct current (HVDC) transmission and distribution systems. The new MMC can be used to generate high frequency near sinusoidal voltages across a high-frequency transformer (HFT) to reduce its size and weight. The approach is unique as it provides the ability to scale to high voltages and make these power electronic transformers (PET) more reliable. In the study by Mohan and Sahoo [51] small scale for PET application was tested with nearly sinusoidal HFT voltages. The proposed hybrid modulation technique combines both phase-shifted and level-shifted carriers, while a proposed simple voltage balancing algorithm based on unequal capacitor sharing may result in proper balanced capacitor voltages over the entire modulation range and output power factor. These results are shown by MATLAB/Simulink simulations. The suggested hybrid modulation technique offers several benefits and features [51]:

- Size and cost halved: requires half the number of submodules to reach the same number of output voltage levels.
- Reduced conduction (33%) and switching losses (75%) relative to standard submodules.
- Improves efficiencies, reduces losses, and increases system reliabilities.
- Robust value proposition.
- Reduces complexities in gate drive circuitry.

The technique can be applied to alternative energy in distributed systems taking and delivering energy from multiple sources, high-voltage, high power usages (e.g., HVDC transmission and distribution, static compensator (STATCOM), multi-megawatt industrial drives), and MV drives or wind turbines.

REFERENCES

1. Bouley, D., Torell, W., and Neal, S., Benefits and Drawbacks of Prefabricated Modules for Data Centers, White Paper 163, Revision 3. A website report by Schneider Electric Data Center Science Center dcsc@schneider-electric.com www.apc.com/support/contact/index.cfm; www.apc.com/salestools/WTOL-7NGRBS/WTOL-7NGRBS_R3_EN.pdf (2017).

Electrical/Electronic Applications 255

2. Data Center Management | Schneider Electric Solutions. A website report www.schneider-electric.us/ (2015).
3. Prefabricated Data Center Modules | Schneider Electric. A website report www.schneider-electric.com/en/.../7550-prefabricated-data-center-modules/ (2015).
4. Dell® Modular Infrastructure | for a Modern Data Center | DellEMC.com. A website report www.dellemc.com/OfficialSite (2013).
5. Dell EMC Data Center Solutions | from the Experts at CDW | CDW.com. A website report www.cdw.com/Dell/EMC (2013).
6. Dell EMC Modular Data Centers | Dell. A website report www.dell.com/learn/us/en/05/.../documents~modular-data-centers.aspx. (2011).
7. Modular data centers, Wikipedia, The free encyclopedia, last visited 16 June 2019 (2019).
8. Containerized Data Center: DELTA. A website report www.deltapowersolutions.com/.../data-center-solutions-containerized-datacente... (2014).
9. Modular Data Center: DELTA. A website report www.deltapowersolutions.com/en/.../data-center-solutions-modular-datacenter... (2014).
10. Sun modular datacenter, Wikipedia, The free encyclopedia, last visited 25 May 2019 (2019). https://en.wikipedia.org/wiki/Sun_Modular_Datacenter.
11. Sun's Data Center on Wheels | Network World. A website report www.networkworld.com/article/2299963/sun-s-data-center-on-wheels.html (2006).
12. Sun's Data Center in a Box | ZDNet. A website report www.zdnet.com/article/suns-data-center-in-a-box/ (17 October 2006).
13. Uninterrupted Power Supplies | UPS Solutions | vertiv.com. A website report www.vertiv.com/ (2013).
14. Why Modular UPS: UPS Systems | ABB. A website report https://new.abb.com › Offerings › UPS and power conditioning › UPS systems (2013).
15. Conceptpower DPA (Modular): UPS Systems | ABB. A website report https://new.abb.com › Offerings › UPS and power conditioning › UPS systems (2013).
16. Uninterruptable power supply, Wikipedia, The free encyclopedia, last visited 6 August 2019 (2019).
17. Fuel Cell-Powered Data Centers: Fuel Cell & Hydrogen Energy... A website report by Fuel Cell & Hydrogen Energy Association, Washington. www.fchea.org/in-transition/2018/11/12/whs20dthibvg3pvhfjekribqhs0bbt (12 November 2018).
18. Hammond, S., The Hydrogen Data Center Challenge. A website report by National Renewable Energy Laboratory, Golden, CO. www.energy.gov/sites/prod/.../fcto-data-center-workshop-2019-hammond.pdf (20 March 2019).
19. Fuel Cells in the Data Center: Gigaom. A website report https://gigaom.com/report/fuel-cells-in-the-data-center/ (2018).
20. Mannel, B., NASA Achieves Optimal Energy Efficiency with its First Modular Computer. A website report www.hpcwire.com/.../nasa-achieves-optimal-energy-efficiency-with-its-first-mo. (18 March 2018).
21. Electra Supercomputer: NASA Advanced Supercomputing Division. A website report www.nas.nasa.gov/hecc/resources/electra.html (3 July 2019).
22. Feldman, M., Atos Broadens HPC Portfolio with Modular Supercomputer. A website report https://atos.net/en.../2018/...2018.../atos-broadens-hpc-portfolio-modular-supercompu (28 March 2018).
23. Intel® HPC Products Portfolio | Breakthrough Results, Faster | intel.com. A website report www.intel.com/HPC (2018).
24. Computer architecture, Wikipedia, The free encyclopedia, last visited 8 July 2019 (2019).
25. Shah, Y.T. (2019). *Modular Systems for Energy and Fuel Recovery and Conversion.* CRC Press, New York.

26. Schimmoller, B.K., Power Plants go Modular, Power Engineering. A website report www.power-eng.com/articles/print/volume.../power-plants-go-modular.html (1 January 1998).
27. Modular Power Plant: MWM. A website report www.mwm.net/mwm-chp...cogeneration/...power-plants/modular-power-plant/ (2013).
28. Siemens Power Plant Solutions | Innovative Power Plants | Create value. A website report new.siemens.com/power-plants (2013).
29. Buecker, B. and Mieckowski, C., Kiewit Power Engineers, Turbine Inlet Air Cooling Cutting Edge Technology. A website report in Power Engineering www.power-eng.com/.../turbine-inlet-air-cooling-cutting-edge-technology.html (01 April 2012).
30. Darrell Proctor, Report: Technology, Renewables will Grow Turbine Market. A website report in Power www.powermag.com/report-technology-renewables-will-grow-turbine-market/ (25 April 2018).
31. Gotlieb, J., Stringfellow, T., Rice, R., and Carter & Burgess Inc., Power Plant Design Taking Full Advantage of Modularization. A website report by Power Engineering www.power-eng.com/.../power-plant-design-taking-full-advantage-of-modulari... (01 June 2001).
32. Stellar Energy Selected by General Electric Co. to Supply a Modular Chiller Plant for Turbine Inlet Air Chilling Application on a 540 MW Gas-Fired Power Plant in Mexico City. A website report by Steller Energy News, Des. 9, New Modular Chiller Plant for GE's Gas-Fired Power Plant in Mexico... www.stellar-energy.net/about-us/news/gas-fired-power-plant-in-mexico-city.aspx (2013).
33. Energy News | Efficient Energy Systems: Stellar Energy News. A website report stellar-energy.net/about-us/news.aspx?p=1&v=30 (2013).
34. Orion Energy Systems Compact Modular (TM) Technology Cooling Properties for Lighting System, Manitowoc, WI (Globe Newswire via COMTEX News Network). (10 December 2009).
35. Orion Energy Systems Awarded Patent for Compact Modular (TM... A website report https://globenewswire.com/news.../2009/.../10/.../Orion-Energy-Systems-Awarded-Patent... (10 December 2009).
36. Jenoptik Modular Energy Systems for Military Platform, Generators and Alternators for On-Board Systems | Jenoptik. A website report www.jenoptik.com/...energy-systems-military-platforms/alternators-and-generat... (2015).
37. Innovative Systems for Security and Defense | Jenoptik USA. A website report www.jenoptik.us/markets/defense-and-security (2015).
38. Electric Power Supply and Drive Systems | Jenoptik. A website report www.jenoptik.com/products/power-supply-systems (2015).
39. Wang, Z.L. and Wu, W. (2012). Nanotechnology-enabled energy harvesting for self- powered micro-/nanosystems. *Angewandte Chemie International Edition*, 51, 11700–11721, Wiley-VCH Verlag GmbH & Co. KGaA, Weinheim, Angewandte Reviews. doi: 10.1002/anie.201201656.
40. Yang, W.M., Chou, S.K., Pan, J.F., Li, J., and Zhao, X. (2010). Comparison of cylindrical and modular micro combustor radiators for micro-TPV system application. *Journal of Micromechanics and Microengineering*, 20(8), 5003. doi: 10.1088/0960-1317/20/8/085003.
41. Kokert, J., Beckedahl, T., and Reindl, L.M. (2016). Evaluating micro-power management of solar energy harvesting using a novel modular platform. *Journal of Physics: Conference Series*, 773, conference 1, 2042. doi: 10.1088/1742-6596/773/1/012042.
42. Lee, P., Ang Eu, Z., Han, M., and Tan, H. (2011). Empirical modeling of a solar-powered energy harvesting wireless sensor node for time-slotted operation. *2011 IEEE Wireless Communications and Networking Conference*, Cancun, Quintana Roo, Mexico, pp. 179–184. doi: 10.1109/WCNC.2011.5779157.

43. Kokert, J., Beckedahl, T., and Reindl, L.M. (2016). Development and evaluation of a modular energy management construction kit. *18. GMA/ITG-Fachtagung Sensoren und Messsysteme 2016*, pp. 84–91.
44. Honsberg, C. and Bowden, S. (2014). A Collection of Resources for the Photovoltaic Educator. Available: www.pveducation.org/pvcdrom/short-circuit-current; PowerMEMS 2016, IOP Publishing, *Journal of Physics: Conference Series* 773 (2016) 012042. doi: 10.1088/1742-6596/773/1/012042.
45. Park, C. and Chou, P.H. (2006). Ambimax: Autonomous energy harvesting platform for multi-supply wireless sensor nodes. *2006 3rd Annual IEEE Communications Society on Sensor and Adhoc Communications and Networks*, San Diego, CA, Volume 1, pp. 168–177, 2007.
46. Bader, S. and Oelmann, B. (2013). Short-term energy storage for wireless sensor networks using solar energy harvesting. *10th IEEE International Conference on Networking, Sensing and Control (ICNSC)*. 10–12 April, IEEE, Evry, France.
47. Varley, J., Martino, M., Poshtkouhi, S., and Trescases, O. (2012). Battery and ultracapacitor hybrid energy storage system and power management scheme for solar-powered Wireless Sensor Nodes. *IECON 2012: 38th Annual Conference on IEEE Industrial Electronics Society Date of Conference*, 25–28 October, IEEE, Montreal, QC, Canada. doi: 10.1109/IECON.2012.6389000.
48. Oletic, D., Razov, T., and Bilas, V. (2011). Extending lifetime of battery operated wireless sensor node with DC-DC switching converter. *2011 IEEE International Instrumentation and Measurement Technology Conference*, 10–12 May, IEEE, Binjiang, China. doi: 10.1109/IMTC.2011.5944247.
49. Smith, T.C. and Lyshevski, S.E. (2016). Nanotechnology for portable energy systems: Modular photovoltaics, energy storage and electronics. *2016 IEEE 36th International Conference on Electronics and Nanotechnology (ELNANO)*, 19–21 April, IEEE, Kiev, Ukraine. doi: 10.1109/ELNANO.2016.7493077.
50. Schienbein, L.A., Hammerstrom, D.J., Droppo, G.W., and Harris, B.E. (18 May 2004). High Voltage Levels, Reduced Voltage Stress and Lower Semiconductor Losses. US20030036806A1, Assignee: Sustainable Energy Technologies, Alberta, CA.
51. Mohan, N. and Sahoo, A.K. (2014). High Power Application Modular Multilevel Converter Patent Protection. US Patent US20150124506.

6 Modular Systems for Energy Usage in District Heating

6.1 INTRODUCTION

The concept of district energy (DE) dates back to ancient Rome, where hot water was used to heat public baths and other buildings. Urban steam systems first became common about 100 years ago (the first North American system was built in 1877 in Lockport, New York), and modern hot water systems have been used extensively in Europe since the 1970s. Today, as modern DE rapidly gains acceptance, systems are being built in increasing numbers in cities and communities across North America. District heating (DH) systems can provide space heating and domestic hot water for large office buildings, schools, college campuses, hotels, hospitals, apartment complexes, and other municipal, institutional, and commercial buildings. Systems can also be used to heat neighborhoods and single-family residences.

There is much debate about the topic of DH as an increasingly important means of generating cheap sustainable energy for future generations. DH (or cooling) or energy is a heat-generating system that is located and distributed centrally, satisfying residential and commercial heating needs such as hot water and space heating. The heat is often generated via fossil fuel plants, but increasingly these are being replaced by "greener" methods, such as geothermal heaters, biomass boilers, and solar arrays. Nuclear power is also a controversial but useful method. Wind energy and industrial waste heat are also considered. The benefits of DE plants (see Figure 6.1) is that they have the ability to provide more efficient heating (and cooling) and better control of pollution when compared to local boilers. Research suggests that DH, when used with combined heat and power (CHP), is the cheapest way to cut carbon emissions and produces the lowest carbon footprint. These types of power plants are being developed in Denmark and other countries as stores for renewable energies, such as those generated from the wind and sun. These renewable energies are generally operated in a modular form [1–3].

There are different types of district heat systems, and sometimes these are simply heat-only plants. Others are full power-and-heat combined plants, producing electricity as well as heat. These can be very efficient, with the more advanced facilities providing nearly 80% of heat efficiency. Such DH plants are being built in Ukraine, Hungary, Slovakia, Switzerland, Brazil, and Sweden. The substations distribute the generated heat to the customer network via a series of insulated pipes, comprising of feed and return routes. These pipes are usually laid underground and the heat is usually distributed by water or steam. Steam is the less common method used, but it

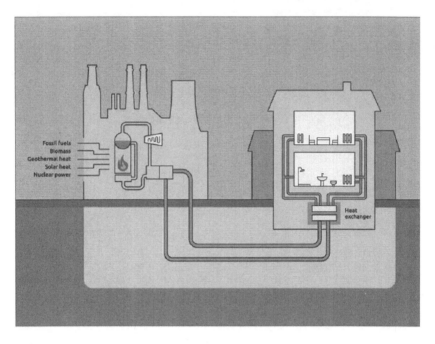

FIGURE 6.1 Animated image showing how DH works [1].

has the advantage of being available for industrial processing at higher temperatures. The heating network of pipes will then be connected to dwellings by heat substations. A system in Norway is currently operating successfully with this method, losing only 10% of thermal energy through its highly efficient distribution system. Once the heat is within the customer's building, it is usually metered to encourage customers to use energy only when necessary.

DH is currently experiencing a renaissance in the UK. Implemented across Europe during the postwar period, DH remains popular on the continent in places such as Germany, Scandinavia, and much of eastern Europe. DH in Denmark, for example, currently heats over 60% of homes with that number rising to 95% in Copenhagen. In contrast, the UK, which saw significant growth in DH with the council housing boom in the 1950s–1970s, fell out of love with DH when the North Sea natural gas network was established in the 1980s. The tide is turning, however, and the UK's energy future with regards to DH looks to be falling in line with the rest of Europe's. The last government funded, via Department of Energy and Climate Change (DECC), over 140 DH feasibility studies to the tune of over £6m. The Government's Heat Strategy published in 2013 firmly placed DH as the preferable source of heating in urban areas by 2050. Today's figure of 2% of domestic demand in the UK being fed by DH is predicted to rise to a figure of 20% by 2030.

The Canadian District Energy Association (CDEA) surveyed 118 owners and operators of DE systems in Canada in 2008 and subsequently reported the survey results [1–4]. These results indicate the scope and growth of the technology. The distribution of DE systems in Canada is described by the building floor space served.

Ontario, with 43%, has the highest share of DE in Canada. The next highest shares are in Alberta and British Colombia with 12% and 10%, respectively. The survey results also show that 27 million square meters of floor space is heated by DE in Canada. Ontario has the largest percentage of DE-heated floor space in Canada with 44%; Alberta and Quebec have the next greatest portions of the DE-heated floor space, at 17% each. There have been two major periods of growth for DE in Canada [1–4]. The first was in the 1970s when energy prices drastically increased. Not only was an expansion experienced in DH but also in CHP plants for industrial process applications. The second significant growth occurred in the late 1990s when the Canadian government encouraged the application of DE to foster sustainable energy and community planning.

DH is defined as "a pipe network that allows centralized heat sources to be connected to many heat consumers". A DH network comprises three main components: "one or more energy center(s), the pipe network itself, and connections to heat customers." It allows heat to be used from sources, which would not be normally possible within individual dwellings or buildings. Most DH systems are kept small in order to avoid excessive heat losses by long pipelines. Renewable energy and small modular nuclear reactors supported DH are generally modular in nature [1–3].

6.1.1 DH Advantages and Drawbacks

There are various advantages of DH systems over individual heat systems. The former tends to be more efficient and the larger-sized combustion units have more advanced cleaning systems in their gas flues. They can often utilize heat outputs from industry for efficient energy "'recycling." However, these systems do require long-term investment, and government policy must focus on the long-term gains. The systems are also best suited for areas with high population densities, particularly those with apartment blocks. For example, in UK, there is a large EnviroEnergy plant in Nottingham, which now heats over 4,600 homes and business premises. Scotland also has a range of district heat schemes, and Southampton has a significant scheme [2].

The key advantages of DH are that, compared to individual gas-fired boilers, DH can deliver heat in a more efficient manner, more cheaply and with lower carbon emissions. It does this mainly by capturing thermal energy or heat that would otherwise be wasted. This wasted heat is primarily created as by-product of generating electricity in power stations. Usually this is ejected into the atmosphere (via cooling towers) or to rivers. However, by generating electricity in CHP plants the wasted heat can be recovered and used in a DH network. CHP plants range in size from small 50 KWe plant up to very large ones at 250 MWe. They can be fueled by natural gas (either in a combustion engine or using gas turbines) or they can be fueled by biomass. Biomass CHP plants usually consist of boilers burning biomass, waste wood, or straw to raise steam to feed steam turbines. This produces both electricity and heat for DH networks.

Another types of CHP plants are waste incinerator plants that are increasingly being used around the world as an alternative to landfill sites for waste disposal. Again these can be used to generate both electricity and heat. Waste heat might also be utilized from industrial or refrigeration processes. This is often at temperatures (typically below 40°C) too low to be used directly for heating dwellings. However if this heat is connected to heat pumps then it can be upgraded to a higher temperature (typically over 80°C). This can then be fed to a DH system. Suitable low-grade heat sources include sewage plants, canals, high-voltage transformers, data centers, etc. As the heat pumps are electrically driven, they provide a true low-carbon source of future heat when coupled with electricity generated from renewables such as photovoltaic solar cells or wind turbines [1–3].

Another source of heat (only) is biomass boilers which can produce heat with low-carbon content. Biomass fuel has exacting handling and storage requirements. This makes biomass boilers more economic at a large scale (typically greater than 1 MW) and thus more suitable for connection to a DH system. DH can use heat from a number of these different sources. In this way, as we progressively move towards a low or zero carbon future, new low-carbon sources of heat can be connected to these thermal networks as new technology and waste heat sources become available. Depending on the exact nature of the DH system and its exact heat source the cost of delivering heat to an individual customer can be as little as 3 c/kWh. This compares to an equivalent figure of 5 c/kWh for gas heating (i.e., including fuel cost and the cost of replacing/installing the necessary capital equipment). Similarly the carbon emissions from heat generated in a DH system can be as little as 60 kg CO_2/kWh compared to the equivalent figure of 240 kg CO_2/kWh for gas heating system using individual boilers [1,2].

The main barrier to DH is its initial capital outlay. This includes the cost of extracting or upgrading the heat from the central CHP plant or waste heat source, the cost of installing the pipework network within the ground (usually DH pipes are run under roads alongside other utilities) and the cost of connecting the system into individual buildings. Again costs vary from scheme to scheme, but typically the cost of running DH pipework in an urban area can be about $2,000/m with a typical connection cost to a two-bedroom flat at approximately $2,000 each. The key to realizing a successful DH scheme, therefore, is to minimize the capital cost but maximize the number of customers connecting to the network. This is ideally done by keeping the central heat source as close as possible to the maximum number of dwellings. One simple way of analyzing this is by looking at the density of dwellings in any one urban area. If it is above eight dwellings per hectare then a DH system could be viable. A further consideration is the cost of operating the DH system. Many lessons have been learned since the first DH systems were installed in the mid-20th century and modern networks. Costs can be kept down by effective design of the system hydraulics, keeping the volume of water circulated around the system to a minimum to satisfy the actual (as opposed to the maximum) overall heating demand. Variable speed pumps linked to multiple heat sources, which are only bought online when the overall system demand requires them, are key components in a successful DH system design [1–3].

The other key consideration is to minimize heat loss from the pipework distribution. Again, this can be achieved by good design and should not exceed 5% of the total system heating capacity. Modern DH pipework is now installed with prefabricated insulation, minimizing the likelihood that this will be damaged during construction. Most DH networks start as relatively small systems serving up to 100–200 dwellings.

Typically these modular systems will consist of a single heat source (often a gas-fired CHP plant), a set of distribution pumps, the DH pipework itself consisting of flow and return pipes, and connections to individual dwellings. These final connections usually consist of a plate heat exchanger combined with a heat meter. The amount of heat delivered to each dwelling is controlled according to its combined hot water and heating demand. These smaller modular networks can then be expanded over time to serve much larger communities up to 10,000 dwellings or more. On these larger systems it is possible to have a number of different heat sources which all feed into the same system. These heat sources can be activated to match the fluctuating overall demand of the system. Typically the base or summer load of a DH network is that required to meet the hot water demand of all the dwellings or buildings. In winter this will rise to meet both the hot water and the heating demand. CHPs and heat pumps are usually used to meet the base load with gas-fired boilers only brought on when the heat demand peaks in times of very cold weather. In addition, large thermal stores (which can be up to 20,000 m^3 in volume) can be used to store surplus heat to be used to smooth out and match the fluctuating heat demand pattern [1–3]. The length of the DH network again typically starts at 1–2 km. As the system network grows these distances grow accordingly. One of the largest systems at the moment is that in Copenhagen, Denmark, which extends up to 20 km in length. As technology advances DH systems are becoming more and more efficient and provide a real tool by which to move to a zero carbon future. With the DECC pushing for their development all those involved in the renewable energy industry should start analyzing their potential as part of an integrated renewable energy mix. The DH is most efficient when it is modular in nature because very large DH system requires long-distance pipelines to serve a larger area of community. This is generally accompanied by larger heat losses. CHP approach is also more efficient in smaller and modular systems. This is particularly true when renewable or nuclear energy is used to supply heating and power needs.

DE systems (DES) use one or more central plants to provide thermal energy to multiple buildings. This approach replaces the need for individual, building-based boilers, furnaces, and cooling systems. Underground pipelines from the heating (or cooling) plant to each of the connected buildings distribute thermal energy in the form of hot water, steam, or chilled water. Energy is then extracted at the buildings and the water is brought back to the plant, through return pipes, to be heated or cooled again. When there are more than one central plants to provide thermal energy, each of them can be operated with different sources like biomass, solar, waste heat, etc. In these situations, DH is called hybrid source DE. A DH system can vary in all sizes from covering a large area, as for instance the Greater Copenhagen DH system, to a small area or village consisting of only few houses. The capacities of a DH grid can be of all sizes, depending on the size of the area. In large DH systems, the DH grid may consist of both, a transmission grid (transporting heat at high temperature/pressure over long distances) and a distribution grid (distributing heat locally at a lower temperature/pressure) [3].

Small DH grids are local concepts to supply households as well as small and medium industries with renewable heat. In some cases, they may be combined with large-scale DH grids, but the general concept is to have an individual piping grid that connects a relatively small number of consumers. Often, these concepts are implemented for villages or towns. They can be fed by different heat sources, including solar collectors, biomass systems, and surplus heat sources (e.g., heat from industrial processes or a biogas plant that is not yet used). Fossil fuel boilers could be installed for peak loads and as a backup in order to increase the economic feasibility of the overall system. Small grids do normally have commercial operators and are larger than microgrids. Micro heating grids are usually installed for fewer customers, e.g., 2–10. An advantage of microgrids is that these systems could be built easier and faster, because of the small amount of customers, without long public procedures. The customers agree on a suitable accounting for the used heat and on who is the operator of the system. Independent of the grid size, it is important not to oversize the grid during planning. Large dimensions cause higher heat losses and higher investment costs. There is a characteristic factor called "heat density" of the grid, which is calculated by the annual sold heat (MWh/a), divided by the length of the grid (in meter, length of pipeline). A common rule of thumb says that this factor should be at least 900 kWh/m per year. The goal should be to sell a high amount of heat at a grid with a short length. In case that the heat density of a potential grid is too low, individual household heating systems may be preferred.

6.2 SMALL MODULAR RENEWABLE HEATING AND COOLING GRIDS

The heating and cooling demand in Europe accounts for around half of European Union's final energy consumption. Renewable energy policies often mainly focus on the electricity market, whereas policies for renewable heating and cooling are usually much weaker and less discussed in the overall energy debate. Therefore, it is important to support and promote renewable heating and cooling concepts, the core aim of the CoolHeating project undertaken in Europe. The objective of the CoolHeating project, funded by the EU's Horizon 2020 program, is to support the implementation of "small modular renewable heating and cooling grids" for communities in southeastern Europe. This is achieved through knowledge transfer and mutual activities of partners in countries where renewable DH and cooling examples exist (Austria, Denmark, Germany) and in countries which have less development (Croatia, Slovenia, Macedonia, Serbia, Bosnia-Herzegovina). Small modular DH and cooling grids are local concepts to supply households and small and medium industries with renewable heat and/or cooling. In some cases, they may be combined with large-scale DH grids, but the general concept is to have an individual piping grid, which connects a smaller number of consumers. Often, these concepts are implemented for villages or towns. They can be fed by different heat sources, including solar collectors, biomass systems, and surplus heat sources (e.g., heat from industrial processes or from biogas plants that is not yet used, but wasted) [3].

Especially the combination of solar heating and biomass heating is a promising strategy for smaller rural communities due to its contribution to security of supply, price stability, local economic development, local employment, etc. On the one hand, solar heating requires no fuel, and on the other hand, biomass heating can store energy and release it during winter when there is less solar heat available.

Thereby, heat storage (buffer tanks for short-term storage and seasonal tanks/basins for long-term storage) needs to be integrated. The main advantages of a biomass/solar heating concept are reduced demand for biomass, reduction of heat storage capacity, and lower maintenance needs for biomass boilers. With increasing shares of fluctuating renewable electricity production (photovoltaic, wind), the Power-to-Heat conversion through heat pumps can furthermore help to balance the power grid. If the planning process is done in a sustainable way, small modular DH/cooling grids have the advantage that, at the beginning, only one part of the system can be realized and additional heat sources and consumers can be added later. This modularity requires good planning and appropriate dimensioning of the equipment (e.g., pipes). It reduces the initial demand for investment and can grow steadily [3,5].

Besides small DH, small district cooling is an important technology with multiple benefits. With increased temperatures due to global warming, the demand for cooling gets higher, especially in southern Europe in which the target countries are located. In contrast to energy demanding conventional air conditioners, district cooling is a good and sustainable alternative, especially for larger building complexes. However, experiences and technologies are much less applied than for DH. The CoolHeating includes both heating and cooling. Especially countries in southern Europe with high solar irradiation need both heating and cooling. The combination of small DH and cooling in the same planning step saves cost and efforts, even if some consumers will demand only either heating or cooling. Thereby, also technical synergies are created (piping, the use of heat pumps). Small modular DH/cooling grids have several benefits. They contribute to increase the local economy due to local value chains of local biomass supply. Local employment as well as security of supply is enhanced. The comfort for the connected household is higher: in the basement of the buildings only the heat exchangers are needed and no fuel storage tank or boiler. Furthermore, no fuel purchase has to be organized. Due to all these benefits, the objective of the CoolHeating project is to support the implementation of small modular renewable heating and cooling grids for communities (municipalities and smaller cities) in southeastern Europe [3,6,7].

In addition to retrofitting residential buildings in the neighborhood, a biomass DH network is deployed. A two-step implementation process has been developed for the DH network in order to maximize the cost-effectiveness of the investment in the distribution network. This approach emphasizes working towards reducing the distribution length and maximizing the connection rate in a smaller area. Accordingly, two locations within the neighborhood have been selected to host the necessary generation equipment, serving small range, compact distribution networks that can be interconnected in a modular approach. This way, further growth of the network is possible, and distributed generation as well as polygeneration can be more easily integrated. For the first phase that is covered in the SmartEnCity project, the estimated installed power is 5,300 kW. The primary fuel for this generation is biomass (wood chips), and natural gas is used as a backup for high-demand scenarios, with an estimated demand coverage of 5%–10%. Preliminary studies have been carried out for both the network layout (first phase) and location of the generation facilities (for both the first and second phase). The estimated space needed for this room in phase 1 is 500 m^2, and future expansions are also foreseen.

The biomass boiler rooms are equipped with biomass boilers, additional backup gas boilers as previously defined, and primary network pumping systems. A resource management center (RMC) enables optimal management of the DH production and distribution system. A website is set up for online consultations and energy consumption information. For the distribution network, a two-pipe system has been chosen. Preinsulated piping with leak detection monitoring systems is used. In the design of DH system, several parameters are important. These include selection of suitable temperature levels that include high-temperature system, medium-temperature system, and low-temperature system. The selection of return temperature and monitoring of temperature levels are also important. The design also requires selection of pipe details such as diameter, length, materials, process for its selection, method of installation, and methods to reduce heat losses. The labor cost for pipes installation and medium for heat transfer also play an important role in the efficiency of a DH system. All of these parameters for DH coolHeating project are described in detail by Rutz et al. [3].

The overall economy of a small DH project depends on many factors at different levels. For example, on the local level, costs for energy from renewable energies are always compared in the society to the costs of energy from fossil fuels. If the renewable energy system is cheaper than a fossil energy system, it will be implemented; if the costs are higher, the project will be often not realized. On the local level, the economy is influenced for instance by local support schemes. Renewable sources that can be used are biomass, solar energy, geothermal energy, and waste heat from various sources. Since the share of fluctuating renewable electricity production constantly increases, another way of producing heat could be by utilizing renewable electricity by the means of heat pumps and electric boilers. These systems, with further help from heat storage, can also help balancing the power grid.

Modularity of these systems enables that only a part of the system can be realized at the beginning with additional heat sources and consumers being added later. Therefore, the initial demand for investment can be reduced and the project can grow steadily. Using renewable energy from DH systems goes in line with the heating and cooling strategy from EU. It helps taking environmental issues of the current EU heating sector, which occurs for half of the EU annual overall energy consumption, with 75% of the fuel it uses coming from fossil fuels. DH is heat analogous of microgrids for electricity. Just like electricity from renewable sources in isolated areas is best managed by microgrids, heat transport in district is best managed by DH network. Both microgrids and DH network using renewable sources are generally modular in nature [3].

6.3 MODULAR DH BY BIOMASS

Prior to the introduction and general implementation of the use of fossil fuels (carbon, oil, and gas) **biomass** was used for heating buildings. Wood was collected from forests in order to burn it in open fires, heaters, and heating stoves, whereby the heat from the combustion could be used to cook, bake, heat the room, and also to heat water for district heating water services and, in certain cases, to send heat to radiators in other parts of the home. The biomass boilers similar to the oil boilers, however, are somewhat larger. Similarly, the volume of fuel is also greater. The biomass boilers produce ash, depending upon the fuel used, of between 0.5% and 2% of the burned fuel. This ash can be used as fertilizer or can be discarded as domestic waste. The initial investment with biomass heat is somewhat higher than fossil fuel heat; however, the savings in fuel costs provide for the timely amortization of the increased initial investment. Furthermore, public subsidies and tax benefits exist (Personal Income Tax refunds) for the promotion of installation [3].

A modular DES based on biomass has the following advantages for both system customers and the surrounding community [1–3,5–13]:

A. **Low, Predictable Energy Costs:** Biomass incurs costs are lower than conventional energy costs. Higher fuel usage provides access to the lower costs associated with bulk purchasing. The use of locally grown biomass as a portion of the fuel mix further enhances the cost-stabilizing benefit of DE. The price of wood fuel is not linked to world energy markets or unstable regions, but instead determined by local economic forces. For this reason, biomass systems do not experience the price instability of conventional fuel systems.

B. **Fuel-Type Flexibility:** Because a central heating plant can have boilers that burn different fuels, the option exists to use whichever biomass fuel is the most economical at any given time. Municipalities can incorporate DE into the infrastructure of their downtown business districts or encourage its use in new developments such as office building complexes and industrial parks. When local biomass fuels, such as wood chips, are used instead of oil or gas, the benefits of renewable energy can be brought to many buildings.

C. **Combined Heat and Power:** DE plants can be designed to produce not only heating and cooling but also electrical power (i.e., cogeneration or CHP). CHP plants are able to get more usable energy out of the input fuel than the one producing electricity only. This subjects of biomass heating and cogeneration with biomass are discussed in more detail later in this section.

D. **Better Air Quality:** Biomass is a source of clean and unlimited energy. It hardly contaminates the environment and does not contribute to the destruction of the ozone layer. Air quality improves—as does community livability—when emissions from a single, well-managed plant replace uncontrolled stack emissions from boiler plants in many individual buildings. The result is magnified when DESs, as they often do, replace multiple systems that use conventional fossil fuels. If the central system uses wood fuel, the emissions of sulfur dioxide (which contribute heavily to acid rain) will decrease, while emissions of particulates and certain toxic air contaminants will increase. The emissions increases, however, do not result in a higher concentration in the air because of changes in the location of the emissions and the improved dispersion of pollutants resulting from a single tall stack. Yet another advantage in improving air quality with DH is the ability to install best available technology emissions control equipment, which may not be affordable in individual building heating plants. More local Jobs Conventional energy systems require labor in fuel extraction, processing, delivery, operation, and maintenance as well as in system construction and installation. Fossil fuel supply is based on energy resources outside the community; thus, all jobs associated with extraction and processing are outside the local and regional economies. By contrast, jobs and most of the raw materials associated with wood fuel extraction, reforestation, and fuel transport are within the local and regional economies.

E. **Dollars Remain in the Local Economy:** The use of biomass promotes the creation of local jobs. Unlike fossil fuels, which generally comes from the outside region, wood fuel is a local and regional resource. The businesses associated with wood supply (logging operations, trucking companies, and sawmills) tend to be locally owned, so that profits are retained in the regional economy. These activities contribute to the state and local tax base. Conversely, the use of fossil fuels creates a net economic drain on a community and state. The Vermont Job Gap Study found that Vermonters spend more than $1 billion annually for fuel and energy imported from outside the state.

F. **Revitalized Communities:** DE infrastructure and stable rates improve a community's business climate, make local businesses more competitive, help to revitalize downtowns and urban core areas so they can better compete with suburban sprawl, and, using biomass as the fuel source, help build a sustainable infrastructure. Reliable Equipment DESs have an unparalleled record of reliable service. They achieve this by well-managed central plant operation, by using multiple fuels, by having backup boilers in one or more locations, and by having standby power at the central plant. The use of biomass helps to clear mountains and promotes the recycling of other industrial wastes.

G. **Use of a Plentiful and Renewable Resource:** Biomass is a renewable resource that can continue to replenish itself when managed and harvested sustainably. Wood-fired heating systems provide a market for lower-grade wood not suitable for furniture or other high-profit products. These markets can be especially critical for restoring commercial and biological quality to harvested forests. In addition, the use of waste wood for energy can reduce the need for and costs of disposal.

H. **Reduced Environmental Risks:** DESs can help to mitigate environmental risks by consolidating fuel storage to one or a very few locations compared to numerous onsite storage tanks that serve individual buildings. Conventional onsite fuel storage includes underground and above-ground storage tanks. Above-ground tanks can pose ire hazards as well as the risk of dislodging in the event of a flood. Failing underground tanks can pose a threat to ground and surface waters.

I. **A Meaningful Way to Address Global Climate Change:** Carbon dioxide (CO_2) is the major greenhouse gas implicated in global warming. When fossil fuels are burned, carbon that was sequestered underground (as oil, gas, or coal) is converted to CO_2 and released into the atmosphere. While CO_2 is a major component of the combustion emissions of both fossil and biomass fuels, burning biomass for energy, over time, adds no net CO_2 to the atmosphere. For biomass energy to be an effective climate change mitigation strategy, however, the biomass must be harvested in a fashion that sustains the forest resource and increases its vitality and productivity over time. If a forest is clear cut and does not regenerate, there will be no trees to sequester, and carbon and CO_2 levels in the atmosphere will increase. A study by the Oak Ridge National Laboratory found that "using part of the forest harvest residue for DH in Vermont has a positive impact on reducing the amount of carbon discharged to the atmosphere [3]."

The use of biomass, however, also carries some disadvantages:
1. The heat produced through biomass boilers is somewhat lower than the heat produced by liquid fossil fuel or oil boilers.
2. Biomass has a lower energy density, which means that the storage systems must be larger.
3. The fuel storage systems and ash removal systems may represent higher maintenance costs.
4. The biomass distribution channels are not as developed as the fossil fuel distribution channels: oil and natural gas.

6.3.1 DH with Modular Biomass CHP

A heat-demand driven CHP unit generates only the amount of heat that is actually needed. If less heat is needed, also less electricity is generated. Ideally, this concept is used, when there is a constant heat demand and 7,500 up to 8,760 full load hours per year. If the heat demand is varying, or if it is decreasing during certain periods, the CHP unit is operated at partial load, according to this definition. This leads to less full load hours (2,000–3,000 h) for DH systems, in which just domestic consumers for space heating are connected [6,7].

A power-demand driven CHP unit only generates the amount of power that is actually needed or that can be fed into the power grid. Most installed biomass CHP units are designed to generate green electricity according to a guaranteed feed-in tariff. Thus, nearly all power-demand driven systems are either operated at maximum full load hours or at grid-related demand. In some countries, such as in Germany, dedicated incentives were introduced to double the capacity at peak power load (e.g., during the day) and to stop operation at low power load (e.g., during night). Thus power-demand driven CHP units will play an increasingly important role in balancing the power grid. If this power-demand driven concept is applied, the heat supply may not match a potential heat demand, or may be too much. In this case, the surplus heat is often wasted. This led to the situation, that units have been installed, which wasted up to 70% of primary energy. After a few years nearly all countries reacted via their legislation and since then the heat utilization of 40%–50% is mandatory for plants that apply for a feed-in tariff. This increases the overall efficiency of the biomass CHP unit to approximately 70%. Hence, the installation of a CHP just makes sense if most of the heat is utilized and if a minimum revenue is gained [1–3,5–15].

Historically, biomass technologies for CHP generation were selected according to the thermal and electric capacity of the system. At this time, Organic Rankine Cycle (ORC) systems were selected for small- to medium-scale systems and steam turbines for large-scale systems. Both are thermodynamic processes based on the principle of the Rankine Cycle. The development of highly efficient steam turbines has been driven by large-scale coal or nuclear power plants with an electrical power of a few hundred MW. For smaller scale ORC processes have been developed, which provide some advantages. The main difference between a steam process and an ORC process is the working media. Water, respectively steam, is replaced by an organic fluid for an ORC process, which has different condensation and evaporation temperatures. With these properties the process can be designed according to the needs of the heat consumer and the heat source. Thus, ORC processes have been optimized for a lower temperature level for the produced thermal energy of 85°C–95°C and a temperature level for the heat source (biomass boiler) of 250°C–350°C. With these parameters, the ORC process is slightly more efficient than a steam cycle as a whole. Another practical reason for choosing this technology is the low effort for operation and maintenance.

Some ORC manufacturers produced standardized ORC modules with complete long-term maintenance contracts. This increased the reliability in such a way, that operation with minimum human resources was achieved easily. Another important condition for choosing ORC system was the need for trained operation personnel. In most EU countries a special education was required for the operation of a steam boiler. Due to lower pressures, temperatures, and different fluid conditions of the organic working media of an ORC process, no such special education was necessary to operate these plants. Finally, the entire ORC plant performed slightly better within the overall economic life cycle assessment compared to different types of steam. Due to this faster market readiness, today ORC plants are widespread. When the market developed rapidly in Europe between 2002 and 2010 due to the provided feed-in tariff for green electricity some steam turbine manufacturers started to develop small-scale steam turbines, and nowadays these two technologies, ORC and steam cycle, are performing similar from an economic point of view [12].

Biomass gasification systems are known for more than 100 years, but became mature after 2002 for mid- to large-scale systems and after 2012 for small-scale systems. Based on some demonstration and commercial plants gasification is nowadays applied to many heating and cooling projects. Gasification is a process that converts the solid biomass into a usable gas consisting mainly of hydrogen, methane, carbon monoxide, and carbon dioxide. The biomass reacts at high temperatures ($>600°C$) with a controlled amount of oxygen ($0<\lambda>1$) to give producer gas. This step is similar to the first step of a combustion process, where gas is converted to gaseous products. On the contrary to a combustion process this gas is not incinerated in situ. Therefore up to 80% of the chemical energy of the biomass is contained in the producer gas. This gas is used in a gas engine to generate power and heat in the case of a CHP. If another gasification agent other than air is used, syngas is produced, but for the utilization in a gas engine, it is sufficient to use air as a gasification agent. The mass production led to this price decrease, thus gasification is nowadays also a good option at small scale [14,15]. All aforementioned technologies have been assessed within different studies. One report combining the summary of some previous studies updated with latest development shows that the discussed processes follow a common economy of scale, whereas just small differences between the discussed CHP processes occur [13]. Use of CHP for DH is rapidly expanding both at small and medium scales. Modular approach has facilitated this momentum.

6.3.2 Examples of Modular Biomass Based DH Schemes

FVB energy Co. is a Swedish company with branches in USA and Canada specializing in DH. FVB corporation is a world leader in biomass-DH systems. They have installed numerous such systems in many parts of the world. Here we illustrate their few success stories for modular biomass-based DH schemes [3,16,17].

Dalhousie University Agricultural Campus 5.4 MWth Biomass Combustion Boiler Wood Waste Fuel is in Operation since 2018. FVB was retained by Dalhousie University to advise on the most economical and sustainable option for executing on their community feed in tariff (COMFIT) contract while simultaneously designing the renewal of their existing biomass steam Energy Center which had reached end of life. It was ultimately decided that the preferred option involved converting the facility and the campus from steam to hot water and installing a biomass combustion boiler. This included converting one of the existing oil-fired steam boilers to hot water, retrofitting campus building mechanical and controls systems, and replacing the distribution piping system to utilize hot water. FVB chose a new 5.4 MWth biomass combustion boiler that generates hot oil that is circulated to an ORC Generator, producing 1.0 MW of electricity and 4.4 MW of heat. The 1.0 MW of power is sold to the local electrical grid year round while the excess heat will be used to provide their campus with medium-temperature hot water for space heating and domestic hot water. A steam to hot water conversion was completed on the campus in 2017; the campus is now served by a hot water DES, which this Energy Center feeds.

In 2008, FVB performed an initial concept study for the new residential development university, located at Simon Fraser University (SFU) in Burnaby, BC, which recommended biomass as a base load renewable energy source for a new hot water DH system to serve the development. Since 2011, FVB has provided design and construction support for nine connected Energy Transfer Stations, 1,150 additional trench meters (3,800 trench feet) of piping, and two interim Energy Centers for the University System. All of this was completed while FVB supported Corix pursuing approvals for the future biomass energy center, and a base load interconnect to the SFU Campus System. The interconnect includes a 10 MWth heat exchanger, 830 trench meters (2,700 trench feet) to the university, and 560 trench meters (1,800 trench feet) to SFU.

FVB has supported Corix in the development and design of a permanent Energy Center: 15 MWth biomass module, a 10 MWth natural gas module, and associated equipment to provide a base load renewable energy to both the SFU campus and university developments, and peaking/backup capacity for the University District Energy Utility. This includes provisions for fuel storage, a biomass combustion system, thermal energy exchange, and flue gas heat recovery. The biomass module also includes an emergency power generator capable of running the natural gas module and allowing for the safe shutdown of the biomass module. A 6 MWth heat pump was designed to dissipate thermal energy in a safe and controlled manner when the biomass module needs to be shut down in emergency situations. The use of thermal oil creates a safer working environment with reduced operating pressures over steam, and also allows for the future addition of power production equipment, turning the operation into a CHP facility [3,16,17].

The biomass-based Prince George DES provides heating for several landmark buildings in downtown Prince George, reducing greenhouse gas emissions by 1,900 tons/year. FVB was involved in the concept development of the modifications at Lakeland necessary to access this "excess and available" biomass energy. The state-of-the-art DES takes waste heat from the Lakeland sawmill and transfers it via insulated piping to heat the downtown core of the city, a move that makes sense financially, environmentally, and socially. The mill produces heat by burning biomass waste for the purpose of drying sawn lumber. FVB's system captures the waste heat from this process and uses it to heat water for the DES. FVB provided feasibility studies, business case analysis, marketing, design, construction, and operations support for the DES, the 7.5 MWth heat recovery energy center, the 5 MWth capacity peaking/backup energy center (expandable to 10 MWth in future), 2,500 trench meters of piping (82,000 trench feet), and 9 building ETS connections. FVB helped the city overcome the challenge of locating the Energy Center by presenting the city with cost estimates based on piping requirements and the Energy Center footprint. FVB also drafted energy service agreements and assisted the city in a marketing campaign to present DES as an attractive alternative to conventional forms of thermal energy.

For Sala-Heby Energi AB, FVB provided the initial concept design analysis and detail design for the central heating plant, distribution piping system, system hydraulic modeling, and building connections. Based on FVB's feasibility study, Sala-Heby Energi constructed a biomass-fueled CHP plant. The biomass CHP plant produces 10 MWe and 22 MWth. From 1998 to the present FVB continues to assist Sala-Heby Energi in expanding their system and connecting new customers. FVB has also prepared the detail design and tender documents for an additional flue gas condensation plant. FVB continues to provide support for combustion optimization, control systems, and other miscellaneous system enhancements.

For Katrineholm Energi AB, FVB provided the initial concept design analysis and detail design for the biomass-fueled central heating plant, DH distribution system, and customer connections. The fuel preparation process included cleaning dirt and rocks from tree residue, chipping the residue into smaller pieces, transporting and storing the fuel, and finally conveying the wood chips into the boiler. FVB also provided engineering services during the project construction and evaluated the feasibility of generating electricity sequentially with DH production (combined heat and power). FVB was later hired to study the feasibility of adding a biomass-fueled CHP plant. At the time the optimum size was 14 MW of power and 31 MW of thermal energy. The DH demand is 210 GWh/year and the peak demand is close to 80 MW.

FVB was hired to study the feasibility for a biomass-fueled CHP Plant installation to supply the existing DH system with thermal energy. FVB optimized the plant size and configuration. The combustion system is a bubbling fluidized bed

(BFB). The CHP plant is sized to supply 34.8 MW of power and 58.6 MW of thermal energy from the condenser and 11.5 MW thermal from the flue gas condenser unit. The BFB boiler can be operated on wood waste, peat, coal, or oil. The 97.5 MW boiler is rated at 144 ton/h at 140 bar, 540°C. The DH demand is 300 GWh/year and the peak demand is 115 MW. The study included conceptual design and performance calculations for each option as well as capital cost estimates and financial analysis [3,16,17].

FVB has prepared the tender documents for the plant, tender evaluation, purchase, and project management and delivery inspection for the owner. The plant consists of a complete automatic biomass fuel handling and combustion system. The boiler is a BFB producing hot water for the DH system in Arvika. The plant has a production capacity of 17 MW and an additional load of 4 MW from the flue gas condensation plant. The project was executed in 2000–2002 [3,16,17].

6.3.3 MODULAR BIOMASS BASED DH IN EUROPE

A. **Denmark, Austria, Sweden, and Ireland:** Biomass, including waste wood and plant material, is extensively used for DH as well as CHP production in Denmark, Austria, and Sweden. It has also found application in cofiring with coal for power generation. However, the relatively high cost and limited availability of local biomass, the availability and lower cost of imports, and the inconvenience of handling biomass compared with competing fuels such as oil and gas has prevented use of biomass for DH in Ireland. The greatest potential for DH in Ireland is in large-scale schemes that target apartments and services (in Dublin) as well as small-scale high-occupancy buildings in the rest of the country located where more expensive existing heating sources prevail (i.e., that are not connected to the gas network). Many Central and Northern European countries have DH systems where the community's needs for hot water and space heating are met by metered heat piped from a central source such as a boiler or CHP plant. Such systems are highly suitable for a centralized biomass-fired cogeneration of electricity and heat [1–3,5,8–15,18–20]. The Spittelau incineration plant to provide DH in Vienna is illustrated in Figure 6.2.

B. **Spain:** In Torrelago district in Laguna de Duero, Valladolid (ES), Spain, district residents have seen a complete retrofitting including renovation of the buildings' façade and an upgrade to a DH system that now integrates renewable energy and smart control solutions. This integrated approach leads to 50% energy savings in the district and a general empowerment regarding energy management issue. Torrelago's DH transition was integrated in the CITyFiED project. Torrelago is a residential district located in Laguna de Duero, near Valladolid, formed by 31 buildings that respond to three different typologies, containing a total of 1,488 dwellings where more than 4,000 residents live. Every building has ground plus 12 floor levels, occupied by an entrance hall and a total of 48 dwellings each with a surface of about 80 or 95 m^2, representing over 140.000 m^2 in total.

FIGURE 6.2 The Spittelau incineration plant is one of the several plants that provide DH in Vienna [1].

Prior to retrofitting, the district had two different DH systems, composed by two independent gas-fired boiler rooms. The system was controlled by an analogical system that managed the flow temperature from the boiler and the temperature of storage domestic hot water tank. A shift to a biomass boiler was considered as the best option. Both DH systems were joined into one in order to harmonize the district overall consumption. A new building was built to place the biomass boiler room and biomass silo. The underground area occupied by the biomass silo is 72.4 m^2 and height 5.9 m, with a useful volume of approximately 400 m^3. Using a digital control system and modulating regulation, boilers reach a performance of 90%. Due to the refurbishment in the DH system, the share of energy savings is increased from 40% to 50%, and the system saves 3,392 tCO$_2$/year compared to the initial gas-fired system, presenting a CO$_2$ reduction of 94% [3,15,18–20].

C. **Bulgaria:** In Varna, Bulgaria, the heart of the DH is a biomass boiler of 6.500 kW, which works with untreated wood (such as wood chips, bark, and sawdust) from sawmills and forestry workings from the area. In support of this boiler there are two auxiliary boilers of natural gas, each with a thermal power of 7,500 kW, which are used to cover the peaks of demand and as a reserve in case of breakdown or failure of the biomass boiler. The annual production of heat power station is of 30,000,000 kWh. The contribution of the auxiliary boilers of natural gas is approximately 2%.

The central of Varna also produces power due to a group of ORC turbogenerators. The heat produced by the biomass boiler is used inside the closed loop of the ORC to evaporate silicone oil. When the high-pressure gaseous state is being reached, the fluid expands in the turbine, generating electricity and is then condensed, while giving up heat to the circuit of the DH network. Due to this highly efficient technology in Varna, with approximately 5,000 kW of thermal power, 980 kW of electrical power can be generated and a thermal power of about 3,800 kW is retrieved. With the use of the ORC group the biomass power central can produce 5,000,000 kWh of electricity annually. The DH central is managed by the Department of Public Works of Bressanone, and its network of heat distribution has been linked to the DH network of the adjacent municipality of Bressanone. This way, the DH network serving Varna and Bressanone reached the length of 120 km, allowing to deploy more than 70 million kWh/year of thermal energy and to serve 1,700 users, including 1,400 in Bressanone and 300 in Varna [3,15,18–20].

D. **England:** In England, Euroheat supplies the Energy Cabin; a biomass heating system housed inside a purpose-built box that simply has to be plumbed and wired-in on site—taking days instead of weeks to install. Prefabrication also helps keep costs down—by up to 30%—due to reduced man hours, plus there is no need to build a dedicated plant room. Designed for projects where the requirement is needed quickly, particularly those where planning may be an issue, prefabrication is suitable for industrial estates and manufacturing facilities, where premises can't be closed, in some instances, at all. Another benefit is that the "box" is considered a nonpermanent structure and can avoid planning legislation all together. Energy Cabins can negate planning permission, making them suitable for listed properties. In general, historic sites are often very difficult to work on, especially where the trappings of the 21st century must remain largely out of view, another factor that modular DH can help with.

A Euroheat DH system is currently being installed by "Northdown Wood & Heat" on an estate of over 20 properties; comprising a mansion, let homes, and a church. With its own managed woodland, biomass, in the form of wood chip, is the ideal fuel. The carbon footprint of the whole scheme is kept to a bare minimum due to the extremely short distance "from tree to radiator." The project is one of the first outings for Euroheat's newest and largest boiler, HDG M400; a 400 kW chip boiler, of which there are two on this site. Combined with an 8,000 L accumulator in the purpose-built boiler room and another 5,000 L accumulator within the village itself, extensive care has been taken to ensure that the presence of biomass is hardly felt by residents and visitors. Each home just has a small heat exchanger—small enough to fit under a sink.

With one, purpose-built site, fuel will need to be delivered less frequently than if each building was individually powered by biomass. A large hopper means this fuel can be stored for months at a time and, if designed correctly, access to the remote plant room will be lorry friendly and, if necessary, out of end users' line of vision [3,6,7,15,18–20].

6.4 MODULAR GEOTHERMAL DH

Geothermal DH was used in Pompeii and Chaudes-Aigues since the 14th century. Direct use geothermal DH systems (GDHS), which tap geothermal reservoirs and distribute the hot water to multiple buildings for a variety of uses, are uncommon in the United States, but have existed in America for over a century. In 1890, the first wells were drilled to access a hot water resource outside of Boise, Idaho. In 1892, after routing the water to homes and businesses in the area via a wooden pipeline, the first GDHS was created. As of a 2007 study, there were 22 GDHS in the United States. As of 2010, two of those systems have shut down [1,3,21–23].

Small modular GDHSs include distribution of heat from a central source to individual customers by means of hot pressurized water flowing through distribution pipes. This heat is not only used for space heating and domestic hot water preparation but can also be used for low-temperature industry needs. Renewable district cooling systems work on similar principles. Those systems use heat gained from renewable sources to generate cold by means of absorption. Using geothermal energy for DH systems goes in line with the Heating and Cooling Strategy from EU. It helps tackling environmental issues of the current EU heating sector, which accounts for half of the EU's annual overall energy consumption with 75% of the fuel it uses coming from fossil fuels.

Main advantages of modular geothermal energy DH and cooling systems are:

- Local economy is increased due to local value chains of local geothermal supply
- Local employment and security of supply are enhanced
- Comfort of the connected households is increased
- Security risk due to fuel combustion in households is eliminated, and the usable space in buildings is increased
- Environmental pollution is reduced and air quality is improved.
- Chilled water is provided to the customer by distribution pipes.

Since the share of fluctuating renewable electricity production constantly increases, another way of producing heat could be by utilizing renewable electricity by means of heat pumps and electric boilers. These systems, with further help from heat storage, can also help balancing the power grid. Modularity of these systems enables that only a part of the system can be realized at the beginning with additional heat sources and consumers being added later. Therefore, the initial demand for investment can be reduced and the project can grow steadily [1,3,21–23].

The main benefits of geothermal heating and cooling are the provision of local and flexible renewable energy for base load, diversification of the energy mix, and protection against volatile and rising fossil fuels prices. With advanced recovery techniques geothermal energy is now accessible in many parts of the world. For example, in Europe, 25% of the population lives in areas directly suitable for geothermal energy for DH (GeoDH [23]). Currently, there are around 250 geothermal DH systems (including cogeneration systems) in operation in Europe, with a total installed capacity of about 4,400 MWth and an estimated annual production amounting to some 13,000 GWh/y [1,3,21–23].

There has been an increase in development of GDHSs in the past few years, in particular in France, Germany, and Hungary. There are 200 planned projects (including upgrading of existing plants), which will imply that the capacity will grow from 4,500 MWth installed in 2014 to at least 6,500 MWth in 2018. Hence, geothermal heat is exploited in many parts of Europe, and there is a potential for further utilization of this renewable energy source. Since geothermal heat is available in many parts of Europe, DH is a method by which geothermal heat can be economically distributed on to buildings at large and small scales.

A key characteristic for geothermal energy is its relatively high investment costs, in particular in areas, where the reservoir is deep underground. Thus, geothermal energy is best feasible in areas with relatively high temperature levels at relatively shallow depths and if it can be supplied as base load capacity to a relatively large DH system. Another key characteristic, in particular regarding the deep reservoirs, is the risk associated with drilling of boreholes of 2–3 km depth. Depending on the accessible temperature level, it may be reasonable to combine geothermal energy with heat pumps in order to increase the temperature levels. These can be either electrical or absorption heat pumps, which can be driven by other renewable energies such as biomass boilers. Hence, the utilization of geothermal energy sometimes implies considerable additional inputs such as biomass or electricity. This not only affects the operation costs, which are relatively low for the geothermal energy itself (pumping costs), but also includes costs for electricity and/or biomass in case of application of heat pumps.

The pumping costs increase with the depth. From experiences in Denmark, it is thus economically more attractive to use heat pumps and extract heat from shallower reservoirs, typically at 1,000–3,000 m depth, where temperatures are 30°C–90°C. This geothermal gradient of 30°C for each 1,000 m depth is a general rule of thumb [24]. When planning geothermal plants, the annual energy production should therefore be relatively large since it must be able to pay back and write off the cost of wells and surface facilities. Based on data from the Danish Energy Agency, a DH system should have an annual sale of at least 400–500 TJ before the geothermal heating prices are competitive with a current price ratio (experiences from Denmark). This may vary from country to country, depending on the geothermal potential. The potential of deep geothermal is significant. However, geothermal DH is at present poorly developed. Four key areas have been identified as important to improve this situation [23]:

- Consistent energy strategies aiming to decarbonize the heat sector
- The removal of regulatory and market barriers, and simplified procedures for operators and policy makers

- The development of innovative financial models for GeoDH projects, which are capital intensive
- The training of technicians, civil servants, and decision-makers from regional and local authorities in order to provide the technical background necessary to approve and support projects.

There are several support projects organized by CoolHeating project concerning the use of geothermal for DH and cooling, such as the modular GDHS in Ozaij, Croatia managed by the University of Zegreb, in the City of Ozalj (Croatia), by the Municipality of Ljutomer (Slovenia), Visoko (Bosnia and Herzegovina), and Karposh (Macedonia), and in the City of Sabac (Serbia). These projects are making good progress [1,3,21–24].

Along with efforts under CoolHeating project, in Stockholm, the first heat pump was installed in 1977 to deliver DH sourced from IBM servers. Today the installed capacity is about 660 MW heat, utilizing treated sewage water, seawater, district cooling, data centers, and grocery stores as heat sources. Another example is the Drammen Fjernvarme DH project in Norway which produces 14 MW from water at just 8°C, and industrial heat pumps are demonstrated heat sources for DH networks. Among the ways that industrial heat pumps can be utilized are [1]:

1. As the primary base load source water from a low-grade source of heat, e.g., a river, fjord, data center, power station outfall, sewage treatment works outfall (all typically between 0°C and 25°C), is boosted up to the network temperature of typically 60°C–90°C using heat pumps. These devices, although consuming electricity, will transfer a heat output three to six times larger than the amount of electricity consumed. An example of a district system using a heat pump to source heat from raw sewage is in Oslo, Norway that has a heat output of 18 MW(thermal) [25].
2. As a means of recovering heat from the cooling loop of a power plant to increase either the level of flue gas heat recovery (as the DH plant return pipe is now cooled by the heat pump) or by cooling the closed steam loop and artificially lowering the condensing pressure and thereby increasing the electricity generation efficiency.
3. As a means of cooling flue gas scrubbing working fluid (typically water) from 60°C postinjection to 20°C preinjection temperatures. Heat is recovered using a heat pump and can be sold and injected into the network side of the facility at a much higher temperature (e.g., about 80°C).
4. Where the network has reached capacity, large individual load users can be decoupled from the hot feed pipe, say 80°C and coupled to the return pipe, at e.g., 40°C. By adding a heat pump locally to this user, the 40°C pipe is cooled further (the heat being delivered into the heat pump evaporator). The output from the heat pump is then a dedicated loop for the user at 40°C–70°C. Therefore, the overall network capacity has changed as the total temperature difference of the loop has varied from 80°C–40°C to 80°C–x (x being a value lower than 40°C).

Concerns have existed about the use of hydroflurocarbons as the working fluid (refrigerant) for large heat pumps. Whilst leakage is not usually measured, it is generally reported to be relatively low, such as 1% (compared to 25% for supermarket cooling systems). A 30-MW heat pump could therefore leak (annually) around 75 kg of R134a or other working fluid. Given the high global warming potential (GWP) of some hydrofluoro carbons, this could equate to over 800,000 km (500,000 miles) of car travel per year. However, recent technical advances allow the use of natural heat pump refrigerants that have very low GWP. CO_2 refrigerant (R744, GWP=1) or ammonia (R717, GWP=0) also have the benefit, depending on operating conditions, of resulting in higher heat pump efficiency than conventional refrigerants. An example is a 14 MW (thermal) DH network in Drammen, Norway which is supplied by seawater-source heat pumps that use R717 refrigerant, and has been operating since 2011. Water at 90°C is delivered to the district loop (and returns at 65°C). Heat is extracted from seawater (from 60-ft (18 m) depth) that is 8°C–9°C all-year round, giving an average coefficient of performance (CoP) of about 3.15. In the process the seawater is chilled to 4°C; however, this resource is not utilized. In a district system where the chilled water could be utilized for air conditioning, the effective COP would be considerably higher.

In the future, industrial heat pumps will be further decarbonized by using, on one side, excess renewable electrical energy (otherwise spilled due to meeting of grid demand) from wind, solar, etc. and, on the other side, by making more of renewable heat sources (lake and ocean heat, geothermal, etc.). Furthermore, higher efficiency can be expected through operation on the high-voltage network [1,3,21–24].

6.5 DECARBONIZING DH WITH MODULAR SOLAR THERMAL ENERGY

DH is a network providing heat, usually in form of hot water. In a DH system, the heat is generated on a larger scale. Therefore, solar thermal, as other technologies, can be scaled up to provide large quantities of hot water. Hence, solar DH (SDH) plants are a large-scale application of conventional solar thermal technology. These plants are integrated into local DH networks for both residential and industrial use. During warmer periods they can wholly replace other sources, usually fossil fuels, used for heat supply. Due to developments in large-scale thermal storage, it is now also possible to store heat in summer for winter use. Solar thermal can also meet a share of the heating demand during the winter. These systems are applicable wherever there are DH networks. Usually such networks exist in cold climates, where there is a substantial demand for heat during autumn and winter. These systems consist of solar thermal plants, made up of hundreds of solar thermal collectors. Considering the requirements of such large systems, larger collectors working with bigger loads have been designed specifically for such application. For smaller systems (block heating), normal solar thermal collectors, either flat plate, evacuated tube or even concentrating, can be used. Economic and environmental benefits derived from the acknowledged reliability of this solar thermal application, combined with the technical expertise gained over decades, have contributed to the growing interest in its commercial operation, and currently there are many plants in operation in Sweden, Denmark, Germany, and Austria. All solar energy powered DH systems are modular in nature [1,25–33].

These solar thermal plants supply heat to a DH network. It can consist of a centralized supply, where a large collector field delivers heat to a main heating central. It can also provide, directly or indirectly, a large seasonal heat store that will contribute to increasing the input of solar thermal plant to the whole system. The temperature requirements highly depend on the currently used temperature in the grid and just follow the demand. The other possible configuration is a decentralized supply or distributed SDH. In this case, solar collectors are placed at suitable locations (buildings, parking lots, small fields) and connected directly to the DH primary circuit on site. This solution can also be interesting for smaller DH networks or block heating networks. A system is considered as very large when it is over 350 kWth (500 m^2), but SDH systems can reach sizes 100 times bigger, i.e., 35,000 kWth. The benefits of solar thermal systems, in particular for such large systems, cover environmental, political, and economic aspects. Environmental benefits relate to the capacity of reducing harmful emissions. The reduction of CO_2 emissions depends on the quantity of fossil fuels replaced directly or indirectly, when the system replaces the use of carbon-based electricity used for water heating. Depending on the location, a system of 1.4 MWth (2,000 m^2) could generate an equivalent of 1.1 GWhth/year, a saving of around 175 kg of CO_2 [1,25–33].

Political and economic benefits are associated with the potential savings in energy costs and the possibility of improving energy security by reducing energy imports, while creating local jobs related to manufacturing, commercialization, installation, and maintenance of solar thermal systems. Regarding energy costs, and potential savings, there are three main aspects to consider that have a bigger impact on the comparable costs of the energy produced by a solar thermal system. These are the initial cost of the system, the lifetime of the system, and the system performance. According to the International Energy Agency, for large systems in Europe, the investment costs can go from 350 to 1,040 USD/kWth (315 to 936 EUR/kWth). In terms of energy costs, it can range from 20 to 70 USD/MWhth (18 to 63 EUR/MWhth) in Southern United States and between 40 and 150 USD/MWhth (36 and 135 EUR/MWhth) in Europe [1,25–33].

Europe

In Europe, the largest progress for the use of solar energy for DH is made in Denmark, Germany, and Austria. In the European Union there are close to 300 systems over 350 kWth in size, feeding into DH. The total capacity installed amounts to 1,100 MW. The combination of solar thermal and DH is a "very good solution" for reducing CO_2 emissions as well as for reaching the EU goal that calls for an 80% reduction in CO_2 emissions by 2020 compared to 1990 (Heating and Cooling strategy). Solar thermal energy is CO_2-free, solar energy is available everywhere, and the heat generation costs are predictable for the coming 25 years. Furthermore the technology is fully developed and mature.

There are, however, challenges to overcome when implementing such projects. The biggest hurdle stems from the space required for renewable energy such as solar thermal. In order to keep costs to a minimum, they need to be installed close to the heat consumers. Yet, land in or close to urban areas is usually limited and expensive. It is still a niche technology and the knowledge about it is not yet widespread. In addition, there is a cost competition with other heat generation technologies, for example with cheap natural gas. Nevertheless, solar thermal plants can nowadays achieve a solar fraction of up to 50% in DH. The generation costs are around 35–60 $/MWh. For this year market researchers predict that solar thermal plants in Europe will already contribute one terawatt hour (=1 billion kWh) to the DH supply. Use of solar heat for DH has been increasing in Denmark and Germany in recent years. The systems usually include interseasonal thermal energy storage (TES) for a consistent heat output day to day and between summer and winter. Good examples are in Vojensat at 50 MW, Dronninglund at 27 MW, and Marstal at 13 MW in Denmark. These systems have been incrementally expanded to supply 10%– 40% of their villages' annual space heating needs. The solar thermal panels are ground-mounted in fields. The heat storage is pit storage, borehole cluster, and the traditional water tank. The scale up of these systems is possible because of their modular nature [1,25–33].

Denmark is way ahead in the use of solar energy for DH. By now there are more than 110 systems with about 700 MW of thermal capacity. The reason why Denmark was able to take the leading position lies also with the political and infrastructural circumstances. On the one hand there are high taxes on fossil fuels like oil and gas. On the other hand DH is very widely spread in Denmark. The Danish examples show very clearly how renewable energy and CHP can be combined in a "smart" way on a local scale by deploying large heat storage units. They allow CHP plants and power-to-heat systems to operate optimally. As for example, since 2012, solar-assisted DH located in Vojens, Denmark has met all the expectations of an SDH installation. The pilot scheme with a 17,000 m^2 (11.9 MWth) large collector field convinced the municipal utility from southern Denmark to add another 52,500 m^2 (36.75 MWth) to the field, as well as seasonal storage, which should increase the annual solar share from the 14% measured in 2014 to an expected 45%. Grid temperatures have been set low to match the solar feed-in—they are 75°C–77°C in summer and 37°C–40°C in winter, respectively. Solar heat prices, including seasonal storage, should be around 42 EUR (52 USD)/MWh, compared with 57 EUR (63 USD)/MWh for natural gas supply. Hence, the plant is purely a commercial venture, which has not been supported by direct subsidies.

On the whole in Denmark, more than 1.3 million square meters of solar collectors are connected to DH. In the last 10 years in Denmark, the annual growth in the total collector area has been 42% in average. Similarly, the average increase in the number of DH plants (see Figure 6.3) with solar heating has been 29%. A number of plants choose to expand their solar capacity after a few years, and so far 16 plants have increased their existing solar thermal capacity—some even more than once. As always there are some significant uncertainties in the planned development. The new plan development, which is extended to mid-2019 (though with an upper limit of 8,000 MWh), which could boost the number of installations again [1,25–33].

While Denmark is the pioneer country in terms of SDH, Graz, Austria has become a pioneer city for solar thermal energy in DH. The project "BIG SOLAR Graz" Austria) achieved a major milestone: the land for the construction of a large-scale solar thermal storage with a technical building as well as a relevant part of a

FIGURE 6.3 Central solar heating plant at Marstal, Denmark. It covers more than half of Marstal's heat consumption [1].

future 450,000 m² solar collector field have been secured. Graz is also home to the largest solar thermal plant in central Europe (5.4 MWth), which has been operating for a while now as well as to new flagship projects such as, for example, the "HELIOS project"; a large-scale thermal storage built on a former domestic refuse landfill and fed by three different heat sources: a solar thermal plant, a power-to-heat module, and a CHP plant powered by landfill gas [1,25–33].

One best practiced example is the HELIOS plant, a project by Energie Graz. During the first construction phase last year, 2,000 m² of flat plate collectors with a 2,500 m³ heat storage were installed on a domestic refuse landfill in the city area of Graz. Heat is also generated by a power-to-heat module with a capacity of 90 kW as well as a 170 kWth CHP plant powered by landfill gas. The feed-in capacity to DH amounts to up to 10 MW. An intelligent storage management system ensures that peak loads in the heating network are diverted, so that heat from renewable sources is prioritized. The system has been in trial operation since few years. Energie Graz is planning to expand the collector area to 10,000 m². Another practical example is the DH Puchstraße, where already 7,750 m² of solar collectors are fed into the DH system of Graz. About 5,000 m² of collectors were installed as roof-mounted systems in 2007, and 2,750 m² were added on the ground by the heating plant in 2014. In this system, where the collectors are also ground-mounted, collectors from different manufacturers are tested for their application with DH. This expansion is possible because of the modular approach [1,25–33].

There is also the community Eibiswald, where a 1,250 m² solar collector field and a 105 m³ heat storage have been supplementing the biomass heating plant since 1997. At that time the solar thermal system was able to cover 90% of the local heating network's demand in summer. By adding new customers the yearly heating demand increased to 8 GWh in 2012 and the network grew to 10,000 m in length. This is why an additional 1,200 m² of solar thermal and a 70 m³ buffer storage were installed. The solar fraction is now 12%. The majority of the heating demand is covered by two wood chip boilers with 2.3 and 0.7 MW capacity. In total there are about 35 MWth (ca. 50,000 m² of collector area) of solar thermal feeding into DH systems in the region around Graz and Styria. The city of Graz is planning to completely decarbonize its heat supply in the medium term and has chosen solar thermal to be one of the main technologies in order to achieve this. In its final expansion stage the plant "BIG SOLAR Graz" will provide 20% of the DH demand and will entirely heat about 4,400 buildings.

There is also progress in Germany. There are about 25 large-scale solar thermal plants with integration into DH in operation. More systems with a total capacity of ca. 40 MWth are in the planning and preparation stage. Currently the strongest market segment is made up of energy villages ("Energiedörfer"); five plants are starting in Randegg and Liggeringen (both in Baden-Wuerttemberg), Mengsberg (Hesse), Ellern (Rhineland-Palatinate), and Breklum (Schleswig-Holstein). Utilities in urban areas are also actively involved, as is evident by the biggest network-integrated solar thermal plant in Germany, Senftenberg, Brandenburg and a new pilot plant by the municipal utility in Duesseldorf. The collector area of large-scale systems in Germany is expected to double within the next years. The majority of the currently planned systems are in the segment of urban DH. Once again, all growth is modular in nature [1,25–33].

The above-mentioned project examples show trends that new generation in DH has begun. DH networks will be the platform for different heat sources: solar thermal, biomass, industrial waste heat, waste incineration, and geothermal and heat pumps. There are, however, various options for installing solar thermal collectors. There is a lack of willingness to provide agricultural land for large-scale solar thermal plants. At the "Energy Bunker" in Hamburg-Wilhemsburg, for example, the solar collectors are mounted on an old bunker. They can also be installed on car park buildings, greenhouses, industrial or multifamily buildings, decommissioned landfills next to wastewater treatment plants, and on noise protection walls. Other options include installations along streets or elevated above agricultural land, a concept already being tested in so-called agrophotovoltaic systems.

North America
Solar Community, Okotoks, Alberta's the Drake Landing Solar Community project came on stream in June 2007. The Drake Landing Solar Community is a prime example of the successful combination of energy efficient technologies with an unlimited renewable energy of the sun. Okotoks, in Alberta, is ideally situated for solar energy collection as it has many days of sunshine per year, and it can get almost as much solar energy as Italy and Greece. However, as solar radiation is lower there during the winter months, the Drake Landing Solar Community uses a central DH system that stores solar energy in abundance during summer months and distributes the energy to each home for space heating needs during winter months. The distribution temperature varies throughout the year, based on the outside air temperature, and the

flow is regulated to match the homeowners' demand. Because of the lower water temperature used in the DH system, each home is equipped with a specially designed air-handler unit for adequate heat distribution. The Drake Landing Solar Community has achieved a world record 97% annual solar fraction for heating needs, using solar thermal panels on the garage roofs and thermal storage in a borehole cluster [1,25–33].

Middle East

The University in Riyadh, Saudi Arabia Operating since mid-2011, the 25.4 MWth solar district water heating plant provides heat for the Princess Nora Bint Abdul Rahman University, which has a campus for 40,000 students, lecturers, and other staff. Its amenities include accommodation, research facilities, and a hospital. The collector surface area is 36,305 m^2 (25.4 MWth), which is equivalent to five football pitches. The solar panels are mounted on the roof of the university building. These are large-surface collectors, 5 m long and 2 m wide (10 m^2). They are used for high-capacity solar thermal systems of over 60 m^2 and are adapted to suit the Arabian deserts with their heavy sandstorms. Each solar collector is made from special solar glass and equipped with a modified mounting system to withstand unfavorable weather conditions, and it has 95% absorption capacity and weighs 170 kg. Throughout its life, the system will save approximately 52 million liters of diesel and reduce the carbon footprint by 125 million kg of CO_2 [1,25–33].

Above discussion indicates that average investment costs for solar thermal systems can vary greatly from country to country and between different systems. The benefits of solar thermal systems, in particular for such large systems, cover environmental, political, and economic aspects. Environmental benefits relate to the capacity of reducing harmful emissions. The reduction of CO_2 emissions depends on the quantity of fossil fuels replaced directly or indirectly, when the system replaces the use of carbon-based electricity used for water heating. Depending on the location, a system of 1.4 MWth (2,000 m^2) could generate an equivalent of 1.1 GWhth/year, a saving of around 175 kg of CO_2 [1,25–33]. Regarding energy costs and potential savings, there are three main aspects to consider that have a bigger impact on the comparable costs of the energy produced by a solar thermal system. These are the initial cost of the system, the lifetime of the system, and the system performance.

6.6 MODULAR DH WITH WIND ENERGY

Although there is a large body of literature on DH and solar energy, biomass, or geothermal units, not much has been done so far on interaction between wind energy and DH grids. The fact that using electricity to produce heat is inefficient discourages researchers to study it as an alternative. However, the idea that curtailed wind power (surplus) can contribute towards that direction, or even can put the system into operational maintenance during summertime, offers another perspective to the whole concept [34–36].

Partly, this idea of increasing local renewable energy production in order to convert surplus electricity into thermal energy was presented by Niemi et al. [37]. The analysis they did was based on multicarrier urban energy systems. It was found that, for the city of Helsinki, wind power production could be increased by 40%–200% by adding the electricity-to-heat conversion option for using surplus wind energy into the heating network. Long et al. [38] tried to balance wind power variability using electric heat pumps and CHP units. Maegaard [39] presented the Danish example with increased integration of wind energy into the grid through the use of CHP facilities. Hong et al. [40] studied scenarios using EnergyPLAN on wind energy integration in parallel with the heat demand into the existing energy system of Jiangsu province, China. It was revealed that, according to the needs of the province, the wind power production could range from 0% to 42% of the total energy demand. The EnergyPLAN model was also used in studies focused mainly in the Danish market aiming at maximizing renewables grid integration [41–43]. Fitzgerald et al. [44] tried to study how power system efficiency can be improved by integrating wind power through an intelligent electric water heating system.

In general, there are few studies that confront the issue, despite the fact that, in many cases, and in a great number of countries, the profitability of wind farms without the support of a scheme based on subsidization is questionable. Therefore, for this reason there are few studies in the literature linking the need for wind energy for DH networks to achieving greater integration shares. Three countries that have examined linkages between excess wind power for heat pump and electrical heating elements for water in DH schemes are Denmark, Greece, and Scotland. In all cases excess wind power has been sold to grid or neighboring countries like Germany or Norway for Denmark. One possibility that has also been examined is to use CHP for conversion of electricity to heat and back to electricity. Greece study was centered around the city of Kozani and Scotland study was done for island of Orkney. All of these studies were theoretical. In all cases, surplus wind energy was expected. The general conclusion was that, while the concept has some merit, the connection between wind energy and DH requires government policy and subsidy to justify capital investment and long-term viability of the general concept [34–36].

6.7 DH WITH SMALL MODULAR NUCLEAR REACTORS

Nuclear DH is mostly viable in very cold climate, i.e., parts of North America, Europe, and Asia. Vital preconditions for the viability include access to nuclear technology and at least a basic nuclear infrastructure along with the public acceptance. Besides internationally accepted safety precautions, some additional design features must be adopted for nuclear plants intended for DH to prevent the ingress of radioactive substances into water. For the reactors in the small modular reactor (SMR) size range in cogeneration mode, the possible share of process heat supply would be larger, and heat could even be the predominant product. This would affect plant optimization and could present more attractive conditions to the potential process heat user. SMRs are more suitable for countries with small or medium electric grids and could be better adapted to cogeneration-generate electricity and process heat than large reactors [45–56].

Historically, nuclear energy sources have been used mainly to produce electricity. The Soviet Union's power generation industry has also followed the path of primarily expanding its nuclear electricity generation capacity. At present, the installed capacity of the 45 nuclear power plant units in the Soviet Union amounts to some 35 gigawatts-electric (GWe). Nuclear power generation is centered, in the main, around the use of water-moderated, water-cooled, pressure-vessel-type reactors (water-water energetic reactor [VVERs]) (these series-produced power units have an electrical output of 440 and 1,000 MW), and channel-type, water-cooled, graphite-moderated reactors of the Reaktor Bolshoy Moshchnosty Kanalnyy (RBMK) type (these series-produced power units have an electrical output of 1,000 and 1,500 MW). The VVER-type reactor units developed by Soviet specialists have also served as a basis for the development of power generation in member countries of the Council for Mutual Economic Assistance (CMEA) [45].

Even when the nuclear power generation industry was at an early stage of development in the USSR, it was clear that focusing solely on the production of electricity would not adequately solve the basic problem which besets the industry, namely, supplanting scarce organic fuel in the country's fuel and energy economy. The only way this problem can be solved is by the extension of nuclear energy into the highly fuel-intensive area of heat production for community and domestic heating and for industrial consumers (1.5 times more fuel is used for this purpose than for the production of electricity). In CMEA countries, which are less well-endowed with fossil fuel resources than the USSR, this problem is even more acute. The plan for developing "nuclear" DH worked out by Soviet specialists to meet the requirements of the country's fuel and energy sector provides for the combination of four different approaches:

- the use of unregulated steam extraction from the turbines in condensing power plants;
- the construction of mixed, DH and condensing nuclear power plants with high back pressure (TK) type turbines (DH and condensing with regulated steam extraction);
- the construction of single-purpose nuclear plants that produce only thermal energy for community and domestic heating purposes (DH atomic power plant (DHAPP));
- the construction of specialized nuclear plants for industrial heating purposes which, due to new technical features incorporated in the design, could be located in the immediate vicinity of points of consumption and be used to produce heat and electricity, or heat only.

At this point, the first three of these approaches are the most developed technically. The use of unregulated steam extraction from the turbines of nuclear power plants in operation and those under construction holds a special position among these various approaches. In practice, this is the only form of nuclear heating, which has been implemented to date. It was started over 20 years ago when a system was set up to deliver heat from the Beloyarsk plant to supply heat and hot water to the buildings and structures on the plant site itself and to the adjacent living areas. Subsequently, this approach was introduced in other plants as well. The total output of the heating systems in plants already operating is fairly impressive; in excess of 3,000 megawatts-thermal (MWth) at the beginning of 1989. The various turbines in use at present in such nuclear power plants, and those in the process of being manufactured for plants under construction, have varying capacities for the output of heat for DH purposes [45].

The design features of heat delivery systems from nuclear power plants are based on current radiation safety requirements for thermal energy consumers. Thus, the heating water is circulating in a tertiary (in relation to the reactor core) circuit. Pressure in this circuit is kept higher than the maximum possible pressure of the highest unregulated steam extracted, which prevents radioactive products from getting into the heating water, should there be a loss of integrity in the heat exchange surface of the boilers. In plants with RBMK reactors, the heating systems have an intermediate coolant circuit between the turbine extraction and the heating water. Pressure in the intermediate circuit is kept higher than the steam pressure but lower than the pressure in the DH circuit. In plants with VVER reactors, the grid water is heated in grid heaters by the steam extracted from the turbine. The highest pressure extraction steam used is lower in pressure than both the pressure in the reactor circuit and the pressure in the grid circuit. To prevent radioactive contamination of the heating water in an accident situation, the heat exchanger is cut off both for the heating steam and for the heated grid water. There is also constant monitoring of radioactivity levels in the heating water. It is important, and an economically attractive idea, that the fullest possible use be made of the existing capacity of DH systems in nuclear power plants. Even allowing for some reduction in electricity output due to the extraction of steam for DH purposes, the dual-purpose (generation of electrical and thermal energy) use of nuclear power plants with VVER-440 power units decreases the volume of inorganic fuel used by 30,000 tons coal equivalent per year for each nuclear unit. Even greater savings are possible when plants with VVER-1000 power units take part in centralized DH. The volume of organic fuel saved by each power unit of this type amounts to 130,000–750,000 tons of coal equivalent per year, depending on the type of turbine installed [45].

Despite this important incentive, in the majority of nuclear power plants, maximum use is not being made of this capacity to provide heat. The reason for this is that the location of the plant is not always ideal in relation to potential heat consumers. The "guaranteed" consumers (within the plant and plant living quarters) usually use only a small part of the total potential heat output from a DH unit. The calculated total heat consumption of users on site amounts, as a rule, to approximately 30 MWth for one power unit with an electrical output of 1,000 or 1,500 MWe. Likewise, the total calculated heat consumption of users in the associated plant living quarters does not exceed 260 MWth. In community and domestic heating schemes (heating, hot water, ventilation), nuclear sources are base-loaded and operate jointly with peak sources using organic fuel. As a result, for a four-unit plant with series-produced VVER-1000 plus K-1000–60/1500–2 turbine, the optimal " l o a d " of the nuclear part of the DH plant due to on-site requirements (within the compound and in living quarters) is not greater than about 230 MW, whereas the total capability is about four times greater. Thus, in modern plants working at design power level, there is a fairly large capacity for heat production, which could be harnessed [45,46].

The most sensible course would be to use this available capacity to provide centralized heating for consumers in the adjacent region; i.e., industrial and residential complexes. But there are a number of limitations to such schemes. Technical and economic considerations dictate a specific area of coverage for centralized heating systems for each nuclear power plant. Beyond that area, the total cost of compensating for underproduction of electricity (owing to the fact that the plant is working in a heat-producing regime) and of transporting the heat to the point of consumption, exceeds the economies to be made from supplanting organic fuel. The size of this coverage area is determined by climatic conditions in the region (which influence the heat output regime), the cost in relation to the organic fuel supplanted, the amount of heat being delivered by the nuclear power plant, the real conditions involved in the laying of the heat transport line from the plant to the industrial and residential complexes where the heat is to be used, the type of generating equipment installed in the plant, and a few other factors. Another significant consideration is the question of which alternative modes of DH can actually be used in the area to provide a realistic comparison with heating delivered by nuclear power plants. It is not a very easy to take into account all of these factors when nuclear power plants are to be built since the closeness of a potential heat consumer is not the only, and frequently not the most important, criterion in the authorization process for a construction site. Current radiation safety requirements in the USSR regulate the minimum permissible distance from a nuclear power plant site to major populated areas. Thus, for example, in the case of a plant with an electrical output of 4,000 MWe, this distance varies from around 25 km (if the population density in the area is 100,000–500,000) to 100 km (if the population density is over 2 million). For a large number of reasons, however, nuclear power plant construction sites have to be located far beyond the standard distances mentioned above. Hence, for the majority of nuclear power plants currently operating, there are significant heat output reserves that have not been utilized [45–50].

6.7.1 Global Assessment of Modular Nuclear Heat Based DH

There are several examples of effective modular utilization of nuclear heat both in Soviet Union and the rest of the world. The Balakovo nuclear power plant, for instance, provides the town of Balakovo with heating (the town is 12 km from the plant, the heat requirement is over 1,000 MWth), the Rostov plant provides heating for the town of Volgodonsk (13 km from the plant, heat requirement of over 1,000 MWth), the Tatarsk plant supplies heat to the town of Nizhnekamsk (40 km from the plant, heat requirement up to 2,000 MWth), and the Bashkir plant to the town of Neftekamsk. Supplying heat to consumers at some distance from nuclear power plants is also now carried out or is being planned in CMEA countries. In the German Democratic Republic, the Bruno Leuschner nuclear power plant has been providing DH for the town of Greifswald since 1984 (22 km from the plant, heat output—260 MWth). A heating system for the town of Magdeburg, based on the Stendhal plant now under construction, is in the planning stage (distance from the plant—95 km).

In Czechoslovakia, it is planned to equip all plants with VVER-440 reactors (12 power units in all) for DH: the Bohunice plant will supply the town of Trnava with heat, the Dukovany plant the town of Brno, and the Mochovice plant the town of Levice. There is a plan to organize heating for the town of Ceske Budejovice from the Temelin plant. The heat transport grids from these plants are very extensive and heat output is fairly high (for example, the heat pipeline from the Dukovany plant to Brno is 40 km long and the calculated heat output is 500 MWth). In Bulgaria, a project for a DH system centered on the Belene nuclear power plant to serve the towns of Pleven (58 km), Svishtov, and Belene is in the development stage. The total calculated heat output for these towns would be around 700 MWth. It is also planned to use the Kozloduj plant to supply heat for the town of Kozloduj. The technical features intended for inclusion in the heat output systems of these plants are similar in many respects to those already mentioned as being used in the USSR for the delivery of heat from plants with VVER reactors. It should be noted that CMEA countries cooperate closely on the solution of nuclear DH problems. This cooperation is coordinated under a comprehensive program for scientific and technical progress for the period up to the year 2000, which has been adopted by those countries [45].

Along with the use of plants now in operation and those under construction for DH purposes, as indicated above, technical decisions have been reached to set up specialized nuclear heating sources—both dual-purpose CHP plants and single-purpose nuclear DH plants (DHAPP). A CHP design was developed for the European part of the USSR, based on VVER-1000 reactor facilities and new turbine units (TK-450/500-60/3000 turbines) with regulated steam extraction and a high thermal output. A two-unit CHP plant of this type can ensure a heat output of up to 2,100 MW. The plan was to bring at least three plants of this type into service by the year 2000, in Odessa, Minsk, and Khar'kov. It should be pointed out that these plants are not simply DH plants, but mixed heating and condensation units (owing to the fact that the turbines used in them have a large "added" condensing capacity). However, this is inevitable where reactors with a large unit output are involved [45–52].

The review of technical policy with respect to nuclear power generation following the Chernobyl accident brought about a change in attitude to the CHP station designs which had already been developed. As the construction sites for the plants had been chosen in the late 1970s and early 1980s (i.e., when less rigorous requirements were in force than at present), the Odessa and Minsk CHP plants did not satisfy the new approach to safety, and their construction was stopped. Design work on the Khar'kov plant was also halted. The design of a single-purpose DHAPP was developed in the USSR simultaneously with the CHP project. A water-cooled, water-moderated reactor with a unit output of 500 MWth was specially designed for the DHAPP. This single-purpose type plant, the AST-500 can be located in the immediate vicinity of population centers; i.e., up to 5 km from the city limits. The AST-500 project is being implemented at present in the towns of Gorky and Voronezh. There had been plans to build a number of DHAPPs in the European part of the Soviet Union; however, the negative attitude of the public towards nuclear power after the Chernobyl accident has delayed (and in certain cases cast doubt upon) the implementation of some nuclear power projects that had previously been scheduled for construction [45–52].

There is definite interest in single-purpose nuclear heating plants in CMEA countries. It should be noted, however, that owing to the specific nature of urban construction in these countries, they require facilities with a significantly smaller capacity than those commonly used in the USSR power generation system. In the light of this, the USSR has developed a design for a reactor facility with a unit output of 300 MWth, having technical features similar to those used in the AST-500. In addition, work is being carried out in cooperation with specialized organizations in CMEA countries for building facilities with an even smaller output. In addition to the work on improving the designs for specialized nuclear heating plants and the development of new types of reactors for them, the search for ways of utilizing more fully the heat output potential of existing nuclear power plants is of great practical significance. Analysis shows that this type of source will dominate the nuclear power infrastructure of the USSR and CMEA countries for a long time into the future [45–52].

Two approaches, in particular, seem highly promising:

- using the available capacity of the heating facilities of nuclear power plants for long-distance industrial heat supply;
- conversion of nuclear power plants that have outlived their standard service life to a DHAPP system (with the reactor facility operating at reduced power and in a less intensive regime).

The implementation of the first of these approaches will allow the pool of potential customers for thermal energy from nuclear power sources to be extended significantly, and it will introduce nuclear energy into the extremely fuel-intensive area of industrial heat supply. Developmental work done in the USSR and Czechoslovakia has shown that, in the matter of long-distance heat supply to industrial consumers, it makes economic sense to employ a system in which heat is transported in the form of pressurized hot water from the nuclear power plant to the point of consumption. The hot water would then be used to produce steam, with the required parameters in the customers' equipment (preferably thermocompressors). The second approach is, at present, at an earlier stage of development, but it is extremely important that an answer be found since it would help solve two problems at once: first, prolonging the operational life of the main equipment in a nuclear energy source, and second, establishing large-scale centralized heat supply sources over a short period of time and at comparatively low cost.

In China, China National Nuclear Corporation (CNNC) launched its Yanlong reactor (referred to as the DHR-400) for DH in November 2017. The move came shortly after the "49-2" pool-type light-water reactor developed by the China Institute of Atomic Energy continuously supplied heat for 168 h. CNNC indicated that the Yanlong reactor—which an output of 400 MWth—has been developed based upon the company's safe and stable operation of pool-type experimental reactors over the past 50 years. It said that Yanlong is a "safe, economical and green reactor product targeting the demand for heating in northern cities." The reactor can be operated under low temperatures and normal pressures. It can be constructed near urban areas due to the zero risk of a meltdown and lack of emissions. In addition, the reactor is easy to decommission [47]. The use of nuclear energy for DH improves China's energy resource structure. Nuclear energy heating could also reduce emissions, especially as a key technological measure to combat haze during winter in northern

China. Thus, it can benefit the environment and people's health in the long run. It can be constructed either inner land or on the coast, making it an especially good fit for northern inland areas, and it has an expected lifespan of around 60 years. In terms of costs, the thermal price is far superior to gas, and is comparably economical with coal and combined heat and power (CHP)."

Two types of nuclear heating reactors—one a deep pool type, the other a vessel type—are developed by Institute of Nuclear and New Energy Technology (INET). The vessel-type reactor was selected as the main development direction. Construction of a 5 MWth experimental nuclear heating reactor (NHR5) at INET began in 1986 and was completed in 1989. The larger, demonstration-scale NHR200-II was developed from this. A feasibility study on constructing China's first nuclear plant for DH would use the domestically developed NHR200-II low-temperature heating reactor technology. SMRs will be used in the future not just for electrical generation but also for providing heating [47].

There is potential for nuclear DH in multiple regions including Finland and Poland. According to energy for humanity, a number of Finnish cities have received political initiatives to evaluate the feasibility of using SMRs instead of fossil fuels to provide DH. A recent study looked at completely decarbonizing electricity, transport, and heating in Helsinki through the use of small, modular advanced reactors. Most of the DH in Finland is supplied by burning coal, natural gas, wood fuels, and peat. It should be noted that while many Finnish cities have progressive climate policies and goals, they have struggled to decarbonize heating and liquid fuels. More than half of the greenhouse emissions of all of Helsinki come from DH, mainly run by fossil fuels. The Society and Energy for Humanity published a report that anticipates that, for Helsinki area, future annual energy use in DH at 8 TWh, electricity at 12 TWh, and 4 TWh of hydrogen for transportation fuels [46,48].

The initial result seems promising. The scenario modeling assesses the investment in 300 MWth of new DH capacity in the Helsinki Metropolitan area in 2030, either as a CHP plant or as a heat-only boiler. According to Värri [50] and Varri and Syri [51], the SMR with data based on the NuScale reactor fared well against most other options. The preliminary results indicate that a heat-only boiler using the technology would be a fairly profitable while investment in a CHP plant would rely heavily on the form that future electricity markets take. In both cases, the potential investments will still rely on the development of multiple factors, especially the cost of capital. DH can potentially rise to a more prominent role in the future as at least in the EU, and it has been recognized as one of the sources of increased energy efficiency and decarbonization [52]. In the Nordics, DH systems are fairly commonplace already comprising 43% of the total heating market [52]. Even so, the decarbonization of these systems has proven challenging as the efforts so far have been heavily based on increasing the share of biomass.

In Poland, the idea of using the Zarnowec nuclear power plant to provide heating for the Gdansk-Gdynia area is being considered (the plant is 75 km from Gdynia). There is also a plan to provide the town of Poznan with heating from the Warta plant, which is scheduled to be built [52]. A significant part of the Polish governments plan to diversify the country's energy system and increase its security of supply is the inclusion of nuclear power in the Polish energy mix. The original plans approved in 2009 included the target of two 3 GW plants, the first of which would be online in 2022. While this timetable did not hold, the idea has not been abandoned. The revised timetable and plan include a total of 6 GW of capacity by 2035. Significant progress has been made in reinforcing the necessary institutions and acquiring further knowledge, but further delays seem likely. While new nuclear plants have had issues with their economy elsewhere, it would seem like the strong political backing might partly override the issue. The strong support is also shared by the public as the approval rate near two possible sites has ranged from 60% to 80% [53,54]. Poland also seems highly interested in high-temperature gas-cooled reactor (HTGR) technology. While the light-water reactors planned could be useful for DH, the country also has a large market for industrial heat with the 13 largest chemical plants already requiring 6.5 GW of heat at the temperature of 400°C–550°C. The initial report on HTGR plans by the Polish Ministry of Energy calls HTGRs "the only practical alternative to replace fossil fuels for industrial heat production" [55]. A potential agreement with the Japanese Atomic Energy Agency and multiple Japanese companies to build a 10 MW experimental reactor by 2025 and a proper 160 MW facility by 2030, most likely based on the HTGTR300C mentioned in Section 2.2.3, is expected to be reached at the beginning of 2018 [56].

For SMR deployment in Europe, Poland seems a likely candidate. Beyond the positivity towards nuclear and the need for cleaner energy, the country also has one of the largest DH networks in Europe that covers approximately half of the population. The DH system supplies around two thirds of the overall heat demand, of which the share of CHP has ranged between 60% and 65%, with the rest powered mostly by often-inefficient heat only boiler (HOB). Partly due to this, the Polish government has plans to double the current share of 14% overall electricity production by CHP plants by 2030. The need to reinvest also encompasses the networks themselves as estimated 35% of the Polish DH networks require investments. The system could be developed by not only fixing the old networks but also by linking the smaller systems into existing larger networks. This work is supported by the government and EU-level programs offering funding and loan support for investments in modernization and development of CHP plants and DH networks. Poland also had a certificate system in place for CHP production that is being phased out in 2018. A replacement system has been under discussion with no concrete developments seen yet [53,54].

The SMR technology itself seems suitable for DH. The small scale means that a single module can fit into a fairly small DH network if the base load is large enough. At the same time, the modular nature of SMRs also means that the plants can be built for a variety of production configurations, and they do not necessarily have to be tied down to a single production type. Similarly the advances in safety seem promising with regard to the emergency planning zone (EPZ) considerations and the plants could hopefully be built fairly close to the networks as long transmission pipelines would bring the costs of the investments up and increase heat losses. The actual economics of the NuScale SMR also show quite a bit of promise, especially for HOB plants. The SMR HOB has the second lowest levelized cost of heat (LCOH) after the MSW CHP in the modeling, and the levelized cost is fairly resilient to any changes to the base values based on the sensitivity analysis. Due to the low operating costs of the SMR plants and their nature as base load production, they would also be bringing down the overall cost of heat production and increasing the competitiveness of DH against individual heat pumps and other sources of heating. As the electrification of heating will not only be happening through heat pumps in DH networks, SMRs could be an important tool for energy companies to keep their DH price at a competitive level against heat pump installations for individual buildings [51,52]. The potential SMR CHP deployment seems more unlikely but, based on the results gained, the same applies to all forms of CHP production outside of MSW CHP. There is some initial promise to the numbers and deployment might be a possibility, but it will be highly dependent on the route that electricity markets take. The levelized cost of electricity (LCOE) and net present value (NPV) values for the SMR CHP are also clearly more sensitive to any modifications to the base data, especially the cost of capital [45–56].

6.8 MODULAR DH BY INDUSTRIAL WASTE HEAT

Xia et al. [57] investigated a method **for integrating low-grade industrial waste heat into a DH network**. Low-grade industrial waste heat could be a considerable potential energy source for DH, on the condition that the heat from different industrial waste heat sources is integrated properly. This study considers a method for integrating low-grade industrial waste heat into a DH system and focuses on how to improve the outlet temperature of heat-collecting water by optimizing the heat exchange flow for process integration. The pinch analysis concept was considered, and a newly developed thermal theory called entransy analysis was introduced. By using entransy dissipation to describe the energy quality loss during the heat integration, this study analyzed how heat exchange flows influence the final outlet water temperature and attempts to provide an efficient method to optimize the heat exchange flow. Finally, the effectiveness of the proposed methodology was demonstrated by testing it in a project involving the recovery of waste heat from a copper plant for DH [57–67].

Industrial efficiency does not stop at the boundaries of a factory. Applying industrial waste heat in DH systems—including heat from electricity generation—is often mentioned as a promising option for energy savings and CO_2-emission reduction. Currently, despite a large amount of available waste heat, application of industrial waste heat in the Netherlands is very limited. So, if waste heat is to fulfill an important role in future emission reductions, there is a major implementation gap to be bridged. The study by Daniëls et al. [60,61] presents an analysis of potentials and costs for industrial waste heat utilization in the Netherlands, for low-temperature application in households, services, and greenhouse horticulture. It starts with identifying the availability of waste heat and the demand for heat. To estimate the technical potential, it evaluates the match between supply and demand: heat should be available at the right temperature level, the right moment, and the right location. As to provide realistic potentials, the study makes a comparison with alternatives, both with regard to alternative application of waste heat on the supply side and with regard to alternative sources of heat on the demand side. It concludes with an assessment of a realistic role of industrial waste heat utilization in DH in the medium to long term, and the roles of major alternatives with regard to their performance in terms of energy savings, emission reductions, and costs. As the analysis points out, this realistic potential is rather limited: an estimated 10–25 PJ of net energy savings may be realized by the utilization of 25–45 PJ of waste heat in DH.

Large quantities of low-grade waste heat are discharged into the environment, mostly via water evaporation, during industrial processes. Putting this industrial waste heat to productive use can reduce fossil fuel usage as well as CO_2 emissions and water dissipation. The purpose of the study by Daniëls et al. [60,61] proposed a holistic approach to the integrated and efficient utilization of low-grade industrial waste heat. Recovering industrial waste heat for use in DH can increase the efficiency of the industrial sector and the DH system, in a cost-efficient way defined by the index of investment vs. carbon reduction (ICR). Furthermore, low-temperature DH network greatly benefits the recovery rate of industrial waste heat. Based on data analysis and in situ investigations, the study discusses the potential for the implementation of such an approach in northern China, where conventional heat sources for DH are insufficient. The universal design approach to industrial-waste-heat based DH is proposed. Through a demonstration project, this approach is introduced in detail. This study finds three advantages to this approach: (a) improvement of the thermal energy efficiency of industrial factories; (b) more cost-efficient than the traditional heating mode; and (c) CO_2 and pollutant emission reduction as well as water conservation.

Heating and cooling in the future will utilize energy gained from waste heat, which will be distributed at low temperature using DH and cooling networks. It will thus make use of the heat wasted by cooling systems in supermarkets and fruit storage facilities, which up to now has simply been released untapped into the atmosphere. South Tyrol's EURAC Institute for Renewable Energy is exploring this new technology in the "FLEXYNETS" project, which is financed to the tune of two million euros by the European research program "Horizon 2020."

At present DH grids run via high temperatures of around 90°C. To heat individual buildings, the networks have to connect to sizeable thermal plants, such as block thermal plants or waste incinerating plants. The technology which will now be researched by the South Tyrol EURAC Institute for Renewable Energy on the other hand runs at temperatures between 10°C and 20°C. This means that the DH grids can be supplied with energy from sources running at much lower temperatures. Space heating, generated for example from a waste incinerating plant, is intended to be supplemented by heat generated in various everyday processes and which is currently wasted. By using low temperatures when distributing heat, one can reduce the huge heat loss in the underground distribution pipelines, which will make the whole grid much more efficient in the future. According to the experts, the energy consumption for heating and hot water could be reduced by 80%, and for cooling buildings by 40%. Across Europe, this would amount to a reduction of 5 million tons of CO_2 emissions by 2030 [52,58,59].

Energy-intensive industries need to adapt in order to play an important role in the low-carbon economy. Efficient use of energy resources and the minimization of wasted heat will be important. The role of the steel industry in recovering recycled metals means these industries are important for a sustainable economy, providing employment and supporting manufacturing. The aim of the study by Raine [67] was to investigate the potential uses for heat storage in Sheffield, UK for capturing heat, which is produced intermittently at a steelworks for both reuse of heat on site at various temperatures and for heat supply to a citywide heat network. Site visits were followed by calculations using data provided. Heat storage options were investigated to ensure that waste heat could be reused effectively. The feasibility of using the DH network in the city to carry low-grade heat away from the plant was considered. Around 4.7 MW of useful heat could be generated from two steelwork sites in Sheffield and a further 10.9 MW from a site in nearby Rotherham, and 22,500 tons of CO_2 could be saved per year by fully exploiting this waste heat resource.

Sheffield has for decades been a pioneering city in the UK in terms of developing DH to provide heat at lower cost and environmental impact to the city. The city-center network is supplied with up to 60 MW of heat from an energy from waste CHP station. In addition, a new CHP power station using biomass is being constructed, and this will supply heat through a new DH network to sites closer to Sheffield city center. The new network will be extended to supply new customers and may connect to the city-center network. The new network also offers the prospect of recovering heat from the city's industrial sector. Finney et al. [68,69] identified potential for at least 10 MW of industrial waste heat for the city's expanded DH system. Sheffield and nearby Rotherham have an estimated annual output of 1.07 million tons of steel produced each year from four electric arc furnaces. Approximately 500 kWh of electricity is used in the furnace for each ton of steel produced [9], meaning an average of around 60 MW of electricity consumption and indirect CO_2 emissions of the order 280,000 tons/year from electricity alone. These processes are necessarily heat-intensive in order to melt the scrap metal, and the use of an electric arc allows the process to be carried out in batches of around 100 tons in four compact furnaces on these three sites. The furnace flue gases from steelworks are very hot, in the range 600°C–1,500°C and, although they are dust-laden, flue gas heat recovery systems in industry are increasingly common, with one example at Port Talbot steelworks [67]. Tenova Group has developed waste heat boilers for electric arc furnace flue gas heat recovery at sites in Germany and South Korea; both examples include the generation of steam. One of the heat recovery systems uses steam to supply heat to a 2.5 MW ORC generator. Higher temperature (higher energy) energy could be used for power generation or through work processes such as vacuum creation in a steam jet ejector. High-pressure steam accumulators are widely used to balance supply and demand on sites for steam and can be used for storing recovered heat. Regenerator materials mainly made from ceramics are already used on many gas-fired furnaces to recover heat; these materials capture the heat of the exhaust air leaving the furnaces and, when the flow direction periodically changes, use that heat to preheat incoming combustion air. Recuperator systems are an alternative, allowing increased efficiency by exchanging heat from exhaust air with incoming air; this involves a heat exchanger, and the heat is not stored.

It is possible to recover heat from cooling water on industry sites, for example in the water that cools the electric arc furnace walls or the water that cools the hot gas extraction ducts. Residual heat may also be available from steam processes that operate typically at 200°C–250°C. In Graz, Austria heat is recovered from a gas-fired reheat furnace along with high-temperature cooling water at around 90°C from two electric arc furnaces [52,58,59]. Buffer tanks could be used in this instance to store water at temperatures in the range 70°C–110°C which are suitable for DH. If the water is above its atmospheric pressure boiling point, then the tank needs to be pressurized, usually including a cushion of steam or nitrogen at the top of the tank to allow for thermal expansion and contraction processes. Many industry sites have relatively low-temperature cooling systems that are adapted from, or in some cases still use, river water cooling. Circulation of water to cooling towers reduces the need to import cool water. In some cases, cooling circuits with water treatment are necessary to prevent contaminants in the water from escaping. This water is often not hot enough for building heating, but heat pumps could be used to draw useful heat from that water. Water tanks can be used to store the cooling water, ranging from ambient temperatures up to possibly 70°C if the cooling systems are adjusted to run at higher temperatures, but large capacities are needed for lower temperature stores. Underground TES is another option that uses the thermal capacity of the ground and can be useful for storing large volumes of low-temperature heat for long periods.

Heat pumps can be used to draw heat from the atmosphere, ground, or bodies of water and supply it at a higher temperature suitable for heating. The CoP for the heat pump describes its performance. The emissions factor for electricity in a given country determines the environmental impact of a heat pump using electricity. One recent example in Norway draws heat from the fjords in order to supply a DH system at 90°C [71–78]. Sheffield's DH operates at 110°C, and delivery of heat from heat pumps at these temperatures is difficult and suitable technology is at an early stage. The Norwegian heat pump uses ammonia as a refrigerant which has its critical point at 132°C, and pressures of 41 bar are required for the ammonia heat pump condensers to work [71–78]. In steelworks, the waste heat sources are primarily flue gases from furnaces (at high temperatures) and the cooling water from the electric furnace (at low temperatures). The electric arc furnace melts scrap metal at very high temperatures and removes impurities. It produces hot and dusty off-gas, which must be extracted and filtered before release. Typically, one extraction duct captures gas from close to the furnace, while a canopy duct captures any fugitive emissions from around the furnace. Some furnaces preheat scrap metal in the extraction duct allowing energy to return to the furnace when the scrap is loaded; this reduces electricity consumption, but the nature of processes at the sites in Sheffield makes this approach difficult. There are also gas-fired furnaces and cooling water where heat could be recovered [71–78].

Industrial sites typically have a range of heating and cooling needs depending upon the stage of the industrial process being carried out. Finding ways to pass heat effectively between parts of the process can save a lot of heating or cooling energy. This method is termed process integration, and the practicality of such steps was considered. The use of DH opens up opportunities for using waste heat streams for a new purpose, and this could become more widespread in future, particularly in the UK. The many options for system arrangement, including how components such as heat storage are integrated, create uncertainty over the optimal way to achieve economic and environmental goals. The method used here comprises a model built in the C++ programming language to investigate how system components interact. This modeling environment was chosen in order to give high flexibility over component arrangement as well as providing a means to adjust appropriate parameters such as the heat storage capacity and the prices of energy under different future scenarios. It is these cost and environmental benefit estimates that will guide decisions. Understanding the potential for using heat pumps is important, as they give the potential to extract heat from cooling water at the steelworks. If an ammonia heat pump is being used then it may be more effective to only raise the water to 90°C, as high pressures are needed for temperatures above that. The output temperature could then be topped using heat from the flue gases [71–78].

High-temperature heat is already stored on sites using regenerator materials and in a steam accumulator. The accumulator matches steady production of steam from boilers to the short discharge needs of the vacuum processes. If more heat is to be recovered then the thermal capacities of the ceramic regenerators or the steam accumulator could be used to balance supply and demand of waste heat, although modifications will be needed to make such connections. If flue gas heat is recovered for generating steam then using existing steam storage facilities may be sufficient. Medium- to low-temperature heat could be stored using hot water tanks which can be linked to DH networks. If the water in the store is circulated in the network then this prevents temperatures losses through heat exchangers. However, the water quality needs to be sufficiently high and the temperatures and pressures of operation need to be compatible with the network. In the case of Sheffield's DH, the store would need to operate between 70°C and 110°C. For low-temperature waste heat, underground TES could be used and would give potentially a high heat capacity at low cost. Storing the cooling water is an option, but low temperatures mean a large volume of storage would be needed, and this increases expense. Any heat recovery project needs be evaluated, not only in economic terms but also in terms of its environmental impacts. This can help industry meet environmental objectives [71–78].

Modern industry recognizes the need to monitor, control, and minimize emissions of particles and harmful gases from their processes. One example of an emission that is closely monitored is dioxins, and while over 90% of human exposure is from dioxins present in food industry is regulated to minimize dioxin emissions. High temperatures of over 850°C are required to destroy these particles, and there is a chance of reformation between 500°C and 250°C. Many steelworks use quenching of gases with water spray to pass this critical temperature zone [52,58,59,67–69]. With flue gas heat recovery, heat exchanger design should account for this issue. Another important issue for heat recovery is the flue gas level of sulfur oxides, if these gases combine with moisture they create acid and this can corrode any heat exchanger increasing maintenance needs. In some instances, lime can be used to counter acidic gases, and activated carbon can be used to capture dioxins and heavy metals. The amount of water consumption at a steelworks is another issue with environmental implications. Approximately 14–28 m^3 of water is required per ton of steel produced in electric furnaces [52,58,59]. For some processes, the water is limited to a certain temperature rise in the cooling circuits to prevent corrosion, and high water velocities are needed to prevent particles from settling in cooling systems. The adjustment of cooling systems is a complex issue and will require detailed work by engineers to consider consequences before proceeding; however, use of high-temperature cooling has been achieved in some instances, and this would carry benefits in terms of transferring heat to DH as well as reducing water use. Even the current low-temperature cooling could provide water suitable for a heat pump [59].

The amount of heat that can be recovered depends upon how that heat is to be used, for example, if the heat is to be used to generate high-pressure steam, then only heat above a certain temperature will be useful. A two-part heat exchanger could be used in order to generate hot water for district heating from the partly cooled flue gases, allowing for a greater level of heat recovery. These would be aspects to consider in the detailed design of heat recovery equipment. For the contribution of heat to DH a new heat store would be required. If the heat is just used when demand is available and discarded when demand is not, then this avoids the need for heat storage; however; it also reduces the environmental benefits. A hot water store could be off-site if connected via DH, although the feed-in temperature and rate need to be carefully controlled if feed into the heat network is instantaneous. If the heat store is operational as part of the network then it can also provide services to the DH operator [71–78]. To effectively use industrial waste heat the operational principles need to be established early with the DH operator, including what charges and payments apply for supplied heat as well as when and how much heat can be fed-in at various times of day. There will be knock-on effects for other heat sources on the network. For example a flexible CHP unit can be switched from electricity to heat production, and therefore the injection of industrial heat increases the capacity to add electricity production to the grid. If electricity is used for a heat pump and then heat is injected to DH, this could have a negative effect on emissions associated with delivered heat, and the way this is accounted will be important. Altering operational temperatures on the heat network would make the recovery of heat more feasible, but the temperatures need to be sufficient to satisfy the needs of all the customers. If the new network connects to the old one then the water needs to be hot enough to be used by an absorption chiller unit in the city center. In the long term, lower temperatures could assist the integration of geothermal and solar energies too [71–78].

Overall, it is quite probable that the energy spent in running a heat pump is inhibitive to the economics, and therefore only the flue gas heat which is much hotter can be recovered. However, if significant charges are associated with river water use for cooling then saving water by recovering heat may have better economics. The amount of heat that is recovered will depend upon the economics of the project and the sale price which heat can achieve through the network. Recovering heat from high-temperature industrial heat sources can boost the overall system efficiency and give environmental advantages; however, there are barriers to making this feasible. In particular, the upfront cost of constructing DH networks means that connections to industry can be capital intensive. High-temperature heat pumps are a quickly developing technology, but some designs have practical limits on delivery temperatures which may limit their applications. In the UK, the high carbon intensity of grid electricity reduces the environmental advantages of electricity-driven heat pumps. Adjusting DH networks to run at lower temperatures in future will increase both efficiency and volume of recoverable industrial waste heat. Working to maximize heat recovery will help energy-intensive industry contribute to reducing the economy's carbon intensity [52,57–70].

6.9 DH BY MODULAR HYBRID SOURCES

6.9.1 Hybrid Modular Geothermal Heat Pump for DH

Accurate calculation of transient subsurface heat transfer is critically important in sizing ground heat exchangers (GHX) in geothermal heat pump (GHP) systems. The size of the GHX is a complicated function of a number of design variables that include the thermal properties of the subsurface, and the dynamics of short- and long-term building heating and cooling loads. The so-called hybrid GHPs have received considerable attention in recent years [79], because they have been shown to significantly improve the economics and energy use of GHP systems. Hybrid GHP systems couple a supplemental heat extraction or rejection subsystem to a conventional GHP system to handle some portion of the building or the ground loads, and as such, permit the use of a smaller, lower-cost GHX. Hybrid GHPs are especially effective in applications that have large peak loads and/or have highly imbalanced loads over the year (i.e., heavily heating or heavily cooling-dominated buildings).

Hybrid GHP systems are more complex in their design than conventional GHP systems due to the transient nature of the supplemental component. Further, recent research on hybrid GHPs identifies more than one method to design a hybrid GHP. For example, Chiasson and Yavuzturk [80,81] describe a method for designing hybrid GHP systems based on annual ground load balancing. Xu [82] and Hackel et al. [83] describe hybrid GHP system design based on lowest life cycle cost, while Kavanaugh [84] describes a method based on designing the GHX for the nondominant load and the hybrid component for the balance of the load. Cullin and Spitler [85] describe yet another method based on minimizing first cost of the system, while designing the GHX to supply both the minimum and maximum design entering heat pump fluid temperature over the life cycle of the system.

Recent research on hybrid GHP systems highlights the complexity of their design. Published research mainly deals with single building applications with one hybrid component. Much less research, if any, has dealt with hybrid GHP systems with more than one hybrid component in a district application. The objective of the literature studies [79–87], therefore, was to describe a system simulation approach to examine the feasibility of a multisource hybrid GHP system for an actual proposed DH application in a cold climate, where conventional GHP systems were deemed to be infeasible and impractical.

6.9.2 The Multisource Hybrid Concept

Here, the term multisource hybrid is used to describe a hybrid GHP system with multiple heat sources. The basic design concept takes advantage of a modular "plug-and-play" structure such that heat sources or sinks can be added as practical.

The concept is centered around a common low-temperature supply pipeline that serves to distribute energy in the form of an aqueous antifreeze solution to the sources and sinks. A low-temperature distribution loop was conceived in this design so that lower-grade heat sources could be rejected to the loop. A lower temperature fluid distribution loop typically requires larger diameter pipe relative to that used in a high-temperature loop, but the added advantage of larger pipe diameter means more fluid volume in the loop and correspondingly more thermal mass (or thermal inertia) of fluid in the pipe, which helps to damp large fluid temperature excursions during peak load times. Amplification of the low-temperature source loop to useful temperatures for space heating is accomplished with water-to-air or water-to-water heat pumps distributed throughout the district in the buildings they serve. The minimum heat pump supply temperature of 0°C was chosen because of the low ground temperatures in cold climates [88].

An integral component of the DES is the GHX, which could consist of one central array or multiple decentralized arrays. The GHX acts to provide a base load heat source for heat pumps, supplemented by a peaking boiler during extreme cold periods. In addition, the GHX acts as a short-term and long-term (i.e., seasonal) storage medium for various waste and other available heat sources, which help to improve the GHX thermal performance during times when heat is needed. The waste and other heat sources considered in this study were limited to solar energy and heat recovered from sanitary sewers. General options exist for "other heat sources and sinks," which could conceivably include heat rejection from refrigeration systems (i.e., ice rinks) or any other source deemed practical. This box could also represent another modular GHX as the district system expands and/or additional GHX's are incorporated at decentralized locations. Each of the individual components of the district GHP systems is described in further detail below.

6.9.2.1 The Ground Heat Exchanger (GHX)

Two types of loads are important in sizing a GHX to meet intended loads: (a) the peak hour load and (b) the annual load. Sizing a GHX therefore differs from sizing conventional heating and cooling equipment (i.e., boilers and chillers) because the earth does not respond instantaneously to heat rejected to and extracted from it; long-term temperature changes take place in the underground GHX storage volume if annual loads are not balanced (i.e., the same amount of energy added approximately equals the amount taken out). When annual loads are not balanced, the GHX must be increased in size to accommodate long-term underground temperature changes over years and decades. Annual loads are sometimes naturally balanced by a distinct heating and cooling season, but this is not the case in a sub-Arctic environment, and the size of the GHX to meet all of the intended loads is excessive and unnecessary, which led to this concept of a hybrid system. Supplemental systems are used to offset peak loads and annual loads on the ground. The GHX design consists of a closed network of vertical boreholes drilled to approximately 100 m deep. Each borehole would be completed with a high density polyethylene plastic u-tube heat exchanger grouted in place with standard bentonite-based grout. The heat transfer fluid consists of an aqueous solution of 20% propylene glycol [88].

6.9.2.2 Geothermal Heat Pumps

The DE concept presented here involves heat pumps distributed throughout the district, located in the buildings they serve. GHPs would simply replace conventional furnaces or boilers in buildings, and would be installed during construction of each individual building. Individual heat pumps would be sized to meet the intended loads of the building; the hybrid components are only designed to assist the GHX in providing source energy to the heat pumps. Therefore, no supplemental heating is necessary within individual buildings, and emergency backup heating in buildings would be up to the preference of the individual building owner. The concept of providing low-temperature source water to customer buildings allows for customer flexibility to choose their preferred type of heating system, either ducted forced air or radiant floor heating [88].

6.9.2.3 Peaking Boiler

For consistent delivery temperature of source fluid from which thermal energy can be extracted by heat pumps, a peaking boiler is added to the district system concept. A peaking boiler system serves to offset peak loads on the ground, thus reducing unnecessary GHX size and cost. The optimum size of the boiler depends on the economic trade-off between the avoided GHX cost and the annual operating cost of the boiler. The fuel source for the boiler could be natural gas, biomass, heating oil, or combined fuel. The design concept involves only operating the boiler during times when the fluid temperature exiting the GHX falls below 0°C, which will occur during peak heating load hours. The boiler will therefore contribute very little to operational costs. The boiler operates on a temperature control, set to maintain a minimum supply temperature of 0°C to the district loop. An additional benefit of a peaking boiler system is that it could be used as a backup in the event of sewage heat recovery interruption (sewer heat exchanger maintenance) or GHX maintenance [88].

6.9.2.4 Solar Thermal Recharge and Sewer Heat Recovery

The role of solar thermal and sewer heat recovery is to offset annually imbalanced loads on the ground by recharging the GHX with thermal energy. Balancing ground loads allows for further reductions in the GHX size and cost. As with a peaking system, the optimum size of the solar collector array and sewer heat recovery system depends on the economic trade-off between the avoided GHX cost and the capital and operating cost of the load balancing systems.

In a DH concept, solar energy would be the "first" energy source added to the district loop, mainly because the most strategic location for solar collector location is on customer roof tops. Solar energy would therefore be added to the district loop immediately downstream of the heat pumps. This would be accomplished with existing off-the-shelf, at-plate solar technology. Solar collectors are typically operated using a differential set point control, meaning that the collector must be warmer than the district loop by a set amount (typically at least 5°C) for useful heat transfer to occur. Thus, useful heat can be collected beginning at low solar collector temperatures (i.e., less than 5°C). Similar to the solar recharging concept, sewer heat would be added to the loop at a strategic location as the fluid returns to the GHX. This allows heat to be collected at any time during year and stored underground in the GHX to improve its thermal performance. Useful heat could only be transferred to the district loop when the wastewater temperature exceeds the district loop temperature.

The hybrid district GHP system described here is intended to serve approximately 124,000 m^2 of mixed residential and commercial floor space in a new subdivision in Whitehorse, Yukon, Canada. Weather conditions at the subject site are sub-Arctic, with a heating design temperature of −37°C, and 12,447°F-day (6,915°C-day) heating degree days. The underground earth temperature is approximately 3°C (37.4°F). The peak heating load is estimated at 5,840 kW (19.9 million Btu/h), and the annual heating energy load is estimated at 9.4 MWh. The choice among various options described above should be based on parametric analysis with optimization with the goal to provide whether feasible combinations exist. The methodology in conducting the optimization analysis involved examining numerous combinations of hybrid system component sizes. In making a choice, a life cycle analysis should also be carried out, which includes capital cost and other economic data [88].

6.9.3 Hybrid Modeling and Cosimulation of DH Networks

While there is intense global research activity addressing problems underpinning the reengineering of the electrical power grid, thermal energy grids are likely to play an increasing role in energy systems. The study by Vesaoja et al. [89] described an ongoing effort in hybrid modeling and cosimulation of the physical and control domains of DH networks. The focus was on modeling each domain using semantics and tools natural to each; the study also described the challenges of, and a method for, integration and synchronization of the simulation models in each domain. Here, the dataflow model of computation used by Simulink provides a natural environment for modeling physics-based dynamics of heat energy flows, and the IEC 61499 automation architecture facilitated distributed systems modeling and enabled rapid deployment to field hardware. At the application level, the study showed how this framework enables study of energy flows within a producer/consumer (prosumer) and the analysis of the economic value of integrating distributed solar thermal generation and storage into a prosumer participating in a DH network.

6.9.4 Hybrid Renewable Energy Systems for Buildings

Primary energy use in buildings accounts for almost 40% of the total energy consumption in the EU40. In residential buildings, approximately 80% of the energy used is required for space heating and cooling and sanitary hot water. A significant number and variety of energy supply technologies can be integrated into the built environment. Many of these technologies can be combined in highly efficient hybrid heating (and/or cooling) systems. Hybrid systems, and in particular systems using two or more renewable energy sources, have a huge potential to reduce CO_2 emissions in the building sector through a wide range of applications, depending on the technology chosen, its overall efficiency, and the avoided environmental impact of the relevant fossil fuel alternative. Until recently, the most common hybrid application was the combination of a fossil fuel burner (mainly gas or oil) and solar thermal collectors. Small-scale systems using a combination of two renewable energy sources have gained market share in recent years. The main examples of hybrid renewable energy systems are

Electrically driven heat pumps and solar photovoltaics.
Electrically driven heat pumps and solar thermal.
Thermally driven heat pumps in combination with solar thermal.

One major challenge of hybrid systems is to maximize the combined efficiency of the energy sources employed and, at the same time, to minimize the operating cost and the environmental impact. This can be only achieved if the system as a whole is considered and not just its various components in isolation. The trade-off between system performance and cost (both related to complexity) is the key thing to understand about hybrid system technology. Improving the relative performance of the individual components is necessary to achieve highly efficient hybrid systems; however, it is not sufficient. In the past decade several European projects and international collaborations have been conducted to improve the global efficiency of hybrid systems, albeit with limited resources. Additional research and development activities are therefore required to optimize these systems and realize their commercial potential [45]. To successfully implement scientific research and technological development in small-scale hybrid systems, it is highly important to establish close links with the building and construction industry. In the near future, the built environment needs to be designed, built, and renovated with a clear vision of integrating multiple renewable energy sources and energy efficiency measures [3,90].

Within the 160 million residential and commercial buildings in Europe, the current housing stock can be divided into three categories:

- Single-family houses which include individual houses inhabited by one or two families. Terraced houses are included in this group.
- Multifamily houses, which contain more than two dwellings in the house. The distinction from the third group varies from country to country. Buildings with eight or fewer stories are regarded as multifamily buildings.
- High-rise buildings, which are defined as buildings that are higher than eight stories.

According to housing statistics, nearly 30% of these buildings were constructed between 1946 and 1970. The annual growth rate of new buildings added to the housing stock is currently estimated at around 1%–1.5% of the housing stock. The number of buildings removed from the stock is about 0.2%–0.5% of the housing stock per year. It is assumed that this trend will continue in the period ahead. Refurbishments affect roughly 2% of the housing stock per year. **Hybrid systems are replaced at a rate that is faster than the rate housing stock is renewed or buildings are refurbished. The systems in about 5% of buildings are replaced each year.** A similar phenomenon is observed in nonresidential buildings with the replacement rate depending on the type of building.

Nonresidential buildings account for the remaining 25% of Europe's building stock and together make up a more heterogeneous group, used for a great variety of functions, each with different energy demands per unit and each built according to different standards. Retail and wholesale buildings account for most of the nonresidential stock while office buildings are the second biggest category, with floor space corresponding to one quarter of the total nonresidential floor space. Variations in usage pattern (e.g., warehouse versus schools), energy intensity (e.g., surgery rooms in hospitals versus to storage rooms in retail), and construction techniques (e.g., supermarket versus office buildings) are some of the factors adding to the complexity of the sector. One peculiar characteristic of this typology of buildings is the higher cooling loads in comparison to housing, which is due to specific appliances (such as computers) and on-average higher comfort requirements. While new buildings can be constructed with high performance levels, it is the older buildings, representing the vast majority of the building stock, which are predominantly with low energy performance and in need of renovation work. The energy performance of each building defines the technical options that can be used to achieve comfortable temperature levels [3,90]. In order to make hybrid systems more attractive, future research should be focused on cost, thermal efficiency, simplification of process implementation, and making the process more user friendly. More efforts should be put on process prefabrication and integration, automation and control, development of new standards and procedures, integration of building components in hybrid systems more easy and making hybrid systems 100% renewables [3,90].

6.9.5 SMALL-SCALE HYBRID PLANT INTEGRATED WITH MUNICIPAL ENERGY SUPPLY SYSTEM

The study by Bakken [91] describes a research program that started in 2001 to optimize environmental impact and cost of a small-scale hybrid plant based on candidate resources, transportation technologies, and conversion efficiency, including integration with existing energy distribution systems. Special attention was given to a novel hybrid energy concept fueled by municipal solid waste (MSW). The commercial interest for the model was expected to be more pronounced in remote communities and villages, including communities subject to growing prosperity. To enable optimization of complex energy distribution systems with multiple energy sources and carriers a flexible and robust methodology must be developed. This will enable energy companies and consultants to carry out comprehensive feasibility studies prior to investment, including technological, economic, and environmental aspects. Governmental and municipal bodies will be able to pursue scenario studies involving energy systems and their impact on the environment, and measure the consequences of possible regulation regimes on environmental questions. This paper describes the hybrid concept for conversion of MSW in terms of energy supply as well as the methodology for optimizing such integrated energy systems.

Due to environmental issues, there is a growing interest for small-scale conversion of renewable sources and waste. Examples are local cogeneration using commercial technologies, and more sophisticated solutions employing fuel cells or

Rankine cycles with organic fluids. These new alternatives are not only expected to offer an increased flexibility to system design and new possibilities to optimize an energy system but will also result in more complex systems. Examples of situations with complex problems related to optimal coordination between different alternative energy carriers are development of a new suburb (including school, kindergarten, shopping center, medical center, etc.), design of new energy-efficient office buildings, or development of modern industrial areas.

When introducing new energy sources and technologies into the existing electricity distribution system it is necessary to take into account the interaction between the new source and the existing system. Dispersed electricity production raises a number of questions based on the needs of the consumer; both technical, economic, and administrative. There is little dispersed generation installed in Norway today, but the amount is increasing, especially in connection with wind farms. It is expected that a larger amount of the electricity consumed will be produced locally in the future. So far technical issues like reliability of supply, power/voltage quality, protection, and safety have been addressed only for traditional electricity supply systems where the generation capacity is centralized. If a larger amount of cogeneration units are installed, the amount of space heating by direct use of electricity will probably go down, leading to a shift in types of equipment connected to the network. Thus, new local energy technologies have to be carefully integrated into the existing energy/electricity distribution system.

During development of new methodology and models it is important with regular testing and verification with realistic data. The study by Bakken [91] presents one of the selected test cases in the project; a small-scale waste fueled cogeneration plant. The study describes the cogeneration plant and the local energy system in general and the waste treatment plant in more detail. It also gives the outline of the analytic methodology under development and describes the hybrid optimization algorithms to be used. In the county of Melhus 20 km south of the city of Trondheim, Norway, a small-scale cogeneration plant was being planned together with a local DH network. The cogeneration plant was fueled by MSW, and supplied 17 GWh/year heat to the DH system and 6 GWh/year electricity to the local grid. The DH network had a total length of approximately 2 km, with the major customers of municipal administration and office buildings, local industry, schools, and health care institutions within a radius of 500 m from the waste plant. Currently, electricity is the major energy source in the area. The county is supplied with a meshed 22 kV grid from four supply stations to the regional grid with a total installed transformer capacity of 67 MVA. The current electricity consumption in the county center is approximately 50 GWh/year plus some oil-fired space heating. The new cogeneration plant is planned with a heat capacity of 2 MW and a gas engine of 700 kVA electric capacity. Assuming some new industrial customers to be connected to the DH in the future, the cogeneration plant is expected to substitute 25%–30% of current electricity consumption in the area. Thus, the installation will not have any major influence on the existing energy system.

As a part of the initial phase to evaluate the possibilities for a local utilization of MSW for the production of heat and power, the local energy distribution company Melhus Energi conducted a study were the available amount of MSW was found to be in the range of 5,000 tons/year. Commercial technologies for energy from waste for such small installations are scarce. The solutions are often based on gasification and pyrolysis, while larger plants are based on combustion. In addition, manufacturers offering small-scale solution often have limited experience with long-term commercial operation of their technology. Based on a total evaluation of cost, available MSW, and possible technologies, Melhus Energi chose to implement the so-called Pyroarc process. The Pyrorac process is based on an updraft gasifier in combination with a decomposition reactor that also includes the use of plasma technology. Solid waste is fed into the gasifier where the organic material is devolatilized and gasified to produce a combustible gas. The inert materials in the waste, ashes, and metals are melted in the high-temperature reaction zone.

The product gas is introduced into a mixing zone in the decomposition reactor just in front of the plasma generator. The dynamic forces of the plasma jet give an effective mixing of the plasma jet and the product gas. The product gas is partially oxidized in the decomposition reactor by the addition of air or oxygen. From the decomposition reactor the gas is led to a gas cooling and cleaning step, which includes removal of particulates, heavy metals, and acid components. The gas cleaning system is designed according to local regulations for emissions [91].

The clean gas can be utilized in suitable combustion processes (boiler, gas engine, gas turbine, etc.) according to local requirements. At Melhus, a gas engine is used for the production of heat and power. The environmental benefits for the Pyroarc process compared to combustion are significant. As the gasifier operates at slagging conditions, the solid residues are very stable in terms of leaching compared to bottom ash from a combustion process. The solid residues from the gasifier can in fact be utilized as a product and not as a waste material that must be stored on landfills. The product gas leaving the gasifier contains tars, chlorinated hydrocarbons, and gaseous nitrogen components (NH_3, HCN). These components can cause severe environmental emissions like dioxins and NOx and the tars can cause operational problems in combustion processes. However, the use of a plasma generator and the conditions in the decomposition reactor (high temperatures, good mixing, and residence time) decompose the components that can form these emissions. Measurements show that there is *no* recombination of halogenated hydrocarbons, and also the NOx emissions are low due to reduced contribution of fuel NOx. The residues from the gas cleaning system can contain zinc and lead concentrations at levels that make recovery of these metals economically feasible. The overall thermal efficiency for the Pyroarc process is in the range of 90%–94%. The chemical energy in the product gas is typically in the range of 70%–80% while the sensible heat of the product gas is in the range of 20%–30%. For CHP production in a gas engine, a net power efficiency of 20% can be achieved. This figure is based on an efficiency of the gas engine of 35% and subtraction of the power needed to operate the plasma generator. For larger units a steam cycle can be added, increasing the net power efficiency to about 35% [91].

Generally, energy systems consist of three types of processes: Energy transport over a geographical distance (AC or DC lines, gas pipelines, liquid natural gas transport, DH, etc.), conversion between different energy carriers (gas power plant, CHP, heat pumps, etc.), and storage of energy (batteries, LNG/gas tanks, heat storage, etc.). The general approach in planning an optimal energy system was a multicriteria decision problem where the objective was to find an optimal network of processes, based on the properties of different processes. To be able to do comprehensive analyses of complex local energy systems with several different energy resources, carriers, and technologies, a robust optimization methodology had to be developed. The main idea of this methodology was based on the knowledge and experience among electrotechnical specialists on complex network structures, load flow models, and linear programming. In this project, however, the concept was further developed from flow of electric current to generic flow of energy. Specialists from other fields are involved to model the different processes and components (Thermal energy, Refrigeration and Air Conditioning, etc.). A major objective was to handle different components at different geographical locations, connected by an energy distribution system. In a hybrid operation, MSW from county and business offices, institutions, companies and households, gas from old landfills, and biomass and waste from forestry and farming can be included. These energy resources had to be transported, processed, and stored; at different locations and in different forms before being converted to end user energy like electricity and heat. Often a choice had to be made between building larger centralized CHP units feeding electricity and DH networks, or remote mini-CHP installations in single buildings (like offices, schools, health care centers, etc.). The model treated energy transport by pipeline and power line as well as by road [91].

The following methodology was used:

- Based on a library of available components, the user builds a model of the distributed energy system with the alternative solutions to be optimized.
- Each component was internally modeled with the necessary mathematical details to account for the specific properties of that technology.
- The connection to the geographic energy network, however, was made by a simple and unambiguous set of linear variables like cost, energy efficiency, and energy quality/environmental aspects.
- The superior network analysis and optimization was made on a generic nodal model without specific knowledge of which components were involved. At this level, the optimization algorithms see only a linear network with nodes and branches where energy flows. This generalization occurred internally in the optimization and was not noticeable by the user.
- For presentation to the user, the results were "translated" back to the component-specific system model.

A major challenge when analyzing such complex energy systems was to combine the multicriteria objective with the modeling demands of a variety of different energy processes. Adding the complexity and time span of the investment analysis created an optimization problem not easily solved using conventional methods. In this approach an important goal was to reduce the number of manual assumptions by separate modeling of each energy technology in sufficient detail. It is easy to argue against such an approach because some simplifying assumptions have to be made in any case. It is impossible to account for all physical aspects in one model due to the different properties of the processes involved. The option of combining different optimization methods allowed this approach to be able to obtain these goals without compromising the main physical characteristics of the processes involved [91].

A combination of different optimization methods adds new possibilities to the modeling of energy-related problems. It is not necessary to account for everything in one large model, as input from other models can be used in the areas where "all-in-one models" meet limitations. An example of such a successful hybrid approach is a model that combines the long-term hydropower operation strategy calculated with stochastic dynamic programming with a detailed deterministic subproblem within 1 week [1]. It is not possible to account for every aspect of the hydropower system when calculating the long-term strategy. In order to calculate a long-term strategy one needs to aggregate in time such that start-up cost, time delays, and hydraulic couplings are not properly accounted for. This makes the results less useful if the modeled system does not meet certain assumptions. Typical assumption for aggregated hydropower models are zero start-up cost, intermittent operation allowed for pumps, limited system size, and a nonsequential time description. In many problems, however, these properties can be accounted for when another model is used for implementing the strategy. Adding the results from the long-term strategy as boundary conditions to a deterministic linear model makes it possible to account for the properties that cannot be included in stochastic optimization. This combination of methods makes it possible to handle details in a proper way despite of the inability of the strategy calculation to handle every hydropower detail.

This approach can be useful also in the case of energy distribution systems with multiple energy carriers. Detailed process models are created to account for properties that are difficult to combine without simplifying assumptions. A transmission model of either AC or DC power including security constraints is used for the local electricity distribution system. Special models for truck transportation to and from waste plants are used, as well as a DH model which can optimize operation of the DH system, taking nonlinear elements into account. An adequate model for hydropower to account for stochastic elements can also be used if there is hydropower in the region. Results from the component models are afterwards used in the linear system model to calculate the optimal operation plan for the selected time period (day/week) for a given topology alternative [89,91–93].

The optimal operation planning kernel must be integrated with an investment analysis scheme to choose the best possible expansion plan over the planning horizon. So far dynamic programming has been used to find the best expansion plan according to the given alternatives and possible introduction times. The overall concept can then be outlined. In this concept the operation planning kernel is used for calculation of the running expense for the different alternatives. Results from each alternative are added to the dynamic programming table and the best route through alternatives (size and type) and time is calculated. Sensitivity and robustness of the alternative can be just as important as the profit of the investment, so a combination of criteria must be used for finding the best solution. To be able to handle a hybrid model like this, it is important that the operation planning kernel is fast, because this calculation is the most time consuming. Also, the number of alternative topologies of the local energy system (new or expanded components, processes, transport channels, etc.)

will influence greatly on the calculation time, hence it is important that the number of alternatives as specified by the user is limited. In most cases it will be possible to rule out the most unrealistic alternatives before optimization. The design of the Graphical User Interface (GUI) was not yet specified, but integration with the existing Geographic Information Systems (GIS) was a possible alternative [91].

The study thus outlined the development of a new methodology for analysis of complex energy distribution systems with multiple energy carriers. The methodology was based on two main levels of modeling. The lower level was used to calculate optimal operation of the system and the upper level handled the investment decisions. In the GUI specific component modules with a standard interface were combined in alternative compositions of the energy system. Each alternative was then generalized to a nodal network with generic energy flow. To enable a multicriteria optimization with a minimum of simplifying assumptions, which might limit the validity of the results, hybrid optimization techniques was implemented, e.g., combinations of stochastic dynamic programming and deterministic short-term optimization. Each energy technology was modeled separately with sufficient detail, supplying the superior linear system model with a simple and unambiguous set of variables like cost, energy efficiency, and environmental impact. The methodology will enable energy companies to carry out comprehensive analyses of their energy supply systems, and governmental bodies will be able to do comprehensive scenario studies of energy systems with respect to environmental impacts and consequences of different regulating regimes for preserving environmental values [89,91–93].

6.9.6 Modular Hybrid Solar and Biomass System for DH

Biomass may be cheap and carbon-neutral, but a solar upgrade of biomass-fired DH could further improve efficiency and reduce local emissions. For example, solar heat helps avoid having to start up and shut down wood-chip boilers or operate them at partial load. It can even replace backup fossil fuel systems, which provide DH networks with energy in summer. Biomass boiler in Sweden benefits from a 10% solar fraction. Larger biomass systems, for instance, between 1 and 100 MWth, have the advantage that their boilers run on wood chips. They take more time to be shut down and restarted, while a partial load results in reduced efficiency, for example, below 20% of nominal power. To prevent partial loading, the systems are usually equipped with a smaller fossil fuel boiler or buffer storage. Of course, an alternative solution is the integration of a solar thermal field, which can provide a significant boost to boiler efficiency, reducing emissions and costs. One such biomass-solar system in Sweden is the DH network set up in 2010 in Ellös, in the Västra Götaland county. It consists of 4 MWth of biomass boilers, a 1,000 m^2 solar array and 200 m^3 of buffer storage. The average annual solar fraction is around 10%, and the main benefit of solar integration is the option to switch off the biomass boiler in summer, except for a few rainy or cloudy days in a row. The size of the solar system is planned to be scaled up to 2,000 m^2 soon [3].

Solar-assisted biomass DH is quite common in Austria. Sixteen of the 32 SDH plants in operation across the country have been combined with a bioenergy system. Their solar arrays range from 100 to 7,000 m^2. One factor encouraging these kinds of developments is the national subsidy program for large solar thermal plants above 100 m². Launched in 2010, it has recently been extended to include plants of up to 10,000 m². It provides 40% of the investment cost, and both small and medium size and designers of innovative systems can get another 5% on top. A biomass-solar DH system in Mürzzuschlag, a town in the northeast of Austria's Styria region. The plant has solar-based production costs of 35 EUR/MWh, which is slightly below the biomass values of 37–38 EUR/MWh.

Over the last 20 years, Austria's countryside has seen the installation of more than 2,000 biomass DH networks, which makes it quite difficult to find a suitable location for a new grid. Instead, solar collector suppliers are focused on solar upgrades of existing biomass DH. Moreover, some of them have 20- to 25-year-old boilers, which should be replaced to meet new efficiency requirements. Some also no longer benefit from CHP feed-in tariffs All these means a great potential for integrating solar arrays into existing networks [3].

Austrian Institute AEE INTEC indicated that a survey among DH utilities analyzed the main reasons for integrating solar thermal into biomass DH. The survey emphasized that solar could replace backup fossil fuel boilers used exclusively in summer to avoid partial load operation. The DH utilities added that solar thermal could also reduce local emissions, such as dust and nitrogen oxides, created during biomass combustion, which would increase public acceptance of biomass DH. Additionally, Austria had a program to recover up to 35% of the investment in a biomass DH installation. The incentive required that overall network efficiency exceeded 75%—not an easy objective for small networks in areas of low population density, where long pipes are necessary. Solar thermal could help meet the target. In that case, the biomass incentive could serve as an indirect support for SDH systems. Hybrid solar-biomass DH systems are also very popular in Romania [94] and China [95].

6.10 ROLE OF THERMAL ENERGY STORAGE IN DH

TES is of assistance to energy suppliers in a DE system by allowing [18,26,42,96–117]

1. the accumulation of thermal energy from off-peak periods for use when demands are high;
2. the storage of excess thermal energy when it is available, for subsequent release to the DE distribution system when thermal demands increase (especially during periods when suppliers are not able to satisfy thermal energy demands with existing facilities). In this way, TES saves thermal energy that would otherwise be wasted;
3. more effective utilization of renewable thermal energy sources like solar than is otherwise possible due to the intermittent nature of the resource supply;
4. Less energy losses in DE systems. Incorporating TES in DE systems allows the reduction of thermal losses, resulting in energy savings and increased efficiency for the overall thermal system. Large seasonal TES

Energy Usage in District Heating

systems have been built in conjunction with DE technology [100,111,112]. Besides improved efficiency and economics, environmental benefits are another reason for the expansion of TES technology in general and with DE.

Many beneficial applications of TES with DE exist or can be developed. For example, Andersson [101] reports that the application of ATES in one DE system reduced energy use by 90%–95% compared to conventional systems not incorporating TES. DE with hybrid systems can advantageously combine sources of energy like natural gas, waste heat, wood wastes and MSW [18,26,96–102]. TES can increase the benefits provided by such systems. Lund et al. [42,102] point out that "low-energy" buildings can be operated with industrial waste heat, waste incineration, power plant waste heat and geothermal energy in conjunction with a DE network. Incorporating TES can improve designs of such systems. The types of energy sources that can be used with DE and TES systems vary, and systems designs are often tailored to best adapt to the energy source.

6.10.1 Advantages and Disadvantages of Using TES in DE

Tanaka et al. [103] state the use of TES in conjunction with DE decreases energy consumption compared to a reference system. They also found that seasonal TES is more effective than short-term TES. Andrepont [114] and others [18,98,99] assessed the economic benefits of using TES technologies in DE systems. They noted that cool TES is applied widely in heating, ventilation and air-conditioning systems by shifting the cooling load from peak periods during the day to the off-peak periods at night. This time shift significantly reduces operating costs, and this benefit is particularly noteworthy in large-scale DE systems. Following additional advantages of using TES in DE systems are also mentioned in the literature [18,26,96–103]:

1. Preventing inefficient operation of chillers and auxiliary equipment during low-level operation
2. Enhancing system reliability and flexibility
3. Balancing electrical and thermal loads in CHP for better economy
4. Lowering accident risks and insurance by enhancing fire protection (since the stored chilled water or other storage fluid in the TES is in the vicinity of the DE and may be utilizable as a reserve firefighting fluid in the event of a fire)

The benefits of TES in facilitating the use of renewable energy, especially solar thermal energy for use in heating and cooling buildings, have been pointed out by several investigators [18,26,96–101]. Demand is growing for facilities that utilize TES, as they are more efficient and environmentally friendly, exhibiting reductions in (a) fossil fuel consumption, (b) emissions of CO_2 and other pollutants and (c) Chlorofluorocarbon (CFC) emissions.

Several investigators have reported [6,7,21–99,103] that buildings with TES systems in DE are environment friendly. The U.S. Green Building Council (USGBC) did not discourage the use of TES in the first version of the Leadership in Energy and Environmental Design (LEED) [115] criteria, but also did not deal with the use of TES in DH systems; however a building with TES is eligible to earn more point for its lower electrical power use. Building a TES requires initial capital for land, construction, insulation and other items. Determining the appropriate location, designing the proper structure and insulation, and executing the design are important steps in installing a TES, which involve significant costs. The high initial cost is a disadvantage of TES.

The use of TES complements the use of solar energy in the DE system, allowing surplus solar heat in the spring and summer to be stored for subsequent use in the fall. Without TES, this surplus energy would be wasted. TES thereby enhances the benefits of using solar energy in the DE system. It is observed that TES reduces annual fuel use and fuel costs by 30% in a typical DE system, reducing the use of natural gas boilers. The use of TES is also advantageous environmentally, reducing emissions to the atmosphere like CO_2 by 46% in the DE system. The advantages of incorporating TES in the DE system, in terms of enhancing the use of solar thermal energy, suggests TES is likely going to become increasingly important and utilized in industry, power generation and DE as use of renewable energy expands. Lund et al. [42,102] indicated that renewable energy is the focus of many countries for improving energy security and mitigating climate change. The use of renewable energy in a sustained form for DH is heavily facilitated by the TES. The role of TES for use of renewable energy in DH can be briefly outlined as follows [18,26,96–99]:

A. **Solar:** Solar energy can be used for space and water heating. While the advantages of using solar energy are significantly enhanced with the help of TES in the thermal system, the key benefit of TES is in its ability to integrate the solar thermal application, i.e., the energy stored by TES can be made available when thermal energy is in demand and solar availability is uncertain. TES increases the impact of solar collectors by avoiding the loss of solar energy when it exceeds demand. Annual TES is desirable if the excess solar energy is stored for a season or longer, while short-term storage is more appropriate if the solar energy is stored for hours or days. In both cases, however, TES provides a beneficial alternative to the use of conventional fuels as a backup energy source. Short-term TES applications in conjunction with solar energy in DH systems have been tested in pilot projects and are now used in several countries [100,111,112].

B. **Geothermal:** The ground source heat pump is an efficient device and can be used advantageously with DE for HVAC purposes. When a ground source heat pump is the source of energy in DE system, TES can help the energy system store extracted heat from the earth for use when in demand. According to Lund et al. [42,102], usually, BTES is an appropriate TES for geothermal energy in conjunctions with ground source heat pumps.

Energy Usage in District Heating 313

C. **Recovered Waste Heat:** Industry can be a supplier of waste heat for a DE system and/or a heat consumer from the DE system. When industry has excess heat or waste heat, it can supply to a DE system, or it becomes a consumer when it requires heat. Holmgren and Gebremedhin [105] point out that cooperation between industry and DE can allow technical and economic factors to be appropriately addressed for both parties in terms of the thermal energy quality and quantity as well as profitability. When the energy for a DE system is supplied by waste heat, the key parameters to be considered in designing DE and TES include (a) heat supply temperature, (b) heat consumption temperature, (c) heat supply time and (d) heat demand time.

The literature [18,98–100] also suggested underground thermal storage such as ATES as an appropriate storage for waste heat. This type of system allows stored thermal energy to be made available for consumers and also permits load leveling in the DE system. DE systems can supply waste heat from an existing boiler which has waste heat or from other industrial processes. DE systems using such energy sources are generally more clean, economic and efficient than DEs using conventional fuels. Holmgren and Gebremedhin [105] also note that the integration of DE with industry reduces not only the cost of heat production but also CO_2 emissions. When a DE system directly operates using fuel, TES is not always needed. Nonetheless, TES provides the possibility of using smaller equipment in designs, by reducing energy need from external energy sources. Electricity is significant energy source in a DE system. TES helps reduce electricity costs during peak demand periods and allows the system to operate during off-peak periods, e.g., running chillers during the night and storing the cold medium for use in cooling the next day. Short-term cold TES is used in such applications.

According to Holmgren and Gebremedhin [105], waste incineration is a useful technology for recovering the energy content of waste. Alternatively, waste that can be disposed of in landfill sites and accelerated processes (e.g., gasification, pyrolysis), can be used to produce a combustible fuel gas [106]. Holmgren and Gebremedhin [105] believe waste incineration is a preferred option compared to others for supplying heat for DE. Also, **TES eliminates energy waste in existing DH network**. When waste incineration supplies the thermal energy for a DE system, several parameters need to be considered in DE and TES design, including (a) capacity of the waste incineration process, (b) availability of the waste and (c) time profile of the heat demand.

6.10.2 ENERGY CENTRAL

The Energy Central is the result of a cooperation between Grundfos [107] and the local DH company. Several years ago, Grundfos headquarters and the local DH company inaugurated a joint system to store the surplus heat from the Grundfos factories in obsolete groundwater boreholes, and use it in the DH network when needed. The synergy between the cooling demand in Grundfos' production plants and the heat demand of Bjerringbro DH company has virtually eliminated energy waste. The remarkable reduction of energy consumption, CO_2 emission and operational costs now benefit both companies and the DH customers in the area. As part of the agreement, a new energy central with state-of-the-art cooling compressors and heating pumps was established. The new setup was based on three elements:

1. Exploitation of surplus heat from cooling in the factories.
2. Indirect storage of heat in an underground aquifer.
3. Heat pumps supply additional heat, when required

Using small, individual refrigeration compressors to cool down machines in production facilities is a very inefficient, but nevertheless very common method. To make matters worse, the heat energy developed in the process is virtually always wasted. With the new shared Energy Central, Grundfos productions facilities are cooled by cold water from the DH network. Also, when the cold water needs to be colder, three large cooling compressors with a total cooling capacity of 2.85 MW and thermal power of 3.65 MW handle the job from a central facility. During the heating season, the cold water used to cool the machines at Grundfos, becomes hot in the process, and along with the excess heat from the cooling compressors in the energy central, the heat is recycled and sent directly to the DH network. Main features of this waste heat-DE-TES partnership are:

- During summer, when the heating demand is minimal, the surplus heat is stored in underground energy storage—a so-called ATES-stock (Aquifer TES) located 80 m underground.
- During the four summer months, a total cooling output of 3,500 MWh is accumulated. And more than 80% of the stored energy during summer will be supplied to the DH network during the heating season. In fact, the energy central covers more than 15% of the town's annual heat requirement, of which 1/3 comes from the hot water storage.
- Pressure drops across valves in a heating system are costly and waste energy. This system however, is frequency controlled and designed for completely open valves. Accordingly, all thermal adjustments are carried out, by speed controlled pumps. As a result, pressure drop across the valves have been completely eliminated and replaced by massive energy savings.
- The ATES system reduces Grundfos' energy consumption for cooling by up to 90%. Combined with significantly reduced heating costs, the result is cheaper heating for the citizens and massive energy savings for Grundfos and the DH plant.
- The carbon emission is reduced by some 3,700 tons a year, equivalent to 1.5 tons per household connected to the heating plant.
- In total, Grundfos and Bjerringbro DH company have invested 4.5 million euro in the new system. Without this joint investment, two separate systems

Energy Usage in District Heating   315

would have been the solution. Compared to these reference systems, annual savings of 400,000 euro was initially expected, but because the system seems to be more efficient than estimated, the payback time will most likely be shorter than the projected 11.25 years.
- With the new setup Bjerringbro DH company seizes the opportunity to store energy and become less dependent on natural gas. As a result, the plant emit, less CO_2 and be able to offer more sustainable heating supply even lower price.

Many European countries aim to be free from the use of fossil fuels in the near future, and this type of system is a big step in that direction. The system is able to store solar heat and thus, in principle, low fossil heat as it runs only on electricity. In the future, this may be supplied by wind turbines or any other renewable source. Many elements of the project can be used globally.

6.10.3 USE OF TES IN LTDH

TES systems are technologies with the potential to enhance the efficiency and the flexibility of the coming fourth generation low-temperature DH (LTDH). Their integration would enable the creation of smarter, more efficient networks, benefiting both the utilities and the end consumers. The study by Rossi Espagnet [117] was aimed at developing a comparative assessment of both latent and sensible heat-based TES systems. First, a technoeconomic analysis of several TES systems was conducted to evaluate their suitability to be integrated into LTDH. Then, potential scenarios of TES integration were proposed and analyzed in a case study of an active LTDH network. This was complemented with a review of current DH legislation focused on the Swedish case, with the aim of taking into consideration the present situation, and changes that may support some technologies over others.

The results of the analysis showed that sensible heat storage was still preferred to latent heat when coupled with LTDH: the cost per kWh stored was still 15% higher, at least, for latent heat in systems below 5 MWh of storage size; though, they require just half of the volume. However, it is expected that the cost of latent heat storage systems will decline in the future, making them more competitive. From a system perspective, the introduction of TES systems into the network results in an increase in flexibility, leading to lower heat production costs by load shifting. It is achieved by running the production units with lower marginal heat production costs for longer periods and with higher efficiency, and thus reducing the operating hours of the other more expensive operating units during peak load conditions. In the case study, savings in the magnitude of 0.5k EUR/year are achieved through this operational strategy, with an investment cost of 2k EUR to purchase a water tank. These results may also be extended to the case when heat generation is replaced by renewable, intermittent energy sources; thus increasing profits, reducing fuel consumption, and consequently emissions. This study represents a step forward in the development of a more efficient DH system through the integration of TES, which will play a crucial role in future smart energy system.

6.10.4 TES-Solar Energy-DH Partnership

DESs are a technology for using energy effectively and efficiently, at local or municipal scales, especially for thermal energy. Combining other beneficial energy technologies with a DES can significantly improve the efficiency of the combined system. The benefits of using TES as a complementary system for solar energy collectors was examined by Andrepont [114], Reed et al. [98], and Rezaie et al. [104]. Rezaie et al. [104] examined this within the context of the Friedrichshafen DH system in Germany, which utilizes renewable energy. The TES in the Friedrichshafen DE system has enhanced the DE's performance. In particular, TES has improved the performance of the solar energy system, reducing annual natural gas consumption and fuel cost by 30% and harmful environmental emissions by 46%. DH and/or cooling systems can be augmented through incorporation of TES. Two examples of such integration are [98,101,104]

1. Andrepont [114] determined that chilled water TES and low-temperature fluid TES, used in large-scale DE systems, significantly lowers installation costs per ton compared with equivalent conventional non-TES chiller plants.
2. In some DE designs, TES is incorporated to store solar energy that would otherwise go to waste during periods when heating is not required. The Friedrichshafen DH system in Germany, for example, uses TES with DE to enhance performance and efficiency [108,109].

The focus of the study by Rezaie et al. [104,110,113,116] was to demonstrate the role and benefits of TES in DE through a case study, via an energy analysis for Friedrichshafen DE system.

Finally, Romanchenko et al. [96] showed that heat load variations in DH systems lead to increased costs for heat generation and, in most cases, increased greenhouse gas emissions associated with the marginal use of fossil fuels. They investigated the benefits of applying TES in DH systems to decrease heat load variations, comparing storage using a hot water tank and the thermal inertia of buildings (with similar storage capacity). A detailed technoeconomic optimization model is applied to the DH system of Göteborg, Sweden. The results show that both the hot water tank and the thermal inertia of buildings benefit the operation of the DH system and have similar dynamics of utilization. However, compared to the thermal inertia of buildings, the hot water tank stores more than twice as much heat over the modeled year, owing to lower energy losses. For the same reason, only the hot water tank is used to store heat for periods longer than a few days. Furthermore, the hot water tank has its full capacity available for charging/discharging at all times, whereas the capacity of the thermal inertia of buildings depends on the heat transfer between the building core *and* its indoor air and internals. Finally, the total system yearly operating cost decreases by 1% when the thermal inertia of buildings and by 2% when the hot water tank is added to the DH system, as compared to the scenario without any storage.

REFERENCES

1. District Heating, Wikipedia, The free encyclopedia, last visited 17 July 2019 (2019).
2. District Heating, SteepWiki. A website report https://tools.smartsteep.eu/wiki/District_heating (2015).
3. Rutz, D., Doczekal, C., Zweiler, R., Hofmeister, M., and Jensen, L.L. Small Modular Renewable Heating and Cooling Grids, A Handbook, a report by Cool Heating.eu ISBN: Translations: 978-3-936338-40-9, © 2017 by WIP Renewable Energies, Munich, Germany. www.coolheating.eu.
4. An Action Plan for Growing District Energy Systems across Canada. A website report by CDEA/ACRT, Ottawa, Canada https://static1.squarespace.com/static/.../t/.../CUIPublication.GrowingDistrictEnergy.pd... (June 2011).
5. Danish Energy Agency, Energinet.dk. (2015). Technology Data for Energy Plants—Generation of Electricity and District Heating, Energy Storage and Energy Carrier Generation and Conversion. May 2012 (certain updates made October 2013, January 2014 and March 2015); ISBN: 978-87-7844-931-3.
6. Ericssona, K. and Wernerb, S. (April 2009). The introduction and expansion of biomass use in Swedish district heating systems. *Energy Institute, Biomass and Bioenergy*, 33(4), 659–678. doi: 10.1016/j.biombioe.2016.08.011.
7. Vallios, I., Tsoutsos, T., and Papadakis, G. (November 2013). Design of biomass district heating systems. *Energy Policy*, 62, 236–246. doi: 10.1016/j.biombioe.2008.10.009.
8. Dimitriou, I. and Rutz, D. (2015). *Sustainable Short Rotation Coppice, A Handbook*. WIP Renewable Energies, Munich, Germany; ISBN: 978-3-936338-36-2; www.srcplus.eu.
9. Laurberg, J.L., Rutz, D., Doczekal, C., Gjorgievski, V., Batas-Bjelic, I., Kazagic, A., Ademovic, A., Sunko, R., and Doračić, B. (2016). Best Practice Examples of Renewable District Heating and Cooling. Report of the Cool Heating Project, PlanEnergi, Denmark, www.coolheating.eu.
10. Rutz, D. and Janssen, R. (2008). Biofuel Technology Handbook. 2nd Version; Biofuel Marketplace Project funded by the European Commission (EIE/05/022); WIP Renewable Energies, Germany, p. 152. www.wip-munich.de/images/stories/6_publications/books/biofuel_technology_handbook_version2_d5.pdf (10 November 2016).
11. Rutz, D., Mergner, R., and Janssen, R. (2015). *Sustainable Heat Use of Biogas Plants—A Handbook*, 2nd edition. WIP Renewable Energies, Munich, Germany; Handbook elaborated in the framework of the BiogasHeat Project; ISBN: 978-3-936338-35-5 translated in 8 languages; www.wip-munich.de/images/stories/6_publications/books/Handbook-2ed_2015-02-20-cleanversion.pdf (10 November 2016).
12. Zweiler, R., Doczekal, C., Paar, K., and Peischl, G. (2008). Endbericht Energetisch und wirtschaftlich optimierte Biomasse-Kraft-Wärmekopplungssysteme auf Basis derzeit verfügbarer Technologien, Energiesysteme der Zukunft, bmvit, FFG-Projektnummer 812771, www.get.ac.at.
13. Zweiler, R. (2013). ToughGas (Entwicklung eines innovativen Wirbelschichtvergasungssystems kleiner Leistung zur Nutzung biogener Reststoffe) (Endbericht No. 834621).
14. Vladimir, N. (April 2014). Biomass Combined Heat and Power (CHP) for Electricity and District Heating, Ph.D. Thesis, Norwegian University of Science and Technology (NTNU), Trondheim.
15. District Heating and Cooling, Combined Heat and Power and Renewable Energy Sources, BASREC—Best Practices Survey, COWI, NUORKIVI Consulting, Denmark, C:\Users\CHOE\Desktop\Dokumenter\BASREC\DHC_CHP_RES_survey_BASREC_Countries.docx (January 2014).

16. FVB Projects | District Energy | FVB Energy Inc. www.fvbenergy.com/projects/ (8 April 2019).
17. FVB Energy Inc. District Energy, District Heating. www.fvbenergy.com/ (2019).
18. Galatoulas, F., Frere, M., and Ioakimidis, C. An Overview of Renewable Smart District Heating and Cooling Applications with Thermal Storage in Europe. *Proceedings of the 7th International Conference on Smart Cities and Green ICT Systems (SMARTGREENS 2018)*, pp. 311–319. ISBN: 978-989-758-292-9. Copyright © 2018 by SCITEPRESS – Science and Technology Publications, Lda. All rights reserved.
19. District Heating System—An Overview | ScienceDirect Topics. A website report www.sciencedirect.com/topics/engineering/district-heating-system (2011).
20. Euroheat and Power. (2015). District Heating and Cooling country by country Survey 2015. Indicator codes: ENER 019, www.euroheat.org/wp-content/uploads/2016/03/2015-Country-by-country-Statistics-Overview.pdf; www.eea.europa.eu/policy-documents/euroheat-and-power-2013-district (19 June 2019).
21. Fjernvarme, D. (2016). Technology. www.geotermi.dk/english/deep-geothermal-energy-in-denmark/technology (11 November 2016).
22. Danish Geothermal District Heating. (2016). The Geothermal Concept. www.geotermi.dk/english/deep- geothermal-energy-in-denmark/technology (9 November 2016).
23. GeoDH. (n.d.).Developing Geothermal District Heating in Europe. www.geodh.eu, https://ec.europa.eu/energy/intelligent/projects/sites/iee-projects/files/projects/documents/geodh_final_publishable_results_oriented_report.pdf (10 November 2016).
24. Frederiksen, S. and Werner, S. (2013). District Heating and Cooling. Studentlitteratur, p. 205.
25. Decarbonising District Heating with Solar Thermal Energy—Solar. www.solar-district-heating.eu/solar-district-heating-on-the-roof-of-the-world-4/ (7 June 2018).
26. Central Solar Heating, Wikipedia, The free encyclopedia, last visited 26 June 2019 (2019).
27. Kempener, R. (2015). Solar Heating and Cooling for Residential Applications: Technology Brief. IEA-ESTAP and IRENA Technology Brief E21—January 2015. www.irena.org/documentdownloads/publications/irena_etsap_tech_brief_r12_solar_thermal_residential_2015.pdf (4 August 2016).
28. Schrøder, P.A., Elmegaard, B., Christensen, C.H., Kjøller, C., Elefsen, F., Bøgild Hansen, J., Hvid, J., Sørensen, P.A., Kær, S.K., Vangkilde-Pedersen, T., and Feldthusen Jensen, T. (2014). Status and Recommendations for RD&D on Energy Storage Technologies in a Danish Context. www.energinet.dk/SiteCollectionDocuments/Danske%20dokumenter/Forskning%20-%20PSO-projekter/RDD%20Energy%20storage_ex%20app.pdf (9 November 2016).
29. Solair Project. (2009). Increasing the Market Implementation of Solar-Air-Conditioning Systems for Small and Medium Applications in Residential and Commercial Buildings (SOLAIR). Project website www.solair-project.eu/142.0.html. Accessed 4/8/2016.
30. Pauschinger, T. Solar Thermal Energy for District Heating—ScienceDirect. A website report www.sciencedirect.com/science/article/pii/B9781782423744000057 (2016).
31. Trier, D., Skov, C.K., Sørensen, S.S., and Bava, F. Solar District Heating Trends and Possibilities, Characteristics of Ground-Mounted Systems for Screening of Land Use Requirements and Feasibility Technical Report of IEA SHC Task 52, Subtask B—Methodologies, Tools and Case studies for Urban Energy concepts prepared by PlanEnergi, Copenhagen, June 2018 for IEA, Paris France. iea-shc.org/.../SDH-Trends-and-Possibilities-IEA-SHC-Task52-PlanEnergi-20180619....
32. Solar District Heating—An Overview | ScienceDirect Topics. A website report www.sciencedirect.com/topics/engineering/solar-district-heating (2013).

33. Hopkins, A.S., Takahashi, K., and Melissa, D.G. Whited, Decarbonization of Heating Energy Use in California Buildings Technology, Markets, Impacts, and Policy Solutions, A report by Synapse, Energy economics Inc. This report was prepared for the Natural Resources Defense Council (NRDC), Decarbonization of Heating Energy Use in California Buildings, A website report www.synapse-energy.com/.../ Decarbonization-Heating-CA-Buildings-17-092-1... (October 2018).
34. Xydis, G. Wind Energy Integration through District Heating. A Wind Resource Based Approach. 4, 110–127. doi: 10.3390/resources4010110, ISSN 2079-9276. www.mdpi.com/journal/resources (2015).
35. Wang, J., Zong, Y., You, S., and Træholt, C. (2017). A review of Danish integrated multi-energy system flexibility options for high wind power penetration. *Clean Energy*, 1(1), 23–35. doi: 10.1093/ce/zkx002. Advance Access Publication Date: 24 November 2017 Homepage: https://academic.oup.com/ce.
36. Wind Power and District Heating, A website report www.pfbach.dk/firma_pfb/forgotten_flexibility_of_chp_2011_03_23.pdf.
37. Niemi, R., Mikkola, J., and Lund, P.D. (2012). Urban energy systems with smart multi-carrier energy networks and renewable energy generation. *Renewable Energy*, 48, 524–536.
38. Long, H., Xu, R., and He, J. (2011). Incorporating the variability of wind power with electric heat pumps. *Energies*, 4, 1748–1762.
39. Maegaard, P. Balancing Fluctuating Power Sources. *Proceedings of 2010 World Non-Grid-Connected Wind Power and Energy Conference (WNWEC)*, 5–7 November 2010, Nanjing, China, pp. 4–7.
40. Hong, L., Lund, H., and Möller, B. (2012). The importance of flexible power plant operation for Jiangsu's wind integration. *Energy*, 41, 499–507.
41. Mathiesen, B.V., Lund, H., and Connolly, D. (2012). Limiting biomass consumption for heating in 100% renewable energy systems. *Energy*, 48, 160–168.
42. Lund, H., Möller, B., Mathiesen, B.V., and Dyrelund, A. (2010). The role of district heating in future renewable energy systems. *Energy*, 35, 1381–1390.
43. Mathiesen, B.V. and Lund, H. (2009). Comparative analyses of seven technologies to facilitate the integration of fluctuating renewable energy sources. *Renewable Power Generation*, 3, 190–204.
44. Fitzgerald, N., Foley, A.M., and McKeogh, E. (2012). Integrating wind power using intelligent electric water heating. *Energy*, 48, 135–143.
45. Losev, V.L., Sigal, M.V., and Soldatov, G.E. (1989). Nuclear district heating in CMEA countries, *IAEA Bulletin*, 3, 46–49.
46. Margen, P. (1978). The use of nuclear energy for district heating. *Progress in Nuclear Energy*, 2(1), 1–28. doi: 10.1016/0149-1970(78)90010-0.
47. CNNC Completes Design of District Heating Reactor—World Nuclear News. A website report www.world-nuclear-news.org/.../CNNC-completes-design-of-district-heating-reactor. A model of the Yanlong reactor (Image: CNNC) (7 September 2018).
48. Csik, B.J. and Kupitz, J. (2007). Nuclear cogeneration: Supplying heat for homes and industries, *IAEA Bulletin*, 39/2 ecolo.org/documents/documents_in_english/cogeneration-nuc-csik-07.html.
49. Tuomisto, H. Nuclear District Heating Plans from Loviisa to Helsinki Metropolitan Area. A paper at Joint NEA/IAEA Expert Workshop on the "Technical and Economic Assessment of Non-Electric Applications of Nuclear Energy" OECD Headquarters, Paris, France www.oecd-nea.org/ndd/.../3_Tuomisto_Nuclear-District-Heating-Plans.pdf (4–5 April 2013).
50. Värri, K. (2018). Market Potential of Small Modular Nuclear Reactors in District Heating, M.S. Thesis, Aalto University, Finland.

51. Varri, K. and Syri, S. (2019). The possible role of modular nuclear reactors in district heating: Case Helsinki Region, Fortum, Keilalahdentie, 2–4, 02150 Espoo, Finland, *Energies*, 12(11), 2195. doi: 10.3390/en12112195. www.mdpi.com/1996-1073/12/11/2195.
52. Connolly, D., Lund, H., Mathiesen, B., Werner, S., Mäller, B., Persson, U., Boermans, T., Trier, D., Ästergaard, P., and Nielsen, S. (2014). Heat roadmap Europe: Combining district heating with heat savings to decarbonise the EU energy system, *Energy Policy*, 65, 475–489 [Online]. Available: www.sciencedirect.com/science/article/pii/S0301421513010574.
53. International Energy Agency. Energy Policies of IEA Countries—Poland-2016 Review (2017).
54. International Energy Agency. Poland: Electricity and Heat for 2015 (cited 9 January 2018) [Online]. Available: www.iea.org/statistics/statisticssearch/report/?country=POLAND=&product=electricityandheat.
55. Sobolewski, J. HTR Plans in Poland. Ministry of Energy, Poland (September 2017).
56. Nikkei Asian Review. Japan to Export Safer Nuclear Reactor to Poland, cited 9 January 2018. [Online]. Available: /https://asia.nikkei.com/Business/Companies/Japan-to-export-safer-nuclear-reactor-to-Poland?page=2 (2017).
57. Xia, J., Zhu, K., and Jiang, Y. (April 2016). Method for integrating low-grade industrial waste heat into district heating network. *Building Simulation*, 9(2), 153–163. First Online: 3 December 2015. doi: 10.1007/s12273-015-0262-3.
58. Papapetrou, M., Kosmadakis, G., Cipollina, A., Commare, U.L., and Micale, G. (25 June 2018). Industrial waste heat: Estimation of the technically available resource in the EU per industrial sector, temperature level and country. *Applied Thermal Engineering*, 138, 207–216. doi: 10.1016/j.
59. Sohani, A.A. Waste Heat Recovery from SSAB's Steel Plant in Oxelösund Using a Heat Pump. M.S. Thesis KTH School of Industrial Engineering and Management Energy Technology EGI_2016–082 MSC Division of ETT SE-100 44 STOCKHOLM.
60. Daniëls, B., Wemmers, A., and Wetzels, W. Dutch Industrial Waste Heat in District Heating: Waste of Effort? Panel: 3. Matching Policies and Drivers: Policies and Directives to Drive Industrial Efficiency. This is a peer-reviewed paper. Also 2016 Dutch Industrial Waste Heat in District Heating: Waste of Effort?—ECEEE. A website report www.eceee.org/…Industrial…/3-matching-policies-and-drivers-policies-and-dir… (2011).
61. Daniëls, B.W., Wemmers, A.K., Tigchelaar, C., and Wetzels, W. Restwarmtebenutting. Potenti.len, besparing, alternatieven, ECN-E-11-058 (November 2011).
62. Heating and Cooling with Waste Heat from Industry. A website report European Academy of Bozen/Bolzano (EURAC) https://phys.org/news/2015-07-cooling-industry.html#jCp (8 July 2015).
63. Raine, R., Sharifi, V., Swithenbank, J., Hinchcliffe, V., and Segrott, A. Sustainable Steel City: Heat Storage and Industrial Heat Recovery for a District Heating Network. *The 14th International Symposium on District Heating and Cooling*, 7–9 September 2014, Stockholm, Sweden.
64. Fang, H., Xia, J., Zhu, K., Su, Y., and Jiang, Y. Industrial waste heat utilization for low temperature district heating. *Energy Policy*, 62(C), 236–246. www.sciencedirect.com/science/article/pii/S0301421513006113. doi: 10.1016/j.enpol.2013.06.104.
65. Tong, K., Fang, A., Yu, H., Li, Y., Shi, L., Wang, Y., Wang, S., and Ramaswami, A. (11 December 2017). Estimating the potential for industrial waste heat reutilization in urban district energy systems: method development and implementation in two Chinese provinces. © 2017. Published by IOP Publishing Ltd. *Environmental Research Letters*, 12(No. 12).
66. Fang, H., Xia, J., and Jiang, Y. (2015). Key issues and solutions in a district heating system using low-grade industrial waste heat. *Energy*, Elsevier, 86, 589–602.

67. Raine, R.D. (September 2016). Sheffield's Low Carbon Heat Network and its Energy Storage Potential, Ph. D. Thesis, University of Sheffield, Sheffield, UK
68. Finney, K. (2011). Sheffield University Waste Incineration Centre (SUWIC), University of Sheffield, Sheffield Heat Mapping and Feasibility Study of Decentralised Energy, http://research.ncl.ac.uk/pro- tem/components/pdfs/Sheffield_EPSRC_progress_ report_3_Sheffield_heat_maps_July2011.pdf. Accessed 15/4/15.
69. Finney, K.N., Sharifi, V.N., Swithenbank, J., Nolan, A., White, S., and Ogden, S. (2012). Developments to an existing city-wide district energy network—Part I: Identification of potential expansions using heat mapping, *Energy Conversion and Management*, 62, 165–175.
70. Doračić, B., Novosel, T., Pukšec, T., and Duić, N. (2018). Evaluation of excess heat utilization in district heating systems by implementing levelized cost of excess heat. *Energies*, 11, 575. doi: 10.3390/en11030575. www.mdpi.com/journal/energies.
71. Koh, S.C.L., Acquaye, A.A., Rana, N., Genovese, A., Barratt, P., Kuylenstierna, J., Gibbs, D., and Cullen, J. Supply Chain and Environmental Analysis, Centre for Low Carbon Futures. www.shef.ac.uk/polopoly_fs/1.153241!/file/SCEnAT-Report.pdf, last accessed 31 July 2014 (2011).
72. Remus, R., Monsonet, M.A.A., Roudier, S., and Sancho, L.D. Best Available Techniques Reference Document for Iron and Steel Production, European Commission, 2013, *The 14th International Symposium on District Heating and Cooling*, 7–9 September 2014, Stockholm, Sweden. http://eippcb.jrc.ec.europa.eu/reference/BREF/IS_ Adopted_03_2012.pdf, last accessed 31 July 14.
73. Dixon, J. and S. Bramfoot. Design of Waste Heat Boilers for the Recovery of Energy from Arc Furnace Waste Gases. Commission of the European Communities (1985).
74. Jones, J. Understanding electric Arc Furnace Operations, RI enter for Materials Production http://infohouse.p2ric.org/ref/10/09047.pdf, last accessed 31 July 2014 (1997).
75. Zuliani, D., Scipolo, V., and Maiolo, J. Opportunities for Increasing Productivity, Lowering Operating Costs and Reducing Greenhouse Gas Emissions in EAF and BOF Steel Making, *Proceedings of AISTech*, 2010: www.millennium-steel.com/articles/ pdf/2010%20India/pp35–42%20MSI10.pdf, last accessed 16 July 2013.
76. World Health Organisation. (2010). Dioxins and Their Effects on Human Health, Fact Sheet No. 225. www.who.int/mediacentre/factsheets/fs225/en/, last accessed 31 July 2014.
77. Born, C. and Granderath, R. (February 2013). Benchmark for Heat Recovery from the Offgas Duct of Electric arc Furnaces, MPT International, pp. 32–35.
78. World Steel Association. (2011). Water Management in the Steel Industry. Article, www.worldsteel.org/media-centre/press- releases/2011/water-management-report. html, last accessed 31 July 2014.
79. Hackel, S. (2008). Development of Design Guidelines for Hybrid Ground-Coupled Heat Pump Systems. ASHRAE Technical Research Project 1384, American Society of Heating, Refrigerating, and Air Conditioning Engineers (ASHRAE), Atlanta, GA.
80. Chiasson, A.D. and Yavuzturk, C., (2009a). A design tool for hybrid geothermal heat pump systems in heating-dominated buildings. *ASHRAE Transactions*, 115(2), 60–73.
81. Chiasson, A.D. and Yavuzturk, C. (2009b). A design tool for hybrid geothermal heat pump systems in cooling-dominated buildings. *ASHRAE Transactions*, 115(2), 74–87.
82. Xu, X. (2007), Simulation and Optimal Control of Hybrid Ground Source Heat Pump Systems, Ph.D. Thesis, Oklahoma State University, Stillwater, Oklahoma.
83. Hackel, S., Nellis, G., and Klein, S. (2009). Optimization of cooling-dominated hybrid ground-coupled heat pump systems. *ASHRAE Transactions*, 115(1), 565–580.
84. Kavanaugh, S.P. (1998). A design method for hybrid ground- source heat pumps. *ASHRAE Transactions*, 104(2), 691–698.

85. Cullin, J. and Spitler, J.D. (2010). Comparison of Simulation-based Design Procedures for Hybrid Ground Source Heat Pump Systems. *Proceedings of the 8th International Conference on System Simulation in Buildings 2010*, Liege, Belgium.
86. Numerical Logics, Inc. (2008). Solar Domestic Hot Water System Sizing for Whitehorse, YT and Dawson, YT, Waterloo, ON.
87. SEL. (2000). TRNSYS, A Transient Systems Simulation Program, Version 15. Solar Engineering Laboratory (SEL), University of Wisconsin-Madison, Madison, WI, USA.
88. Chiasson, A. (May 2011). A feasibility study of a multi-source hybrid district geothermal heat pump system. *GHC Bulletin*, 9–14. https://oregontechsfcdn.azureedge.net/...source/geoheat.../art3e476ee4362a663989f6f....0
89. Vesaoja, E., Yang, C.-W., Nikula, H., and Sierla, S. Hybrid Modeling and Co-Simulation of District Heating Systems with Distributed Energy Resources. Conference Paper (PDF Available) April 2014 Conference: *Workshop on Modeling and Simulation of Cyber-Physical Energy Systems*, 2014, Berlin, Germany.
90. Nakomcic-Smaragdakis, B. and Dragutinovic, N. (January 2015). Hybrid renewable energy system application for electricity and heat supply of a residential building. *Thermal Science*, 20(00), 144–145. 237 Reads. doi: 10.2298/TSCI150505144N.
91. Bakken, B.H., Fossum, M., and Belsnes, M.M. Small-Scale Hybrid Plant Integrated with Municipal Energy Supply System MSc SINTEF Energy Research, N-7465 Trondheim, Norway, bjom.bakken@energy.sintef.no Ossiach, International Energy symposium www.iaea.org/inis/collection/NCLCollectionStore/_Public/33/009/33009933.pdf?... (2001).
92. Holmgren, K. A System Perspective on District Heating and Waste Incineration, Division of Energy Systems Department of Mechanical Engineering Linköpings universitet, Linköping, Sweden 2006, Linköping Studies in Science and Technology Dissertation No. 1053,ISBN: 91-85643-61-0. ISSN 0345-7524 Printed in Sweden by LiU-Tryck, Linköping https://pdfs.semanticscholar.org/b2e8/167a72eb12a853b1254a7e4a498948e2313a.pdf (2006).
93. Ulloa, P. and Themelis, N.J. Doubling the Energy Advantage of Waste-to-Energy:, District Heating in the Northeast U.S. Copyright © 2007 by ASME. *15th North American Waste to Energy Conference*, 21–23 May 2007, Miami, FL NA EC15–3201 https://pdfs.semanticscholar.org/de71/f5322ef03abcb2cd5d48ae8a2134337ca593.pdf.
94. Ilie, A. and Visa, I. (2019). Hybrid solar-biomass system for district heating. *E3S Web of Conferences*, Volume 85, 04006. doi: 10.1051/e3sconf/20198504006. EENVIRO 2018 www.e3s-conferences.org/articles/e3sconf/pdf/.../e3sconf_enviro2018_04006.pdf.
95. Zhang, C., Sun, J., Ma, J., Xu, F., and Qiu, L. (June 2019). Environmental assessment of a hybrid solar-biomass energy supplying system: A case study. *International Journal of Environmental Research and Public Health*, 16(12), 2222. Published online 24 June 2019. doi: 10.3390/ijerph16122222. PMCID: PMC6617335, PMID: 31238546.
96. Romanchenko, D., Kensby, J., Odenberger, M., and Johnsson, F. Thermal energy storage in district heating: Centralized storage *vs.* storage in thermal inertia of buildings. *Energy Conversion and Management*, 162, 26–38. doi: 10.1016/j.enconman.2018.01.068.
97. Kiviluoma, J., Heinen, S., Qazi, H., Madsen, H., Strbac, G., Kang, C., Zhang, N., Patteeuw, D., and Naegler, T. (2017). Harnessing flexibility from hot and cold. *IEEE Power & Energy Magazine*, 15(No. 1), 25–33. Contributiontojournal›Journalarticle–Annualreportyear:2017›Research›peer-reviewhenrikmadsen.org/wp.../harmenessing_flexibility_from_heating_and_cooling.pdf.
98. Reed, A.L., Novelli, A.P., Doran, K.L., Ge, S., Lu, N., and McCartney, J.S. (2018). Solar district heating with underground thermal energy storage: Pathways to commercial viability in North America. *Renewable Energy*, 126, 1–13.
99. Schuchardt, G.K. (neé BESTRZYNSKI). (December 2016). Integration of decentralized thermal storages within district heating (DH) networks, *Environmental and Climate Technologies*, 18, 5–16. doi: 10.1515/rtuect-2016-0009. www.degruyter.com/view/j/rtuect.

100. Dincer, I. and Rosen, M. (2010). *Thermal Energy Storage: Systems and Applications*, 2nd Edition. doi: 10.1002/9780470970751.
101. Andersson, O. (1997). ATES Utilization in Sweden: An Overview. *Proceedings of MEGASTOCK'97 7th International Conference on Thermal Energy Storage*, Volume 2, pp. 925–930.
102. Lund, H., Werner, S., Wiltshire, R., Svendsen, S., Thorsen, J.E., Hvelplund, F. and Mathiesen, B.V. (2014). 4th Generation District Heating (4GDH): Integrating smart thermal grids into future sustainable energy systems. *Energy*, 68, 1–11.
103. Tanaka, H., Tomita, T., and Okumiya M. (2000). Feasibility study of a district energy system with seasonal water thermal storage. *Solar Energy*, 69(6), 535–547.
104. Rezaie, B., Reddy, B.V., Rosen, M.A. (June 2017). Assessment of the thermal energy storage in Friedrichshafen district energy systems. *Energy Procedia*, 116, 91–105. doi: 10.1016/j.egypro.2017.05.058.
105. Holmgren, K. and Gebremedhin, A. (2004). Modelling a district heating system: Introduction of waste incineration, policy instruments and co-operation with an industry. *Energy Policy*, 32, 1807–1817. doi: 10.1016/S0301-4215(03)00168-X.
106. Sahlin, J., Knutsson, D., and Ekvall, T. (2004). Effects of planned expansion of waste incineration in the Swedish district heating systems. *Resources, Conservation and Recycling*, 41, 279–292.
107. Energy Central Bjerring bro District Heating and Grundfos. A website report www.districtenergy.org/HigherLogic/.../DownloadDocumentFile.ashx?...0 (2007).
108. Schmidt, T. and Mangold, D. (2006). Seasonal thermal energy storage in Germany. *Structural Engineering International*, 14. doi: 10.2749/101686604777963739.
109. Schmidt, D. (2009). Low exergy systems for high-performance buildings and communities. *Energy and Buildings*, 41, 331–336.
110. Rezaie, B., Reddy, B.V., and Rosen, M.A. (2011). Role of thermal energy storage in district energy systems. In: Rosen, M.A. (ed) *Energy Storage*. NOVA Publishers, Hauppauge, NY.
111. Dincer, I. and Rosen, M.A. (2011). *Thermal Energy Storage: Systems and Applications*. Wiley, West Sussex.
112. Rosen, M.A., Le, M.N., and Dincer, I. (2005). Efficiency analysis of a cogeneration and district energy system. *Applied Thermal Engineering*, 25, 147–159.
113. Rezaie, B. and Rosen, M.A. (2012). District heating and cooling: Review of technology and potential enhancements. *Applied Energy*, 93, 2–10.
114. Andrepont, J.S. (2006). Developments in thermal energy storage: Large applications, low temps, high efficiency, and capital savings. *Energy Engineering*, 103, 7–18.
115. Griffin, T. (2010). LEED® district energy. The new version 2.0 guideline: It's here!" *District Energy*, 96, 55.
116. Rezaie, B., Reddy, B.V., and Rosen, M.A. (2015). Exergy analysis of thermal energy storage in a district energy application. *Renewable Energy*, 74, 848–854.
117. Rossi Espagnet, A. Techno-Economic Assessment of Thermal Energy Storage Integration into Low Temperature District Heating Networks, M.S. Thesis KTH School of Industrial Engineering and Management Energy Technology EGI-2016-068 Division of Heat and Power Technology, SE-100 44 STOCKHOLM (2016).

7 Modular Systems for Energy Usage for Desalination and Wastewater Treatment

7.1 INTRODUCTION

Freshwater is a necessity in everyday life, and is vital for the survival of human beings. The ability to get freshwater in dry areas, or in times of shortage, comes from the process of desalination, where the salt in seawater is displaced and it becomes drinkable water [1–25]. Desalination is a water treatment process that separates salts from saline water to produce potable water or water that is low in total dissolved solids (TDS). Globally, the total installed capacity of desalination plants was 61 million m^3/day in 2008 [20]. Seawater desalination accounts for 67% of production, followed by brackish water at 19%, river water at 8%, and wastewater at 6%. The most prolific users of desalinated water are in the Arab region, namely, Saudi Arabia, Kuwait, United Arab Emirates, Qatar, Oman, and Bahrain [21].

Desalination plants consume a lot of energy, though, and they are not as green as they could be. Modern day facilities that desalinate water uses high-energy consuming process of reverse osmosis (RO). Not only does the environment suffer from the energy expenditures of desalination plants, but so does the economy. The desalination plant in San Diego cost more than $1 billion, and it only provides about 7% of drinking water to the city of San Diego. Of the world 71% of water (29% land), only 4% is drinkable. The areas that are most affected by the shortage of freshwater are arid and semiarid areas in Asia and North Africa. In 2002, according to a United Nations Educational, Scientific and Cultural Organization report, the deficit of drinking water in the world was around 230 billion m^3/year, and this amount was expected to rise to 2,000 billion by 2025. In fact, a report from the World Economic Forum published in 2015 highlighted the problem and indicated that the lack of freshwater could be the great global threat for the next decade.

According to this report, there are approximately 19,000 desalination plants throughout the world, providing water to municipal and industrial users. Almost half of the global installed desalination capacity is in the Middle East, followed by the European Union with 13%, the United States with 9%, and North Africa with 8.5%.The largest desalination plant—the $3.8 billion Al-Jubail 2 in Saudi Arabia—has 948,000 m^3/day multiple-effect distillation–thermal vapor compressor

(MED-TVC) capacity, plus 2,745 MW power generation using gas turbines. The Saudi Saline Water Conversion Corporation (SWCC) takes about 62% of output to supply Riyadh. China is building a 1 million m^3/day RO plant to supply potable water for Beijing.

SWRO desalination stands for saltwater RO. This high-pressure system used to desalinate saltwater requires a high amount of energy. The amount of energy consumed from a desalination plant, which supplies water to 300,000 people, is equivalent to one jumbo jet's power. Energy consumption is one of the biggest hurdles desalination faces. It consumes an average of 10–13 kWh per every thousand gallons. Desalination is also viewed as one of many factors contributing to climate change and global warming. The ocean is home to many creatures, and desalination poses a threat to ocean biodiversity and marine habitats. If fossil energy is used to provide energy for desalination process, carbon emission can be significant [1–25].

As shown later, desalination can be done by a number of different techniques besides RO. In recent years, modular desalination operation is preferred to lower the upfront costs, improve efficiency, reduce the time for plant installation, and make the process more flexible. The use of renewable energy like solar, geothermal, wind, and water along with the use of small modular nuclear reactors also reduce energy consumption and improve its return on investment. The use of renewable energy also makes the process more environment friendly. Unlike fossil energy, nuclear energy is clean. Freshwater is a major priority in sustainable development. It is generally obtained from streams and aquifers, desalination of seawater, mineralized groundwater, or treatment of urban wastewater. A study in 2006 by the UN's International Atomic Energy Agency (IAEA) showed that 2.3 billion people lived in water-stressed areas, and 1.7 billion of them have access to <1,000 m^3 of potable water per year. With population growth, these figures will increase substantially. Water can be stored, while electricity at utility scale cannot. This suggests two synergies with base-load power generation for electrically driven desalination: undertaking it mainly in off-peak times of the day and week, and load shedding in unusually high peak times.

Cumulative investment in desalination plants reached about US$21.4 billion in 2015 and is expected at least to double by 2020 according to a 2016 report by market analyst, Research and Markets. The report, Seawater and Brackish Water Desalination, includes a prediction that investment by 2020 should top $48 billion, showing a compound annual growth rate of 17.6%. The report assesses the market for large industrial or municipal facilities with a capacity >1,000 m^3/day. It highlights a growing gap between freshwater resources and demand from all sectors. Most desalination today uses fossil fuels, and thus contributes to increased levels of greenhouse gases [1–10].

7.1.1 Process Alternatives

Desalination can be achieved by using a number of techniques. Total world capacity in 2016 was 88.6 million m^3/day of potable water, in almost 19,000 plants. Of this, 73% is membrane desalination, and 27% thermal, though in the year to July 2016,

93% of new capacity contracted was membrane based. Industrial desalination technologies use either phase change or involve semipermeable membranes to separate the solvent or some solutes. Thus, desalination techniques may be classified into two main categories [3,25]:

- Phase-change or thermal processes—where base water is heated to boiling. Salts, minerals, and pollutants are too heavy to be included in the steam produced from boiling and therefore remain in the base water. The steam is cooled and condensed. The main thermal desalination processes are multistage flash (MSF) distillation, MED, and vapor compression (VC), which can be thermal (TVC) or mechanical (MVC).
- Membrane or single-phase processes—where salt separation occurs without phase transition and involves lower energy consumption. The main membrane processes are RO and electrodialysis (ED) or ED reversal. RO requires electricity or shaft power to drive a pump that increases the pressure of the saline solution to the required level. ED also requires electricity to ionize water, which is desalinated by using suitable membranes located at two oppositely charged electrodes.

Three other membrane processes that are not considered desalination processes, but that are relevant, are microfiltration (MF), ultrafiltration (UF), and nanofiltration (NF). The ion-exchange process is also not regarded as a desalination process, but is generally used to improve water quality for some specific purposes, e.g., boiler feed water [5,25]. The range of applicability of these different technologies is illustrated in Table 7.1. In the past 15 years, membrane technology has developed to a mature and reliable technique, which is usable for (drinking) water purification. Within this technology, RO has been widely used. This method requires that the pressure difference between two sides of membrane is higher than osmotic pressure in order for water to permeate through the membrane. The osmotic pressure of seawater is around 26 bar [1–10,25].

All processes require a chemical pretreatment of raw seawater—to avoid scaling, foaming, corrosion, biological growth, and fouling—as well as a chemical posttreatment. The two most commonly used desalination technologies are MSF and RO systems. As the more recent technology, RO has become dominant in the desalination industry. In 1999, about 78% of global production capacity comprised MSF plants and RO accounted for a modest 10%. But by 2008, RO accounted for 53% of worldwide capacity, whereas MSF consisted of about 25%. Although MED is less common than RO or MSF, it still accounts for a significant percentage of global desalination capacity (8%). ED is only used on a limited basis (3%) [4].

7.1.1.1 Hybrid Thermal-Membrane Desalination

Hybrid thermal-membrane plants have a more flexible power-to-water ratio, efficient operation even with significant seasonal and daily fluctuations of the electricity and water demand, less primary energy consumption, and an increase in plant efficiency, thus improving economics and reducing environmental impacts. MSF plus

TABLE 7.1
Range of Applicability of Various Desalination Processes

Technology	Salt Concentration in Water (ppm)
Ion exchange	10–800
RO	50–50,000
ED	200–10,000
Distillation	20,000–100,000

RO or MED-TVC plus RO hybrid plants exploit the best features of each technology for different quality products or a blended product. Several thermal distillation processes capable of using waste heat from power generation are in use: MSF distillation process using steam was earlier prominent. It works by flashing a portion of the water into steam in multiple stages of what are essentially countercurrent heat exchangers, and it accounted for 23% of world capacity in 2012 [1–10]. It is more energy-intensive than MED, but it can cope with suspended solids and any degree of salinity. The Japan Atomic Energy Agency (JAEA) has designed a 600 MW high temperature reactor called the GTHTR300, which produces 300 MW and uses the waste heat in MSF desalination, the projected water cost being half that of using gas-fired CCGT. An increasing number of plants use MED with 8% world capacity in 2012, or multieffect VC (MVC or VCD) distillation or a combination of these, e.g., MED-TVC with thermal VC [11].

MED is a low-temperature thermal process of obtaining freshwater by recovering the vapor of boiling seawater in a sequence of vessels (called effects), each maintained at a lower temperature than the last. Because the boiling point of water decreases as pressure decreases, the vapor boiled off in one vessel can be used to heat the next one, and only the first one (at the highest pressure) requires an external source of heat, such as that from the condenser circuit of a power plant. It is more expensive than RO but can cope with any degree of salinity. MD is an emerging process that is thermally driven. Desalination is energy-intensive. RO needs up to 6 kWh of electricity per cubic meter of water (depending on both process and its original salt content), though the latest RO plants such as in Perth, Western Australia, use 3.5 or 4 kWh/m^3 including pumping for distribution. Hence 1 MW continuous power will produce about 4,000–6,000 m^3/day from seawater. MSF and MED require heat at 70°C–130°C and use about 38 kWh/m^3 thermal input, plus 3.5 kWh/m^3 electrical energy for MSF and 1.5 kWh/m^3 for MED-TVC. A variety of low-temperature and waste heat sources may be used, including solar energy (especially for MED). The choice of process generally depends on the relative economic values of freshwater and particular fuels, and whether cogeneration is a possibility. Thermal processes are more capital-intensive. Forward osmosis (FO) may be used in conjunction with a subsequent process for desalination. The FO draws water through a membrane from a feed solution into a more concentrated draw solution, which is then desalinated without the problems of fouling, such as often encountered with simple RO. FO plants operate in Gibraltar and Oman [11,25].

The cost of water desalination varies depending on water source access, source water salinity and quality, specific desalination process, power costs, concentrate disposal method, project delivery method, and the distance to the point of use. Power costs in water desalination may account for 30%–60% of the operational costs; thus, slight variations in power rates directly impact the cost of treated water. The major technology in use and being built today is RO driven by electric pumps, which pressurize water and force it through a semipermeable membrane against its osmotic pressure. This accounted for 63% of 2012 world capacity, up from only 10% in 1999. With brackish water, RO is much more cost-effective, though MSF gives purer water than RO. RO relies on electricity to drive the actual process and requires clean (filtered) feedwater. The energy efficiency of seawater RO depends on recovering the energy from the pressurized reject brine. In large plants, the reject brine pressure energy is recovered by a turbine; commonly a Peloton wheel turbine recovers 20%–40% of the consumed energy [1–10].

RO can be carried out in different types of membrane modules. A typical RO plant is illustrated in Figure 7.1. More common modules are hollow fiber, spiral, tubular, and flat plate. The natural osmotic pressure in mineral water are the same for spiral modules and hollow fiber modules. Pretreatment costs for the purification of surface water are higher when hollow fiber membranes are used. These membranes require a more specific pretreatment, because they are more susceptible to fouling. The use of tubular modules and plate & frame modules may be more expensive than that of hollow fiber modules or spiral modules. The costs of the use of tubular modules and plate & frame modules are approximately equal. The required system space for the modules is very high when hollow fiber modules or spiral modules are used. Tubular modules take up much less space. The literature indicates that for RO systems, for the desalination of seawater, spiral membranes are most often used [11]. About three-quarters of Israel's water is desalinated, and one large RO plant provides water at 58 c/m^3, claimed to be the world's cheapest. Until 2013 it also claimed to have the world's largest seawater RO plant at Soreq, producing 627,000 m^3/day. In 2015 Israel and Jordan signed a $900 million agreement for a new desalination plant at Aqaba on the Red Sea, supported by the World Bank and based on a 2013 agreement. The new agreement involves desalination of 80 million m^3/year (220,000 m^3/day) at the Aqaba plant, with Israel buying half of that amount for use in its southern port town of Eilat and the Arava region—both desert areas with chronic water shortages. Jordan will get half the water for the arid southern part of that country. As part of the deal, Israel will supply an additional 50 million m^3 of water for the central and northern parts of Jordan from its Lake Kinneret. In addition to the desalination, over 100 million m^3 of concentrated brine will be pumped 180 km north to replenish the Dead Sea [1–10].

Singapore in 2005 commissioned a large RO seawater desalination plant supplying 136,000 m^3/day—10% of its needs, at 49 c/m^3, and in 2013 commissioned a 318,500 m^3/day RO plant on a build-own-operate basis, costing $700 million, to provide water at 36 c/m^3. The UAE is heavily dependent on seawater desalination, much of it with cogeneration plants. Algeria in mid-2013 had 2.1 million m^3/day capacity and another 400,000 m^3/day is envisaged. In February 2012 China's state council announced that it aimed to have 2.2–2.6 million m^3/day seawater desalination

FIGURE 7.1 RO desalination plant in Barcelona, Spain [26].

capacity operating by 2015, and early in 2015, the target under the government's special plan for seawater utilization was 4 million m³/day. The Kwinana desalination plant near Perth, Western Australia, has been running since early 2007 and produces about 140,000 m³/day of potable water, requiring 24 MW of power (4.1 kWh/m³), and about 3.7 kWh/m³ across the membranes [11,25].

The discussion here on the use of modular operation for desalination here is broken into three parts. First, the use of modular operation and their benefits are articulated with a number of industrial examples. Second, the effectiveness of renewable energy like solar, geothermal, wind, and water energy for various types of modular desalination processes are articulated. Finally, the worldwide modular desalination processes by small modular nuclear reactor (both stationary and floating) are examined.

Desalination and Wastewater Treatment 331

7.2 MODULAR DESALINATION PLANTS

7.2.1 IDE Chemical-Free River Osmosis Desalination Plant

IDE's modular solution simplifies the installation of the plant while reducing installation and startup time and cost. Modular SWRO desalination design is an excellent solution for remote locations due to the use of prefabricated modules with minimal maintenance requirements [12–14].

The Key Benefits of IDE's Membrane Desalination Plants are [12–14]

- **Consistent quality standards:** IDE's small- to large-scale plants produce high-quality drinking and industrial water from both seawater and brackish water, while meeting strict health and safety standards. The plants provide ~3 million m^3/day of clean water to dozens of millions of people globally for domestic and industrial use.
- **Industry-leading efficiency:** With the most experience in megasized thermal and membrane desalination, IDE consistently improves cost efficiencies.
- **Proven project management track record:** Close to 50 years of desalination success has proven IDE's ability to deliver complex projects on time, within budget and above specifications.

IDE's modular Sea/Brackish Water Reverse Osmosis (SWRO/BWRO) units are ideal if one needs a desalination solution that is flexible, able to grow, and completely self-contained. IDE utilizes the most recent RO technologies for small to midsize desalination projects. These are designed to provide high performance and reliability, together with ease of operation and low energy consumption [12–14].

Starting with capacities of 500 m^3/day, the systems can very easily be expanded to capacities in excess of 30,000 m^3/day. The equipment and materials of construction are selected from leading suppliers in order to maximize system reliability and minimize maintenance cost. The produced water conforms to World Health Organization regulations/requirements for potable water. The supply of modular SWRO units eases transportation to site, reduces on-site erection activities, and saves on required infrastructure and buildings, resulting in significant reduction in project implementation cost and risks.

The Key Benefits of IDE's Modular SWRO Systems for brackish water are [12–14]:

1. Increased operational efficiency and reliability
2. Frequency converter (variable frequency drive)
3. Post shutdown product water flush
4. Increased energy efficiency
5. Lower operating expense
    a. Standard pretreatment with low velocity, pressurized, multimedia filters
    b. High-quality materials of construction

6. Simple operation
    a. Panel-mounted instruments and indications
    b. Product TDS monitor
    c. High TDS shutdown

IDE also provides optional features of, supervision of installation and commissioning, supply of turnkey services, maintenance program, supervision of operation and maintanence contract, boron removal system, and posttreatment and polishing.

7.2.1.1 IDE PROGREEN Plant

IDE PROGREEN™ is a unique, eco-friendly, cost-effective SWRO desalination "plant in a box" that eliminates the need for chemicals. Bringing high-quality clean water to a variety of resorts and industries, IDE PROGREEN™ provides an end-to-end solution (pretreatment, RO, and posttreatment) in one simple to use safe package [12–14].

IDE PROGREEN™ is a modular, RO desalination plant that delivers affordable clean water without the use of chemicals, and with reduced energy consumption. It provides a complete solution in one neat compact package, in a wide range of capacities from 500 to 20,000 m^3/day. It is easily transported, keeps O&M costs to a minimum, and eliminates the need to invest in expensive infrastructure. The system is designed to increase return on investment. It is [12–14]

1. Eco-Friendly
    a. Entire process is chemical-free
    b. Reduced energy consumption
2. Cost-Effective
    a. Minimal infrastructure investment required
    b. Modular units for fast & easy installation
    c. Operational costs reduced by 10%
    d. Fully customizable to a variety of water sources and customer requirements
3. Safe & Easy
    a. Plug & Play installation
    b. No need for chemical handling or disposal permits
    c. Easy operation and maintenance (O&M)
    d. Its key features include top quality materials of construction, highly efficient energy recovery systems, end-to-end modular programmable logic controller with 15″ touch screen and
    e. remote performance monitoring.

Its operating parameters are [12–14]

1. Capacity range 500–20,000 m^3/day
2. Seawater TDS 35,000–41,000 ppm^3
3. Temperature 15°C–32°C (59°F–90°F)
4. Recovery rate of 45%

Systems can be provided, containerized, or skid mounted for different feedwater parameters—"green" or conventional, upon customer demand. It is specifically designed to be flexible, and it is suitable for a diverse range of applications, including resorts and islands, municipalities, industries, and agriculture sector.

IDE Technologies also designed and supplied the desalination plant for the Quebrada Blanca Phase 2 (QB2) project in Chile. The desalination plant is located at the port site for the QB2 copper mine in the Tarapacá region and is the second largest seawater desalination plant in Chile. IDE's experience encompasses more than 400 desalination plants worldwide, with more than 20 successfully operating desalination plants in Chile. The SWRO desalination plant for the QB2 project produces high-quality water for use in the copper concentrator. IDE's design of the QB2 desalination plant is in accordance with the stringent environmental requirements and safety regulations in Chile.

7.2.2 MIT Study on Modular Desalination Plants

Massachusetts Institute of Technology (MIT) study highlights benefits of smaller, modular desalination plants. The study looked at the city of Melbourne, where a 12-year drought from 1997 to 2009 led to construction of the $5 billion Victorian Desalination Plant. The facility was approved in 2007 and opened in 2012, at which time the drought had already passed. As a result, the plant has barely been used. As an alternative, the study suggested smaller, modular desalination plants that could have met Melbourne's needs at a lower price. The study does not argue that a single solution applies to all cases, but presents a new method for pinpointing the best plan. It notes that, in many cases, "moderate investment increases", together with flexible infrastructure design, can mitigate water-shortage risk significantly [15].

The new paper by MIT on "**Water Supply Infrastructure Planning: Decision-Making Framework to Classify Multiple Uncertainties and Evaluate Flexible Design**," was recently published in the *Journal of Water Resources Planning and Management*. The MIT team's new framework for water-supply analysis incorporates several uncertainties that policymakers must confront in these cases, and runs large numbers of simulations of water availability over a 30-year period. It then presents planners with a decision tree about which infrastructure options are best calibrated to their needs. The significant uncertainties include climate change and its effects on rainfall, as well as the impact of water shortages and population growth [15].

7.2.2.1 Melbourne Case Study

In the Melbourne study, the researchers looked at six infrastructure alternatives, including multiple types of desalination plants and a possible new pipeline to more-distant sources, and combinations of these things. The results highlight a vexing problem in water-access planning: shortages can be acute, but they may last for relatively short periods of time. The team ran 100,000 simulations of 30-year conditions in Melbourne and found that in 80% of all years, there would be no water shortages at all. For the years where drought conditions did hold, large water shortages were more common than minor water shortages. As a result, when costs were factored into the analysis, simply building no new infrastructure was the best option around 50% of the time. However, doing nothing was also the "worst-performing alternative" around 30% of the time. This is why building smaller desalination plants can make sense. The Melbourne desalination plant can produce 150 million m^3 of water per year. But in MIT's simulations, building a desalination plant half that size usually works well. It was the best-performing option in 20% of the simulations, and in the top three of 90% of the simulations [15].

Building smaller at first also gives planners the ability to bring a new plant online quicker and then scale up if needed. One needs to build a certain number of modules in the beginning, and one can add a certain number later. One can proactively plan to adopt in the future. The researchers acknowledge that the results of their study would likely vary from region to region, but they think the framework could help planners make the case that building on a smaller scale may position cities and countries best in the long run [15].

7.2.3 NexGen Desal Modular Desalination Plant

Complex System Manufacturing NexGenDesal™ Plants are designed as multiples of 1,000 m^3/day "Units," assembled as standardized, skid-mounted subsystem "Modules." Upon order, the modules can be assembled and tested in a factory setting, and shipped to the plant site to be integrated in a "Plug & Play" fashion. In comparison to the prevailing "Engineer, Procure, Construct" (EPC) method of project fulfillment, Complex System Manufacturing [16,17]:

1. *Speeds Projects to Market* (e.g., manufacturing can commence prior to permitting, more activities can occur in parallel)
2. *Minimizes Disruption of Operations during Plant Expansions* (e.g., lowers the risk of unplanned downtime from construction-related incidents, maximizes revenue stream by minimizing shutdown duration)
3. *Minimizes and Controls Risk* (e.g., protects against permitting delays, minimizes weather impacts, eliminates craft labor shortages, ensures predictable schedule performance)
4. *Reduces Project Costs* (e.g., minimizes project-specific engineering, maximizes value engineering and supply chain management savings, produces persistent economies of scale, enables multiplant spares packaging that reduces anchor inventory)
5. *Improves Project Performance* (e.g., produces a single, multiplant experience curve, optimizes ISO 9000 quality management processes, enables continuous improvement of product design and operator training)

For Small, Distributed, Community-Based seawater desalination plants to effectively compete with the prevailing large, regional water treatment model, a number of scale economies must be addressed. NexGenDesal plants can compete with large regional systems on a CapEx basis by securing equal or greater economies of scale through standardization and manufacturability. Its OPEEX is also competitive because of lowest specific power consumption. Lights out operability allows OPEX to compete with larger plants by spreading OPEX over multiple plants. Predictive diagnostics reduces spare inventory while standardization allows reduction of spare parts inventory even further by spreading the spares package over multiple plants. Multiple plants also allow spreading of legal and financial risks and technical and administrative costs [16,17].

Modular desalination units of NextGenDesal are based on main principles of the RO technology engineered in mobile design. The unique features of this modular system are

1. turnkey solution for water desalination
2. small size
3. no internal pipes
4. possibility of simple connection between modules
5. low price
6. no release of solid waste into the surrounding environment
7. units are self-contained and mobile
8. low energy consumption
9. can be installed in existing buildings with no industrial zone or waste area required

With appropriate use and replacement of modules, the system can be used for both sea and brackish water.

7.2.4 KSB Modular Desalination Plant

The growing demand for smaller desalination systems has led to a number of companies developing containerized systems. KSB claims its entry to the market can achieve up to 75% energy saving and does not require two electric motors to drive booster and high-pressure pumps. Research from the International Desalination Association (IDA) has forecast the market value for containerized desalination plants in 2013 to be US$830 million, with an annual growth rate of 15%.

The benefits of containerized systems are clear compared to large-scale fixed infrastructure: they are designed to provide a method of producing low-cost drinking water, quickly and efficiently. Large plants, in comparison, can take longer in planning, construction, and require considerable upfront investment costs. The phrase of the "tourist dollar" has been used in connection with many industries—restaurants, hotels, famous landmarks—and is actually the lifeblood of many of these businesses. Yet, commonly unnoticed is the strain the influx of tourists can put on a town, city, or country's water sector. KSB believes this is where there lies a huge opportunity here for containerized filtration [18,19].

The global market for containerized desalination plants is fragmented and obtaining accurate data is difficult. Taking the tourism industry as an example, there are many holiday resorts around the world where freshwater has to be constantly available, but the costs of building and operating large desalination plants are prohibitive. With containerized desalination plant, hotels, resorts, and manufacturing sites can be independent of regional water supply companies and have total control over their requirements and costs. As a result, the pump and valve manufacturer have developed a compact high-pressure pump unit for RO seawater desalination. Called the SALINO Pressure Center, it has been developed for RO systems with the capacity to produce of up to 1,000 m³/day of permeate. The unit operates by receiving pretreated seawater via a feed pump with the salt concentration typically being in the region of 38,000 ppm. The seawater enters the pressure center at 2.5 bar and is raised to 70 bar and transferred into the RO membrane where the brine is separated from the water to produce permeate. The resulting high-pressure brine produced in the membrane is returned to the pressure center where its mechanical energy is recovered, contributing to raising the pressure of the incoming pretreated seawater [18,19].

The Pressure Center is a hydraulic system that is designed to provide pressure boosting and energy recovery. KSB claims it is down to combining all the elements in a single unit and reducing the number of components, thereby reducing the complexity of pipework and installation costs. By removing the need for a separate booster pump and motor, it is claimed that the number of components associated with other types of systems is reduced, and energy savings of up to 75% are made compared with systems that do not use an integral energy recovery device. Because there is no fluid exchange between the brine and the feedwater, energy-consuming mixing is avoided, according to the manufacturer. The system combines the four elements required in the RO process for seawater desalination: a high-pressure pump, an electric drive, together with a booster pump and an energy recovery device. The container solution employs a single lubricant-free axial piston pump and single axial piston motor, the latter serving as an integrated energy recovery device. As a result, three of the elements are fulfilled by one and the same unit: creating high pressure, compensating pressure losses and recovering energy. The entire system runs on a single electric motor and frequency inverter and the electric drive has a rating of 29 kW, not requiring a separate booster pump. KSB says the fluctuating salt content in seawater can be processed by a response through the integrated control system. In tests the unit desalinated one cubic meter of seawater with a salt content of 35,000 ppm at a power input of ~2 kW/h [18,19].

7.3 USE OF MODULAR RENEWABLE ENERGY SYSTEMS FOR DESALINATION

Using desalination technologies driven by renewable energy resources is a viable way to produce freshwater in many locations today. As the renewable technologies continue to improve—and as freshwater and cheap conventional sources of energy become scarcer—using renewable energy technology in desalination will become even more attractive. The selection of the appropriate renewable energy desalination technology depends on a number of factors, including plant size, feedwater salinity, remoteness, availability of grid electricity, technical infrastructure, and the type and potential of the local renewable energy resources [27–114].

Using renewable energy sources in water desalination has many advantages and benefits. The most common advantage is that they are renewable and cannot be depleted. They are a clean energy, not polluting the air, and they do not contribute to global warming or greenhouse gas emissions. Because their sources are natural, operational costs are reduced and they also require less maintenance on their plants. Using these resources in water desalination in remote areas also represents the best option due to the very high cost of providing energy from the grid. And implementing renewable energy in these areas fosters socioeconomic development. Since most renewable energy in remote areas tend to small and modular in nature, they will be more efficient, less costly, and flexible. Renewable energy can be used for seawater desalination either by producing the thermal energy required to drive the phase-change processes or by producing electricity required to drive the membrane processes. The major sources of alternative energy discussed here are solar, wind, Geothermal, and water [27–114].

Proper matching of stand-alone power-supply desalination systems has been recognized as being crucial if the system is to provide a satisfactory supply of power and water at a reasonable cost. Stand-alone renewable energy systems for electricity supply are now a proven technology and economically promising for remote regions, where connection to the public electric grid is either not cost-effective or feasible, and where water scarcity is severe. Solar thermal, solar photovoltaic (PV), wind, geothermal, and even water (if possible) technologies could be used as energy suppliers for desalination systems. Solar energy—both solar thermal and solar PV—can be used to drive MSF, MED, RO, and ED. Wind energy can drive VC, RO, and ED. Geothermal energy reservoirs with moderate temperature can drive MSF and MED units, while geothermal high-pressure reservoirs can be utilized to drive mechanically driven desalination units by shaft power or by producing electricity to drive VC, RO, and ED units. As shown later, modular wave energy can be used for RO [27–113].

7.3.1 SOLAR-ASSISTED MODULAR DESALINATION SYSTEMS

Solar energy can drive the desalination units by either thermal energy and electricity generated from solar thermal systems or by PV systems. The cost distribution of solar distillation is dramatically different from that of RO and MSF. The main cost is in the initial investment. However, once the system is operational, it is extremely inexpensive to maintain, and the energy has minimal or even no cost. Solar-assisted desalination systems are divided into two parts: solar thermal-assisted systems and solar PV-assisted systems.

7.3.1.1 Solar Thermal-Assisted Modular Systems

Solar thermal energy can be harnessed directly or indirectly for desalination. A simple illustration of this process is described in Figure 7.2. Collection systems that use solar energy to produce distillate directly in the solar collector are called direct-collection systems, whereas systems that combine solar energy collection devices with conventional desalination units are called indirect systems. In indirect systems, solar energy is used either to generate the heat required for desalination and/or to generate electricity used to provide the required electric power for conventional desalination plants such as MED and MSF plants. Direct solar desalination requires large land areas and has a relatively low productivity. However, it is competitive with indirect desalination plants in small-scale production due to its relatively low cost and simplicity.

FIGURE 7.2 Desalination assisted by solar energy [115].

7.3.1.1.1 Direct Solar Thermal Desalination

Direct systems are those where the heat collection and distillation processes occur in the same equipment. Solar energy is used to produce the distillate directly in the solar still. The method of direct solar desalination is mainly suited for small production systems, such as solar stills, and it is used in regions where the freshwater demand is low. This device has low efficiency and low water productivity due to the ineffectiveness of solar collectors to convert most of the energy they capture, and to the intermittent availability of solar radiation. For this reason, direct solar thermal desalination has so far been limited to small-capacity modular units, which are appropriate in serving small communities in remote areas having scarce water. Solar still design can generally be grouped into four categories: (a) basin still, (b) tilted-wick solar still, (c) multiple-tray tilted still, and (d) concentrating mirror still. The basin still consists of a basin, support structure, transparent glazing, and distillate trough. Thermal insulation is usually provided underneath the basin to minimize heat loss. Other ancillary components include sealants, piping and valves, storage, external cover, and a reflector (mirror) to concentrate light. Single-basin stills have low efficiency, generally below 45%, and low productivity (4–6 L/m^2/day) due to high top losses. Double glazing can potentially reduce heat losses, but it also reduces the transmitted portion of the solar radiation [30]. On a much smaller scale, a solar microdesalination unit [31] may be used in remote areas and is capable of producing about 1.5 L/day.

A tilted-wick solar still uses the capillary action of fibers to distribute feedwater over the entire surface of the wick in a thin layer. This allows a higher temperature to form on this thin layer. Insulation in the back of the wick is essential. A cloth wick needs frequent cleaning to remove sediment built-up and regular replacement of wick material due to weathering and ultraviolet (UV) degradation. Uneven wetting of the wick can result in dry spots that reduce efficiency [32]. In a multiple-tray tilted still, a series of shallow horizontal black trays are enclosed in an insulated container with a transparent glazing on top. The feedwater supply tank is located above the still, and the vapor condenses and flows down to the collection channel and finally to the storage. The construction of this still is fairly complicated and involves many components that are more expensive than simple basin stills. Therefore, the slightly better efficiency it delivers may not justify its adoption [33].

The concentrating mirror solar still uses a parabolic mirror for focusing sunlight onto an evaporator vessel. The water is evaporated in this vessel exposed to extremely high temperature. This type of still entails high construction and maintenance costs [34].

7.3.1.1.2 Indirect Solar Thermal Desalination

Indirect solar thermal desalination methods involve two separate systems: the collection of solar energy by a solar collecting system, coupled to a conventional desalination unit. Processes include HD, MD, solar pond-assisted desalination, and solar thermal systems such as solar collectors, evacuated-tube collectors, and concentrating collector (concentrating solar power (CSP)) systems driving conventional desalination processes such as MSF and MED.

7.3.1.1.2.1 HD Process These units consist of a separate evaporator and condenser to eliminate the loss of latent heat of condensation. The basic idea in HD process is to mix air with water vapor and then extract water from the humidified air by the condenser. The amount of vapor that air can hold depends on its temperature. Some advantages of HD units are the following: low-temperature operations, able to combine with renewable energy sources such as solar energy, modest level of technology, and high productivity rates. Two different cycles are available for HD units: HD units based on open-water closed-air cycle, and HD units based on open-air closed-water cycle. These two options are described below. In an open-water closed-air cycle, seawater enters the system, is heated in the solar collector, and is then sprayed into the air in the evaporator. Humidified air is circulated in the system and, when it reaches the condenser, a certain amount of water vapor starts to condense. Distilled water is collected in a container. Some of the brine can also be recycled in the system to improve the efficiency, and the rest is removed [35].

An open-air closed-water cycle is used to emphasize recycling the brine through the system to ensure high utilization of the saltwater for freshwater production. As air passes through the evaporator, it is humidified. And by passing through a condenser, water vapor is extracted [36].

7.3.1.1.2.2 Membrane Distillation MD is a separation/distillation technique where water is transported between "hot" and a "cool" stream separated by a hydrophobic membrane, permeable only to water vapor, which excludes the transition of liquid phase and potential dissolved particles. The exchange of water vapor relies on a small temperature difference between the two streams, which results in a vapor pressure difference, leading to the transfer of the produced vapor through the membrane to the condensation surface. In the MD process, the seawater passes through the condenser usually at about 25°C and leaves at a higher temperature, and then it is heated to about 80°C by an external source such as solar, geothermal, or industrial waste [35]. The main advantages of MD lie in its simplicity and the need for only small differentials to operate. However, the temperature differential and the recovery rate determine the overall efficiency for the process. Thus, when it is run with a low-temperature differential, large amounts of water must be used, which adversely affects its overall energy efficiency.

Membrane desalination is a promising process, especially for situations where low-temperature solar, geothermal, waste, or other heat is available. MD was introduced commercially on a small scale during the 1980s, but it has not demonstrated large-scale commercial success due to the high cost and problems associated with membranes. Therefore, more intensive research and development is needed, both in experimentation and modeling, focusing on key issues such as long-term liquid/vapor selectivity, membrane aging and fouling, feedwater contamination, and heat-recovery optimization. Scale-up studies and realistic assessment of the basic working parameters on real pilot plants, including cost and long-term stability, are also considered to be necessary [37].

7.3.1.1.2.3 Solar Pond-Assisted Desalination Salinity-gradient solar ponds are a type of heat collector, as well as a mean of heat storage. Hot brine from a solar pond can be used as a heat source for MSF or MED desalination units. Solar ponds can store heat because of their unique chemically stratified nature. A solar pond has three layers: (a) upper or surface layer, called the upper convection zone, (b) middle layer, which is the nonconvection zone or salinity-gradient zone, and (c) lower layer, called the storage zone or lower convection zone. Salinity increases with depth from near pure water at the surface to the bottom, where salts are at or near saturation. Salinity is relatively constant in the upper and lower convection zones, and increases with depth in the nonconvection zone. Saline water is denser than freshwater; therefore, the water at the bottom of the pond is more dense (has a higher specific gravity) than water at the surface. The solar pond system is able to store heat because circulation is suppressed by the salinity-related density differences in the stratified water. Convection of hot water to the surface is repressed by the salinity (density) gradient of the nonconvection zone. Thus, although solar energy can penetrate the entire depth of the pond, it cannot escape the storage zone [38].

7.3.1.1.2.4 Concentration Solar Thermal Desalination Concentrating solar thermal power technologies are based on the concept of concentrating solar radiation to provide high-temperature heat for electricity generation within conventional power cycles using steam turbines, gas turbines, or Stirling and other types of engines. For concentration, most systems use glass mirrors that continuously track the position of the sun. The four major CSP technologies are parabolic trough, Fresnel mirror reflector, power tower, and dish/engine systems. Debate continues as to which of these is the most effective technology [39].

   a. **Parabolic trough:** Parabolic trough power plants consist of large parallel arrays of parabolic trough solar collectors that constitute the solar field. The parabolic collector is made of reflectors, each of which focuses the sun's radiation on a receiver tube that absorbs the reflected solar energy. The collectors track the sun so that the sun's radiation is continuously focused on the receiver. Parabolic troughs are recognized as the most proven CSP technology, and at present, experts indicate the cost to be 10 US cents/kWh or less.

b. **Fresnel mirror reflector:** This type of CSP is broadly similar to parabolic trough systems, but instead of using trough-shaped mirrors that track the sun, flat or slightly curved mirrors mounted on trackers on the ground are configured to reflect sunlight onto a receiver tube fixed in space above these mirrors. A small parabolic mirror is sometimes added atop the receiver to further focus the sunlight. As with parabolic trough systems, the mirrors change their orientation throughout the day so that sunlight is always concentrated on the heat-collecting tube.

c. **Dish/Stirling engine systems and concentrating PV (CPV) systems:** Solar dish systems consist of a dish-shaped concentrator (like a satellite dish) that reflects solar radiation onto a receiver mounted at the focal point. The receiver may be a Stirling or other type of engine and generator (dish/engine systems) or it may be a type of PV panel that has been designed to withstand high temperatures (CPV systems). The dish is mounted on a structure that tracks the sun continuously throughout the day to reflect the highest percentage of sunlight possible onto the thermal receiver. Dish systems can often achieve higher efficiencies than parabolic trough systems, partly because of the higher level of solar concentration at the focal point. Dish systems are sometimes said to be more suitable for stand-alone, small power systems due to their modularity. Compared with ordinary PV panels, CPV has the advantage that smaller areas of PV cells are needed; because PV is still relatively expensive, this can mean a significant cost savings.

d. **Power tower:** A power tower system consists of a tower surrounded by a large array of heliostats, which are mirrors that track the sun and reflect its rays onto the receiver at the top of the tower. A heat-transfer fluid heated in the receiver is used to generate steam, which, in turn, is used in a conventional turbine generator to produce electricity. Some power towers use water/steam as the heat-transfer fluid. Other advanced designs are experimenting with molten nitrate salt because of its superior heat-transfer and energy-storage capabilities. Power towers also reportedly have higher conversion efficiencies than parabolic trough systems. They are projected to be cheaper than trough and dish systems, but a lack of commercial experience means that there are significant technical and financial risks in deploying this technology now. As for cost, it is predicted that with higher efficiencies, 7–8 cents/kWh may be possible. But this technology is still in its early days of commercialization.

7.3.1.1.2.5 CSP Systems Coupled with Desalination Plant The primary aim of CSP plants is to generate electricity, yet a number of configurations enable CSP to be combined with various desalination methods. When compared with PVs or wind, CSP could provide a much more consistent power output when combined with either energy storage or fossil-fuel backup. There are different scenarios for using CSP technology in water desalination [40], and the most suitable options are described below.

7.3.1.1.2.5.1 PARABOLIC TROUGH COUPLED WITH MED DESALINATION UNIT   In this coupling steam generated by the trough (superheated to around 380°C) is first expended in a noncondensing turbine and then used in a conventional manner for desalination. The steam temperature for the MED plant is around 135°C; therefore, there is sufficient energy in the steam to produce electricity before it is used in the MED plant. It is important to emphasize that water production is the main purpose of the plant—electricity is a byproduct. Although conventional combined-cycle (CC) power plants can be configured in a similar manner for desalination, a fundamental difference exists in the design approach for solar and fossil-fuel-fired plants. The fuel for the solar plant is free; therefore, the design is not focused primarily on efficiency but on capital cost and capacity of the desalination process. In contrast, for the CC power plant, electricity production at the highest possible efficiency is the ultimate goal [41].

7.3.1.1.2.5.2 PARABOLIC TROUGH COUPLED WITH RO DESALINATION UNIT   In this case, as in MED, the steam generated by the solar plant can be used through a steam turbine to produce the electric power needed to drive the RO pumps. As an alternative for large, multiunit RO systems, the high-pressure seawater can be provided by a single pump driven by a steam turbine. This arrangement is similar to the steam-turbine-driven boiler feed pumps in a fossil-fuel power plant. Often, MED and RO are compared in terms of overall performance, and specifically for energy consumption. Based on internal studies by Bechtel [42], one can conclude that, in specific cases, the CSP/RO combination requires less energy than a similar CSP/MED combination.

However, an analysis presented in [43] suggests that, for several locations, CSP/MED requires 4%–11% less input energy than CSP/RO. Therefore, before any decision can be made on the type of desalination technology to be used, it is recommend that a detailed analysis be conducted for each specific location, evaluating the amount of water, salinity of the input seawater, and site conditions. It appears that CSP/MED provides slightly better performance at sites with high salinity such as in closed gulfs, whereas CSP/RO appears to be more suitable for low-salinity waters in the open ocean.

One additional advantage of the RO system is that the solar field might be located away from the shoreline. The only connection between the two is the production of electricity to drive the RO pumps and other necessary auxiliary loads.

7.3.1.2 Modular Solar Thermal Applications

Although the strong potential of solar thermal energy to seawater desalination is well recognized, the process is not yet developed at the commercial level. The main reason is that the existing technology, although demonstrated as technically feasible, cannot presently compete, on the basis of produced water cost, with conventional distillation and RO technologies. However, it is also recognized that there is still potential to improve desalination systems based on solar thermal energy.

Among low-capacity production systems, solar stills and solar ponds represent the best alternative in low freshwater demands. For higher desalting capacities, one needs to choose conventional distillation plants coupled to a solar thermal system, which is known as indirect solar desalination [44]. Distillation methods used in indirect solar desalination plants are MSF and MED. MSF plants, due to factors such as cost and apparent high efficiency, displaced MED systems in the 1960s, and only small-size MED plants were built. However, in the last decade, interest in MED has been significantly renewed, and the MED process is currently competing technically and economically with MSF [45]. Recent advances in research of low-temperature processes have resulted in an increase of the desalting capacity and a reduction in the energy consumption of MED plants providing long-term operation under remarkable steady conditions [46]. Scale formation and corrosion are minimal, leading to exceptionally high plant availabilities of 94%–96%.

Many small systems of direct solar thermal desalination systems and pilot plants of indirect solar thermal desalination systems have been implemented in different places around the world [47]. Among them are the de Almería (PSA) project in 1993 and the AQUASOL project in 2002. Study of these systems and plants will improve our understanding of the reliability and technical feasibility of solar thermal technology application to seawater desalination. It will also help to develop an optimized solar desalination system that could be more competitive against conventional desalination systems. On a commercial basis, CSP technology will take some years until it becomes economic and sufficiently mature for use in power generation and desalination.

7.3.2 MODULAR PV-BASED DESALINATION

A PV or solar cell converts solar radiation into direct current (DC) electricity. It is the basic building block of a PV (or solar electric) system. An individual PV cell is usually quite small, typically producing about 1 or 2 W power. To boost the power output, the solar cells are connected in series and parallel to form larger units called modules. Modules, in turn, can be connected to form even larger units called arrays. Any PV system consists of a number of PV modules, or arrays. The other system equipment includes a charge controller, batteries, inverter, and other components needed to provide the output electric power suitable to operate the systems coupled with the PV system. PV systems can be classified into two general categories: flat-plate systems and concentrating systems. CPV system has several advantages compared with flat-plate systems: CPV systems increase the power output while reducing the size or number of cells needed; and a solar cell's efficiency increases under concentrated light.

PV is a rapidly developing technology, with costs falling dramatically with time, and this will lead to its broad application in all types of systems. Today, however, it is clear that PV/RO and PV/ED will initially be most cost competitive for small-scale systems installed in remote areas where other technologies are less competitive. RO usually uses alternating current (AC) for the pumps, which means that DC/AC inverters must be used. In contrast, ED uses DC for the electrodes at the cell stack, and hence, it can use the energy supply from the PV panels without major modifications. Energy storage is again a concern, and batteries are used for PV output power to smooth or sustain system operation when solar radiation is insufficient. PV-powered RO is considered one of the most promising forms of renewable-energy-powered desalination, especially when it is used in remote areas. Therefore, small-scale PV/RO has received much attention in recent years, and numerous demonstration systems have been built. Two types of PV/RO systems are available in the market: BWRO and SWRO PV/RO systems. Different membranes are used for brackish water and much higher recovery ratios are possible, which makes energy recovery less critical [48].

Brackish water has a much lower osmotic pressure than seawater; therefore, its desalination requires much less energy and a much smaller PV array in the case of PV/RO. Also, the lower pressures found in BWRO systems permit the use of low-cost plastic components. Thus, the total cost of water from brackish water PV/RO is considerably less than that from seawater, and systems are beginning to be offered commercially [49]. Many of the early PV/RO demonstration systems were essentially a standard RO system, which might have been designed for diesel or mains power, but powered from batteries charged by PV. This approach generally requires a rather large PV array for a given flow of product because of poor efficiencies in the standard RO systems and batteries. Large PV arrays and the regular replacement of batteries typically make the cost of water from such systems rather high.

The osmotic pressure of seawater is much higher than that of brackish water; therefore, its desalination requires much more energy, and, unavoidably, a somewhat larger PV array. Also, the higher pressures found in seawater RO systems require mechanically stronger components. Thus, the total cost of water from seawater PV/RO is likely to remain higher than that from brackish water, and systems have not yet passed the demonstration stage [50–54].

ED uses DC for the electrodes; therefore, the PV system does not include an inverter, which simplifies the system. Currently, there are several installations of PV/ED technology worldwide. All PV/RD applications are of a stand-alone type, and several interesting examples are outlined below. In the city of Tanote, in Rajasthan, India, a small plant was commissioned in 1986 that features a PV system capable of providing 450 peak watts (W_p) in 42 cell pairs. The ED unit includes three stages, producing $1 m^3$/day water from brackish water (5,000 ppm TDS). The unit energy consumption is 1 kWh/kg of salt removed [55]. A second project is a small experimental unit in Spencer Valley, New Mexico (USA), where two separate PV arrays are used: two tracking flat-plate arrays (1,000 W_p power, 120 V) with DC/AC inverters for pumps, plus three fixed arrays (2.3 kW_p, 50 V) for ED supply. The ED design calls for 2.8 m^3/day product water from a feed of about 1,000 ppm TDS. This particular feedwater contains uranium and radon, apart from alpha particles. Hence, an ion-exchange process is required prior to ED. Unit consumption is 0.82 kWh/m^3 and the reported cost is 16 US\$/m^3 [56,57]. A third project is an unusual application in Japan, where PV technology is used to drive an ED plant fed with seawater, instead of the usual brackish water of an ED system [46]. The solar field consists of 390 PV panels with a peak power of 25 kW_p, which can drive a 10 m^3/day ED unit. The system, located on Oshima Island (Nagasaki), has been operating since 1986. Product-water quality is reported to be below 400 ppm TDS, and the ED stack is provided with 250 cell pairs.

7.3.3 DESALINATION SYSTEMS DRIVEN BY WIND

Remote areas with potential wind energy resources such as islands can employ wind energy systems to power seawater desalination for freshwater production. The advantage of such systems is a reduced water production cost compared with the costs of transporting the water to the islands or to using conventional fuels as power source. Different approaches for wind desalination systems are possible. First, both the wind turbines as well as the desalination system are connected to a grid system. In this case, the optimal sizes of the wind turbine system and the desalination system as well as avoided fuel costs are of interest. The second option is based on a more or less direct coupling of the wind turbine(s) and the desalination system. In this case, the desalination system is affected by power variations and interruptions caused by the power source (wind). These power variations, however, have an adverse effect on the performance and component life of certain desalination equipment. Hence, backup systems, such as batteries, diesel generators, or flywheels might be integrated into the system. Main research in this area is related to the analysis of the wind plant and the overall system performance as well as to developing appropriate control algorithms for the wind turbine(s) as well as for the overall system. Regarding desalinations, there are different technologies options, e.g., ED or VC. However, RO is the preferred technology due to low specific energy consumption [80].

Wind turbines can be used to supply electricity or mechanical power to desalination plants. Like PV, wind turbines represent a mature, commercially available technology for power production. Wind turbines are a good option for water desalination, especially in coastal areas, presenting a high availability of wind energy resources. Many different types of wind turbines have been developed. A distinction can be made between turbines driven mainly by drag forces versus those driven mainly by lift forces. A distinction can also be made between turbines with axes of rotation parallel to the wind direction (horizontal) and with axes perpendicular to the wind direction (vertical). The efficiency of wind turbines driven primarily by drag forces is low compared with the lift-force-driven type. Therefore, all modern wind turbines are driven by lift forces. The most common types are the horizontal-axis wind turbine (HAWT) and the vertical-axis wind turbine (VAWT). Wind-driven desalination has particular features due to the inherent discontinuous availability of wind power. For stand-alone systems, the desalination unit has to be able to adapt to the energy available; otherwise, energy storage or a backup system is required. Wind energy is used to drive RO, ED, and VC desalination units. A hybrid system of wind/PV is usually used in remote areas. Few applications have been implemented using wind energy to drive an MVC unit. A pilot plant was installed in 1991 at Borkum, an island in Germany, where a wind turbine with a nominal power of 45 kW was coupled to a 48 m^3/day MVC evaporator. A 36-kW compressor was required. The experience was followed in 1995 by another larger plant at the island of Rügen. Additionally, a 50 m^3/day wind MVC plant was installed in 1999 by the Instituto Tecnologico de Canarias (ITC) in Gran Canaria, Spain, within the Sea Desalination Autonomous Wind Energy System (SDAWES) project [47]. The wind farm is composed of two 230-kW wind turbines, a 1,500-rpm flywheel coupled to a 100-kVA synchronous machine, an isolation transformer located in a specific building, and a 7.5-kW uninterruptible power supply located in the control dome. One of the innovations of the SDAWES project, which differentiates it from other projects, is that the wind generation system behaves like a mini power station capable of generating a grid similar to conventional ones without the need to use diesel sets or batteries to store the energy generated.

Regarding wind energy and RO combinations, a number of units have been designed and tested. As early as 1982, a small system was set at Ile du Planier, France [48], which as a 4-kW turbine coupled to a 0.5-m^3/h RO desalination unit. The system was designed to operate via either a direct coupling or batteries. Another case where wind energy and RO were combined is that of the Island of Drenec, France, in 1990 [60]. The wind turbine, rated at 10 kW, was used to drive a seawater RO unit. A very interesting experience was gained at a test facility in Lastours, France, where a 5-kW wind turbine provides energy to a number of batteries (1,500 Ah, 24 V) and via an inverter to an RO unit with a nominal power of 1.8 kW. A 500 L/h seawater RO unit driven by a 2.5-kW wind generator (W/G) without batteries was developed and tested by the Centre for Renewable Energy Systems Technology (CREST) UK. The system operates at variable flow, enabling it to make efficient use of the naturally varying wind resource, without need of batteries [61].

The ED process is interesting for brackish water desalination since it is able to adapt to changes of available wind power and it is most suitable for remote areas than RO. Modeling and experimental tests results of one of such system installed at the ITC, Gran Canaria, Spain is presented in [63]. The capacity range of this plant is 192 to 72 m/day.

Excellent work on wind/RO systems has been done by ITC within several projects such as AERODESA, SDAWES, and AEROGEDESA [62]. In addition in Coconut Island off the northern coast of Oahu, Hawai, a brackish water desalination wind-powered RO plant was analyzed. The system directly couples the shaft power production of a windmill with the high-pressure pump; 13 L/min can be maintained for a wind speed of 5 m/s [84].

Additionally, a wind/RO system without energy storage was developed and tested within the JOULE Program (OPRODES-JORCT98-0274) in 2001 by the University of Las Palmas. The RO unit has a capacity of 43–113 m^3/h, and the W/G has a nominal power of 30 kW [85]. The European Community Joule III project funded different research programs and demonstration projects of wind desalination systems on Greek and Spanish islands. In addition, an excellent job on combining wind/RO was done by ENERCON, the German wind turbine manufacturer. ENERCON provides modular and energy-efficient RO desalination systems driven by wind turbines (grid-connected or stand-alone systems) for brackish and seawater desalination. Market-available desalination units from ENERCON range from 175 to 1,400 m^3/day for seawater desalination and 350–2,800 m^3/day for brackish water desalination. These units in combination with other system components, such as synchronous machines, flywheels, batteries, and diesel generators, supply and store energy and water precisely according to demand [64]. Several industrial wind/RO installations for sea water include (a) Ile de Planier, France with 500 L/h capacity (commission in 1983), (b) Fuerteventura island, PUNTAJANDIA project with a capacity of 2,333 L/h (commission in 1995), (c) Therasia island, Greece with a capacity of 200 L/h (commission in 1997), (d) Pozo Izquierdo, Gran Canaria,

AEROGEDESA project with a capacity of 800-L/h (commission in 2003), and (e) CREST,UK with a capacity of 500L/h (commission in 2004) [25].

Besides the ones mentioned above, other wind-driven RO systems are as follows: (1) An RO system driven by a wind power plant, in the Island of the County Split and Dalmatia, as reported in [73]. (2) An RO plant in the Middle East, which is a $25\,m^3$/day plant connected to a hybrid wind–diesel system [89]. (3) A wind-powered RO system [90] in Drepanon, Achaia near Patras (Greece). Finally, European Commission (1998) presents other facilities at

- Island of Suderoog (North Sea), with $6-9\,m^3$/day;
- Island of Helgoland, Germany ($2.480\,m^3$/h);
- Island of St. Nicolas, West France (hybrid wind–diesel); and
- Island of Drenec, France (10 kW wind energy converter).

For general information on wind desalination research, see [81–83]. For information on large stand-alone wind desalination systems, see [84,85]; for small systems, see [86]; and for an overview of the research activities in North America, see [87].

7.3.4 DESALINATION BY BIOMASS AND GEOTHERMAL ENERGY

The use of biomass in desalination is not in general a promising alternative since organic residues are not normally available in arid regions and growing of biomass requires more freshwater than it could generate in a desalination plant. Also, even though geothermal energy is not as common in use as solar (PV or solar thermal collectors) or wind energy, it presents a mature technology that can be used to provide energy for desalination at a competitive cost. Furthermore, and comparatively to other renewable energy technologies, the main advantage of geothermal energy is that the thermal storage is unnecessary, since it is both continuous and predictable [65].

The earth's temperature varies widely, and geothermal energy is usable for a wide range of temperatures from room temperature to well over 300°F. The main advantage of geothermal energy is that thermal storage is unnecessary in such systems. Geothermal reservoirs are generally classified as being either at low (<150°C) or high temperatures (>150°C). Generally speaking, high-temperature reservoirs are suitable for, and sought out for, commercial production of electricity. Energy from the earth is usually extracted with ground heat exchangers, made of a material that is extraordinarily durable but allows heat to pass through efficiently. The direct use of moderate and high temperatures is for thermal desalination technologies. The direct use of geothermal fluid of sufficiently high temperature in connection to thermal desalination technologies is the most interesting option [94]. The main advantage of geothermal energy compared to other renewable energy resources is that thermal storage is not necessary since it is both continuous and predictable [65]. A high-pressure geothermal source allows the direct use of shaft power on mechanically driven desalination, while high-temperature geothermal fluids can be used to power electricity-driven RO or ED plants. The availability and/or suitability of geothermal energy, and other RE resources for desalination are given in reference [73].

The first geothermal energy-powered desalination plants were installed in the United States in the 1970s [65–69], testing various potential options for the desalination technology, including MSF and ED. An analysis [70] discussing a technical and economic analysis of an MED plant, with a capacity of 80 m^3/day, powered by a low-temperature geothermal source and installed in Kimolos, Greece showed that high-temperature geothermal desalination could be a viable option. A study [71] presented results from an experimental investigation of two polypropylene-made HD plants powered by geothermal energy [72]. Recently, a study [73] discussed the performances of a hybrid system consisting of a solar still in which the feedwater is brackish underground geothermal water. Finally, the availability and/or suitability of geothermal energy and other renewable energy resources for desalination are given in [74].

7.3.5 Advantages and Disadvantages of Technologies

There are advantages and disadvantages when comparing membrane with thermal technologies, and many factors need to be considered depending on the purpose and objectives for considering a particular desalination process. Advantages of membrane processes over thermal processes include [114]:

- Lower capital cost and energy requirements;
- Lower footprint and higher space/production ratio;
- Higher recovery ratios;
- Modularity allows for up- or downgrade and minimal interruption to operation when maintenance or membrane replacement is required;
- Less vulnerable to corrosion and scaling due to ambient temperature operation; and
- Membranes reject microbial contamination.

Advantages of thermal processes over membrane processes include:

- Very proven and established technology;
- Higher quality product water produced;
- Less rigid monitoring than for membrane process required;
- Less impacted by quality changes in feedwater; and
- No membrane replacement costs.

7.3.6 Economic Analysis for Renewable Energy Desalination Processes

Several factors affect desalination cost. In general, cost factors associated with implementing a desalination plant are site specific and depend on several variables. The cost estimation procedures are described in [95]. The major cost variables are as follows: (a) Quality of feedwater, where, the low TDS concentration in feedwater (e.g., brackish water) requires less energy for treatment compared with high TDS feedwater (seawater). (b) Plant capacity, where it affects the size of treatment units, pumping, water storage tank, and water distribution system. Large-capacity plants require high initial capital investment compared with low-capacity plants. But, due to the economy of scale, the unit production cost for large-capacity plants can be lower [96,97]. (c) Site characteristics, where it can affect water production cost such as availability of land and land condition, the proximity of plant location to water source, and concentrate discharge point is another factor. Pumping cost and costs of pipe installation will be substantially reduced if the plant is located near the water source and if the plant concentrate is discharged to a nearby water body. (d) Costs associated with water intake, pretreatment, and concentrate disposal can be substantially reduced if the plant is an expansion of an existing water treatment plant as compared to constructing a new plant. (e) Regulatory requirements, which are associated with meeting local/state permits and regulatory requirements [98]. It is difficult to compare the costs of desalination installations at an aggregated level because the actual costs depend on a range of variables specific to each site [99].

Desalination plant implementation costs can be categorized as construction costs (starting costs) and O&M costs. Construction costs include direct and indirect capital costs. The direct cost includes land, production wells, surface water intake structure, process equipment, auxiliary equipment, buildings, and concentrate disposal (type of desalination technology, plant capacity, discharge location, and environmental regulations). The indirect capital cost is usually estimated as percentages of the total direct capital cost. Indirect costs may include freight and insurance, construction overhead, owner's costs, and contingency costs. The O&M costs consist of fixed and variable costs. Fixed costs include insurance and amortization costs. Usually, insurance cost is estimated as 0.5% of the total capital cost. Typically, an amortization rate in the range of 5%–10% is used. Major variable costs include the cost of labor, energy, chemicals, and maintenance. For low TDS brackish water, the replacement rate is about 5% per year. For high TDS seawater, the replacement could be as high as 20%. The cost for maintenance and spare parts is typically <2% of the total capital cost on an annual basis [98].

It can be observed from these data that

1. The fixed costs are a major factor for both brackish and seawater
2. The major difference in cost between desalination of brackish water and seawater is energy consumption, while the remaining factors are decreased proportionally, but remain about the same; and costs associated with membrane replacement, maintenance & parts, and consumables are relatively small. These costs depend on the status of technology and may be further reduced as technology evolves but will not have a significant impact on the overall cost of desalination.

Ghoneyem and Ileri [101] estimated that a production-size solar still can produce water for $20/m^3 (1994 dollars), while Madani and Zaki [102] estimated solar distilled water production costs as low as $2.4/m^3. According to Bouchekima et al. [103], recent improvements in solar distillation technology make it an ideal technology for remote isolated areas with a water demand <50 m^3/day. All other technologies are more expensive at this small scale. Fath [104] believes solar stills are the technology of choice for water production needs up to 200 m^3/day. The dominant competing process is RO that has an energy requirement of between $22 \times 10^6 J/m^3$ and $36 \times 10^6 J/m^3$

(6 and 10 kWh/m^3) of water treated and investment costs of between US$600 and $2,000/m^3 of production capacity [105]. The most commonly used solar distillation technology is a single-effect, single-basin still characterized by a relatively large thermal mass, i.e., the water basin [106].

The economics of operating solar desalting units tend to be related to the cost of producing energy with these alternative energy devices. Presently, most of the renewable energy systems have mature technology, but despite the free cost of renewable energy resources, their collecting systems tend to be expensive, although they may be expected to decline as further development of these devices reduces their capital cost.

We first look at the cost distribution of both conventional and renewable energy-operated desalination units. For the renewable systems, the investment costs are the highest and the energy costs are the lowest. One study has considered the technoeconomic viability of solar desalination using PV and low-grade thermal energy using solar ponds [66]. The results show that the cost of water produced by a conventional RO system is less than that by a conventional MSF system. However, for solar-based systems, the partial solar-based MSF system gives the lowest cost of water production. Because of limited capacity of solar units, the capital costs and operating costs are not as well established as for the other processes. For solar stills, the cost of water production is high due to the low productivity of these stills. However, this type of desalination is only used in remote areas where there is no access to conventional energy resources. The results show that the water costs for multieffect solar stills are much lower than for simple stills [108–113].

One study [78] showed that solar-pond desalting systems have considerable potential to be cost-effective if favorable site conditions exist. The cost comparison of solar-pond-powered desalination with conventional SWRO for two production capacities (20,000 and 200,000 m^3/day) indicate that the unit water-cost difference is relatively small. However, investment costs and specific investment cost for solar-powered systems are still higher compared with the SWRO systems, where the difference decreases as the capacity increases. Cost figures for desalination have always been difficult to obtain. The total cost of water produced includes the investment cost, as well as the operating and maintenance cost. In a comparison between seawater and brackish water desalination, the cost of the first is about three to five times the cost of the second for the same plant size. As a general rule, a seawater RO unit has low capital cost and significant maintenance cost due to the high cost of membrane replacement. The cost of the energy used to drive the plant is also significant. The major energy requirement for RO desalination is for pressurizing the feedwater. Energy requirements for SWRO have been reduced to about 5 kWh/m^3 for large units with energy recovery systems, whereas for small units (without energy recovery system), this may exceed 15 kWh/m^3. For brackish water desalination, the energy requirement is between 1 and 3 kWh/m^3. The product water quality ranges between 350 and 500 ppm for both seawater and brackish water units. According to published reports [50–54], the water cost of a PV seawater RO unit ranges from 7.98 to 29 US$/m^3 for product-water capacity of 120 to 12 m^3/day, respectively. Also for a PV/RO brackish water desalination unit, a water cost of about 7.25 US$/m^3 for a product-water capacity of 250 m^3/day has been reported in the literature [62–66].

In general, ED is an economically attractive process for low-salinity water. EDR has greater capital costs than ED because it requires extra equipment (e.g., timing controllers, automatic valves), but it reduces or almost eliminates the need for chemical pretreatment. In ED applications, the electricity from a PV system can power electromechanical devices such as pumps or to DC devices such as electrodes. The total energy consumption of an ED system under ambient temperature conditions and assuming product water of 500 ppm TDS would be about 1.5 and 4 kWh/m^3 for a feedwater of 1,500–3,500 ppm TDS, respectively. The water cost of a PV-operated ED unit ranges from 16 to 5.8 US$/m^3 [57,58]. The main advantage of PV desalination systems is the ability to develop small-scale desalination plants.

Wind energy could be used to drive RO, ED, and VC desalination units. A hybrid system of wind/PV was also used in remote areas. Few applications have been implemented using wind energy to drive an MVC unit, and a number of wind/RO combinations systems have been designed and tested. ENERCON provides modular and energy-efficient RO desalination systems driven by wind turbines for brackish and seawater desalination. The estimated water cost produced from the installed wind/RO unit ranges from 7.2 to 2.6 US$/m^3 of freshwater. According to a published report [55], the water cost of a wind brackish water RO unit (capacity of 250 m^3/day) is of the order of 2 Euro/m^3, whereas for the same feedwater salinity and size, the water cost of a wind/ED unit is around 1.5 Euro/m^3. For stand-alone wind-powered MVC units with a capacity range between 5 and 12.5 m^3/h, the mean water cost varies between 3.07 and 3.73 Euro/m^3 [79].

7.3.7 MODULAR SOLAR DESALINATION—GRID-FRIENDLY WATER TECHNOLOGY

The process of desalination is now less energy-intensive compare to what it was just a decade ago, due to new reverse-osmosis plant designs and nanostructured membranes that have drastically improved the energy efficiency of turning saltwater into freshwater. As shown above, pairing *solar* wind and other renewable energy with desalination is a potential next step, with applications from the industrial scale to the village-microgrid level. The most efficient desalination technologies produce poor quality water when they vary the pressures or throughputs, which means they need a steady and predictable supply of electricity. Solar panels can't provide that, absent diesel generators or batteries, which add cost and complexity [28,107].

A desalination system that can follow the ups and downs of local green power could allow more projects to flourish without those constraints—and taken to bigger scales, it could turn constant-energy desalination plants into "demand response desalination" systems. Between 2000 and 2004, Sisyan LLC developed the first continuous-production, photovoltaic, seawater reverse-osmosis (PVSWRO) system. In simple terms, Kunczynski of Sisyan LLC has broken the problem down through a "multimodule design concept"—that is, lots of smaller reverse-osmosis systems, each of which can be turned on or off to match the local electricity supply, whether it's coming from intermittent solar or reliable grid power [28,107].

It's a similar concept to using lots of smart thermostats, water heaters, commercial refrigerators, and other on-or-off electricity loads to follow the ups and downs of intermittent supply of wind and solar power in a stepwise fashion. In this case,

Sisyan has put 3-hp modules together, each using a variable frequency drive that can maintain consistent pressure on each unit within a certain operating range at about 22 kW. If total power supply falls below that minimum range, Sisyan turns off one of the units and directs the remaining power to the still-operating units, keeping them running at optimal capacity. The Sisyan La Paz system has been running for more than 12 years, first in a configuration that used a battery to smooth power flow, and more recently in a solar-only setting. This system can be used in resorts and small communities. But beyond that, the same modular concepts could be applied to the larger-scale desalination plants that are now being built in Israel and the Middle East, where they're a critical source of freshwater [28,107].

California is also the epicenter of distributed solar power, as well as host to a range of initiatives meant to incorporate this intermittent resource into the grid. That includes demand response—the ability to ramp power consumption up and down at a site to help the grid manage peak loads or variations in supply—as well as new flexible, fast-ramping resources to help the state manage the "duck curve" distortions of its daily grid demand-supply balance. Sisyan's white paper also lays out the concept of a "demand response desalination system," which could consist of anywhere between 5 and 500 of its 22-kW modules tied together. That's more than fine-grained enough to manage most demand response programs today, and good enough to meet many of the flexibility requirements of matching the ups and downs of local solar power production [28,107]. There are limits on the cost-effectiveness of this approach—it's more capital-intensive and less efficient than a massive central desalination plant that's collocated with a big thermal plant, But the energy recovery mechanisms used are on par with any other small or medium-sized system [28,107].

7.3.8 Modular Wave Energy System for Desalination

The invention by McCormick and Washington (U.S Patent No. 8784653 B2, July 27, 2013) is generally directed to the generation of potable water using attenuator type wave energy converters (AWECS) and RO membranes. The U.S. Department of Interior (DOI) funded the Subfloor Water Intake Structure System (SWISS), currently utilized in desalination desalinization plants in California and Japan [116–121]. The SWISS approach is to install a permanent subfloor well/intake system for the source water for the traditional shore structures. The in situ sand provides the filtration media [119].

Ocean wave energy conversion is directed to the exploitation of ocean wave energy to produce energy in one or more of four forms, those being hydraulic, pneumatic, mechanical, or electrical. The articulated-barge wave energy conversion system dates back to the 1970s when both Sir Christopher in the United Kingdom and Glen Hagen of the United States suggested the system. The system was studied in the late 1970s by Haren [116,120,121] at MIT. He found that the optimum articulated-barge configuration was a three-barge system. In the 1980s, McCabe showed that the efficiency of the three-barge system could be substantially improved by suspending an inertial-damping plate below the center barge. McCabe, then, produced a prototype of the system, coined the Wave Pump (MWP). The MWP was primarily designed as a producer of potable water.

Ocean Energy Systems (OES) is in the business of designing and manufacturing articulated-barge systems to produce potable water by RO desalinization of seawater. U.S. Patent Publication No. 2009/0084296 (McCormick) describes a system directed to a wave-powered device having enhanced motion making use of an AWECS (also U.S. Patent Publication No. 2010/0320759, Lightfoot et al.). The AWECS basically comprises a forward barge, a rear barge, and an intermediate or center barge, all of which arranged to float on a body of water having waves. The barges are hingedly coupled together so that they can articulate with respect to each other in response to wave motion. The AWECS also includes high-pressure pumps that straddle and pivotably connect the barge pairs, e.g., at least one pump connects the forward barge and the intermediate barge, and at least another pump connects the rear barge and the intermediate barge. The pumps are designed to draw in the water through a prefilter, pressurize the water, and deliver the water to an onboard RO desalinization system. That system includes an RO membrane. As an incoming wave makes contact with the forward barge first, the hydraulic fluid in the pump(s) coupled between the forward barge and the center barge are driven in a first direction; as the wave continues, the hydraulic fluid in the pump(s) coupled between the rear barge and the center barge are driven in a second opposite direction. The end results are bidirectional hydraulic pumps.

In U.S. Provisional Patent Application Ser. No. 61/707,206, filed on September 28, 2012, there is disclosed an AWECS arranged for producing electrical energy from the wave energy. To that end, it makes use of an AWECS similar to that described above, except that it also makes use of a commercially available rotary-vane pump to drive a generator to produce the electricity. In particular, the invention of that Provisional Application entails a floating device having a first portion (e.g., a first barge) movably coupled (e.g., hinged) to a second portion (e.g., a second barge); at least one hydraulic or pneumatic pump (e.g., a linear pump) coupled between the first portion the said second portion, the hydraulic pump driving a hydraulic fluid therein when the first portion moves with respect to the second portion due to wave energy. A fluid rectifier is provided in the AWECS and is in fluid communication with at least one hydraulic or pneumatic pump that generates a unidirectional hydraulic or pneumatic fluid flow. A rotary vane pump is coupled to the fluid rectifier.

The rotary vane pump uses the unidirectional flow to generate a rotational motion via a drive member. A rotating electrical generator (e.g., a DC generator) is coupled to that drive member, so that the drive member causes the rotating electrical generator to generate electricity when the drive member is rotating. A filter anchor is provided that includes a filter housing for filtering seawater prior to entry into a water desalination system for placement on a sea floor. The filter housing has an exterior and an interior chamber, at least one inlet for providing the seawater to the interior chamber, and at least one outlet for providing filtered water to exit the interior chamber. A sand filter is disposed in the filter housing, separating the exterior from the interior chamber. The filter housing has at least one water conduction outlet conduit for allowing filtered water to exit the interior chamber to provide filtered water.

The inlets for providing seawater may provide for a surface intake velocity of <0.5 ft/s to restrict incursion of fish larva and macro or microvertebrae. The filter anchor may be of a size to permit container transportable via truck transportation.

The interior chamber of the filter anchor may be substantially filled with clean, washed, coarse sand, from either a local beach or shoreline source or from sand obtained from a commercial sand source. The filter housing may have hatches between the exterior and the interior chamber which, when opened, provide for submersion of the filter housing via flooding of the interior chamber and controlled sinking of the filter anchor to the sea floor. The filter anchor, prior to use as a filter, may be floatable and towable to a deployment site in the sea. At least one submersible pump and submersible air snorkel may be included such that the filter anchor is refloatable when the hatches are in a closed position, wherein the interior chamber is substantially filled with air, wherein the submersible pump and air snorkel can be activated to float the filter anchor.

A wave energy conversion system is also provided that includes an articulated-barge system for converting wave energy into energy used to pump water to a desalination system which generate potable water. At least one filter anchor is included. Each filter anchor includes a filter housing and a filter disposed therein for filtering seawater prior to entry into a water desalination system for placement on a sea floor. The filter housing has an exterior and an interior chamber, at least one inlet for providing the seawater to the interior chamber, and at least one outlet for providing filtered water to exit the interior chamber. The filter is disposed in the filter housing, separating the exterior from the interior chamber. The filter housing has at least one water conduction outlet conduit providing for filtered water to exit the interior chamber to provide filtered water to the desalinization system on the articulated barge. A mooring buoy is attached to each filter anchor by a mooring line. The desalinization system may include an RO membrane. The filter may be a sand filter. The filter anchor may include at least one feed line in the interior chamber to provide the filtered water to the water conduction outlet conduit. The filter housing may be constructed using a steel sheet. At least one inlet may be a manually controlled hatch or an automatically controlled hatch.

A method of anchoring a wave energy conversion system and providing filtered water to a desalination system is also provided. The method includes towing an articulated barge for converting wave energy into energy used to pump water to generate potable water to a location in a sea and towing at least one filter anchor to the location and sinking all of the at least one filter anchors to the sea bed. Each filter anchor includes a filter housing and a filter disposed therein for filtering seawater prior to entry into a water desalinization system for placement on a sea floor. The filter housing has an exterior and an interior chamber, at least one inlet for providing the seawater to the interior chamber, and at least one outlet for providing filtered water to exit the interior chamber. The filter is disposed in the filter housing, separating the exterior from the interior chamber, the filter housing has at least one water conduction outlet conduit for filtered water to exit the interior chamber to provide filtered water to the desalination system on the articulated barge.

The method also includes the steps of providing a mooring buoy for each filter anchor at the location, attaching each mooring buoy to one of the filter anchors by a mooring line, attaching each filter anchor to the articulated-barge system, and supplying filtered water to the articulated-barge system. The desalinization system may use a reverse osmosis membrane. The method may include the step of intaking seawater having an intake velocity of <0.5 ft/s to restrict incursion of fish larva and macro or microvertebrae. The method may include the step of transporting the filter anchor via a highway. The method may include opening a plurality of filter hatches located between the exterior and the interior chamber to submerse of the filter housing via flooding of the interior chamber and controlled sinking of the filter anchor to the sea floor. The step of towing the filter anchor to a deployment site in the sea may be included. The steps of closing the hatches and activating at least one submersible pump and submersible air snorkel to fill the filter anchor with air, to refloat the filter anchor can be included.

Besides the patent by Murtha, McCormick, and Washington [116,120,121], a new company called Atmocean Inc. [122] hopes to bring to market an ocean-wave-powered process that supplies potable water—cutting costs and limiting pollution, as compared to traditional desalination systems that require power plants. Atmocean Inc., Santa Fe, N.M., has conducted 29 sea trials since its founding in 2006. In collaboration with Sandia National Laboratory and Reytek Corp., both of Albuquerque, its recent trials were in the Pacific Ocean, off Peru, in 2015. Following modifications, the company tested a one-eighth-scale model at Texas A&M in February. Reytek is preparing a full-scale test unit for deployment in 2017 off Marystown, Newfoundland.

7.4 DESALINATION WITH SMALL MODULAR NUCLEAR REACTORS

There are several reasons why use of nuclear heat from small modular nuclear reactors for desalination makes sense [123–162]:

1. the overall cost of fossil heat generation is being dominated by the cost of the fuel itself
2. the current trends in fossil-fuel prices and supply uncertainties
3. concerns about green house gas emissions
4. the new generation of modular nuclear power plant (NPP) systems have highly enhanced safety levels and competitive economics.

The coupling of desalination with nuclear systems is not technically difficult [127,128,130,135–147] but needs the following considerations:

- avoiding radioactivity cross-contamination
- providing backup heating energy sources in case the nuclear system is not in operation (e.g., for refueling and maintenance)
- incorporating certain design features in case the thermal desalination option is used.

At present most desalination plants use fossil fuels and to a lesser degree clean energy. While for some time nuclear energy has been used for desalination processes; the enormous potential of nuclear energy is as yet not fully explored. It can generate much more freshwater than what is being currently produced, and also it is more affordable and does not release greenhouse gasses. Nuclear desalination is generally cost-competitive compared with the desalination using fossil fuels.

The IAEA [127,128] says that only nuclear reactors are capable of delivering copious quantities of energy which will be required for desalination projects of the future. Along with desalination of brackish or seawater, treatment of urban wastewater will be increasingly important and undertaken. The processes used to desalinate water have been described earlier. Any process that produces freshwater requires a lot of energy, which raises costs. The cost of water, however, varies significantly in different areas of the world. According to the United States Environmental Protection Agency (EPA), it is estimated that tap water costs an average of $2 per 1,000 gallons. However, the desalination project at Coquina Coast in Florida calculates the cost at $6.27–7.74 per 1,000 gallons. The difference in cost depends on the source of energy used in the process. With nuclear energy it is possible to achieve an enormous economy of scale, which lowers the cost. A study carried out in Tunisia discovered that the costs of nuclear desalination were from one-third to less than half of those related to desalination via fossil fuels, depending on the desalination technology that is used [123–162].

At present, desalination plants mostly use fossil fuels, which contributes to increase the levels of greenhouse effect gases. In December 2015, the initiative "Global Clean Water Desalination Alliance—H_2O minus CO_2" was presented at the COP21 climate talks in Paris, with a call to its 17 member countries to use clean energy sources in their new desalination plants. Several countries have implemented nuclear desalination, including India, Japan, and Kazakhstan. The latter operated a 750-MW thermal power plant for over 25 years. This plant not only generated desalinated water but also produced heat and electricity. Desalination of water with nuclear energy is an established, proven, and known technology with thousands of men hours.

As mentioned earlier, there are a number of processes that have been demonstrated for producing clean water from seawater; however, global experience is dominated by three primary processes: two distillation-based technologies [MED and MSF] and one membrane-based technology [RO] [6]. All seawater desalting processes—MSF, MED, and SWRO—consume significant amounts of energy and materials. In view of the rising fuel costs, the amount and cost of fuel consumed to desalinate seawater are some of the main factors determining the operational cost of water desalination. Similarly, the materials selected and the increased cost of materials for desalination have a significant impact on the capital cost. These rising costs, in turn, become a major factor in choosing the method and technology to be used [159]. The RO process employs a pressure-driven separation technique where water is forced under high pressure through a water-permeable membrane. No heating or phase change takes place. The main energy is the electricity required for the initial pressurization of the feedwater, 5–7 MPa for seawater or 2–3 MPa for brackish water. The advantages are simple processing, low installation, and maintenance costs. The drawbacks are the necessary pretreatment of the feedwater, the short lifetime of the membranes, and the comparatively high content (1%–2%) of salt passing through the membrane [158,159].

Another key distinction in the three methods is the way that they couple with a power source. The RO plant has the most straightforward coupling since it can operate using only electricity, which is needed to run the high-pressure pumps. Therefore, it is not essential to colocate the desalination plant with the power plant so long as a grid connection is available. However, there may be an advantage for colocation of the power and RO desalination plant in terms of shared infrastructure and protection against grid disruption. Also, low-grade steam or warm wastewater from the power plant can be used to preheat the saline feedwater of the RO plant to improve its clean water production efficiency, although the quality of the distillate may be adversely impacted. Both MED and MSF plants require a thermal heat source, such as a steam line from the secondary side of the nuclear plant. This steam is typically extracted from a low-pressure turbine stage, which results in a commensurate decrease in the electrical output of the power plant and may have implications on the reliability and flexibility of operations for both the power plant and the desalination plant. Also, the use of a tertiary heat transport loop is typically required to ensure that no radionuclides such as tritium are carried over from the reactor's secondary loop to the distillation plant [123–162]. The choice of desalination method(s) is determined primarily by the characteristics of the source water and the water quality required by the end user. For example, RO technology typically has a lower capital cost but is less effective with feedwater that contains high level of organic materials that can foul the membranes or that have high salinity levels and can only produce potable water without further treatment. The two thermal distillation processes are much more tolerant of "dirty" or "salty" feedwater and produce high purity water.

7.4.1 GLOBAL LANDSCAPE OF MODULAR NUCLEAR REACTORS WITH DESALINATION

Small nuclear reactors, known as Small Modular Reactors, are the most suitable kind for desalination, often with electricity cogeneration with low-pressure steam from the turbine and hot seawater feed from the final cooling system. A small reactor can produce between 80 and 100,000 m^3 a day. As seawater desalination techniques improve and more countries opt to build double-purpose electric plants (cogeneration), nuclear plants will need more advanced technologies plants, as well as more economical and efficient desalination systems.

A mini review of small/medium nuclear reactors (SMRS) for potential desalination applications is recently published by Ahmed et al. [125]. This review suggests that SMRs are a promising alternative for powering large-scale desalination plants. The modern generations of these systems manifest cost-effectiveness and built-in safety features. The compatibility with geological and topological challenges is an added advantage. Moreover, funding opportunities and packages could be easily arranged for SMR. In U.S. modular NuScale reactor is preferred to be used with various desalination technologies. The NuScale small modular reactor design is, especially well suited to support water desalination due to its high degree of modularity, enhanced safety and robustness, and flexible plant design. The NuScale plant can easily and effectively couple to a variety of desalination technologies, and can be economically competitive for simultaneously producing clean electricity and clean water [135–147].

Worldwide, the accumulated operating experience of nuclear desalination has exceeded 200 reactor-years. All nuclear reactor types can provide the energy required by the various desalination processes. However, SMRs may offer, when available, the largest potential as coupling options to nuclear desalination systems in developing countries [127,128,132,157,161]. The development of innovative reactor concepts and fuel cycles with enhanced safety features as well as their attractive economics are expected to improve public acceptance and further the prospects of nuclear desalination. As mentioned above, nuclear heat assisted modular desalination plants have been operated in Kazakhstan and Japan for many years. In Aktau, Kazakhstan, the liquid metal cooled fast reactor BN-350 has been operating as an energy source for a multipurpose energy complex since 1973, supplying regional industry and population with electricity, potable water, and heat. The complex consists of a nuclear reactor, a gas and/or oil fueled thermal power station, and MED and MSF desalination units. The seawater is taken from the Caspian Sea. The nuclear desalination capacity is about 80,000 nvVd. A part of this capacity has now been decommissioned. In Japan, all of the NPPs are located at the seaside. Several NPPs of the electric power companies of Kansai, Shikoku, and Kyushu have seawater desalination systems using heat and/or electricity from the nuclear plant to produce feedwater for the steam generators and for on-site supply of potable water. MED, MSF, and RO desalination processes are used. The individual desalination capacities range from about 1,000–3,000 nvVd. The experience gained so far with nuclear desalination is encouraging. There are nine nuclear units in Japan and one in Kazakhstan as mentioned above. In Japan, the desalination plants are constructed on-site at the NPPs with the aim of supplying the required makeup cooling water to these NPPs. Such desalination plants have clear-based desalination plant in the world [160]. The optimization of water desalination using nuclear reactors has been analyzed [162].

Small and medium-sized nuclear reactors are suitable for desalination, often with cogeneration of electricity using low-pressure steam from the turbine and hot seawater feed from the final cooling system. The main opportunities for nuclear plants have been identified as 80–100,000 m³/day and 200–500,000 m³/day ranges. The U.S. Navy nuclear powered aircraft carriers reportedly desalinate 1,500 m³/day each for use onboard. A 2006 IAEA report based on country case studies showed that costs would be in the range 50–94 c/m³ for RO, 60–96 c/m³ for MED, and $1.18–1.48/m³ for MSF processes, with marked economies of scale. These figures are consistent with later reports. Nuclear power was very competitive at 2006 gas and oil prices. A French study for Tunisia compared four nuclear power options with CC gas turbine and found that nuclear desalination costs were about half those of the gas plant for MED technology and about one-third less for RO. With all energy sources, desalination costs with RO were lower than MED costs. At the April 2010 Global Water Summit in Paris, the prospect of desalination plants being colocated with NPPs was supported by leading international water experts. As seawater desalination technologies are rapidly evolving and more countries are opting for dual-purpose integrated power plants (i.e., cogeneration), the need for advanced technologies suitable for coupling to NPPs and leading to more efficient and economic nuclear desalination systems is obvious [127,128,135–147].

The IAEA Coordinated Research Program (CRP) New Technologies for Seawater Desalination using Nuclear Energy was organized in the framework of a Technical Working Group on Nuclear Desalination that was established in 2008. The CRP ran over 2009–2011 to review innovative technologies for seawater desalination, which could be coupled to main types of existing NPP. The CRP focused on low-temperature horizontal tube MED, heat-recovery systems using heat pipe based heat exchangers, and zero brine discharge systems.

An IAEA preliminary feasibility study on nuclear desalination in Algeria was published in 2015, for Skikda on the Mediterranean coast, using cogeneration. The nuclear energy option was very competitive compared with fossil fuels. The feasibility of integrated nuclear desalination plants has been proven with over 150 reactor-years of experience, chiefly in Kazakhstan, India, and Japan [127,128,135–147].

Large-scale deployment of nuclear desalination on a commercial basis will depend primarily on economic factors. Indicative costs are 70–90 c/m^3, much the same as fossil-fueled plants in the same areas. One strategy is to use power reactors which run at full capacity, but with all the electricity applied to meeting grid load when that is high and part of it to drive pumps for RO desalination when the grid demand is low. The BN-350 fast reactor at Aktau, in Kazakhstan, successfully supplied up to 135 MW of electric power while producing 80,000 m^3/day of potable water over 27 years, about 60% of its power being used for heat and desalination (see Figure 7.3). The plant was designed as 1,000 MW but never operated at more than 750 MW, but it established the feasibility and reliability of such cogeneration plants. (In fact, oil/gas boilers were used in conjunction with it, and total desalination capacity through ten MED units was 120,000 m^3/day.) In Japan, some ten desalination facilities linked

FIGURE 7.3 The Shevchenko BN-350, a former nuclear-heated desalination unit in Kazakhstan [26].

to pressurized water reactors operating for electricity production yield some 14,000 m³/day of potable water, and over 100 reactor-years of experience have accrued. MSF was initially employed, but MED and RO have been found more efficient there. The water is used for the reactors' own cooling systems [127,128,135–147,157].

In 2002 a demonstration plant coupled to twin 170 MW nuclear power reactors (pressurized heavy-water reactor [PHWR]) was set up at the Madras Atomic Power Station, Kalpakkam, in southeast India. This hybrid Nuclear Desalination Demonstration Project () comprises an RO unit with 1,800 m³/day capacity and an MSF plant unit of 4,500 m³/day costing about 25% more, plus a recently added barge-mounted RO unit. This is the largest nuclear desalination plant based on hybrid MSF-RO technology using low-pressure steam and seawater from a nuclear power station. In 2009, a 10,200 m³/day MVC plant was set up at Kudankulam to supply freshwater for the new plant. It has four stages in each of the four streams. A low-temperature nuclear desalination plant uses waste heat from the nuclear research reactor at Trombay has operated since about 2004 to supply makeup water in the reactor. Pakistan commissioned a 4,800 m³/day MED desalination plant in 2010, coupled to the Karachi NPP (KANUPP, a 125 MW PHWR) near Karachi, though in 2014 it was quoted as 1,600 m³/day. South Korea has developed a small nuclear reactor (SMART) design for cogeneration of electricity and potable water. The 330 MW SMART reactor (an integral PWR) has a long design life and needs refueling only every 3 years. The main concept has the SMART reactor coupled to four MED units, each with TVC (MED-TVC) and producing 40,000 m³/day in total, with 90 MW. Argentina has designed an integral 100 MW PWR (CAREM) suitable for cogeneration or desalination alone, and a prototype is being built next to Atucha. A larger version is envisaged, which may be built in Saudi Arabia.*NHR-200:* China's Institute of Nuclear and New Energy Technology (INET) has developed this, based on a 5 MW pilot plant [127,128,133,135–147,155,157].

In the Middle East, a major requirement is for irrigation water for crops and landscapes. This need not be potable quality, but must be treated and with reasonably low dissolved solids. In Oman, the 76,000 m³/day first stage of a submerged membrane bioreactor (SMBR) desalination plant was opened in 2011. Eventual plant capacity will be 220,000 m³/day. This is a low-cost wastewater treatment plant using both physical and biological processes, which produces effluents of high-enough quality for some domestic uses or reinjection into aquifers. In Australia AGL plans to install a 2,000 m³/day RO desalination plant to treat water from fracking in its Gloucester coal seam gas project. This will be used for irrigation, rather than being potable quality. The Australian Commonwealth Scientific and Industrial Research Organization has found that the addition of nutrients could make desalinated water more financially attractive to farmers, who normally pay 20 c/kl for irrigation water, compared to most desalinated groundwater which costs more than $1/kl. In South Africa, Veolia is building a 1,700 m³/day seawater desalination plant at Lamberts Bay, upgradable to 5,000 m³/day. This will be the seventh plant along the west and south Cape coasts. A 450,000 m³/day plant costing $1.23 billion is planned for Koeberg, near Cape Town. Spain is building 20 RO plants in the southeast to supply over 1% of the country's water. Spain has 40 years of desalination experience in the Canary Islands, where some

1.1 million m³/day is provided. Tunisia is looking at the feasibility of a cogeneration (electricity-desalination) plant in the southeast of the country, treating slightly saline groundwater [127,128,133,135–147,155,157].

For seawater desalination and electricity generation, Russian design organizations have developed and can supply to customers floating nuclear power/desalination stations based on the small modular KLT-40 type reactor plant derived from Russian icebreakers. ATETs-80 is a twin-reactor cogeneration unit using KLT-40 and may be floating or land-based, producing 85 MW plus 120,000 m³/day of potable water. The small ABV-6 reactor is 38 MW thermal, and a pair mounted on a 97 m barge is known as Volnolom floating NPP, producing 12 MW plus 40,000 m³/day of potable water by RO.

The industrial enterprises of Russia are working on the development of a floating nuclear cogeneration plan for the country's northern regions. This can serve as a prototype for the floating nuclear power station (FNPS) desalination complex. The electric energy generated by the FNPS is partially transmitted to the ship for seawater desalination and its excess is used for supply to coastal users. The KLT-40 reactor has been successfully operated for many years in Russian nuclear ships and has been modified for each new ship generation based on accumulated experience. The KLT-40 reactor plant was the winner among plants of the same power level at a competition "Small Nuclear Power Stations-91" held by the Russian Federation Nuclear Society. Depending on customer requirements, the power/desalination station may be a one-reactor (APWS-40 type) or a two-reactor (APWS-80 type) design; hence, the reactor thermal power is within 80–160 MW. Advantages of floating power/desalination complexes are [134,156]

- convenient maintenance by a floating base at a mooring site and decommissioning by tugging to the Supplier's country;
- commercial production and long-term confirmation of service life characteristics of the KLT-40-type reactor plants and desalination units;
- possibility of installation in different coastal regions of the world;
- high fabrication quality at a shipyard and "turnkey" delivery to the customer in a short period of time.
- For any of the options, the supply of electrical energy to users within the plant is provided by the shipboard electric power station

Excess electricity can be used either for the production of additional desalinated water using RO or for sale. The cost of desalinated water can be materially reduced by compensating part of the production costs out of the profits from sale of electric energy. Construction of desalination plants using distillation plants with film-type horizontal evaporators seems to be the most practical at this time. The use of back pressure turbines as the source of thermal energy results in some excess electric energy in relation to the power consumed for the station's own needs. In order to reduce the probability of radioactive contamination of the desalinated water, two intermediate circuits are provided in the station design. One of these is pressurized to a higher pressure than the reactor side circuit. As previously stated, the most economical approach is seawater desalination using RO. Therefore, in parallel with the APWS-80 development, a floating complex for the production of electricity and seawater desalination using RO was developed. It is proposed that this system uses high thermal efficiency condensing turbines in conjunction with the desalination units. The complex includes two floating structures: an FNPS and a ship for seawater desalination using RO. The FNPS is a special non-self-propelled ship for the production of electricity. It is designed for use in a protected aquatorium or harbor [134,135–147,156] (Figure 7.4).

The arrangement, which separates power generation and desalinated water production, has certain advantages over an arrangement in which they are combined on one floating structure. It simplifies a solution to the problem of preserving a high efficiency of desalinated water production from the complex when the reactors are shutdown by supplying the desalination plant with electric energy from the external grid. The scheme seems to be sufficiently flexible since it allows the optimal ratio for production of the required amounts of water and electric energy. It is intended for use in a protected water area, together with a complex of external servicing structures. The reactor operates at a reduced power of ~80 MWth. If each reactor was to be operated at its full power of 160 MWth, desalinated water capacity and electric power generation would be doubled. Besides the reactor plant, the station includes the desalination plant, a drinkable water production plant, and ship general systems. As the desalination plant, four distillation plants with film-type horizontal-tube evaporators are used in APWS-80. Many years of experience exist using analogous plants for an industrial complex in Aktau (Kazakhstan).

These plants meet international safety requirements for marine NPPs and the requirements of Russian regulatory codes for NPPs, including IAEA recommendations [127,128]. The rapid maneuvering characteristics of the ship's reactor plant

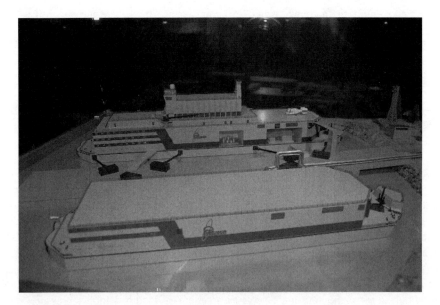

FIGURE 7.4 Model of the Project 20870 (back) with a desalination unit (front)—Russian floating nuclear desalination [163].

Desalination and Wastewater Treatment 363

allow it to closely follow power demand, and thereby to ensure highly economical operation. The ship for RO desalination is a non-self-propelled structure housing systems and equipment providing for the supply of seawater, its pretreatment, desalination, supply of desalinated water to users, and cleaning of the desalination units. With the objective of optimizing the technical and economic characteristics of the complex, some variations with different desalinated water output were considered. The desalinated water production plant can also be arranged on shore in the vicinity of the FNPS. The advantage of having the desalination plant arrangement on a ship is that there is the possibility of plant manufacture and testing at a shipyard in the Supplier country [135–147,155,157].

At present new technologies are being developed for seawater desalination using RO. For example, in the Canadian CANDESAL desalination program the use of RO technology is accompanied by preheating of the seawater in the turbine condenser. Preheating allows considerable reduction in the specific power consumption for desalination and in the cost of desalinated water. In this connection it seems beneficial to develop a joint Canada–Russian project for floating nuclear power/desalination complex, with the FNPS based on the KLT-40 shipboard reactor plant and the new application of RO seawater desalination. For floating nuclear reactors has high potentials due to possibility of installation in different coastal regions of the world; high fabrication quality at a shipyard and "turnkey" delivery to the customer in a short period of time. The supply of electrical energy to users within the plant is provided by the shipboard electric power station [127,128,135–147,155–157,164].

7.5 MODULAR SYSTEMS FOR WASTEWATER TREATMENT

For many municipal governments, drinking water and wastewater plants are typically the largest energy consumers and account for 30%–40% of total energy consumed. Nationwide, potable water and wastewater systems account for ~2% of energy use in the United States. This adds over 45 million tons of greenhouse gases to our atmosphere annually. A 2013 study by the Electric Power Research Institute and Water Research Foundation reported that treatment plants consumed about 30 billion kWh of electricity per year, or about 0.8% of the total electricity used in the United States. In California, it takes ~475–1,400 kWh of energy to treat 300,000 gallons of wastewater [165–167]. A typical wastewater treatment plant is illustrated in Figure 7.5 [168].

Wastewater treatment involves four main steps: Primary treatment, Secondary treatment, Anaerobic digestion, and the Disinfection phase. Primary treatment separates solids from liquid waste, while Secondary treatment has bacteria break down dissolved waste that contains ammonia and other pollutants. The remaining solids are then separated from the treated liquid. An anaerobic step occurs when solids from the primary and secondary steps are digested by microorganisms in a sealed tank without oxygen. Then finally, the Disinfection phase. Bacteria plays a key role in breaking down our sewage and industrial wastewater by consuming organic pollutants and inorganic nutrients such as ammonia. However, they require certain conditions to thrive: optimal temperature, food, and oxygen. Getting bacteria, the oxygen they need requires one-quarter to half of the energy used by a wastewater

FIGURE 7.5 Wastewater treatment plant in Cuxhaven, Germany [168].

treatment plant. Traditional plants pump air into the tank where the bacteria do their work, diffusing it through small holes to create oxygen bubbles the bacteria can access. This process wastes a lot of energy because most the bubbles rise to the top and pop without the bacteria using them [165–167].

Ironically, the organic matter found in wastewater contains up to five times as much energy as the treatment plants use. Despite all the energy seemingly there for the taking, reducing fossil-fuel demand of treatment plants is challenging and requires myriad approaches. Utilizing energy-efficient solutions could save money and reduce those greenhouse gas emissions. This subject was treated in detail in Chapter 2. Today, the industry is experimenting with new technologies and evaluating them as energy and cost reduction measures. For example, AOS PredictEnergy Analytics Software allows effective visualization for plant operators, enabling the process of wastewater at the lowest cost and creates wastewater energy cost savings. PredictEnergy's analytics platform with near real-time data, provides a level of visibility into plant health like never before and further reducing costs by eliminating unnecessary downtime. Predict Energy is using a variety of technologies to reduce the electricity they use through energy efficiency and to generate electricity on-site to offset [165–167].

Wastewater treatment energy consumption varies considerably throughout the country. There are several reasons for this, including variations in climate, equipment efficiency, types of pollutants, and specific energy practices. There are many benefits of limiting energy use and ways you can reduce your energy consumption. It's important that your facility is accountable for its water treatment energy consumption. Preliminary treatment, sedimentation, chlorination, and processing sludge are some of the processes that each takes different amounts of power. There are also other miscellaneous uses of energy to consider, such as light usage. Energy Star reports that energy usage can range from over 50k British thermal unit/gallon each day to <5. Energy consumption of wastewater treatment plants is often determined by the type of pollutants in the water, how much of these pollutants are present, and the methods used to remove them. For example, trickle filtration will normally use less energy, while nutrient removal will use higher amounts of energy. Nutrient removal, however, is often a necessary process.

7.5.1 Benefits of Reducing Water Treatment Plant Energy Consumption

According to the EPA, there are several benefits of reducing energy consumption in a wastewater treatment plant [165–167,169].

- **Increased Life of Equipment and Infrastructure**—By installing more energy-efficient methods, energy consumption can be reduced and therefore extend the life of the equipment. The need for maintenance will be limited as well.
- **Reduced Pollution and Emissions**—Energy reduction can reduce GHG emissions and limit the amount of air pollution that your facility creates. The release of pollutants can be limited by reducing the use of fossil fuels.
- **Reduced Energy Costs**—Wastewater facilities use a significant portion of a community's energy resources. Overall costs can be reduced when your facility runs more efficiently and uses less energy.
- **Improved Security**—Greater energy efficiency can also help prevent blackouts and brownouts by limiting the electrical demand in a community. It can also reduce the potential for a water shortage.
- **Improved Public Health**—Limiting energy consumption can reduce levels of water and air pollution. Increasing overall efficiency can also reduce the risk of water-related illnesses.

The National Renewable Energy Laboratory (NREL) states that there are several ways to reduce the amount of energy consumption in wastewater treatment facility.

Conduct an Energy Audit—The first step toward reducing energy use is to do an audit to find out exactly what the energy expenditures are for your plant.
Create a Plan—It's important to create an industrial water treatment plan that is unique to your specific facility. After identifying objectives, it is time to put together an action plan that will need the approval of management and

the implementation of training programs. The following are several steps the plan could include:

- **Install New Equipment**—Older equipment is more expensive to repair and usually doesn't operate as efficiently as newer models. It's essential to have the most energy-efficient equipment possible and to keep up a schedule of preventative maintenance.
- **Improve Pump Optimization**—Pumping processes normally use a lot of energy and provide opportunities for savings. It's important to know which blowers or pumps to use, and how often to use them when conserving energy.
- **Implement Renewable Energy**—Employing sources of on-site renewable energy is an excellent way to save on energy consumption while protecting the environment. Solar, water, and wind are potential options.
- **Monitor and Maintain Improvements**—You'll need to evaluate what's working and what's not on a regular basis, making improvements where needed.

There are a number of modular systems in the market that can treat wastewater cost-effectively. Here we examine few of them.

7.5.2 Modular Systems for Energy Usage-Water Treatment-AdEdge Technology

EPA regulations for contaminant removal have placed an unprecedented challenge upon small public water systems to cost-effectively meet safe drinking water compliance objectives. Many small potable water systems such as schools, day care centers, housing developments, and manufacturing facilities need options that are simple, affordable, and effective. Additionally, solutions are needed that require less engineering by adapting to existing space—and infrastructure.

AdEdge Water Technologies' answer to this market need is a line of modular water treatment systems designed specifically for these applications [169–171].

AdEdge Water Technologies has successfully deployed many systems in the field, accumulating an unparalleled track record of performance. Similar to the skid-mounted AdEdge Package Units (APUs), these compact modular systems can use various AdEdge media types to remove arsenic, iron, manganese, uranium, radium, nitrates, and other contaminants. Systems are shipped ready for field assembly and installation by a qualified contractor or installer. Modular systems are available in various flow rates and configurations.

The AdEdge modular system for wastewater treatment offers the following benefits [169–171]:

- AdEdge offers a complete prepackaged modular system with automated valves, flowmeter, pressure gauges, and flow controls for simple permitting and installation.
- AdEdge will work closely with customer to size, design, and offer value-added engineering assistance and permitting support.

Desalination and Wastewater Treatment

- Performance guarantee resulting in reduced liability; AdEdge's extensive network of field installations (now over 2,000 installations) allows one to predict and stand behind the performance of our systems and media.
- Small footprint and portability to fit in tight spaces where limited room is available.

The AdEdge modular system can be used for [169–171] nontransient, noncommunity water systems, schools, churches, day care centers, retirement homes, manufacturing facilities, housing developments, rest stops, environmental remediation, industrial process applications, and industrial parks.

7.5.3 Newterra Advantages for Wastewater Treatment

Newterra modular wastewater treatment plant offers the following benefits [172,173]:

1. The modular "train" designs of newterra portable water treatment and sewage systems allow additional treatments units to be added or removed from a system quickly and efficiently to match actual requirements. This ability to scale a system up and down provides flexibility and cost efficiencies.
2. Newterra modular systems are factories built in ideal conditions by an experienced production team. Every portable water treatment system undergoes rigorous quality assurance process, including comprehensive hydraulic, electrical, and controls testing.
3. Newterra water treatment system has go-anywhere portability. ISO shipping containers are the building blocks of newterra portable water treatment systems. These rugged enclosures protect advanced treatment equipment from the elements and simplify transportation. The modular systems can be transported by land or sea or even on ice roads of north.
4. By engineering and prebuilding "universal" components of the system, newterra focuses on the site-specific elements of modular treatment solutions and compress delivery times.
5. On-site construction is costly and vulnerable for delays. In remote areas, work crew availability and logistics are significant challenges. Newterra eliminates these concerns. The modular solutions arrive on-site prewired, preplumbed, and pretested for quick setup and operation in days.
6. Newterra system is easily relocatable. At the conclusion of a project or when capacity requirements warrant, the full system or treatment modules can be redeployed to other sites.

7.5.4 Envirogen Modular Wastewater Treatment

Envirogen Technologies, Inc.'s (Envirogen's) modular, transportable process alternative for the treatment of high-strength organic waste streams at the source combines low cost with high performance removal of contaminants. Called TIMBR™

(Transportable Internal Membrane Biological Reactor), the system combines features that include the following [174,175]:

- Low capital and operating costs.
- Small footprint, allowing the system to be installed at the source of the waste generation.
- Tolerant of influent variations, making it ideal for treating wastewater from either continuous or batch processes.
- Modular, transportable, prefabricated design minimizes installation cost and time of installation. Additional modules are available for separate metals precipitation, nutrient removal, biosolids treatment, and disinfection.

7.5.4.1 Membrane Biological Reactors

TIMBR is a modular configuration of Envirogen's Membrane Biological Reactor (MBR) technology. An MBR system consists of a suspended-growth bioreactor combined with a membrane liquids/solids separation unit. In an MBR, contaminated water is first fed to the suspended growth reactor for degradation of the organics in the stream. The TIMBR system maintains a high biomass concentration (typically 10,000–30,000 mg/L), which yields unusually high destruction efficiencies, even for recalcitrant compounds. The membrane is the key to maintaining high biomass, allowing treated water to pass from the reactor, while keeping solids in the reactor for a longer, controlled length of time [176].

TIMBR was developed in response to a need by the industry for compact, easily relocated, on-site wastewater treatment systems. In the TIMBR configuration of the MBR technology, Envirogen uses a membrane that is internal to the reactor; a proprietary feature that allows for cleaning the membrane without removing it from the bioreactor. The first TIMBR system was an 80,000-gallon-per-day plant designed to treat a combined municipal and industrial waste stream. The system was designed for the removal of high-strength organics, nutrients and metals. It was equipped with UV disinfection, allowing it to discharge to either surface or groundwater. The TIMBR system is the treatment of choice regardless of whether the discharge will be receiving a body of water or to a treatment facility.

The bioreactor includes a chamber that contains one or more internal membranes as required to meet plant conditions. Water from the bioreactor is drawn through the membranes separating the permeate from the mixed liquor volatile suspended solids. Compressed air is injected into the bioreactor to not only treat the organics but also to continuously flex the membranes and prevent them from becoming fouled. Fitted with a modem and an automated control system, the TIMBR system can be operated from a remote location, ensuring safe, reliable disposal of the treated wastewater, leachate, or groundwater stream. The system can be fitted with additional modules for the treatment of nutrients (phosphorous and nitrogen). Metals can be removed with the biosolids, or if it is necessary to have metal-free biosolids, a pretreatment module can be provided. The basic system, shown below, will treat 10,000 gallons per day of leachate containing 3,000 ppm organics. It consists of a self-contained $40' \times 8' \times 10'$ container requiring minimal site preparation and includes all tanks, pumps, blowers, and controls [174–176].

7.5.5 MODULAR TUBULAR MICROBIAL FUEL CELLS FOR ENERGY RECOVERY DURING SUCROSE WASTEWATER TREATMENT AT LOW ORGANIC LOADING RATE

In the study by Kim et al. [177] energy recovery while treating low organic loads has been investigated using longitudinal tubular microbial fuel cell (MFC) reactors. Duplicate reactors, each consisting of two modules, were operated with influent sucrose organic loading rates (OLRs) between 0.04 and 0.42 g chemical oxygen demand/l/day. Most soluble COD (sCOD) removal occurred in the first modules with predominantly volatile fatty acids reaching the second modules. Coulombic efficiency (CE) in the second modules ranged from 9% to 92%, which was three to four times higher than the first modules. The maximum energy production was 1.75 W h/g COD in the second modules at OLR 0.24 g/L/day, up to ten times higher than the first modules, attributable to a nonfermentable substrate. A simple plug flow model of the reactors, including a generic nonelectrogenic reaction competing for acetate, was developed. This modular tubular design can reproducibly distribute bioprocesses between successive modules and could be scalable, acting as a polishing stage while reducing energy requirements in wastewater treatment.

7.5.6 MODULAR PACKAGE, FIELD-ERECTED WASTEWATER TREATMENT PLANTS

As the need for improved effluent quality increases to meet regulatory requirements, Global Treat, Inc provides budgetary pricing and treatment selection, as well as design, execution, and startup of wastewater treatment plants. The use of modular, expandable plants to provide treatment of current wastewater flows is a cost-efficient solution for wastewater treatment. The company provides field-erected and modular package biological treatment plants. The package and modular biological wastewater plants are fabricated with all of the necessary components, including screening, aeration, mixing, clarification, activated sludge, recirculation, and aerobic sludge digestion [178].

Aeration is the first phase of Global Treat's Sludge-Activated Treatment process. Air diffusers are used to mix wastewater with microorganisms. These microorganisms, known as "sludge," remove contaminants through absorption of organic matter. After aeration is complete, the mixture of sludge and water move into the next zone called the Clarifier. In the Clarifier, microorganisms and organic matter settle, allowing leftover water to flow into the *chlorination tank*. Sludge is recycled and used to treat additional wastewater. During the disinfection phase, chlorine or UV light exposure disinfect wastewater before it is introduced back into the environment. Global Treat provides UV, chlorine gas, chlorine tablets, and both automatic and manual chlorine liquid feed systems [178].

7.5.7 EVOQUA MODULAR WATER TECHNOLOGIES

An industrial wastewater often requires treatment to remove solids, oil and grease, bacteria, regulated chemicals, and other substances before it can be reused or discharged to the environment [177,179–181]. Evoqua Water Technologies provides a complete range of modular solutions and services for treating industrial wastewater to meet process needs as well as environmental regulations. Their products include filtration, clarifying systems, biological treatment (anaerobic and aerobic), dewatering, disinfection, ion exchange, carbon filtration, water reuse, RO, and other technologies for removing impurities from water. They also provide customers with the engineering and knowhow that helps them choose the right approach, and support our solutions through North America's largest service network. They serve chemical processing, food and beverage, hydrocarbon processing, life science, metals and mining, microelectronics, pharmaceutical, and power industries.

7.5.7.1 Evoqua Innovative Modular Desalination Technology

Though brackish water salinity levels aren't as high as those of seawater, they're still too high for most applications as well as human consumption. In fact, a large percentage of groundwater is too salty for most applications. Therefore, treatment is necessary to reduce the saline levels so it's safe to use. Unfortunately, conventional desalination methods and treatments use hazardous chemicals to remove the tunable dissolved solids (TDS) [177,179–181]. Evoqua system, on the other hand, uses electrochemistry. They have reimagined ED and the process, creating the NEXED® module—a high-tech brackish water desalination system. Launched in January of 2016, this module has revolutionized the typical membrane desalination process with electrochemical desalination and, therefore, met and continues to meet the growing demands of brackish salt removal.

ED is a membrane desalination process in which ions are transported through selective ion-permeable membranes from one solution to another under the influence of an electrical potential gradient. Alternating ion selective membranes (anionic and cationic) create separated streams of concentrated and diluted feedwater that can be influenced by the electrical potential gradient and resulting current. The NEXED® module, for electrochemical desalination, uses membrane-based desalination technology to remove salt from brackish water more efficiently and effectively. The module is designed to cost, meaning it reduces the overall cost of desalination of brackish water (500–15,000 ppm NaCl) in a variety of real-world applications [177,179–181].

Commercially used since 1960, EDR is an effective membrane desalination process. During the process, an electric current transfers dissolved solids through an ED stack. The stack has an alternate layer of anionic and cationic ion exchange membranes that work together to separate naturally occurring dissolved solids from the water. EDR is commonly used for the desalination of brackish water with limited dissolved solids as well as water with high scaling potential. The NEXED module, an innovation in the world of EDR, offers a variety of advantages over conventional EDR, in addition to the modules modular and stackable design. Other benefits include (a) less silt fouling, (b) fewer operating costs, (c) less biofouling potential, (d) resistance to process upsets, (e) minimal space requirements, (f) real-time adjustment of salt removal, (g) optimization for low energy, high recovery, and small footprint, (h) reduced clean in place and maintenance, (i) cost-driven design and low

Desalination and Wastewater Treatment

lifecycle costs, (j) optimized membrane for electrochemical desalination, (j) tunable system, and (k) low-pressure operation with system only requires low-pressure piping, reducing the installment costs.

One of the biggest benefits of the NEXED module is its unique modular design. The design optimized for membrane utilization has a multitude of advantages. First the construction lends itself to automated manufacturing process. In turn this increases consistency and reduces cost simultaneously. Additionally, this design is more flexible than others. This flexibility allows one to create optimized solutions for every customer, regardless of their needs. This groundbreaking brackish water desalination system gives users the capability to reduce salinity levels without the use of chemicals. In addition, it requires less energy to operate. In fact, it operates at up to 30% less than other desalination methods. The NEXED module is the next generation of water desalination. In fact, it has been used for more than electrochemical desalination. What makes the module so special is the unique, innovative membrane, which is now used in applications other than brackish water desalination [177,179–181].

The technology used in the NEXED module provides a TDS removal capability. This adjustable output provides cost-effective treatment options, such as consistent water quality with variable feedwater parameters or partial removal of contaminants without the need for blending. Because output quality can be manipulated by input power adjustments, this tunable feature also allows for options to minimize footprint and provide for optimized energy consumption. NEXED modules were designed as desalting engines and have many potential applications including brackish water TDS reduction, RO reject recovery, small footprint bulk TDS removal, water reuse, and variety salinity applications.

Evaqua produces a HydroDI™ module, which is a chemical-free and environmentally friendly alternative to traditional water softening. Unlike RO systems, the HydroDI system can be adjusted to balance both hardness and salt removal. It uses Evoqua's groundbreaking NEXED electro desalination crossflow membrane for small flow-rate requirements as a salt-free softener and water conditioner. In fact, the typical flow rate is only 1 gpm (gallon per minute) and is used in a process designed to treat a whole household. Modules using the 7″ membrane can also be applied to commercial applications such as coffee shops to ensure consistent water quality. Evaqua also produces a module using 12″ membranes, which are optimized for higher salinity levels. Designed for municipal and industrial applications, these modules can be combined in parallel for larger flow rates in applications such as RO reject recovery [177,179–181].

7.5.8 Package Systems from ClearBlu

Treatment ponds for wastewater treatment are a common solution for facilities that have ample space available, but this solution is not realistic for many industrial wastewater generators. Space-constrained facilities need small footprint options that allow them to meet discharge requirements or enable water reclamation if desired. This is where ClearBlu Environmental's packaged wastewater treatment systems shine. The company implement our highly effective microbubble aeration technology into self-contained systems that are customized to the needs. These systems are highly desirable in situations where there is very little space available to dedicate to wastewater treatment, or if the facility has land, but simply cannot justify giving up valuable space that would otherwise generate revenue (planting additional vines at a winery, for example) [182,183].

Packaged wastewater treatment systems are highly customizable to fit the varying needs of facilities. The system may consist of some combination of biodigester(s) and/or polytank(s). A biodigester is a marine grade aluminum or stainless steel tank constructed with an over–under weir system and is packed with a fixed film media. They are the heavy lifters of a package system, as they have been shown to process waste up to ten times faster than an open body of water. Their unique design provides the ultimate environment for bacteria to thrive and process organic waste efficiently. The coalescing media provides a surface for bacteria to adhere to, which has been proven to increase their efficiency versus free floating in water. The design of the media provides thousands of square feet of surface area in a very small footprint. The weir system ensures a treatment path, guaranteeing that when the water is pumped through the system it passes by the maximum amount of BOD reducing bacteria possible. The biodigester is also supplied with microbubble aeration to provide the highly aerobic environment required for bacteria to efficiently digest the organic waste. In addition to biodigesters, polytanks are used to increase retention time in package systems. They are fitted with microbubble aerators, and often recirculation pumps are implemented to give the wastewater more than one pass through the system to maximize BOD reduction.

The benefits of ClearBlu packaged wastewater treatment systems are multifaceted [182,183]:

- **Small footprint:** Free up space for revenue-generating components or enable space-constrained facilities the option for treatment, where the possibility may not have existed otherwise.
- **Odor free:** ClearBlu offers robust automatic pH balancing systems that can be designed into the package system.
- **Easily expandable:** Reduce initial capital investment upfront by designing a system that meets the needs of current production. Systems can easily grow with the facility by adding tanks and/or biodigesters to meet the growing demand.
- **Relocation ability:** If the business relocates to a new facility in the future, the system can be drained, dismantled, and moved to the new location, protecting the capital investment for the long haul.

7.5.9 PCS Package Wastewater Treatment Plants

Pollution Control Systems (PCS) is recognized worldwide for its products and technologies to treat wastewater to a safe sanitary water effluent quality discharge, meeting and/or exceeding effluent standards recommended by the EPA [184,185].

PCS selection of a packaged treatment plant offers the user a pre-engineered and prefabricated method of treating wastewater with an aerobic process. The final effluent can be released safely into the environment, such as receiving streams, rivers, etc. Treated nonpotable water is also being used as a new source of water to promote agricultural and aquaculture production, industrial uses, water sustainability, and reclamation uses such as irrigation, wash down, and/or artificial recharge [184,185]. PCS technologies include Clarification, Chlorination/Disinfection, Extended Aeration (EA), Activated Sludge Process, Nitrification/Denitrification, Oil/Water Separation (OWS), and Tertiary Treatment. Its typical applications include Land development/housing/subdivisions; Small and medium-sized communities and cities; Mobile home parks; Remote mining, logging, and construction sites; Recreational areas such as parks, campgrounds, and marinas; Manufacturing facilities, power plants, military bases; Schools and other educational campuses; Government compounds; Low flow/high strength and high flow/low strength applications; and Biological wastewater treatment for industrial wastewater flows.

7.5.9.1 Advantages
- Pre-engineered, prefabricated structures result in lower cost
- Unit is easily transported to the customer's project site
- Design allows for quick turnaround time for delivery and installation
- Treatment system is simple to operate and requires low manpower
- Effective EA principle
- User friendly—low and easy maintenance
- Regulatory compliant
- Custom design/application-specific systems
- Long service life

PCS technologies is a modular system where pre-engineered modular components such as diffused air blowers, flow equalization tanks, aeration tanks, sludge holding tanks, wastewater clarifiers, and disinfection units allow for the package plants to be sized specifically for the customer's application. They can be designed to handle a variety of influent flow rates and BOD loadings to meet discharge requirements. A decentralized wastewater treatment system, commonly referred to as a "packaged plant," utilizes the biological EA principle of operation, which is a variation of the activated sludge treatment process. This system functions by creating an environment with sufficient oxygen levels and agitation to allow for bio-oxidation of the wastes to suitable levels for discharge.

Waste material in domestic wastewater is generally organic (biodegradable) which means that microorganisms can use this matter as their food source. A biological wastewater treatment system makes use of bacteria and other microorganisms to remove up to 95% of the organic matter in the wastewater. Biological wastewater treatment systems and processes were actually developed by observing nature. As waste enters the stream, the dissolved oxygen content in the water decreases and bacteria populations increases. As the waste moved downstream, the bacteria would eventually consume all of the organic material. Bacterial populations would then decrease, the dissolved oxygen in the stream would be replenished, and the whole process would be repeated at the next wastewater discharge point.

The influent wastewater enters the wastewater treatment package plant by passing through a comminutor and/or bar screen for gross solids removal. This step provides for the mechanical reduction of solids prior to aeration. Once the wastewater has entered the aeration chamber, the untreated flow is mixed with an active biomass in a rolling action that takes place throughout the length and width of the chamber in a slow forward progression. This rolling mixing action is the result of air originating from air diffusers located along one side of the bottom of the tank. This ensures that adequate mixing is maintained in the tank. The chambers are filleted on each side along the bottom to assure and enhance the rolling motion of the water and to eliminate any "dead zones" in the tank. The oxygen transfer achieved with the diffused air passing through the wastewater coupled with the rolling action provides sufficient oxygen supply allowing microorganisms to oxidize treatable wastes into carbon dioxide, water, and stable sludge.

After aeration, the wastewater flows to the clarifier that typically has a hopper bottom configuration. The wastewater clarifiers are sized to provide the required retention time based on an average 24-h design flow. During the settling period, solids settle at the bottom of the clarifier. Airlift pumps with adjustable pumping capabilities are used to return these solids, as activated sludge, to the aeration chamber to maintain the maximum efficiency of the biological process. When necessary, excess sludge is wasted to an aerated sludge digestion tank for additional treatment and reduction. A skimmer airlift pump is used to return floatable solids and scum to the aeration chamber for further processing. The treated water flows from the clarifier to a disinfection chamber for treatment via chlorination or UV disinfection prior to discharge to complete the treatment process. Tertiary filters may also be used where a higher quality of effluent is required.

REFERENCES

1. Global Water Desalination Market 2018–2025 Current Trends. A website report www.reuters.com/brandfeatures/venture-capital/article?id=48863 (28 August 2018).
2. Seawater Desalination Costs, by Water Reuse Association, White Paper by Water Reuse Association, September 2011; Revised January 2012, Seawater Desalination Costs – WateReuse. A website report https://watereuse.org/wp-content/uploads/.../WateReuse_Desal_Cost_White_Paper.pdf (2012).
3. Walton, M., Commentary: Desalinated Water Affects the Energy Equation in the Middle East, WEO Energy Analyst, IEA Paris, France. www.iea.org/.../desalinated-water-affects-the-energy-equation-in-the-middle-ea (21 January 2019).
4. Shahzad, M.W., Burhan, M., Ybyraiymkul, D., and Ng, K.C. (2019). Desalination processes' efficiency and future roadmap. *Entropy*, 21, 84. doi: 10.3390/e21010084, www.mdpi.com/journal/entropy.
5. Krishna, H.J. (2004). Introduction to Desalination Technologies: Texas Water Development. www.twdb.texas.gov/publications/reports/numbered_reports/doc/.../C1.pdf.
6. Desalination, Wikipedia, The free encyclopedia, last visited 23 July 2019 (2019).
7. Desalination Technology: An Overview | ScienceDirect Topics. A website report www.sciencedirect.com/topics/earth-and-planetary.../desalination-technology (2016).

8. Ghalavand, Y., Hatamipour, M.S., and Rahimi, A. (2014). A review on energy consumption of desalination processes. *Desalination and Water Treatment*. doi: 10.1080/19443994.2014.892837.
9. Lozier, J.C., Desalination Technology Overview. A website report by *Water Resources Research Center Conference*, Yuma, AZ. https://wrrc.arizona.edu/sites/wrrc.arizona.edu/files/programs/conf2011/.../Lozier.pdf (2011).
10. Chaudhry, S., (2013). An overview of industrial desalination technologies ASME Industrial Demineralization (Desalination). *Best Practices and Future Directions Workshop*, January 28–29, Washington, DC. https://community.asme.org/.../An-Overview-of-Industrial-Desalination-Technologies.
11. Hamed, O.A. (2005). Overview of hybrid desalination systems: Current status and future prospects. *Desalination*, 186 (1–3), 207–214; *Presented at the International Conference on Water Resources and Arid Environment*, 5–8 December 2004, Riyadh, Saudi Arabia. doi: 10.1016/j.desal.2005.03.095.
12. Modular Desalination System. A website report rustrade.org.uk/eng/wp-content/uploads/PROMTECHEXPORTDesalination.pdf (2019).
13. Chemical-Free Desalination Technology | IDE Technologies. A website report www.ide-tech.com › Home › Solutions › IDE PROGREEN™ Plants (2011).
14. IDE PROGREEN™ Plants | IDE Technologies. A website report www.ide-tech.com › Home › Solutions (2011).
15. Jacob, A., MIT Study Highlights Benefits of Smaller, Modular Desalination Plant. A website report www.filtsep.com/desalination/.../mit-study-highlights-benefits-of-smaller-modu (15 August 2017).
16. Modular, Scalable, Factory Tested Desalination Plants: CAP Holdings. A website report www.nexgendesal.com/modular-scalable-factory-tested-desalination-plants.html.
17. Small, Scalable, Community-Based Seawater Desalination: CAP. A website report www.nexgendesal.com/small-scalable-community-based-seawater-desalination.html (2012).
18. Modular Desalination: Small Packages, Big Energy Gains? | WaterWorld. A website report www.waterworld.com/.../desalination/.../modular-desalination-small-packages-b (18 July 2013).
19. Fresh Water for the Future Desalination Made Efficient: KSB. A website report www.ksb.com/blob/52744/.../fresh-technologie-en-data.pdf (2013).
20. Economic and Social Commission for Western Asia. (2001). Energy Options for Water Desalination in Selected ESCWA Member Countries. New York, United Nations.
21. CORDIS Database. (2006).
22. UNEP (United Nations Environment Program). (2003). Key Facts about Water. www.unep.org/wed/2003/keyfacts. Accessed 15/01/2006.
23. http://desaldata.com/.
24. World Health Organization. (1984). *Guidelines for Drinking Water Quality*, Vol. I. World Health Organization, Geneva.
25. Al-Karaghouli, A.A. and Kazmerski, L.L. (2011). Renewable Energy Opportunities in Water Desalination, National Renewable Energy Laboratory Golden, Colorado, USA.
26. Desalination, Wikipedia, The free encyclopedia, last visited 12 August 2019 (2019).
27. Thermal Solar Energy System Technology: An Overview. A website report www.sciencedirect.com/topics/.../thermal-solar-energy-system-technology (2015).
28. Ullah, I. and Rasul, M.G. (2019). Recent developments in solar thermal desalination technologies: A review. *Energies*, 12, 119. doi: 10.3390/en12010119, www.mdpi.com/journal/energies.
29. John, J.St., Can Modular Desalination Provide Cheap, Clean Water and Demand. A website report www.greentechmedia.com/.../can-modular-desalination-provide-cheap-clean-w (26 August 2014).

30. Al-Hayek, I. and Badran, O.O. (2004). The effect of using different designs of solar stills on water distillation. *Desalination*, 169, 121–127.
31. Economic and Social Commission for Western Asia. (2009). ESCWA Water Development Report 3. Role of desalination in addressing water scarcity, New York, United Nations.
32. Sodha, M., Kumar, A., Tiwari, G., and Tyagi, R. (1981). Simple multiple wick solar still: Analysis and performance. *Solar Energy*, 26, 127–131.
33. Fernandez, J.L. and Chargoy, N. (1990). Multi-stage, indirectly heated solar still. *Solar Energy*, 44(4), 215–223.
34. Al-Hinai, H., Al-Nassri, M.S., and Jubran, B.A. (2002). Effect of climatic, design and operational parameters on the yield of a simple solar still. *Energy Conversion and Management*, 43, 1639–1650.
35. Mathioulakis, E., Belessiotis, V., and Delyannis, E. (2007). Desalination by using alternative energy: Review and state-of-the-art. *Desalination*, 203, 346–365.
36. Bohner, A. (1989). Solar desalination with a higher efficiency multi effect process offers new facilities. *Desalination*, 73, 197–203.
37. Banat, F.A. (1994). Membrane distillation for desalination and removal of volatile organic compounds from water, Ph.D. Thesis, Mcgill University.
38. Qiblawey, H.M. and Banat, F. (2008). Solar thermal desalination technologies. *Desalination*, 220, 633–644.
39. Price, H. (2003). Assessment of Parabolic Trough and Power Tower Solar Technology Cost and Performance Forecasts, Sargent & Lundy LLC Consulting Group, National Renewable Energy Laboratory, Golden, CO. www.nrel.gov/solar/parabolic_trough.html.
40. Trieb, F. and Müller-Steinhagen, H. (2007). Concentrating Solar Power for Seawater Desalination in the Middle East and North Africa, Submitted to Desalination.
41. García-Rodríguez, L. (2003). Renewable energy applications in desalination: State of the art. *Solar Energy*, 75, 381–393.
42. Zachary, J. and Layman, C. (2010). Bechtel power corp. Adding desalination to solar hybrid and fossil plants. *Electric Power Journal*, 154(5), 1–6.
43. Al-Shammiri, M. and Safar, M. (1999). Multi-effect distillation plants: State of the art. *Desalination*, 126, 45–59.
44. Quteishat, K. and Abu-Arabi, M. (2006). Promotion of Solar Desalination in the MENA Region. Middle East Desalination Centre, Muscat, Oman. www.menarec.com/docs/Abu-Arabi.pdf. Accessed 28/03/2006.
45. Collares Pereira, M., Carvalho, M.J., and Correia de Oliveira, J. (2003). A new low concentration CPC type collector with convection controlled by a honeycomb TIM Material: A compromise with stagnation temperature control and survival of cheap fabrication materials. *Proceedings of the ISES Solar World Congress 2003*, 14–19 June, Göteborg, Sweden.
46. Water Corporation. (September 2000). A Strategic Review of Desalination Possibilities for Western Australia.
47. Zarza, E. and Blanco, M. (1996). Advanced M.E.D. solar desalination plant: Seven years of experience at the Plataforma Solar de Almería. *Proceedings of the Mediterranean Conference on Renewable Energy Sources for Water Production*, Santorini, Greece, p. 182.
48. Thomson, M., Gwillim, J., Rowbottom, A., Draisey, I., and Miranda, M. (2001). Battery Less Photovoltaic Reverse Osmosis Desalination System. Technical Report, S/P2/00305/REP, ETSU, DTI, UK.
49. Fiorenza, G., Sharma, V., and Braccio, G. (2003). Technoeconomic evaluation of a solar powered water desalination plant. *Energy Convers Manage*, 44, 2217–2240.

50. Kehal, S. (1991). Reverse osmosis unit of 0.85 m³/h capacity driven by photovoltaic generator in South Algeria. *Proceedings of the New Technologies for the Use of Renewable Energy Sources in Water Desalination Conference*, Session II, Athens, Greece, pp. 8–16.
51. Tzen, E. and Perrakis, K. (1996). Prefeasibility Study for Small Aegean Islands, RENA-CT94-0063, CRES, Greece.
52. Thomson, M. and Infield, D. (2002). A photovoltaic powered seawater reverse-osmosis system without batteries. *Desalination*, 153, 1–8.
53. Hafez, A. and El-Manharawy, S. (2003). Economics of seawater RO desalination in the Red Sea region, Egypt. Part 1: A case study. *Desalination*, 153, 335–347.
54. Perrakis, K., Tzen, E., and Baltas, P. (2001). PV-RO Desalination Plant in Morocco, JOR3-CT95-0066, CRES, Greece.
55. Finken, P. and Korupp, K. (1991). Water desalination plants powered by wind generators and photovoltaic systems. *Proceedings of the New Technologies for the Use of Renewable Energy Sources in Water Desalination Conference, Session II*, Athens, Greece, pp. 65–98.
56. Adiga, M.R., Adhikary, S.K., Narayanan, P.K., Harkare, W.P., Gomkale, S.D., and Govindan, K.P. (1987). Performance analysis of photovoltaic electrodialysis desalination plant at Tanote in the Thar desert. *Desalination*, 67, 59–66.
57. Kuroda, O., Takahashi, S., Kubota, S., Kikuchi, K., Eguchi, Y., Ikenaga, Y., et al. (1987). An electrodialysis seawater desalination system powered by photovoltaic cells. *Desalination*, 67, 161–169.
58. Lichtwardt, M. and Remmers, H. (1996). Water treatment using solar powered electrodialysis reversal. *Proceedings of the Mediterranean Conference on Renewable Energy Sources for Water Production*, Santorini, Greece.
59. Carta, J.A., Gonzalez, J., and Subiela, V. (2004). The SDAWES project: An ambitious R&D prototype for wind powered desalination. *Desalination*, 161, 33–48.
60. Maurel, A. (1991). Desalination by RO using RE (solar and wind): Cadarache Center Experience. *Proceedings of the New Technologies for the Use of RE Sources in Water Desalination*, 26–28 September, Greece, pp. 17–26.
61. Peral, A., Contreras, G.A., and Navarro, T. (1991). IDM—Project: Results of One Year's Operation. *Proceedings of the New Technologies for the Use of RE Sources in Water Desalination*, 26–28 September, Greece, pp. 56–80.
62. Miranda, M. and Infield, D. (2002). A wind powered seawater RO system without batteries. *Desalination*, 153, 9–16.
63. De La Nuez Pestana, I., Francisco Javier, G., Celso Argudo, E., and Antonio Gomez, G. (2004). Optimization of RO desalination systems powered by RE: Part I. Wind energy. *Desalination*, 160, 293–299.
64. ENERCON. (2006). Desalination Units. www.enercon.de.
65. Barbier, E. (2002). Geothermal energy technology and current status: An overview. *Renewable and Sustainable Energy Reviews*, 6, 3–65.
66. Barbier, E. (1997). Nature and technology of geothermal energy. *Renewable and Sustainable Energy Reviews*, 1(1–2), 1–69.
67. Awerbuch, L., Lindemuth, T.E., May, S.C., and Rogers, A.N. (1976). Geothermal energy recovery process. *Desalination*, 19, 325–336.
68. Boegli, W.J., Suemoto, S.H., and Trompeter, K.M. (1977). Geothermal desalting at the East Mesa test site. (Experimental results of vertical tube evaporator, MSF and high-temperature ED. Data about fouling, heat transfer coefficients, scaling.) *Desalination*, 22, 77–90.
69. Ophir, A. (1982). Desalination plant using low grade geothermal heat. *Desalination*, 40, 125–132.

70. Karytsas, K., Alexandrou, V., and Boukis, I. (2002). The Kimolos geothermal desalination project. *Proceeding of International Workshop on Possibilities of Geothermal Energy Development in the Aegean Islands Region*, Milos Island, Greece, pp. 206–219.
71. Bourouni, K., Martin, R., and Tadrist, L. (1997). Experimental investigation of evaporation performances of a desalination prototype using the aero-evapocondensation process. *Desalination*, 114, 111–128.
72. Bouchekima, B. (2003). Renewable energy for desalination: A solar desalination plant for domestic water needs in arid areas of South Algeria. *Desalination*, 153, 65–69.
73. Belessiotis, V. and Delyannis, E. Renewable energy resources, in Encyclopedia of Life Support Systems (EOLSS). *Desalination: Desalination with Renewable Energies*. Available: www.desware.net/DeswareLogin/LoginForm.Aspx.
74. Zejli, D., Bouhelal, O.-K., Benchrifa, R., and Bennouna, A. (2002). Applications of solar and wind energy sources to sea-water desalination: Economical aspects. *International Conference on Nuclear Desalination: Challenges and Options*, October 2002, Marrakech, Morocco.
75. Al-Hallaj, S. and Selman, J.R. (2002). A Comprehensive Study of Solar Desalination with a Humidification: Dehumidification Cycle. The Middle East Desalination Research Center, Muscat, Sultanate of Oman.
76. Suri, R.K., Al-Marafie, A.M.R., Al-Homoud, A.A., and Maheshwari, G.P. (1989). Cost-effectiveness of solar water production. *Desalination*, 71, 165–175.
77. Templitz-Sembitsky, W. (2000). The Use of Renewable Energy for Sea-Water Desalination; a Brief Assessment Technical Information, W16e, Gate Information Services, Germany.
78. Glueckstern, P. (1995). Potential uses solar energy for seawater desalination. *Desalination*, 101, 11–20.
79. Damm, W., Gaiser, P., Kowalczyk, D., and Plantikow, U. (1996). Wind powered MVC desalination plant: Operating results and economical aspects. *Proceeding of the Mediterranean Conference on Renewable Energy Sources for Water Production*, Santorini, Greece, pp. 143–146. ISBN 960-905777 0-2.
80. Ackermann, T. and Soder, L. (2002). An overview of wind energy-status 2002. *Renewable and Sustainable Energy Reviews*, 6, 67–128.
81. Rodriquez, L. (2002). Sea water desalination by renewable energy – a review, *Desalination*, 143(2), 103–113.
82. C.R.E.S. Centre for Renewable Energy Sources. (1998). Desalination Guide Using Renewable Energies. Maxibrochure under THERMIEDG XVII, Pikermi-Attiki, Greece.
83. Feron, P. (1985). Use of windpower in autonomous reverse osmosis seawater desalination. *Wind Engineering*, 9(3), 180–199.
84. Rahal, Z. and Infield, D.G. (October 1997). Wind powered stand alone desalination. *Proceedings of the European Wind Energy Conference*, Dublin, Ireland. Available: http://info.lboro.ac.uk/departments/el/research/crest/publictn.html.
85. Rahal, Z. and Infield, D.G. (August 1997). Computer modelling of a large scale stand alone wind-powered desalination plant. *Proceedings of the British Wind Energy Conference*, Stirling, UK. Available: http://info.lboro.ac.uk/departments/el/research/crest/publictn.html.
86. Infield, D. (1997). Performance analysis of a small wind powered reverse osmosis plant. *Solar Energy*, 61(6), 415–421.
87. Manwell, J.F. and McGowan, J.G. (1994). Recent renewable energy driven desalination system research and development in North America. *Desalination*, 94(3), 229–241.
88. Vujcic, R. and Krneta, M. (2000). Wind-driven seawater desalination plant for agricultural development on the islands of the County of Split and Dalmatia. *Renewable Energy*, 19, 173–183.

89. Stahl, M. (1991). Small wind powered RO seawater desalination plant design, erection and operation experience. *Seminar on New Technologies for the Use of Renewable Energies in Water Desalination*, 26–28 September, Commission of the European Communities, DG XVII for Energy, CRES (Centre for Renewable Energy Sources), Athens.
90. Kostopoulos, C. (1996). *Proceedings of the Mediterranean Conference on Renewable Energy Sources for Water Production*. European Commission, EURORED Network, CRES, EDS, Santorini, Greece, 10–12 June, pp. 20–25.
91. ITC (Canary Islands Technological Institute). (2001). Memoria de gestion (in Spanish).
92. Lui, C.C.K., Park, J.-W., Migita, R., and Qin, G. (2002). Experiments of a prototype wind-driven reverse osmosis desalination system with feedback control. *Desalination*, 150, 277–287; *Twelfth International Water Technology Conference*, IWTC12 2008 Alexandria, Egypt, 37.
93. Veza, J., Penate, B., and Castellano, F. (2001). Electrodialysis desalination designed for wind energy (on-grid test). *Desalination*, 141, 53–61.
94. MEDRC R&D Report, Matching Renewable Energy with Desalination Plants, IT Power Ltd (2001).
95. USBR. (2003). Cost estimating procedures. In: *Desalting Handbook for Planners (Chapter 9)*, 3rd Edition. Desalination and Water Purification Research and Development Program Report No. 72. United States Department of Interior, Bureau of Reclamation, Technical Service Center, Washington, DC, pp. 187–231.
96. LBG-Guyton Associates. (2003). Brackish Groundwater Manual for Texas Regional Water Planning Groups. Prepared for Texas Water Development Board. 31 p.
97. Younos, T. (2004). The Feasibility of Using Desalination to Supplement Drinking Water Supplies in Eastern Virginia. VWRRC Special Report SR25–2004. Virginia Water Resources Research Center, Virginia Tech, Blacksburg, VA, 114 p.
98. Younos, T. (2005). The economics of desalination, Universities Council on Water Resources. *Journal of Contemporary Water Research and Education*, 132, 39–45.
99. NRC Review of the Desalination and Water Purification Technology Roadmap. National Research Council; The National Academies Press, Washington, DC.
100. Miller, J.E. (2003). Review of Water Resources and Desalination Technologies. Sandia National Laboratories, Albuquerque, NM, 49 p. www.sandia.gov/water/docs/MillerSAND2003_0800.pdf.
101. Ghoneyem, A. and Ileri, A. (1997). Software to analyze stills and an experimental study on the effects of the cover. *Desalination*, 114, 37–44.
102. Madani, A.A. and Zaki, G.M. (1995). Yield of stills with porous basins. *Applied Energy*, 52, 273–281.
103. Bouchekima, B., Gros, B., Oahes, R., and Diboun, M. (1998). Performance study of the capillary film solar distiller. *Desalination*, 116, 185–192.
104. Fath, H.E.S. (1997). High performance of a simple design, two effects, solar distillation unit. *Energy Conversion and Management*, 38, 1895–1905.
105. Komgold, E., Korin, E., and Ladizhensky, I. (1996). Water desalination by pervaporation with hollow fiber membranes. *Desalination*, 107, 121–129.
106. Aboabboud, M.M., Horvath, L., Mink, G., Yasin, M., and Kudish, A.I. (1996). An energy saving atmospheric evaporator utilizing low grade thermal or waste energy. *Energy*, 21(12), 1107–1117.
107. SISYAN. The Off-Grid Facility Operates on 60kWp of Solar PV Arrays and Desalinated Water. Over 800... between 2007 and 2015. A website report sisyan.com/ (2015).
108. Water Desalination Using Renewable Energy, Insights for Policymakers Energy Technology System Analysis Programme IEA-ETSAP and IRENA© Technology Policy Brief I12 (January 2013). www.etsap.org - www.irena.org, https://iea-etsap.org/E-TechDS/PDF/I12IR_Desalin_MI_Jan2013_final_GSOK.pdf.

109. Renewable Energy Desalination: An Emerging Solution to Close the Middle East and North Africa's Water Gap. p. cm. (MENA Development Report) Includes Bibliographical References. ISBN 978-0-8213-8838-9 (alk. paper); ISBN 978-0-8213-9457-1, The World bank report, Washington, DC (2012). siteresources.worldbank.org/.../Resources/Renewable_Energy_Desalination_Final_Repor.
110. Shatat, M., Riffat, S., and Ghabayen, S. (2012). State of the desalination technologies using conventional and sustainable energy sources, Copyright© 2012 IUG. *The 4th International Engineering Conference: Towards Engineering of 21 Century*, pp. 1–16. https://pdfs.semanticscholar.org/0860/82e5913fd692ff598914868cf4ce0eeb52cb.pdf.
111. Qiblawey, H.M. and Banat, F. (2007). Solar thermal desalination technologies. *Presented at the Conference on Desalination and the Environment*. Sponsored by the European Desalination Society and Center for Research and Technology Hellas (CERTH), Sani Resort, 22–25 April, Halkidiki, Greece. doi: 10.1016/j.desal.2007.01.059.
112. Reif, J.H., Solar Power Technologies for Desalination. A website report Solar Power Technologies for Desalination: Duke Computer Science, https://users.cs.duke.edu/~reif/paper/solar/solardesalchapter/solardesalchapter.pd (2015).
113. A Company in France has Developed Standalone Solar-Powered Desalination Technology that doesn't Require Batteries: A World First, they Say. A website report Groundbreaking Solar-Powered Desalination that doesn't Require.... https://landartgenerator.org/blagi/archives/5948 (24 May 2018).
114. ASIRC (Australian Sustainable Industry Research Centre Ltd.). (2005). Overview of Treatment Processes for the Production of Fit for Purpose Water: Desalination and Membrane. Technologies Report No. R05-2207, 25 July, p. 7.4.
115. Solar Desalination | Department of Energy. A website report www.energy.gov/eere/solar/solar-desalination (19 June 2018).
116. Murtha, R., McCormick, M.E., and Washington, M.K. (2016). Modular Sand Filtration-Anchor System and Wave Energy Water Desalination System and Methods of Using Potable Water Produced by Wave Energy Desalination. United States Patent Application 20160236950, Application Number: 15/023791.
117. Franzitta, V., Curto, D., Milone, D., and Viola, A. (2016). The desalination process driven by wave energy, a challenge for the future. *Energies*, 9, 1032. doi: 10.3390/en9121032, www.mdpi.com/journal/energies.
118. NikWB., W., Olanrewaju, S., Rosliza, R., Prawoto, Y., and Muzathik, A. (2011). Wave energy resource assessment and review of the technologies. *International Journal of Energy and Environment*, 2, 1101–1112.
119. Lovo, R. (May 2001). Initial Evaluation of the Subfloor Water Intake Structure System (SWISS) vs. Conventional Multimedia Pretreatment Techniques, Assistance Agreement No. 98-FC-81-0044. Desalination Research and Development Program Report No. 66, U.S. Department of Interior.
120. McCormick, M. (1981). *Ocean Wave Energy Conversion*. Wiley-Interscience, New York. Reprinted by Dover Publication, Long Island, New York in 2007.
121. Murtha, R., McCormick, M.E., and Washington, M.K. (2014). Modular Sand Filtration: Anchor System and Wave Energy Water Desalination System Incorporating the Same. Assignee Murtech Inc Publication of US20140158624A1, 2014-07-15; Publication of US8778176B2 2019-07-25.
122. Armistead, T., Wave Energy Desalination Is Small, Modular and Cheap | 2017-07-12. A website report by Engineering News Record, www.enr.com/.../42354-wave-energy-desalination-is-small-modular-and-cheap (12 July 2017).
123. IAEA. (2007). Economics of Nuclear Desalination: New Developments and Site Specific Studies, Vienna, Austria, IAEA-TECDOC-1561. ISBN 978-92-0-105607-8, ISSN 1011-4289, www-pub.iaea.org/MTCD/Publications/PDF/te_1561_web.pdf.

124. Nuclear Desalination: World Nuclear Association. A website report www.world-nuclear.org/information.../non...nuclear.../nuclear-desalination.asp (2019).
125. Ahmed, S.A., Hani, H.A., Al Bazedi, G.A., El-Sayed, M.M.H., and Abulnour, A.M.G. (2014). Small/medium nuclear reactors for potential desalination applications: Mini review. *Korean Journal of Chemical Engineering*, 31(6), 924–929. doi: 10.1007/s11814-014-0079-2.
126. Jones, E., Qadir, M., van Vliet, M.T.H., Smakhtin, V., and Kang, S. (2019). Review, the state of desalination and brine production: A global outlook. *Science of the Total Environment*, 657, 1343–1356. doi: 10.1016/j.scitotenv.2018.12.076.
127. IAEA. (1997). Nuclear Desalination of Sea Water. *Symposium Proceedings*, Vienna, Austria.
128. IAEA. (1998). Nuclear Heat Applications: Design Aspects and Operating Experience. IAEA-TECDOC-1056.
129. Konishi, T. and Misra, B.M. (2001). Freshwater from the Seas. IAEA Bulletin 43.
130. Methnani, M. (2003). Coupling and thermodynamic aspects of seawater desalination using high temperature gas cooled reactors. *Proceedings of IDA World Congress on Desalination and Water Reuse*, Bahamas.
131. United Nations World Water Development Report. (2003). www.un.org/esa/sustdev/publications/WWDR_english_129556e.pdf.
132. Kim, Y.M., Kim, S.J., Kim, Y.S., Lee, S., Kim, I.S., and Kim, J.H. (2009). Development of a package model for process simulation and cost estimation of seawater reverse osmosis desalination plant. *Desalination*, 238, 312.
133. Khamis, I. (2010). Nuclear Desalination.
134. Cobb, J. (January 2015). New Technologies for Seawater Desalination Using Nuclear Energy, TecDoc 1753, World Nuclear Association, press@world-nuclear.org, International Atomic Energy Agency.
135. International Atomic Energy Agency. (1990). *Use of Nuclear Reactors for Seawater Desalination*. IAEA-TECDOC-574, Vienna.
136. International Atomic Energy Agency. (1992). *Technical and Economic Evaluation of Potable Water Production through Desalination of Seawater by Using Nuclear Energy and other Means*. IAEA-TECDOC-666, Vienna.
137. International Atomic Energy Agency. (1996). *Options Identification Programme for Demonstration of Nuclear Desalination*. IAEA-TECDOC-898, Vienna.
138. International Atomic Energy Agency. (1996). *Potential for Nuclear Desalination as a Possible Source of Low Cost Potable Water in North Africa*. IAEA-TECDOC-917, Vienna.
139. International Atomic Energy Agency. (1997). Nuclear desalination of sea water. *Proceeding Series*, STI/PUB/1025, Vienna.
140. International Atomic Energy Agency. (2000). Introduction of Nuclear Desalination: A Guidebook. IAEA Technical Report Series No. 400, Vienna.
141. International Atomic Energy Agency. (2000). Safety of Nuclear Power Plants: Design: Safety Requirements. IAEA Safety Standard Series No. NS-R-1, Vienna.
142. International Atomic Energy Agency. (2000). *Status of Non-Electric Nuclear Heat Applications: Technology and Safety*. IAEA-TECDOC-1184, Vienna.
143. International Atomic Energy Agency. (2001). *Safety of Nuclear Plants Coupled with Seawater Desalination Units*. IAEA-TECDOC-1235, Vienna.
144. International Atomic Energy Agency. (2002). *Status of Design Concepts of Nuclear Desalination Plants*. IAEA-TECDOC-1326, Vienna.
145. International Atomic Energy Agency. (2006). *Basic Infrastructure for a Nuclear Power Project*. IAEA-TECDOC-1513, Vienna.
146. International Atomic Energy Agency. (2006). *Potential for Sharing Nuclear Power Infrastructure Between Countries*. IAEA-TECDOC-1522, Vienna.

147. International Atomic Energy Agency. (2007). *Economics of Nuclear Desalination: New Developments and Site Specific Studies*. IAEA-TECDOC-1561, Vienna.
148. International Atomic Energy Agency. (2007). *A Global Overview on Nuclear Desalination 327, Managing the First Nuclear Power Plant Project*. IAEA-TECDOC-1555, Vienna.
149. International Atomic Energy Agency. (2007). *Status of Nuclear Desalination in the IAEA Member States*. IAEA-TECDOC-1542, Vienna.
150. International Atomic Energy Agency. (2008). *Advanced Application of Water Cooled Nuclear Power Plants*. IAEA-TECDOC-1584, Vienna.
151. International Atomic Energy Agency. (2008). *International Status and Prospects of Nuclear Power*. IAEA, Vienna, Austria.
152. International Atomic Energy Agency. (2008). *Nuclear Technology Review 2008*. IAEA, Vienna.
153. International Atomic Energy Agency. (2008). *Power Reactor Information System (PRIS) Database*. IAEA, Vienna.
154. International Atomic Energy Agency. (2009). Issues to Improve the Prospects of Financing Nuclear Power Plants. IAEA Nuclear Energy Series No. NG-T-4.1, Vienna.
155. International Atomic Energy Agency. (2009). Non-electric applications of nuclear power: Seawater desalination, hydrogen production and other industrial applications. *Proceeding of the International Conference*, 16–19 April, Oarai, Japan, Vienna.
156. Zverev, K.V., Polunichev, V.I., and SerGeev, Yu.A. (1997). Floating Nuclear Power Station of APWS-80 Type for Electricity Generation and Fresh Water Production.
157. Khamis, I. (2009). A global overview on nuclear desalination. *International Journal of Nuclear Desalination*, 3(4), 311. citeseerx.ist.psu.edu/viewdoc/download?doi=10.1.1.61 2.4613&rep=rep1.
158. Spadaro, J., Langlois, L., and Hamilton, B. (2000). Greenhouse gas emissions of electricity generation chains: Assessing the difference. *IAEA Bulletin*, Vienna, 42(2), 19–24.
159. Tewari, P.K. and Rao, I.S. (2002). LTE desalination utilizing waste heat from a nuclear research reactor. *Desalination*, 150, 45–49.
160. Tewari, P.K. and Khamis, I. (2007). Non-electrical applications of nuclear power. *Proceedings of International Conference*, 16–19 April, Oarai, Japan.
161. Wade, N.M. (2001). Desalination plant development and cost update. *Desalination*, 136, 3–12.
162. Lacomte, M. and Bandelier, P. (2002). Optimization of water desalination by high temperature reactor using electricity/heat cogeneration. *Proceedings of the International Conference on Nuclear Desalination: Options and Challenges*, 16–18 October, Marrakech, Morocco.
163. Russian floating nuclear power station, Wikipedia, The free encyclopedia, last visited 11 August 2019 (2019).
164. KLT-40 reactor, Wikipedia, The free encyclopedia, last visited 4 December 2017 (2019).
165. How Much Energy Does a Wastewater Treatment Plant Use? Written by AOS Treatment Solutions on (1 March 2018), info@aosts.com.
166. How do you Reduce Wastewater Energy Consumption? By Nicolette l (05 October 2017).
167. Energy Analytics by PredictEnergy® Software to measure Energy Data. A website report https://predictenergy.com/energy-analytics-by-predictenergy/, Predict Energy platform (2015).
168. Wastewater treatment, Wikipedia, The free encyclopedia, last visited 10 August 2019 (2019).
169. Lytle, D., Cost Effective Treatment Technologies for Small Community Drinking Water Systems. EPA Report, Cost-Effective Treatment Technologies for Small Drinking Water... – EPA. A website report www.epa.gov/sites/production/files/2018-04/.../dw_small_communities_slides (2018).

170. AdEdge Water Technologies: Water Treatment Solutions. A website report www.adedgetech.com/ (2018).
171. AdEdge Technologies Arsenic Removal Systems | Water Softeners. A website report www.water-softeners-filters.com/adedge-technologies (2018).
172. Advantages of Modular Water Treatment Systems | Newterra. A website report https://newterra.com/modularity-advantages (2019).
173. Sewage Treatment Solutions | Domestic Wastewater | Newterra. A website report https://newterra.com/applications/sewage-treatment (2019).
174. Modular Water Treatment System: Envirogen Technologies. A website report www.envirogen.com/files/files/ETI-TIMBR-MBR.pdf (2019).
175. Wastewater Treatment: Envirogen Technologies. A website report www.envirogen.com/pages/applications/wastewater-treatment/ (2019).
176. Envirogen Technologies Membrane Biological Reactor: Water Online. A website report www.wateronline.com/doc/envirogen-membrane-biological-reactor-0001 (2019).
177. Kim, J.R., Premier, G.C., Hawkes, F.R., Rodríguez, J., Dinsdale, R.M., and Guwy, A.J. (2010). Modular tubular microbial fuel cells for energy recovery during sucrose wastewater treatment at low organic loading rate. *Bioresource Technology*, 101(4), 1190–1198. doi: 10.1016/j.biortech.2009.09.023, www.ncbi.nlm.nih.gov/pubmed/19796931.
178. Field Erected Wastewater Plant: Modular Wastewater...: Global Treat. A website report www.globaltreat.com/wastewater-plants.php (2019).
179. MODULAB® High Flow Water Purification System: Evoqua Water. A website report www.evoqua.com/en/brands/IPS/Pages/MODULAB.aspx (2015).
180. Mobile and Temporary Water Treatment Services: Evoqua Water. A website report www.evoqua.com/en/brands/IPS/Pages/mobile-temporary-water-services.aspx (2015).
181. Evoqua Launches the Next Generation in Electrochemical Desalination. A website report www.evoqua.com/en/about/newsroom/Pages/nexed-announcement.aspx (10 June 2015).
182. Packaged Wastewater Treatment Systems | ClearBlu Environmental. A website report www.clrblu.com/package-systems/ (2017).
183. Clearblue® Introduces the First and Only Ovulation Test System to... A website report https://news.pg.com/press.../clearblue-introduces-first-and-only-ovulation-test-system-... (10 October 2017).
184. Modular Waste Water Treatment, Package Wastewater Treatment. A website report www.pollutioncontrolsystem.com/contact-us (2016).
185. Pollution Control Systems: Industrial Filtration and Collection. A website report www.pollutioncs.com (30 April 2016).

8 Modular Systems for Energy and Fuel Storage

8.1 METHODS FOR ENERGY AND FUEL STORAGE

Electric energy storage (also simply called "energy storage") encompasses a wide range of technologies that are capable of shifting energy usage from one time period to another. These technologies could deliver important benefits to electric utilities and their customers, since the electric system currently operates on "just-in-time" delivery. Generation and load must be perfectly balanced at all times to ensure power quality and reliability. Strategically placed energy storage resources have the potential to increase efficiency and reliability, to balance supply and demand, to provide backup power when primary sources are interrupted, and to assist with the integration of intermittent renewable generation. Energy storage is capable of benefiting all parts of the system—generation, transmission, and distribution—as well as customers.

Electric energy storage resources are usually described in terms of their nameplate power rating and their energy storage capacity. For example, a 10 MW/20 MWh storage system is capable of delivering 10 MW of AC power for 2 h, for a total of 20 MWh of energy delivered to the grid (10 MW × 2 h = 20 MWh). Systems can be as large as pumped hydropower facilities that provide hundreds of megawatts of power for many hours or as small as off-grid battery systems that support electric service for small, remote residences and facilities. This flexibility is one of their attractive qualities [1–12].

8.1.1 Energy Storage Technologies

Energy storage encompasses a wide range of technologies and resource capabilities, and these differ in terms of cycle life, system life, efficiency, size, and other characteristics. Although battery technology has attracted a great deal of industry attention in recent years, pumped hydro technology still supplies the vast majority of grid-connected energy storage (95%). The remaining categories combined comprise only 5% of installed capacity, as the Table 8.1 below illustrates [1–12].

TABLE 8.1
Energy Storage Technologies and Their Percentage Contribution

Technology Class Examples

Electrochemical Storage: Batteries (1%), capacitors, supercapacitors (0.5%)
Mechanical Storage: Flywheels (1%), compressed air (0.5%)
Thermal Storage: Ice, molten salt, hot water, PCMs, physico-chemical storage (2%)
Bulk Gravitational Storage Pumped hydropower (95%), gravel

8.1.1.1 Electrochemical Storage (Batteries)

This class of energy storage includes the following: advanced lead–acid, lithium-ion, sodium-based, nickel-based flow batteries and electrochemical capacitors. Technologies are further divided into subcategories based on the specific chemical composition of the main components (anode, cathode, separator, electrolyte, etc.). Each class and subcategory is at a different stage of commercial maturity and has unique power and energy characteristics that make it more or less appropriate for specific grid support applications [1–12].

8.1.1.1.1 Advanced Lead–Acid Batteries

Invented in the 19th century, lead–acid batteries are the most fully developed and commercially mature type of rechargeable battery. They are widely used in both mobile applications like cars and boats and stationary consumer applications like uninterruptible power supply (UPS) units and off-grid PV. However, several issues have prevented widespread adoption for utility-scale grid applications. These include short cycle life, slow charging rates and high maintenance at power rating >1 MW. These perform a variety of services including peak shaving, on-site power, ancillary services, ramping, and renewables capacity firming. Lead–acid batteries rely on a positive, lead-dioxide electrode reacting with a negative, metallic lead electrode through a sulfuric acid electrolyte. Ongoing research and development have produced several proprietary technologies in two categories: advanced lead–acid and lead acid–carbon batteries [1–12].

8.1.1.1.2 Lithium Ion

First commercialized in 1991, lithium-ion batteries have experienced tremendous research and development investment and publicity in the last few years due to their high energy density, voltage ratings, cycle life, and efficiency. They have been the preferred battery technology for portable electronic devices and electric vehicles (EVs), and now they are being scaled up and deployed for utility grid services. Approximately 70 systems with power rating of 1 MW or greater are currently in operation around the world. Because it can adapt to a range of power and energy ratings, this technology can perform a wide variety of services. Grid-scale units range from small, regulation pilot projects of 1 MW/0.5 MWh (30 min duration) to large lithium-ion technologies and are also classified based on cell shape: cylindrical, prismatic, or laminate. Cylindrical cells have high potential capacity, lower cost, and good structural strength [1–12].

8.1.1.1.3 Sodium Sulfur

Sodium sulfur (NaS) battery technology was invented by Ford Motors in the 1960s, but research, development, and deployment by Japanese companies like NGK Insulators and Tokyo Electric Power Company over the past 25 years have established NaS as a commercially viable technology for fixed, grid-connected applications. Sodium sulfur batteries use a positive electrode of molten sulfur, a negative electrode of molten sodium, and a solid beta alumina ceramic electrolyte that separates the electrodes. Batteries require charge/discharge operating temperatures between 300°C and 350°C, so each unit has a built-in heating element. High operating temperatures and hazardous materials require the systems to include safety features like fused electrical isolation, hermetically sealed cells, sand surrounding cells to mitigate fire, and a battery management system (BMS) that monitors cell block voltages and temperatures. Typical units are composed of 50 kW modules. The advantages of sodium sulfur are its high power and long duration, extensive deployment history, and commercial maturity. Downsides include risk of fire, round-trip efficiencies of 70%–90%, and potentially high self-discharge/parasitic load values of 0.05% cycling applications because the internal heating element continually consumes energy. In sodium–nickel chloride batteries, the cathode is composed of nickel chloride instead of sulfur. These require operating temperatures between 260°C and 350°C and therefore must have internal thermal management capability. Able to withstand limited overcharging, they are potentially safer than sodium sulfur, and they have a higher cell voltage [1–12].

8.1.1.1.4 Nickel-Based Technology

The two main subtechnologies in the nickel-based family are nickel–cadmium (NiCd), which has been in commercial use since 1915, and nickel–metal hydride (NiMH), which became available around 1995. Nickel-based batteries are primarily used in portable electronics and EVs due to their high power density, cycle life, and round-trip efficiency. Only two operational projects have energy ratings >1 MWh. One of them provides electric supply reserve capacity in Alaska, and the other performs renewable capacity firming on Bonaire Island. Although Sandia laboratory states that Nickel-cadmium and nickel metal hydride deployments exist, in general, nickel-based technology is not yet competitive with other battery types. All nickel-based batteries employ a cathode of nickel hydroxide. Subcategories are classified according to anode composition: nickel–cadmium, nickel–iron, nickel–zinc, nickel–hydrogen, and nickel–metal hydride. The first three use a metallic anode; the last two have anodes that store hydrogen.

Nickel–cadmium chemistry is a low-cost, mature technology with high energy density, but the toxicity of cadmium necessitated a search for alternatives. Nickel–metal hydride was developed in response. While, the metal hydride chemistry is safer and has a higher specific energy, and nickel–metal hydride's safety made it the battery of choice for electric and hybrid vehicles, lithium-ion technology is challenging its market potential. Other nickel chemistries are in the research and development phase [1–12].

8.1.1.1.5 Flow Batteries

Flow batteries are fundamentally different than other types of electrochemical storage because the systems' power and energy components are separate. This feature allows flow systems to be tailored to specific applications and constraints. A number of megawatt-scale demonstration projects are testing the deep discharge ability, long cycle life, and easy scalability that characterize flow batteries. Some chemistries have been more extensively developed and deployed than others; maturity ranges from development stage (for iron–chromium and zinc–bromine) to precommercial (for vanadium). Projects in operation range from 5 MW/10 MWh (2 h duration) to 3 kW/8 kWh (2 h, 40 min duration). One or both of a flow battery's active materials is in solution in the electrolyte at any given time. In traditional flow batteries, the electrolyte solution is stored in separate containers and pumped to the cell stack and electrodes where an oxidation-reduction reaction occurs.

Many forms of storage, particularly batteries and ice energy, are more flexible when it comes to sizing and siting. Battery resources can be sized from 20 kW to 1,000 MW and sited at the customer's location or interconnected to the transmission system. Other factors may also limit the siting of storage, such as space availability, permitting, and interconnection upgrade requirements. Battery or flywheel storage projects can move from concept to commissioning in 2–3 years. Smaller systems (in the 1–5 MW range) have been commissioned in less than 2 years. Modular containerized systems can be installed in the field and be brought online within months, assuming the systems become standardized and the interconnection process is streamline. The distributed solar systems are well suited for modular operations [1–12].

8.1.1.2 Electrochemical Capacitors and Super Capacitors

8.1.1.2.1 Capacitors

A **capacitor** is a passive two-terminal electronic component that stores electrical energy in an electric field. The effect of a capacitor is known as capacitance. While some capacitance exists between any two electrical conductors in proximity in a circuit, a capacitor is a component designed to add capacitance to a circuit. The capacitor was originally known as a **condenser** or **condensator** [1]. The original name is still widely used in many languages but not commonly in English.

The physical form and construction of practical capacitors vary widely, and many types of capacitor are in common use. Most capacitors contain at least two electrical conductors often in the form of metallic plates or surfaces separated by a dielectric medium. A conductor may be a foil, thin film, sintered bead of metal, or an electrolyte. The non-conducting dielectric acts to increase the capacitor's charge capacity. Materials commonly used as dielectrics include glass, ceramic, plastic film, paper, mica, air, and oxide layers. Capacitors are widely used as parts of electrical circuits in many common electrical devices. Unlike a resistor, an ideal capacitor does not dissipate energy. When two conductors experience a potential difference, for example, when a capacitor is attached across a battery, an electric field develops across the dielectric, causing a net positive charge to collect on one plate and net negative charge to collect on the other plate. No current actually flows through the dielectric. However, there is a flow of charge through the source circuit. If the condition is maintained sufficiently long, the current through the source circuit ceases. If a time-varying voltage is applied across the leads of the capacitor, the source experiences an ongoing current due to the charging and discharging cycles of the capacitor.

Capacitance is defined as the ratio of the electric charge on each conductor to the potential difference between them. The unit of capacitance in the International System of Units (SI) is the farad (F), defined as one coulomb per volt (1 C/V). Capacitance values of typical capacitors for use in general electronics range from about 1 picofarad (pF) (10^{-12} F) to about 1 millifarad (mF) (10^{-3} F). The capacitance of a capacitor is proportional to the surface area of the plates (conductors) and inversely related to the gap between them. In practice, the dielectric between the

Systems for Energy and Fuel Storage

plates passes a small amount of leakage current. It has an electric field strength limit, known as the breakdown voltage. The conductors and leads introduce an undesired inductance and resistance.

Capacitors are widely used in electronic circuits for blocking direct current while allowing alternating current to pass. In analog filter networks, they smooth the output of power supplies. In resonant circuits, they tune radios to particular frequencies. In electric power transmission systems, they stabilize voltage and power flow [2]. The property of energy storage in capacitors was exploited as dynamic memory in early digital computers [1–12].

8.1.1.2.2 Super Capacitors

Also called electrochemical double-layer capacitors and ultracapacitors, this technology class bridges the gap between batteries and traditional capacitors; it stores energy electrostatically. Supercapacitors are characterized by low internal resistance, which allows rapid charging and discharging, very high power density (but low energy density), and high specific energy (30 Wh/kg) and corresponding high cost per kWh. A **supercapacitor** (**SC**) (also called a **supercap**, **ultracapacitor,** or **Goldcap** [2]) is a high-capacity capacitor with capacitance values much higher than those of other capacitors (but lower voltage limits) that bridge the gap between electrolytic capacitors and rechargeable batteries. They typically store 10–100 times more energy per unit volume or mass than electrolytic capacitors, can accept and deliver charge much faster than batteries, and tolerate many more charge and discharge cycles than rechargeable batteries.

Supercapacitors are used in applications requiring many rapid charge/discharge cycles rather than long-term compact energy storage: within cars, buses, trains, cranes, and elevators, where they are used for regenerative braking, short-term energy storage, or burst-mode power delivery [3]. Smaller units are used as memory backup for static random-access memory (SRAM). Unlike ordinary capacitors, supercapacitors do not use the conventional solid dielectric, but rather, they use electrostatic double-layer capacitance and electrochemical pseudocapacitance [4], both of which contribute to the total capacitance of the capacitor, with a few differences [1–12]:

- **Electrostatic double-layer capacitors** (**EDLCs**) use carbon electrodes or derivatives with much higher electrostatic double-layer capacitance than electrochemical pseudocapacitance, achieving separation of charge in a Helmholtz double layer at the interface between the surface of a conductive electrode and an electrolyte. The separation of charge is of the order of a few ångströms (0.3–0.8 nm), much smaller than in a conventional capacitor.
- Electrochemical pseudocapacitors use metal oxide or conducting polymer electrodes with a high amount of electrochemical pseudocapacitance in addition to the double-layer capacitance. Pseudocapacitance is achieved by Faradaic electron charge-transfer with redox reactions, intercalation or electrosorption.
- Hybrid capacitors, such as the lithium-ion capacitor, use electrodes with differing characteristics: one exhibiting mostly electrostatic capacitance and the other mostly electrochemical capacitance.

The electrolyte forms an ionic conductive connection between the two electrodes which distinguishes them from conventional electrolytic capacitors where a dielectric layer always exists, and the so-called electrolyte (e.g., MnO_2 or conducting polymer) is in fact part of the second electrode (the cathode or, more correctly, the positive electrode). Supercapacitors are polarized by design with asymmetric electrodes or, for symmetric electrodes, by a potential applied during manufacture [1–12].

8.1.1.3 Mechanical Storage
Mechanical storage technologies use compressed air and flywheels to store energy [1–12].

8.1.1.3.1 Compressed Air
Compressed air energy storage (CAES) resources compress air and store it in a reservoir, typically underground caverns or above-ground storage pipes or tanks. Underground facilities are considered less expensive than those above ground and can operate for between 8 and 26 h; however, siting underground compressed air storage facilities requires detailed research and time-consuming permission requirements.

8.1.1.3.2 Flywheels
The flywheels is the other mechanical energy storage technology. In flywheels, a rotor (flywheel) is accelerated to a very high speed in a very-low-friction environment. The spinning mass stores potential energy to be discharged as necessary. Flywheels are modular and can range from 22 kW (Stornetic's EnWheel) to 160 kW (Beacon Power) in size. Flywheels are best for short-duration, high-power, high-cycle applications. They also have a much longer cycle life than other storage alternatives. Flywheels are less heat sensitive than batteries, and they last longer (up to 20 years guaranteed performance). Power grid uses include voltage/voltage-ampere reactive (VAR) support and frequency regulation. Primary competitors to flywheels are supercapacitors or ultracapacitors [1–12].

8.1.1.4 Thermal Storage
Thermal storage comes in many forms; the most well-known bulk thermal storage solution is molten salt. Paired with solar thermal generation plants, molten salt thermal storage is used to improve the dispatchability of concentrated solar power (CSP) facilities. The stored energy powers steam turbines to continue generation after the solar day has ended. Other forms of thermal storage are more distributed in nature. These primarily interact with building heating and cooling systems and support demand-side services such as demand response. Some technologies, such as direct load control of water heaters, have already demonstrated deployment in electrical and heating networks. SCE and PG&E recently awarded contracts to IceEnergy for distributed thermal storage to reduce air conditioning loads. Although promising, many of these technologies are aimed at reducing peak loads during high-temperature periods.

Energy storage also involves storage of heat and cold. Besides thermal storage outlined above, storage of heat and cold also involves sensible heat storage such as molten salt, ice chillers, ice box, phase change materials, and thermochemical heat storage. Modular systems used for these types of storage are also described in this chapter [1–12].

8.1.1.5 Bulk Gravitational Storage

Bulk gravitational storage includes technologies such as pumped hydro and gravel in railcars.

8.1.1.5.1 Gravitational Potential Energy Storage with Solid Masses

Changing the altitude of solid masses can store or release energy via an elevating system driven by an electric motor/generator. Potential energy storage or gravity energy storage was under active development in 2013 in association with the California Independent System Operator. It examined the movement of earth-filled hopper rail cars driven by electric locomotives from lower to higher elevations. Methods include using rails and cranes to move concrete weights up and down, using high-altitude solar-powered buoyant platforms supporting winches to raise and lower solid masses, using winches supported by an ocean barge for taking advantage of a 4 km (13,000 ft) elevation difference between the surface and the seabed [13]. Efficiencies can be as high as 85% recovery of stored energy [1–12].

8.1.1.5.2 Pumped Hydro

Pumped hydro is a mature technology used throughout North America and the world. Off-peak power is used to pump water from a lower reservoir to a higher reservoir; then the water is released to generate electricity during peak periods. Because pumped hydro facilities require above-ground reservoirs, specific land configurations are needed. Pumped hydro projects are rarely located close to urban centers, and permitting can take many years due to their large environmental impact.

The siting of an energy storage resource is an important consideration for development feasibility; it affects both costs and benefits. Some resources, like pumped hydro, must be located in areas with specific geology, water access, and transmission lines. Natural gas combustion turbines have similar constraints, plus they face air emissions constraints in many locations as well. Pumped hydro and CAES facilities, due to their size and environmental impacts, require significantly longer development timelines for analysis, design, and extensive permitting activity than many storage resources. It can take 5–10 years (or more) to complete one of these projects, depending on public support or opposition for a particular project, the ability to negotiate environmental impact studies, and other necessary approvals. Their large size and often remote location also may mean that new transmission is needed; obtaining the necessary permits and regulatory approvals required to start transmission construction can also take years, although this activity may take place concurrently with storage facility planning. At present time, large-scale pumped hydro is not suitable for modular operation. Small modular hydropower is, however, gaining heavy acceptance in rural areas [1–12].

8.1.1.5.3 Gravel/Railcar

The gravel/railcar storage method operates in a similar manner to pumped hydro. Off-peak power is used to move rail cars filled with gravel or another heavy material up a slope. When power is needed, the railcar moves down the slope, converting gravitational energy into electricity as it moves down. Unlike pumped hydro, railcar/gravel energy storage does not require reservoirs to function. Rather, it requires a long slope of existing or new railroad track. This makes it potentially easier to site than pumped hydro, although it is still not suitable for urban areas, nor is it suitable for railroad segments where there is existing traffic [1–12].

8.1.1.6 Fuel Storage

Besides energy storage, various types of fuel storage like biomass, hydrogen, liquid fuel, compressed natural gas, and liquid natural gas and propane also use a variety of modular systems. These are also described in some detail in this chapter.

8.2 MODULAR BATTERY STORAGE

8.2.1 Methods for Integrating and Controlling Modular Battery-Based Energy Storage Systems

Mechanisms for electric energy storage can mitigate supply–demand imbalances, by storing electric energy during periods of excess supply and returning it to the electric power system during periods of excess demand. Hence, devices or systems to store electric energy, including those based on batteries, may become more prevalent within the electric power system. In particular, battery energy storage systems (BESSs) composed of modular subunits may become more prevalent because of their inherent scalability. Such modular subunits may be constructed from a wide variety of battery cell types and chemistries, including lithium-ion, nickel–cadmium, nickel–metal hydride, lead–acid, zinc–air, and others both currently available and emerging.

The patent by Kaplan et al. [14] includes systems, apparatuses, methods, and programs for enabling user installation, removal, exchange, maintenance, monitoring, control, charging, and discharging of energy storage equipment, such as battery modules or power conversion modules, within an energy storage system (ESS). An example of such systems is a modular battery energy storage system (MBESS). The patent also describes maintenance, monitoring, control, charging, and discharging of an ESS comprised of user-accessible energy storage equipment, such as battery modules or power conversion modules. In particular, a modular energy storage system (MESS) includes one or more energy storage units, one or more power conversion units that can be coupled to an external system, and a power transfer network. The power transfer network is coupled to the one or more energy storage units via a first interface. The power transfer network is also coupled to the one or more power conversion units via a second interface. The power transfer network includes one or more conductors that can be coupled to the first interface and to the second interface for transferring power between the one or more energy storage units and the one or more power conversion units. The power conversion units are configured to convert an electrical parameter of power transferred to and from the energy storage units in accordance with an electrical requirement of the external system or the one or more energy storage units. The power conversion units can also be coupled to a power grid operated by an electric utility.

The ESS further includes a controller in communication with the power transfer network. The controller is configured to selectively cause one or more of the conductors in the power transfer network to electrically connect one or more of the energy storage units to one or more of the power conversion units based at least in part on power or energy supply or demand of the external system or an amount of energy stored in the one or more energy storage units. An energy storage unit in the ESS may differ from another energy storage unit with regard to an electrical characteristic, such as energy capacity, power capacity, current capacity, or voltage, or with regard to a physical characteristic. Similarly, a power conversion unit in the ESS may differ from another power conversion unit with regard to an electrical characteristic, such as power capacity and current. In some embodiments, the ESS may comprise more energy storage units than power conversion units.

The controller may be configured to communicate with the power transfer network using a standardized, publicly available protocol. Further, the controller may be configured to communicate with one or more of the energy storage units and/or with one or more of the power conversion units. The controller may also be configured to communicate with a controller of another ESS. Aspects of the ESS may be distributed. For example, at least one of the energy storage units can be housed physically separate from another energy storage unit in the ESS. Similarly, at least one of the power conversion units is housed physically separate from another power conversion unit. The power transfer network includes one or more switches that are responsive to the controller to selectively electrically connect a conductor in the power transfer network to an energy storage unit and to a power conversion unit. At least one of the switches may be a multiway switch. There are multiple power transfer network interfaces with modular points of connection. An energy storage unit or a power conversion unit may be electrically connected to a conductor in the power transfer network via a DC–DC converter. The power transfer between an energy storage unit and a power conversion unit may be controlled using pulse-width modulation. The disclosure also includes a power transfer control system for a modular energy storage network. The patent also disclosed a computer-readable storage medium having computer-executable instructions stored thereon. The instructions, in response to execution by a computing device, cause the computing device to undertake actions. Such actions include receiving information indicating a power or energy supply or demand of an external system or information indicating an amount of energy stored in one or more energy storage units. More details on the workings of battery storage units are described in the patent.

In recent years, significant efforts have been made by industries to lay out the details of battery-controlled storage system for stable power generation from renewable sources like solar and wind. The increased role of renewable and distributed energy in overall power generation mix requires several issues to be addressed. Power obtained from photo voltaic (PV) plants and wind parks is intermittent by nature. The availability of the primary energy carriers (being wind and sunlight) is dependent on the weather conditions, which are predictable over a limited time horizon only. BESSs can help to mitigate these effects. When traditional power plants are replaced by distributed generation, the new plants are required to contribute equally well to grid stabilization [2]. The grid code thus requires small- and medium-sized generation plants to take part in static and dynamic voltage control as well. This can be achieved by using BESSs. Also, traditional thermal power stations such as coal-fired power plants or nuclear reactors generate electricity with the help of turbines, which have a high rotating mass. In case of a power plant outage, the energy stored as mechanical inertia helps to keep the grid frequency stable until the primary control of the network stabilizes the grid. When the share of power produced with traditional generators decreases, transient disturbances on the system frequency may become more pronounced. The short ramp-up times of BESSs may counteract these effects [3]. Many of these issues may be addressed by building a stronger grid to level out power flows across large geographical areas [4]. However, depending on the degree of capacity utilization, a strongly interconnected grid may be unfavorable in terms of return on investment. On the other hand, (battery) ESSs can reduce power fluctuations in the region where they occur. The buffering energy locally will in general lead to uneconomically large system sizes and poor utilization as well [5]. As a consequence, the future electricity grid is expected to show a combination of both approaches. The success of the storage technologies applied will depend on their technical performance and economic attractiveness.

The power electronic converter systems play an important role in the performance of ESS. The aim of the project "Power Electronic Converter Systems for Modular Energy Storage Based on Split Batteries" by the Swiss Federal Office of Energy (SFOE) was to increase the efficiency, compactness, and reliability of BESSs connected to the medium-voltage grid. Different approaches based on the modular multilevel converter (MMC) were compared. The single-star bridge-cell (SSBC), the single-delta bridge-cell (SDBC), and the double-star chopper-cell topologies were identified as the most suitable candidates for next-generation BESSs: As opposed to conventional BESSs, solutions based on these topologies can be directly connected to the medium voltage grid without a transformer, increasing the overall efficiency while at the same time cutting down on system volume. Due to the modular architecture, redundancy can be added to the systems which increases their availability.

Since the presented MMC topologies do not feature a low-voltage DC link, the batteries are integrated into the modules. Different possibilities to realize this connection were compared. It was found that a direct connection to the modules might decrease the battery lifetime. Consequently, the analyzed solutions included DC–DC converters in each module to decouple the power flowing in and out of the modules from the charging process of the batteries. In order to identify the best solution, the performance of the candidate topologies was compared in a systematic way. An optimal design methodology was developed to identify the theoretical limitations when designing for low power losses and small size. In addition, the control of these new systems in case of a fault was analyzed thoroughly. Of the compared candidate systems, the SSBC was proven to be the most attractive one: The overall energy efficiency of the proposed solution was around 86%, which compares favorably to the 85% overall energy efficiency reported for state-of-the-art systems. Both figures included a round-trip efficiency of 94% of the batteries. A hardware prototype was designed and commissioned to validate the results of the analyses. The system was designed with maximum versatility in mind to support future research effort on modular architectures at the Laboratory for High Power Electronic Systems [15–59].

As an alternative, the study by Yang et al. [22] proposes a reverse-blocking MMC for a BESS (RB-MMC-BESS). Besides integrating distributed low-voltage batteries to medium- or high-voltage grids, with the inherited advantages of traditional MMCs, the RB-MMC-BESS also provides improved DC fault handling capability. This study analyzes such a new converter configuration and its operating principles. Control algorithms were developed for AC side power control and the balancing of battery state of charge. The blocking mechanism to manage a DC pole-to-pole fault was analyzed in depth. Comprehensive simulation results validated both the feasibility of the RB-MMC-BESS topology and the effectiveness of the control and fault handling strategies. The modular structure and the comprehensive IT management system provides a system with nearly unlimited scaling.

Zhang et al. [19] also proposes a novel topology of PV-BESS based on MMC, where the batteries are connected to the submodules through DC–DC converters. It is necessary to analyze the control strategies of both DC–DC converters and MMC. Specifically, the capacitor voltage balancing and the modulation mode are important. The sorting method is proposed to balance the capacitor voltages in this study. Moreover, carrier phase shift-square wave modulation with the highest voltage utilization ratio and the highest power transfer capability is proposed to generate pulse-width modulation (PWM) singles for MMC-PV-BESS. In order to verify the availability of the proposed control strategy, a simulation model was built, and related experiments were carried out. The simulation results and experimental results were consistent with the theory. As a result, the voltages of all capacitors could be balanced, and the MMC-PV-BESS could reliably work.

Helling et al. [26] showed that the dependency of energy systems on battery storage systems is constantly increasing, but current battery systems are inflexible; only cells with the same electrical parameters can be combined, and cell defects cause a high reduction of the overall battery lifetime or even a system blackout. In addition, the maximum usable capacity and the maximum charging current are limited by the weakest cell in the system. Current BMSs can increase the usable battery capacity to some extent and are able to enlarge the maximum usable charging current. With the battery modular multilevel management system (BM3) presented in this study, a very flexible, fault tolerant, and cost-efficient battery system can be implemented. With this system, it is possible to establish either serial or parallel connections between neighboring cells or to bypass a cell. Thus the cells can be operated according to their needs and their state of charge (SOC). Separate balancing means for balancing the SOC of cells, however, have become obsolete.

Thus, ESSs require battery cell balancing circuits to avoid divergence of cell SOC. In the study by Evzelman et al. [42], a modular approach based on distributed continuous cell-level control is presented that extends the balancing function to higher-level pack performance objectives such as improving power capability and increasing pack lifetime. This is achieved by adding DC–DC converters in parallel with cells and using state estimation and control to autonomously bias individual cell SOC and SOC range, forcing healthier cells to be cycled deeper than weaker cells. The result is a pack with improved degradation characteristics and extended lifetime. The modular architecture and control concepts are developed, and hardware results are demonstrated for a 91.2-Wh battery pack consisting of four series Li-ion battery cells and four dual active bridge (DAB) bypass DC–DC converters.

In order to solve the inconsistency of the battery pack in the traditional BESS, a new type of battery module ESS topology and control strategy based on flexible grouping was proposed by Li et al. [23]. The study presents a modular BESS based on one integrated primary multiple-secondaries transformer. Different from commonly used full-power control mode, the part-power control mode is adopted in this study. The current flowing through the switches is the difference of each battery module current and the average current of all battery modules. By applying part-power control method, the system volume, cost, and loss is effectively reduced. Based on the DAB converter and one integrated primary multiple-secondaries transformer, phase shifting control strategy is implemented to control part of the charging and discharging current of each battery module independently, which makes each battery module work under the best condition and helps in improving the energy utilization. Based on analyses of the data of the passenger car battery after 3 years of actual operation, simulation results show the correctness and superiority of this topology and its control method. In recent years, hybrid systems (particularly for automobiles) are gaining prominence. Hybrid drives offer significant potential for reducing fuel consumption and emissions in off-highway applications due to the recovery and storage of kinetic or potential energy. Systems supplier MTU is developing a modular hybrid system comprising standardized components. One of the key subcomponents in this system is the ESS, which stores the produced energy for use when it is actually needed.

In the LiANA+ project funded by the German Federal Ministry of Economics and Technology, Akasol GmbH, with partners, developed a modular high-performance ESS based on lithium-ion cells for use in a wide range of applications and power classes. Akasol tested nine different cell types based on criteria such as power density, safety, price, cycle stability, quality, and availability and then selected two of these cell types for prototype development. In terms of detailing and time prioritization, attention was focused on the larger 46 Ah cell, since the simulations indicated the prospect of a promising solution.

The dynamics, power, and efficiency of an electrical accumulator depend on the SOC, state of health (SOH), and temperature of the battery. To cater to various off-road applications in the future, a modular approach was adopted for the battery system. The existing solution at Akasol has had its design optimized to allow better cooling properties and, therefore, higher continuous power with a longer life. The module includes 12 lithium-ion cells and ensures their thermal and electrical connection. It also contains the measuring system for monitoring the state of the cells. Several modules are grouped together in a battery housing. Various parallel and serial connections allowed these battery banks to be arranged into energy accumulators with different power and energy values. The LiANA+ project created banks of 15 serially connected modules creating the nominal voltage of 666 V. Three of these battery banks are connected in parallel to produce a total energy content of about 92 kWh and a peak output of 552 kW [29–31].

The risk assessment produced in the LiANA+ project stipulated the safety objectives for the electronic safety equipment, determined the safety functions and their requirement levels up to SIL2, and defined the shutdown of a battery bank as a safe state. The BMS realizes these functions, and the manufacturer of this system, Sensor-Technik Wiedemann GmbH, is able to provide verification in accordance with the applicable standards. Each of the modules installed in the battery includes a cell supervision circuit (CSC). Each CSC further includes a 16-bit microcontroller plus a redundant highly accurate measuring device for individual cell voltages and temperatures. Each battery is assigned two contactors which allow two-pole switching of the respective battery. These are supplemented by a highly accurate, shunt-based current measurement and insulation monitoring as well as a precharging unit that allows controlled charging of circuit capacitance [15–17,31].

In the case of parallel operation, one BMS can be assigned a master function. This BMS assumes the coordination role and makes the ESS appear to be an individual battery with correspondingly higher capacity. There is a 32-bit microcontroller with floating-point processor. Its basic software is already decoupled from the functionally safe components and performs the more complex algorithms determining the state of the ESS. Testing and validation of the functions of the ESS in interaction with the other hybrid drive components was performed on a hybrid test bench adapted for this purpose. The tests confirmed the functions of the system and were followed by an endurance run in order to gain long-term experience with lithium-ion ESSs of this size.

In order to evaluate the fuel-saving potential of the selected rail vehicle compared to conventional railcars, an optimum operating strategy was devised based on a control-oriented simulation model at the University of Rostock. The main components were modeled in terms of the longitudinal dynamics of a two-part local railcar converted into a hybrid vehicle. In accordance with the basic idea of the Bellman optimality principle—each end of an optimum sequence of decisions is optimum per se—the sequence of decisions is built up successively from an end state in the direction opposite to the simulation direction. Here the battery SOC is selected as the dynamic state variable and the load distribution between electric motor and diesel engine as the control variable. The result for the speed profile is a fuel saving of 18.1% compared to the conventional diesel vehicle. The corresponding progression of the energy accumulator SOC meets the requirement for a balanced SOC and also displays a maximum charge stroke of 4%, superimposed by numerous micro cycles <1%. Useful life simulations gave reason to expect that about 125,000 of these load cycles are possible. Superimposed with the calendric aging of the cells, an operating time of over 10 years can be expected.

The invention by Ibok [27] addresses the limitation of singular energy storage devices by integrating multiple modular storage units interfaced with each other such that the electrical energy content of one unit can be throttled into a connecting unit when energy content of said unit is detected to be below a predetermined threshold.

Specifically, the invention outlines a method of integrating power density intensive components (batteries) and energy density components (capacitors) into single power supply wherein the electrical energy content in the power density intensive component is replenished from the energy density intensive component when the electrical energy content drops below a predetermined level. Additionally, two or more energy density intensive components are integrated so that each component can be charged or discharged to a predetermined level in the presence of a voltage gradient across the matrix. An integrated modular storage scheme is also disclosed wherein stackable modular storage units are connected to portable energy generation units such as solar vanes. Such storage modules are energized passively for subsequent energy utilization. A method of EV power fabrication wherein the power supply unit comprises an integrated battery and storage unit implies that the storage unit can be located anywhere in the car displaced from the battery unit. The power to the battery unit is throttled from the storage unit when it drops below a predetermined level. Since the storage unit is standardized, the battery of an EV does not have to be swapped out being integrated with the storage unit. The storage unit can be replaced or swapped out thus increasing the travel range of the car [59].

8.2.2 MODULAR BATTERY SYSTEM FOR COMMUNITY ENERGY STORAGE

The concept of community energy storage (CES) [20,21] has captured the imagination of the growing ranks of stakeholders with interest in electricity storage (storage). It involves electric-utility-owned storage that is distributed, being located at the periphery of the utility distribution system, near end users. The potential benefits of distributing the storage capacity rather than using one or a few large units can be significant. Although it is not a value proposition per se, CES embodies many attractive facets of the broader storage value proposition for the electricity grid and marketplace of the future, one that is smarter, more sustainable, more diverse, and more distributed and modular. CES is especially important as an example of grid-connected and utility-owned-and-operated distributed energy storage systems (DESS). DESSs are modular storage systems that are located at or near end-user homes and businesses.

The genesis of the CES concept was investigated by American Electric Power (AEP), who after starting with a larger battery storage settled on the concept involving numerous much smaller units—rated at 25 kW for 3 h or 75 kWh—that are distributed and located at or near end-user sites. So, instead of deploying one or two large battery systems with a power output of 2 MW at the utility substation, the alternative is to deploy 80 individual systems at or near end-user homes and businesses whose power output is 25 kW [20,21].

AEP describes the approach as a fleet of small distributed energy storage units connected to the secondary of transformers serving a few houses controlled together to provide feeder-level benefits. Special design attention was given to making the CES resemble conventional utility equipment. One notable advantage of using many smaller units is "unit diversity." Because there are so many units, it is unlikely that a substantial amount of CES power will be out-of-service at any time. That is helpful if reliability is especially important. CES is designed to "island" which means that when a localized portion of the distribution system becomes electrically isolated from the rest of the grid, CES can "pick up" the end-user demand and can serve that demand while there is stored energy. So, CES functions autonomously to provide "backup" power in case of, for example, a traffic accident or a fallen tree.

Important elements of the rationale for CES include the following: (a) it can provide numerous benefits; (b) it is a flexible solution for many existing and emerging utility needs; and (c) to one extent or another, eventually, utility engineers will include modular distributed storage as a standard alternative in their growing toolkit of solutions and responses. CES is expected to provide numerous benefits in many possible combinations. It can serve as a robust, fast-responding, and flexible alternative to generation. It can store low-priced energy and use that energy when the price is high. CES can also be used to provide most types of "ancillary services" that are needed to keep the electrical grid stable and reliable. Depending on the location, CES may reduce the need for transmission and distribution (T&D) capacity because CES provides power locally, so less T&D equipment is needed to serve the local "peak demand." CES can also improve the local electric service reliability and power quality. Of particular interest is CES used to maintain a stable voltage in the distribution system.

CES can play an important role in the integration of renewable energy (RE) generation into the grid, including large scale/remote wind generation and distributed (e.g. rooftop) photovoltaics. CES addresses two notable RE generation integration challenges. First, CES can be charged with wind generation output, much of which occurs at night when the energy is not very valuable. In some circumstances, demand for energy is less than the amount being generated, so wind generation is "curtailed" (turned off) or the system operator must pay someone to take the energy. By charging at night, CES takes advantage of the time when transmission systems are less congested and more efficient. Second, CES can be used to manage localized challenges related to "power quality" posed by wide penetrations of photovoltaic (PV) systems, especially in residential areas. Of particular note are undesirable voltage fluctuations that occur such as those associated with rapid variations of output due to passing clouds [20,21].

INCHARGE POWER [52], a business of INRADA GROUP out of the Netherlands, is the first company in the world to take a refugee camp off-grid, providing independence from the grid utility provider and reducing dependency on costly, maintenance intensive, dirty, noisy generators. The system powers the most recognized refugee camp in the world, Camp Kara Tepe on the Island of Lesvos Greece. INCHARGE's Power Scout system is a complete flat pack modular kit that is easy to transport, simple, quick, and easy to install in under 8 h and provides much needed power to the 1,000 residents of the camp. The system utilizes solar power during the day and stores the excess energy produced through their Power Scout power management system and into their unique Power Safe crystal energy storage modules (the greenest, safest alternative to lithium ion and acid battery technology "99% recyclable" with high performance under high and low temperatures). The Power Scout system requires no maintenance and will auto-restart itself if the system runs completely empty, meaning it requires no manual intervention like other systems. Combined systems produce 24,000 Wh on solar energy and store a total of 96,000 Wh; the system produces enough excess energy to provide residents power for lights, fans, and phone charging throughout the night and enough backup power to support emergency crisis situations as and when needed.

INRADA indicates [52] that modular battery storage technology can provide off-grid generator replacement which is cost effective and maintenance free. Systems can be immediately adopted by all agencies who operate across the world in remote camp environments; they are safe, simple, easy to install, and modular, meaning they can be upgraded without modification to the existing installation as and when needed should more power be required. The benefit of adding crystal batteries is simple; they are safe, store energy from the sun faster, and operate fantastically well under elevated heat conditions, whereas acid batteries lose up to 50% of their expected life with every 7°C (44°F) increase over 25°C (77°F). These installed systems generate and store enough renewable energy to provide power to over 1,000 residents and utilize the skills of the refugees on the camp for the installation. The Power Scout system is anticipated to deliver power for the expected lifetime of the camp. The system requires no infrastructure; they operate completely stand-alone; one may call it a remote energy island, drop-ship the kit into any environment anywhere, and within 8 h, one is completely power self-sufficient, no grid tie challenges, no network constraints, no long distance power line infrastructure. The return on investment for this system versus the use of generators is very high.

In order to provide CES more reliably, the Modular Energy Storage Architecture (MESA) Standards [35–38] alliance and the SunSpec alliance have jointly released the first open, non-proprietary ESS specification. The draft specification, referred to as SunSpec Energy Storage Model Specification, incorporated in MESA specifications as a "MESA Device," was developed through a joint effort and proposed standards for how the different components of an ESS (power meters, power conversion systems, and batteries) communicate with one another. As utilities upgrade their operations to effectively manage and integrate the next generation of distributed assets, standards like SunSpec and MESA will be essential to successfully achieving the smart-grid vision. A technical storage working group have met weekly over the past 9 months to draft the specification [35–38].

Snohomish County Public Utility District (PUD) brought online its second ESS at a utility substation. The battery storage systems aimed to transform the marketplace and how utilities manage grid operations. They also were designed to improve reliability and the integration of renewable energy sources. The Snohomish county PUD systems are built using an innovative approach known as Modular Energy Storage Architecture (MESA). It offers a non-proprietary and scalable approach to energy storage. What sets MESA [35–38] apart from other energy storage efforts is it uses standard interfaces between equipment components, such as the power conversion system, batteries, and control system. These provide utilities more choices, reduce the complexity of projects, and, ultimately, lower costs.

An industry consortium of electric utilities and technology suppliers developed MESA for the purpose of clearing barriers to growth in energy storage to serve the community. More than 30 organizations have joined the MESA alliance as members or partners, including Snohomish County PUD, Seattle City Light, Pacific Northwest National Laboratory (PNNL), Duke Energy, Doosan GridTech, and General Electric. The PUD's first ESS includes two large-scale lithium-ion batteries, one manufactured by GS Yuasa International Ltd. and supplied by Mitsubishi Electric and a second manufactured by LG Chem. Parker Hannifin manufactured the power conversion systems and battery containerization. The system is located at a substation near the utility's main operations center.

Systems for Energy and Fuel Storage   401

8.2.3 CONTAINERIZED ENERGY STORAGE POWER STATIONS

Energy storage power station has been widely used in peak load and power regulation, cooperation with the thermal power plant output, as well as to meet the requirements of emergency power supply within a small range. Container-type modular storage system as a form of energy storage power station, with high efficiency space utility, convenient installation and transportation, short station completion cycle, strong environment adaptability, high intelligence, and many other advantages, has been widely used in various industrial and commercial electric power applications. The station has the following features:

- High modular degree, simple structure and convenient transportation, installation, and maintenance;
- According to the properties of the selected batteries, we can choose the specialized Intelligent BMS;
- The system is equipped with local and remote monitor system, implementing real-time online control;
- We can adopt air cooling mode, optimize the air distribution, implement intelligent control and temperature control, and save energy consumption; temperature field is uniform;
- Complete safety protection measures to ensure operation personnel and ESS work safely and reliably;
- As per the customer requirement and load character, it provides the custom design solution;
- Based on the system requirement, we can adopt the specialized energy storage battery—lead–acid, gel, lead–carbon, or lithium-ion battery—and configure energy storage units flexibly.

There are numerous containerized energy storage power systems offered by the industry. Kratos' [24,25] prefabricated enclosures provide end users with a secure, scalable, and rapidly deployable environment for housing mission critical electrical, power, and cooling equipment. These systems are custom designed and purpose built to meet the requirements of each project. Whether the need is for a modular walk-in enclosure, externally accessible module, or ISO container-based platform, Kratos can provide a proven solution. Kratos' specialized modular electronic enclosure systems are in use by major commercial enterprises and government organizations worldwide. Kratos' rapidly deployable solutions are offered with high levels of equipment integration, electromagnetic interference (EMI) shielding, power distribution, environmental controls, and various security features [24].

Just like Kratos, DropBox Inc. [25] has been able to design, engineer, build, and deliver portable containerized power solutions to customers worldwide. From the U.S. to Germany, Russia, and beyond, DropBox Inc. has worked with customers to create battery storage, inverter housing, and motor control systems that are turnkey, portable shipping container modifications. Over the past few years, DropBox Inc. has modified ISO containers to house battery storage, inverter storage, power control systems, and motor control systems. The containerized ESSs built by DropBox Inc.

are designed to tie into a grid, store electricity when the demand on the grid is low, and redistribute the stored energy during peak demand. These portable power solutions from DropBox Inc. may be delivered with already installed inverters, battery systems, motor control, and PC systems for monitoring integrated transformers and switchgear stations [25].

Siemens has also developed SIESTORAGE [57,58], a sustainable, containerized, and modular stationary energy storage and power flow management system that combines fast-acting power regulation function and lithium-ion batteries. The system can reach a performance of up to 20 MW at a capacity of 20 MWh. The modular concept can be adapted to specific demands, covering any storage power need or capacity, and provides a wide range of applications for utilities, network operators, industries, cities, and infrastructure. SIESTORAGE enables them to save potential with asset optimization.

As mentioned before, the use of renewables on a large scale leads to new challenges for grid stability: Short-circuit power is a measure for grid stability which producers using wind and solar energy can usually not provide. The infeed of energy from distributed sources can cause a reversed load flow. In distribution grids not designed for this event, damages and power outages can be the result. Since conventional ESSs cannot readily ensure stable grid operation on the lower distribution grid levels today, there is a high demand for energy storage solutions that provide balancing power for primary reserve power. SIESTORAGE is able to deliver available power with next to no delay. Indeed it improves the voltage and supply quality by providing active and reactive power on demand, thus compensating for low-voltage fluctuations in generation within milliseconds [57,58].

SIESTORAGE also ensures the reliability of electrical grids for isolated sites and areas where access to power is limited. In this case, the system represents a sustainable solution combined with renewable energy sources also suited for microgrids supplied with diesel gensets. When no balancing power is available to improve the gensets' efficiency, SIESTORAGE can serve as a "range extender" at higher loads for smaller gensets, optimizing the size and efficiency of the generators. During low-load periods, energy can be taken from the grid and stored for peak-load periods since exceeding the maximum output agreed to with the utility just once can cause high costs. This way, SIESTORAGE helps to avoid expensive peak loads and provides a sustainable solution for industrial processes, infrastructure businesses, and energy-efficient buildings [57,58]. Another field of application for SIESTORAGE is the continuous power supply of sensitive industrial production processes, data centers, hospitals, etc. The system is able to guarantee energy reliability even in the case of an outage. The black-start capability of SIESTORAGE makes the startup of a grid possible when the main supply is not available. The energy stored is sufficient to start a gas turbine, for example, and bridge the grid's power requirements. The variation between power generation and actual load is compensated with the help of spinning reserve. SIESTORAGE reliably provides balancing power within milliseconds, guaranteeing a constant energy supply and cost savings for power generation and the provision of additional reserve power. Fluctuation in feed puts strain on low- and medium-voltage grids. SIESTORAGE compensates for this by storing excess energy and transferring it into the grid later. The amount of energy to be transmitted remains constant and the load on the distribution grid even, avoiding grid extensions and protecting the grid's low voltage (LV) and medium voltage (MV) components. With SIESTORAGE, the size of generators can be optimized, since SIESTORAGE functions as "range extender" to smaller gensets. SIESTORAGE is able to reduce the runtime of diesel generators (switch off at lower loads), thus providing lower fuel consumption and gas emissions for a better environmental footprint [57,58].

Due to the parallel connection of the inverters on the AC side, the very high redundancy of the SIESTORAGE system is an advantage in case of a single point of failure, which has no influence on the availability of the storage system. This leads to the highest availability of power and a high reliability. Through individual balancing of the battery cabinets, the installed battery capacity is optimized at the maximum and provides more reliability by lowest maintenance. The modularity of SIESTORAGE enables the highest design flexibility. The system can be combined and adapted to suit any customer's needs. It comprises an inverter cabinet, a control cabinet, a grid connection cabinet, and up to five battery cabinets per inverter, depending on the battery supplier. It is scalable from ~0.1 to 20 MW and can be integrated into a standard container. SIESTORAGE can thus be adapted to meet any storage power and capacity needs for any application. SIESTORAGE maximizes return and optimizes energy consumption. The results are integrated solutions with state-of-the-art components ranging from storage components, including power electronics and Li-ion batteries, to LV and MV switchgear, transformers, and energy automation, all of which ensure grid integration [57,58].

BYD Company Ltd. [40], the world's leading manufacturer of rechargeable batteries, is targeting the Australian renewable energy market by bringing a profusion of innovative ESSs based on its proprietary lithium–iron phosphate battery. BYD exposed its MINI-ES, B-Box—in high-voltage and low-voltage versions—and containerized energy storage solutions to provide the market with increased efficiency, economy, and flexibility. The BYD B-Box storage system comes in two versions: low-voltage B-Box LV and high-voltage B-Box HV. The B-Box LV system operates with a Goodwe S-BP inverter, making it cost effective, and the B-Box HV system operates with an SMA Sunny Boy Storage inverter, which raises the system's level of efficiency. Both are aimed at the retrofit market that is booming in Australia, that is, a market in which people already have solar panels installed and want to install an ESS. Another advantage of the BYD B-Box systems is that they are modular and scalable from 2.5 to 442 kWh—users can increase the capacity of parallel battery rack to meet different requirements of energy storage—whereas major competitors reach only 13.5 kWh with a large-sized system. A variety of configurations of the two versions of the BYD B-Box can also be used for commercial purposes [40].

The MINI-ES is a small-sized, all-in-one residential ESS that is easy to install and maintain, with an original capacity of 3 kW/3 kWh that can be expanded to 6 kWh. This system is fully certified by Australia's SAA, in line with the latest Australian AS/NZS 4777.2: 2015 standard, and has become very popular with a number of Australian local power grid companies and electricity retailers. To date, over 500 Australian households have chosen to install the BYD MINI-ES and have shown great satisfaction with the product. The BYD containerized energy storage has a modular design to meet different large-scale energy storage projects demands and features extreme safety, reliability, and efficiency. It can be used for peak load and frequency regulation to stabilize renewable energy or as a backup system to prevent outages. Current installed capacity of BYD ESS exceeds 400 MWh globally [40].

Solarwatt [47–49] launched an MESS, which can be easily adapted to the needs of different customer groups. The "MyReserve Matrix" energy storage consisted only of the control module command and the battery modules, and it has a modular design. The system was almost unlimited in terms of capacity and performance, making it suitable for any number of different applications: from single-family homes to industrial use. The MyReserve storage system consists only of two elements the size of a shoebox: the MyReserve Pack battery module and the MyReserve Command power electronics. All connectors, sensors, and software are completely contained in this small module. If users need more storage capacity or performance, an unlimited number of MyReserve Pack or Command modules can be combined. This allows application-oriented configurations as small as 2.2 kWh or as large as 2 MWh.

Another advantage for installers and consumers was easy replacement of defective parts due to the modular design. This also allows easy recycling. The new MyReserve is the first Li-ion battery system that can be fully non-constructive when dismantled. The customizable size of the storage system makes profitability even more achievable for end customers. The consistent modularity makes it easier to introduce the MyReserve successfully in the various foreign markets. If the battery needs to be adapted to specific conditions on site, only one of the building blocks needs to be modified, instead of the entire product. This results in significant scale effects.

In the spring of 2017, the PUD [13,44–46] completed deployment of multiple advanced vanadium-flow batteries, built by UniEnergy Technologies, at a second PUD substation. By capacity, it is the world's largest containerized vanadium-flow battery storage system. Together with the lithium-ion batteries, these systems will store enough energy to power nearly 1,000 homes for 8 h. The vanadium-flow batteries and control systems are housed in 20 shipping containers, each 20 ft (6 m) in length, packed with tanks of liquid electrolyte solution. While this technology requires a larger footprint than lithium-ion systems, it can store 70% more energy and has a longer lifetime. To support the PUD projects, PNNL (pacific northwest national laboratory) has developed use cases, or detailed descriptions, of the many ways energy storage can increase renewable energy use as well as how it can improve overall grid efficiency and resiliency. The PUD is actively employing these use cases as it implements and evaluates its energy storage projects. PNNL also is providing analytical and technical support, including conducting benefits analysis, designing test plans, and enhancing control strategies [13,44–46].

UniEnergy Technologies (UET) recently installed a 2 MW/8 MWh vanadium-flow battery system at a Snohomish PUD substation near Everett, Wash. Snohomish PUD concurrently is operating a modular, smaller (1 MW/0.5 MWh) lithium-ion battery energy storage installation. The PUD explains: The utility is managing its energy storage projects with an Energy Storage Optimizer (ESO), a software platform that runs in its control center and maximizes the economics of its projects

Systems for Energy and Fuel Storage

FIGURE 8.1 A rechargeable battery bank used in a data center [13].

by matching energy assets to the most valuable mix of options on a day-ahead, hour-ahead, and real-time basis. Through standardization, MESA accelerates interoperability, scalability, safety, quality, availability, and affordability in energy storage components and systems. Application of the MESA standards should permit future system upgrades and module replacements as energy storage technologies mature [13,35–38]. A typical rechargeable battery bank used in data center is illustrated in Figure 8.1.

8.2.4 MODULAR FLOW CELL BATTERIES

Flow cell [15–17,56] battery (also known as redox flow cell battery) technology is being applied by the Swiss firm nanoFlowcell AG for use in automotive all-electric power plants. This flow cell battery doesn't use rare or hard-to-recycle raw materials and is refueled by adding "bi-ION" aqueous electrolytes that are "neither toxic nor harmful to the environment and neither flammable nor explosive." Water vapor is the only "exhaust gas" generated by a nanoFlowcell®. The e-Sport limousine and the QUANT FE cars successfully demonstrated a high-voltage electric power automotive application of nanoFlowcell® technology. Since its production in 2015, flow cell batteries have not made significant inroads as an automotive power source; however, the firm now named nanoFlowcell Holdings remains the leader in automotive applications of this battery technology [15–17,56,60,61].

In contrast to most other electric car manufacturers, nanoFlowcell Holdings has adopted a low-voltage (48 V) electric power system for which it claims the following significant benefits. "The intrinsic safety of the nanoFlowcell® means its poles can be touched without danger to life and limb. In contrast to conventional lithium-ion battery systems, there is no risk of an electric shock to road users or first responders even in the event of a serious accident. Thermal runaway, as can occur with lithium-ion batteries and lead to the vehicle catching fire, is not structurally possible with a nanoFlowcell® 48 V drive. The bi-ION electrolyte liquid—the liquid "fuel" of the nanoFlowcell®—is neither flammable nor explosive. Furthermore, the electrolyte solution is in no way harmful to health or the environment. Even in the worst-case scenario, no danger could possibly arise from either the nanoFlowcell® 48 V low-voltage drive or the bi-ION electrolyte solution [60,61]."

In comparison, the more conventional lithium-ion battery systems in the Tesla, Nissan Leaf, and BMW i3 electric cars typically operate in the 355–375 V range, and the Toyota Mirai hydrogen fuel cell electric power system operates at about 650 V. In the high-performance QUANT 48 V "supercar," the low-voltage application of flow cell technology delivers extreme performance [560 kW (751 hp), 300 km/h (186 mph) top speed] and commendable range [>1,000 km (621 miles)]. The car's four-wheel drive system is comprised of four 140 kW (188 hp), 45-phase, low-voltage motors and has been optimized to minimize the volume and weight of the power system relative to the previous high-voltage systems in the e-Sport limousine and QUANT FE.

A version of the QUANTiNO [61] without supercapacitors currently is being tested. In this version, the energy for the electric motors comes directly from the flow cell battery, without any buffer storage in between. These tests are intended to refine the BMS and demonstrate the practicality of an even simpler, but lower performance, 48-V power system. The power from energy storage system can be used to maintain grid stability. We need dispatchable grid storage systems because of the proliferation of grid-connected intermittent generators and the need for grid operators to manage grid stability regionally and across the nation.

Flow cell battery technology has entered the market as a utility-scale energy storage/power system that offers some advantages over more conventional battery storage systems, such as the sodium sulfur (NaS) battery system offered by Mitsubishi, the lithium-ion battery systems currently dominating this market, offered by GS Yuasa International Ltd. (system supplied by Mitsubishi), LG Chem, Tesla, and others, and the lithium iron phosphate ($LiFePO_4$) battery system being tested in California's GridSaverTM program. Flow cell battery advantages include the following [15–17,56,59]:

1. Flow cell batteries have no "memory effect" and are capable of more than 10,000 "charge cycles." In comparison, the lifetime of lead-acid batteries is about 500 charge cycles and lithium-ion battery lifetime is about 1,000 charge cycles. While a 1,000 charge cycle lifetime may be adequate for automotive applications, this relatively short battery lifetime will require an inordinate number of battery replacements during the operating lifetime of a utility-scale, grid-connected ESS.
2. The energy converter (the flow cell) and the energy storage medium (the electrolyte) are separate. The amount of energy stored is not dependent on the size of the battery cell, as it is for conventional battery systems. This allows better storage system scalability and optimization in terms of maximum power output (i.e., MW) versus energy storage (i.e., MWh).
3. No risk of thermal runaway, as may occur in lithium-ion battery systems.

The firm UET offers two MESSs based on flow cell battery technology: ReFlex and the much larger Uni.System™, which can be applied in utility-scale dispatchable power systems. UET describes the Uni.System™ as follows: "Each Uni.System™ delivers 600 kW power and 2.2 MWh maximum energy in a compact footprint of only five 20′ containers. Designed to be modular, multiple Uni.System can be deployed and operated with a density of more than 20 MW per acre, and 40 MW per acre if the containers are double-stacked."

8.2.5 Other Notable Modular Battery Storage Systems

In 2017, Zyskowski and Neroutsos [13] reported that Mercedes-Benz Energy and Vivint Solar are partnering to bring the Mercedes-Benz home ESS to solar customers in the U.S. The ESS will consist of modular 2.5 kWh batteries that can be combined to create a system as large as 20 kWh, making it easy to scale and tailor to individual customer profiles. Vivint Solar will customize systems based on the customer's individual energy consumption needs, providing consumers with a tool to help them manage their energy costs while using solar. The pairing will offer the home solar plus storage system in California first.

Li et al. [18] not only examined the vehicle battery technology in terms of energy consumption and the environment, but also highlighted nanotechnologies and systems design. The study included current situation and future development trends of batteries. Graphene batteries have a higher specific energy, higher specific capacity, and lower cost than conventional batteries. The study pointed out that if nanographene technology enhances battery power and prolongs life span, it will have attractive prospects in the field of EVs. The study summarized the recent progress in graphene nanobatteries regarding structural models, nanoscale effects, and system design. The study also proposed future directions for research on battery EV technology as well as the prospect of graphene battery applications.

GNB's Restore 500 [51] series is a turnkey solution for easy transportation and installation. The planning and execution schedules can be significantly reduced due to the modular and standardized assembly. GNB's integrated BMS continuously detects and evaluates relevant battery data to operate the battery in partial state of charge—this ensures a significant reduction of total cost of ownership of the overall system. Restore 500 is the right choice for applications such as Hybrid & Green Deployment, Grid & Power Quality, Renewable Energy Management, or Back-up Power (UPS). GNB® Industrial Power has developed Restore 500, a modular "Plug & Store" ESS that helps to control energy from renewables and stabilizes power generation and consumption [51].

Ioakimidis et al. [59] indicated that the lithium iron phosphate ($LiFePO_4$) battery is becoming the favored choice of electrochemical energy storage for EVs and hybrid vehicles (HVs) due to its high energy density and low self-discharge. However, battery operating temperature plays a vital role in the reliability, lifespan, safety, and performance of EVs and HVs. Battery thermal management system (BTMS) must keep the operating temperature of the battery pack between 20°C and 40°C in order to achieve good performance and long lifespan. To this day, this task remains a challenging subject for the EV development. BTMS consumes energy from the onboard battery pack, thus reducing the range of the vehicle. In order to reduce this adverse impact, this paper presents a novel approach that takes advantage of the non-uniform surface distribution of Li-ion battery cell, which results from complex reactions inside the cell. First, Li-ion hotspots were identified and found next to the positive and negative tabs. Then, thermoelectric coolers (TECs) are mounted next to the tabs and in the center of the Li-ion battery. The control circuit is designed to turn on and off TECs in order to reduce the parasitic power feeding the BTMS. Experimental results show the feasibility of this system.

KREISEL ELECTRIC [33,34] developed MAVERO, *a highly efficient and flexible ESS for private and commercial application. Due to the modular design, each unit can be tailored precisely to individual conditions. The flexible home storage system will be available in four different sizes. The battery packs are based on Li-ion technology and the usable capacity ranges from 8 kWh to 22 kWh. The elegantly designed casing is available in two colors and communicates all charge and discharge activities by means of LED visualization* [33,34]. ABB's modular approach [30] enables a wide range of system sizes and layouts to accommodate the requirements specific to each energy storage project. The modular system design offers a range of leading-edge power conversion systems, integrated system controls, state-of-the-art protection systems for AC and DC equipment, primary interconnecting equipment such as transformers and switchgear, and the latest in lithium-ion technology to optimize the energy usage and density of the overall system. ABB's energy storage module (ESM) is a **single- or three-phase system in arc-proof enclosures up to 5 MW/4 h with output voltage ranging from 120 V to 40.5 kV.**

ABB's ESM portfolio offers a range of modular products that improve the reliability and efficiency of the grid through storage. In addition to complete ESSs, ABB can provide battery enclosures and Connection Equipment Modules (CEMs) as separate components. The ESM portfolio maintains the balance between generation and demand, benefitting the grid in a number of ways [30]:

1. Provides smooth grid integration of renewable energy by reducing variability
2. Stores renewable generation peaks for use during demand peaks
3. Flattens demand peaks, thereby reducing stress on grid equipment
4. Supports infrastructure as loads increase with EV use.

Rothgang et al. [28] showed that with large-scale battery systems being more and more used in demanding applications regarding lifetime, performance, and safety, it is of great importance to not only utilize cells with a high cyclic and calendric lifetime but also optimize the whole system architecture. The aim of this work was therefore to highlight the benefits of a modular system architecture allowing the use of hybrid battery systems combining high-power and high-energy cells in a multitechnology system. The study showed that in order to achieve an optimized performance, efficiency, and lifetime for an EV, the complete drive train topology has to be taken into account instead of optimizing one of the components individually. Consequently, the topic was analyzed from the system's point of view, addressing in particular the modularization of the battery as well as the power electronics needed to do so. The study showed that a highly flexible battery system can be realized by DC–DC converters between a modular, hybrid battery system and the drive inverter. By the DC–DC converters, the battery output voltages and the inverter input voltages are decoupled. Hence, the battery's topology can be chosen unrestrictedly within a wide range and easily be interconnected to a common DC link of a different voltage. This flexibility helps the lifetime of battery and its impact on system weight (28).

The MESS from RCT Power [39] can be upgraded and adapted to fit user requirements. The high-voltage battery guarantees efficiency and autonomy for the ESS. The RCT storage solution utilizes solar energy to make the home more independent of energy providers. Stored solar energy is available in the evenings, when the sun doesn't shine, or during power cuts. The modular system can be easily upgraded and adapted to fit user requirements. The battery storage has a particularly space-saving design and is easy to install. The "Fits-all" technology allows us to combine all types of solar panels with the storage system. The systems from RCT Power come with three to six modules, the nominal capacity starts with 5.7 kWh up to 11.5 kWh. The high-voltage battery guarantees efficiency and autonomy. All batteries are delivered in the elegant design of RCT Power [39].

Powerstorm [54,55] Company produced its commercial-scale MESS which is a hybrid "off-grid" system that uses solar and wind energy, along with a generator and lithium-ion batteries for energy storage. Powerstorm stores its energy in its space-saving lithium-tray batteries. The systems are environmentally friendly, and they afford energy density, rapid charging, and other characteristics needed for various demanding storage applications. Powerstorm systems are designed to generate electricity on demand to keep mission-critical systems running during power outages. On-site storage makes it easier to add on-site generation from solar and wind by making it possible to accommodate both peak load and peak production. No facility is too large or too small to benefit from Powerstorm. With the many benefits afforded by an intelligent energy storage and management system, the investment usually pays for itself in 5–7 years. The payback can be even shorter by participating in the local utility's demand response programs—an opportunity that energy storage makes possible by eliminating the potential adverse impact on business operations. The ultimate future of smart buildings and campuses is a microgrid that consumes "net zero" energy from the grid and has a minimal or zero carbon footprint. Powerstorm ESS can make that future possible [54,55].

The world's first modular large-scale battery storage system, a 5 MW device is being built at a research university in Aachen, Germany. The modular, multimegawatt, multitechnology medium voltage battery storage system, handily abbreviated to M5BAT [50], is being built at the technical institute RWTH Aachen University. It will assess and demonstrate the appropriateness of combining a number of different battery types into one system. The project seeks to further the integration of renewable energy into electricity networks, while it is also hoped that increased modularity will allow the system to be applied to a broader range of uses than existing large-scale battery storage devices. The project involves researchers from RWTH Aachen University and the research center of multinational utility company E.ON. Also involved are the GNB Industrial Power unit of battery maker Exile Technologies, E.ON itself, and solar inverter manufacturer SMA Solar Technology. GNB Industrial

Power and SMA supply technical components. Modularity of battery systems is becoming an ever-bigger question for the industry. Sizing batteries to meet the different needs of each region and application can be both expensive and tricky. One German manufacturer known primarily for its ESSs, ASD Sonnenspeicher, claims that with a new system that it has developed called Pacadu, batteries of different type and size can be neatly put into parallel circuits [50].

The significant literature on modular battery storage [15–59,60,61] described here shows wide acceptance of modular approach by industry for battery storage systems. Many of these systems are containerized, and some of them can handle even large-scale operations.

8.3 MODULAR ELECTRICAL AND ELECTROMAGNETIC ESSs

8.3.1 MODULAR ULTRACAPACITORS OR SUPERCAPACITORS

Ultracapacitors and batteries are highly complementary technologies within a hybrid system. The ultracapacitors fulfill short-duration, high-power application needs such as renewables intermittency smoothing and frequency response, while the batteries are well suited for longer-duration applications requiring more energy, such as time shifting. Following points should be noted about ultracapacitor cell technology [62–73]:

- Ultracapacitors and batteries can be combined to increase the overall value of the energy storage system.
- Ultracapacitors help to keep peak power loads off the batteries.
- Like batteries, ultracapacitors are scalable.
- Ultracapacitors offer the advantages of longer lifetime and wider operating temperature range.

Figure 8.2 illustrates basic diagram for supercapacitor. As shown in Figure 8.3, there are different types of supercapacitors. Capacitors have numerous applications in electrical and electronic applications. The use of ultracapacitor as an energy storage module has following advantages and disadvantages.

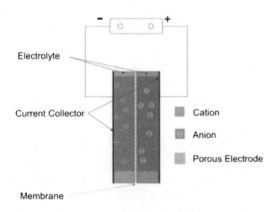

FIGURE 8.2 Schematic illustration of a supercapacitor [66].

Systems for Energy and Fuel Storage

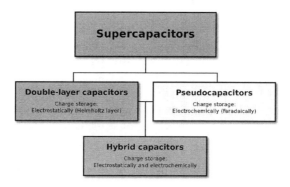

FIGURE 8.3 A diagram that shows a hierarchical classification of supercapacitors and capacitors of related types [66].

Advantages:

1. It can charge and accumulate energy quickly (see Figure 8.4).
2. It can deliver the stored energy quickly.
3. Losses are small compared to other storage medium, long service life, and low (or no) maintenance.

Disadvantages:

1. Low energy capacity compared to batteries.
2. Limited energy storage per dollar cost.
3. Stored energy will eventually deplete due to internal losses.

Some novel designs are being developed to overcome disadvantages. *For example, Shanghai is experimenting with supercapacitor buses, called the Capabus. Siemens is developing a hybrid storage utilizing both capacitors and batteries to be used in the powering of light rail systems. In both these applications, capacitors are*

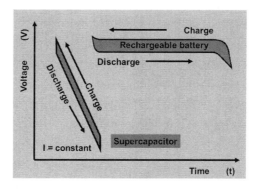

FIGURE 8.4 The voltage behavior of supercapacitors and batteries during charging/discharging [66].

quickly charged at stops or along the route by connecting to overhead charging points. Ultracapacitors from Maxwell technologies [63] are 48 V general purpose modules. They are extensively used for propulsion and energy recuperation in bus; propulsion, voltage stabilization, and locomotive engine starting in rail; and energy recuperation in forklifts and cranes as well as for some industrial storage and power delivery purposes. The rugged, environment-resistant design makes the modules ideal for hybrid mining, construction, train, and bus applications. The modules are specifically engineered to provide cost-effective solutions for hybrid bus and construction equipment and are also widely utilized in UPS, telecommunications, and other industrial electronics applications. These come in two models, BMOD0083 P048 B01 and BMOD0165 P048 BXX, CXX. They are used in hybrid vehicles, rail, industrial heavy equipment and UPS systems. Their other features include DC life up to 1,000,000 duty cycles or 10 years, 48 V DC working voltage, active cell balancing temperature output, availability of over voltage outputs, high power density, compact, rugged, fully enclosed, splash-proof design [62,63].

8.3.2 MODULAR DOUBLE-LAYER CAPACITORS

Many energy storage modules will use electric double-layer capacitors [64–66], often referred to as supercapacitors. Supercapacitors use a liquid electrolyte and charcoal to form what is known as an electrical double layer. This greatly increases the capacitance. Capacitors with large Farad rating and small size can be obtained. Note: As they have a large internal resistance, double-layer capacitors are not suitable for AC circuits. The greater the capacitance or the voltage, the more the energy it can store. When connecting capacitors in series, the total capacitance reduces but the voltage rating increases. Connecting in parallel keeps the voltage rating the same but increases the total capacitance. Either way, the total energy storage of any combination is simply the sum of the storage capacities of individual capacitors. By connecting two modules in series (doubling the voltage, halving the capacitance), the energy storage can be doubled. These are safe designs. Capacitors store energy and will remain charged when disconnected from any supply. Before working on any capacitive systems which have been isolated from the power supply, be careful to take all necessary steps to ensure the capacitors are fully discharged.

Hybrid energy systems (HESs) employ capacitors in conjunction with batteries to benefit from the advantages of both technologies while minimizing the disadvantages. Within a hybrid energy system, the capacitors can charge and deliver energy more quickly, thereby reducing strenuous duty of the batteries. The batteries can achieve greater energy densities and provide for longer running duration. The use of this type of HESs is becoming more popular, particularly in transportation applications [64–66].

8.3.3 MODULAR INTEGRATED ENERGY STORAGE DEVICE AND POWER CONVERTER BASED ON SUPERCAPACITORS

With the increasing energy demand and emergence of renewable energy, large amounts of energy must be stored and later very quickly restored, generating high power [62]. A promising approach is to use supercapacitor banks which have a shorter reaction time and larger power output than batteries. To reach a high enough voltage, supercapacitors need to be connected in series. The problem is that since the supercapacitors all have a slightly different capacitance, they cannot be simultaneously fully charged and the available storage maximized, unless an equilibrating circuit is used (at the expense of additional losses). Special care must also be taken to avoid over voltage on individual capacitors. Moreover, supercapacitor banks require a power converter to be used alongside, which increases the size of the system and causes EMI due to high-frequency switching [62–66].

This new design integrates the storage unit and the power converter in the same compact modular device made of a number of basic cells connected in series comprising only low-cost, low-voltage components. The basic cell contains two supercapacitors as well as switches protecting against overvoltage by supplying an alternative current path and allowing to make use of the full storage capacity by controlling the voltage of each capacitor individually. The switches are controlled by a programmable microcontroller capable of performing any desired conversion (DC–DC, AC–DC, DC–AC, AC–AC) by putting a variable number of cells in series. The innovative integrated power converter hence does not rely on high-frequency switching which eliminates the EMI problem. The device could be applied for braking energy recovery in trains, subways, cars, motorcycles, bicycles, etc. In addition to providing energy storage, the device replaces standard stand-alone power converters with the advantage that the storing device provides short-term autonomy in case of power outage. For renewable energy, it could also be useful to level the power output of wind turbines and solar panels in which the energy production and consumption are quite irregular [62].

Its **competitive advantages** are as follows [62–66]:

1. One compact device for both energy storage and power conversion
2. Maximized stored energy by controlling the voltage on each capacitor
3. Low-frequency operation and thus no electromagnetic interference
4. Suitable for any type of power conversion (AC–DC, AC–AC, DC–AC, or DC–DC)
5. Quick and easy to repair modular arrangement of low-cost, low-voltage components.

Jiang et al. [62] outlined a bifunctional converter module for supercapacitor energy storage based on an input-series-output-series (ISOS) circuit. Compared to the existing topologies, the proposed circuit acts both as a supercapacitor cell voltage equalizer and as an output voltage regulator. During the charging process, series-connected flyback converters transfer energy from fully charged cells to the string via a bypass. Therefore, the fully charged cells maintain the rated terminal voltage, and the charging speed of the remaining cells increases. During the discharging process, the output voltage of the module can be regulated with the same set of converters. The working principles during the discharging and charging processes are analyzed. The designs of the flyback converter and the control system are presented. A prototype consisting of three submodules is implemented. Experimental results verify the effectiveness of the proposed topology [62].

8.3.4 SUPERCAPACITOR MODULE SAM FOR HYBRID BUSES

As a prototype, the TOHYCO-Rider Bus was first introduced to the public in 2002. The feasibility of this unique concept, combining supercapacitors and inductive power transfer of energy, was then demonstrated. After that the TOHYCO-Rider bus became semi-industrialized for public service. The challenge was to achieve a safe and reliable performance of this bus. Supercapacitor module SAM for hybrid buses used in this system is an advanced energy storage device. Because of the rapid improvement of the supercapacitor technology and due to the fact that these components are delivered only with axial connector pins, the energy storage of the bus had to be completely redesigned to the Maxwell types BCAP0008 with 1800F [67–73]. Therefore, a complete modification of the vehicle, including new cabling and improvements regarding electromagnetic compatibility (EMC), took place [67–73].

The control tasks and the energy management was improved and tested by simulation with MATLAB-Simulink and SIMPLORER. Also the MMI (man–machine interface) and diagnostic system were finished. Supercapacitors have become very important energy storage devices during the last several years. Today supercapacitors are also important components for hybrid cars and buses. Since 1998, The Competence Center Integral, Intelligent and Efficient Energy Systems (CC IIEE) has been working on hybrid bus systems with supercapacitors and has realized the TOHYCO-Rider bus. The concept consists of a serial hybrid drive, a supercapacitor energy storage of 1.5 kWh, and the fast and contactless charging station inductive power transfer (IPT). The basic principle is the following: quickly recharge the bus at every bus stop by the IPT charging station. For short rides without IPT, use an emergency traction battery. The TOHYCO-Rider bus project shows two very important features: the supercapacitor component SAM and the hybrid bus system with the control strategies. The supercapacitor itself cannot be integrated in a drive system like batteries. It is necessary to have well-developed modules that control overvoltages, fulfill charge balancing, and have adequate supervision circuits. The advantages of supercapacitors often have to be combined with those of batteries and therefore this combination is often used. Furthermore, the mounting and packaging has to be done in a way for optimal maintenance possibilities. All this is combined in the concept of SAM [6,8,17]. For the TOHYCO-Rider, a second generation of SAM could be realized independently of the pilot service bus project [6]. For the bus, SAM consists of only supercapacitors.

The combination of a drive-chain system by means of supercapacitors and IPT is a quite unique public mobility concept: only little energy is charged in a short time and holds on only for reaching the next station and charging place. Ideally, a whole fleet of buses share the infrastructure. The experiences of the mentioned projects showed that the future research for projects with the improved storage technologies should concentrate on a SAM with advanced functions that is even cheaper and an energy management system that plays very well together with SAM. While passengers leave or get into the bus, it is recharged in a few seconds. As a result, only little weight has to be carried around instead of tons of installed batteries. Another important aspect that could be solved by the CC IIEE is the hybrid bus energy management which is much more complicated than that in a traditional EV. Due to the higher number of power sources and drains in the DC link, the stability of regulation becomes quite complex. The developed controller area network (CAN)-bus system can be transferred to other hybrid vehicles and general drive applications very easily.

The basic concept of a SAM consists of a combination of batteries, supercapacitors, and intelligence depending on the application. SAM can also consist only of supercapacitors as is the case in the TOHYCO-Rider project. But SAM also includes a development methodology from simulation to the energy management algorithm and further on to the production of commercial packages. A supercapacitor module like SAM has to fulfill at least the following basic specifications [67–73]:

1. Overvoltages of >2.5–2.7 V have to be avoided.
2. For serial combination of supercapacitors a charge balancing circuit as well as temperature control have to be installed.
3. The power dissipation has to be under control and may be ventilated away.
4. The total voltage U is not constant but a function of the changing energy W (C = capacity): Therefore the voltage has to be transferred to the DC-link voltage.
5. The behavior of the supercapacitor stack has to be communicated to the superior application control.
6. The costs have to be very low due to the competition situation with batteries. Circuitries for each supercapacitor have to be <1 US$.
7. All the mentioned points above have already been realized in the first- and second-generation SAM (Blue-Angel hybrid car and TOHYCO-Rider bus). For a generally usable SAM with a great market potential, the advanced specification for a third-generation SAM can be defined by the following additional points. SAMs should be integrated as easily as board batteries to a system. Therefore the SAM should be combined and connected together very easily by a special plug.

Every single SAM should be easily exchangeable for maintenance reasons. The SAMs should be able to communicate with the most popular communication systems, such as CAN-bus or others in the same way. A combination with different types of power converters should be possible. The supercapacitor packs should be combined with SAMs with specific sizes according to customers wishes. SAM should fulfill the relevant standards, especially for EMC. A very important issue is the kind of balancing circuits for supercapacitors. There are a lot of different possibilities to control overvoltage and balance the charges between the single supercapacitors [67–73]. An analysis of the mentioned references showed that a lot of applications could be realized with only the VP (virtual parallel) circuitry of CC IIEE [67–73], the resistor ladder network (shortly named R) and the zener-diode principle (shortly named Z).

The VP circuitry has a charge equalizing bypass which is connected by a PWM with every supercapacitor (SCAP). Charge is flowing from the supercapacitor with the highest voltage to the one with the lowest voltage. This works automatically without measuring the individual supercapacitor voltages. This solution is quite expensive, but it can move charge very rapidly and has a high supervision standard. A very cheap and broadly used solution is the R network. Resistors are connected directly over the supercapacitors. Active R circuitries have additional switches that work dependently on the supercapacitor voltages. Normally the charge equalizing speed and efficiency are low. An easy and passive solution works with zener diodes connected in parallel over the supercapacitors. Charge equalizing works only at maximum voltages of the supercapacitors. One way to achieve this is to use advanced SAM of this configuration. For the selection of the type of circuitry three factors are relevant:

1. **Number of duty cycles:** With a lot of duty cycles, the speed of the balancing process is important, and the VP solution should be applied instead of R or Z networks.
2. **Compensation current:** For R networks, the compensation current is rather limited. For the Z network, it can be quite high but then produces losses. With the VP solution, the current can be chosen depending on the type of application. The higher the current, the deeper the efficiency.
3. **Costs:** A compromise has to be found between the costs and the requirements. The VP circuitry is more expensive than an R or Z network. But for the costs, also aspects of reliability, lifespan, and maintenance have to be considered. For example, due to the long-time parametric drift of the supercapacitors, higher balancing currents may be useful. Therefore R and Z networks are not adequate.

All the above-mentioned specifications will be realized in a third-generation SAM. Additionally industrial use of these new SAMs demands the following two points [67–73]:

1. Due to different types of applications, it should be possible that different types of balancing circuits (VP, R, Z) can be used on the same PCB without additional development efforts.
2. For a stack consisting of many supercapacitors, the balancing circuit has to be equipped with a very good self-check utility,

8.4 MODULAR MECHANICAL STORAGE SYSTEMS

8.4.1 Modular Compressed Air ESSs

Fargion and Habib [46] considered a novel ESS based on the compression of air through pumped water. Differently from CAES on trial, the proposed indirect compression leaves the opportunity to choose the kind of compression from adiabatic to isothermal. The energy storage process could be both fast or slow leading to different configurations and applications. These novel storage systems are modular and could be applied at different scales for different locations and applications, being very flexible in charge and discharge processes. The system may offer an ideal energy buffer for wind and solar storage with no (or negligible) environmental hazard. There are a lot of different systems proposed for energy storage; right now the ones commercially developed are just hydrodynamic storage, for large systems, and lead-based batteries for medium to small ones. Alternatives for mechanical storage,

CAES systems have been under testing for a long time in some sites like Huntorf in Germany (operating since 1978) and McIntosh, Alabama, USA (operating since 1991). These systems are quite promising as they don't have the geographical limits of hydrodynamic storage, thus being of wider use. Nonetheless they need to provide heat to air before turbine expansion, leading to a delay in activation and fuel consumption. Moreover this requires many ancillary services for the storage plant. These systems have low energy efficiency, rated 40%–75% [11], too, due partly to thermal issues and partly to mechanical issues related to air compression and air expansion with a variable pressure gap [74–83].

Some evolutions of CAES systems have been proposed, trying to overcome the need for heating of air before expansion. Actually, air has to be cooled after compression, too, in order to reduce its specific volume so as to increase stored mass of high pressure air in the vessel. Thus, heat storage has been proposed to avoid fuel consumption for heating in so-called adiabatic CAES [11,76–79]. This will improve energy and exergy efficiency of systems, but it won't be useful to avoid activation delay that limits the kind of service CAES systems could provide to power grid. Moreover, usual compressors and turbines are not suitable to operate with variable back pressure. So during charging phase, air is compressed up to the highest storage pressure. Before being introduced into the turbine, air is expanded in a valve, lowering its pressure to the lower storage pressure. Thus, despite the fact that the exploitation of CAES technology has already started, a lot of improvement is still possible.

In traditional CAES, compression of air takes place in the compressor, and the air is then moved to the storage vessel. Similarly, air is taken from the vessel and introduced into the turbine for expansion. In the proposed system, air is compressed and expands directly in the storage vessel. This is done through a water piston that modifies air volume, reducing it during charge and increasing it during discharge. The water piston is used as heat storage so as to absorb heat during compression and reject it during expansion. The new system is thus a hydraulic compressed air energy storage (HYCAES). It is composed of high-pressure storage vessel, almost full of air when fully out of power, an atmospheric pond for water storage, a water pump and a hydraulic turbine and connecting pipes [80,81]. It is not ever-new, as there are some papers illustrating similar systems [79–82]. In these and other [83], thermodynamic aspects of proposed systems were analyzed to prove their energy feasibility.

Reversible compression of air by water piston can be done through different polytropic transformations, according to heat exchange of air. Rapid compression and high volume to surface ratios provide an almost adiabatic transformation. In order to avoid limiting power to energy ratio in the system, a rapid phenomenon is assumed, so that heat exchange through vessel is negligible. Nonetheless, a perfect mixing of water to air is assumed so as to have an almost infinite contact surface that lets any heat exchange rate be provided to air. Air and water are assumed to be at the same temperature during transformation. This transformation has the lowest possible polytropic index. It is possible to reach an almost isothermal transformation through a sufficient sparkling of water during compression. Final pressure of storage system is limited by its mechanical stability, as it should be as high as possible, as this enhances energy storage per unit mass of air. On the contrary, even though lower initial pressure means more storable energy per unit mass, it also means lower initial mass. CAES has been proved to be cost effective with efficiency comparable to hydrodynamic storage. It should be highlighted that in the proposed system the fully charged storage vessel is mainly filled with water rather than air. As energy is stored in air compression, this means that energy storage cost per unit volume is higher than that in conventional CAES. Nonetheless, in existing systems, compressed air has to be cooled prior to being stored, losing a lot of energy, and it has to be reheated thereafter, before expansion in turbine. These lead to a large amount of thermal energy loss in conventional CAES. Conventional compressed air storage is almost isothermal, as air is cooled after compression and heated before expansion. Actually, temperature increases during storage and decreases during air extraction, thus reducing energy efficiency, but for present calculations, this will be neglected. Stored energy per unit volume is thus given by the difference in air mass as specific internal energy is almost constant. Thus, neglecting the dependence of specific heat on temperature (limited to 7% in the relevant thermodynamic states of air), it could be expressed as being lower than 25% for pressure ratios higher than 2.4.

On the other hand, the optimization of such a system is rather different; thus comparison of the vessel usage ratio of the proposed system with optimal pressure ratio to conventional is more meaningful. Thus, for low-pressure ratios, usage ratio is higher. It should be noted that Huntorf plant has a 1.57 compression ratio that corresponds to a 40.5% usage ratio. Thermodynamic analysis for Huntorf plant, based on available data [11], with unit efficiency compressor, shows that only 31% of compression work is actually gathered in energy storage, as the rest is rejected during cooling of compressed air. The proposed system didn't need any cooling prior to storage. Thus, the new system proposed had a vessel cost about 2.5 times that of conventional CAES, but it did not have any plant cost for cooling heat exchangers or heating systems with related fuel ancillary services. So, the investment cost was estimated to be between 1 and 1.5 times that of conventional CAES. During operation, newly proposed HYCAES will be 3.3 times more efficient in energy storage for thermal issues only; thus break-even will be obtained in a half of the time needed for conventional CAES. In summary, thermodynamic analysis of proposed system showed that isothermal compression of air through a water piston is possible. The proposed HYCAES is suitable for energy storage with the main advantage of having neither fuel nor a heat storage system. Cost analysis has shown that loss in vessel usage due to water piston displacement is well compensated by the reduction of thermal energy loss after compression and of heat demand before expansion. The proposed system is thus an alternative to available large-scale energy storage. In conclusion, the energy buffer is based on the combination of the well-known huge thermal buffer due to the heat capacity of water over the air. This acts as a thermodynamic reserve that avoids most of the energy dispersion of common CAES storable energy per unit volume [74–83].

8.4.2 Low-Pressure, Modular CAES System for Wind Energy Storage Applications

The construction and testing of a modular, low-pressure CAES system is presented in the study by Alami et al. [74]. The low-pressure assumption (5 bar max) facilitates the use of isentropic relations to describe the system behavior and practically eliminates the need for heat removal considerations necessary in higher-pressure systems to offset the temperature rise. The maximum overall system efficiency is around 97.6%, while the system physical footprint is <0.6 m^3 (small storage room). This provides a great option for storage in remote locations that operate on wind energy to benefit from a non-conventional storage system.

The overall size and capacity of the system can be changed by changing the number of active cylinders, which in this case are off-the-shelf, small pressure vessels used for fire protection. Moreover, the system operation is automated and capable of addressing both high-energy and high-power-density applications with an infinite number of charge–discharge cycles by augmenting the capacity with the required number of storage cylinders. The system is eco-friendly and has low maintenance costs compared to chemical storage. Thus, this study presents an experimental evaluation of a modular, low-pressure CAES. The system is flexible and responds to systems requiring high power density and high energy density alike. The system fares well compared with a battery storage system, with less environmental and disposal cost.

8.4.3 Design of a Modular Solid-Based Thermal Energy Storage for a Hybrid Compressed Air ESS

The share of renewable energy sources in the power grid is showing an increasing trend worldwide. Most of the renewable energy sources are intermittent and have generation peaks that do not correlate with peak demand. The stability of the power grid is highly dependent on the balance between power generation and demand. CAES systems have been utilized to receive and store electrical energy from the grid during off-peak hours and play the role of an auxiliary power plant during peak hours. Using thermal energy storage (TES) systems with CAES technology is shown to increase the efficiency and reduce the cost of generated power [75]. In the study by Lakeh et al. [75], a modular solid-based TES system is designed to store thermal energy obtained from conversion of grid power. The TES system stores the energy in the form of internal energy of the storage medium up to 900 K. A three-dimensional computational study using commercial software (ANSYS Fluent) was completed to test the performance of the modular design of the TES. It was shown that solid-state TES, using conventional concrete and an array of circular fins with embedded heaters, can be used for storing heat for a high-temperature hybrid CAES (HTH-CAES) system [75].

8.4.4 Modular Flywheel Energy Storage

One energy storage technology now arousing great interest is the flywheel energy storage systems (FESSs), since this technology can offer many advantages as an energy storage solution over the alternatives. Flywheels have attributes of a high cycle life, long operational life, high round-trip efficiency, high power density, low environmental impact, and can store megajoule (MJ) levels of energy with no upper limit when configured in banks [84–110].

A typical flywheel energy storage used by NASA is illustrated in Figure 8.5. Flywheel applications range from large scale at the electrical grid level, to small

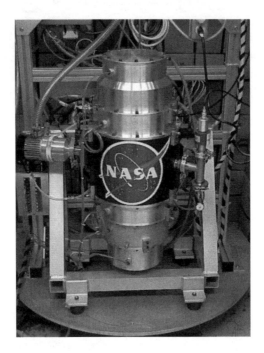

FIGURE 8.5 NASA G2 flywheel [111].

scale at the customer level [89,90]. A high power and capacity is reached by arranging flywheels in banks, rather than by using large machines [98]. The best and most suitable applications of flywheels fall in the areas of high power for a short duration (e.g., 100s of kW/10s of seconds) [87], when frequent charge–discharge cycles are involved [89]. The most common applications are power quality such as frequency and voltage regulation [7,99], pulsed power applications for the military [97], attitude control in space craft [97], UPS [93], load leveling [7], hybrid and EVs [93,97], and energy storage applications [97]. As part of energy storage applications, flywheels perform storage applications both at the grid, as well as at the customer level. FESSs offer the unique characteristics of a very high cycle and calendar life and are the best technology for applications which demand these requirements. A high power capability, instant response, and ease of recycling are additional key advantages. Since this book focuses on modular systems, here we briefly outline in detail the use of flywheels in transportation and renewable industries where modular systems are most applicable.

In transportation, flywheels are used in hybrid and EVs to store energy, for use when harsh acceleration is required or to assist with uphill climbs. In hybrid vehicles, constant power is provided by internal combustion engines to keep the vehicle running at a constant optimum speed, reducing fuel consumption and air and noise pollution and extending the engine life by reducing maintenance requirements [88,89]. At the same time, energy from regenerative braking during vehicle slowdown is stored in flywheels, which is supplied back to provide a boost during acceleration or climbing hills [88,89,96]. The only competitors to flywheels in hybrid vehicle applications are chemical batteries and ultracapacitors. However, ultracapacitors suffer from low energy density and higher cost. Flywheels rank better than batteries based on their longer lifetime, higher power density, higher efficiency, and frequent charge–discharge capability [89]. Furthermore, flywheels are developed for rail applications, both for hybrid and electric systems. They also find a place in gas turbine trains for the same purpose. The desired speed and maximum weight of the train determines the power and energy requirements. It is estimated that 30% of the braking energy could be recovered by this system, due to receptivity issues [89].

In electrical vehicles with chemical batteries as their source of propulsion, flywheels are considered to cope well with a fluctuating power consumption. This will prolong the lifetime of the battery as its charge–discharge cycles become more regular [10]. In train energy recovery systems, flywheels are installed at stations or substations to recover energy through regenerative braking and supply it back into the system for traction purposes. Flywheels are well suited for this application due to the high rate of charge–discharge cycles needed. In addition, they allow voltage sag control for transmission and distribution lines, without increasing the line capacity of the railway [84,110].

In the study by Hedlund et al. [84], a review of flywheel energy storage technology was conducted, with a special focus on the progress in automotive applications. The study found that there are at least 26 university research groups and 27 companies contributing to flywheel technology development. Flywheels are seen to excel in high-power applications, placing them closer in functionality to supercapacitors than to batteries. Examples of flywheels optimized for vehicular applications were found with a specific power of 5.5 kW/kg and a specific energy of 3.5 Wh/kg. Another flywheel system had 3.15 kW/kg and 6.4 Wh/kg, which can be compared to a state-of-the-art supercapacitor vehicular system with 1.7 kW/kg and 2.3 Wh/kg, respectively. Flywheel energy storage is reaching maturity, with 500 flywheel power buffer systems being deployed for London buses (resulting in fuel savings of over 20%), 400 flywheels in operation for grid frequency regulation, and many hundreds more installed for UPS applications. The industry estimates the mass-production cost of a specific consumer-car flywheel system to be 2000 USD. For regular cars, this system has been shown to save 35% fuel in the U.S. Federal Test Procedure (FTP) drive cycle [84].

In the study by Heshmat and Walton [85] the authors present the design and operating test results for a novel energy storage flywheel system. Proof of concept testing of a small flywheel with motor/generator using compliant foil gas bearings to support both generator and flywheel was conducted. Tests were completed from ambient pressures down to about 1 psia to show that flywheel energy storage systems using foil bearings are competitive with magnetic bearing based systems. Due to the shock tolerance inherent in foil bearings the total system size and weight are less than magnetic bearing based systems which require auxiliary/catcher bearings. Comparison of measured and predicted windage loss show that total windage losses for the foil bearing based system are less than that needed to power the magnetic bearings. Tradeoff studies show that power densities of 2000 W/L and Energy Densities of greater than 150 W-Hr/L are possible with the foil bearing system and high strength composite materials at practical/achievable stress levels. The power and energy densities are not just for the flywheel but for the flywheel, generator, bearings and housing [85]. Heshmat and Walton [112] examined lubricant free foil bearings for highly efficient and reliable flywheel storage system.

A number of flywheels for trackside energy recovery systems have been demonstrated by URENCO and Calnetix [103]. VYCON's flywheel, known as Metro's Wayside Energy Storage Substation (WESS), can recover 66% of the braking train energy [105]. The collected data, after 6 months of operation, showed 20% energy savings (~541 MWh), which is enough to power 100 average homes in California [104]. A total of 190 metro systems operating in 9,477 stations and ~11,800 km of track have been reported globally [92]. The introduction of energy storage into rail transit for braking energy recovery can potentially reduce 10% of the electricity consumption, while achieving cost savings of $90,000 per station [82].

Flywheels are also used in roller coaster launch systems to accumulate the energy during downhill movements and then rapidly accelerate the train to reach uphill positions, using electromagnetic, hydraulic, and friction wheel propulsion [107]. Since the late 2000s, the use of flywheel hybrid storage systems in motorsports has seen major developments, beginning with Formula 1 and followed by the highest class of World Endurance Championship (WEC) [92]. In public transport, city buses are an ideal application for electric flywheel hybridization, due to their higher mass and frequent start–stop nature. The technology can save fuel and reduce greenhouse gas emissions by up to 30% [108]. Williams Hybrid Power (WHP) started developing flywheel energy storage for use in buses for the Go-Ahead Group in March 2012.

Flywheels find applications in space vehicles where the primary source of energy is the sun and where the energy needs to be stored for the periods when the satellite is in darkness [88,89]. For the past decade, the NASA Glenn Research Centre (GRC) has been interested in developing flywheels for space vehicles. Initially, designs used battery storage, but now, FESSs are being considered in combination with or to replace batteries [88,89]. The combined functionality of batteries and flywheels improves the efficiency and reduces the spacecraft mass and cost [88]. It has been shown that the flywheel offers a 35% reduction in mass, 55% reduction in volume, and a 6.7% area reduction for solar array [109]. FESS is the only storage system that can accomplish dual functions, by providing satellites with renewable energy storage in conjunction with altitude control [94,95].

In the military, a recent trend has been toward the inclusion of electricity in military applications, such as in ships and other ground vehicles, as well as for weapons, navigation, communications, and their associated intelligent systems. Hybrid electric power is essential for future combat vehicles, based on their planned electrically powered applications. Flywheels appear as an appropriate energy storage technology for these applications. They are combined with supercapacitors to provide power for high-speed systems requiring power in <10 μs.

Flywheels are also likely to find applications in the launching of aircrafts from carriers. Currently, these systems are driven by steam accumulators to store the energy; however, flywheels could replace these accumulators to reduce the size of the power generating systems that would otherwise be sized for the peak power load [89]. A FESS is integrated into a microgrid serving the U.S. Marine Corp in California to provide energy storage applications throughout the entire distributed generation at the base [100]. The purpose of the project was to provide energy security to military facilities using renewable energy. It is a network of interconnected smaller microgrids that are nested into a 1.1 MW bigger-scale microgrid, which include solar PV systems, diesel generators, batteries, and 60 kW, 120 kWh FESSs [110]. Flywheel storage is intended to decrease the dependency on diesel generators by about 40% and provide peak shaving applications by mainly supplying high power loads such as in elevators. In addition to extending the lifespan of the batteries, the FESS is estimated to work for 50,000 cycles and have a lifespan of 25 years [110]. Flywheels can assist in the penetration of wind and solar energy in power systems by improving system stability. The fast response characteristics of flywheels make them suitable in applications involving renewable Energy Sources (RES) for grid frequency balancing. Power oscillations due to solar and wind sources are compensated for by storing the energy during sunny or windy periods and are supplied back when demanded [90,91]. Flywheels can be used to rectify the wind oscillations and improve the system frequency, whereas, in solar systems, they can be integrated with batteries to improve the system output and elongate the battery's operational lifetime [90].

Pena-Alzola et al. [93] indicate that the formation of a hybrid system by the addition of wind turbines and PV panels could not result in fuel savings, as expected. This is because diesel generators, even unloaded, will consume up to 40% fuel. Diesel generators should only be started when demanded and shut down most of the time. Therefore, FESSs can reduce frequent start/shutdown cycles of the diesel generators, thus reducing fuel consumption and bridging the power fluctuations [93]. Amiryar and Pullen [86] see a great benefit of flywheels backing up solar PV, since they can cope with the high cycles due to the cloud passing, yet provide ride through, as long as standing losses are kept low. There has been a wide range of flywheel systems developed for the penetration of renewable energy systems. For example, ABB's PowerStore, Urenco Power, Beacon Power and VYCON technology have all provided flywheel-based systems for wind and solar applications [86]. On a larger scale, the world's first high-penetration solar PV diesel power stations were installed in 2010, supplying the towns of Nullagine and Marble Bar in Western Australia. A FESS is operated as a UPS system, to allow maximum solar power injection during sunshine and ramp up diesel generators when the sun is obscured. This enables a saving of 405,000 L of fuel and 1,100 metric tons of greenhouse gas emissions each year. Moreover, the integration of flywheels in the system has helped the PV system to supply 60% of the average daytime energy for both towns, generating 1 GWh of renewable energy per year [100].

The overall objective of the Mohawk innovative technology [43] was to design, test, and demonstrate the ability of a 500 kW test bed prototype energy storage module to withstand a shipboard environment. The Phase I trade-off design studies sized and identified a shock tolerant flywheel energy storage modular system configuration that can yield the desired power and energy densities while having minimal maintenance requirements. Under Phase II, a technology demonstrator will be designed, built, and tested. The overall goal was to demonstrate and verify the power and energy density gains and reduced footprint possible through effectively integrating the generator, bearing, and flywheel components. To achieve the desired power and energy densities, a composite flywheel and very-low-loss bearings were necessary. University of Texas center for electromechanics (UT-CEM) as subcontractor established the composite flywheel structure and manufacturing layup, while Ministry of international trade and industry-Japan (MiTi) designed, fabricated, and tested the demonstrator module. Testing under Phase II assessed system dynamics, thermal management, and electrical performance with successful results.

Beacon Power [101], a manufacturer of grid-scale FESSs installed its first 20-MW flywheel energy storage plant at the Pennsylvania site. Beacon's carbon fiber composite flywheels are designed to store excess energy during periods of low energy demand and quickly discharge it during demand peaks. Frequency regulation is a grid reliability service that is performed to correct short-term unpredictable imbalances in electricity supply and demand. On the power grid, supply of electricity must match demand to maintain frequency at 60 Hz. Beacon's 20 MW flywheel plant provides frequency regulation services by absorbing electricity from the grid when there is too much and storing it as kinetic energy. When there is not enough power to meet demand, the carbon fiber composite flywheels inject energy back into the grid. These cycles can occur multiple times in time periods as short as 1 min. Since its first installation, the plant has installed 200 flywheel modules. The first 4 MW of energy storage entered commercial operation in the project management (PJM) Interconnection grid system, with the full 20-MW plant operational during second quarter of 2014. The Beacon flywheel [101] facility provides a fast, accurate, and reliable response to grid changes that system operators need to increase system efficiency and power quality. Furthermore, flywheels offer a long asset life with no degradation of performance, as well as the ability to move energy in and out of the grid many more times than other technologies, which contributes to low life-cycle cost and high-quality service. To date, Beacon's flywheels have accumulated more than 3.5 million operating hours.

8.5 MODULAR THERMAL STORAGE

8.5.1 TES Module Using Phase Change Composite Material

HVAC (heating, ventilation, and cooling) accounts for ~15%–30% of a commercial building's electricity cost and about 25% of electricity consumption. Electricity costs charged by the utilities companies for commercial building customers includes consumption charges ($/kWh) and demand charges ($/kW) depending on the time of use (TOU) charges [113]. Thus, there is great interest to reduce the electricity consumption charges and demand charges by using new energy-efficient technologies and products. One such product is a TES system using a phase change material (PCM) to offset the cooling costs associated with air-conditioning systems by reducing the electricity consumption during peak time periods. A novel phase change composite (PCC) TES material was developed that has very high thermal conductivity and favorable operating temperature than ice with fast charge/discharge rate capability.

Compared to ice, the PCC TES system is capable of very high heat transfer rate and has lower system and operational costs. Proof of concept demonstration and technical feasibility were successfully completed on a 4.5 kWh PCC TES prototype unit. Performance results show that a PCC TES system can be designed for a commercial building and can maintain high efficiency at high discharge rates [114]. Latent heat storage is attractive because large amounts of energy can be stored in a relatively small volume with a small temperature change in the media. Latent heat storage is useful in many areas including solar heating systems described by El-Dessouky [115], district cooling as described by Chiu et al. [116], and waste heat recovery by M. Chinnapandiana [117]. Out of the four phase transformation options being solid–liquid, liquid–gas, solid–gas, and solid–solid, solid–liquid is most common and cost-effective of all and applicable to the temperature range considered herein. This study focuses on using a solid–liquid PCM for building cooling applications by utilizing their low temperature latent heat of fusion to reduce the cooling load on an existing air-conditioning unit.

In the U.S., electricity consumption accounts for about 61% of the total energy consumed in the commercial building sector, and cooling and refrigeration alone accounts for more than 25% of the electricity consumption [118]. Most of the electricity consumption for building cooling happens during the peak operating hours between 10 a.m. and 5 p.m. when the solar gain into the building is very high. During these peak operating times, the air-conditioning units typically run close to their full load during hot summer days when their efficiency is low and carbon emissions are very high. When this scenario is aggregated over thousands of commercial buildings, the load on the electricity grid is so significant that utility companies charge three to four times higher electricity rates during peak operating times and also charge additional demand electricity charges in order to lower the demand on the grid. Currently, the utility companies have TOU electricity rates for electricity consumption and also demand charges for peak power consumption. The electricity rates during peak hours between 9 a.m. and 5 p.m. are three to four times higher than those during off-peak hours, and demand charges are about $15–18/kW [119]. This will result in significant electricity costs that will encourage commercial building owners to adopt more energy-efficient technologies and renewable energy sources to partially meet the electricity demands or shift the electricity consumption to off-peak duration. Thus, solar PV ESSs have gained lot of attention in recent years for reducing peak electricity consumption in commercial and residential sectors. Such ESSs typically use battery systems like lead–acid or lithium-ion battery to store the PV electrical energy and shift the building electricity load completely or partially during the peak operating hours [113–125].

Recently, TES systems are becoming very popular in meeting the building cooling needs as demonstrated by Austin Energy utility company. This utility company's central cooling system uses chilled water to meet peak cooling demands of various commercial buildings and thereby shave about 15 MW of its summer peak electrical demand [121]. Ice Energy is another commercial TES company selling 4–20 t ice-based TES systems to shift more than 90% of peak cooling demand by running their ice TES systems during peak summer days [122]. There are some inherent disadvantages in pure ice-based TES system such as the need for a dedicated refrigeration system to make ice which increases system weight and capital and operating costs, poor thermal conductivity of ice that requires significant amount of copper, slow charging duration (>10h), and slow discharging rates. Because of these reasons, the final system energy storage capacity reduces by more than 50% compared to theoretical energy storage capacity of pure ice [123].

The proposed TES system in this study employs an organic PCM compound that is integrated into graphite composite matrix. This combination brings significant advantages over current methods. PCM compounds alone have thermal conductivity of about 0.2 W/m.K. The thermal conductivity of the proposed phase change composite (PCC) material is considerably higher at about 10–20 W/m.K. An additional significant advantage of this PCC is its containment: The PCM will not leak out from the graphite matrix when the PCM is in liquid phase because it is held by capillary force inside the microscopic pores of graphite matrix. This PCC does not require a container, let alone a sealed one. It is possible to choose the PCC phase-change temperature. The PCC can be impregnated with PCM ranging in temperature from −37°C to 151°C all of which are commercially available. An additional advantage of the graphite PCC is the large surface area in contact with the heat exchanger. The PCC can be machined to mate with tubes and flat surfaces as desired, providing very high contact surface area. The proposed TES system can be integrated with existing AC refrigeration without the need for an additional compressor and relative components. The existing cycle is used to charge the TES system during off-peak times when cooling load is lower rather than a dedicated refrigeration system necessary for ice-based TES system [113–115]. The proposed PCC TES system is intended to be used to cool commercial buildings in addition to existing air-conditioning units [126]. The proposed technology relies on existing compressor and refrigerant system to charge the system during off-peak times, resulting in lower equipment cost. During the night, the AC unit charges the PCC TES system when electricity prices are very low and the cold energy is stored in the PCC material. During the peak hours of the day, the TES system will run in parallel with the AC unit and cools the warm air when the cooling demand exceeds a peak threshold. In this manner, the AC unit can be downsized, compressor efficiency can be improved, and peak consumption charges and demand charges can be significantly reduced [113–125].

8.5.1.1 PCM Modular System

Acciona Solar [127,128], under the Thermal Storage funding opportunity announcements (FOA), plans to design and validate a prototype and demonstrate a full-size (800 MWth) TES system based on PCMs. Acciona's PCM module is designed to be the building block of a TES system that can be deployed at costs in line with the benchmark established by the Department of Energy (DOE) for the year 2020. The goal is to develop a reliable, unsophisticated, modular, and scalable TES system that can be mass-manufactured, utilizing the most advanced automated fabrication and assembly processes, which can be installed in the field in the most cost-effective configuration. Achieving this goal would facilitate attainment of levelized cost of electricity (LCOE) of $0.07/kWh by 2015. Acciona Solar intends to demonstrate a full-size, 800 MWhth TES system with costs <$15/kWhth with a round-trip efficiency >93% [114].

Jacobs and Sokoloff [129] present a modular, adjustable TES system for industrial/commercial HVAC applications, thermal process applications, and power production applications. The system uses a combination of sensible TES and latent TES to provide a higher storage capacity per unit volume. A heat transfer fluid (HTF) flows in both directions between a tank within a TES unit and a source heat exchanger and between the tank and a load heat exchanger. At microscale, each TES unit is made adjustable and modular through the use of interchangeable PCM cartridges submerged in the HTF in the tank. The most appropriate PCM cartridge can be utilized for a given application. At macroscale, each TES unit is housed in a standardized shipping container storage tank, allowing for system-level scalability and modularity. The units can be assembled in any combination for scalable capacity or variable temperature gradients based on the application design specifications.

8.5.2 Modular IceBrick™ and Chillers Thermal Energy Storage Cell

Nostromo [130], the pioneer in encapsulated ice energy storage solutions, has announced its **IceBrick™ TES** cell. The **IceBrick™** is designed to be the core element of the most cost-effective, behind the meter storage system available and consists of plain water and a proprietary nucleate. Each cell will be able to store and discharge an amount of energy which is equivalent to 25 kWh of electricity consumed by cooling systems at peak demand hours. Weighing only 1,000 kg (2,200 lb) and measuring $50 \times 50 \times 400$ cm ($20'' \times 20'' \times 157''$) the **IceBrick™** is the first ever modular cell that can be installed on rooftops and along exterior walls, taking up minimal amounts of space. **IceBrick's™** [130,131] revolutionary design ensures the system has ZERO recycling issues, and with an expected cost of $250/kWh, it reduces the cost per kilowatt hour by 80% when compared to the leading ice storage systems.

Nostromo has developed a system that will cost half as much as Li-ion systems and won't be using any rare or poisonous materials. The only known and environmentally sustainable way to shift the grid's peak demand from high noon to the dead of the night is to store energy when there is a surplus in power production capabilities and discharge it when the demand is rising and resources are scarce. With **IceBrick™**, Nostromo has redesigned ice storage and enabled, for the first time, the use of water as a viable energy capacitor for commercial users [130,131].

Tandem Chillers Inc. [131] manufactures ice making chillers for TES systems. Off-peak cooling (OPC) is the process of making and storing ice at night (using TES), when demand for cooling and electricity prices are at their lowest, and using the ice to provide cooling for air-conditioning during the day. This simple concept of electrical peak load shifting can lower your monthly hydro costs by thousands of dollars. Tandem Chillers designed its first modular chiller in 2003 with a limited number of scroll compressor chillers and screw compressor chillers. Tandem chillers are (a) of a sound refrigeration design, (b) serviceable and (since they are small, modular, compact without space on either side or front and back for servicing the chiller) of Tandem-made design so that (c) the chillers could be removed for service **without shutting down the rest of the chilling system**. Tandem chillers are modular in **Assembly** and modular in **Disassembly**. This means that each of our chillers operates as a stand-alone chiller in a group of up to 12 chillers. Advancements in chiller technology have provided a number of chiller-related components that can help them to achieve higher operating efficiencies. There are seven main reasons that Tandem ice making chillers are more energy efficient and environmentally friendly. Some key features are:

1. **Oversized Condensers and Evaporators:** Make for a much longer lasting and more efficient compressor. With a hot gas bypass, a water flow switch, a freeze thermostat, non-rusting water piping, you have the perfect cooling solution for anyone who has to worry about cooling reliability and cost effectiveness.
2. **Highest Energy Efficiency Ratings (EERs) Available:** 18–19 Integrated part load value (IPLV) chillers are rated at full load efficiency, application part load value (APLV), and IPLV. The latter two values are generally more useful because chillers operate primarily at part load.
3. **Scroll Compressors:** Tandem makes sure that the compressor matches the condenser/evaporator for the highest efficiency and lowest consumption of energy.
4. **Cooling Capacity:** They have a cooling capacity of 1,800 kW, with a coefficient of performance (COP) in excess of six, even under part load conditions.
5. **Reliability:** They can last years without any interruption to the process.
6. **Low Operation and Maintenance Costs:** Low energy consumption.
7. **Software:** For chillers to operate at peak efficiency, they need to be monitored constantly. This is now possible with programmable controls like the pCO controller from Carel.

The Tandem chillers are used for computer room, industrial processes, such as plastics processing, air-conditioning, and ice making [130,131].

8.5.3 Modular TES Tanks Using Modular Heat Batteries

This invention by Laverman [132] relates to the storage and extraction of thermal energy. More particularly, this invention is concerned with a TES apparatus which employs a plurality of heat batteries of modular design located in an insulated tank. With the recent interest in energy conservation and efficiency, and with the development of solar energy technology, there has been considerable interest in the application of TES. Many applications, both industrial and commercial, have been evaluated for possible use of TES concepts. The temperature range over which it is desired to store thermal energy varies considerably with these different concepts, from ~200°F to temperatures in excess of 1,000°F.

Many of the ESSs already proposed employ a type of tank to contain the heated material. However, major design problems are involved with such tanks because of the thermal movements associated with placing the storage tank in service and with the normal temperature cycles through which it operates. These temperature variations cause considerable difficulty in the load bearing insulation and foundation of the storage tank. A need accordingly exists for improved TES apparatus. According to the invention by Laverman [132], there is provision for a TES apparatus comprising a thermally insulated tank having a bottom, side wall, and roof; a plurality of spaced-apart modular heat batteries inside the tank supported on load bearing thermal insulation on the tank bottom, each heat battery constituting an enclosed metal shell containing a bed of solid objects around which a liquid can flow; conduit means to feed a hot or cold liquid from outside of the tank to the top of each bed in each battery; and conduit means to withdraw a hot or cold liquid from the bottom of each bed in each battery and deliver it to a destination outside of the tank. Each heat battery can be a vertical circular cylindrical shell with a flat bottom. Furthermore, the solid objects in each battery can be rocks. The tank desirably has a flat bottom and a vertical circular cylindrical side wall which supports the roof. The side wall thermal insulation can comprise a layer of granular insulation supported between the side wall and a thin gauge metal barrier which is suspended from the side wall by a plurality of horizontal rods; and the bottom insulation can comprise concrete load bearing insulation, such as in the form of blocks. The roof thermal insulation can be provided by a layer of granular insulation supported by a metal deck suspended by rods from the roof.

The apparatus can also include means to flood the tank with an inert gas for safety purposes. For the same reason, means can be included to supply each battery with a blanket of an inert gas. Each battery bottom is desirably provided with means which maintains it axially stationary while permitting radial expansion and contraction with temperature change. Each battery is preferably supported on load bearing thermal insulation. According to a second aspect of the invention, there is provided a method of storing thermal energy consisting of distributing a flowing hot liquid to the top of a plurality of heat storage batteries containing a bed of cold solid objects, allowing the hot liquid to flow downwardly in a trickle flow manner over the bed of solid objects contained in the heat storage batteries, allowing a vertical temperature gradient, including a downwardly moving thermocline heat transfer zone, with the flowing hot liquid above the zone and the flowing cold liquid below the zone, to develop; and removing the cold liquid from below the thermocline heat transfer zone from the bottom of the heat storage batteries. According to the invention, there is a provision for recovering stored thermal energy comprising distributing a flowing cold liquid to the top of a plurality of heat storage batteries containing a bed of hot solid objects; allowing the cold liquid to flow downwardly in a trickle flow manner over the bed of solid objects contained in the heat storage batteries; allowing a vertical temperature gradient, including a downwardly moving thermocline heat transfer zone in which the cold liquid is located above the zone and hot liquid is located below the zone, to develop in the heat storage batteries; and removing the hot liquid from below the thermocline heat transfer zone from the bottom of the heat storage batteries [132].

8.5.4 MODULAR MOLTEN SALT TES PLANTS

The invention by Wasyluk et al. [133] primarily relates to a system that produces and stores thermal energy from the sun for processes such as thermal desalination or electricity generation. Generally, a solar receiver is a component of a solar thermal energy generation system whereby radiation from the sun (i.e., sunlight) is used as a heat source. The radiation and heat energy from the sun is concentrated on the solar receiver and is transferred to an HTF flowing through the receiver which can be stored and used to generate steam for a process or for power generation or for both (cogeneration). The receiver is usually a large unit permanently mounted on top of an elevated support tower that is strategically positioned in a field of heliostats, or mirrors, that collect rays of sunlight and reflect and concentrate those rays onto the tube panels of the receiver. An efficient, compact solar receiver for such systems which uses molten salt or a similar HTF and which is simple in design; modular; rugged in construction; and economical to manufacture, ship, and install would be desirable.

Currently wind and solar PV power generators do not have economical energy storage capability. Without energy storage, fluctuations on the grid are inevitable due to changing winds, clouds, and darkness at night. A molten salt solar plant is able to efficiently store the collected solar energy as thermal energy, which allows the process or power generation to be decoupled from the energy collection. The process or power plant can then continue to operate as needed, such as during cloud cover and at night, for some amount of time depending on the number of receiver towers and size of the thermal storage system relative to the amount of energy required by the process or power cycle [133].

The invention primarily relates to solar thermal energy generation systems that use solar receivers to absorb solar energy and certain storage tank structures for storage of the HTF to provide thermal energy for process and/or power generation. Preferably, the systems use molten salt as the HTF and storage fluid. Disclosed in various embodiments of the patent are methods of operating a solar thermal energy generation and storage system. The HTF (e.g., molten salt) is pumped from a set of cold storage tanks to a solar receiver. The HTF is heated to a maximum temperature of about 850°F, and it then flows by gravity to a set of hot storage tanks. The heated fluid is then pumped to steam generation system to provide thermal energy to generate steam for a process and/or to drive a turbine and generate electricity. Molten salt systems designed specifically to produce electricity use higher-temperature molten salt, typically 1,050°F, needed to meet the steam temperatures required by conventional utility-scale steam turbines and to provide a more efficient power cycle. However, processes such as thermal desalination do not require high-temperature working fluids. Therefore, the maximum temperature of the HTF (molten salt) in this disclosure is selected to be less than what could be achieved, so that the solar receiver, hot salt piping, hot salt storage tank, hot salt piping, and steam generation system (SGS) heat exchangers can be made from lower-grade alloys, thus reducing the cost of the plant. Alternately, the thermal energy can be used to generate steam to drive a turbine and produce electricity, but with lower power cycle efficiency due to lower steam temperatures resulting from lower salt temperature.

The patent also discloses solar thermal energy generation and storage systems and steam generation systems that include one or more vertical receiver towers. A solar receiver includes a vertical support structure that supports multiple tube panels, which can be arranged in quadrants. The tube panels are fluidly connected to form at least one flow path. A plurality of heliostats is arranged around the vertical tower. A set of cold storage tanks is configured to supply "cold" HTF to the solar receiver(s). A set of hot storage tanks is configured to receive "hot" HTF from the solar receiver(s). These and other non-limiting aspects and/or objects of the disclosure are more described in details by Wasyluk et al. [133,134].

eSolar developed a tower-mounted molten salt receiver surrounded by a heliostat field utilizing eSolar's small heliostat technology. This basic thermal module can be replicated, without scaling or redesign, as many times as required to create plant sizes from 50 to 200 MW with capacity factors ranging from 20% to 75%. The study by Pacheco et al. [135,136] describes the details of this molten salt reference plant design as developed over the past 2 years, as well as a potential scenario for staged initial commercial deployment that manages both risk and cost. eSolar is a provider of large-scale modular solar power tower systems for both electric power and non-power (e.g., desalination and enhanced oil recovery) applications. While our initial systems rely on water/steam as the working fluid, we (together with our partner Babcock &Wilcox Power Generation Group, Inc. (B&W) and with support from the U.S. Department of Energy) have over the past 2 years completed a reference plant design of a molten-salt-based system that can be scaled to match a broad range of customer requirements without significant redesign. eSolar's modular solar power tower technology has been proven at Sierra SunTower Generating Station located in Lancaster, California [137–140]. This facility, comprising two thermal modules and associated power generation system, utilizes their first-generation technology, including a B&W direct steam receiver and 24,360 of eSolar's 1.1-m^2 heliostats. While this direct steam system is effective for generating power when the sun is shining, storage is difficult and expensive to incorporate in a direct steam system. Recognizing the importance of storage to the future competitiveness of CSP, a project was initiated in 2010 to incorporate proven molten salt technology into eSolar modular approach. Results of the first phase of that work, a conceptual design, have resulted in a heliostat field utilizing eSolar's small heliostat technology.

To minimize risk, the details of the molten salt components are based directly on the lessons learned from the successful Solar Two molten salt pilot plant [133,134]. The unique feature of our technology is the ability to replicate the basic thermal module, without scaling or redesign, as many times as required (typically 2–14), to create plant sizes from 50 to 200 MW with capacity factors ranging from 20% to 75%. (In all cases, the multiple solar modules feed a single power block.) For example, ten modules are in our base 100-MW commercial configuration with 50% capacity factor. Examples of alternative configurations include five modules powering a 50-MW plant with 50% capacity factor and 14 modules powering a base load 100-MW plant with a 75% capacity factor [133–141]. In addition to scalability, the modular design also has a high tolerance to single points of failure. Planned and unplanned outages of a receiver, for example, will only affect one module's output, with the bulk of the plant still functioning as designed. The receiver is lifted by a crane to the top of a 100-m tall steel monopole tower. The hexagonal heliostat field surrounding the receiver and tower is comprised of about 47,000 of eSolar's new 2.2-m^2 SCS5 heliostats, calibrated and controlled by the Spectra software system. Unique to our modular design is the requirement for a field piping system to deliver 288°C "cold" molten nitrate salt from the centrally located storage system to the receivers and return 565°C "hot" salt to the storage system. The thermal storage system, comprised of large cold and hot salt storage tanks, is located in the power block, along with the molten salt steam generator and a conventional reheat steam turbine/generator system. The B&W-designed steam generator includes preheater, evaporator, superheater, and reheater heat exchangers, all designed to accommodate rapid daily startup and assure dynamic stability in all operating conditions. This work has been a concerted effort to reduce cost and improve plant performance. With significant heliostat field cost reductions, optimized performance and O&M, and flexibility to vary plant size and capacity factor with minimal engineering and risk, e-solar has achieved its goal of provider of cost-competitive photovoltaics and other CSP facilities with storage [133–141].

8.5.5 Modular Thermochemical Heat Storage

A 3 kWh thermochemical heat storage (TCS) module was built as part of an all-in-house system implementation focusing on space heating application at a temperature level of 40°C and a temperature lift of 20 K. It has been tested, and measurements showed a maximum water circuit temperature span (released by adsorption) of 20–51 K which is by all means suitable for space heating. Thermochemical heat storage is a promising technology to solve the mismatch between seasonal heat supply and demand as a typical problem for temperate climate zones. Such systems are able to make use of solar irradiation during summer time to cover the heat demand during winter time. Storing the energy is based on reversible thermochemical reactions such as [142–145]

$$\text{Zeolite} + \text{water} \rightarrow \text{zeolite.water} + \text{Heat.}$$

Such reactions release heat by adsorption/absorption which can be used as input for space heating (SH) and domestic hot water (DHW) applications. The TCS system can be charged by desorption taking the excess solar energy. Usually, water is used as a sorbate. A thermochemical reaction has a significant advantage over existing long-term TES technologies such as sensible heat storage in aquifer (ATES), boreholes (BTES), caverns (CTES), pits and water tanks [143], or latent heat storage (LHTES) using PCMs [114]. Heat can be stored in thermochemical material (TCM). On the other hand, TCS systems also have to sufficiently fulfill heat transfer characteristics in terms of heating power and temperature lift for space heating and/or domestic hot water applications. The sorption pair zeolite 5A–water was chosen for its hydrothermal and mechanical stability as well as safety precautions and minimization of corrosion of heat exchanger components. Placed in a cylindrical vessel of stainless steel, eight heat exchanger blocks with sizes of 1,000 × 300 × 33 mm (height × width × depth) were arranged in such a way that the water flow can be heated in parallel by the adsorption process. Further design specifications resulted in a deliverable hydration heating power of 800 W. For evaporation and condensation, a single unit was developed. It consists of a combination of a copper fin connected on one side to a copper spiral and a capillary working material on the other side so that the heat exchanger can serve both processes of evaporation and condensation. The unit contains a heat exchange surface of 1.4 m^2, resulting from 20 spirals with a diameter of 0.3 m, delivering a theoretical power peak of 3,000 W.

An all-in-house system using a 3 kWh thermochemical heat storage module was designed and built, as described by Finck et al. [142,144]. Temperature levels for dehydration and hydration were determined by their heat sources. Regeneration with solar collectors can be performed in a temperature range of 80°C–120°C using a condensation temperature of 20°C–30°C. Hydration takes place at 20°C (return temperature from SH) and has to fulfill a temperature lift of 20 K to obtain an SH inlet temperature of 40°C. The hydration process goes along with evaporation that is designed to be functional in a temperature range of 5°C–15°C. Both evaporation and condensation are strongly influenced by their external heat source/heat sink which can be ambient air or ground storage [142–145]. A decrease of 20 K from 100 to 80 K can reduce heat released by adsorption from 3.5 to 2.5 kWh, almost 30% less.

8.5.6 Modular "Thermal Capacity on Demand" in Rapid Deployment Building Solutions

Smart-plan, observe and debrief (POD) is a unique and innovative research project which provided an alternative to traditional classroom planning. It proposed a rapid deployment building solution, transitory or permanent in its use, modular in design, flexible in setup, and self-sustaining in use, requiring nominal site works and providing for all of its energy demands from renewable energy sources. Its feasibility was tested by Ceranic et al. [146] via a design case study which investigated potential of its novel "thermal capacity on demand" energy performance approach. It combined a modular thermal storage solution capable of balancing heating demand and supply for a low-rise, low-mass superstructure with renewable technologies and the level of backup power/services needed [146–152].

In this study, low-temperature diurnal sensible heat storage was used [146,147], with a loosely packed rock bed as a medium. Faster response rate, lower temperature, lower energy losses, and lower risk of boiling/freezing and leakage make this option an economical alternative to seasonal storage [148]. Furthermore, the medium can often be sourced by recycling the existing waste on the site, giving it an added environmental benefit. In the compact site conditions, the size and thermal performance of diurnal and seasonal store envelope can often be restricted by the available storage space. Hence, to charge the stores to a required temperature level, heat is often added by the heat pumps [148]. Diurnal stores can provide a significant "load shifting capability" and reduced energy losses, but the required storage volumes are large for the small- to medium-size building typologies and will only be fully resolved with improving the effectiveness and reducing the costs of latent or thermochemical heat storage systems [149,150].

Smart-POD was designed as a sustainable, rapid deployment, and potentially autonomous modular building solution, representing the outcome of the investigation undertaken into the combination of the technological processes involved. Principally, research into innovative thermal storage methods was conducted given the apparent lack of thermal mass that lightweight modular building systems suffer from. Furthermore, research into off-site rapid construction methods, passive design techniques and principles, energy-efficient building envelopes, and renewable energy technologies was also undertaken to balance energy requirements against the gains that can be made from its surrounding environment, including energy gains made whilst the POD is not in use. Its capabilities can be defined as follows [151,152]:

1. **Sustainable**—both in terms of its cost and energy performance, with a novel concept of "thermal capacity on demand"
2. **Modular**—allowing for it to be used as a stand-alone unit or as a cluster. Smart-POD is designed to allow schools to build flexibly, without significant changes to space or infrastructure
3. **Autonomous**—meaning it is designed as a self-sufficient unit, with an option of "plug and play" connections to existing school infrastructure and fixed services
4. **Reusable**—with a rapid redeployment to other sites
5. **Transportable**—delivered to site by road and operational within 24 h (post site and foundation preparation).

The proposed single classroom pod comprises the following:

- Teaching area with a flexible furniture arrangement and central teaching point, including IT provision and easy access to outside, as per curriculum requirements in primary schools
- Storage facilities and Approved Document M DDA compliant WC facilities
- An entrance lobby area for coat storage, with double "air lock" to minimize heat gains/losses
- Large area of triple glazing facing south to south-east to maximize solar gains in the winter, with effective insulation.

When the outside air is colder than the inside, fresh air is drawn in through the mechanical ventilation heat exchanger (MVHR) at the top of the building and heat exchanged with the outgoing warm stale air. The indoor air is further warmed by passive solar heat gains, latent heat gains from people, appliances, etc. It fills the space from the top, displacing any stale cool air via the side vents (if needed), near the floor level. The sensor measures the temperature and if within the comfort band range, releases air directly into the classroom. If not, the fan drives air down through the heat store. In this scenario, prewarmed thermal store is now being used to heat the building, diffusing it evenly up through the floor grilles to avoid drafts, hot spots, and convection air currents. The thermal store is precooled and is now being used to cool the building. The incoming warm fresh air is precooled by outgoing air in a contraflow MVHR. The sensor measures the temperature and, if within the comfort band range, releases air directly into the classroom. If not, the fan drives air down through to the heat store and cools it further, diffusing it up evenly through the floor grilles. The fresh cool air will passively "pool" at the bottom of the room, and warmed air will passively rise to the MVHR to be exhausted from the building. Final scenario 6 considers *Night Cooling of Space and Thermal Mass*, by using lower summer night temperatures to precool the store in preparation for a hot day. The building is passively cooled by allowing stale warm air to rise up and convect out of the building, replenished and "pushed up" by cool fresh air, as it enters via the side vents. In addition, the cool air is passively drawn through the thermal store to cool it. As a system backup, the proposed secondary heating/cooling system is a reverse cycle air source heat pump powered by electricity [146–152].

For a typical primary school, CIBSE Energy Benchmarks floor area per year (heating and electricity). More recently, tailored energy benchmarks for offices and schools were analyzed from the sample of decent (DEC) data for 6,686 primary schools [152], giving overall energy profiles (including energy required for lighting, IT equipment, controls, fans, and MVHR). Hourly heat gains of 75 W per pupil and 140 W per adult teacher were used in calculations, with 10 h average daily occupancy assumed. The estimated average daily heat gains are in range of 17.4–26.6 kWh, giving projected temperature rises from 2.8°C to 3.6°C from the thermal store initial temperatures in the 20°C–21.6°C range. Its thermal capacity is such that it would require 8.70 kWh energy to raise its temperature by 1°C. Should temperature in the store deviate from its steady state, the air-to-air heat pump would be used to lift the temperature to a required level. The above calculations are based on a total of 215 school days per year, with assumed average hourly equipment heat gains of 20 W per pupil and 150 W per adult teacher, based on 7 h of daily use.

Based on the model calculations, it is estimated that the Smart-POD would generally be heated entirely by its users and their equipment all year round [153]. Further sensitivity analysis on the extreme winter temperatures (10°C below average) reveals that is still possible to maintain classroom temperatures of over 17°C. In a similar manner, for extreme summer temperatures, (10°C above average) analysis shows that is still possible to maintain classroom temperatures of below 26°C, highlighting the effectiveness of thermal store and its integrated "thermal capacity on demand" approach. Relatively little additional effort of backup heat source would be required on those extreme temperature days to bring ambient temperature within the thermal comfort band range. Hence, the predicted performance of a proposed low-temperature diurnal thermal storage solution indicates an effective climatic adaptability potential, enhanced by integrated passive design strategies and bespoke modes of building control. The research also identified several specific uses for the Smart-POD [146–153]:

- whilst refurbishing and/or retrofitting
- as rapid replacement for fire- or flood-damaged schools or facilities
- to accommodate partial closure due to poorly maintained buildings
- as a quick, or phased, temporary or permanent response to predicted or confirmed increase in pupil numbers
- as a cost-effective, rapid-build alternative to traditional construction methods.

8.5.7 SCALABLE MODULAR GEOTHERMAL HEAT STORAGE SYSTEM

In the study by Nordbeck et al. [154], an innovative modular heat storage system is investigated experimentally and by numerical modeling. A single storage module consists of a helical heat exchanger in a water saturated porous cement matrix. The experiment comprises a 5-day thermal loading stage, followed by 16.5 days of passive cooling, and was especially designed to quantify the thermal insulation efficiency.

An inverse modeling approach was applied to successfully match temperature measurements within the storage by numerical simulation. The thus determined heat loss rates amount to 130 W for the fully loaded storage and to 50 W on average during passive cooling.

8.6 ECONOMIC VIABILITY OF MODULAR HYDROPOWER STORAGE SYSTEM

To date, the vast majority of global and domestic pumped-storage hydro (PSH) developments have focused on the construction of large (generally >300 MW), site-customized plants (see Figure 8.6) The viability of alternative design paradigms for PSH technologies has been actively discussed in industry and the research community, but no reliable determinations on the viability of these concepts have been made. Of particular interest is the development of smaller, distributed PSH systems incorporating elements of modular design to drive down cost and increase the ease of implementation. A small modular PSH could present a significant avenue to cost-competitiveness through direct cost reductions and by avoiding many of the major barriers facing large conventional designs such as access to capital; the long, uncertain licensing process; and the suppression of market prices (and subsequently revenues) caused by adding utility-scale storage to grid. These distributed modular units would typically serve large commercial and industrial loads in regions with adequate topography; examples include large industrial facilities, national laboratories, and data centers [156–169].

However, the cost and design dynamics of this new form of PSH development are not known, and it is ultimately unclear whether the benefits of modularization are sufficient to outweigh the economies of scale inherent in utility-scale development

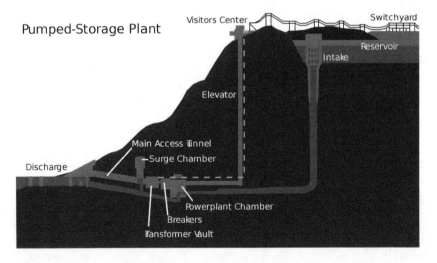

FIGURE 8.6 Diagram of the *Tennessee Valey Authority (TVA)* pumped storage facility at *Raccoon Mountain Pumped-Storage Plant* [155].

or prove superior to alternative distributed-storage technologies (i.e., batteries). The study by Witt et al. [157] and others [156,158] fills portions of this knowledge gap by evaluating the technical feasibility and economic viability of modularizing the design of PSH. Determining feasibility involves both an evaluation of the technology strategies for modularization and the market realities facing alternative PSH designs, including the size, geography, and power market distribution of potential locations and the production economies of scale necessary to reach economic viability. Equipment vendor expertise is utilized to evaluate modularized implementations of PSH components and subsystems to address technical viability. Various configurations and their cost–performance trade-offs were explored, including standardized reversible Francis units, as well as "off-the-shelf" applications of industrial pumps. Additional future research is needed to address civil work cost reductions, including the application of alternative materials (e.g., carbon fiber) to the penstock and manufactured reservoirs.

To systematically explore the cost–performance trade-offs of modularization, the analysis reported in this study focuses on the potential development of a small modular PSH at an abandoned coal mine, with existing upper and lower reservoirs, operating as a closed loop. In spite of intense interest in PSH development and increasing importance of energy storage for integration of variable renewables, new PSH development has been limited to a single plant (only 40 MW) in the last decade [168]. This lack of development has been attributed to the interaction of many complex factors, including improper valuation by markets and extensive permitting and licensing timelines [167]. These issues can be partially alleviated by a more direct approach involving a new type of PSH capable of bypassing many of the market and regulatory issues currently inhibiting new project deployment through prohibitive project designs, implementation schedules, and associated risks. Smaller and simpler units could enjoy streamlined regulatory treatment and be better suited for design standardization and replication, in turn reducing market prices as a larger plant would. The use of smaller and simpler pumping and generating units would allow the equipment manufacturers to focus standardization around particular head and flow ranges, similar to what is occurring in small hydro application. As such, this new PSH framework could be applicable to a wide variety of situations, including but not limited to locations with [156–169]

- existing upper and lower reservoirs,
- existing waterways, tunnels, or pipelines connecting existing reservoirs,
- suitable head differential but without existing reservoirs (closed-loop),
- and existing hydroelectric generation where only new turbines and/or a pump house is required.

In a recent report [165], Idaho National Laboratory performed an assessment to identify locations across the U.S. that may be suitable for new PSH development. Based on a minimum capacity of 10 MW, the assessment found that over 2,500 sites are suitable for new PSH development, including 31 hydroelectric plant sites, 7 non-powered dam sites, 97 greenfield sites, and 2,370 paired water body sites. When the screening requirement was reduced to include all sites with at least 1 MW of potential, a significant number of additional sites were introduced, including 44 hydroelectric plant sites, 20 non-powered dam sites, and 1,829 paired water body sites. This assessment demonstrates the unique opportunity for PSH development, though the number of sites which may be suitable for modularized development is likely to be much lower.

Even if a site is physically suitable for this scope of PSH development, economic feasibility must be considered. As described, a more direct approach to PSH focused on simplifying the project development process, shortening the delivery cycle from concept to commissioning, and increasing the reliability and predictability of project success would provide numerous financial benefits. Under the existing paradigm of custom site layouts and unit design, smaller plants are typically more expensive on a per kilowatt basis. However, the standardization and modularization of very small PSH units may enable significant cost reduction potential. Development of this new PSH mode, referred to as modular PSH (m-PSH), is a currently a major focus for the DOE [159]. To investigate the feasibility of developing m-PSH units, DOE's Wind and Water Power Technologies Office has tasked Oak Ridge National Laboratory [168] with assessing the cost and performance trade-offs of modularizing small PSH plants and the potential for cost reduction pathways. To assess the feasibility of developing small m-PSH, it is important to first define "small" to define the research and design space. Compared to larger projects, smaller and simpler PSH may be deployable at a higher number of potential locations and reduce the overall development schedule and life-cycle cost. Smaller, distributed PSH reduces the need for transmission upgrades and new transmission lines because it may enable integration in the distribution system. Smaller PSH concepts can be generally classified into three types based on use and size [156–169]:

- **Utility-Scale (20–200 MW):** The function of these units is similar to larger custom plants providing general support to the grid, but the smaller size may allow for standardization and modularization of design and make alternative market arrangements (i.e., direct support of variable renewable energy installations) economically feasible.
- **Municipal, Industrial, Commercial (1–20 MW):** PSH plants of this size would generally serve dedicated loads from high-demand facilities or address their associated localized transmission issues. Candidate locales include large industrial plants, national laboratories, and data centers.
- **Distributed (<1 MW):** These microsized PSH plants could potentially support isolated communities (such as remote villages or mining installations) or high-congestion areas of load by balancing the local microgrid.

However, the cost, implementation schedules, and design dynamics of these potential new forms of PSH development are not known, and it is ultimately unclear in each case whether the benefits of modularization are sufficient to outweigh the economies of scale inherent in large-scale development or to prevail against alternative storage technologies competing in similar markets (e.g., batteries, flywheels, CAES). To evaluate these trade-offs, different size and technology configurations of modular

PSH plants were considered. To capture major market and cost drivers, the following aspects of PSH development were addressed [156–169]:

- Project size
- Adjustable versus Single-speed technology
- Site features
- Market location.

Typical periods of generation would occur during peaking hours and last from 6 to 10 h, while pumping could last from 14 to 18 h. The generating-to-pumping ratio is largely location and system dependent, though pumping time can be reduced to take advantage of cheaper off-peak energy production. Using equipment and civil cost estimates provided by manufacturers, contractors, and consultants for various modular designs, Oak Ridge National Laboratory (ORNL) is evaluating individual project viability by simulating revenue streams from various competitive energy and ancillary service markets across the country. To illustrate this evaluation process, the following two sections detail an example cost estimate and simulated market revenue for a 5 MW single-speed modular unit in the PJM energy market.

The first m-PSH case study involved utilization of an existing dam as an upper reservoir and an old coal mine as a lower reservoir. The dam was located in Kentucky, owned by a coal company and covers ~520 acres. The total volume of water stored in the mine is ~770,000,000 gallons, with a net head of about 500 ft. For a specific configuration, the assumed design parameters were as follows [156–169]:

- Rated power = 5 MW
- Rated net operating head = 150 m (492.2 ft)
- Nominal turbine runner size = 1.0 m with a nominal flow of $4 m^3/s$ (141.3 cfs). Economy of scale from a volume order was assumed, and some non-hardware costs, such as engineering, project management, transportation, etc. were included.

To provide a complete cost estimate for supply of the entire power plant hardware, additional components such as the unit governor, controls, protection system, switchgear, interconnecting wiring, interconnecting piping, cooling water systems, and bearing oil systems would also be required. The study allowed an additional $600,000–$800,000 for these pieces of equipment and systems and addition of a setup medium voltage transformer at an existing substation. The civil works preconcept estimate included some modifications to the existing switchyard. In addition to these hardware costs, other costs are omitted from the calculation that would be required, such as the powerhouse structure, penstock, and tailrace costs, as well as labor costs for the installation of civil works components. Other soft costs such as permitting, environmental studies, and other consulting engineering services would be required.

There are many other necessary configurations and parameters that could be optimized later through more detailed analysis. Below are a few design options to be considered:

1. The required submergence for the pump and turbine for the centerline of the turbine or pump wheel.
2. Machinery speed for both the pump and the turbine.
3. Vertically or horizontally arranged equipment layout. The preferred initial approach for this analysis is a simple, low-cost arrangement.
4. Two conduits to the lower and upper reservoir, with a bifurcation after the butterfly valve (BFV) and governor control valve (GV). Alternatively, the lower reservoir may only require a suction chamber and draft tube, as well as a gate (draft tube gate or stop logs). This arrangement is typical in pump wells and at draft tube ends.
5. Direct drives between the motors and pumps and turbines and generators.
6. Single-stage pumps compared to other designs.

What was desired in the preliminary project layout was simplicity and modularization. Early concept work had demonstrated that both were achievable, but more engineering work was needed to complete the preliminary project layouts based on manufacturer information. Preliminary design required a 4 ft. diameter penstock varying in thickness from 3/8 to 1/2 inch. The turbine-generator assembly was proposed in one module, and the pump-motor assembly was proposed in a second module. The modular approach allows for assembly and testing of a completed module prior to arriving on site. From upper surface to the lower reservoir, there will be a winch hoist and steel stairs for access. A construction crane would be used to place the modules at the lower reservoir level. To the extent practical, electrical and control equipment will be located at ground surface in a prefabricated metal building. A new reinforced concrete intake would be constructed at the upper reservoir. The intake structure would be furnished with a trash rack and a vertical lift steel gate with hoist. The penstock would be fully vented downstream of the intake structure. Preconcept civil works estimates are in the range of $5.0–$7.0 million [156–169].

Given the small size of the PSH units considered in this study, if the power was sold into a wholesale electricity market, the facility would be a price taker (i.e., it would not influence the market price). The mathematical programming model outlined in the study by Witt et al. [157,169] and Bradbury et al. [158] maximizes r in the discount rate. This model makes the following assumptions:

1. Generation can only take place during peak hours (7 a.m. to 10 p.m.).
2. Pumping can only take place during off-peak hours (11 p.m. to 6 a.m.).
3. The rated capacity of the pump is the same as the rated capacity of the turbine-generator.
4. Spinning reserves can be provided when the unit is either in generation mode or in pumping mode.
5. With a single-speed turbine, regulation can only be provided when the facility is in generation mode (single-speed turbine) and operating at partial load (because PJM does not have separate regulation-up and regulation-down products).
6. With an adjustable-speed turbine, regulation can be provided when the unit is either in generation mode or in pumping mode.

7. Ability to switch from pumping at full volume to generating at full volume within 1 h.
8. No startup costs after idle periods.

The first two assumptions should be relaxed in markets where negative prices happen frequently. In PJM, in 2011, the day-ahead prices were positive for the 8,760 h in the year. As for the real-time market, it cleared at a negative price only 1 h in the whole year. Simulated revenues using this approach should be interpreted as an upper bound to potential revenues due to two reasons. First, the model assumes that the plant owner has perfect foresight of the price levels for the entire year and that the system operator would accept the bids from the plant owner 100% of the time. Secondly, the assumed annual unit availability factor is 100%.

Initial results for the PJM market (in which Kentucky is located):

1. 5 MW PSH
2. 75% turnaround efficiency
3. 10 h of storage
4. Single-speed turbine
5. 2011 day-ahead energy and ancillary service prices.

Based on initial results for the PJM market, total annual revenues from participation in energy and ancillary service markets were estimated. The benefit–cost ratio (BCR) and internal rate of return (IRR) are two standard metrics useful in evaluating the economic feasibility of a project. The BCR is calculated as the ratio of the net present value of life-cycle benefits to the net present value of life-cycle costs. This means that not only the level but also the timing of revenues versus expenditures matters for determining the feasibility of a project. The IRR is the annual rate of return for which the net present value of life-cycle net benefits (i.e., benefits minus costs in each period) equals zero. The levelized cost of energy (LCOE) can be interpreted as the minimum price at which a project owner must sell the electricity generated by a project to make the project economically feasible. It is a measure of the long-term cost for all resources and assets used in the operation of an energy project. Based on revenue numbers, these parameters appeared attractive.

In summary, the concept of modular PSH is technically feasible using conventional pumping and turbine equipment presently available and may offer a path to reducing the project development cycle from inception to commissioning. When applied to an existing site where there are existing waterworks and reservoirs, the actual installation cost may be competitive with other energy storage options. The tariff availability, project capital cost, and development and licensing uncertainty have been an industry challenge for privately funded, large-scale PSH. Smaller size, modular PSH may offer an opportunity to overcome some of these challenges and may avoid the large transmission costs associated with large-scale pumped storage. Modular PSH is not intended to replace conventional large economy of scale pumped storage but may offer a possible alternative for wider energy storage deployment. The preliminary analysis of the first case study indicates promise in terms of the overall costs, and through the use of modular approach, much of the actual manufacturing, fabrication, assembly, and testing can be done prior to on-site delivery. The substation modifications, mechanical hook-up of penstocks, and electrical/control wiring [156–169] can then be done onsite.

8.6.1 USE OF WIND-POWERED MODULAR PUMPED HYDROPOWER IN CANARY ISLANDS

A significant number of islands have found themselves obliged to place restrictions on the penetration of renewable sourced energy in their conventional electrical grid systems. In general, this has been due to certain energy-related characteristics often connected to their very nature as islands. These limitations attempt to prevent the appearance of problems that might affect the stability and safety of the electrical system. The restrictions imposed on the direct penetration of wind-sourced energy in the conventional grids of the Canary Islands are an obstacle to meeting the renewable energy objectives set out by the European Union. As a partial solution to the problem, the study by Bueno and Carta [170] proposes the installation on Gran Canaria island (Canarian Archipelago) of an appropriately administered wind-powered PSH system. The results obtained from the application of an optimum-size economic model of such a system indicates that penetration of energy from renewable sources can be increased by 1.93% (52.55 GWh/year) at a competitive cost for the unit energy supplied. These results are obtained on the hypothesis that two of the largest existing reservoirs on the island (with a difference in height of 281 m between the two and a capacity of some 5,000,000 m^3 used in each) are employed as storage deposits. Investment, operating, and maintenance costs are taken into account, as well as those costs involving health and environmental damage associated with energy production and use (externalities) [170].

The system would consist of a wind farm with a rated output of 20.40 MW; a modular pumping station with a rated output of 17.80 MW, operated so that the variation in the energy demand for pumping is in sympathy with the wind generation; and a hydraulic plant with a rated output of 60.00 MW. The proposed system would have no negative effect on either the reliability of the electrical system or consumer satisfaction. Furthermore, it would mean a fossil fuel saving of 13,655 metric tons/year and a reduction in CO_2 emissions into the atmosphere of 43,064 metric tons/year. For regions that have topographically suitable sites and which suffer energy problems similar to those of the Canary Islands, it is thus suggested that an analysis be made of the technical and economic feasibility of the installation of power systems such as one proposed here. Within the general guiding framework of a policy promoting clean and renewable energy, these systems represent an enormous and as yet barely explored potential [170].

8.7 MODULAR HYDROGEN STORAGE

8.7.1 MODULAR METAL HYDRIDE HYDROGEN STORAGE SYSTEM

Myasnikov et al. [171] describes a metal hydride hydrogen storage unit comprising a pressure containment vessel having a longitudinal axis, a plurality of cells at least partially filled with a hydrogen storage alloy, a plurality of primary modular blocks containing at least a portion of the plurality of cells, and a plurality of fins wherein each of the fins is disposed between two of the primary modular blocks. The plurality of modular blocks and/or the plurality of fins may be radially disposed inside the pressure containment vessel about the longitudinal axis of the pressure containment vessel. The plurality of fins may have a corrugated or grooved configuration. The plurality of cells may have an open top, an open bottom, and a cell wall. The hydrogen storage material may be retained in the plurality of cells via a porous filter material disposed at the top and/or bottom of each of the plurality of cells. The plurality of cells may have a circular configuration or a polygonal configuration. The primary modular blocks preferably have a height less than one-half of the inner diameter of the pressure containment vessel. The pressure containment vessel may be wrapped in a fiber reinforced composite material.

The metal hydride hydrogen storage unit may further comprise one or more heat exchanger tubes at least partially disposed within the pressure containment vessel, the one or more heat exchanger tubes being in thermal communication with the hydrogen storage material. The metal hydride hydrogen storage unit may further comprise an axial channel disposed about the longitudinal axis of the pressure containment vessel. One or more secondary blocks including at least a portion of the plurality of cells may be disposed in the axial channel. The one or more secondary modular blocks may have a cylindrical configuration.

In accordance with the present invention, there is provided herein a modular metal hydride hydrogen storage unit. Through compartmentalization, the metal hydride hydrogen storage unit maintains a substantially uniform metal hydride powder density after repeated cycling. The design of the metal hydride hydrogen storage unit reduces the amount of strain applied on the interior of the hydrogen storage unit as a result of the expansion of the hydrogen storage material upon absorbing and storing hydrogen in metal hydride form. The metal hydride hydrogen storage unit may also be able to absorb a portion of the stress created by the expansion of the hydrogen storage material, thereby further reducing the strain applied on interior of the hydrogen storage unit. The modular design of the metal hydride hydrogen storage unit also allows for assembly of the hydrogen storage unit using prefabricated pressure containment vessels.

8.7.1.1 Modular Heat Exchanger for Metal Hydride Hydrogen Storage

There is great interest in developing hydrogen-powered devices, especially hydrogen-powered automobiles. One prerequisite for this application is that there is enough hydrogen to give driving ranges comparable to those of conventionally fueled automobiles. However, hydrogen poses the problem of very low density. To overcome this obstacle, cryogenic storage containers filled with metal hydrides have been developed to store enough hydrogen in its liquid form to match its convention rival. However, with this technology, there is a requirement to heat the liquid hydrogen to make it available for use.

Pourpoint et al. [172] and Visaria et al. [173] have developed a unique modular heat exchanger for use in hydrogen fuel cells. The internal design optimizes both pellet contact area for increased heat transfer and hydride pellet capacity. Additionally, the modular design allows for easy replacement of defective or malfunctioning modules. These highly efficient heat exchangers can provide the necessary heat to evaporate the liquid hydrogen while remaining compact and durable enough for use over the broad range of temperatures and under the high environmental pressure of a hydrogen storage tank. The design is durable, and the modular design makes repair/replacements easy.

8.7.2 Hydrogen Storage Module Based on Hydrides

Hydrogen is stored conventionally as a gas in steel cylinders at high pressures (e.g., 2,000 psi) and at lower pressures as a liquid in insulated containers. Both methods of storage require comparatively bulky storage containers. In addition to their unwieldy size, such containers are inconvenient due to the high pressure required for gas storage in cylinders and the ever-present danger of gaseous hydrogen evolving from boiling-off of the liquid form.

In recent years, considerable attention has been focused on the storage of hydrogen as a metallic compound, or hydride, or various substances. Metal hydrides can store large amounts of hydrogen at low and even subatmospheric pressures in relatively small volumes. This low-pressure storage of hydrogen is relatively safe and allows the construction of hydrogen containers having forms significantly different than those presently known. Hydridable metals are changed with hydrogen by introducing pressurized gaseous hydrogen into valved containers. The hydrogen gas reacts exothermically with the metal to form a compound. Discharging of the metal hydride is accomplished by opening the valve of the container, to permit decomposition of the metal hydride, an endothermic reaction. It has been found expedient when gas is desired from the storage vessel to heat the storage vessel thereby increasing the flow of hydrogen or providing hydrogen at pressures substantially above atmospheric.

During the adsorption/desorption process, the hydridable metal has been found to expand and contract as much as 25% in volume as a result of hydrogen introduction and release from the metal lattice: Such dimensional change leads to fracture of the metal powder particles into fine particles. After several such cycles, the powder self-compacts causing inefficient hydrogen transfer. Additionally, and of even greater significance, high stresses due to the compaction of the powder and expansion during hydride formation are directed against the walls of the storage container. The stress within the powder has been observed to accumulate until the yield strength of the container is exceeded whereupon the container plastically deforms, buckles, or bulges and eventually ruptures. Such rupture is extremely dangerous since a fine, pyrophoric powder is violently expelled by a pressurized, flammable hydrogen gas. Small, experimental cylinders of the aforedescribed type have indeed been found to burst when subjected to repetitive charging/discharging conditions.

The problem of expansion and compaction has been recognized so that containers are only partially filled with hydridable metal powders. The problem of hydridable metal powder particle breakdown has been addressed in U.S. Pat. No. 4,036,944 wherein a thermoplastic elastomer binder is used to form pellets of the hydridable metal particles. Although this provides a solution to a portion of the problem of hydrogen storage, polymers are notoriously poor heat conductors and are

Systems for Energy and Fuel Storage   445

subjected to thermal deterioration. Since heat is generated during hydrogen charging and since heat may, in many cases, be added during discharging, such polymer-containing pellets appear to be only partially useful under somewhat restrictive operational conditions. Additional problems exist in the storage and transport of hydridable metals. There is a need for a means whereby hydridable metals can be shipped from the maker and loaded into pressure vessels without allowing the metal to react to any significant extent with atmospheric gases and moisture. A more difficult problem arises when it is required to move hydridable metal in the gas-charged condition [174].

It has now been discovered that by means of a novel structure, these difficulties and disadvantages can be avoided. It is an object of the invention by Turrilon and Sandrock [174] to provide a novel hydrogen storage module. Generally speaking, the invention contemplates a hydrogen storage module comprising a fluted tube section of generally circular cross section crimped and closed on both ends over a gas-permeable filter disk. Within the closed tube section is a charge of hydridable metal (or a hydride thereof) occupying no greater than about 78% of the volume of the tube section when the metal is in the hydrogen-free condition. The gas-permeable filter disk at each end of the tube section has metal crimped over it only on the periphery thereof, and that metal of the crimp has depressed, generally radially extending paths therein so as to provide means of gas passage in the event one module presses tightly against another or against a flat surface. The module is advantageously of such a configuration that the ratio of length to diameter is low, for example, the ratio is in the range of about 1 to 10 and advantageously in the range of 1 to 6. Again, advantageously, the ends of the module adjacent the filter disks are of circular cross section with a diameter less than the diameter of the tube section proper. The fluting along the main portion of the tube section is so constructed that the radius of curvature of the rounded grooves of the flutes is less than the radius of curvature of the tube section itself. This is to ensure the availability of gas passage space between undeformed tube sections even when tube section modules tend to nest with one another. Detailed storage options and storage cost and performance are outlined by the patent of Turrilon and Sandrock [174].

8.7.3 Calvera Modular Hydrogen Storage System

Calvera Hydrogen [175,176] offers a complete range of H_2 storage, from 200 to 1,000 bars of working pressure. It analyzes the integration of cylinders, valves, and structures to optimize the device, with the maximum amount of hydrogen in the minimum space, according to each project.

- All systems are modular and extendible, in order to minimize investment in extensions.
- Calvera Hydrogen collaborates with hydrogenation manufacturers (HRS) with complete, optimized solutions for filling vehicles up to 700 bars, in the most efficient manner.
- Scalable storage system to reach the required modular capacity of hydrogen and other gases, at the lowest possible cost. Its chief function is to work as

a buffer at the end customer, filling from the H_2 Tube trailer by means of a cascade process.
- With this solution, the need to maintain a fixed H_2 Tube trailer at the premises of small customers is removed. It leads to a drastic reduction in opex and capex, owing to the reduced number of Tube trailers in the logistics operation.

Several racks can also be connected to reach the required capacity [175,176].

8.7.4 Demonstration of a Microfabricated Hydrogen Storage Module for Micropower Systems

The development of microelectromechanical system (MEMS) sensors and actuators needs an onboard power source with high energy capacity [177]. A number of micropower systems, such as a microfabricated hydrogen–air PEM (Proton-exchange membrane) fuel cell system [178], and a new kind of low-pressure nickel–hydrogen battery that utilizes a metal hydride to store hydrogen, rather than a negative electrode in the nickel–metal hydride battery [179,180], are being developed. PEM fuel cell technology can provide both the steady state and pulse power required by these devices and can be integrated with the MEMS fabrication. Due to low working pressure and limited contact with the corrosive solution, the low-pressure nickel–hydrogen battery is safe and has long cycle life. A durable hydrogen fuel source is needed for these micropower systems.

The electric capacity of a micropower system is determined by the amount of fuel-hydrogen available from the hydrogen source. For the micropower systems, a high volume capacity is critical. Due to the mechanical strength of the micropower systems and the proposed working environment which is near ambient condition, the hydrogen source should absorb or release hydrogen readily under atmospheric pressure at room temperature, and the hydrogen release rate should be fast, so that the micropower system can provide high current. Since the hydrogen source is integrated with the micropower systems, it should be compatible with the microfabrication process and the working environments, such as the high humidity in the fuel cell environment or high concentration KOH in the nickel–hydrogen battery environment. For longer service life and lower cost, the hydrogen source should be reusable [177–192]. Intermetallic hydrogen storage alloys that have high volume capacity higher than that of liquid hydrogen and proper working pressure and temperature can be identified [181]. The hydrogen in intermetallic hydrogen storage alloys can also be reversibly absorbed/desorbed at high rates. However, a common characteristic of the hydrogen storage metals and alloys is that they need high-pressure hydrogen for activation before they can absorb hydrogen readily under the normal working pressure; for example, $LaNi_{4.7}Al_{0.3}$ and $CaNi_5$ have a plateau pressure of around 0.05 MPa at room temperature, but in order to activate $LaNi_{4.7}Al_{0.3}$ and $CaNi_5$ in short time, a pressure of 1–3 MPa is needed. These pressures are far greater than desired working range of the micropower systems.

By modifying the surface composition/structure of the hydrogen storage alloys, the activation on pressure can be lowered. The surface modification can be chemical treatment [182,183], coating with Pt group metals by electrodeless plating [184], or mechanically grinding the intermetallic alloys with catalysts to nanosize particles [185–187]. Based on the hydrogen storage capacity and working pressure, $LaNi_{4.7}Al_{0.3}$ and $CaNi_5$ were selected for the hydrogen storage module for the micropower systems. However, the as-received $LaNi_{4.7}Al_{0.3}$ and $CaNi_5$ could not be activated easily under ambient pressure and temperature. The activation behaviors of $LaNi_{4.7}Al_{0.3}$ and $CaNi_5$ were improved greatly by Pd treatment composed of grinding with small amount of palladium [188,189]. The Pd-treated alloys could absorb hydrogen readily under atmospheric hydrogen at room temperature even after being exposed to air for more than two 2 years [188,189]. The palladium-treated hydrogen storage alloy could also retain the ready activation property and air exposure durability after the microfabrication process which involved mixing the palladium-treated alloy with polymer binder and solvent and baking at high temperature. The pressure composition isotherm (PCI) of the Pd-treated alloys was not affected by microfabrication process. The detailed results on the effect of microfabrication process and the performance of the inks made with the Pd-treated alloys are reported somewhere else [190].

The mechanism of the increased performance and durability is hydrogen spillover and has been presented [191]. In this work, the use of the palladium-treated alloys for hydrogen source along with a new-type nickel–hydrogen battery and a PEM fuel cell was demonstrated. The objective of this work was to demonstrate a hydrogen storage module for an onboard electrical power source for micropower systems. **In summary, the following conclusions are drawn from this study [177–192]:**

- The palladium-treated intermetallic hydrogen storage alloys is compatible with the thin-film ink preparation process. The think-film inks made with the palladium-treated alloys can absorb hydrogen readily under atmospheric hydrogen.
- The contact with 26 wt% KOH solution slightly decreases the absorption rate in the first hydrogen absorption, but the absorption rate can be recovered after a few cycles of absorption and desorption.
- The thin-film ink can keep its structural integrity after 5,000 absorption/desorption cycles. The change of storage capacity with increasing the cycle number is determined by the property of the hydrogen storage alloy and the testing environment. With the existence of water vapor in the hydrogen, the degradation is accelerated.
- Both palladium-treated $LaNi_{4.7}Al_{0.3}$ and $CaNi_5$ can be used as the hydrogen source for the microfabricated power systems. The current provided by these two alloys can be higher than the requirement of the microfabricated PEM fuel cell. The efficiency of hydrogen provided by $LaNi_{4.7}Al_{0.3}$ can be higher than 90% at room temperature.
- Due to the low pressure of the first plateau of $CaNi_5$, not all the hydrogen in $CaNi_5$ can be used by the fuel cell at room temperature. This leads to the low hydrogen efficiency of $CaNi_5$ module compared to the $LaNi_{4.7}Al_{0.3}$ module.

- By monitoring the hydrogen pressure, the remaining amount of hydrogen can be determined. Hence the remaining power capacity of the fuel cell power system can be monitored.

8.7.5 Conception of Modular Hydrogen Storage Systems for Portable Applications

Hydrogen, till now the most prominent candidate as a future sustainable energy carrier, yields a gravimetric energy density three times as high as liquid hydrocarbon. Furthermore it is proven to be the most environmentally friendly fuel. Unfortunately, a few components regarding storage and tank solutions have not yet reached a technology level required for broad use. Thus, Paladini et al. [193] propose solutions and device concepts for everyday use and space applications of both devices. This contribution assesses both state of the art of storage materials and existing technologies of power generation systems for application in portable devices. The aim of the study by Paladini et al. [193] was to define the characteristics of a modular system, being suitable for a wide range of different devices, operating on advanced metal hydrides as the active hydrogen supply component. The concept has been studied and modeled with respect to volume, mass, and power requirements of different devices. The smallest system developed is intended to run, for example, a mobile phone. Minor tuning and straightforward scale-up of this power supply module should make it suitable for general applicability in any portable device.

8.8 MODULAR LIQUID FUEL STORAGE SYSTEM

The DRS Modular Fuel System (MFS) provides the ability to rapidly establish a fuel distribution and storage capability without a bag farm or engineer support [194]. The system can be used at any location without the availability of construction and material handling equipment. Consisting of fourteen 9,464 L/2,500 gallon tank racks and two pump filtration modules, the MFS increases mobility, capacity, and speed in fuel distribution, while decreasing deployment and recovery time. The MFS is mobile, whether it is full, partially full, or empty. It is compatible with the Heavy Expanded Mobility Tactical Truck, Load Handling System (LHS), the Palletized Load System (PLS) truck and trailer, and the United States Marine Corps Logistics Vehicle System Replacement. The MFS tank racks can also be used for line-haul of bulk petroleum throughout the total hauling system. By using two tank racks—one on the truck and one on the trailer—the PLS and LHS can transport up to 18,927 L/5,000 gallons of bulk petroleum per trip. The MFS is a logistics support system for Modular Brigades and provides liquid logistics support for the Army's Future Combat Systems. MFS has two 600 gallon per minute pump, three-stage pump filtration modules, 1,006 m of hose, 116 nozzles, fittings and adapters, two eight-point refuel-on-the-move kits and petroleum test kit, spill control kit, and a slot for additive injector [195,196].

The MFS provides the Army with a rapidly emplaced fuel distribution system that can receive, store, and distribute fuel at a fraction of the time required for a

Systems for Energy and Fuel Storage   449

collapsible tank system; support vehicle, ground equipment, and aircraft fuel requirements; and tank racks compatible with line-haul operations. MFS is a heavy Expanded Mobility Tactical Truck which is accompanied by Load Handling System, Palletized Load System and trailer, United States Marine Corps Logistics Vehicle System Replacement, C-130, and CH-47. Tank rack are low velocity air droppable, ISO/Convention for Safe Containers certified for 192,000 kg/423,000 lb and 10 K material handling equipment compatible. The system was modernized to provide the ability to rapidly establish fuel distribution and storage capability at any location regardless of material handling equipment availability [195,196].

The TRM (tank racks module) is air-transportable with fuel capacity fuel storage tank that can provide unfiltered, limited retail capability through gravity feed or the pump [196]. The TRM also includes hose assemblies, refueling nozzles, fire extinguishers, grounding rods, a NATO slave cable, and a fuel spill control kit. TRM full retail capability is being developed and will include replacing the existing electric pump with a continuously operating electric 20 gpm pump, a filtration system, and a flow meter for fuel accountability. The PRM (pump racks module) includes a self-priming 600 gpm diesel-engine-driven centrifugal pump, filter separator, valves, fittings, hoses, refueling nozzles, aviation fuel test kit, fire extinguishers, and grounding. The PRM has an evacuation capability that allows the hoses in the system to be purged of fuel prior to recovery and is capable of receiving, storing, filtering, and issuing all kerosene-based fuels.

8.9 MODULAR COMPRESSED NATURAL GAS STORAGE

8.9.1 Modular Compressed Natural Gas (CNG) Station and Method for Avoiding Fire in Such Station

The invention by Compo [197] refers to a modular transportable and easy-to-install compressed natural gas (CNG) refueling station for replacing traditional facilities for delivering CNG by a practical, economical solution. This modular station is a compact solution including several technologies already used separately in traditional CNG refueling stations. With the present invention, these solutions are incorporated in a "micropackage" and are used through the application of security engineering techniques or alternative solutions replacing traditional refueling gas station constructions as well as traditional security rules.

Based on the analysis of past work on gas explosion, Compo [197] concluded that the CNG refueling stations need an integral solution avoiding long construction times. The main objective of the invention by Compo [197] was to develop a technological solution for overcoming the possible explosion problems, incorporating all the elements included in a traditional CNG station but in a compact "micropackage" solution. Every CNG station needs a "measurement bridge" for measuring the gas flow the station is taking from the general gas pipeline. A bunker lodging storing vessels, including a compression system, a control, and logical switchboard, is also needed. With the proposed invention, all these elements are incorporated in a single compact "packaged" unit that is a single transportable unit. The most important problem to be solved in this project resides in bringing the "packaged" unit purposed to the same to security level as a traditional concrete bunker of a traditional CNG station.

In summary, the invention by Compo [197] is referred to a modular CNG station, characterized by comprising a modular compact unit with a flat supporting surface, lateral walls, and a detachable roof; at least two different areas are defined inside the said modular compact unit: a first antiexplosive area where a set of gas storage vessels and a compressor unit are lodged and a second area where an engine for driving said compressor unit and a measuring bridge unit are arranged; the station is capable of being transported and includes connectors for receiving gas from the general gas pipeline and connector for gas dispenser units. Said antiexplosive area is defined by fixed walls and at least one door with closing bolts connected to microswitches in turn connected to the PLC. The engine which drives the compressor unit is selected from the following: electrical engine, gas-feed internal combustion engine, gasoil-feed internal combustion engine, and fuel-oil internal combustion engine. Said second area also includes a measuring bridge unit for controlling the following variables of the station through the PLC: vessel's inner pressure, line pressure, line electrical tension, compressor exit pressure, and atmosphere control. The station includes in the second area an electrical switching board. The roof of the station comprises a detachable modular structure capable of being detached from the station structure during an explosion, and said compressor unit is connected to heat exchangers capable of regulating the temperature of each compression stage. The most important area of this compact station resides in the front area.

8.9.2 Modular CNG Tank

Today's natural gas vehicles are fitted with onboard fuel tanks that are too large, difficult to integrate into the vehicle, and expensive to properly facilitate the widespread adoption of natural gas vehicles in the U.S. and globally. Additionally, the low volumetric density of natural gas—roughly 30% less energy by volume than gasoline—limits the driving range of natural gas vehicles and makes cost-effective storage solutions a significant challenge. Dramatic improvements must be made to the capacity, conformability, and cost of onboard storage to accelerate natural gas vehicle adoption.

United Technologies Research Center (UTRC) [198,199] is developing a conformable modular storage tank that could integrate easily into the tight spaces in the undercarriage of natural-gas-powered vehicles. Traditional steel and carbon fiber natural gas storage tanks are rigid, bulky, and expensive, which adds to the overall cost of the vehicle and discourages broad use of natural gas vehicles. UTRC is designing modular natural gas storage units that can be assembled to form a wide range of shapes and fit a wide range of undercarriages. UTRC's modular tank could substantially improve upon the conformability level of existing technologies at a cost of ~$1,500, considerably less than today's tanks. If successful, UTRC's modular natural gas tank would serve as a viable alternative to today's storage tanks, offering a low-cost storage solution without sacrificing driving range [198,199].

Natural gas vehicles produce ~10% less greenhouse gas emissions than gasoline-powered vehicles throughout the fuel life cycle. CNG currently costs half as much per gallon of gasoline equivalent. With the average American spending over $2,000 per year on gas, enabling the use of natural gas vehicles could save drivers $1,000 per year.

The UTRC team has designed a conformable CNG tank. Under its ARPA-E award, the team advanced integrated design and optimization methods for complex structural components and developed manufacturing strategies for lightweight alloys and carbon fiber composites. UTRC is executing a licensing strategy to commercialize its conformable tank technology. After a detailed cost analysis, the team determined a projected $1,700 volume production cost for its aluminum tank design, well within the current natural gas tank market price range. UTRC is also pursuing parallel paths for licensing its technology for high-pressure (3,600 psi) CNG applications and continues to work with automotive OEMs to advance the technology for commercialization [198,199]. UTRC designed its tank through an integrated computational optimization, employing detailed topology optimization and structural analysis. The team's final design is flat, multichambered, and modular. It is adaptable to the wide range of aspect ratios applicable to different vehicle platforms. Composed of two D-shaped chambers on the outer edges and sandwiching a variable number of "stadium"-shaped interior chambers, the conformable UTRC design can provide 30% more gas storage in comparison to cylindrical tanks occupying the same space, at a manufacturing cost that is comparable to cylindrical tanks. Domed end caps complete the chambers, and adjacent chambers can be internally connected so that only one external valve is necessary. UTRC validated its design with scalable prototypes. The team also ensured that the manufacturing processes are viable for volume productions [198,199].

8.10 THE FUTURE OF MODULAR LNG TANKS

Efforts to lower the unit costs of liquefaction of natural gas has seen a move away from very-large-scale bespoke trains to a modular, multitrain approach, based on smaller, midscale 0.5–1.0 MMTPA (million tons per annum) trains, such as Energy World's proposed plant in Sengkang, Indonesia. For those countries with an established gas distribution network, large-scale regasification terminals, in excess of 1.0 bcfd, are appropriate, whereas archipelagoes such as the Caribbean [200], Indonesia, and the Philippines need to consider a hub and spoke solution in which large-scale LNG imports (7.0 MMTPA) can be distributed by smaller LNG carriers (30,000 m^3) directly to the power station [201–217].

For the past 20 years, the traditional solution for LNG storage in excess of 10,000 m^3 has been a stick-built 9% Ni steel single or full containment LNG storage tanks (see Figure 8.7). Most LNG projects have targeted throughputs >1,000 million standard cubic ft. per day (MMSCFD) or 7 MMTPA. The storage volumes for this size of regasification or liquefaction plant have exceeded 160,000 m^3. Indeed as the capacity of LNG carriers has increased up to 266,000 m^3 (Q-Max), the onshore storage tank size has also increased to ensure filling or discharge can be achieved within 24 h. Relatively little work has been done to develop cost-effective storage tank sizes

FIGURE 8.7 Fifteen thousand gallon modular LNG storage tank [218].

for the LNG to power market. Tank sizes >160,000 m³, required to receive a standard export LNG carrier, would provide 10 months of storage for a 100 MW CCGT. Even for a larger power station, it is clear that there is a mismatch between the storage tank and the exporting LNG carrier. Smaller carriers exist, using Type C or membrane technology, but there is a definite requirement for smaller ships to support cost-effective LNG to power delivery. Ships in the range of 10,000–30,000 m³ would allow smaller marine facilities and be compatible with the required onshore storage. Another market that is expected to see significant expansion is the LNG marine fuels business. Eagle LNG has recently completed its project in Maxville, FL, USA, and Conrad Shipyard is building an LNG bunkering barge. The LNG volumes for each ship are suitable for Type C storage containers, but aggregated onshore LNG storage tank volumes in excess of 10,000 m³ are necessary.

The small to midscale LNG market, supplying power stations, or the marine fuels business requires a smaller capacity LNG storage tank, in the range of 10,000–100,000 m³. The traditional solution based on 9% Ni steel technology is stick built on site. It is well known that the unit price of LNG stored reduces as the single tank size increases [207]. However economies of scale can also be achieved by production volume. The modular LNG tank seeks to reduce the unit cost for smaller LNG storage volumes by targeting off-site manufacturing productivity levels. The economies of scale are based on not the volume of a single tank but the number of units produced to achieve the required volume. A good reference case was the production of 25,000 m³ tanks in South Carolina [208]. In this case, the productivity improvements were observed for the initial ten tanks. It should be noted that the first sphere in that project experienced severe component fit-up and some welding issues. Since the basic tank unit can be in the range of 10–40k m^3, larger total volumes can be achieved with multiple tanks, which can also align with project phasing goals. The initial concept considered a maximum volume of 36,000 m^3, and this was considered to be close to the upper bound of what could, or should, be prefabricated and transported, before costs were negatively impacted. However, an opportunity to consider a 40,000 m^3 single containment design on the U.S. GoM coast provided the basis for the next phase of development.

The key technical developments are summarized below [201–217].

1. The tank is elevated above ground to provide both space for the SPMTs and also air flow to eliminate base slab heating.
2. The cellular concrete base slab is replaced with a steel grillage and concrete deck. This reduced weight is a significant issue for the larger tank volume.
3. Nine percent Ni was chosen over membrane based on owner preference and concerns over permitting delays that might arise since membrane tanks have not yet been approved by Federal Energy Regulatory Commission (FERC).
4. Side wall discharge was proposed. This is consistent with National Fire Protection Association (NFPA) 59, and if in-tank shut off valves are provided, the design spill is significantly reduced. The tank elevation also ensures that the pump does not need to be recessed below ground to achieve the minimum Net positive suction Head (NPSH).

The results of the techno-economic evaluation concluded that side wall pump discharge could reduce costs by up to $6 MM for a 2 × 140k m^3 storage tanks (1998 prices). But the prize is even greater for the modular LNG tank. Not only is the pump platform significantly reduced in size, but the tanks can be manifolded reducing the total number of pumps. The pumps can also be located outside of the bunded areas with easy access for maintenance. For larger total volumes, based on multiple units, the modular LNG tank will require individual bunded areas. This area can be optimized based on the work carried out by Coers [210]. Based on these technical developments, execution plans were developed. These plans resulted in favor of the use of modular approach with some additional considerations.

The work carried out on the 40k m^3 modular LNG tank confirmed technical feasibility and schedule advantages over a stick-built solution. It also highlighted the importance of fabrication yard setup costs. When these are spread over many tanks, they are not significant, as for any pre-engineered, manufactured product. To ensure that competitive pricing is achieved from the start, it was recognized that off-site prefabrication should not be delivered on a bespoke design basis for each project. The modular LNG tank concept would be enhanced if standard designs could be offered for any site, anywhere in the world. A standard tank design would permit the fabricator to further improve its fabrication and erection methods. Key site-specific drivers for modular LNG tank design are

1. soil conditions and foundation design
2. seismic conditions and inertial loads on tank and foundation
3. other environmental loading conditions (such as wind and snow loading)

4. temperatures and effect on insulation design
5. tow route, duration, and storm conditions.

The soil conditions will always be site specific, and if settlement criteria are satisfied, then there is no direct impact on the modular tank design, except for seismic response. Other environmental loading conditions are not significant drivers of tank shell and roof quantities, and conservative assumptions could be made to eliminate this variation. Preserving a standardized design is always a compromise. Perlite insulation could be maintained at a constant thickness and heat leak variations addressed by changes in the roof and base insulation thicknesses. This would impact the overall height of the tank and is not necessarily the most efficient solution. Further work is required to understand the sensitivity to this issue, but if insulation properties cannot be easily adjusted for a given thickness, then conservative insulation thicknesses could be appropriate.

Tank response during the tow has been investigated. It is clear that any extreme motions that would impact the basic tank design can be addressed with temporary sea fastenings and strengthening to the outside of the tank which can be ultimately removed and reused. The key driver on tank shell design and quantities is seismic loading. This is the most significant lateral load on the tank and, in areas of moderate to high seismicity, will govern the tank geometry and shell weight. Some tank designs have adopted seismic isolation to reduce the inertial loading and shell quantities. According to Earthquake Protection Systems Inc. (EPS) [211], an 85% reduction in seismic loading was achieved, which reduced the overall cost of the tank construction.

Despite the cost savings on the Peru LNG tanks, seismic isolation is not the default approach for dealing with moderate to high seismic loads. Lowering the tank aspect ratio (H:R), using inner tank straps to prevent uplift, and advanced non-linear dynamic soil structure interaction (DSSI) can be used to lower the inertial load effects on the tanks. Seismic isolation automatically elevates the tank and introduces a second foundation or base slab. This increases schedule and cost, to which the isolator cost is also added. For the modular LNG tank, these costs are already included, and the elevated tank is part of the overall concept to allow for installation using self-propelled modular transporter (SPMTs). In fact, the modular LNG tank is very well suited to adopting seismic isolation because all components are included in the existing design for other reasons.

Initial calculations confirm that tuning the elastomeric bearing will lower the seismic loads to those of the base design. The base design could be chosen utilizing the 33% over stress permitted under the Operating Basis Earthquake (OBE). For areas of high seismicity, friction pendulum bearings of the type provided by EPS may be required. The solution for any specific site requires a detailed analysis of the tank foundation system incorporating isolators. It is important that the foundation system (shallow or deep) is incorporated into the model, because significant reduction in loads can arise due to non-linear response in the soil resulting in longer-period response and higher levels of damping. Seismic isolation results in longer-period response which is accompanied by an increase in tank transient displacements. This will impact the design of incoming pipework, but experience has shown that differential movements can be accommodated in the piping design. If displacements are considered too high, then viscous dampers can be added to the isolation system to reduce peak displacements.

Isolation of vertical ground motions is not as common and has not been proposed for LNG tanks to date. Vertical accelerations will increase the effective weight of the LNG and therefore the hoop stresses. In areas of high seismicity, such as the west coast of the U.S., peak spectral accelerations approaching 1 g can occur, but careful DSSI can mitigate these effects. Long-period ground motions cannot be isolated, and these give rise to sloshing effects on the liquid surface. The codes are clear on the requirements for freeboard under both Order of British Empire (OBE) and Safe shutdown Earthquake (SSE) conditions. As seismic intensity increases, the freeboard height for a given tank aspect ratio increases. To preserve a standard tank design, baffles could be installed on the underside of the roof to disrupt the sloshing wave, but this is a novel approach which might not be acceptable to owners or regulators. Alternatively, it is accepted that the tank height must be increased to address this issue. However it would require only a minor height adjustment to the standard tank design. Further work is required to understand the variations and impact that vertical and horizontal seismic accelerations have on the modular tank design, but initial results are encouraging, and a standardized tank design is possible, which should translate into further reductions in cost and schedule [201–217].

8.10.1 Membrane Modular LNG Tank

Membrane tanks are not new; indeed more than 100 onshore membrane tanks have been built since 1972, and over 85% of all LNG carriers utilize the membrane technology solution [214]. Two membrane tanks are currently under construction for Energy World Corporation at Sengkang, Sulawesi, Indonesia, and Pagbilao, Philippines. In addition, there have been recent developments in international codes to recognize and incorporate design provisions for membrane tanks. Nevertheless, the dominant tank technology for LNG storage remains 9% Ni steel. A description of the membrane technology and comparison with aboveground 9% Ni storage tanks is presented by Ezzarhouni et al. [206]. Whilst this comparison was for a full integrity or full containment design, there are many attributes of the system that are compatible with the objectives of the modular LNG tank and would enhance the overall concept, further lowering the costs and reducing the schedule. These benefits are summarized below:

- Global Telecom and Technology (GTT) has developed a highly modular membrane system based on pre-engineered, manufactured components. This is well aligned with the objectives of a standardized tank design.
- There is only one structural tank and it is located on the outside. The inner 9% Ni and outer A36 shells are replaced with a 1.2 mm stainless steel liner and A537 Class 2 outer shell. Total steel weight and costs reduce significantly.
- Stainless steel and A537 Class 2 have much shorter procurement lead times and will continue to exhibit much lower price volatility.

- The total volume of wall insulation, based on polyurethane foam (PUF) filled plywood boxes, is less. Hence, for the same overall external tank diameter and volume, the corresponding tank height is reduced, further reducing the shell quantities.
- The tank transportation weight is lighter than the 9% Ni steel option, despite having all insulation installed prior to load-out.
- Membrane tanks do not require hydrotesting. Leak tightness is demonstrated through the ammonia leak test. Foundation proof loading is of questionable value even for 9% Ni LNG tanks and is not required for membrane LNG tanks which use PUF bottom insulation.
- No hydrotest means that the tank can leave the fabrication complete with all insulation installed and fully precommissioned. After installation at the project site, the ammonia leak test could be rerun to satisfy the owner and regulator that no damage was sustained during the sea tow.
- The design is fundamentally more robust with respect to transportation loadings. Recalling that 85% of all LNG carriers use the technology, it is a well-proven technology able to accommodate the strains associated with vessel motion. Further, all transportation loads can be designed into the outer tank which can easily accommodate sea fastening and temporary strengthening. There is no thin-walled inner shell to sea-fasten [201–217].

Additional design benefits of the membrane LNG tank are as follows:

1. Thermal cycling of 9% Ni tanks is not recommended because of the inner tank radial movements. However, the membrane tank is not subject to the same constraints as the liner accommodates the thermal strains within the stainless steel configurations.
2. The membrane insulation space is maintained under a nitrogen purge which is continuously monitored. This is considered a more effective method of leak detection than temperature sensors which rely on spill of LNG rather than vapor.
3. The membrane liner permits the use of sumps in the tank bottom thereby increasing the net useable tank volume.
4. In summary, membrane modular LNG tank undergoes important steps toward "plug and play" objective.

The ongoing development work on the modular LNG tank concept has confirmed technical feasibility of both 9% Ni and membrane solutions. The membrane option will offer a more robust design for transportation and also lower costs and shorter schedules.

More importantly, the concept of a cheaper and quicker prefabricated small- to medium-sized tank with "plug and play" capability, based on a standard design that can be installed for any site, anywhere in the world, is achievable. Single containment is not appropriate for all projects and jurisdictions. Full containment options are too heavy to transport cost effectively, but initial work looking at precast wall panels and wire-wound prestressing as used in the water tank industry, combined with the membrane technology should offer cost and schedule savings. Cheaper and faster smaller tanks will greatly assist the developing market. The key modular LNG tank drivers are as follows [214]:

1. Standardized tank design by volume based on site-specific seismic isolation.
2. Off-site tank fabrication in parallel with foundation construction.
3. Dedicated fabrication yard leading to improved productivities and higher quality.
4. Off-site precommissioning of tank.
5. Reduced man hours executed on site.
6. These drivers target "plug and play" capability while reducing costs and schedule compared to the stick-built traditional solution.

    Modular construction of tanks for storage of liquefied gases has been proposed in U.S. Patent 6,729,492 and U.S. Patent Publication 2008/0314908 and in the patent by Lokken [217].

8.10.2 Vacuum-Insulated LNG Modular Storage Systems

While as discussed in the previous section modular LNG storage vessels at scales around 30,000 m^3 and higher are suitable for prismatic membranes and in some cases 9% Ni tanks, at the other end of the scale, the most successful concept for LNG storage vessels has been vacuum-insulated tanks. Vacuum insulation is the best insulation technology available and is likely to remain so. A vacuum is maintained in the annular space between the two inner and outer tanks in order to reduce the convective heat transfer. In addition, the annular space is filled with an absorptive material to reduce the heat transfer due to radiation. With very low boil-off rates, the tank pressure can be easily maintained below the opening pressure of the safety valves for very small storage volumes. Therefore, vacuum-insulated tanks will continue to be the preferred alternative for small LNG modular storage tanks below 240 m^3.

The standard storage system for transporting various liquid hydrocarbons at low temperatures for several decades has been the IGC Code Type-C austenitic steel pressure vessels. In these tanks, insulation of 300–350 mm is applied by polyurethane foam (PUR) sprayed directly on the surface. Vacuum-insulated LNG storage tanks are normally fitted with 250–300 mm annular space, despite the lower thermal conductivity. The actual constraint comes from the need to install the interconnecting pipes in the annular space, rather than from insulation requirements. In installations where fast LNG bunkering time is a necessity, the large bunkering pipes may result in an even bigger annular space. The annular space and pipe design normally undergoes finite element modeling [213,215,216]. A vacuum-insulated tank is, in principle, two pressure vessels, where one pressure vessel is installed inside the other, and a vacuum is applied in the annular space. The inner tank is designed to withstand the internal pressure plus an additional 1 bar of pressure for the vacuum in the annular space. The outer tank needs to withstand only the suction force from the vacuum or the buckling force. Similarly, the saddles for a single-shell tank are simpler in design and can even be incorporated in the ship's hull. It is possible because the saddles cannot come into contact with cryogenic LNG in any damage scenario, thus avoiding becoming brittle. Consequently, the saddles can be made of carbon steel instead of stainless steel. This straightforward single-shell mechanical construction reduces the engineering effort, saving large amounts of material. These cost savings can be directly transferred to the end user, reducing the overall cost of the LNG storage system.

Today Wärtsilä [213,215,216] is recognized as a leader in propulsion solutions for gas-fueled vessels. There are several alternative LNG storage systems available, which have already earned their place in the LNG distribution chain. Due to the better insulation properties, for volumes below 250 m^3, the vacuum-insulated tank will be the preferred choice in order to meet a holding time above 15 days. However, the robustness and simplicity of the single-shell design reduces both engineering and material costs without sacrificing system safety or integrity. It is, therefore, likely that the popularity of this design will increase for gas-fueled vessels in the 300–5,000 m^3 volume range. The business case for gas-fueled vessels continues to become more attractive. Existing and new technologies are being adopted to help drive the total cost of ownership down and make LNG an environmentally and economically sustainable propulsion solution.

Natural gas is stored as LNG at very low temperature (–260°F/–162°C), where it occupies 1/600th the volume of natural gas in gaseous state. Chart offers a complete range of standard and custom-engineered tanks and complete systems for storage and regasification, manufactured at their facilities in North America, Europe, and China, for applications such as vehicle fueling, bunkering, satellite plants for gas distribution in areas not connected to the natural gas grid, peak shaving plants and backup supply during high demand/emergency periods, virtual pipeline solutions, and mobile units. The tanks are available in capacities ranging from 89 to 1,225 m^3; tanks can be oriented horizontally or vertically according to customer requirements and application. Each tank contains enough natural gas to last 500 years for a single-family energy demand.

Chart LNG storage tanks [219,220] utilize a lightweight stainless steel inner vessel, with a durable outer jacket and our proprietary, highly efficient vacuum insulation technology. The design allows for greater thermal performance and extended LNG hold times that result in minimal loss of stored product. Tanks are typically combined with other Chart in-house engineered components for a complete storage solution providing natural gas at the point of use according to customer capacity and withdrawal requirements. Typical elements include vaporizers and vacuum tank, manifolds, and control system.

Chart standard and modular liquefaction plants and associated process technology enable displacement of liquid fuels through small- and mid-scale LNG. Chart small-scale liquefaction plant features standard plant solutions with repeated designs in three standard sizes, proven low-risk standardized technology platform, maximum standardization, and brazed aluminum heat exchanger for optimum operating efficiency. Chart mid-scale liquefaction plant features, modular plant solutions with replicable designs and multiple modules, low-risk standardized technology platform, modular construction with brazed aluminum heat exchanger for optimum operating efficiency [219,220].

REFERENCES

1. Wagner, L. (2014). Chapter 27, Overview of energy storage technologies. In: Letcher, T.M. (ed) *Future Energy: Improved, Sustainable and Clean Options for our Planet*, 2nd Edition. Elsevier Science, Amsterdam, pp. 613–631. doi: 10.1016/B978-0-08-099424-6.00027-2.
2. Energy Storage Technology: An Overview | ScienceDirect Topics. A website report www.sciencedirect.com/topics/engineering/energy-storage-technology (2015).
3. Energy storage, Wikipedia, The free encyclopedia, last visited 25 July 2019 (2019).
4. Energy Storage Technologies | Energy Storage Association. A website report energystorage.org/energy-storage/energy-storage-technologies (2019).
5. Gur, T. (2018). Review of electrical energy storage technologies, materials and systems: Challenges and prospects for large-scale grid storage. *Energy and Environmental Science*, 11, 2696–2767. doi: 10.1039/C8EE01419A.
6. Medina, P., Bizuayehu, A.W., Catalao, J.P.S., Rodrigues, E.M.G., and Contreras, J. (2014). Electrical energy storage systems: Technologies' state-of-the-art, techno-economic benefits and applications analysis. *In Proceedings of the 47th Hawaii International Conference on System Sciences*, 6–9 January, Waikoloa, HI, USA, pp. 2295–2304.
7. Chen, H., Cong, T.N., Yang, W., Tan, C., Li, Y., and Ding, Y. (2009). Progress in electrical energy storage system: A critical review. *Progress in Natural Science*, 19, 291–312.
8. Hadjipaschalis, I., Poullikkas, A., and Efthimiou, V. (2009). Overview of current and future energy storage technologies for electric power applications. *Renewable and Sustainable Energy Reviews*, 13, 1513–1522.
9. Sabihuddin, S., Kiprakis, A., and Mueller, M. (2014). A numerical and graphical review of energy storage technologies. *Energies*, 8, 172–216.
10. EPRI. (2010). Electric Energy Storage Technology Options: A White Paper Primer on Applications, Costs, and Benefits. Palo Alto, CA.
11. SBC Energy Institute. (2013). *Leading the Energy Transition, Factbook, Electricity Storage*. SBC Energy Institute, Gravenhage, Netherlands.
12. Ferreira, H.L., Garde, R., Fulli, G., Kling, W., and Lopes, J.P. (2013). Characterization of electrical energy storage technologies. *Energy*, 53, 288–298.
13. Zyskowski, J., Neroutsos, N. (2017). Snohomish County PUD Launches Second Energy Storage System. A website report mesastandards.org/snohomish-county-pud-launches-second-energy-storage-system/ (12 July 2017).
14. Kaplan, D.L., Kaplan, J.I., and Darlington, G.P. (9 May 2012). Modular Energy Storage System. US 20130113294 A1.
15. Battery storage power station, Wikipedia, The free encyclopedia, last visited 29 July 2019 (2019).
16. Different Types of Batteries and their Applications: Circuit Digest. A website report https://circuitdigest.com/article/different-types-of-batteries (24 July 2018).
17. Types of Batteries | The Rechargeable Battery Association: PRBA. A website report www.prba.org/battery-safety-market-info/types-of-batteries/ (2019).
18. Li, Y., Yang, J., and Song, J. (2018). Nano energy system model and nanoscale effect of graphene battery in renewable energy electric vehicle. *Journal of Renewable and Sustainable Energy*, 69, 652–663.
19. Zhang, D., Li, H.L., and Jiang, J.G. (2018). A novel photovoltaic battery energy storage system based on modular multilevel converter. *Journal of Renewable and Sustainable Energy*, 10, 053508. doi: org/10.1063/1.5045526.
20. Community Energy Storage: A New Revenue Stream for Utilities and… A website report www.nrel.gov/…/community-energy-storage-a-new-revenue-stream-for-utilities… (24 September 2018).

21. Koirala, B.P., van Oost, E., and der Windt H. (2018). Community energy storage: A responsible innovation towards a sustainable energy system? *Applied Energy*, 231, 570–585. doi: 10.1016/j.apenergy.2018.09.163.
22. Yang, X., Xue, Y., Chen, B., Mu, Y., Lin, Z., Zheng, T.Q., and Igarashi, S. (2017). Reverse-blocking modular multilevel converter for battery energy storage systems. *Journal of Modern Power Systems and Clean Energy*, 5(4), 652–662.
23. Li, Q., Liang, H., Diao, W., Li, D., and Jiang, J. (2017). Modular battery energy storage system based on one integrated primary multi-secondaries transformer and its independent control strategy. *IEEE Transportation Electrification Conference and Expo, Asia-Pacific (ITEC Asia-Pacific)*, 7–10 August, IEEE, Harbin, China. doi: 10.1109/ITEC-AP.2017.8080928.
24. Kratos Energy Solutions. A website report https://kratosenergysolutions.com/ (2015).
25. DropBox Inc., Provides Portable Containerized Energy Storage in 2012. A website report www.dropboxinc.com/.../container/.../DropBox-Inc-Provides-Portable-Containerized-... (6 January 2012).
26. Helling, F., Götz, S., and Weyh, T. (2014). A battery modular multilevel management system (BM3) for electric vehicles and stationary energy storage systems. *16th European Conference on Power Electronics and Applications*, 26–28 August, IEEE, Lappeenranta, Finland. doi: 10.1109/EPE.2014.6910821.
27. Ibok, E.E. (19 September 2013). Modular, Portable and Transportable Energy Storage Systems and Applications thereof. US 20130244064 A1.
28. Rothgang, S., Baumhöfer, T., Hoek, H., Lange, T., De Doncker, R.W., and Sauer, D.U. (2015). Modular battery design for reliable, flexible and multi-technology energy storage systems. *Applied Energy*, 137, 931–937. doi: 10.1016/j.apenergy.2014.06.069.
29. Sitras ESM 125: Siemens Download Center. A website report www.downloads.siemens.com/download-center/download?DLA14_51 (2017).
30. Energy Storage Modules (ESM): Modular Systems | ABB. A website report https://new.abb.com › Offerings › Medium Voltage Products › Modular Systems (2017).
31. LiANA+: Development of Large Electrical Storage Systems for use... – ZSW. A website report www.zsw-bw.de/.../projects/batteries/liana-development-of-large-electrical-stora... (2018).
32. Fargion, D. and Habib, E. (2016). A Novel Flexible and Modular Energy Storage System for Near Future Energy Banks, Cornell University Publication, Ithaca, NY, arXiv:1601.03350v1 [physics.flu-dyn] for this version.
33. Lettner, M., Kreisel Electric Presents MAVERO: The Modular Energy Storage System. A website report www.kreiselectric.com/.../kreisel-electric-presents-mavero-modular-energy-storage-s... (27 June 27 2016).
34. MAVERO Energy Storage: KREISEL ENERGY. A website report www.kreiselenergy.com/en/ (2016).
35. Wheeler, D., New Plug and Play Energy Storage Standard Announced by MESA... A website report mesastandards.org/new-plug-and-play/ (1 October 2014).
36. Tansy, D., Energy Storage Models Available: SunSpec Alliance. A website report https://sunspec.org/energy-storage-models-available/ (30 September–1 October 2014).
37. MESA Standards: Open Standards for Energy Storage. A website report mesastandards.org/ (2016).
38. Open Standards for Energy Storage: MESA Standards. A website report mesastandards.org/wp-content/uploads/2018/01/MESA_Brochure_website_v4.pdf (2016).
39. RCT Power: Modular and Upgradable Energy Storage Systems - pv... A website report www.pveurope.eu/.../Storage/.../RCT-Power-modular-and-upgradable-energy-s... (2018).

40. BYD's HV Battery-Box Achieved Top Ranking in Energy Storage Test - pv... A website report www.pveurope.eu/.../Storage/.../BYD-s-HV-Battery-Box-achieved-top-ranking-... (28 February 2019).
41. Muneeb ur Rehman, M., Evzelman, M., Hathaway, K., Zane, R., Plett, G.L., Smith, K., Wood, E., and Maksimovic, D. (2014). Modular approach for continuous cell-level balancing to improve performance of large battery packs preprint. *Presented at IEEE Energy Conversion Congress and Exposition Pittsburgh*, 14–18 September 2014, Pennsylvania, NREL, Golden, CO. www.nrel.gov/publications.
42. Evzelman, M., Muneeb ur Rehman, M., Hathaway, K., Zane, R., Plett, G., Smith, K., Wood, E., and Maksimovic, D. (2014). Modular approach for continuous cell-level balancing to improve performance of large battery packs. *2014 IEEE Energy Conversion Congress and Exposition (ECCE)*, Pittsburgh, PA, USA. doi: 10.1109/ECCE.2014.6953991.
43. Mohawk Innovative Technology, Inc. › Hooshang Heshmat, Ph.D... A website report mohawkinnovative.com › Company › Our Team (2019).
44. Snohomish County PUD Launches Second Energy Storage System... A website report www.tdworld.com/.../snohomish-county-pud-launches-second-energy-storage-... (12 July 2017).
45. Balkundi, S., FIU, Florida, Expandable Modular Energy Storage Regulator: Research. A website report research.fiu.edu › Technology & Innovation › Technologies Available for Licensing (2016).
46. Fargion, D. and Habib, E., MaimAir: A Flexible and Modular Energy Storage System for Tomorrow Energy. A website report https://briefs.techconnect.org › Papers (22 May 2016).
47. Energy Storage: Solarwatt Launches Fully Modular MyReserve Matrix... A website report www.pveurope.eu/.../Energy-Storage/Energy-storage-Solarwatt-launches-fully-... (2018).
48. Solarwatt Launches Europe-Wide Distribution of MyReserve Matrix battery. A website report www.pveurope.eu/.../Energy-Storage/Solarwatt-launches-Europe-wide-distribut... (26 February 2018).
49. Containerized Energy Storage System: Total Solution | Kokam. A website report kokam.com/container-2/ (2019).
50. Colthorpe, A., E.ON Claims Energy Storage World First with Large-Scale Modular Battery M5BAT. A website report www.energy-storage.news/.../e.on-claims-energy-storage-world-first-with-large-... (12 August 2015).
51. KW25 | GNB Presented Energy Storage Products at Intersolar Europe... A website report www.solarserver.com/.../gnb-presented-energy-storage-products-at-intersolar-eur... (2017).
52. InCharge Power: Happy Dutch National Sustainability Day... A website report www.facebook.com/InChargePower/posts/723203437827642 (2015).
53. Containerized Energy Storage Systems for Hybrid Solutions: Ingeteam. A website report www.ingeteam.com › Press room › Corporate (12 February 2018).
54. Powerstorm Announces the Commercial Availability of its Modular... A website report www.businesswire.com/.../Powerstorm-Announces-Commercial-Availability-M... (20 July 2015).
55. Powerstorm ESS Batteries: Power. New Energy. Solar: Powerstorm. A website report powerstormtechnologies.com/product-details/19 (2015).
56. Significant Advances in the Use of Flow Cell Batteries | The Lyncean... A website report https://lynceans.org/all-posts/significant-advances-in-the-use-of-flow-cell-batteries/ (30 March 2017).
57. Siestorage: Digital Asset Management: Siemens. A website report https://assets.new.siemens.com/siemens/.../public.1505828841.78bc9b600465c5165e4 (2017).

58. The Modular Energy Storage System for a Reliable Power Supply. A website report www.windenergy.org.nz/.../9.45SiemensSIESTORAGEPresentationMarcelBzank.pdf (2017).
59. Ioakimidis, C.S., Murillo-Marrodán, A., Bagheri, A., Thomas, D., and Genikomsakis, K.N. (2019). Life Cycle Assessment of a Lithium Iron Phosphate (LFP) electric vehicle battery in second life application scenarios. *Sustainability*, 11, 2527. doi: 10.3390/su11092527, www.mdpi.com/journal/sustainability.
60. Nanoflowcell, Wikipedia, The free encyclopedia, last visited 2 July 2019 (2019).
61. Nano Flowcell Quantino Covers 218,000 Miles on Flow Batteries. A website report www.greencarreports.com › News › Electric Cars (15 April 2019).
62. Jiang, W., Wu, X., and Zhao, H. (2017). Novel bifunctional converter-based supercapacitor energy storage module with active voltage equalizing technology. Institute of Electrical Engineers of Japan, John Wiley & Sons, Inc. doi: 10.1002/tee.22382, https://onlinelibrary.wiley.com/doi/abs/10.1002/tee.22382.
63. Ultracapacitor Modules: Maxwell Technologies. A website report www.maxwell.com/products/ultracapacitors/modules (2017).
64. McFadyen, S., Capacitors: Energy Storage Application: myElectrical. A website report https://myelectrical.com/notes/entryid/223/capacitors-energy-storage-application (29 June 2013).
65. Murata: High Performance Electrical Double-Layer Capacitors. A website report www.mouser.com/pdfDocs/Murata-DMF-DMT_TechnicalGuide.pdf (2013).
66. Supercapacitors, Wikipedia, The free encyclopedia, last visited 28 July 2019 (2019).
67. Haerri, V.V. and Martinovic, D. (2007). Supercapacitor module SAM for hybrid busses: An advanced energy storage specification based on experiences with the TOHYCO-rider bus project. *IECON 2007: 33rd Annual Conference of the IEEE Industrial Electronics Society*, 5–8 November, IEEE, Taipei, Taiwan. doi: 10.1109/IECON.2007.4460395.
68. Kahlen, H. (1999). Energy and power from supercapacitors and electrochemical sources. *Proceedings EVS-16 (el. Vehicle Symposium)*, Tokyo, Japan.
69. Härri, V.V. and Schneuwly, A. (2000) Supercapacitors revolutionize energy storages. *Proceedings EVS-17 (el. Vehicle Symposium)*. www.hta.fhz.ch/iiee.
70. Härri, V.V., Eigen, S., Zemp, B., and Carriero, D. (2003). Minibus TOHYCO-rider with SAM-supercapacitor-storage. *Annual Report 2003 of Swiss Federal Office of Energy SFOE*, Department Mobility and Energy-Storages, 03 December. www.hta.fhz.ch/iiee.
71. Härri, V., Collins, P., and Martinovic, D. (2007). Small bus TOHYCO-rider with super capacitor storage SAM. *Final Report of Swiss Federal Office of Energy SFOE*, Department Mobility and Energy-Storages, May 2007. www.hta.fhz.ch/iiee.
72. Linzen, D., Buller, S., Kardenand, E., and DeDoncker, R.W. (2005). Analysis and evaluation of charge-balancing-circuits on performance, reliability, and-lifetime of supercapacitor systems. *IEEE Transactions on Industry Applications*, 41(5), 1135–1141.
73. Härri, V.V. and Egger, S. (2001). Supercapacitor circuitry concept SAM for public transport vehicles and other applications. *Proceedings EVS-18 (El. Vehicle Symposium)*. www.hta.fhz.ch/iiee/iiee.
74. Alami, A.H., Aokal, K., Abed, J., and Al-Hemyari, M. (2017). Low pressure, modular compressed air energy storage (CAES) system for wind energy storage applications. *Renewable Energy*, 106, 201–211. doi: 10.1016/j.renene.2017.01.002.
75. Lakeh, R.B., Villazana, I.C., Houssainy, S., Anderson, K.R., and Kavehpour, H.P. (2016). Design of a modular solid-based thermal energy storage for a hybrid compressed air energy storage system. *ASME 2016 10th International Conference on Energy Sustainability collocated with the ASME 2016 Power Conference and the ASME 2016 14th International Conference on Fuel Cell Science, Engineering and Technology, Volume 2: ASME 2016 Energy Storage Forum*, 26–30 June, Advanced Energy Systems Division, Solar Energy Division, Charlotte, NC, USA. ISBN: 978-0-7918-5023-7.

76. Fertig, E. and Apt, J. (2011). Economics of compressed air energy storage to integrate wind power: A case study in ERCOT. *Energy Policy*, 39, 2330–2342.
77. Bullough, C., Gatzen, C., Jakiel, C., Koller, M., Nowi, A., and Zunft, S. (2004). Advanced adiabatic compressed air energy storage for the integration of wind energy. *Proceedings of the European Wind Energy Conference, EWEC 2004*, 22–25 November, London, UK.
78. Hartmann, N., Vöhringer, O., Kruck, C., and Eltrop, L. (2012). Simulation and analysis of different adiabatic Compressed Air Energy Storage plant configurations. *Applied Energy*, 93, 541–548.
79. Kim, Y.M., Lee, J.H., Kim, S.J., and Favrat, D. (2012). Potential and evolution of compressed air energy storage: Energy and exergy analyses. *Entropy*, 14, 1501–1521.
80. Kim, Y.M., Shin, D.G., and Favrat, D. (2011). Operating characteristics of constant-pressure compressed air energy storage (CAES) system combined with pumped hydro storage based on energy and exergy analysis. *Energy*, 36, 6220–6233.
81. Wang, H., Wang, L., Wang, X., and Yao, E. (2013). A novel pumped hydro combined with compressed air energy storage system. *Energies*, 6, 1554–1567.
82. Qin, C. and Loth, E. (2014). Liquid piston compression efficiency with droplet heat transfer. *Applied Energy*, 114, 539–550.
83. Raju, M. and Khaitan, S.K. (2012). Modeling and simulation of compressed air storage in caverns: A case study of the Huntorf plant. *Applied Energy*, 89, 474–481.
84. Hedlund, M., Lundin, J., de Santiago, J., Abrahamsson, J., and Bernhoff, H. (2015). Flywheel energy storage for automotive applications. *Energies*, 8(10), 10636–10663. doi: 10.3390/en81010636.
85. Heshmat, H. and Walton, J.F. (2010). An advanced high efficiency modular, flywheel energy storage system. *Presented at The IAPG Mechanical and Electrical Systems Working Groups Meeting*, 3–5 May, Golden, CO.
86. Amiryar, M.E. and Pullen, K.R. (2017). A review of flywheel energy storage system technologies and their applications. *Applied Science*, 7, 286.
87. Vafakhah, B., Masiala, M., Salmon, J., and Knight, A. (2008). Emulation of flywheel energy storage systems with a PMDC machine. *In Proceedings of the 18th IEEE International Conference on Electric Machines*, 6–9 September, Vilamoura, Algarve, Portugal, pp. 1–6.
88. Liu, H. and Jiang, J. (2007). Flywheel energy storage: An upswing technology for energy sustainability. *Energy and Buildings*, 39, 599–604.
89. Hebner, R., Beno, J., and Walls, A. (2002). Flywheel batteries come around again. *IEEE Spectrum*, 39, 46–51.
90. Bolund, B., Bernhoff, H., and Leijon, M. (2007). Flywheel energy and power storage systems. *Renewable and Sustainable Energy Reviews*, 11, 235–258.
91. Sebastián, R. and Alzola, R.P. (2012). Flywheel energy storage systems: Review and simulation for an isolated wind power system. *Renewable and Sustainable Energy Reviews*, 16, 6803–6813.
92. Bender, D.F. (2015). Sandia Report, Sandia National Laboratories, Albuquerque, ME, USA.
93. Pena-Alzola, R., Sebastián, R., Quesada, J., and Colmenar, A. (2011). Review of flywheel based energy storage systems. *In Proceedings of the 2011 International Conference on Power Engineering, Energy and Electrical Drives*, 11–13 May, Malaga, Spain.
94. Babuska, V., Beatty, S., DeBlonk, B., and Fausz, J. (2004). A review of technology developments in flywheel attitude control and energy transmission systems. *In Proceedings of the 2004 IEEE Aerospace Conference*, Big Sky, MT, USA, 6–13 March, Volume 4, pp. 2784–2800.
95. Bitterly, J.G. (1997). Flywheel technology past, present, and 21st century projections. *In Proceedings of the Thirty-Second Intersociety Energy Conversion Engineering Conference (IECEC-97)*, 27 July–1 August, Honolulu, HI, USA, Volume 4, pp. 2312–2315.

96. Faias, S., Santos, P., Sousa, J., and Castro, R. (2008). An overview on short and long-term response energy storage devices for power systems applications. *Proceedings of International Conference on Renewable Energies and Power Quality*, 1, 442–447.
97. Okou, R., Sebitosi, A.B., Khan, A., and Pillay, P. (2009). The potential impact of small-scale flywheel energy storage technology on Uganda's energy sector. *The Journal of Energy in Southern Africa*, 20, 14–19.
98. Beacon Power LCC. Beacon POWER's Operating Plant in Stephentown, New York. Available: http://beaconpower.com/stephentown-new-york/. Accessed 10/2/2017.
99. Whittingham, M.S. (2017). History, evolution, and future status of energy storage. *Proceedings of IEEE 2012*, 100, 1518–1534; *Applied Science*, 7, 286.
100. US Department of Energy Global Energy Storage Database. Available: www.energystorageexchange.org/projects. Accessed 1/2/2017.
101. Beacon Power. (2016). 20 MW Flywheel Frequency Regulation Plant. Final Technology Performance Report, Beacon Power, Tyngsboro, MA, USA.
102. EasyStreet Ramps up Data Center Operations, Deploys Additional VYCON Flywheel Systems to Protect its Green Data Center. Available: www.calnetix.com/newsroom/press-release/easystreet-ramps-data-center-operations-deploys-additional-vycon-flywheel. Accessed 1/2/2017.
103. Tarrant, C., Kinetic Energy Storage Wins Acceptance. Available: www.railwaygazette.com/news/single-view/view/kinetic-energy-storage-wins-acceptance.html. Accessed 12/2/2017.
104. VYCON Technology Allows Los Angeles Metro to be First Transit Agency in U.S. Using Flywheels to Achieve Nearly 20 Percent in Rail Energy Savings. Available: www.calnetix.com/newsroom/press-release/vycon-technology-allows-los-angeles-metro-be-first-transit-agency-us-using. Accessed 1/2/2017.
105. Castro, F., Ng, L.S.B., Dombek, A., Solis, O., Turner, D., Bukhin, L., and Thompson, G. (2014). La Metro Red Line Wayside energy storage substation revenue service regenerative energy saving results. *In Proceedings of 2014 Joint Rail Conference (JRC 2014)*, 2–4 April, Colorado Springs, CO, USA, pp. 1–5.
106. Schroeder, P. and Yu, D.T. (2010). Guiding the Selection and Application of Wayside Energy Storage Technologies for Rail Transit and Electric Utilities, Transit Cooperative Research Program. *Contractor's Final Report for TCRP Project J6/Task 75*, Transport Research Board, Washington, DC, USA.
107. Bleck and Bleck Architects LLC. Available: http://bleckarchitects.com/2014/08/flywheel-launchedcoaster/. Accessed 11/2/2017.
108. Williams Hybrid Power: Advanced Flywheel Energy Storage. Available: www.esa-tec.eu/space-technologies/for-space/williams-hybrid-power-advanced-flywheel-energy-storage/. Accessed 11/2/2017.
109. Truong, L.V., Wolff, F.J., and Dravid, N.V. (2000). Simulation of flywheel electrical system for aerospace applications. *In Proceedings of the 35th Intersociety Energy Conversion Engineering Conference and Exhibition*, 24–28 July, Las Vegas, NV, USA, Volume 1, pp. 601–608.
110. US Marine Corps Utilising Microgrid Energy Storage Project. Available: www.decentralized-energy.com/articles/2015/09/us-marine-corp-utilising-microgrid-energy-storage-project.html. Accessed 1/2/2017.
111. Flywheel energy storage, Wikipedia, The free encyclopedia, last visited 11 August 2019 (2019).
112. Heshmat, H. and Walton, J. (2016). Lubricant free foil bearings pave way to highly efficient and reliable flywheel energy storage system. *ASME 2016 10th International Conference on Energy Sustainability collocated with the ASME 2016 Power Conference and the ASME 2016 14th International Conference on Fuel Cell Science, Engineering and Technology*. doi: 10.1115/ES2016-59350.

113. Khateeb Razack, S., Shabtay, Y., Bhaskar, M., Shabtay, Y., Stilman, H., and Al-Hallaj, S. (2016). Design and performance of thermal energy storage module using high thermal conductivity phase change composite material. *International Refrigeration and Air Conditioning Conference.* http://docs.lib.purdue.edu/iracc/1568.
114. Soares, N., Costa, J.J., Gaspar, A.R., and Santos, P. (2013). Review of passive PCM latent heat thermal energy storage systems towards buildings' energy efficiency. *Energy and Buildings*, 59, 82–103.
115. El-Dessouky, H. and Al-Juwayhel, F. (1997). Effectiveness of a thermal energy storage system using phase- change materials. *Energy Conversion and Management*, 38, 601–617.
116. Chiu, J.N.-W., Martin, V., and Setterwall, F. (2010) System integration of latent heat thermal energy storage for comfort cooling integrated in district cooling network, KTH: Department of Energy Technology, Brinellvägen 68, SE-100 44, Stockholm, Sweden.
117. Chinnapandiana, M., Pandiyarajan, V., and Velraj, R. (2011). Experimental investigation of a latent heat storage system for diesel engine waste heat recovery with and without cascaded arrangement. *International Conference on Mechanical, Automobile and Robotics Engineering*, Barcelona, Spain.
118. United States Department of Energy Information Administration (EIA). (1 May 2016). Retrieved from www.eia.gov/energyexplained/index.cfm?page=us_energy_commercial.
119. Southern California Edison Commercial Time of Usage Electricity Rates. www.sce.com/NR/rdonlyres/6B523AB1-244D-4A8F-A8FE-19C5E0EFD095/0/090202-Business-Rates-Summary.pdf.
120. Sharma, A., Tyagi, V.V., Chen, C.R., and Buddhi, D. (2007). Review on Thermal Energy Storage with Phase Change Materials and Applications, Department of Mechanical Engineering, Kun Shan University, 949.
121. Lazar, J. (2016). *Teaching the "Duck" to Fly*, 2nd Edition. The Regulatory Assistance Project, Montpelier, VT. Available: www.raponline.org/document/download/id/7956.
122. IceEnergy Ice Bear Product. (1 May 2016). Retrieved from www.ice-energy.com/grid/#whitepapers.
123. IceEnergy Ice Bear 30 Product Specification Sheet (1 May 2016). Retrieved from http://ice-energy.com/wp-content/uploads/2013/12/ice_bear_product_sheet.pdf.
124. Mills, A., Farid, M., Selman, J.R., and Al-Hallaj, S. (2006). Thermal conductivity enhancement of phase change materials using a graphite matrix. *Applied Thermal Engineering*, 26, 1652–1661. doi: 10.1016/j.applthermaleng.2005.11.022.
125. Zero Energy Offices. (2011). Phase Change Materials for HVAC Applications. www.pcmproducts.net/Building_Temperature_Control.htm.
126. Indirect, Dual-Media, Phase Changing Material Modular Thermal Energy Storage System. A website report www.energy.gov/.../project-profile-indirect-dual-media-phase-changing-materi... Imperial Journal of Interdisciplinary Research (IJIR).
127. Acciona, Wikipedia, The free encyclopedia, last visited 12 July 2019 (2019).
128. Photovoltaic Solar Energy and its Contribution | ACCIONA. A website report www.acciona.com/renewable-energy/solar-energy/photovoltaic/ (2011).
129. Jacobs, D.E. and Sokoloff, Z. (22 April 2016). Current Assignee Locap Energy LLC, Modular Thermal Energy Storage System. US10072896B2.
130. IceBrick | Nostromo. A website report nostromo.energy/icebrick-2/ (2013).
131. Tandem Chillers. A website report https://tandemchillers.com/ (2013).
132. Laverman, R.J. (25 June 1985). Modular Thermal Energy Storage Tanks Using Modular Heat Batterie. US4524756A.
133. Wasyluk, D.T., Alexander, K.C., Santelmann, K.L., and Marshall, J.M. (07 April 2016). Modular Molten Salt Solar Towers with Thermal Storage for Process or Power Generation or Cogeneration. US20160097376A1.

134. Tyner, C. and Wasyluk, D. (2014). eSolar's modular, scalable molten salt power tower. *Energy Procedia*, 49, 1563–1572.
135. Pacheco, J.E., Moursund, C., Rogers, D., and Wasyluk, D. (2011). Conceptual design of a 100 MWe modular molten salt power tower plant. *In Proceedings of SolarPACES 2011*, 20–23 September, Granada Spain.
136. Pacheco, J.E. (ed) (2002). *Final Test and Evaluations Results from the Solar Two Project*. Sandia National Laboratories, Albuquerque, NM, SAND2002-0120.
137. Campell, M., Newmaker, M., Lewis, N., George, C.T., and Cohen, G. (2011). Design of a modular latent heat storage system for solar thermal power plants. *Proceedings of the 5th Energy Sustainability Conference: ESFuelCell2011*, 7–10 August, Washington, DC, pp. 679–685, ISBN: 978-0-7918-5468-6.
138. Meduri, P.K., Hannemann, C.R., and Pacheco, J.E. (2010). Performance characterization and operation of eSolar's Sierra Suntower Power Tower Plant. *Proceedings of the 2010 SolarPACES Conference*, Perpignan, France.
139. Slack, M., Meduri, P., and Sonn, A. (2010). eSolar power tower performance modeling and experimental validation. *Proceedings of the 2010 SolarPACES Conference*, Perpignan, France.
140. Ricklin, P., Slack, M., Rogers, D., and Huibregtse, R. (2013). Commercial readiness of eSolar next generation heliostat. *Proceedings of the 2013 SolarPACES Conference*, Las Vegas, NV, USA.
141. Castro, P., Selvam, P., and Suthan, C. (2016). Review on the design of PCM based thermal energy storage systems. *Imperial Journal of Interdisciplinary Research (IJIR)*, 2(2), 203–215. ISSN: 2454-1362, www.onlinejournal.in.
142. Finck, C., Henqueta, E., van Soesta, C., Oversloota, H., de Jonga, A.-J., Cuypersa, R., and van't Spijkera, H. (2014). Modular thermochemical heat storage. *SHC 2013, International Conference on Solar Heating and Cooling for Buildings and Industry*, 23–25 September 2013, Freiburg, Germany; *Energy Procedia*, 48, 320–326.
143. Novo, A.V., Bayon, J.R., Castro-Fresno, D., and Rodriguez-Hernandez, J. (2010). Review of seasonal heat storage in large basins: Water tanks and gravel-water pits. *Applied Energy*, 87, 390–397.
144. Finck, C., van't Spijker, H., de Jong, A.J., Henquet, E., Oversloot, H., and Cuypers, R. (2013). Design of a modular 3 kWh thermochemical heat storage system for space heating application. *IC-SES 2 Sustainable Energy Storage in Buildings Conference*, Dublin.
145. De Jong, A.J., Trausel, F., Finck, C., and Cuypers, R. (2013). Thermochemical heat storage: System design issues. *SHC 2013, International Conference on Solar Heating and Cooling for Buildings and Industry*, 23–25 September, Freiburg, Germany.
146. Ceranic, B., Beardmore, J., and Cox, A. (2017). A novel modular design approach to "thermal capacity on demand" in a rapid deployment building solutions: Case study of smart-POD. *Energy Procedia*, 134, 776–786. doi: 10.1016/j.egypro.2017.09.582.
147. Steijger, L.A., Buswell, R., Smedley, V., Firth, S.K., and Rowley, P. (2013). Establishing the zero carbon performance of compact urban dwellings. *Journal of Building Performance Simulation*. doi: 10.1080/19401493.2012.724086.
148. Steijger, L. (2013). Evaluating the feasibility of 'zero carbon' compact dwellings in urban areas. Ph.D. Dissertation, Loughborough University; Pinel, P., Cruickshank, C.A., Beausoleil-Morrison, I., and Wills, A. (2011). A review of available methods for seasonal storage of solar thermal energy in residential applications. *Renewable and Sustainable Energy Reviews*, 15, 3341–3359.
149. Pavlov, G.K. and Olesen, B.W. (2012). Thermal energy storage: A review of concepts and systems for heating and cooling applications in buildings: Part 1 – Seasonal storage in the ground. *HVAC&R Research*, 18(3), 515–528.
150. Kenisarin, M.M. (2010). High-temperature phase change materials for thermal energy storage. *Renewable and Sustainable Energy Reviews*, 14, 955–970.

151. Ceranic, B., Beardmore, J., and Cox, A. (2013). Smart-POD CRD Project Completion Report, Sustainable Construction INet Energy Benchmarks CIBSE TM46: 2008, Page Bros, Norwich. ISBN: 978-1-903287-95-8.
152. Boardass, B., Cohen, R., Burman, E., and Field, J. (2014). Tailored energy benchmarks for offices & schools. *CIBSE ASHRAE Technical Symposium*, Dublin. www.cibse.org/knowledge/cibse-technical-symposium-2014/tailored-energy-benchmarks-for-offices-schools.
153. Sarbu, I. and Sebarchievici, C. (2018). A comprehensive review of thermal energy storage. *Sustainability*, 10, 191. doi: 10.3390/su10010191, www.mdpi.com/journal/sustainability.
154. Nordbeck, J., Beyer, C., and Bauer, S. (2017). Experimental and numerical investigation of a scalable modular geothermal heat storage system. *Energy Procedia*, 125, 604–611. doi: 10.1016/j.egypro.2017.08.217.
155. Pumped storage hydroelectricity, Wikipedia, The free encyclopedia, last visited 12 August 2019 (2019).
156. Modular Pumped Storage Hydropower Feasibility and Economic... A website report by DOE, Washington, DC, www.energy.gov/sites/.../modular-pumped-storage-hydropower-feasibility.pdf (2015).
157. Witt, A., Hadjerioua, B., Uría-Martínez, R., and Bishop, N. (2015). Evaluation of the Feasibility and Viability of Modular Pumped Storage Hydro (m-PSH) in the United States, ORNL/TM-2015/ 559, Environmental Science Division, Oak Ridge, TN.
158. Bradbury, K., Pratson, L., and Patiño-Echeverri, D. (2014). Economic viability of energy storage systems based on price arbitrage potential in real-time U.S. electricity markets. *Applied Energy*, 114, 512–519. doi: 10.1016/j.apenergy.2013.10.010.
159. Department of Energy (DOE). (2013). Grid Energy Storage. December, 2013.
160. Federal Energy Regulatory Commission (FERC). (2014). Preliminary Permit Application Trends. http://ferc.gov/industries/hydropower/gen-info/licensing/pump-storage/trends-pump-storage.pdf. Accessed 1/05/2014.
161. Federal Energy Regulatory Commission (FERC). (2014). Issued Preliminary Permits. http://ferc.gov/industries/hydropower/gen-info/licensing/issued-pre-permits.xls. Updated 23/04/2014.
162. Federal Energy Regulatory Commission (FERC). (2014). FERC Seeking Pilot Projects To Test Two-Year Hydro Licensing Process. www.ferc.gov/media/news-releases/2014/2014-1/01-06-14.asp.
163. HDR. (2011). Oak Ridge National Laboratory Pumped Storage Reconnaissance Study. ORNL, September 2011.
164. HDR. (2011). Quantifying the Value of Hydropower in the Electric Grid: Plant Cost Elements. EPRI, November 2011.
165. Idaho National Laboratory (INL). (2014). Assessment of Opportunities for New United States Pumped Storage Hydroelectric Plants Using Existing Water Features as Auxillary Reservoirs. INL, March 2014.
166. Innovation Reclamation Technologies & Engineering Co., Inc. (IRTEC). (2013). Justus Dam/Refuse Area/Deep Mine Requested Information, Communication, December 2013.
167. NHA. (2012). Pumped Storage Development Council Challenges and Opportunities for New Pumped Storage Development. A White Paper Developed by NHA's Pumped Storage Development Council. July, 2012.
168. Oak Ridge National Laboratory (ORNL). (2014). National Hydropower Asset Assessment Program. Accessed 1/05/2014.
169. Witt, A., DeNeale, S., Papanicolaou, T., Abban, B., and Bishop, N. (2018). Standard Modular Hydropower: Case Study on Modular Facility Design. Technical Report ORNL/TM-2018/915, ORNL, Oak Ridge, TN. doi: 10.2172/1484123.

170. Bueno, C. and Carta, J. (2006). Wind powered pumped hydro storage systems, a means of increasing the penetration of renewable energy in the Canary Islands. *Renewable and Sustainable Energy Reviews*, 10, 312–340.
171. Myasnikov, V., Gerasimov, A., and Sobolev, V. (07 October 2008). Modular Metal Hydride Hydrogen Storage System. US7431756B2.
172. Pourpoint, T.L., Velagapudi, V., Mudawar, I., Zheng, Y., and Fisher, S.T. (2010). Active cooling of a metal hydride system for hydrogen storage. *International Journal of Heat and Mass Transfer*, 53, 1326–1332. doi: 10.1016/j.ijheatmasstransfer.2009.12.028.
173. Visaria, M., Mudawar, I., and Pourpoint, T. (2011). Enhanced heat exchanger design for hydrogen storage using high-pressure metal hydride: Part 2. Experimental results. *International Journal of Heat and Mass Transfer*, 54, 424–432. doi: 10.1016/j.ijheatmasstransfer.2010.09.028.
174. Turillon, P.P. and Sandrock, G.D. (23 January 1979). Hydrogen Storage Module. US4135621A.
175. Calvera: Storage for Hydrogen (H2) High Pressure. A website report www.calvera.es/en/business-lines/hydrogen-h2/storage-for-hydrogen/ (2015).
176. Calvera: Gas Storage CNG Container Tube Trailer Hydrogen and Biogas. A website report www.calvera.es/en/ (2015).
177. Shan, X., Payer, J.H., Wainright, J.S., and Dudik, L. (2011). Demonstration of a microfabricated hydrogen storage module for micro-power systems. *Journal of Power Sources*, 196(2), 820–826. doi: 10.1016/j.jpowsour.2010.07.077.
178. Wainright, J.S., Savinell, R.F., Liu, C.C., and Litt, M. (2003). Microfabricated fuel cells. *Electrochimica Acta*, 48, 2869.
179. Tam, W.G. and Wainright, J.S. (2007). A microfabricated nickel–hydrogen battery using thick film printing techniques. *Journal of Power Sources*, 165, 481.
180. Tam, W.G. and Wainright, J.S. (2006). A micro-fabricated nickel-hydrogen battery, *ECS Transactions*, 1, 1.
181. Sandrock, G. (1999). Panoramic overview of hydrogen storage alloys from a gas reaction point of view. *Journal of Alloys and Compounds*, 293, 877.
182. Wang, X.L., Suda, S. and Wakao, S. (1994).Improved surface properties of Zr-M-Ni hydride electrodes by oxidation treatment*. *Zeitschrift fur Physikalische Chemie*, 183(1–2), 294.
183. Wang, X.L. and Suda, S. (1993). Hydriding and dehydriding reactions of… under quasi-isothermal conditions. *Journal of Alloys and Compounds*, 194, 173–177.
184. Willey, D.B., Pederzolli, D., Pratt, A.S., Swift, J., Walton, A., and Harris, I.R. (2002). Low temperature hydrogenation properties of platinum group metal treated, nickel metal hydride electrode alloy. *Journal of Alloys and Compounds*, 330–332, 806.
185. Zaluski, L., Zaluska, A., and Strom-Olsen, J.O. (1997). *Journal of Alloys and Compounds*, 253–254, 70.
186. Oelerich, W., Klassen, T., and Bormann, R. (2001). Atomic hydrogen adsorption and incipient hydrogenation of the Mg(0001) surface: A density-functional theory study. *Journal of Alloys and Compounds*, 322, L5.
187. Zaluski, L., Zaluska, A., and Tessier, P. (1996). Hydrogen absorption by nanocrystalline and amorphous Fe-Ti with palladium catalyst, produced by ball milling. *Journal of Materials Science*, 31, 695.
188. Shan, X., Payer, J.H., and Wainright, J.S. (2006). Increased performance of hydrogen storage by Pd-treated $LaNi_{4.7}Al_{0.3}$, $CaNi_5$ and Mg_2Ni. *Journal of Alloys and Compounds*, 426, 400.
189. Shan, X., Payer, J.H., and Wainright, J.S. (2007). Improved durability of hydrogen storage alloys. *Journal of Alloys and Compounds*, 430, 262.

190. Shan, X., Payer, J.H., Wainright, J.S. and Dudik, L. (2011). A micro-fabricated hydrogen storage module with sub-atmospheric activation and durability in air exposure. Journal of Power Sources, 196, 827.
191. Shan, X., Payer, J.H., and Jennings, W.D. (2009). Mechanism of increased performance and durability of Pd-treated metal hydriding alloys. *International Journal of Hydrogen Energy*, 34, 363.
192. Brodd, R.J. and Bullock. K.R. (2004). Batteries, 1977 to 2002. *Journal of the Electrochemical Society*, 151, k1.
193. Paladini, V., Miotti, P., Manzoni, G., and Ozebec, J. (2003). Conception of modular hydrogen storage systems for portable applications. *Towards a Greener World: Hydrogen and Fuel Cells Conference and Trade Show*, Canadian Hydrogen Association, Canada, p. 446.
194. Modular Fuel System Flexible, Highly…: Leonardo DRS. A website report www.leonardodrs.com/media/6278/mfs_datasheet.pdf (2007).
195. Finck, J.W. (03 November 2015). Modular Fuel Storage System. US Patent US9174531B2.
196. Modular Fuel System (MFS) Tank Rack Module (TRM)… A website report https://govtribe.com/…/modular-fuel-system-mfs-tank-rack-module-trm-production-w… (2011).
197. Campo, O.D. (11 May 2004). Modular Compressed Natural Gas (CNG) Station and Method for Avoiding Fire in such Station. US6732769B2, United States.
198. United Technologies Research Center Provides Exclusive License to… A website report www.angpinc.com/…/united-technologies-research-center-provides-exclusive-license-t… (4 February 2016).
199. United Technologies Research Center (MOVE): ARPA-E. A website report https://arpa-e.energy.gov/?q=impact-sheet/united-technologies-research-center-move (27 February 2017).
200. Castalia. (2015). Natural Gas in the Caribbean: Feasibility Studies, Revised nal report (Vols I and II). Report to the Inter- American Development Bank, 30 June 2015.
201. International Gas Union (IGU). (2016). 2016 World \Energy Report.
202. Shell. (2017). Shell LNG Outlook 2017. www.shell.com/energy-and-innovation/natural-gas/liqueed-natural-gas-lng/lng-outlook.html.
203. Shiryaevskaya, A. and Burkhardt, P. (2017). Hottest thing in LNG is producing power as record glut looms. *Bloomberg*, 18 January 2017. www.bloomberg.com/news/articles/2017–0118/hottest-thing-in-lng-is-producing-power-as-record-glut-looms.
204. Raine, B. (2014). Onshore mid-scale LNG terminal storage and modularization. *Trinidad Oil and Gas Conference*, May 2014, Port of Spain, Trinidad.
205. Raine, B. and Powell, J. (2015). Onshore mid-scale LNG terminal storage modularization. *Gastech* 2015, 29 October, Singapore.
206. Ezzarhouni, A., Powell, J., and Elliott, S. (2016) Why a membrane full integrity tank? *LNG 18*, 11–15 April, Perth.
207. Long, B. (1998). Bigger and cheaper LNG tanks? Overcoming the obstacles confronting freestanding 9% Nickel Steel Tanks up to and beyond 200,000 m^3. *LNG 12*, 4–7 May, Perth, Paper Session 5.6.
208. Veliotis, P.T. (1977). Solution to the series production of aluminum LNG spheres. *Society of Naval Architects and Marine Engineers Transactions*, 85, 481–504.
209. Antalffy, L.P., Aydogean, S., De la Vega, F.F., Malek, D.W., and Martin, S. (1998). Technical-economic evaluation of pumping systems for LNG storage tanks with side and top entry piping nozzles. *LNG 12*, 4–7 May, Perth, Poster Session B.8.
210. Coers, D. (2005). Transshipping LNG: Downscaling Field- Erected Storage Tanks for Lower Pro le. Presentation with photos provided by CB&I.

211. Peru LNG, Melchoriate, Peru, Triple Pendulum Bearings Protect Critical Storage Tanks. Earthquake Protection Systems Inc. http://www.earthquakeprotection.com/pdf/Peru_LNG_Dec08.pdf.
212. Symans, M.D., Seismic Protective Systems: Seismic Isolation, FEMA, Instruction Material Complementing FEMA 451, Design Examples, Seismic Isolation 15-7-1. www.ce.memphis.edu/7119/PDFs/FEAM_Notes/Topic15-7-SeismicIsolationNotes.pdf.
213. Soren Karlsson, Creating Optimal LNG Storage Solutions. A website report in Watsila Technical Journal, www.wartsila.com/twentyfour7/in-detail/creating-optimal-lng-storage-solutions (20 October 2015).
214. Oh, B.T., Hong, S.H., Yang, Y.M., Yoon, I.S., and Kim, Y.K. (2003). The development of KOGAS membrane for LNG storage tank, document ID, ISOPE-I-03-394. *The Thirteenth International Offshore and Polar Engineering Conference*, 25–30 May, International Society of Offshore and Polar Engineers, Honolulu, Hawaii, USA.
215. LNG to Power: Wärtsilä Energy Solutions, Oil & Gas. A website report www.wartsila.com/energy/lng-to-power (26 December 2018).
216. LNG to Power Solutions: MAN Energy Solutions. A website report www.man-es.com/energy-storage/.../energy-storage.../energy-storage.../man-es... (2016).
217. Lokken, R.T., Modular Membrane LNG Tank. WO2017116548A1; WIPO (PCT) Other languages French Inventor Roald T. Lokken, Worldwide applications, 2016 CN AU US KR CA WO, Application filed by Exxonmobil Upstream Research Company, 06 July 2017.
218. LNG storage tank, Wikipedia, The free encyclopedia, last visited 3 April 2018 (2018).
219. LNG Storage Equipment | LNG Technology | Chart Industries. A website report www.chartindustries.com/Energy/LNG-Solutions-Equipment (2019).
220. LNG Cryogenic Storage Tanks & Regasification | Chart Industries. A website report www.chartindustries.com/Energy/LNG-Solutions-Equipment/Storage (2019).

9 Modular Systems for Energy and Fuel Transport

9.1 INTRODUCTION

In this chapter, we examine modular systems used for energy and fuel transport, which is necessary as a part of energy usage management. In urban environment, electricity is transported by smart and complex grid. This grid operates two ways: from supplier to consumer and vice a versa. While this electricity transport system provides stable electricity to all utility customers, it does not extend to some rural and isolated areas. In these areas where distributed energy, particularly by small renewable sources, is more prominent, small modular microgrids and nanogrids are being developed. These grids can be connected to large complex smart grids, or they can be independent. This means of electricity transport is growing very rapidly, particularly to enhance electricity from renewable resources in overall energy mix. The development of modular microgrids and nanogrids has also led to the development of microenergy internet and the infusion of open energy system for distributed energy. Sections 9.2–9.5 examine microgrids, nanogrids, microenergy internet, and infrastructure for open energy system. All of these are modular in nature.

The remaining five sections of the chapter examine the use of modular approach for various types of fuel transport systems. These sections evaluate modular fuel transport for biomass, liquid fuel, natural gas, liquefied natural gas (LNG), compressed natural gas (CNG), natural gas hydrates (NGH), natural gas liquids (NGL), and hydrogen. In each case, emphasis is placed on the modular systems for fuel transport.

9.2 MICROENERGY SYSTEMS AND MODULAR MICROGRID

Two major trends are driving the transformation of the energy world: digitalization and the shift from centralized to distributed energy systems. These trends as well as prosumers' needs are driving the development of microenergy systems and related modular microgrids. Technically, a grid is any combination of power sources, power users, wires to connect them, and some sort of control system to operate it all. In essence, it is a mechanism for transporting electricity from source (s) of generation to various consumers in a controlled manner. Unlike a large-scale smart grid that spans over large urban area, microgrid is a small, freestanding grid. It can consist of several buildings, one small building (sometimes called a "nanogrid"), or even one person (a "picogrid") with a backpack solar panel, an iPhone, and some headphones.

Several years ago, a research firm GTM indicated that there are 1,900 basic and advanced, operational and planned microgrids in the U.S., with the market expected to **grow quickly**. Most microgrids today are basic, one-generator affairs, but more complex microgrids are popping up all over like in Brooklyn, Alcatraz island, and Sanoma, California [1–97]. Ton and Smith [98] describe the Department of Energy initiative for microgrid project and also review the present global status of microgrids.

Some microgrids stand on their own, apart from any larger grid, **often in remote rural areas**. These off-grid microgrids are a relatively cheap and quick way to **secure some access to power for people who now lack it**. Often these microgrids can be extended more quickly than large grids. Most microgrids, especially in wealthier nations, are grid-connected—they are embedded inside a bigger grid, like any other utility customer. What makes a microgrid a microgrid is that it can flip a switch (or switches) and "island" itself from its parent grid in the event of a blackout. This enables it to provide those connected to it with (at least temporary) backup power. Again, most actually existing microgrids are extremely basic, such as a hospital with a diesel generator in the basement, or a big industrial facility with a combined-heat-and-power (CHP) facility on site that can provide some heat and power during a blackout. Microgrids contain all the elements of complex energy systems; they maintain the balance between generation and consumption, and they can operate on and/or off grid. They are ideal for supplying power to remote regions or locations with no connection to a public network. In addition, more and more industrial operators are using microgrids to produce the electricity they need cost effectively, sustainably, and reliably.

Microgrids use a variety of energy sources, including photovoltaic (PV) and wind-power plants as well as small hydropower and biomass-power plants. Biodiesel generators and emergency power units, storage modules, and intelligent control systems ensure the security of supply. Using sophisticated software, operators can optimize power usage based on demand, utility prices, and other factors. Application areas most impacted are university campuses as well as commercial and industrial sites. Microgrids are designed to provide uninterrupted 24/7 power and to balance load demands for an organization with changing power needs. Relevant applications are critical infrastructures, military institutions, commercial and industrial areas, remote locations, and islands. By using primarily renewable energy, microgrids reduce carbon dioxide emissions, which is often required by government regulations. That makes them especially attractive for campuses, utilities, and islands [1–97].

Microgrids won't be a core part of the clean-energy transition until they serve all three grid needs—to be greener, more reliable, more resilient. Right now, most microgrids around the world rely on diesel generators, which are polluting and loud, so they're not very green. (In the U.S., the primary sources are **CHP and natural gas**). They get turned on only when the grid is down, so they don't help with day-to-day reliability. Of the three grid needs, most serve only resilience, for those lucky enough to be connected to one. As basic as most of them are today, microgrids hold great promise for the future. Technology is rapidly expanding the possibilities. The reasons for rapid growth of microgrids are as follows [1–97]:

- Electricity use is becoming more controllable and adaptable, as **every system and appliance learns to communicate over the internet**.
- Small-scale and community-scale electricity generators are getting cheaper, cleaner, and more diverse; they now include solar panels, small-scale wind, efficient natural gas generators and fuel cells, CHP, and more. (Solar panels, in particular, have become **very cheap**.)
- Energy storage is also becoming cheaper and more diverse, from various kinds of batteries and fuel cells to thermal storage in hot water or ice. (The Stone Edge Farm microgrid in Sonoma boasts **five separate forms of storage**.) Every bit of new storage helps to smooth out the variations in solar and wind, allowing more to be absorbed.
- Software, AI, and machine learning are enabling intelligent integration of all these diverse resources.

Smart design and software can create microgrids specifically designed to integrate distributed renewable energy, microgrids designed to provide 99.9999% reliability, or microgrids designed for maximum resilience. There are even "nested" microgrids within microgrids. Smarter microgrids can communicate on an ongoing basis with their parent grids. By aggregating together distributed, small-scale resources (solar panels, batteries, fuel cells, smart appliances and HVAC (heating, ventilation, and cooling) systems, etc.), a microgrid can present itself to the larger grid as a single entity—a kind of **Voltron** composed of distributed energy technologies. This makes things easier on grid operators. They don't necessarily relish the idea of communicating directly with millions (or **billions**) of discrete generators, buildings, and devices. It's an overwhelming amount of data to assimilate. Microgrids can gather those smaller resources together into discrete, more manageable, and predictable chunks.

Grid operators can put these chunks to good use. A smart microgrid can provide "grid services"—storing energy when it's cheap, providing energy when it's expensive, serving as backup capacity, or smoothing out frequency and voltage fluctuations. A single smart microgrid, aggregating diverse, distributed low-carbon resources, can provide cheap, clean, reliable power to those within it. It can also provide grid services to the larger grid around it. What really tickles the imagination is a grid that contains dozens or hundreds of networked microgrids—even a grid that is someday *composed* of networked microgrids. This kind of "modular architecture," with multiple semi-autonomous nodes operating in parallel, is more secure and efficient than a centralized system with a few, large points of failure. In essence, microgram is a bottom-up modular approach for providing electricity transport from suppliers to consumers. Microgrids may never eliminate the need for large utilities, power plants, and transmission lines, but moving more power generation, management, and consumption under **local control** makes everyone less dependent on them; and it makes the grid greener, more reliable, and resilient. There are other ways of aggregating small-scale distributed energy resources (DERs) that do not involve a physical switch that can island them off from the grid. These "virtual" aggregations can gather together multiple small resources (batteries, solar panels, whatever) and treat the result as a single unit that participates in grid-services markets.

These **"virtual power plants"** offer lots of benefits, to participants and to the grid, but they do not offer the core microgrid value proposition: resilience, i.e., independence from the larger grid in times of need [1–97].

Distributed energy systems are made up of many nodes that are part of a larger system. They tend to be hybrid systems made up of flexible loads, electrical, thermal, and mechanical storage systems, and local conventional or renewable production systems. Specialized software supplies the intelligence necessary for linking various components together to meet specific operational objectives. This software solution approach can establish the basis for standardized configurable microgrids. In order to get to that level of modular and streamlined installations, solution providers need to focus more on standardized energy services delivery rather than on custom-engineered project delivery. The transition to mass adoption requires standardization, simplification, and personalization. These requirements include the flexibility to match the acceptable and appropriate level of risk, as well as the preferred business model of the customer. This transition also requires new tools and processes to train and activate a large workforce already skilled in large-scale deployments in adjacent industries. DER software is making this a reality with plug-and-play device interoperability, data-driven system configuration, and equipment installation and commissioning that are within the capabilities of most licensed electricians [1–97].

Software as a service (SaaS) has matured in most industries and is making significant headway in the electric power industry as well. There are four general requirements for SaaS systems being applied to microgrids: intelligent gateways that host local applications, a backend that hosts system applications, service provider network operations, and customer portals. If solution providers master these critical elements, this microgrid market can grow much faster than anyone could ever predict or forecast. Microgrids empower energy consumers to take active control of their energy systems and operate their interconnection with the grid as a managed resource. They also empower innovative energy services companies to offer energy solutions to consumers in ways not possible before. And they empower communities and corporations to set goals and implement solutions that transcend the narrow limitations of a kilowatt-hour mindset while simultaneously meeting consumer objectives and larger social imperatives such as greenhouse gas reduction and energy security [1–97].

9.2.1 Modular Microgrids

Modular microgrids are on their way in. In fact, they boast so many benefits they could liberate microgrids. Larger-scale microgrids tend to be so complex that they sometimes die in the planning process. They die because they are sold as complex projects and require money spent on complex engineering that doesn't always make economic sense. They also require complex control schemes. Modular microgrids, on the other hand, often involving a number of microgrids under 100 kW in size, are smaller, expandable, and have simpler controls (see Figure 9.1). They don't have to be forced into an existing system in the form of a retrofit. And they're easier for utilities to get installed. Such benefits are the key to liberating microgrids. It's easier to make modular microgrids plug and play. The smaller microgrids can run off of smart inverters, don't need complex control schemes, and can provide more renewables

Systems for Energy and Fuel Transport 475

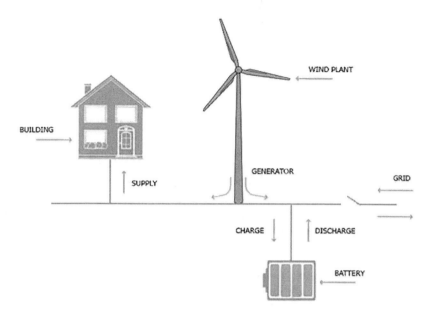

FIGURE 9.1 A typical scheme of an electric based microgrid with *renewable energy* resources in grid-connected mode [99].

and less reliance on fossil fuels. Modular microgrids will lead to the development of more greenfield projects. If people understand that one doesn't need to tie into existing diesel plants, and want to focus on just solar plus storage, those are the kinds of microgrids that will lead to more greenfield projects [1–97].

The key players helping to advance modular microgrids are Tecogen [42,43], Blue Pillar, and Spirae, according to Navigant. Duke Energy's Coalition of the "Willing" is helping move the market, with its goal of developing a common interoperability framework for microgrids. Tecogen, for example, offers CHP units and has embedded in each unit a chip with an inverter, which controls for no cost. Some argue whether Tecogen's unit is really a microgrid. The company is focused on 1 MW power and less. In that category, it's easier to make microgrids plug and play. The Tecogen unit is 100 kW, which can be put together, and then solar and batteries can be added. These types of offerings will take up a growing share of the market. In fact, Tecogen ranked fourth in total installed microgrid projects globally last year in Navigant Research's Microgrid Deployment Tracker. In a separate report of ranking by microgrid developers that offer their own controls platform, Tecogen ranked fifth.

Tecogen produces InVerde, a small natural gas engine often deployed as a modular 100-kW CHP unit. It includes a Consortium for Electric Reliability Technology Solutions' islanding software. The Sacramento Municipal Utility District strung some of these together to produce a microgrid. Islanding is provided by an inverter that can support a number of generators on the same microgrid. Each one can act on its own to respond to load changes and regulate power quality. Blue Pillar [100] offers a modular microgrid with very little engineering. It also has a library of all existing assets, designed to help developers identify where microgrids could be developed.

Spirae [101] offers DER management systems, along with real-time microgrid control systems for electric utilities and energy service providers.

A microgrid is simply a collection of DER that, when managed as a single system with appropriate control software, turns that collection into a virtual system with well-defined properties and capabilities. The key requirement for a microgrid is that it be able to actively manage power and energy flow within some defined ranges. Such systems are now available. The main obstacles to more deployment of modular microgrids are issues related to connecting distributed resources to the grid. A lot of it boils down to the nitty-gritty interconnection of all distributed resources, where utilities are worried about too much solar and wind and have policies about them. However, many utilities are getting more comfortable with solar plus storage and microgrids. Modular microgrids will play an important role along with the bigger projects. Smaller plug-and-play modular microgrids will serve smaller needs, while at the same time, for large cities, huge projects will require a certain amount of complex engineering and smart grids.

Modular microgrids are being considered as a solution for implementing more reliable and flexible power systems compared to the conventional power grid. Various factors, such as low system inertia, might make the task of microgrid design and operation to be non-trivial. In order to address the needs for operational flexibility in a simpler manner, the dissertation by Kim and Mathews [97] discusses modular approaches for design and operation of microgrids. This research investigates Active Power Distribution Node (APDN), which is a storage-integrated power electronic interface, as an interface block for designing modular microgrids. To perform both voltage/current regulation and energy management of APDNs, two hierarchical control frameworks for APDNs are proposed. The first framework focuses on maintaining the charge level of the embedded energy storage at the highest available level to increase system availability, and the second framework focuses on autonomous power sharing and storage management. The detailed design process, control performance, and stability characteristics are also studied. The performance is also verified by both simulation and experiments. The control approaches enable application of APDNs as a power router realizing distributed energy management. The decentralized configuration also increases modularity and availability of power networks by preventing a single point of failure. The advantages of using APDNs as a connection interface inside a power network are discussed from an availability perspective by performing a comparison using Markov-based availability models.

Furthermore, the operation of APDNs as power buffers is explored and the application of APDNs enabling modular implementation of microgrids is also studied. APDNs enable the system expansion process—i.e., connecting new loads to the original system—to be performed without modifying the configuration of the original system. The analysis results show that a fault-tolerant microgrid with an open architecture can be realized in a modular manner with APDNs. APDNs also enable simplified selectivity planning for system protection. The effect of modular operation on microgrids is also studied by using an inertia index. The index not only provides insights on how system performance is affected by modular operation of modular microgrids, but is also used to develop a simpler operation strategy to mitigate the effect of plug-and-play operations [1,2].

9.2.2 Expandable, Modular Microgrid Unit

Several microgrid projects are already in service around the world. The Aichi project described in [13] is an independent power system constructed in 2005 in Japan that employs PV panels and a battery storage system. The Hachinohe project detailed in [14] features a microgrid system that uses a private distribution line to transmit electricity primarily generated by a gas engine system. Although the loads connected to a microgrid continue to grow, the same level of reliability and power quality is expected. This entails expanding the microgrid and providing additional resources. In addition, this requires new studies and new design and commissioning efforts for the entire microgrid. Therefore, expandability is an important factor that must be taken into account when designing a microgrid [12].

The study by Falahati et al. [11] proposes a modular microgrid unit (MMGU) that can work autonomously but is also capable of connecting to other MMGUs to form a larger microgrid. New units can simply be added when the load increases. A real-time automation, control, and protection (RTACP) system provides distributed control and protection logic for each MMGU and can communicate with other MMGUs through an Ethernet network. The study presents the single-line diagram of the proposed MMGU and lists the key power equipment as well as the typical values of the MMGU elements. Various modes of operation are discussed, and possible control and protection strategies are outlined. In addition, the automation scheme for each MMGU and for all of the MMGUs together is discussed in detail.

The MMGU enhances microgrid operation when it is used in conjunction with renewable energy sources (RESs) such as PV units and wind turbines. However, RESs are inherently unreliable and non-dispatchable, and they suffer from low stability and poor power quality. The proposed MMGU can alleviate all of the aforementioned drawbacks because it is equipped with dispatchable resources and a storage unit along with intelligent control algorithms. An MMGU can be applied in any facility that uses a synchronous machine bidirectionally (e.g., to pump and store). Pumped storage power plants, water pump utilities, and offshore facilities are a few applications that can use MMGUs to provide their own power [15]. The system topology and the proximity of generation to loads are two key factors that affect microgrid reliability [12]. In the proposed MMGU, the focus is on developing a reliable unit in which there is minimum distance and interface between the generation and load. The entire microgrid, when MMGUs are connected, can be joined to the main busbar. The main busbar is usually connected to the utility grid at the point of common coupling (PCC). Reliable arrangements, such as double incoming/double busbar, can be deployed to improve the operation reliability. Some of the key components of an MMGU are as follows: (a) Synchronous machine which is the focal point of the MMGU because it is the only distributed generation (DG) in the unit. The machine is operated in both motor and generator modes. (b) Main bus to other switches. A 1,500 Ah battery can store the energy from the generator to energize a typical 30 kW load for about 6 h. For most remote sites, PV panels on the top of the roof are an available option. The MMGU is engineered to be capable of using PV panels to directly connect to the battery storage unit. Intelligent electronic devices (IEDs) have a crucial role in the operation, control, and protection of the MMGU.

For each feeder, IEDs are installed that locally control and protect dedicated zones. All IEDs transfer data to the local RTACP unit based on the IEC 61850 protocol.

Each MMGU is typically designed to provide power for 30 kW of local load and 1 MW of external load. The number of units needed for a microgrid is calculated by dividing the total installed load by the nominal capacity of each unit, and that number grows by increasing the installed external load. A more detailed method of finding the minimum number of MMGUs required is a reliability assessment of the entire microgrid based on the total load and acceptable level of reliability indices (e.g., System Average Interruption Frequency Index [SAIFI] and Customer Average Interruption Frequency Index [CAIFI]). The variable-speed operation of small-scale generation units is a topic of common interest. In the proposed MMGU, variable-speed operation is implemented with a current source converter (CSC) that drives a synchronous machine [15,16].

The main advantage of the proposed MMGU is the flexibility in different operation modes. Each mode needs a certain combination of breaker statuses, which may be achievable either automatically, through operator initiation, or manually. The MMGU storage unit is made up of two connected parts: a battery and a charger. Battery systems store electrical energy in the form of chemical energy. The charger is generally a bidirectional converter that allows energy to be stored in and taken from the batteries. In addition to reserving energy for future demand, the battery stabilizes and permits DG units to run at a constant and volt-ampere reactive (VAR) compensation. It can run in several modes. FSM (Finite State Machine) mode represents the simplest mode of operation in that there is no power electronic equipment between the motor and outgoing feeder, and the motor is energized through the bulk power system [16–21].

Although FSM mode is more affordable and reliable than VSM (Value Stream mapping) mode, it has some important drawbacks such as vulnerability to under voltage and under frequency of the utility bus and voltage stability problems during the VAR deficiency in the system. VSM mode entails more expensive design and more complicated control, protection, and automation implementation. The design is less reliable in comparison with FSM mode. On the other hand, the power electronic interface provides a higher level of controllability and protection. VSM mode can also be used to start FSM mode in order to limit the starting current [19,20]. When the synchronous machine is in generation mode, it generates power and sends it to the main bus. MMGU also provides real-time automation and control. It also has all the protection devices imbedded in it. These are described in detail by Falahati et al. [11,22]. Some of the key advantages of MMGU are expandability, modularity, high reliability, enhanced power quality and stability, awareness, intelligence, agility, and it being a dual application synchronous machine. This machine can be operated in both motor and generator modes.

9.2.3 Modular Microgrids for Renewable Resources

Numerous attempts have been made to develop a microgrid to support the use of renewables for power generation. For example, the HMEP (hybrid microenergy project) [1,2], founded by members of the Alaskan home building industry, is particularly suited for Alaska's rural communities as an economical and sustainable supplement to diesel for producing electricity and heat. A hybrid micropower system designed for rural Alaska that is simple to install and easy to maintain has worldwide potential. HMEP is composed of solar PV, thermal, wind, and biomass systems. Biomass system includes gasification, anaerobic digestion, and bio-oil or biodiesel technologies. Demonstration projects were carried out to confirm the viability of biomass, solar, and wind technologies and the integration of systems prior to large-scale replication in Alaska [1,2].

In general, the transition to 100% renewable energy systems is an important factor in making energy systems sustainable. The transition path toward 100% renewable electricity supply has been analyzed for many countries such as Denmark, Portugal, Macedonia, Croatia, China, and New Zealand. Morvaj et al. [23,24] evaluated the use of modular microgrid for renewable energies. The main scope of this study was to assess the feasibility of using the heat demand of buildings as a function of outdoor temperature for renewable energy in the electricity grid supply. The district of Alvalade, located in Lisbon (Portugal), was used as a case study. The district consists of 665 buildings that vary in both construction period and typology. Three weather scenarios (low, medium, high) and variable renovation scenarios were considered.

The results of the study indicate that the future of power systems likely lies in the implementation of microgrid concepts which will additionally increase the complexity of designing and operating distribution grids. A microgrid is defined as a network structure based on the control of all aspects related to the network operation (distributed generators, storage devices, controllable flexible loads, etc.) at the distribution level, which allows the network to coordinate efficiently all its resources as if it was a single energy system [17]. With customer involvement, it can improve the overall energy efficiency and reliability of the grid and decrease energy consumption [18]. In this study, the impact on energy systems of different levels of renewable energy in the electricity grid supply was analyzed on district scale (at the low-voltage-distribution-grid level). The study examined how urban districts should be optimally designed in the (near) future when the goals of energy roadmaps are achieved. The main goal of the analysis was to compare the differences in the design and operation of distributed multi-energy systems for two approaches—individual and microgrid.

In some countries, generation from mains alone is not sufficient to cover national electricity demand. Microgrid with modern DERs could play an important role to alleviate dependency on the main grid. DER comprises wind turbine, solar photovoltaics, diesel generator, gas engine, micro turbine, fuel cells, etc. The analysis of different DERs is a vital task to judge a reliable microgrid (MG). Due to the gap between typical loads and supply within MGs, larger-scale energy generation could provide a possible solution to balance power demand and supply. Some distributed generation options are not reliable, have large installation and operating costs, and may generate considerable CO emission, such as gas generators [7–10]. Solar and wind are good choices as RESs; however, they are not sustainable enough to support the base load [2]. Diesel generator, coal-fired generation, and gas turbines are significant sources of greenhouse gas emissions. Small modular reactors (SMRs) would come to attention as flexible, reliable, and cost-effective electric power sources for the future world. An SMR can provide stable power output as desired.

The optimization model used for the analysis can simultaneously determine the optimal design and operation of a multi-energy system (with district heating network layout and electrical distribution grid upgrades needed) while ensuring that the solutions are within the distribution grid limits using linearized AC power flow. The considered technologies are CHP, photovoltaics, heat pump (HP), gas boiler, and heat storage. The model was applied to a large urban case study which was based on the IEEE Low Voltage Test Feeder case [21]. It consisted of 55 residential buildings and a radial network with four branches. In order to reduce the computational complexity, buildings that were connected to the same grid-connection point were aggregated. As a result, the number of buildings were reduced to 37 connected by a three-phase low-voltage (0.4 kV) radial distribution grid. The grid was assumed to be balanced, and the mutual impedances were taken as zero.

The optimization model of distributed multi-energy systems was applied to an urban case study for different renewable shares in the electricity grid, and the differences between individual and microgrid approaches were compared. Overall the microgrid approach had better performance than the individual approach. The installed capacity of heat pumps was generally lower in the microgrid approach because PV electricity can be more efficiently used, and smaller HP capacities were needed to transform excess electricity into heat and store it in the heat storage. The installed PV capacity was smaller in the microgrid approach because of the microgrid's opportunity to exchange electricity between buildings. Only in the 100% scenario are capacities roughly the same. The total cost of the microgrid approach was generally lower for the same carbon limits because less PV was needed to achieve the same carbon emissions. The difference was most significant in (near) carbon optimal solutions when this PV benefit was most expressed. As for carbon emissions, the microgrid approach can achieve lower emissions both with and without grid upgrade. Only in the 100% scenario was there no difference because the grid was already carbon neutral.

A number of industries have advanced "plug-and-play" modular microgrid to advance the use of renewables. ABB [3–5] introduced a modular and scalable "plug-and-play" microgrid solution to address the globally growing demand for flexible technology in the developing market for distributed power generation. The cost-efficient, containerized solution is relevant for mature and emerging countries and will help maximize the use of RESs while reducing dependence on fossil fuels used by generator sets. ABB's innovative technology with the PowerStore Battery and the dedicated Microgrid Plus control system as well as cloud-based remote service can not only provide power access to remote areas, but also secure cost-efficient uninterrupted power supply to communities and industries during both planned and unplanned power outages from the main grid supply.

All the pieces of equipment required to run the microgrid—ABB's power converter and dedicated control system, Microgrid Plus, as well as battery storage—have been integrated into a container for faster, easier, and safer deployment. The customer can choose to configure the microgrid to integrate energy from solar, wind, main grid, or diesel generator supply, based on the application and local conditions. ABB's modular microgrid is compact and has four predesigned variants in the range of 50–4,600 kW, to meet varying customer needs. The standard integrated functionalities include grid-connected and off-grid operation with seamless transition. It is a containerized solution designed for easy transportation, fast installation, and commissioning on site. Operations and maintenance is enabled via a cloud-based remote service system. This modular, standardized, and scalable microgrid solution will provide cost-efficient access to reliable power for rural and urban applications, as a plug-and-play solution.

ABB's modular microgrid system aimed at supporting the use of on-site renewables. The containerized unit includes battery energy storage and the firm's microgrid control system. ABB also plans to make a modular microgrid MGS100 that is suited to remote communities, extreme environments, and places that lack access to reliable grid power. The MGS100 is also designed for small commercial and industrial facilities with inconsistent grid supply. It can employ renewable energy, including solar plus storage, prioritizing solar by day, switching to battery mode at night, and using an AC generator if the battery runs out. Encased in a single container, the MGS100 has three power ratings—20, 40, 60 kW nominal load power. Because it's modular and scalable, the microgrid's capacity can be increased when required. The system has islanding capability and can be connected to a central grid [3–5].

Siemens Management systems [9,10] and solutions based on SICAM automation equipment and software solutions are based on the leading Spectrum Power™ platform. Siemens Microgrid Management Systems monitor and control networks with large and small distributed energy generators, renewable assets, conversion plants, storage, and loads. Their scalable system helps to automate, display, alarm, and control all elements in networks, thus assuring the needed quality of supply at all times. It generates schedules, automatically monitors their observance, and readjusts them in real time. This is enabled by automated sequences based on rules or forecasts that draw on a large number of constantly updated parameters—such as weather forecasts, type of plant, or energy price. Siemens Microgrid Management Systems increase the value of any energy plant and can protect any system from outages, whether it is connected to the local grid or running as an island. Siemens solutions are flexible and expandable. While previously the only network was electrical, increasingly there is an opportunity in the distribution of heat and cold and the creation of a holistic, multi-vector energy system. Ownership and management of the underlying electrical network is no longer a necessity for creating a locally optimized system with energy cooperation between different organizations, people, and buildings. With multiple energy vectors, multiple sources, and storage, the options for energy delivery increase, and costs are reduced [9,10].

Green village electricity (GVE) deployed a solar microgrid to deliver power to households that are not connected to a central grid or large, traditional power plants. The project started in a small way in Haiti after the earthquake in 2010. GVE helped deliver 15 solar microgrids in trailers to communities that had never been on a central power grid. The power stations, which they called Sunblazers [102,103], shipped solar panels that produce 4 kWh of electricity per day, plus 40–80 12-V lead-acid battery packs to various communities. Local technicians cut and welded frames and wired the panels. Then they hauled the finished products into remote communities where they powered homes through the battery packs. The kits were composed of almost entirely off-the-shelf parts configured for efficiency and portability.

Today, ten of the units are still running. Now Sunblazer is in its second and third generations, Sunblazer II and Sunblazer Lite. Both new models are lighter and more portable than the first prototype, and their panels are more efficient. Sunblazer II includes six 300 W panels that deliver 9 kWh of electricity per day in equatorial sun. The panels store power in four station batteries that can charge a portable household battery pack in about 4 h. The chassis is steel tubing with a floor of sealed plywood or steel. Sunblazer Lite is about half the size and weight with two panels that deliver nearly 3 kWh/day. It has a stainless steel frame and an enclosure of high-density polyethylene plastic [102,103]. The new models also include mostly off-the-shelf components, and the designs are intentionally fluid, open to adaptation to suit changing conditions in communities and countries around the world. Each station can expand to deploy more panels and charge more batteries. GVE uses a modular generator. In Nigeria, GVE is deploying 24 kW of solar power to three different villages, for a total of 72 kW that powers 200 homes [102,103].

Eaton's microgrid energy solutions help meet demand and keep power flowing despite power disruptions from natural disasters, accidental faults, or intentional attack. In the early 2000s, Eaton engineers began exploring mobile power systems projects. From 2005 to 2013, mobile microgrid projects led to stationary microgrid demonstrations, and each one offered Eaton new insights and lessons learned about how to manage multiple generation sources—natural gas generators, solar, wind, and battery storage—within seamless, islanded grids [7–9]. In 2013, Eaton was selected to help design an innovative Smart Grid for Portland General Electric in Salem, Oregon—one of the largest microgrid demonstrations ever. The project used multiple inverters coupled with lithium-ion batteries to provide 5 MW of energy storage in both grid-connected and islanded modes and demonstrated a unique capability to provide seamless islanding in the event of an unplanned utility outage. Eaton developed a modular Microgrid Energy System design using industry-leading substation hardware that could efficiently scale to meet a wide range of applications and adapt to evolving generation and load assets. At its core is Eaton's Power Xpert Energy Optimizer with advanced control algorithms that help maintain system stability, shave peak demand, shift load, and manage black starts. Coupled with its Cooper Power series SMP family of controllers, servers, and I/O modules and Yukon Visual T&D HMI solutions—all utility proven and cybersecure by design—Eaton is able to help maximize renewable energy contribution, provide utility demand response functionality, and proactively manage generation assets to maximize microgrid performance [7–9].

Lešić et al. [27] applied model predictive control (MPC) to microgrids for buildings. Application of the MPC to building energy management had a great impact on research trends and contributed with experimentally validated building energy efficiency increase by 17% [28] to 28% [29] (theoretically expected up to 70% [30] in particular comprehensive applications), heat pump energy savings of 13% [31], load shifting possibility [32] for power grid balancing, or additional economic savings with hourly-variable electricity prices [33]. Furthermore, introduction of microgrid with distributed power generation and storage units allows buildings to enter the energy market as active participants and ensure revenues [34] with bidding strategies, where increase in profit is achieved with combined microgrid and building climate control (e.g., by about 15% in [35]). With inherent ability to handle constraints, MPC has led to the system reliability and improvement of users comfort at the same time when compared with state-of-the-art adopted control methods. This is also expected to contribute to productivity within commercial buildings. A motivation for integration of all building subsystems toward energy efficiency and cost-effectiveness increase is evident. Simultaneous building thermal comfort satisfaction and smart grid demand response management is addressed in [32,36,37]. The papers illustrate an opportunity of utilizing buildings to resolve the critical issue of power imbalance in the distribution grid. Buildings act as thermoelectric storage systems, which are identified as the main need for integration of large share of RESs [38].

Buildings are complex systems composed of many subunits responsible for maintaining safe and steady operation such as sensors, thermal actuators, zone (room/office) climate controllers, central heating/cooling medium production units, medium supply ducts, lighting, shading, fire alarms, and security systems. Recently, PV systems, wind turbines, energy storage units with corresponding charging curves are also included in the form of a microgrid. These subsystems are all very different not only in dynamics, priorities, means of operation but also implementation aspects such as energy levels, protocols, maintenance services, etc. Rather than having one large control structure to handle all the subsystems at once, it is more natural to separate it into subunits in a hierarchical or distributed way [39–41]. In addition, many of the listed subsystems can be subjected to optimization methods, and many of them already have such algorithms implemented.

The study is focused on integration of several energy systems for major mutual cost saving opportunities, applied to building climate control with integrated microgrid. The following contributions are highlighted:

i. proposition of modular hierarchical MPC method for system coordination through predicted energy consumption and corresponding prices and
ii. detailed realistic simulations performed on the chosen test site of building with integrated microgrid in real (non-averaged) weather and market conditions and numerous scenarios. Results give a reliable estimate of building climate and microgrid operation costs with applied simple and more sophisticated controllers including the proposed coordinated modular MPC.

The proposed method is applied to a particular case of a building with integrated microgrid, where building zone heating/cooling is coordinated with renewable energy production, storage's state of charge, and volatile electricity pricing from the electricity distribution grid. The microgrid-level MPC maintains cost-optimal energy balance of the microgrid, and the zone-level MPC maintains the zone temperature comfort in energy-optimal way. In the proposed coordinated operation, microgrid-level MPC receives predicted energy consumption profiles from the zone-level MPC and issues back the optimized price profile relevant for the predicted consumption.

The optimized price profile includes information on how a zone-level energy consumption change reflects on the overall building energy cost while taking into account the optimized microgrid operation. The approach stands out with its modularity and easy switching between uncoordinated operation of individual modules and coordinated control for mutual economic gain. The modularity is also suitable for coordination extension to new vertical modules, e.g., central HVAC unit level between the current levels or aggregator and district levels on the grid side above the microgrid level. Following the same established methodology, levels are coordinated by information on a predicted consumption in the up direction and issuing a price for the received predicted consumption in the down direction. In case of absent modules, the remaining ones are still coordinated via the matching interfaces for achieving the economical optimum in the current system configuration. The approach is readily applicable for buildings and grids operating in various conditions: different climates, building zone topologies, thermal actuator types, or additional microgrid components, including controllable or non-controllable loads.

Finally, Müller et al. [25] examined the application of Product Service System (PSS) layer method for modular microenergy systems. The study (a) introduces PSS layer method, (b) uses microenergy system as a case study for PSS, and (c) sustainability is emphasized as a driver for PSS and microenergy systems. All three topics are set in relation to an industrial case where the PSS layer method had been applied to a microenergy system which in this case is a solar home system. Such systems are used for energy supply in off-grid installations in rural regions, e.g., in weak infrastructures of developing countries.

9.2.4 Modular Microenergy Grids for Oil and Gas Facilities

Modular microenergy grid (MEG) is an emerging concept in intellectual networks that integrates different energy resources like electricity, heat, hydrogen, and natural gas. MEG needs to be more reliable, secure, economic, eco-friendly, and safer. The main characteristics of MEG include the following [6–9]: (a) Self-healing: Online self-assessment of the grid operation state, able to detect fault quickly without or with little manual intervention, (b) Interaction: ability to incorporate consumer equipment and behavior in the design and operation of the grid, (c) Optimization: ability to optimize its capital assets while minimizing operation and maintenance costs in the whole life cycle, (d) Compatibility: ability to accommodate a wide variety of distributed generation and storage options, and (e) Integration: intelligent decision support system based on the integration of information.

It can be seen from above that safety and sharing are regarded as the core advantage and key technical problem of MEGs to be tackled. However, precise and real-time fault diagnosis has become one of the key requirements in the application of MEG due to its impact on the overall grid safety [10]. The study by Gabbar et al. [26] examined the **power supply** to a **petroleum refinery**. The intention of this research was to create solutions in different ways for a reliable service and minimum power losses in a petroleum refinery. This modular MEG is designed for oil and gas plants. The loads that are used in this project are boilers and furnaces, cooling tower, heat exchanges, and vent/flare. The proposed hybrid MEG consists of PV, wind turbine (WT), fuel cell (FC), battery, micro gas turbine (MGT), AC loads, AC distribution lines, DC distribution lines, DC loads, and DC-AC-DC converters. The energy that is produced by distributed generators (DGs) is stored in the battery.

In order to satisfy the energy supply requirements for oil and gas plants, it is possible to utilize higher generation capacity, such as small **nuclear power plants**. A small nuclear power plant also referred as small modular reactor (SMR) is a new emerging technology which could be compatible as distributed generators. According to the International Atomic Energy Agency (IAEA), a plant having an electrical output power <300 MWe is defined as SMR or small nuclear plant [26]. There are various sizes and technologies of the SMR which might be an alternative solution for independent microenergy grid. Light water reactor (LWR), fast neutron reactor (FNR), and graphite moderator reactor (GMR) are the main categories of SMRs, and the sizes could vary from 10 to 300 MWe. Depending on electricity demand, one or multiple SMR units could be installed in same site in a MEG. As SMR could provide continuous power, the access power would be transferred to the main grid. SMR wouldn't be recommended for distribution like solar or **wind power** because of safety concern. Besides, SMRs would have more installed capacity than gas power plant. Islam and Gabbar [104] also examined the role of SMRs for modern microgrids. Energy supply is one of the most important concerns of every government, energy authority, and researchers to meet the growing regional energy demands. Their target was to ensure low carbon dioxide emission and sustainable, efficient, and cost-effective energy solutions.

ABB provided Woodside, Australia's largest independent oil and gas company, with a PowerStore™ battery storage system that is capable of remote management of operations and service [105]. Oilfield Technology reports that the system was installed on the Goodwyn A platform, reducing carbon emissions and helping to lower cost of operations and maintenance. The Goodwyn A offshore production platform is located about 135 km northwest of Karratha in Western Australia and has been operating since 1995. ABB's containerized, plug-and-play ABB Ability PowerStore battery storage system supports Goodwyn A's existing gas turbine generators. The battery replaces one of the six existing gas turbine generators and also reduces the need for using the emergency diesel generator. Short-term backup can be provided via the new battery energy storage system (ESS) incorporated within the microgrid, to provide a "spinning reserve." A dedicated ABB Ability Microgrid Plus control system acts as the brain of the solution, and it is also possible to remotely operate the microgrid if the need arises or the platform has to be de-manned for any reason [105].

9.3 CONCEPT OF MODULAR MICROENERGY INTERNET

As a next step from modular microgrid, Mei et al. [44] examined the concept, design principles, and engineering practices of microenergy internet. Just like modular approach for microgrid, a similar approach for microenergy internet is needed to save cost. The energy internet is one of the most promising future energy infrastructures that could both enhance energy efficiency and improve its operating flexibility. Analogous to the microgrid, the microenergy internet emphasizes the distribution level and demand side. This study proposes concepts and design principles of a smart microenergy internet for accommodating microgrids, distributed polygeneration systems, energy storage facilities, and associated energy distribution infrastructures.

Since the dispatch and control system of the smart microenergy internet is responsible for external disturbances, it should be able to approach a satisfactory operating point while supporting multiple criteria, such as safety, economy, and environmental protection. To realize the vision of a smart microenergy internet, an engineering game-theory-based energy management system with self-approaching-optimum capability was investigated. Based on the proposed concepts, design principles, and energy management system, the study presents a prototype of China's first conceptual solar-based smart microenergy internet, established in Qinghai University.

Increasing worldwide energy demand has created a global urgency for high-efficiency and environmentally friendly energy utilization systems. It has been validated that the integration of different modular energy distribution systems could bring additional flexibility and reliability to system operation and enhance energy efficiency [45]. Thus, the coupling of energy infrastructures including district heating network [46,47], natural gas network [48,49], the electric vehicle transportation infrastructure [50,51], and the power network, all have been investigated extensively to improve system efficiency and reliability. The integrated energy system [45], energy internet [52], multi-source multi-product system [53], multi-energy system [54], as well as the energy microgrid [55], all provide the possibility for coupling among multi-carrier energy systems, including electricity, natural gas, and heat. The integration of such infrastructures has been viewed as a promising future approach to realize a safe, cost-effective, and environmentally friendly energy system [56,57].

The energy hub is one of the key components in a multi-carrier energy system. It provides the capabilities of conversion, transmission, and storage among multiple energy carriers [58]. The energy hub also helps to build natural linkages among traditional independent infrastructures, including power distribution network (PDN), gas distribution network (GDN), district heating network (DHN), cold distribution network (CDN), and electrified transportation network (ETN). CHP units with thermal energy storage [47], compressed air energy storage (CAES) systems [59,60], and concentrating solar power facilities [61] are typical energy hubs that offer power and heat cogeneration and storage capabilities. These hubs have been investigated at both transmission- and distribution-level coupled energy systems [60–62].

Since the energy distribution level simplifies the implementation of the coupled energy system functions, research on the future form of such coupled infrastructures both at the distribution level and on the demand side are of great interest [63]. One such coupled system is the microenergy internet, which emphasizes the integration of multiple energy networks at the distribution level. Specifically, the microenergy internet is a distributed form of energy internet composed of distributed RESs; energy storage facilities; multi-carrier energy sources; multi-carrier loads; and distribution infrastructures, such as PDN, DHN, GDN, and ETN. Based on this, we have the smart microenergy internet, which is formed by capturing multi-criteria self-approaching-optimum capabilities, as illustrated by Mei and Li [64]. Although extra reliability and flexibility can be achieved by integrating multi-carrier energy devices and different energy distribution infrastructures under the framework of the smart microenergy internet, the energy management of such systems is challenging due to multiple decision makers having competitive or/and cooperative targets and operating in stochastic conditions and with asymmetric market information [65].

In these scenarios, traditional single-agent-based control and decision theory is inadequate. As such, an advanced modeling, analysis, and decision tool is necessary for energy management in a smart microenergy internet.

Game theory has been a fundamental mathematical tool of economists, politicians, and sociologists for decades due to its capability of solving decision-making problems involving multiple agents with cooperative or/and competitive goals [66]. The concepts, theories, and methodologies of game theory can be used to guide the resolution of engineering design, operation, and control problems in a canonical and systematic way [66,67]. Game theory has been widely used in smart grid planning, economic dispatch, and market equilibrium analysis [67–71]. Specifically, engineering game theory has become a powerful tool for the development of advanced dispatch and control schemes for the smart microenergy internet. Thus, in the study by Mei et al. [44], engineering game-theory-based dispatch and control methods were investigated to help in realizing a self-approaching-optimum energy management system of the smart microenergy internet. When developing advanced energy infrastructures, the construction of engineering demonstration sites is important for testing and validation. To this end, a prototype system of a solar-based smart microenergy internet has been established under the guidance of the smart microenergy internet framework in Qinghai University, China. The system comprises three kinds of solar-based power and heat sources, including a multi-function PV station, a solar chimney power station, and a full solar spectrum power generation; an energy hub based on solar-thermal CAES (ST-CAES) system; a smart microgrid for library with building integrated photovoltaics; electricity demands such as carbon fiber recycling system and campus load; multi-carrier demands (heating, cooling, and power), such as solar-based wooden house; and an engineering game-theory-based self-approaching-optimum energy management system.

Solutions such as smart grid, integrated energy system, and energy internet have all been proposed in recent years to satisfy increasing energy demands in a secure and reliable way. However, most of these concepts focus on the transmission-level side with considerably less attention given to the distribution level and demand side of energy systems. Taking into consideration the high-efficiency and flexibility characteristics of distributed cogeneration, we focus here on the distribution and demand side as a way to accommodate microgrids, distributed polygeneration systems, energy storage facilities, and associated energy distribution infrastructures within the framework of the smart microenergy internet. The microgrid is regarded as the distributed way of the smart grid, having two fundamental operation modes: the autonomous mode and the grid-connection mode [72]. The microgrid pays more attention to the distribution level and demand side. The generated power in a microgrid is mainly for self-use in the autonomous mode. Insufficient or surplus power can be regulated with the connected PDN or power utility in the grid-connection operation mode.

The energy internet is an energy management system with the electrical system as its core and the smart grid as its basis. It incorporates renewables and accommodates multi-carrier energy capabilities for improving the overall energy utilization ratio with advanced information, communications technologies, and power electronics technology [73,74]. Inside the energy internet is the energy hub, which is a key facility wherein multiple energy carriers can be converted, conditioned, and stored. Specifically, the energy hub represents an interface between different energy infrastructures and/or load in the energy internet. Energy hubs consume power at their input ports, e.g., electricity and natural gas infrastructures, and provide certain required energy services such as electricity, heating, and cooling at the output ports. Although the energy hub connects multiple infrastructures in an energy internet, the complexity of the energy internet puts barriers on achieving some of its predefined functions. To facilitate the functions of the energy internet at the distribution level, Mei et al. [44,64] investigated the smart microenergy internet as a class of energy networks composed of PDN, DHN, and CDN with energy hubs. Energy infrastructures including electricity, district heating, natural gas, and transportation network are coupled through energy hubs, electric vehicle, heat storage system, and absorption refrigerator. The authors of the study firmly believe that the implementation of the smart microenergy internet can simplify energy internet functionalities, such as multi-carrier energy synergy that incorporates more renewable energy resources with improved reliability. Undoubtedly, the performance of the smart microenergy internet depends on its dispatch and control strategies. A multi-criteria, self-approaching-optimum operation energy management system is required to realize the secure, cost-effective, and environmentally friendly operation of the smart microenergy internet [64]. Microenergy internet is the next level of energy infrastructure that needs to be modular if it is cost effective and widely applied.

9.4 MODULAR NANOGRID

A nanogrid is different from a microgrid. Although some microgrids can be developed for single buildings, they mostly interface with the utility. Some aren't even fully islandable. A nanogrid, however, would be "indifferent to whether a utility grid is present." Rather, it would be a mostly autonomous DC-based system that would digitally connect individual devices to one another, as well as to power generation and storage within the building. A nanogrid is a single domain of power—for voltage, capacity, reliability, administration, and price. Nanogrids include storage internally; local generation operates as a special type of nanogrid. A building-scale microgrid can be as simple as a network of nanogrids, without any central entity [93–96].

The nanogrid is conceptually similar to an automobile or aircraft, which both house their own isolated grid networks powered by batteries that can support electronics, lighting, and internet communications. Uninterruptible power supplies also perform a similar function in buildings during grid disturbances. Essentially, it would allow most devices to plug into power sockets and connect to the nanogrid, which could balance supply with demand from those individual loads. It presumes digital communication among entities, embraces DC power, and is only intended for use within (or between) buildings. Building-scale microgrids are built on a foundation of nanogrids and pervasive communication. The system capacity could be anywhere from a few kilowatts to hundreds of kilowatts [93–96].

According to Nordman [94], there are lots of potential benefits to structuring local DC power distribution in this way. Conversion losses would be cut, investments in inverters and breakers would be reduced, and device-level controls would enable a much more nimble way to match generation or storage capabilities with demand. The building would also theoretically be immune to problems more likely to be encountered with a local microgrid or the broader centralized grid. However, obstacles currently outweigh those benefits by a significant degree. Nordman thinks nanogrids should be universal, meaning they would operate on the exact same communications and voltage standards. But because nanogrids are still mostly conceptual, no organization has attempted to create those standards. Nanogrids can't really scale without them. There's also another structural issue to deal with on the grid. Theoretically, these localized, autonomous systems could be scaled without much interference with the utility. But Nordman [94] also envisions nanogrids being tied to larger microgrids, which are ultimately tied to the central grid. That inevitably brings utilities into the picture, and they likely wouldn't have much incentive to support a system designed to drastically cut their electricity sales. That's why growth in nanogrids will likely occur in developing countries with weak grid services where it makes sense to power buildings in isolation. The concept is compelling. It's also so completely different from the status quo that it currently has little chance of scaling in this grid-centric world.

A building is often only as intelligent as the electrical distribution network it connects with. That's why smart buildings are often seen as an extension of the smart grid. Meters, building controls, intelligent lighting and HVAC systems, distributed energy systems, and the software layered on top are indeed valuable for controlling localized energy use within a building. But in many cases, the building relies on the utility or regional electricity grid to value those services. Some analysts define these technologies as the "enterprise smart grid" because of their interaction with the electricity network. Nanogrid can help when there is no supporting centralized grid [93–96]. Arrival of the concept of smart grids and electricity market introduces the demand-side management (DMS) which leads to controllable loads. Furthermore, by the rapid penetration of RESs, such as PV and wind generation systems, operating as distributed generation systems (DGSs), many issues about utilization of these units have emerged. Smart homes, with the presence of these units and issues about DMS, have gathered so much attention to managing the energy from multiple sources, loads, and the utility grid. The paper by Hagh and Aghdam [106] introduces a multi-port converter to be used in smart homes and hybrid AC/DC nanogrids. Beside energy management among generation units and energy storage devices, this converter is capable of power management of appliances of the home using various communication networks. Furthermore a comparison between optimally tuned proportional integral derivative (PID) and fractional order proportional integral derivative (FOPID) controllers has been carried out. Genetic algorithm (GA) is used for optimizing the parameters of the controllers.

The business case for nanogrids echoes many of the same arguments used on behalf of microgrids. These smaller, modular, and flexible distribution networks are the antithesis of the bigger is better, economies of scale thinking that has guided energy resource planning over much of the past century. Nanogrids take the notion of a bottom-up energy paradigm to extreme heights. In some cases, nanogrids help articulate a business case that is even more radical than a microgrid; in other cases, nanogrids can peacefully coexist with the status quo. Many believe that the linking of batteries to distributed solar PV systems is a game changer. The recently published

Solar PV plus Energy Storage Nanogrids report shows that this technology represents a <$1 billion market worldwide today, but this number is expected to grow to $14 billion by 2024, with residential customers leading the market. When it comes to resilience, states such as Massachusetts have already signaled their preference for building-level solar PV plus energy storage nanogrids, since they face less regulatory hurdles than community resilience microgrids. Ironically enough, the potential loss of federal investment tax credits for solar PV in the United States as early as 2016 and growing utility opposition globally to traditional solar PV support mechanisms such as net metering and feed-in tariffs only help to build the business case for solar PV plus energy storage nanogrids. In order to extract the greatest value from solar PV in the absence of subsidy—whether that be utility demand charge abatement or greater reliability and resilience—will require linking this variable distributed generation to an ESS, either in the context of a microgrid or a nanogrid [93–96].

The intermittency of solar PV, which can be more extreme than wind on a second-by-second basis, has long been viewed as a drawback to widespread deployment as a substitute for 24/7 fossil-fuel generation. Rooftop solar PV in particular can feature capacity factors as low as 20%. If such small systems—whose primary advantage for residential applications is providing financial benefits (offsetting expensive peak grid power)—are coupled with ESSs, the value of solar energy is magnified. In essence, it can be stored and then discharged during time periods most advantageous to asset owner. These same storage systems can also offer resiliency benefits when the larger grid goes down. While the decline in solar PV pricing has been underway for quite some time, it is only recently that batteries—particularly lithium ion—have begun to match solar PV with a similar downward momentum, thereby increasing the appeal of this technology pairing [93–96]. The most radical interpretation of this solar PV plus energy storage nanogrid vision is at the residential level, the application where the nanogrid model is likely to meet opposition from utilities—that is, unless utilities begin offering nanogrid services. So far, utilities in Ontario, Australia, and New Zealand are doing just this (Powerstream, Vector, and Ergon, respectively). It is safe to say the size of the microgrid market is larger than that of nanogrid due to sheer scale. But microgrids also incorporate CHP and wind, as well as other resources. If we narrowed the comparison to total capacity of just solar PV plus energy storage microgrids versus nanogrids, it is the smaller nanogrid that would likely come out on top today and perhaps over the long term.

With microgrids gaining popularity for deployment in front of and behind the meter, it's only a matter of time before the unique strategy receives a spin-off technology of its own. Enter the nanogrids, DERs with capacity around or below 100 kW. Despite bringing less capacity to the table, nanogrids are capable of delivering much of the same functionality as their larger cousins. By serving as an independent source of power or a supplement to a larger electrical system, nanogrids can provide resiliency to commercial facilities during outages, offer demand response capabilities, and help companies to reduce energy costs. The main differentiator between micro- and nanogrids is the size of the area they are expected to serve. While microgrids have been introduced as a means of helping businesses with campuses featuring multiple buildings to become more energy dependent, nanogrids are suitable for a single commercial building owned by an organization interested in reaping the same rewards. When combined with solar or another form of on-site generation, nanogrids can even help commercial buildings to become completely self-sustaining with regard to electricity [93–96].

Nanogrids offer improved flexibility over the power grid's traditional limitations. **Demand for nanogrids behind the meter is on the rise.** Nanogrids will be most useful in developing countries where grid services are weak. However, businesses globally have already caught on to the usefulness of nanogrids for supplementing solar generation. One of the reasons that nanogrids have caught on is because the technology offers improved flexibility over the power grid's traditional limitations. Nanogrids, small, modular distribution networks, are considered the antithesis of the bigger-is-better economies of scale thinking that has guided energy resource planning over much of the past century. The research data shed light on the advantages of "distributed solar PV installations in behind-the-meter building-level applications for residential and commercial customers [93–96]."

Technology trends typically occur in waves. The experiences of first adopters act as the proving ground for new technology, and entire industries can quickly adopt solutions if these strategies come with a proven track record or obvious opportunities to reduce costs and maintenance. This has been the case for traditional microgrids. While the technology was initially looked over by some utilities and businesses managing large campuses, the modern demand for these solutions has put microgrids at the forefront of energy innovation. For example, California utility San Diego Gas and Electric recently debuted a solar-plus-storage microgrid designed to serve the needs of 2,800 customers too remote for efficient connections to local energy infrastructure. One of the first microgrid setups to take advantage of solar power and storage in the nation, the SanDiego gas and electric (SDGE) deployment is likely to become a model for utility microgrids in similar climates. Additionally, the continued success of independent, utility-run microgrids will eventually usher in greater acceptance of nanogrid technology.

The shape of the nation's energy infrastructure is undergoing a rapid evolution. **Regulatory simplicity provides extra incentive for nanogrid deployment.** Businesses fueling the demand for nanogrids aren't just interested in the technology as an offshoot of traditional microgrids. Nanogrids, especially those capable of turning a commercial facility into its own self-sufficient and renewable power station, have more to offer than performance scaled to suit the needs of a single building. Utility Dive emphasized that because nanogrids are only assigned to a single building, they are far less likely to incur regulatory scrutiny or rouse the ire of local utilities than traditional microgrids. In addition to helping companies to avoid suits and fines, the reduced complications of integrating nanogrids into a facility's energy profile means that acquiring permissions and completing installation will take place over a far shorter time frame than that of a traditional microgrid [93–96].

9.5 MODULAR OPEN ENERGY SYSTEM

Microgrids are promising solutions for integrating large amounts of microgeneration by reducing the negative impact to the utility network [76]. In general terms, microgrids can be defined as structures that combine DG units, ESSs, and loads [77].

Microgrids including batteries allow to shift peak demand and flatten the consumption pattern. While their architecture may vary greatly depending on the type or number of building blocks as well as the application context [77–79], a clear distinction can be drawn between AC- and DC-based microgrids. Justo et al. [77] concluded that even though AC microgrids are now predominant, the number of DC microgrids is expected to increase in the coming years as they will soon be the right candidates for the future energy system. As outlined above, nanogrids can be seen as smaller and technologically simpler microgrids, typically serving a single building or a single load. As they face less technical and regulatory barriers than their microgrid counterparts, substantial deployment is already underway [80]. This is why, compared to microgrids, nanogrids are often seen as a bottom-up approach, well suited also for off-grid areas and with a clear preference for DC solutions [80].

However, to prevent both power outages and wasting generated electricity, most microgrids/nanogrids include a utility grid connection. In Europe, where the utility grid is advanced and DERs are widespread, feed-in tariffs have been more attractive compared to purchasing a storage unit, but they are decreasing yearly [77] because the higher the intermittency of power sources, the higher are the costs for upgrading the conventional grid and including energy storage [81]. Interconnections and energy storage facilities are required to reduce the stress that intermittent renewables cause on primary generation such as nuclear and thermal [82,83]. Thus, the future energy grid in those areas is predicted to be based on the various DG units, storage devices, and controllable loads that are connected with advanced information and communication devices such as automated meter infrastructure. In those systems, DG units, ESSs, loads, and also microgrids are aggregated in clusters and can be seen as VPPs [84] (virtual power plants) that can then be treated as single entity. VPP can be considered as top-down approach that taps into the existing grid via smart meters and software to add the intelligence necessary to manage demand–response [77]. The aim is to remotely and automatically dispatch and optimize DG via an aggregation platform linking retail to wholesale markets [85]. For supporting both a wide range and flexible number of DERs, VPPs must be both loosely coupled and generally adopted by all players requiring standardization of communication [86].

Altogether, nanogrids and microgrids are usually bound to fixed building blocks, and VPPs are essentially software solutions bound to the existing utility grid infrastructure. Werth et al. [75] propose Open Energy Systems (OESs) as a new type of scalable and bottom-up distribution network that shares some characteristics of all three approaches: building blocks are a flexible number of DC nanogrids, interconnected via a local DC power grid and controlled in a distributed way. The general concept may be seen as a multi-level DC grid system whose two-level implementation needs to be investigated. It provides both hardware and software for exchanging energy in between DC nanogrids of a local community so that one can spread fluctuations over the community without needing to feed in energy to the utility grid. Each house is equipped with one subsystem, a DC nanogrid including batteries, that is connected to a dedicated, shared DC power bus as well as a communication line allowing power exchanges within a community. The concept is an application of open systems science [87,88] to energy distribution: instead of the conventional top-down approach requiring a grid infrastructure and large-scale power transmissions on the top, OES is based on independent subsystems that are interconnected to share power from unstable distributed sources. It is called open because the connections can vary freely and energy can be exchanged both among subsystems as well as with the environment, for instance, by harnessing solar energy.

Werth et al. [75] extend these nanogrids with a bidirectional DC–DC converter and a network controller so that power can be exchanged between houses over an external DC power bus. In this way, demand–response fluctuations are absorbed not only by the local battery, but can be spread over all batteries in the system. By using a combination of voltage- and current-controlled units, we implemented a higher-level control software independent from the physical process. A further software layer for autonomous control handles power exchange based on a distributed multi-agent system, using a peer-to-peer architecture. In parallel to the software, Werth et al. [75] made a physical model of a four-node OES on which different power exchange strategies can be simulated and compared. First results show an improved solar replacement ratio and thus a reduction of AC grid consumption, due to power interchange. Werth et al. [75] describe the OESs architecture, that is, the components of a stand-alone nanogrid as well as the interconnections of such a subsystem to make an OES. They also introduce the layered control scheme (from the lowest physical layer of energy transfer to the highest system control layer).

9.5.1 OES as Multi-Level DC Grid System

The concept and advantages of interconnecting microgrids have been analyzed by Falvo and Martirano [89] and Brenna et al. [90], who suggested sustainable energy microsystems as a type of multi-level grid systems. They classified the level of integration among the subsystems and showed promising results in their preliminary analysis [91]. While the main concepts and theoretical analysis are valid for the study by Werth et al. [75] as well, they propose a system where there is no direct connection to the utility grid. Indirectly, subsystems may be internally connected to the utility grid that would then serve as an auxiliary power source as it is the case for common commercially available residential systems, but this does not impact OES. Werth et al. [75] do not feed in electricity to the utility grid, meaning that there is no power trading with the utility grid. Power trading and load shedding are exclusively done within the OES community over its own system, a DC-based distributed grid. This allows one to develop a parallel, utility-independent grid without compromising on network reliability for residents, a crucial requirement for the real platform in Okinawa. The OES is built on two DC layers: (a) the DC nanogrids installed in each house and (b) the DC microgrid that connects the nanogrids over a higher voltage DC power bus line. Werth et al. [75] chose DC for both levels because of the following reasons:

1. More efficient transmission lines and improved system stability because of the absence of reactance and external disturbances [77]
2. Wide variety of "raw" electricity output from renewable energy can be easily connected to a DC power bus line and batteries without requiring

AC–DC conversions. The same is true for a vast variety of consumer equipment that could be seen as DC loads (~80% of loads in commercial and residential structures are now DC [91]) which results in an increasing attention for DC distribution [77].

Progress of DC conversion technologies has led to efficiencies above 90% making it practical to convert voltage sources to current sources (over 94% efficiency for 48–380 V conversion at 2 kW using a resonant-type bidirectional DC–DC converter [92]). The core nanogrid includes the following modules:

1. PV panel and PV charger (PVC = a DC–DC converter including maximum power point tracking control);
2. three battery units (3.6 kWh) and a battery management unit;
3. an inverter allowing to use AC power for input as well as to supply energy from the battery to AC loads—both functions are combined in an uninterruptible power supply module;
4. a nanogrid controller that controls all internal modules over an internal communication line.

Internally, all modules on the rack are connected together over the same DC bus whose voltage is determined by the battery voltage—nominally 51.2 V. The system can therefore be considered as a DC microgrid with bipolar configuration and including an ESS as described in [77]. This basic stand-alone nanogrid can thus be converted to a full OES unit—which is a local system able to exchange energy with other systems over a designated, external DC power bus by adding following two modules:

1. a current- and voltage-regulated bidirectional DC–DC converter (resonant-type);
2. a network controller that communicates with the internal nanogrid controller and that can monitor and control the DC–DC converter. It is also responsible to establish the network communication and serves as interface to all hardware units. This logical unit is the basis of building robust software for distributed power exchange.

The advantage of such a modular design that separates the core nanogrid from the interconnecting modules is that, in principle, any kind of nanogrid that fulfills the previously mentioned requirements can be used as long as the interface between nanogrid controller and network controller is adapted so that a minimum set of monitoring/control functionalities can be accessible over the network. The nanogrid controller must guarantee that under no circumstances the internal safety is compromised. The DC interconnection shall be cut and the subsystem shall fall back to stand-alone mode if anything goes wrong. All higher-level networking logic is designed to be completely independent of the lower-level software and the internal functioning of the microgrid, allowing a clear layering and thus freeing the conceptualization of a power exchange strategy from low-level physical processes.

As the number of subsystems connected to the same network increases, distances between houses become bigger, and resistive losses on the power bus line (380 V DC) reduce efficiency. It gets more complicated to keep the bus voltage stable, and also distributed control becomes more complex and increases delays. Thus, rather than increasing the size of one OES by multiplying the number of connected units, we can increase the system size by connecting several OES systems together by using the same type of strategy as for connecting subsystem but one layer higher (meaning higher voltages and conversion capacities). An entire OES community could share energy with another neighboring OES community using a dedicated high-voltage DC power bus and appropriate DC–DC converters. The study is conducted bottom-up starting from the practical feasibility and aiming at directly testing assumptions on real appliances. A three-subsystem experimental prototype in the Sony Computer Science Laboratories (CSL) in Tokyo, another three-subsystem prototype at the Okinawa Institute of Science and Technology (OIST) as well as our full-scale platform at OIST are briefly presented by Werth et al. [75]. In terms of software, Werth et al. [75] have built real-time visualization of the energy flow within subsystems and in between subsystems. An anonymous full system visualization version can be shown to the community including a subsystem view for the residents (restricted to their own subsystem). A full system and individual visualization including a manual control interface can be accessed by researchers or administrators. Combined with previously collected data, these interfaces are the basis for overall system evaluation.

In summary, as the demand for sustainable energies continues to increase, it is important to find ways not only to generate but also to distribute the power coming from inherently distributed and unstable power supplies such as renewable resources. This study analyzes a new type of DC-based, distributed interconnection of DC nanogrids. The study proposes a new concept, both in terms of hardware and software architecture, and shows the benefits on four-node simulations using physical model. The study further demonstrates the feasibility on a full-scale platform, which is one way of putting the concept in practice. Note that the research is still ongoing and some parts of the concepts still need to be studied further. In the future, this study can constitute the basis of higher-level intelligent exchange strategies using weather forecasts, predictions for peak cutting, or even further implementing mechanisms such as monetary control. It explores such an alternative grid system that can not only develop alongside with the existing grid system but also work without it. Because of its open architecture, it can develop gradually, one subsystem at the time, thus requiring gradual investment. This is particularly interesting for areas that are currently off-grid. On a theoretical level, it provides an application model for open systems science in practice as well as explores the limitations of decentralization. A more detailed evaluation of OES is given by Werth et al. [75].

9.6 MODULAR TRANSPORT OF BIOMASS

9.6.1 MODULAR SYSTEM FOR LOW-COST TRANSPORT AND STORAGE OF HERBACEOUS BIOMASS

A conceptual system adopting features from cotton, silage, and container shipping systems was evaluated between 2008 and 2011 by Searcy et al. [107]. The evaluation included both simulations of the anticipated full-scale system and field trials of forming, transporting, and storing biomass modules containing energy sorghum. The simulations utilized Integrated Biomass Supply Analysis and Logistics (IBSAL) and incorporated the anticipated module forming machine that would operate in partnership with a forage harvester, as well as a machine to load the modules onto a flatbed semi-trailer. When compared to the DOE target for logistics costs of $38.59/Mg, the estimated cost was lower for distances up to 80 km. Field results were promising, with biomass modules of up to 5.2 Mg formed, stored for up to 12 months, loaded on a truck in 2 min or less, and transported for 96 km with no significant change of shape or size. Difficulty in field drying of energy sorghum was consistent over 3 years of harvest, as was the ability to use field drying in windrows without increasing the ash content of the biomass.

The manually formed module packages did not maintain an anaerobic environment throughout the storage period, and excessive biomass degradation occurred. In addition, the dry matter density in the modules was ~180 kg/m^3 rather than the 240 kg/m^3 targeted in the simulation. Despite the conceptual evaluation not achieving all the desired features, these studies demonstrated the economic and logistical advantages of a system based upon large packages of chopped bio.

9.6.2 Novel Modular Timber Transport Projects in Sweden

Timber transports must be made more energy efficient if Swedish forestry is to contribute to a cleaner environment. Consequently, in 2006, Skogforsk initiated a project aimed at lowering the total number of timber transports needed in Sweden and reducing the associated diesel consumption and emissions of carbon dioxide and other substances. This was to be attained through advances in transport technology and by using vehicles with higher gross weights but with no negative effects in terms of road wear or road safety. A literature study indicated that there would be no negative impact on road safety. Current conventional timber transport vehicles have a gross weight of 60 metric tons and are 24 meters long. They generally comprise a three-axle truck with space for one stack of wood and a four-axle trailer that carries two stacks. Over 2,000 wood and chip vehicles are in operation in Sweden [108–110].

9.6.2.1 ETT Project

The project was given the name One More Stack ("En Trave Till," ETT). The aim was to use longer rigs with higher gross weights than conventional rigs in order to accommodate an additional 6-m stack of wood, i.e., four stacks instead of the usual three. Because the stacks on the ETT vehicles are also bigger, two ETT vehicles could replace three conventional timber vehicles. In 2006, Skogforsk initiated the project "ETT Modular System for Timber Transport." The ETT project evaluated the effects of transporting timber on a vehicle that is 30 m long and with a gross weight of 90 tons in a normal road traffic environment. When the project started, the expectation was that, in comparison with conventional timber vehicles, fuel consumption, carbon dioxide emission, and transport costs could be reduced by 20%–25%.

The basic premise was that gross weight can be increased without any negative effect on road safety. The argument is that fewer vehicles are needed to transport the same volume of timber and that road wear would not increase because the weight is

Systems for Energy and Fuel Transport

distributed evenly over the vehicle and over more axles. However, the load-bearing capacity of bridges with long spans will need to be calculated, and bridges may need to be reinforced. The vehicle was loaded at a timber terminal in Överkalix, and each trip comprised a journey of 160 km to a sawmill in Munksund outside Piteå [108–110].

9.6.2.2 Results from Studies of the ETT Project
Based on the above-described ETT project study, following conclusions were made:

- Diesel consumption was reduced by just over 20%, and carbon dioxide emissions were reduced accordingly. Emissions of other environmental contaminants were also reduced accordingly.
- The number of vehicles needed to transport a given amount of wood was ~35% less than if the same wood volume had been transported using conventional timber vehicles.
- 700 runs using the ETT rig found no negative reactions when overtaking other road users.
- The Swedish National Road and Transport Research Institute (VTI) carried out four different studies to determine how road safety is impacted by heavier and longer lorry combinations compared with current conventional heavy vehicles. The emphasis was on overtaking scenarios. The conclusion was that no definite demonstrable differences could be shown and that more studies were needed.
- Transport costs were reduced by just over 20%.
- The maneuverability, stability, and braking capacity of the ETT rig were comparable to those of a conventional 60-ton timber vehicle.
- No increase in road wear was shown. This is because the greater gross weight is distributed over more axles. However, the load on long bridges is increased.
- During the test period, the ETT rig drove a total of 800,000 km and transported ~150,000 m^3 solid volume.

9.6.2.3 ST Project
In Sweden they were permitted to run the vehicle, which had a gross weight of 90 metric tons, on public roads during the test period. The ETT vehicle was driven with 65 tons of timber between Överkalix and Piteå. After 6 months, the ETT project was supplemented with a secondary project known as Bigger Stacks ("Star Travar," ST), in which timber vehicles were compiled in a way that increased the payload, while complying with applicable vehicle length and axle pressure regulations. "ETT Modular System for Timber Transport" serves as the collective name for both projects. In the project, we used the loading units (modules) of the European Modular System (EMS), a standardized design for load units currently used in the vehicle industry. The modules were dolly, link, and trailer.

The ST vehicles comprised two types of rig: a four-axle crane truck with dolly and trailer and a tractor with link and trailer. They have been tested both as part of a staging system and as separate vehicles. In the staging system, the crane truck loads the wood in the forest and transports it to a staging area, and then the tractor transports the wood from the staging area to the mill. This allows optimization of the various advantages offered by the two vehicles, thereby increasing the payload and reducing fuel consumption [108–110].

As separate vehicles, the ST crane truck and ST tractor carried timber from forest or terminal directly to the mills. Both the crane truck and the tractor had gross weights of 74 tons, so, once again, dispensation was required from the Swedish Transport Administration before the system could be driven on public roads. The ST vehicles were tested in the counties of Dalsland, Bohuslän, and Värmland inwestern Sweden. All the vehicles tested in the project were fitted with air suspension system with loading scale indicator, alcohol ignition interlock devices (alcolocks), and computer systems to enable real-time analysis of transport performance. Productivity, fuel consumption, etc. were monitored via studies and measurements during the project period [108–110]. The project involved extensive collaboration between around 30 companies, organizations, and public agencies and yielded three new modular forest vehicles that effectively reduce environmental impact and lower the number of timber vehicles needed to transport wood [108–110].

9.6.2.4 Results from Studies of the ST Rigs
Based on ST project study, following conclusions were made:

- When vehicles were used in the staging system, fuel consumption was reduced by up to 8%, and emissions of carbon dioxide and other environmental contaminants were also reduced accordingly.
- Estimates indicate that transport costs were reduced by up to 5%–10%.
- Staging can pay off under certain conditions, but the logistics pose a challenge.
- No increase in road wear was shown. This is because the greater gross weight is distributed over more axles. However, the load on long bridges is increased.
- The maneuverability of the ST rigs was comparable to that of a conventional 60-ton vehicle.
- The ST rigs were driven over 500,000 km in total, and they transported 190,000 m^3 solid volume.

9.7 MODULAR FUEL SYSTEM (MFS)

The purpose of Modular Fuel System (MFS) is to provide the ability to rapidly establish fuel distribution and storage capability at any location regardless of material handling equipment availability. The MFS, formerly known as the Load Handling System Modular Fuel Farm (LMFF), is transported by the Heavy Expanded Mobility Tactical Truck-Load Handling System (HEMTT-LHS) and the Palletized Load System (PLS). It is composed of 14 tank rack modules (TRMs) and two each of the pump and filtration modules, commonly known as pump rack modules (PRMs). The TRM can be used with the MFS PRMs, the HEMTT Tankers, or as a stand-alone system. TRM when used with the HEMTT Tanker doubles the HEMTT

Tanker's capacity. The TRM is air-transportable with fuel and includes a baffled, 2,500-gallon-capacity fuel storage tank that can provide unfiltered, limited retail capability through gravity feed or the 25-gallon per minute (gpm) electric pump. The TRM also includes hose assemblies, refueling nozzles, fire extinguishers, grounding rods, a NATO slave cable, and a fuel-spill control kit [111–113].

TRM full retail capability is being developed, and it includes replacing the existing electric pump with a continuously operating electric 20 gpm pump, a filtration system, and a flow meter for fuel accountability. The PRM includes a self-priming 600 gpm diesel engine-driven centrifugal pump, filter separator, valves, fittings, hoses, refueling nozzles, aviation fuel test kits, fire extinguishers, grounding rods, flow meter, and NATO Connectors. The PRM has an evacuation capability that allows the hoses in the system to be purged of fuel prior to recovery and is capable of refueling both ground vehicles and aircraft. MFS is capable of receiving, storing, filtering, and issuing all kerosene-based fuels. The MFS rapidly establishes fuel distribution and storage capability at any location regardless of material handling equipment availability. The MFS enables retail operation for the warfighter by storing, transporting, and issuing fuel [111]. Besides MFS army has also developed intelligent and modular intelligent track system (ITRAK) modular transport system. This system is described in references [114,115].

9.7.1 DRS MFS

The DRS MFS provides the ability to rapidly establish a fuel distribution and storage capability without a bag farm or engineer support. The system can be used at any location without the availability of construction and material handling equipment. Consisting of fourteen 9,464 L/2,500 gallon tank racks and two pump filtration modules, the MFS increases mobility, capacity, and speed in fuel distribution, while decreasing deployment and recovery time. The DRS MFS is mobile, whether it is full, partially full, or empty. It is compatible with the HEMTT-LHS, PLS truck and trailer, and the United States Marine Corps Logistics Vehicle System Replacement. The MFS tank racks can also be used for line haul of bulk petroleum throughout the theater. By using two tank racks—one on the truck and one on the trailer—the PLS and LHS can transport up to 18,927 L/5,000 gallons of bulk petroleum per trip. The MFS is a logistics support system for Modular Brigades and provides liquid logistics support for the Army's Future Combat Systems [112,113]. The Modular Fuel Storage and Distribution System is designed for rapid deployment and mobility to provide refueling capabilities to front-line combat units. The system consists of four identical stand-alone units that can be attached together quickly to form a single 20′ ISO container for easy shipment by land, sea, or air. The modular containers act as standalone 5,000 USG/20,000 L fuel storage and distribution systems that can be deployed by ground or air to forward tactical locations as individual systems or used together to form a larger 20,000 USG/80,000 L system at a larger fuel depot. Modular containers include helicopter lifting points and forklift pockets [112,113].

The system includes a diesel and electric discharge pump system with mechanical totalizing flow meter, fuel filtration (aviation or ground), and an inlet connection to refuel the system from bulk fuel sources as well as a hose reel with 100 ft/30 m of hose, grounding reel, and fuel nozzles to refuel military vehicles. All components remain inside modular containers for protection. Each modular container includes a 5,000 USG/20,000 L Terra Tank TM collapsible fuel bladder with Insta-Berm TM secondary containment, Rain Drain TM rain water management system, Sun Shade, and connection hoses [112,113].

9.7.2 ISO TANK RACK DESIGN

The Army's decision to use a separate International Organization for Standardization (ISO) tank rack and pump rack configuration established the foundation for the flexible MFS. A surprisingly large amount of equipment was incorporated into the two racks as standard equipment, so a separate vehicle no longer will be required to follow the tanker to carry essential equipment. The standard equipment setup includes 3,300 ft of hose and the full complement of nozzles needed to support multiple retail, bulk, and aviation distribution points. The system can support two refuel-on-the-move configurations for a total of 16 distribution points. A full range of nozzles are provided with the MFS, including the D-1, 1, 1.5 inch, closed-circuit refueling, in line with agreement of North Atlantic Treaty Organization. The MFS is compatible with all military fuel mixtures, and an additive fuel injector can be installed [116,117]. Redundancy is critical in today's fast-paced battlefield environment. The MFS achieves redundancy by using ISO-certified, self-contained tank and pump racks. Built-in test equipment with manual override and bypass tubing also has been installed for all critical functions. The independence of each pump or tank rack provides the needed redundancy and ensures that a failure in any single component will not compromise mission support [116–117].

Safety and environmental concerns were addressed early in the development of the system, ensuring that exceptional safety standards were set. For example, the common "dead man" switch on the fuel manager control panel is backed up by multiple emergency shut-off switches, and the remote-controlled tank valves are backed up by manual safety shut-off valves. In addition, a new Fuel Tank Sealant System (FTSS) is being considered for incorporation into the system. The FTSS would provide secondary containment without adding significant weight to the MFS's streamlined weight configuration.

9.7.3 MODULAR CANISTERS FOR SPENT NUCLEAR FUEL

The discharge burnups of spent fuel from nuclear power plants keep increasing with plants discharging or planning to discharge fuel with burnups in excess of 60,000 MWD/MTU. Due to limited capacity of spent fuel pools, transfer of older cooler spent fuel from fuel pool to dry storage, and very limited options for transport of spent fuel, there is a critical need for dry storage of high-burnup, higher-heat-load-spent fuel so that plants could maintain their full core offload in a concrete storage over pack (HSM-H). These canisters are designed to meet all the requirements of both storage and transport regulations. They are designed to be transported off-site either directly from the spent fuel pool or from the storage over pack in a suitable transport cask [118].

9.8 MODULAR SYSTEMS FOR NATURAL GAS TRANSPORT

LightSail Energy, a developer of sustainable energy technologies, announced the launch of its innovative LightStore® high-capacity gas transport module that incorporates state-of-the art carbon composite pressure vessels that are lightweight, yet low cost, allowing breakthrough economics in gas storage and transport—the lowest gas transport cost per mile ever achieved. LightSail Energy received the U.S. Department of Transportation (DOT) special permit for the lightweight module (below 80,000 lb. GVW) capable of transporting close to half a million standard cubic foot of CNG each truck trip [119,120]. The LightStore gas transport module comprises a modular 53-ft ISO 1496-3 compliant container; convenient one-touch operational controls; remote data capability; and numerous safety features including innovative pressure vessel mounting systems, fire safety, and emergency shut-offs. The LightStore module is ideal for economic storage of natural gas for vehicle refueling and power generation and for transportation of large quantities of gas to locations where pipelines do not exist. The modularity and high capacity of LightStore can positively impact the economics of natural gas projects since each module allows transportation of up to half a million standard cubic foot of natural gas per truck load, which is up to five times more capacity than a steel tube trailer—enough fuel to refuel a fleet of Class 8 trucks or keep a thousand homes warm in a cold winter day [119,120].

LightStore vessels have been rigorously tested to ensure safety, performance, and reliability to achieve ASME and DOT approvals. Tests included those prescribed by ASME Section X, which is the gold standard for stationary storage; ISO 11119-3, which is the established international standard for composite pressure vessels; and ISO 11515, which is the latest global standard for large capacity vessels for bulk gas transportation. LightStore vessels are the first high-capacity vessels in the world to meet the requirements of ISO 11515 for all-composite vessels, which includes very rigorous impact testing. LightSail currently manufactures LightStore vessels at their ASME-certified facility in Berkeley, CA. LightSail Energy develops grid-scale renewable compressed air energy storage solutions and manufactures low-cost, lightweight, composite pressure vessels and systems for storage and transportation of various gases [119,120].

9.8.1 Gas Compressor Modules

Gas compressors play a vital role in gas storage and transport. ABB offers new modular gas compressor with reduced costs with digital gas compression packages. The gas compressor modules integrate ABB control and electrical systems with reciprocating gas compressors to ensure predictable and efficient delivery and operation. The benefits of this compressor are as follows [121]:

1. The gas compressor modules integrate highly reliable ABB technologies and an Ariel Corporation gas compressor.
2. Gas compressor modules are fully designed, assembled, and commissioned packages based on ABB advanced technologies and premium reciprocating gas compressors.

3. A modular approach that accelerates delivery, optimizes compressor performance, and reduces costs.
4. Modules arrive on site fully assembled, pretested, validated, and commissioned to help reduce your risk. Module commissioning includes rigorous testing of mechanical, piping, and automation and startup by ABB experts to help ensure consistent quality, predictable performance, and compliance with stringent industry requirements.
5. After implementation, ABB provides advanced monitoring and optimization services to ensure machines remain in good health and are working optimally.
6. Because they are the world's first digital gas compressor modules, these solutions provide operators with real-time insights about the status of the compressor to ensure more accurate management of equipment.
7. Enables more predictable and efficient operations.
8. Reduces capex and increases operational efficiencies by giving you a single source from which to purchase, deploy, service, and maintain equipment.
9. Reduces risk with fully designed, pretested, validated, assembled, and commissioned solutions.

The main features of the compressor are [121] maximum rated power up to 7,500 kW; maximum discharge pressure up to 420 bar; maximum speed up to 1,800 rpm; low to moderate speed, 150 up to 1,800 rpm; and ABB advanced technologies, such as Direct Torque Control. The compressor can be used for pipeline transportation and CNG transportation.

9.9 MODULAR TRANSPORT OF COMPRESSED LIQUID AND SOLID NATURAL GAS

Natural gas is transported on land and sea by various techniques. On land, as shown above, the most common form is the natural gas pipeline. However, it is also transported by trucks in smaller quantities. Natural gas is also transported as compressed natural gas, liquid natural gas, and natural gas hydrates. In this section, we examine the modular nature of all of these different forms and types of natural gas transport.

9.9.1 THE TRANSPORT OF CNG

The transport of natural gas in the form of CNG represents one of the first alternative technologies evaluated in the past. Currently, CNG is internationally recognized as an alternative fuel, with good performance and low emission of pollutants into the atmosphere. The CNG chain consists of the following stages: (a) treatment of the gas (not always required), (b) compression and cooling (optional), (c) loading and transport with tankers, and (d) receipt and offloading for decompression. The chain is thus extremely simple and does not present the need for plants with special properties, with the exception of a significant compression capacity, which is nevertheless within current technological limits. The working pressures involved during

Systems for Energy and Fuel Transport

storage are in the order of 200–250 bar at ambient temperature or slightly lower. The development of advanced engineering technologies has allowed the transport system to become more efficient and safer: gas storage systems with a higher degree of intrinsic safety have been developed using composite materials. Furthermore, if the gas is cooled slightly (to about 30°C), the pressure can be lowered to half the storage pressure at ambient temperature. Storage can thus be optimized by reducing pressure and its associated hazards, obtaining a storage capacity equal to or greater than that at ambient temperature.

CNG technology presents a volume reduction factor ranging from 200 to 250 times the original volume, slightly over a third of that which can be obtained with the LNG transport system. Currently, the CNG system differs only in terms of the technologies developed to contain natural gas which are adopted on specially designed vessels (see Figure 9.2).

9.9.2 Carriers for the Transport of CNG

Over the past several years, numerous technologies for CNG storage and transport have been developed. Here we examine few of them.

9.9.2.1 Coselle Technology

From the development of other transport technologies (LNG), and the evolution of the oil and gas market, Cran & Stenning designed and developed a new type of pressurized tank named Coselle (from the words "coil" and "carousel"); the potential generated by this innovation has renewed interest in the marine transport of CNG. The development of Coselle modules has opened up the possibility of using CNG technology to transport natural gas across oceans [123]. CNG is considered cheaper than both LNG and the gas pipeline in the case of modest volumes of production, such as production of between 5 and 15 million ft^3 over distances of between 800 and

FIGURE 9.2 Jayanti Baruna—CNG Cargo Carrier, *Indonesia* [122].

3,000 km or 250 and 5,000 km, respectively. The possibility of making intermediate stops and reusing infrastructure makes this technology even more attractive [124].

The central idea on which the Coselle technology is based is the creation of capacious but compact transport vessels. Gas containment systems, together with control and safety systems account for a large portion of the costs of a CNG carrier. The costs of the Coselle storage system, assuming an equivalent safety level, are lower than those for a system using pressure bottles. An example of a transport vessel using the Coselle technology is a double-hulled tanker with a gross tonnage of 60,000 tons; to insulate it from potential sources of fire hazard, the holds are saturated with nitrogen. However, this transport system requires a pretreatment of the gas to dehydrate it, in order to avoid the formation of hydrates and other deposits which might obstruct the pipes and reduce the capacity and efficiency of transport, as well as compromise safety [123–125].

9.9.2.2 VOTRANS Technology

Coselle development technology in turn has led to the emergence of other technologies, such as EnerSea Transport's Volume Optimized TRANsport and Storage (VOTRANS), Knutsen OAS's Pressurized Natural Gas (PNG), TransCanada's Gas Transport Module (GTM), and Trans Ocean Gas's Composite Reinforced Pressure Vessel (CRPV). The latter two technologies are based on the same fundamental principle, in other words the use of containers made of mixed structures in steel and composites. The VOTRANS system, developed by EnerSea Transport of Houston, is an innovative transport system. This system is based on an optimization of the volumes occupied, on specially designed transport carriers, on loading and offloading systems similar to other CNG systems, but at lower pressure and temperature.

The vessels are designed with horizontal or vertical tanks of carbon steel (API standard), giving a very large total storage capacity. The largest carriers allow the CNG technology to be used at the production centers which deliver the highest daily flow rates of gas over particularly large distances. An individual VOTRANS storage tank consists of a series of 6–24 tanks, connected to one another to form a single storage system. There is also the option of converting existing single-hulled vessels to the VOTRANS system, with the aim of speeding up the time required for facilities to become operational and lowering costs. As far as safety is concerned, EnerSea has conducted numerous studies to demonstrate that the proposed system does not present greater risks than other gas transport systems.

In addition to the VOTRANS transport system, EnerSea is developing a storage system for onshore installation with horizontal or vertical tanks, named VOLANDS (Volume *GTM technology*). The GTM system is based on a newly designed carrier of new design for the transport of natural gas, using the technology patented by NCF Industries. This technology is based on pressurized tanks in a reinforced composite material, consisting of large diameter pipes in High Strength Low Alloy (HSLA) steel, reinforced with high-performance composites. This material has a high resistance to corrosion and a mechanical resistance of over 650 MPa. If GTM tanks are compared to equivalent tanks made of steel alone, it becomes evident that the former are about 35%–40% lighter, thus allowing for applications which were previously impossible, at a lower cost [126,127].

As already mentioned, the GTM system is based on large diameter pipes in HSLA steel with both ends welded. The pipework thus obtained undergoes the patented reinforcement process with composite materials based on glass fiber, increasing resistance whilst minimizing the increase in weight. The glass fiber increases the circumferential resistance, whilst the steel, which contributes only partly to circumferential resistance, absorbs all longitudinal loads. A typical storage tank is about 24 m long, with a diameter of 1–1.5 m. The working pressure is about 200 bar (maximum allowed pressure 250 bar). This technology is not innovative, but it is applied to a new process and dimensions never reached before. Since 1973, NCF, the owner of the patent, has developed numerous applications demonstrating its versatility and effectiveness [126,127].

9.9.2.3 CRPV Technology

Trans Ocean Gas (TOG) proposes the CRPV technology for the transport of natural gas, based on the use of tanks in a composite material grouped into modules and inserted vertically one inside the other within the vessel's hull. The tanks were designed in collaboration with Lincoln Composites, which has applied this technology in the aerospace industry and currently produces tanks for LNG plants in the automotive sector. Tanks in a composite material (CPVs, composite pressure vessels) are lighter and safer than their steel counterparts, in addition to being corrosion resistant.

Each individual element has a diameter of about 1 m and a length of 12 m and is designed to withstand a pressure of 250 bar. The CPVs are of a plastic material reinforced with fibers (FRP, fiber reinforced plastic): the body of the tank is in high-density polyethylene (HDPE) and is reinforced with a cladding of glass fiber or carbon fiber The ends are in stainless steel; these form the mandrel during the process of covering the tank with fibers and allow it to be welded to conventional piping. Trans Ocean Gas believes that it is preferable to use glass fiber rather than carbon fiber for CNG transport, with the aim of containing costs at the expense of lightness; a CPV clad in glass fiber still only weighs one-third of a conventional steel tank, and it is thus possible to use a carrier with a larger storage capacity and higher sailing speed. The modular system developed by TOG consists of a framework containing about 18 CPVs arranged vertically and linked to one another at both ends. These modules, known as cassettes, can be arranged in several tiers depending on the size of the vessel; for example, a ship with a tonnage of 60,000 tons has two tiers of cassettes [128].

The transport system is completed with valve systems placed on the main deck and a conventional cooling system, used to maximize storage capacity and inhibit the formation of hydrates during the loading and offloading phases. The compression unit placed onboard can be used for loading and offloading at an offshore mooring terminal.

9.9.2.4 PNG Technology

Knutsen OAS has developed PNG carriers for the transport of CNG. The design scheme is based on the use of cylindrical steel tanks in a vertical arrangement, grouped to form storage units. The tanks always have a diameter of 1 m and a thickness of 33.5 mm, whereas their length depends on the capacity of the vessel. Knutsen has developed three different cylinders respectively.

For loading and offloading operations, the possibility of connecting through the keel of the vessel has been studied. This allows for both direct loading from subsea satellites at a depth of between 50 and 500 m and the use of a specially designed mooring system for safe loading/offloading operations from the coast. In this context, Knutsen has developed an offloading terminal using the same storage technology as the vessel, in other words a series of vertical tanks. This type of terminal has the purpose of shortening the time required to offload the vessel and allowing the delivery of natural gas to the network to be regulated [128].

9.9.3 The Modular Transport of LNG

Since the 1960s, when the transport of gas in the form of LNG began, high safety, reliability, and environmental protection standards have been met in liquefaction and regasification processes and during transport using purpose-built tankers (methane tankers). Technological innovation has had considerable impact over the years, leading to a gradual reduction of costs in all stages of the production chain. Given the growing need for energy and the increasing gas consumption, this has contributed to the sector's growth and the consequent success of LNG as an alternative source of supply. However, the strong competitiveness of pipeline transport has relegated LNG to a niche market, restricted to countries excluded from or not reached by the transport network, and to countries which, being heavily dependent on imports, have wished to diversify sources of supply [128].

9.9.3.1 Tankers for the Transport of LNG

The transport of LNG using methane tankers began in the 1960s and took off in the 1970s. Two different designs were developed, based on two different concepts still used today: in the first design, the tanks containing the LNG are structurally integrated into the double hull of the tanker, over which the loads exerted by the cargo are spread; in the second design, the tanks are independent of the ship's structure and must therefore be self-supporting [128].

9.9.3.2 Tankers with Integrated Tanks

The two most widespread integrated tank technology systems are that developed by Technigaz and that belonging to Gaztransport; these two companies merged in 1994 to form Gaztransport & Technigaz, which was later acquired by Eni group [128].

In the Technigaz system, the tanks consist of an elastic membrane made of ribbed stainless steel, which rests on the hull by means of a thermal insulation layer, and a secondary barrier, which has the task of protecting the hull of the tanker from any LNG leaks. Only a few special steels are compatible with the low temperatures caused by contact with LNG. The primary barrier consists of a ribbed membrane made from welded sheets of special steel; these sheets are orthogonally ribbed in order to reduce thermal stresses caused by the considerable temperature differences; the double ribbing system also allows bending stresses to be absorbed by the ribs themselves and tensile stresses by the flat parts of the sheets. The thermal insulation between the tank and the hull, originally made of balsa wood, has been replaced with polyurethane foam reinforced with glass fibers. The secondary barrier is made of a composite "triplex" material, consisting of a sheet of aluminum in a fiber glass sandwich. The membrane's resistance to temperature variations and its low thermal inertia allow the tanks to be cooled rapidly during the loading of LNG. The return trip of the empty methane tankers can thus be undertaken without the need to keep the tanks at low temperature [128].

The Gaztransport system involves two independent barriers. The primary and the secondary barriers are both made of invar (steel alloy containing 36% nickel) in the form of flat-welded strakes 0.7 mm thick, supported by thermal insulation boxes in balsa wood filled with silicone-treated perlite. The membranes are liquid- and gas-proof.

9.9.3.3 Tankers with Self-Supporting Tanks

In the second design scheme, the tanks must resist the stresses induced by the weight of the LNG which they contain. The Norwegian company Moss Rosenberg has tackled this problem by building methane tankers with four or six spherical tanks. The spheres are thermally insulated using suitable insulation materials; a gap is maintained between the tank and the insulation, filled with dry air or inert gas (nitrogen) to increase the system's insulation capacity and ensure the elasticity of the primary barrier. Each sphere is supported by a cylindrical jacket resting on the tanker's hull; the latter is protected from potential LNG leaks with a secondary barrier placed at the base of the spheres [128].

9.9.3.4 The LNG Transport Fleet

In 2003, the LNG transport fleet consisted of 145 carriers; within the fleet, 50% were carriers of the Moss Rosenberg type, 37% the Gaztransport type, 11% the Technigaz type, and the remaining 5% other minor typologies. The capacity of methane tankers has evolved over time. The costs of LNG transport vessels increased until 1998, reaching around \$2,600/m^3 of capacity, subsequently declining to \$1,200/m^3 in 2002. Forecasts up to 2007 predict slightly declining costs. To deliver LNG to customers in north-west Europe, Shell built and recently took delivery of Cardissa, a state-of-the-art LNG bunker vessel with capacity to hold around 6,500 m^3 of LNG fuel. The vessel will deliver fuel from the Gate terminal in Rotterdam and in locations throughout Europe. Shell has also finalized a long-term agreement to charter an LNG bunker barge with a capacity to carry 3,000 m^3 of LNG fuel. Operating out of Rotterdam, the LNG bunker barge will be able to refuel vessels operating on Europe's inland waterways. The Gate terminal in Rotterdam will boost the availability of LNG as a marine fuel [128].

In the U.S., Shell has finalized a similar agreement for an LNG bunker barge with a 4,000 m^3 fuel capacity. This is the first ocean-going barge of its kind to be based there. It will supply LNG to marine customers along the southern East Coast and support growing cruise-line demand for LNG marine fuel. Corban Energy Group provides its customers access to the most cost efficient, reliable, and technically advanced LNG equipment available in today's marketplace. The company LNG equipment, such as ISO tank containers, storage tanks, and specialized gas transportation products and services have been thoroughly tested to meet all safety standards [128].

9.9.3.5 LNG ISO Containers

Corban LNG ISO containers are optimized specifically for transporting LNG worldwide by rail, sea, or road and are also ideal for short-term storage. They are designed for performance, ease of operation, and safety. Made with the highest quality materials available, ISO containers guarantee years of operation and service, due to their construction. Having complied to all of the codes and approvals listed below, the containers are some of the highly performing, most cost efficient, and fully functional methods of transporting LNG resources today. The company provides perfect turnkey solution to their LNG transportation problems [129].

Features and Benefits of LNG 40ft ISO containers include intermodal transportation through road, rail, and sea; custom design availability; depending on the needs of the business, they are passed and approved by CSC, IMDG, DOT, RID, ADR, TPED, ASME (global regulation and intermodal transportation approval; low boil-off rate (BOR), at 0.15%–0.30% per day; long holding time (55–81 days); easy accessibility of valves and gauges; and customizable valve cabinets and instruments [129].

9.9.3.6 LNG Transport Trailers

These transport trailers are also available for transportation applications. Corban Energy Group has manufactured batches of trailers for transportation and fueling purposes. Safety features and designs prevent spillage and leaks from the transfer of LNG from terminal to the truck's tank. The tank has a double-walled design that protects the LNG from evaporating at ambient temperatures too quickly, as well as a boil-off gas management system that uses any vaporized gas as fuel for the transport trailer [128,129].

9.9.4 THE TRANSPORT OF NGH WITH GTS TECHNOLOGY

Recently, interest has been shown in the transport of natural gas using gas to solid (GTS) technology, in other words, natural gas transformed into hydrates. To illustrate the feasibility of this project, we can compare the LNG chain to the GTS chain, estimating both production and regasification costs, as well as transport costs [130–132]. Despite the results of numerous studies which have analyzed the entire transport chain in detail, proposing different production and storage methods, and despite designs of vessels for marine transport, the transport of natural gas in the form of hydrates has still not been used commercially.

Like LNG, the GTS transport chain comprises the following main stages: (a) treatment and transport by pipeline to the coast, (b) treatment of the gas to meet the specifications required for the solidification process, (c) transformation into methane hydrates, (d) storage and loading of hydrates, (e) transport of hydrates using tankers, (f) receipt and storage, and (g) regasification. The hydrate formation plants represent the heart of this chain and are the main item of investment expenditure [130–132].

9.9.4.1 Properties of Methane Hydrates and Options for Hydrate Storage and Transport

The formation process of methane hydrates has been known for long time, but the systematic study of processes suitable for inclusion in a natural gas transport chain is recent and has intensified since 2000. The formation process consists of making natural gas react with water in a purpose-built reactor in order to generate hydrate in two forms: simple ice crystals (dry hydrates) and a semi-liquid mixture (slurry). Studies on the formation of hydrates aim to identify possible processes for their production and to determine the stability of the product. The stability of hydrates depends heavily on the composition of the natural gas used. It has been shown that hydrates formed by methane alone are more unstable than those formed by a mixture also containing ethane, propane, and butane. Using the stability diagram, it is possible to identify the potential types of transport systems. The possibilities involve using either pressurized transport or transport at atmospheric pressure. If a transport system at atmospheric pressure is used, the storage temperature must be kept below 40°C. By contrast, if transport is pressurized, the storage temperature can be selected, for example 0°C.

The study conducted by Mitsui Engineering & Shipbuilding (MES), with the aim of promoting the development of a complete transport chain for natural gas using hydrates, has identified an improvement in the efficiency of transport as a result of the form which the hydrate is given. In practice, after pretreating the natural gas to eliminate acid gases and transforming it into hydrates, it takes the form of a coarse powder, with grains ranging from a few tens of microns to a few millimeters in size. Handling the hydrate in this form is fairly difficult, because it involves a high degree of sensitivity to the temperature fluctuations which always accompany both the storage and transport phases, due to low density and ease of dissociation. To this end, various forms in which to transport the hydrates have been researched, including large and small rectangular blocks, pellets, and hydrate powder; these have all been compared with slurry. Pellets offer the greatest advantages in terms of the volume of gas transported, the efficiency of self-preservation and flow, and the constancy of the properties of the mass. Pellets are made by compressing the hydrate powder and compacting it to form pellets of the requisite size. For the production and transport of natural gas by sea, NGH is 36% and 25%, respectively, cheaper than LNG. The total production, transport, and regasification cost for NGH is also 26% lower than that for LNG [130–132].

The self-preservation property is defined as the capacity to halt the dissociation of the hydrate under pressure. Storage and transport systems for natural gas hydrates can basically be divided into pressurized systems and refrigerated systems at atmospheric pressure. A particularly important aspect of transport is the ability to fill the available volumes, since methane hydrates are in the solid phase. Gas hydrates field formation plant produces hydrate storage tanks and shipping storage tanks. In the case of pressurized systems, one refers to the transport of slurry, in other words hydrate suspensions in the pseudoliquid phase, with the same filling capacities as any other fluid. The slurry is transferred to pressurized tanks (about 10 bar) and loaded onto tankers which allow the temperature to be kept around 2°C. The alternative is to use the holds of tankers for direct loading; however, these must ensure thermal insulation and the potential for pressurization.

In the case of hydrate production in the form of powder, one has the problem of a cargo with void spaces. As already specified, the optimal form for transport seems to be that of pellets. In order to increase the efficiency of filling the tanks used for transport, the use of pellets of differing sizes has been suggested, allowing a higher cargo density to be obtained. The pellets also slide easily, and during the loading phase, there is therefore no need to level their surface; during unloading, it is sufficient to place the hopper at a suitable angle for them to slide outside of their own accord. The high volumetric efficiency of storage thus remains unaltered, and the offloading phase is brief. Since the required temperatures are between 15°C and 50°C, depending on the type of hydrate, it is unnecessary to use particularly high-performance materials; standard merchant ships can therefore be used, taking particular care in terms of cargo containment [130–132].

By cargo containment, one refers to the various expedients and techniques employed to store and conserve the cargo on board, limiting or preventing the absorption of heat from the outside. It is considered technically and economically advantageous to insulate as much as possible the cisterns destined to receive the cargo and to use as a means of propulsion the gas freed by the portion of hydrate which dissociates during the voyage. As far as "physiological" losses resulting from transport are concerned, optimization is probably one of the parameters which contributes most to the economic feasibility of transport. Whereas on the one hand it would be preferable to have no losses, on the other this would increase the cost of the carrier (and thus of transport), closely linked to the efficiency of heat insulation. The insulation of the cisterns containing the hydrates has a direct impact on the typology of the carrier and its safety requirements. This is because either the inner or outer surface of a tank may be insulated. The former solution implies the creation of a tank which is independent of the tanker, contained and supported by the vessel itself, but whose structures do not contribute to the tanker's overall robustness. This solution may also be used for pressurized transport and is similar to that used for double-hulled merchant ships. The second solution, involving the internal insulation of the ship's hulls, which in this case act as tanks, obviously leads to a reduction in costs; standard merchant ships can be used.

The final point to be analyzed regarding the transport and storage system is that of the processes for shifting the cargo. The studies carried out involve the use of mechanical movement systems. For the loading phase, a horizontal conveyor belt which stows the natural gas hydrate has been proposed. For the offloading phase, an angled conveyor belt is used, which moves the hydrate to the deck; it is then brought onshore for storage, again with conveyor belts. Alternative methods have also been proposed, such as a pumping system for slurry or a pneumatic system using pressurized gas; in the first analysis, all these systems seem suitable, but a detailed study is still required in order to identify any advantages of one compared to another [130–132]. Studies to design a carrier for the transport of natural gas hydrates have been carried out by various groups, including Mitsui Engineering, Transmarine, Three Quays, and ELP (Emerging Leaders Program). However, none of these have gone beyond the project phase. One possibility is to use a double-hulled ship with several holds insulated by the internal hull and ballasts placed in the space between the internal and external hull [130–132].

9.10 TRANSPORT MODULES FOR HYDROGEN (H_2)

Hydrogen is transported in a number of different ways: by transport modules, tube trailers, mobile transport units, cryogenic liquid trucks, rails, barges, and ships. We briefly examine few of these modular operations here.

X-STORE transport modules [133] with type 44 composite lightweight design vessels are among the most efficient gas-transport systems available worldwide. They are being successfully used for the transport of hydrogen. Currently, there are modules with 250, 300, and 500 bar in use. The modules are available in a container-type design and also in the form of a battery vehicle in sizes from 10 to 45 ft. These lightweight design high-pressure vessels are arranged vertically, and they optimally utilize the container space. The extremely light and at the same time rugged design of modules in combination with the optimal use of space, provides the world's largest transport volumes. This in turn reduces the operational costs of the customers. The modules are manufactured in a container-type construction and thus comply with international safety standards for maritime containers. They are mounted on a conventional ISO-standard chassis, which offers significant advantages when service and maintenance work is required on the module or the chassis. Hydrogen **tube trailers (see Figure 9.3)** are semi-trailers that consist of 10–36 cluster high-pressure hydrogen tanks varying in length from 20 ft (6.10 m) for small tubes to 38 ft (11.58 m) on jumbo tube trailers. They are part of the hydrogen highway and usually precede a local hydrogen station.

Modular tube trailers carry 8–54 tubes [134]. **Intermediate trailers carry** intermediate tubes that are assembled in banks of five tubes in lengths of 19 and 38 ft (5.79 and 11.58 m) and provide mobile or stationary storage. **Jumbo tube trailers carry** 10 tubes and a 44-ft (13.41 m) chassis, operating with pressures in excess of 3,200 psi (220.63 bar; 22.06 MPa). As of September 2013, the Linde Group has introduced a tube trailer operating at 500 bar (50.0 MPa; 7,250 psi) utilizing new, lighter storage materials to more than double the amount of compressed gaseous hydrogen to 1,100 kg (2,425 lb) or a normal 13,000 m^3 (17,000 cu yd) of hydrogen gas. The new trailers can be filled and emptied in less than 60 min. As on July 2016, Nishal Group had multiple cascade configurations in the form of cascade banks operating at 200 bar. Powertech has pioneered the use of lightweight carbon fiber composite

FIGURE 9.3 Compressed hydrogen tube trailers [134].

tanks for the high-pressure bulk transportation of hydrogen, for mobile hydrogen fueling station applications, and for portable self-contained hydrogen fueling units. Transportable compressed hydrogen units can be custom-designed to meet customer needs, including transport trailers, mobile fuelers, and portable filling stations. Mobile station designs can either contain only high-pressure hydrogen storage for a limited number of fills or can include a high-speed compressor for multiple high-pressure fueling using a medium-pressure hydrogen source. Powertech has provided mobile hydrogen fueling at numerous fuel cell vehicle events across North America, including the Alaska/Canada (AlCan) Highway drive by Toyota in 2007. Portable fueling units have also provided various automotive OEMs with hydrogen fuel supply in support of cold temperature fuel cell electric vehicle (FCEV) tests in remote locations [134].

9.10.1 Tube Trailers, Cryogenic Liquid Trucks, Rail, Barges, and Ships

The majority of today's transportation fuels are transported to local terminals over a network of pipelines and then distributed locally to the points of use via tanker trucks. Barge, rail, and truck transport are also used. Similarly, hydrogen fuel is transported today by three modes: regional pipeline networks, on commercial roadways using cryogenic liquid cargo trailers, and on commercial roadways using high-pressure gaseous tube trailers. Rail, barge, and ship travel are also potential transport modes, but they are not in commercial use today. High-pressure cylinders and tube trailers at 182 bar (2,640 psi) are commonly used to distribute gaseous hydrogen within 320 km (200 miles) of the source. Higher-pressure, 250-bar composite tube trailers have recently received U.S. Department of Transportation (DOT) certification and can carry 560 kg of hydrogen onboard. Hydrogen can be economically distributed within 600 miles of the source using liquid hydrogen tanker trucks that have capacities of 3,000–4,000 kg of hydrogen.

Successful widespread use of hydrogen will require a delivery infrastructure that accommodates diverse means of distribution. Although the most economical means of transporting hydrogen in the future may be by a larger pipeline network similar to that used for natural gas, other modes of transport may be more efficient for outlying areas or dense urban settings. Rail and barge transport offer higher load-carrying capacities and higher weight limits than over-the-road trailers. Trucks, railways, and barges may also play a key role during the transition phase, when hydrogen demand is low and economic incentives for building hydrogen pipelines are not yet in place. Hydrogen is currently shipped overseas using tube skids or a high-efficiency liquid storage container similar in size to over-the-road trailers. In the future, it is conceivable that liquid hydrogen tanker ships (similar to LNG tankers) may be used to transport large volumes of hydrogen between U.S. ports and overseas.

9.10.1.1 Tube Trailers

Tube trailers (see Figure 9.4) are currently limited by DOT regulations to pressures of less than 250 bar. Further development and testing of types II, III, or IV higher-pressure composite vessels for hydrogen, with the development of appropriate codes

Systems for Energy and Fuel Transport

FIGURE 9.4 Hydrogen tube trailers [135].

and standards, will eventually allow the use of higher-pressure hydrogen tube trailers that also comply with federal truck weight limitations. Other approaches being researched for more cost-effective stationary gaseous hydrogen storage may also be applicable for transportation. This includes the use of cryo or cold gas and possibly the use of solid carriers in the tube vessels. With sufficient technology development to minimize capital cost, high-pressure composite tube trailers could dramatically decrease the cost of hydrogen transport via tube trailer by significantly increasing the carrying capacity.

Hydrogen leak detection, in the absence of odorizers, is a challenge. Currently, commercially available leak detection equipment is handheld. Ideally, an online leak detector (direct or indirect measurement) would be a desirable addition to a tube trailer. Improved monitoring and assessment of the structural integrity of tubes and appurtenances may be called for in the presence of higher containment pressures. Some examples of potentially novel methods include in-situ strain monitoring and acoustic emission monitoring. Codes and standards will need to address integrity management for the operating envelope [135].

9.10.1.2 Liquid Hydrogen Trailers

Cryogenic liquid hydrogen trailers (see Figure 9.5) can carry up to 4,000 kg of hydrogen and operate at near atmospheric pressure. Some hydrogen boil-off can occur during transport despite the super-insulated design of these tankers, potentially on the order of 0.5% per day. Hydrogen boil-off of up to 5% also occurs when unloading the liquid hydrogen on delivery. If cost effective, a system could be installed to compress and recover the hydrogen boil-off during unloading if warranted. Based on the economics of off-loading liquid hydrogen into a customer's tank (distance from source, driver hours, losses), most organizations plan deliveries to serve up to three customer sites. It is estimated that merchant liquid hydrogen suppliers possess more than 140 liquid hydrogen trailers. Current markets include food processing; refineries; chemical processes; oil hydrogenation; and glass, electronics, and metals manufacturing.

FIGURE 9.5 Liquid hydrogen delivery [136].

9.10.2 Modular Station with Hydrogen Produced On-Site

A modular station with an electrolysis unit or a steam reformer attached to produce hydrogen is often used. Having on-site production using an electrolyzer or steam reformer and modular fueling station components only requires a small footprint for installation. The addition of a steam methane reformer or an electrolyzer to a hydrogen refueling station, however, increases the capital cost significantly, with the steam methane reformer estimated to have the installed cost (which includes site preparation, engineering and design, permitting, component capital, and installation costs), and generally results in hydrogen cost breaking even in about 7 years [137,138]. On-site production does allow for a more compact station footprint and does not require consideration for tube-trailer delivery routes. With current technology and projected utility prices, operating a steam methane reformer costs slightly less than operating an electrolyzer. These offsetting differences in capital versus operational costs cause the price of hydrogen to be very similar for a station with SMR-produced hydrogen and electrolysis-produced hydrogen. In both cases, the cost of hydrogen is $6–$10/kg more for the stations that produce hydrogen on-site than the stations that have hydrogen delivered. This estimate does not include the startup and shutdown inefficiencies (and associated costs) for either the SMR or electrolyzer. The penalty for cycling an SMR is likely greater than for cycling an electrolyzer, potentially giving an electrolyzer an economic advantage for on-site production [137,138].

Electrolyzers also have the potential to produce less carbon dioxide than SMRs. By definition, SMRs require methane, which contains carbon, to produce hydrogen, releasing the carbon molecules as carbon dioxide and contributing to climate change. A reduced carbon footprint could be achieved by using biogas or implementing carbon capture and storage (although this would incur an efficiency penalty). Electrolyzers on the other hand require only electricity and water, giving this technology the potential to be completely carbon free, given the right source of electricity, such as solar, wind, hydroelectric, or nuclear. The key to making on-site production more cost competitive with delivered hydrogen is reducing the utility costs. For electrolysis, this can be achieved through efficiency improvements (although around 50%–70% efficiencies are achieved with current technology), through the purchase of low-cost off-peak electricity or by using electricity generated behind the meter. Of course, reducing capital cost and maintenance requirements of on-site production units will also improve the economics of this technology [137,138].

According to literature [137,138], current modular fueling stations, where components are assembled and tested at a central production facility, have installed costs no higher than those of conventional stations. Using a high estimate [137,138] of $1.5 M for the modular unit, the installed cost of the 300 kg/day modular station is nearly the same as that of a 300 kg/day conventional station. Using a lower cost estimate of $1 M for the modular unit, the installed cost of a 100 kg/day conventional station is 11% higher than that of the modular station. With greater industry experience and higher volume production, one anticipates the cost of modular stations to reduce in the near future. Developing stations in this manner may lead to better quality control and reduced maintenance requirements for hydrogen fueling stations. It is also possible to achieve more compact station footprints by modularizing the station components by building in appropriate fire-rated barrier walls. Because of these potential benefits, it is anticipated that the modularization and standardization would increase in the future [137,138].

REFERENCES

1. Hybrid Micro Energy Project (HMEP). A website report by Cold Climate Housing and Research Center, Fairbanks, AK www.cchrc.org/sites/default/files/docs/HMEPHandout2.pdf (2019).
2. HMEP Guidance Resources | PHMSA. A website report www.phmsa.dot.gov/grants/hazmat/hmep-guidance-resources (25 February 2019).
3. ABB Pioneers Microgrid Solution for Installation on Offshore Platform. A website report, Zurich, Switzerland www.abbaustralia.com.au/cawp/seitp202/95582016ef0b47 8448258286000e2f9b.aspx (12 December 2017).
4. ABB to Pioneer Microgrid Solution for Installation on Offshore Platform. A website report www.abbaustralia.com.au/cawp/seitp202/b2ba69e22a266d0c482581f4002539 21.aspx (12 December 2017).
5. Williams, D., ABB Microgrid Technology to Power Offshore Gas Platform. A website report (12 December 2017).
6. Buck, E., Microgrid and Energy Storage Systems- Eaton's Intelligent Grid Solution Series. A website report www.gulfcoastenergynetwork.org/wp-content/.../Energy-MicroGrids-Eaton-Buck.pdf (3 June 2015).

7. Microgrid Energy Systems: Eaton. A website report www.eaton.com/FTC/utilities/MicrogridEnergySystems/index.htm (2015).
8. Eaton Microgrid. (2016). *A Paper Presented at the 2016 IEEE PES Innovative Smart Grid Technologies Conference*, September 2016, IEEE, Minneapolis, MN.
9. Advanced Microgrid Management System. A website report by Siemens https://w3.usa.siemens.com/smartgrid/.../microgrid/.../SP7_%20Microgrid_Manageme... (2015).
10. Spectrum Power 7™ Microgrid Management System (MGMS). A website report by Siemens https://w3.usa.siemens.com/smartgrid/.../microgrid/.../Spectrum_Power_7_MGMS_Ap... (2015).
11. Falahati, B., Chua, E., and Kazemi, A. (2016). A novel design for an expandable, modular microgrid unit. *2016 IEEE Power and Energy Society Innovative Smart Grid Technologies Conference (ISGT)*, 6–9 September, IEEE, Minneapolis, MN, USA. doi: 10.1109/ISGT.2016.7781186.
12. Khodayar, M.E., Barati, M., and Shahidehpour, M. (2012). Integration of high reliability distribution system in microgrid operation. *IEEE Transactions on Smart Grid*, 3(4), 1997–2006.
13. Kroposki, B., Lasseter, R., Ise, T., Morozumi, S., Papatlianassiou, S., and Hatziargyriou, N. (2008). Making microgrids work. *IEEE Power and Energy Magazine*, 6(3), 40–53.
14. Kojima, Y., Koshio, M., Nakamura, S., Maejima, H., Fujioka, Y., and Goda, T. (2007). A demonstration project in Hachinohe: Microgrid with private distribution line. *Proceedings of the Second Annual IEEE International Conference on System of Systems Engineering*, April 2007, San Antonio, TX.
15. Suul, J.A., Uhlen, K., and Undeland, T. (2008). Variable speed pumped storage hydropower for integration of wind energy in isolated grids: Case description and control strategies. *Proceedings of the Nordic Workshop on Power and Industrial Electronics*, June 2008, Espoo, Finland.
16. Galasso, G. (1991). Adjustable speed operation of pumped hydroplants. *Proceedings of the International Conference on AC and DC Power Transmission*, September 1991, London, England.
17. Suul, J.A., Undeland, T., and Uhlen, K. (March 2008). Pumped Storage for Balancing Wind Power Fluctuations in an Isolated Grid. Available: http://en.escn.com.cn/Tools/download.ashx?id=110.
18. Yokoyama, R., Niimura, T., and Saito, N. (2008). Modeling and evaluation of supply reliability of microgrids including PV and wind power. *Proceedings of the IEEE Power and Energy Society General Meeting*, July 2008, Pittsburgh, PA.
19. Barnes, M. (2003). *Practical Variable Speed Drives and Power Electronics*. Elsevier, Newnes.
20. Schlunegger, H. and Thöni, A. (2013). 100 MW full-size converter in the grimsel 2 pumped-storage plant. *Proceedings of the Hydro 2013: International Conference and Exhibition*, October 2013, Innsbruck, Austria.
21. Kargarian, A., Raoofat, M., and Mohammadi, M. (2011). Probabilistic reactive power procurement in hybrid electricity markets with uncertain loads. *Electric Power Systems Research*, 82(1), 68–80.
22. Falahati, B. and Fu, Y. (2012). A study on interdependencies of cyber-power networks in smart grid applications. *Proceedings of the IEEE PES Innovative Smart Grid Technologies (ISGT) Conference*, January 2012, Washington, DC.
23. Morvaj, B., Evins, R., and Carmeliet, J. (2017). Comparison of individual and microgrid approaches for a distributed multi energy system with different renewable shares in the grid electricity supply. *Energy Procedia*, 122, 349–354. doi: 10.1016/j.egypro.2017.07.336.

24. Morvaj, B., Evins, R., and Carmeliet, J. (2016). Optimization framework for distributed energy systems with integrated electrical grid constraints. *Applied Energy*, 171, 296–313. doi: 10.1016/j.apenergy.2016.03.090.
25. Müller, P., Kebir, N., Stark, R., and Blessing, L. (2009). PSS layer method: Application to microenergy systems. In: Sakao, T. and Lindahl, M. (eds) *Introduction to Product/Service-System Design*. Springer, London. doi: 10.1007/978-1-84882-909-1_1.
26. Gabbar H. A., Honarmand N., Abdelsalam A.A. (2014). Resilient micro energy grids for continuous production in oil and gas facilities. *Advances in Robotics and Automation*, 3, 125. doi: 10.4172/2168-9695.1000125
27. Lešić, V., Martinčević, A., and Vašak, M. (2017). Modular energy cost optimization for buildings with integrated microgrid. *Applied Energy*, 197, 14–28. doi: 10.1016/j.apenergy.2017.03.087.
28. Olesen, B.W. (2007). Indoor environmental input parameters for design and assessment of energy performance of buildings addressing indoor air quality, thermal environment, lighting and acoustics. EN 15251, January 2007.
29. IBM ILOG CPLEX Optimization Studio: CPLEX Users Manual, Ibm Corp, IBM Corp., Version 12.6. (2015).
30. Comodi, G., Giantomassi, A., Severini, M., Squartini, S., Ferracuti, F., Fonti, A., Nardi Cesarini, D., Morodo, M., and Polonara, F. (2015). Multi-apartment residential microgrid with electrical and thermal storage devices: Experimental analysis and simulation of energy management strategies. *Applied Energy*, 137, 854–866.
31. Gulin, M., Martinčević, A., Lesic, V., and Vasak, M., Multi-Level Optimal Control of a Microgrid-Supplied Cooling System in a Building. A website paper www.fer.unizg.hr/_.../ISGT_Europe_%5BGulin_Martincevic_Lesic_Vasak%5D.pdf (October 2016).
32. Gulin, M., Pavlovic, T., and Vasak, M. (2017). A one-day-ahead photovoltaic array power production prediction with combined static and dynamic on-line correction. *Solar Energy*, 142, 49–60.
33. Fiorentini, M., Wall, J., Ma, Z., Braslavsky, J.H., and Cooper, P. (2017). Hybrid model predictive control of a residential HVAC system with on-site thermal energy generation and storage. *Applied Energy*, 187, 465–479.
34. Ogunjuyigbe, A.S.O., Ayodele, T.R., and Monyei, C.G. (2015). An intelligent load manager for PV powered off-grid residential houses. *Energy for Sustainable Development*, 26, 34–42. doi: 10.1016/j.esd.2015.02.003.
35. Gulin, M., Vasak, M., and Baotic, M. (2015). Analysis of microgrid power flow optimization with consideration of residual storages state. *European Control Conference (ECC)*, pp. 3126–3131. doi: 10.1109/ECC.2015.7331014.
36. Gulin, M., Pavlovic, T., and Vasak, M. (2016). Photovoltaic panel and array static models for power production prediction: Integration of manufacturers' and on-line data. *Renewable Energy*, 97, 399–413.
37. Pao, L.Y. and Johnson, K.E. (2009). A tutorial on the dynamics and control of wind turbines and wind farms. *Proceedings of American Control Conference*, June 2009, St. Louis, MO.
38. Korkas, C.D., Baldi, S., Michailidis, I., and Kosmatopoulos, E.B. (2016). Occupancy-based demand response and thermal comfort optimization in microgrids with renewable energy sources and energy storage. *Applied Energy*, 163(C), 93–104.
39. Koehler, S., Danielson, C., and Borrelli, F. (2016). A primal-dual active-set method for distributed model predictive control: A primal-dual active-set method for distributed model predictive control. *Optimal Control Applications and Methods*, 38. doi: 10.1002/oca.2262.
40. Brandstetter, M., Schirrer, A., Miletic, M., and Kupzog, F., Hierarchical Predictive Load Control in Smart Grids. A website report www.semanticscholar.org/.../Hierarchical-Predictive-Load-Control-in-Smart-Grids... (September 2015).

41. Ma, Y., Kelman, A., Daly, A., and Borrelli, F. (2012). Predictive control for energy efficient buildings with thermal storage. *IEEE Control Systems Magazine*, 32, 44–64.
42. Tecogen Selected for 12 Unit Microgrid Order: Tecogen, Inc. (TGEN). A website report www.tecogen.com/news.../press.../tecogen-selected-for-12-unit-microgrid-orde... (December 11 2018).
43. InVerde e+. Tecogen, Inc. (TGEN). A website report www.tecogen.com/chp-cogeneration/inverde (2018).
44. Mei, S., Li, R., Xue, X., Chen, Y., Lu, Q., Chen, X., Ahrens, C.D., Li, R., and Chen, L., Paving the Way to Smart Micro Energy Internet: Concepts, Design Principles, and Engineering Practices. A website report from Cornell University https://arxiv.org › math (2016).
45. Geidl, M. and Andersson, G. (2007). Optimal power flow of multiple energy carriers. *IEEE Transactions on Power Systems*, 22(1), 145–155.
46. Li, J., Fang, J., Zeng, Q., and Chen, Z. (2016). Optimal operation of the integrated electrical and heating systems to accommodate the intermittent renewable sources. *Applied Energy*, 167, 244–254.
47. Nuytten, T., Claessens, B., Paredis, K., Bael, J.V., and Six, D. (2013). Flexibility of a combined heat and power system with thermal energy storage for district heating. *Applied Energy*, 104, 583–591.
48. Wang, C., Wei, W., Wang, J., Liu, F., Qiu, F., Correa-Posada, C.M., and Mei, S. (2016). Robust defense strategy for gas-electric systems against malicious attacks, arXiv:1603.00832.
49. Wang, C., Wei, W., Wang, J., Liu, F., and Mei, S. (2016). Strategic offering and equilibria in coupled gas and electricity markets, arXiv:1607.04184.
50. Wei, W., Mei, S., Wu, L., Wang, J., and Fang, Y. (2017). Robust operation of distribution networks coupled with urban transportation infrastructures, *TPWRS*, 32, 2118–2130.
51. Wei, W., Mei, S., Wu, L., Shahidehpour, M. and Fang, Y. (2017). Optimal traffic-power flow in urban electrified transportation networks. *IEEE Transactions on Smart Grid*, 8(1), 84–95.
52. Huang, A.Q., Crow, M.L., Heydt, G.T., Zheng, J.P., and Dale, S.J. (2010). The future renewable electric energy delivery and management (FREEDM) system: The energy internet. *Proceedings of the IEEE*, 99(1), 133–148.
53. Hemmes, K., Zachariah-Wolf, J.L., Geidl, M., and Andersson, G. (2007). Towards multi-source multi-product energy systems. *International Journal of Hydrogen Energy*, 32(S10–S11), 1332–1338.
54. Mancarella, P. (2014). MES (multi-energy systems): An overview of concepts and evaluation models. *Energy*, 65, 1–17.
55. Jin, X., Mu, Y., Jia, H., Jiang, T., Chen, H., and Zhang, R. (2016). An optimal scheduling model for a hybrid energy microgrid considering building based virtual energy storage system. *Energy Procedia*, 88, 375–381.
56. Geidl, M. (2007). Integrated modeling and optimization of multi-carrier energy systems, Ph.D. Dissertation, ETH Zurich.
57. Lund, H., Werner, S., Wiltshire, R., Svendsen, S., Thorsen, J.E., Hvelplund, F., and Mathiesen, B.V. (2014). 4th Generation District Heating (4GDH): Integrating smart thermal grids into future sustainable energy systems. *Energy*, 68(4), 1–11.
58. Geidl, M., Koeppel, G., Favre-Perrod, P., Klöckl, B., Andersson, G., and Fröhlich, K. (2007). Energy hubs for the future. *IEEE Power and Energy Magazine*, 5(1), 24–30.
59. Mei, S., Wang, J., Tian, F., Chen, L., Xue, X., Lu, Q., Zhou, Y., and Zhou, X. (2015). Design and engineering implementation of non-supplementary fired compressed air energy storage system: TICC-500. *Science China Technological Sciences*, 58(4), 600–611.

60. Li, R., Chen, L., Yuan, T., and Li, C. (2016). Optimal dispatch of zero-carbon emission micro energy internet integrated with non-supplementary fired compressed air energy storage system. *Journal of Modern Power Systems and Clean Energy*, 4(4), 566–580.
61. Li, R., Chen, L., Wei, W., Xue, X., Mei, S., Yuan, T., and Zhao, B. (2017). Economic dispatch of integrated heat-power energy distribution system with concentrating solar power energy hub. *Journal of Energy Engineering*, 143(5), 1–11.
62. Li, Z., Wu, W., Wang, J., Zhang, B., and Zheng, T. (2016). Transmission- constrained unit commitment considering combined electricity and district heating networks. *IEEE Transactions on Sustainable Energy*, 7(2), 480–492.
63. Lasseter, R.H. (2011). Smart distribution: Coupled microgrids. *Proceedings of the IEEE*, 99(6), 1074–1082.
64. Mei, S. and Li, R. (2016). Smart micro energy internet and its engineering implementation. *Chinese Association for Artificial Intelligence Communication*, 6(10), 1–5.
65. Peters, M., Ketter, W., Saar-Tsechansky, M., and Collins, J. (2013). A reinforcement learning approach to autonomous decision-making in smart electricity markets. *Machine Learning*, 92(1), 5–39.
66. Morgenstern, O. and von Neumann, J. (1944). *Theory of Games and Economic Behavior*. Princeton University Press, Princeton, NJ.
67. Mei, S., Wei, W., and Liu, F. (2017). On engineering game theory with its application in power systems. *Control Theory and Technology*, 15(1), 1–12.
68. Mei, S., Liu, F., and Wei, W. (2016). *Foundation of Engineering Game Theory and its Applications to Power System*. Science Press, Beijing.
69. Mei, S., Wang, Y., Liu, F., Zhang, X., and Sun, Z. (2012). Game approaches for hybrid power system planning. *IEEE Transactions on Sustainable Energy*, 3(3), 506–517.
70. Wei, W., Liu, F., and Mei, S. (2016). Charging strategies of EV aggregator under renewable generation and congestion: A normalized Nash equilibrium approach. *IEEE Transactions on Smart Grid*, 7(3), 1630–1641.
71. Mei, S., Zhang, D., Wang, Y., Liu, F., and Wei, W. (2016). Robust optimization of static reserve planning with large-scale integration of wind power: A game theoretic approach. *IEEE Transactions on Sustainable Energy*, 5(2), 535–545.
72. Lopes, J.A.P., Moreira, C.L., and Madureira, A.G. (2006). Defining control strategies for microgrids islanded operation. *IEEE Transactions on Power Systems*, 21(2), 916–924.
73. Huang, A.Q., Crow, M.L., Heydt, G.T., and Zheng, J.P. (2010). The future renewable electric energy delivery and management (FREEDM) system: The energy internet. *Proceedings of the IEEE*, 99(1), 133–148.
74. Dong, Z., Zhao, J., Wen, F., and Xue, Y. (2014). From smart grid to energy internet: Basic concept and research framework. *Automation of Electric Power Systems*, 38(15), 1–11.
75. Werth, A., Kitamura, N., and Tanaka, K. (2015). Conceptual study for open energy systems: Distributed energy network using interconnected DC nanogrids. *IEEE Transactions on Smart Grid*, 6(4), 1621–1630. doi: 10.1109/TSG.2015.2408603.
76. Barnes, M., Kondoh, J., Asano, H., Oyarzabal, J., Ventakaramanan, G., Lasseater, R., Hatziargyriou, N.D., and Green, T.C. (2007). Real-world microgrids: An overview. In *Proceedings of the IEEE International Conference on System of Systems Engineering*, April 2007, San Antonio, TX, USA, pp. 1–8.
77. Justo, J.J., Mwasilu, F., Lee, J., and Jung, J.-W. (2013). AC-microgrids versus DC-microgrids with distributed energy resources: A review. *Renewable and Sustainable Energy Reviews*, 24, 387–405.
78. Mariam, L., Basu, M., and Conlon, M.F. (2013). A review of existing microgrid architectures. *Journal of Engineering*, 2013, 1–8.
79. Asmus, P. and Lawrence, M. (2013). *Microgrids*. Navigant Research, Boulder, CO.

80. Asmus, P. and Lawrence, M. (2014). *Nanogrids*. Navigant Research, Boulder, CO.
81. Gross, R. and Green, T. (2006). *The Costs and Impacts of Intermittency*. Technol. Pol. Assess. Funct., UK Energy Research Centre, London, UK.
82. Abe, R., Taoka, H., and McQuilkin, D. (2011). Digital grid: Communicative electrical grids of the future. *IEEE Transactions Smart Grid*, 2(2), 399–410.
83. Jin, C., Loh, P.C., Wang, P., Mi, Y., and Blaabjerg, F. (2010). Autonomous operation of hybrid AC-DC microgrids. *In Proceedings of the IEEE International Conference on Sustainable Energy Technologies (ICSET)*, Kandy, Sri Lanka, pp. 1–7.
84. Kok, J.K., Scheepers, M.J.J., and Kamphuis, I.G. (2010). Intelligence in electricity networks for embedding renewables and distributed generation. In: Negenborn, R.R., Lukszo, Z., and Hellendoorn, H. (eds) *Intelligent Infrastructures (Intelligent Systems, Control and Automation: Science and Engineering)*. Springer, Amsterdam, The Netherlands, Volume 4, pp. 179–209.
85. Asmus, P. and Lawrence, M. (2014). *Virtual Power Plants Demand*. Navigant Research, Boulder, CO.
86. Nikonowicz, L. and Milewski, J. (2012). Virtual power plants: General review: Structure, application and optimization. *Journal of Power Technologies*, 92(3), 135–149.
87. Tokoro, M. (2010). *Open Systems Science*. IOS Press, Amsterdam, The Netherlands.
88. Tokoro, M. (2014). Sony CSL-OISTDC-based open energy system (DCOES). *In Proceedings of 1st International Symposium Open Energy System*, Tokyo, Japan, pp. 64–67.
89. Falvo, M.C. and Martirano, L. (2011). From smart grids to sustainable energy microsystems. *In Proceedings of 10th International Conference on Environment and Electrical Engineering (EEEIC 2011)*, May 2011, Rome, Italy, pp. 1–5.
90. Brenna, M., Falvo, M., Foiadelli, F., Martirano, L., and Poli, D. (2012). Sustainable energy microsystem (SEM): Preliminary energy analysis. *Proceedings of the 2012 IEEE PES Innovative Smart Grid Technologies*, January 2012, Washington, DC, USA, pp. 1–6.
91. Asmus, P. and Lawrence, M. (2013). Direct Current Distribution Networks. Technical Report, Navigant Research, Boulder, CO, USA.
92. Miyawaki, S., Itoh, J., and Iwaya, K. (2012). Comparing investigation for a bi-directional isolated DC/DC converter using series voltage compensation. *In Proceedings of the 27th Annual IEEE Power Electronics Conference and Exposition (APEC)*, February 2012, Orlando, FL, USA, pp. 547–554.
93. Peter Asmus Nanogrids vs. Microgrids: Energy Storage a Winner in Both Cases. A website report www.forbes.com/sites/pikeresearch/2015/10/26/nanogrids-vs-microgrids/ (21 October 2015).
94. Nordman, B., Networked Local Power Distribution with Nanogrids. A website report, Lawrence Berkeley National Laboratory https://eta-intranet.lbl.gov/sites/default/files/microgridMAY2013.pdf (29 April 2013).
95. Burmester, D., Rayudu, R., Seah, W., and Akinyele, D. (2016). A review of nanogrid topologies and technologies. *Renewable and Sustainable Energy Reviews*, 67, 760–775. doi: 10.1016/j.rser.2016.09.073.
96. Burmester, D. (2018). Nanogrid topology, control and interactions in a microgrid structure, Ph. D. Thesis in Computer Science, Victoria University of Wellington, New Zealand. https://researcharchive.vuw.ac.nz/xmlui/bitstream/handle/10063/.../thesis_access.pdf?
97. Kim, S.-Y. and Mathews, J.A., Korea's Greening Strategy: The Role of Smart Microgrids. A website report in The Asia Pacific Journal www.microgridresources.com/about-microgrids97/what-is-a-microgrid (December 2016).
98. Ton, D.T. and Smith, M.A. (2012). The U.S. Department of Energy's Microgrid Initiative, The Electricity Journal, Elsevier Inc. doi: 10.1016/j.tej.2012.09.013, www.energy.gov/.../The%20US%20Department%20of%20Energy's%20Micro...
99. Microgrid, Wikipedia, The free encyclopedia, last visited 12 August 2019 (2019).

100. Witter, B., Blue Pillar Enables Next-Gen Microgrids with Energy IoT; Part I. A website report blog.bluepillar.com/blue-pillar-enables-next-gen-microgrids-with-energy-i... (12 September 2017).
101. John, J.S., Spirae Launches the Operating System for Distributed Energy... A website report www.greentechmedia.com/.../spirae-launches-the-operating-system-f... (30 October 2014).
102. IEEE Smart Village Launches SunBlazer IV. A website report https://smartvillage.ieee.org/news/ieee-smart-village-launches-sunblazer-iv/ (11 March 2019).
103. Goodier, R., Microgrids where the Big Grids Don't Go: ASME. A website report www.asme.org/topics-resources/content/microgrids-big-grids-dont-go (29 March 2019).
104. Islam, M.R. and Gabbar, H.A. (2014). Study of small modular reactors in modern microgrids. *International Transactions on Electrical Engineering Systems*, 25(9), 1943–1951. doi: 10.1002/etep.1945.
105. ABB to Provide Woodside with Battery Storage System: World Oil. A website report www.worldoil.com/.../abb-to-provide-woodside-with-battery-storage-system (12 December 2017).
106. Hagh, M.T. and Aghdam, F.H. (2016). Smart hybrid nanogrids using modular multiport power electronic interface. *2016 IEEE Innovative Smart Grid Technologies: Asia (ISGT-Asia)*, 28 November–1 December, IEEE, Melbourne, VIC, Australia. doi: 10.1109/ISGT-Asia.2016.7796456.
107. Searcy, S.W., Hartley, B.E., and Thomasson, J.A. (2014). Evaluation of a modular system for low-cost transport and storage of herbaceous biomass. *Bioenergy Research*, 7, 824. doi: 10.1007/s12155-014-9427-7.
108. Löfroth, C., Larsson, L., Enström, J., Cider, L., Svenson, G., Aurell, J., Johansson, A., and Asp, T., ETT: A Modular System for Forest Transport, a Three-Year Roundwood Haulage Test in Sweden. A website report https://pdfs.semanticscholar.org/3667/a1aa8e03d54abc3f146775a9d4b0fcc2db46.pdf (2006).
109. Löfroth, C. and Svenson, G., ETT: Modular System for Timber Transport: Skogforsk. A website report www.skogforsk.se/cd_48e588/.../ett---modular-system-for-timber-transport.pdf (2006).
110. Lena Larsson Increased Weight and Dimensions for Transport of Timber Future Weights and Dimensions on Heavy Commercial Vehicles, Elmia Jönköping. A website report www.nvfnorden.org/lisalib/getfile.aspx?itemid=4013 (25 August 2010).
111. Modular Fuel System (MFS). A website report https://fas.org/man/dod-101/sys/land/wsh2013/244.pdf (2013).
112. Modular Fuel System (MFS) | Leonardo DRS. A website report www.leonardodrs.com/products-and-services/modular-fuel-system-mfs/ (2013).
113. Modular Fuel System Flexible, Highly... Leonardo DRS. A website report www.leonardodrs.com/media/6278/mfs_datasheet.pdf (2013).
114. iTrak Intelligent Track Systems | Rockwell Automation. A website report www.rockwellautomation.com/en_NA/.../overview.page?...iTrak...System (2015).
115. iTRAK Modular Transport System Increases Flexibility and Drives... A website report mepca-engineering.com/itrak-modular-transport-system-increases-flexibility-and-driv... (2 June 2015).
116. ISO Container Loading Platforms | Carbis Solutions Fall Protection. A website report www.carbissolutions.com/product/iso-container-loading-platforms/ (2012).
117. ISO Tanks, Containers & Intermodal Equipment | CS Leasing. A website report www.cslintermodal.com/equipment (2012).
118. Bondre, J. (2004). A complete NUHMOS solution for storage and transport of high burn up spent fuel. *14th International Symposium on the Packaging and Transportation of Radioactive Materials (PATRAM 2004)*, 20–24 September, Berlin, Germany. www.iaea.org/inis/collection/NCLCollectionStore/_Public/37/088/37088556.pdf.

119. LightSail Energy Launches LightStore(R) Gas Transport Module... A website report https://sports.yahoo.com/news/lightsail-energy-launches-lightstore-r-200000203.html (11 October 2016).
120. LightSail Energy Launches LightStore® Gas Transport Module Targeting Breakthrough Economics in Natural Gas Transportation. A website report, Berkeley, CA (Marketwired) www.efe.com › English edition › News release (11 October 2016).
121. ABB Gas Compressor Modules Reduce Costs with Digital Gas... A website report https://library.e.abb.com/.../Gas%20Compressor%20Modules%20Data%20Sheet.pdf (2017).
122. CNG carrier, Wikipedia, The free encyclopedia, last visited 24 June 2019 (2019).
123. Stenning, D., Coselle CNG: Economics and Opportunities Develop Stranded Gas... A website report - The New Way to Ship Matural Gas by Sea www.ivt.ntnu.no/ept/fag/tep4215/innhold/LNG%20Conferences/.../Stenning.pdf (2019).
124. Wagner, J.V. and van Wagensveld, S. (2002). Marine Transportation of Compressed Natural Gas A Viable Alternative to Pipeline or LNG. doi: 10.2118/77925-MS.
125. Coselle CNG System: Wärtsilä. A website report www.wartsila.com/encyclopedia/term/coselle-cng-system (2019).
126. White, C.N. and Dunlop, J.P. (2005). VOTRANS CNG Provides Transport Solutions for Deepwater Associated Gas. doi: 10.4043/17492-MS.
127. White, C.N., Britton, P.S., Terada, Y., Doi, N., and Murata, M. (2005). Technical advancements: VOTRANS large-scale CNG marine transport, International Society of Offshore and Polar Engineers. *The Fifteenth International Offshore and Polar Engineering Conference*, 19–24 June, Seoul, Korea.
128. Claudio Alimonti Dipartimento di Ingegneria Chimica, dei Materiali, delle Materie Prime e Metallurgia Università degli Studi di Roma 'La Sapienza' Roma, Italy, Transporting natural gas by sea Encyclopedia of Hydrocarbons, pp. 855–878. A website report, Hydrocarbon gas transport and storage, VOLUME I/EXPLORATION, PRODUCTION AND TRANSPORT Transporting natural gas by sea – Treccani www.treccani.it/export/sites/default/Portale/sito/altre.../pag855-878Ing3.pdf (2004).
129. LNG ISO Containers: Corban Energy Group. A website report www.corbanenergygroup.com/lng-iso-containers/ (2019).
130. Atilhan, M., Aparicio, S., Benyahia, F., and Deniz, E. (2012). *Natural Gas Hydrates*. doi: 10.5772/38301.
131. Gudmundsson, J.S. and Børrehaug, A. (1996). Frozen hydrate for transport of natural gas. *2nd International Conference on Natural Gas Hydrates*, 2–6 June, Toulouse, France. www.ipt.ntnu.no/ngh/library/paper3.html.
132. Gudmundsson, J., Mork, M., and Graff, D. (2002). Hydrates non-pipeline technology. *Fourth International Conference on Gas Hydrates*, 19–23 May, Yokohama, Japan. A website report citeseerx.ist.psu.edu/viewdoc/download?doi=10.1.1.562.5324&rep=repl...
133. X-STORE CNG Type 4 Storage System: Hexagon Purus. A website report www.hexagonxperion.com/.../transport-modules.../transport-modules-for-comp... (2011).
134. Compressed hydrogen tube trailers, Wikipedia, The free encyclopedia, last visited 23 September 2018 (2018).
135. Hydrogen Tube Trailers | Department of Energy. A website report www.energy.gov/eere/fuelcells/hydrogen-tube-trailers (2017).
136. Liquid Hydrogen Delivery. Liquid Hydrogen Delivery | Department of Energy, Division on Energy Efficiency and Renewables, Washington, DC. A website report www.energy.gov/eere/fuelcells/liquid-hydrogen-delivery (2015).
137. Shah, Y.T. (2017). *Chemical Energy from Natural and Synthetic Gas*. CRC Press, Taylor and Francis Group, New York.
138. Shah, Y.T. (2019). *Modular Systems for Energy and Fuel Recovery and Conversion*. CRC Press, Taylor and Francis Group, New York.

10 Modular Approach for Simulation, Modeling, and Design of Energy Systems

10.1 INTRODUCTION

In the previous volume on modular systems [1] and in the previous nine chapters, we examined modular systems for energy and fuel recovery and conversion (front end of energy industry) and energy usage management (back end of the energy industry). We demonstrated that the use of modular approach is not only advantageous, but also commonly accepted as an industrial practice across the entire spectrum of energy industry. The use of modular approach for energy systems will not only expand and penetrate further deeper in the energy industry, but will also change how the energy industry will be managed in the future. Just like many other industries, building, vehicle, and computer, the management of energy industry will be altered with the infusion of modular systems to make the industry more flexible, innovation infusion ready, and cost effective [2–42].

In this last chapter, we briefly examine the use of modular approach for the simulation, modeling, optimization, and design of energy systems. Unlike many other systems, energy systems are very complex with multiple suppliers, users, stakeholders, and endgames. Simulation, modeling, optimization, and design of energy systems require good understanding of their end purpose. For example, simulation of an equipment or process generally involves performance prediction and optimization. On the other hand, simulation of economics and social effects of energy systems require understanding of politics, social issues, global economics, and multi-sector dynamics. While the former modeling can be more scientific, the latter modeling will be more statistics and probability based. In both cases, simulation, modeling, and optimization can be done with different mathematical principles and algorithms. In some cases, simulation process can also be standardized to make modular effort simpler and cost effective. Process-based scientific modeling is generally carried out bottom-up while policy based socio-economic model is generally carried out top-down.

In the scientific community, thus two contrasting streams of modeling techniques for analyzing energy system-related issues have been developed: the bottom-up, sectoral, or engineering approach and the top-down, macroeconomic approach [2,3]. The sectoral approach aims at developing bottom-up models through descriptions of technologic aspects of the energy system. In contrast, the traditional macroeconomic approach is to describe the economy as a whole and to emphasize the possibilities to substitute different production factors in order to optimize social welfare. The two approaches differ considerably in their identification of the relevant system and may therefore produce different guidance for policymakers. Comprehensive evaluations of transitions in energy systems are subject to a central obstacle: When modeling the system with much technical precision and details, i.e., bottom-up, the changes in, and feedback from, macrovariables of the broader economic system are not accounted for. However, when modeling from a top-down perspective to describe the macrovariables well, technical details are abstracted away, and the model cannot be understood anymore as a consistent representation of the (technical) energy system.

However, as tools to support decision-making and policy development, these modeling approaches complement each other. Therefore, linking bottom-up sectoral (engineering) models with (macroeconomic) top-down models can be an important contribution for designing energy systems compatible with sustainable economic growth. Hybrid modeling is used to indicate a mix of both modeling perspectives with the expectation that "the whole" should exceed the sum of its parts: integrating aspects and functionality from top-down and bottom-up modeling approaches results in "hybrid" models, which may provide more insight than the individual models could on their own. There are two types of linking. Soft-linking models means that the "user" controls the data exchange between the models. Hard-linking models means that the data exchange is automated. Integrated, hybrid models combine perspectives and characteristics, modeling functionality, for both bottom-up and top-down approaches.

The idea of hybrid modeling attempts to close the gap between the two approaches and to provide a framework that allows us to evaluate overall economic developments with a sound engineering basis. A first step in hybrid modeling is soft-linking; i.e., models are run separately and the exchange of data is controlled manually. Hard-linking refers to automated data processing and exchange. Lastly, model integration is the strongest degree of model-linking; i.e., the models are united into a single framework and solved as a whole, either in a single common run or using a decomposition algorithm. The advantages of soft-linking can be summarized as practicality, transparency, and learning, while the advantages of hard-linking are efficiency, scalability, and control. Challenges when linking or integrating models concern the negotiation of common or overlapping input data and results and scalability issues (especially relevant for complementarity problems). Applying a decomposition approach may prove an efficient solution in such cases. Furthermore, the integration of models complicates the general handling, such as the model maintenance, which is why the degree of linking between models should be a conscious decision by model developers and users. The chapter shows that in the ultimate case of large-scale energy model, such as NEMS (National Energy Management System) model developed by Energy Information Administration of the U.S. Department of Energy, is a linked modeling system consisting of several demand, supply, and market modules.

In this chapter, with the help of case studies, we examine the role of modular approach for bottom-up modeling, top-down modeling, and hybrid modeling with clear emphasis on the usefulness of modular approach for each type of modeling. We also briefly review the literature on different types of modular simulation, modeling, optimization, and design carried out over the years. We will demonstrate how modular approach can facilitate not only simulation and modeling of energy systems but also design of new systems. As shown before, the use of modular approach makes the simulation, modeling, optimization, and design of energy systems more flexible, cost effective, and adaptable for new innovations in energy technology and political–socio-economic conditions [2–43].

10.2 SCIENTIFIC MODULAR SIMULATION AND MODELING (BOTTOM-UP APPROACH)

An individual system like a product (e.g., gas turbine, fuel cell, etc.) or process (e.g., combustion, gasification, etc.) can be technically analyzed and optimized, generally by scientists and engineers, for its performance, efficiency, and cost. These are localized analyses with no considerations of its global economic, environmental, or social effects. This type of localized scientific analysis is aimed toward understanding and optimizing the performance of the particular energy system. Modular approach has been widely used for scientific simulation, modeling, and optimization of simple and complex energy systems. The module, a building block of any system can be separated from the overall system for independent analysis and coding. Individual modules can be modeled in any level of detail, from lumped parameter to a detailed finite element level. Numerous simple and complex mathematical techniques can be used. The modules link together through predetermined parameters requiring a module to accept and generate certain inputs and outputs. Given sufficient flexibility in the connection, interaction between a finite difference or finite element model is possible. Through a graphic interface, an overall system model may be developed from a library of modules without need for extensive coding or debugging. Individual refinement from in-depth analysis and formulation of various modules can take place offline and can be focused specifically on the issue at hand without regard to managing its integration in the greater model.

In a modular method, a set of equations, including differential and algebraic, are divided into subsets. Those subsets are the so-called modules of the model. In order to illustrate basic conceptual understanding of scientific modular modeling, here we take a compressor of a gas turbine as a case study. Gas turbine or a micro gas turbine is a very important energy system for power generation across the entire energy industry, and the compressor plays a vital role in its performance. The differential equation representing the motor in a compressor that calculates the motor magnetic torque can be lumped together as a motor module. It takes the applied torque as input and calculates the motor speed and power consumption as output. Once this is running, a more advanced module that considers, e.g., voltage and torque variation and temperature may be developed offline. In general, modules can be regarded as subroutines or data files in a complete computer program. Division of the system into subsets can be arbitrary, but a module generally has three characteristics.

First, it represents either a particular part, a mechanism, or an aspect of the system. For example, for a compressor, bearing modules represent the physical parts of bearings. The compression chamber represents more a process than a physical part, because it depends on interactions with other modules to define volume, pressure, mass, heat transfer, etc. The *geometry module* represents the kinematics of a particular compressor design, generating information about chamber geometry, port states, loads and load transfer, etc. Second, a module should be as independent as possible. It can run with minimum specified inputs and is truly a subprogram or "mini model" that is dedicated to modeling a particular aspect. Third, it can be reconfigured through data input to adapt to any specific model. For example, the same bearing module can be used in reciprocating, rotary, scroll, or any other compressor type simply by specifying individual geometry. To maintain flexibility and clarity for the users, the classification of subsets needs special attention. The present approach divides modules into several types and standardizes the connections among those modules.

A simple model can be started with a only a few modules and gradually expanded and refined by adding more modules. Different levels of complexity would satisfy different needs of analysis. For instance, a basic model to study valve action in a compressor may not require a motor or bearings, so those modules may be excluded: A refinement to study detailed fluid flow losses may then add flow passage modules between the suction line and suction valve. This can all be accomplished with a minimum number of modules for the rest of the system or by working with an existing model without modification of the rest of the model except in areas of interest. Several modules that accomplish a certain function can be further grouped together into a supermodule. A basic example is a *compressor main body*, a supermodule that consists of the geometry, chamber, port, valve, bearing, and other modules that make up a fundamental compression device, such as the mechanism of a single cylinder of a reciprocating compressor. This allows more complicated assemblies, such as multi-cylinder compressors, to be constructed without generating each cylinder separately. Thus users at different levels of understanding or with different design expertise can work with the same set of modules. An expert in bearings can work on the bearing modules without looking into the details of the compression main body supermodule, while an engineer or designer can go into the compression main body supermodule to investigate the effects of each individual design parameter on the efficiency and capacity of the whole system.

10.2.1 Module Types

For a compressor, the modules can be divided into five types: fluid flow, mechanical, electromechanical, compression chamber, and geometry. The compressor main body is a supermodule, in which generic modules are assembled together in a specific way so that a specific type of compression process, such as reciprocating, rotary, scroll, screw, or other, is represented. It contains the information of geometry configuration, compression process (compression chamber), flow and leakage paths, and the load characteristics specific for a particular type of compressor. The fluid flow modules include valves, flow passages (channels or pipes), and suction and discharge ports and lines. The suction and discharge line can be defined as a constant pressure and temperature or specified as a function of time. The suction gas from the gas sources goes to the suction valve though several flow passages and then is fed into the compression chamber in the compressor main body module. The discharge gas coming out of the compression chamber passes though the discharge valve and goes to the other gas source though a gas passage.

The mechanical modules include a shaft and bearings. The shaft module takes the mechanical load and torque information from the compressor main body and distributes the load to the associated bearings and to the motor. The torque balance between motor and the sum of compressor and bearing loads occurs in the shaft module and determines the shaft angular velocity and angular position. The motor module is an electromechanical module. It converts electrical energy into mechanical work and at the same time generates heat, which is rejected into the fluid. The motor and other electronic control mechanisms can be combined into another supermodule where connections with fluid passages or other modules provide pressure and temperature input to the controls.

The compression chamber module converts mechanical energy from the motor into internal energy of the fluid. Therefore, it has both mechanical connections and fluid flow connections as well as heat transfer connections. However, the compression chamber module also needs information about its geometry and behavior depending on the compressor type being simulated. This information is provided by the geometry module, which contains information about the basic compression mechanism geometry and performs the calculations regarding compression chamber behavior and load generation. These include the traditional kinematic calculations regarding the compression mechanism as well as information on fixed port openings and closings.

10.2.2 STANDARD CONNECTIONS

Standard connections are like sockets and plugs, so modules can be connected by "plugging in" to another module with the same type of connection. Standardized connections offer several advantages. First, the modules of the same type can be easily connected without assigning separate data exchanges. Second, modules of the same type can be easily exchanged. For example, a bearing module with a different degree of complexity or detail can be exchanged for another in a model as the focus of the analysis changes. Third, an attempt to connect incompatible modules, such as a mechanical connection between a bearing and gas passage would be rejected by the model. Finally, standard connections make it possible for a user without deep understanding of the underlying computer code to build or modify a model.

10.2.3 CONNECTION TYPES

Major categories of connections are fluid flow, mechanical, electrical, geometry, and heat transfer. A basic fluid flow connection includes mass flow rate, vapor quality, and two thermodynamic parameters that determine the thermodynamic state of the fluid. Mechanical connections can be further divided into two subcategories of linear motion and rotating motion. The former includes acceleration, velocity, position, force, and position of the force. The latter includes angular acceleration, angular velocity, angle, force, position of the force, and torque. Heat transfer connections include heat transfer coefficient, heat transfer area, temperature, and heat transfer rate. While the connection type is related to its module type, a module can have more than one type of connection. The suction and discharge ports have fluid flow connections but might also have geometry connections in fixed port machines such as scroll or screw compressors. Bearings have mechanical connections but can also have thermal connections to account for heat generation and rejection. Thus, adoption of modular techniques takes advantage of the best features of all modeling approaches. Module development, as in the anatomic approach, can focus on the behavior and analysis of individual subsystems without regard to their interaction with the rest of the compressor. However, the model, once constructed of linked modules, takes full advantage of interaction analysis as provided by the system approach. In addition, the prepackaged nature of the modules and preconfigured nature of the connections draw effort away from repetitive individual analysis of components and redirect the focus to details of the compressor design and interaction, which is after all the ultimate interest of the system designer.

In general, an energy system generally tends to be a process with various modules interconnected. Chemical engineers have long thought of process models in terms of the physical systems they represent. Consequently, the most common approach to process- or plant-wide model description is sequential modular. Here, individual module models relate to distinct physical/chemical/biological processes and link together according to flow sheet process topology. The calculation flow is thus rigidly fixed by the flow sheet. This procedure is generally reliable, easy to assemble, and usually robust. However, it often lacks the flexibility to perform design and optimization tasks simultaneously.

On the other hand, the equation-based approach for system or process modeling offers complete flexibility in specifying design constraints, solving optimization problems, and deriving a solution procedure. However, since the entire equation set bears little resemblance to the process or system flow sheet, much more work is required to set up and test the process or entire system model. One can, however, use simultaneous or sequential modular methods to capitalize the advantages of the above approaches. In this case, simultaneous modular seeks the flexibility of equation-based systems while working with "black box" process modules. In a sequential modular method, the results for one module are used in the subsequent analysis. As shown later, these approaches have been demonstrated to be useful in several examples articulated in the literature. Thus, bottom-up modular approach can not only simulate and model but can also simultaneously optimize the system. Since (convex) optimization problems (of the bottom-up engineering models) are a subclass of the complementarity problem class, the cost minimization problem of the energy system models can be incorporated in an extended version of a computable general equilibrium model [5,12].

10.3 STANDARDIZED MODULAR DESIGN— ASPEN PLATFORM AND OTHERS

All industries, including the energy industry, are required to continually invest in new processes, address new markets, innovate, and meet evolving demand for products. This has created a pressure point on the capital and lifecycle cost of assets and on the projects designed to create those assets. Project overruns in the oil, gas, and petrochemical industries have cost impacts that extend over the lifetime of the delivered asset, pressuring companies to deliver on schedule to maximize the profitability

Simulation, Modeling, and Design of Systems

of the completed asset. Owners have become convinced that standardized or modular designs are a big answer to this conundrum. By using standardized designs, whenever and wherever possible, a number of cost and risk issues are addressed. Upstream players such as Chevron, ExxonMobil, and Marathon Oil have expressed this as a key business initiative; and consultants such as EY Consultants, Accenture, and Douglas-Westwood, as a key implementation strategy.

Implementing standardized designs and/or adopting a modular approach to process units reduces design, schedule, and cost uncertainty and, as a result, saves significant amounts of time and money and may potentially help achieve a faster start-up. Owners are trading off the best possible design ("gold-plated design"), which has higher CAPEX for predesigned processes known to fulfill the function and deliver a lower lifecycle cost. Standardized designs improve front-end engineering efficiency and execution, helping to get projects to the construction phase more quickly and with a lower engineering cost. Additionally, standardized designs improve construction management by increasing the proportion of fabrication work performed in the shop versus in the field, especially when combined with modularization and simplified construction management. With the Aspen ONE-integrated engineering workflow, based on model-based applications, process designs, and cost estimates, a "best practices" reference design can be created, complete with engineering documentation and ready for reuse in a modular fashion to rapidly complete similar conceptual engineering projects. This includes quickly and reliably modifying the process based on varying locations, applications, and scale, thereby reducing engineering, construction, and cost risk.

Breaking the habit of reinventing solutions associated with traditional engineering methods and organizational structures is difficult, organizationally. Engineers embrace creativity and continuous improvement, with their training and experience telling them to optimize the design for the production objectives, feedstock, and location, biasing them toward one-of-a-kind designs. However, some EPCs (engineering, procurement, and construction) have seen the inexorable trend and have begun to change their engineering processes by using standardized design as a potential strategic advantage. Some examples are Fluor's "Third Generation Modular Design" solutions, Chart Industries's small packaged liquefied natural gas (LNG) plants, and Technip's standardized design corporate strategy. The changes that are required are on the one hand organizational, requiring engineering disciplines to collaborate on definition of standardized designs, but they also must be coupled with an evolution of the underlying software tools: the right set of modular, model-based design, workflow-supporting integration tools that can go a long way toward enabling and promoting change. When the right tools are available, the engineers can focus on the value-added engineering project at hand.

When an organization is dependent on traditional tools to create templates, such as Microsoft Excel spreadsheets, they can be limited to scaling or factoring them to different project parameters, throughputs, or locations, as well as enumerating their project schedules, costs, and risks. That approach is sufficient when working in known locations and sizing, but it is hampered by many limitations when planning a project that is significantly larger or smaller than existing designs or is in a new location. It does not promote a rigorous approach to capturing best practices or employ a sophisticated approach of adapting a standardized approach to a new situation. It also introduces high maintenance requirements for the customized Excel spreadsheets, which end up being dependent on a few individuals.

In contrast, the modeling tools from Aspen Tech can capture both process units and entire designs as templates, showing that the reference design is easily usable as the starting point for the next project. And, by extension, the entire integrated ("activated") workflow supports capturing those multiple models (across disciplines), such as heat exchanger and energy models, as aspects of these standardized design templates. Having the right tools available supports the strategy in which the engineer can consider a plant as being composed of modules—some requiring previously designed and used modules, while others are individually designed for the project. Samsung Heavy Industries, e.g., has demonstrated that by using this approach for floating liquid natural gas (FLNG) projects, EPCs can reduce front end engineering (FEED) costs by 50% or more and EPC delivery can be significantly expedited and reduced by 10% or more.

Project design for upstream and midstream oil and gas is the first key area to embrace a modular approach and reuse standardized design modules. Many oil and gas companies, who have in the past designed and built customized projects to specific locations and hydrocarbon characteristics and the nature of the existing infrastructure, are exploring and driving forward standardization. Reusing existing templates for repeatable process units helps reduce risk of the work, because many of these units are fundamentally the same from project to project. This includes dehydration, NGL separation, and gas purification but also includes more specific equipment modules such as large compressor systems, acid gas removal strippers, and subsea modules. The precedent is the process licensor workflow and business model in which the capture of designs in software has proven to be highly successful by licensors, such as UOP, DuPont Clean Fuels, Technip Stone and Webster Process Technology, and Bechtel's LNG groups.

The concept of off-site fabrication and modularization in engineering and design can be scalable from small to large projects, such as floating production, storage and offloading vessels, which are scaled to the oil and gas flow characteristics and the size of a particular producing formation. Compressor modules can be standardized because the same equipment design and layout can be reused in various settings, accounting for throughput, contaminants, viscosity, weight and size limits, and other factors and verified using dynamic models. For larger facilities, such as LNG plants, the focus moves to replicating modules that make up the plant. The key risks that need to be traded off against stick-built and one-of-a kind design are the logistics involved in ship or land transport from the fabrication site and the lead times required to assemble modules in fabrication yards. This is not only a logistical and construction cost trade-off; it also relates to factors such as safety risk in remote construction workforces, engineering quality achievable in fabrication yards, and the availability of peak on-site workforces.

Many companies have successfully adopted modular standardization to apply common design specifications and guidelines for process units across projects (i.e., a refinery or production platform). Dow Chemical Company, for instance, has discussed in public forums their use of standardized engineering reference designs for certain process equipment and units. The use of libraries containing design templates, which include data sheets, equipment, and line lists, is a powerful way of avoiding unnecessary duplication of data entry and copying, helping to minimize engineering time and reduce costly overruns. A key to this strategy is aligning the engineering stages from conceptual design through basic engineering to detailed design. Collaboration across the project teams is essential to leverage important documentation.

Off-site modular assembly is becoming a strong alternative for construction in process plant development. This highly efficient process alleviates the challenges typically associated with tight project schedules, changing site conditions and the availability of skilled field labor, while minimizing variability in the quality of the finished product. The safe and correct assembly of equipment, such as columns and reboilers, is critical to performance and reliability. Units derived from fabrication workshops, such as steel casings, stacks and ducts, burners and piping, can be preassembled for shipping anywhere around the world, and modular construction can be more easily executed with available on-site skills. Safety is often a driving concern, especially in remote or dangerous areas (such as the arctic).

As modular design and construction projects become more prevalent, powerful, and integrated, engineering tools can help engineers complete data sheets faster and communicate with all stakeholders working on the project. Many engineering and construction (E&Cs) have standardized on the Aspen ONE Engineering software suite by Aspen Tech, which contains process modeling analysis and design tools that are integrated and accessible through process simulators. Engineers can optimize process designs for energy use, capital and operating costs, and product yield through the use of activated energy, economics, and equipment design during the modeling process. E&Cs continually seek ways to improve workflow and streamline processes. Aspen HYSYS and Aspen Plus are the tools of choice for engineers using a modular approach to design. Process units targeted for reuse can be captured as templates, kept in an organized library and quickly accessed when a design is being modeled. The tool helps deliver faster project execution, meeting increasing demands, and minimizing performance degradation, while complying to strict environmental and product quality standards.

In addition to process modeling, concurrent development of accurate CAPEX estimates is a key advantage of standardized design. To achieve that, it is imperative to deliver accurate cost estimation early in the concept design and basic design stages. Implementing standard practices and methods enterprise-wide ensures design quality, reduces maintenance costs, and meets safety compliance. Costs can be equivalently captured in Aspen Capital Cost Estimator (ACCE) as template costs. These can be easily accessed during process design via the Activated Economics workflow from the process modeling tools, which transfers the scope from process engineering teams to estimating teams. ACCE is a powerful tool for evaluating the efficacy of modules for projects. Its modeling approach enables costs to be evaluated based on the specific sizing and location parameters of a new instance of a standardized design. The software provides estimators with an early look at resource constraints, such as craft, labor, and fabrication equipment, and then enables them to easily evaluate and quickly shop versus field fabrication, including a whole host of trade-off scenarios.

It is also important to capture design knowledge to improve the ability of less experienced engineers in delivering high-quality designs. Aspen Basic Engineering (ABE) is an industry-leading process engineering solution that enables global organizations to seamlessly and accurately bring together and template all aspects of front-end engineering design and basic engineering. Now it is possible to achieve a huge competitive advantage by delivering process data packages for licensed technologies and other repeatable designs in half the time that's currently required. Through capturing process technologies and best practice designs in reusable templates, engineers can apply them repeatedly in future projects for dramatic time savings. In addition, time-consuming data sheet and equipment list development can be automated.

10.3.1 Customer Success: Sadara and DSM

A prominent example of the effective application of modularization is Sadara Chemical Company's Petrochemical Complex Green field project. During prefeed, the company used Aspen Capital Cost Estimator to identify an unattainable large temporary workforce requirement at the project location in Saudi Arabia, which was based on the resource loading predicted by the model-based estimating methodology. At that point, modular construction of many process units was compared with on-site stick-built construction, and it was determined that the work force requirement could be reduced to feasible levels at a reasonable cost trade-off. That project was nearing startup in 2016 and is tracking close to the original FEED estimate.

Another interesting example of modularization is the global science-based company, DSM. The company has entered a mode of rapid introduction of new health and science products. To support that business strategy, the engineers at DSM developed concept designs and costs for reusable process building-block modules, amenable to the style of batch manufacturing that supports this new generation of products. Using these reusable building blocks, process engineers can support the business by rapid development of conceptual designs with associated lifecycle-cost estimates, enabling better commercialized decision-making [1].

In summary, with business leaders seeking lower capital project investments, while still driving toward business growth, standardized designs and modularization are two engineering strategies that respond to those business pressures. Standardized designs reduce project risk and therefore project cost in multiple ways. Modularization increases project management efficiency and presents opportunities for trade-offs between on-site fabrication and shop modular fabrication. When modular construction is considered, lead times can be improved, helping the shop fabricator efficiently fabricate and then ship. Therefore, early and accurate conceptual design becomes even more important when trying to achieve fast-tracked designs.

A key for both strategies is more focus on front-end design and the supporting engineering workflows. When better engineering tools are put in place, an organization can evolve toward a more integrated and collaborative workflow in which standardized designs can be used to achieve lower costs and, in many cases, more successful projects. Using these standardized designs, it can reduce the time spent on repetitive design tasks and increase the time available for value-added engineering work, such as being able to optimize the design for life cycles capex and opex, reduce energy consumption, and increase sustainability. Aspen ONE Engineering is an ideal platform to achieve this, with a focus on Aspen HYSYS, Aspen Plus, ACCE, and ABE.

Standardized modular design gives EPCs the opportunity to gain a competitive position and take advantage of the unique characteristics of integrated engineering modeling and analysis software tools. This supports the concept of repeatable designs, which saves time when re-entering data and enables the optimization of a design across the feasibility study, conceptual engineering, and front end engineering and design (FEED) workflows. The software tools also help knowledge sharing across the organization and allow efficient access for project delivery teams to streamline and deliver accurate engineering solutions that meet deadlines. In essence, modularization expedites project execution by compressing project schedules and integrating global design teams for faster on-time delivery. Companies that are embracing this approach are seeing significant success in the marketplace.

10.4 MODULAR GLOBAL (TECHNO-ECONOMETRIC/SOCIAL) ENERGY MODELS (TOP-DOWN APPROACH)

Unlike technically and scientifically oriented energy system analysis outlined above, global energy models are used to project the future energy demand and supply of a country or a region in a multi-sector arena. They are mostly used in an exploratory manner assuming certain developments of boundary conditions such as the development of economic activities, demographic development, or energy prices on world markets. They can also be used to evaluate the effects of energy technologies on environmental issues such as carbon release and resulting global warming effects. They are also used to simulate policy and technology choices that may influence future energy demand and supply and hence investments in energy systems, including energy efficiency policies.

Policy and technology choices sometimes induce a dilemma in the choice of global energy model [5]. Detailed techno-economic (or process-oriented) models were first developed in the early 1970s, particularly after the first oil crisis in 1973, when analysts started to examine the options of oil use and the more efficient use of final energies. Modern macroeconomic energy models have their origin in the late 1950s, when energy supply companies and energy administrations had to make decisions about the future energy supply to meet the rising energy demand.

Global energy system econometric/social models represent a more or less simplified picture of the real energy system, its effect on environment and the real economy; at best they provide a good approximation of today's reality. Nevertheless, it would be impossible to answer very specific questions on energy technologies or economic implications without making some cutbacks and approximations, with an uncertain reliability on quantitative figures used by those models. A large diversity of modeling approaches has been developed over time depending on their target group, intended use, regional coverage, conceptual framework, and the information available. Some of these are outlined in subsequent sections. Even though they are often regarded as "macroeconomic" models, computable general equilibrium (CGE) models are based on a microeconomic framework: Consumers demand goods in order to maximize utility, and producers supply goods under profit maximization. The first successful implementation of a CGE model was made in 1960 by Leif Johansen [8].

Most contemporary CGE models belong to the class of complementarity problems. This class includes not only various non-linear and non-convex problems but also the simpler linear and quadratic programming problems. As such, a linear or quadratic program can be cast as a complementarity problem and thereby easily integrated in another complementarity problem. The TIMES model [2] is a generic model for the energy sector developed through an International Energy Agency (IEA) implementing agreement, the Energy Technology Systems Analysis Program (ETSAP). The generic model is tailored by the input data to represent a specific region, ranging from global models down to single-city models. The planning period is usually 20–50 years. The predecessor of TIMES is the MARKAL model. MARKAL is used for the IEA's Energy Technology Perspectives reports (ETP)—published every second year. Much of the model structure is similar between MARKAL and TIMES, but TIMES allows the user to define more flexible time periods [4–16].

The MESSAGE model is another energy systems model, developed at the International Institute for Applied Systems Analysis (IIASA). It is part of IIASA's Integrated Assessment Scenario Analysis Framework. MESSAGE is a time-dependent linear programming model which provides an optimal allocation of fuels and energy carriers to meet a given demand. MESSAGE has a reference energy system (RES) that represents the most important energy carriers and conversion technologies. Energy demands are exogenous to the model. The general model characteristics are therefore very similar to TIMES and MARKAL [16–35].

10.4.1 TOP-DOWN VERSUS BOTTOM-UP APPROACHES

One major advantage of top-down energy models is their application of feedback loops to welfare, employment, and economic growth. This endogenous assessment of economic and societal effects results in higher consistency and facilitates a comprehensive understanding of energy policy impacts on the economy of a country or region. On the other hand, top-down models suffer from the lack of technological detail and deliver rather generalized information. Consequently, they might not be able to give an appropriate indication of technological progress, non-monetary barriers to energy efficiency, or specific policies for certain technologies or branches. Especially in the long run, when substantial technological change, saturation, and intrasectoral structural change can be expected and has to be included in a plausible model, top-down models are not suited to show in a transparent manner credible technology futures.

Furthermore, driven by the assumption of efficiently allocating markets, top-down modeling approaches tend to underestimate the complexity of obstacles and their non-monetary form like lack of knowledge, inadequate decision routines, or group-specific interests of technology producers or of whole sales. CGE models [44–53] assume that any policy implies additional cost, although highly profitable (but unrealized) investments in energy efficiency may reduce cost and increase profits and tax income. Transaction costs are only implicitly covered and cannot be changed by relevant policies such as technical standards or energy efficiency networks. Finally, as they are focused on monetary terms, they consequently tend to favor monetarily related policies.

In contrast to top-down macroeconomic modeling, bottom-up modeling approaches incorporate a high degree of technological detail which enables them to present very detailed pictures of energy demand and energy supply technologies, as well as plausible technology futures. Bottom-up models can also give detailed evaluations of sector- or technology-specific policies [38]. However, this high degree of detail means that bottom-up modelers are heavily dependent on data availability and credibility with regard to their many assumptions on technology diffusion, investments and operating cost. There are also criticisms of bottom-up modeling concerning the neglect of program costs, the feedback of energy policies as well as the lack of macroeffects of the presumed technological change on overall economic activity, structural changes, employment, and prices.

The imperfections of both bottom-up and top-down models outlined above have led to the development of hybrid energy system models, which integrate bottom-up and top-down models in a variety of ways to seek the desired answers from the overall modeling effort.

10.5 HYBRID GLOBAL ENERGY SYSTEM MODELS

Obviously, both top-down and bottom-up techno-economic models have specific advantages and limitations, of which modelers, users of the results, and policymakers are, however, often not sufficiently aware. As outlined earlier, bottom-up models are generally constructed and used by engineers, natural scientists, and energy supply companies, whereas top-down models tend to be developed and used by economists and public administrations. The understanding of the two approaches has increased substantially over the last decade [3,11,12], and this has resulted in the development of hybrid global energy system models which combine both top-down and bottom-up approaches [3,38,40–42]. These hybrid models have shaped the recent discussions about greenhouse gas emission targets (and target sharing), phasing out nuclear energy after Fukushima, and the speed of introducing renewable energies and realizing energy efficiency potentials.

There was insufficient understanding of the strengths, weaknesses, and limitations of the top-down and bottom-up models used by the different disciplinary communities, but with the increasing interdisciplinary competence of energy research teams over the last decade, there has also been a growing demand to combine the two approaches and, hence, their advantages. To overcome the above-mentioned weaknesses and limitations of conventional top-down and bottom-up energy models, energy modeling is currently moving in the direction of hybrid energy system modeling combining at least one macroeconomic model with at least one set of bottom-up models for each final energy sector and the conversion sector. According to Hourcade et al. [3] and Bataille [36], a high-quality hybrid model system should incorporate at least three properties: (a) technological explicitness, (b) microeconomic realism, and (c) macroeconomic completeness. Top-down modeling on its own provides energy modelers with a high degree of macroeconomic completeness through the feedback loops for economy, welfare, etc. combined with microeconomic realism, e.g., the decision- making processes of the different agents, etc. Pure bottom-up modeling, on the other hand, offers a high level of technological explicitness and a low level of macroeconomic completeness. To address these issues, as shown in next several sections, several methods for linking bottom-up and top-down models have been investigated. As shown, many of these linkages are tailored toward the desired objectives and usage of the model results. Many new hybrid models are constantly being evolved to satisfy emerging needs of various nations and global environmental and economical priorities.

10.6 HYBRID MODELS WITH SOFT LINKS

Merging three properties mentioned above into one hybrid system can take place in several different ways. The simplest form of linking top-down and bottom-up approaches is called "soft-linking," which is the manual transfer of data, parameters, and coefficients. A linking of models is achieved through iterations with information exchange between the models. The first example of linked energy-economy models was reported by Hoffman and Jorgenson [14]. They linked the Brookhaven Energy System Optimization Model (BESOM) with a general equilibrium model and, later, with an input–output model. During the following decades, several studies have linked economic and systems engineering models; initially these links were "soft," i.e., the information transfer between the models was directly controlled by the user. Soft-linking is a necessary starting point in order to test different modeling and linking approaches. By soft-linking, we mean that the user evaluates results from the models and decides if and how the inputs of each model should be modified to bring the two sets of results more in line with each other, i.e., how to make the input assumptions and results of the models consistent with each other, i.e., converge.

The advantages of soft-linking can be summarized as practicality, transparency, and learning. Soft-linking seems the most reasonable starting point for linking models based on different approaches. Initial investments in computer programming are kept low, and the modelers can obtain results for evaluation and learning fairly quickly. Generally, the knowledge achieved from soft-linking results in the formation of modular, autonomous "hard-linking" between the models. As mentioned earlier, the advantages of hard-linking are efficiency, scalability, and control. For the reasons of efficiency, hard-linking is the preferred end product: As the number of model runs increases, and more model users become involved, more resources are needed to retain the quality of soft-linked models than for hard-linked ones. The energy system models TIMES/MARKAL and MESSAGE have frequently been used in model-linking exercises. Most of these examples involve soft-linking. A large proportion of this literature describes applications for specific geographical regions. Based on the report by Holz et al. [35], in the following, we briefly present some of examples.

10.6.1 MARKAL–MSG

MSG is a CGE model of the Norwegian economy developed in the early 1990s. MARKAL–MSG is an example of a soft-linked system [9]. This combination can be used in three different ways:

1. Data for useful energy demand from macroeconomic model is soft-linked with energy mix data transferred from energy systems model.

2. Data for demand for energy services and initial electricity production from macroeconomic model is soft-linked with energy mix and electricity price data, respectively, from energy systems model.
3. Data for production and consumption quantities from macroeconomic model is soft-linked with data for technology choices from energy systems model. Tax increase giving higher present value of cost is transformed and allocated to sectors in MSG.

Johnsen and Unander [9] found that the feedback from MARKAL in energy demand in the Norwegian residential sector had little impact on the general economy. Later, Martinsen [18] soft-linked MARKAL and MSG to analyze technology learning. Bjertnæs et al. [17] also soft-linked MARKAL and MSG to analyze the effects on Norway of a national CO_2 tax and international CO_2 quota prices.

10.6.2 MARKAL–EPPA

Schafer and Jacoby [19,20] developed a hybrid CGE-MARKAL model focusing on transport. The CGE model is the Emission Prediction and Policy Analysis (EPPA) model. EPPA works on a macroscale, with a rougher, more aggregate sector classification than MARKAL. Being a CGE, EPPA is constructed on a social accounting matrix basis stated in (monetary) value terms, while MARKAL uses physical flows. A third model, of modal splits, is applied to connect the aggregate transport sector of the EPPA model to the technology detail in MARKAL. Thus, in this soft-linking, prices, taxes, and transport demand from macroeconomic model is soft-linked with adjustment of CES elasticities and autonomous energy efficiency variable (AEEI) in EPPA from energy systems model.

10.6.3 TIMES–EMEC

EMEC [54] is a static CGE model, which has been developed and maintained at the Swedish National Institute of Economic Research (NIER) for over 10 years. EMEC and TIMES-Sweden have been soft-linked by Berg et al. [15]. Some changes have been necessary to make the soft-linking possible. In the production functions, the energy mix is modeled with Leontief functions instead of general CES functions, thereby fixing the proportions of energy based on the TIMES results. Also the household utility from energy to heating is changed to a Leontief representation. The mix is adjusted during the iterations, based on TIMES results. In short, in this linkage, energy demand (converted from monetary units to corresponding mass and energy units) from macroeconomic model is soft-linked with energy mix (to Leontief functions) from energy systems model.

10.6.4 MESSAGE–MACRO

As pointed out by Holz et al. [35], Wene [2] soft-links MESSAGE and MACRO. MACRO receives costs/prices for energy supply from MESSAGE. From these, MESSAGE supplies the quadratic demand functions for MACRO, and the overall energy demand is adjusted. MESSAGE is rerun with these adjusted demands to give adjusted prices. Messner and Schrattenholzer [32] brought the linkage between MESSAGE and MACRO further and introduced a hard-linked system.

10.7 HYBRID MODELS WITH HARD LINKS

If the transfer between bottom-up models and top-down macro models is further evolved using automatic routines, a "hard-link" is established between the different models. **This "hard-link" is usually created in a modular form.** The first example of hard-linking of energy-economy models was reported by Manne and Wene [16]. By hard-linking, we mean that all information processing and transfer is formalized and usually handled by computer programs and transformation modules. In areas where the models overlap, an algorithm may be used to negotiate input data and results. Usually one model is given control over certain results, and the other model is set up to reproduce the same results, typically with a different aggregation level. Some examples of hard-links are given below.

10.7.1 THE DECOMPOSITION METHOD IN THE COMPLEMENTARITY FRAMEWORK

As pointed out by Holz et al. [35], the examples of soft-linking in earlier section all involve energy system models based on an optimization approach (often linear programming). However, the development of specific and powerful solvers for complementarity models (i.e., PATH for GAMS) pushed the development of (mixed) complementarity (MCP) models not only for CGE models but also for bottom-up market equilibrium models. Initially Böhringer and Rutherford [13] developed an integrated MCP model for an energy market equilibrium within a top-down CGE model. However, by addressing real-world problems with large-scale models, a problem of dimensionality, or numerical tractability, quickly arose. Hence, Böhringer and Rutherford [13] presented a decomposition method for the formulation that permits a convenient combination of top-down equilibrium models and bottom-up energy system models. They combined energy demand of macroeconomic model with net energy output and inputs of non-energy goods to the energy system from energy system model by a hard-link. More precisely, they use the complementarity method to solve the top-down general equilibrium model and quadratic programming to solve the bottom-up energy supply model. This formulation remains challenging to implement, and subsequently Abrell and Rausch [22] applied this decomposition algorithm and develop a multi-country multi-sector general equilibrium model for Europe, integrating a high-frequency electricity dispatch and trade decisions.

10.7.2 MARKAL–MACRO

There also exist hard-linking examples involving the MARKAL model. Both MARKAL and MACRO are dynamic and assume perfect foresight. MACRO is an aggregate macroeconomic model for the entire global economy and is solved by non-linear optimization. Input factors for production are capital, labor, and different energy carriers. MACRO transfers energy demand to MARKAL. For this demand level, MARKAL determines the composition of the energy system that minimizes the energy expenditure. This sets up further data exchange between MACRO and MARKAL. Such data exchange is fully automated, so it can be labeled as a hard-linked system. The linking ensures consistency between energy supplies, demands, and prices. Because MACRO includes substitution possibilities between energy, labor, and capital, the solution space increases. As a consequence, final energy prices are lower in MARKAL–MACRO compared to MARKAL stand-alone. MARKAL–MACRO hard-linked model has been used to analyze energy policies in China, Italy, and UK [16–42].

10.7.3 MESSAGE–MACRO

Messner and Schrattenholzer [32] brought a hard-linked system between MESSAGE and MACRO. MACRO defines an intertemporal utility function to be maximized for a representative producer–consumer in each of the eleven world regions for electricity and non-electricity energy. The main variables are production factors such as capital, labor, and energy inputs, which determine the total output of an economy. The optimal quantities of the production factors are determined by their relative prices. Actual demands are determined by MACRO in a way that is consistent with projected GDP. MACRO also disaggregates total production into macroeconomic investment, overall consumption, and energy costs. More recently, Abrell and Rausch [22] also presented a hard-link between electricity demand and price from macroeconomic model and electricity dispatch from energy system model.

10.8 CHALLENGES AND FURTHER PERSPECTIVES ON HYBRID MODELS

The next step for connecting top-down and bottom-up models is to apply partial model elements (top-down or bottom-up) in their modeling counterparts. Catenazzi [38] defines two hybrid energy model systems in this context: "macroeconomic models with bottom-up energy supply models" and "bottom-up models with some limited macroeconomic sub-models." The MARKAL model family is an example of the latter. A similar approach is used by Hourcade et al. [3] who define the following three categories of hybrid energy models: "bottom-up models with macroeconomic feedbacks," "bottom-up models with microeconomic behavioral parameters for technology choices," and "top-down models with more technological explicitness or parameters for endogenous technological change."

A slightly different classification of approaches—based on different model types instead of linking types—is given in Böhringer and Rutherford [12,13]. They define three broad categories of hybrid modeling efforts that aim to combine bottom-up engineering models with top-down economic models:

a. Linking between individual, "equally important," stand-alone models (typically soft-links).

b. Linking where one of the models dominates and is complemented by the other one. The subdominant model is usually implemented in a reduced form. This is a typical model constellation that enables hard-links between the models (although hard-linking can also be implemented between individual stand-alone models).
c. Combining bottom-up and top-down characteristics directly in an integrated model, possibly belonging to a more general model class.

While future is in the development of hybrid models, there are various types of hybrid models. Some combine different modules—which may be optimization or equilibrium models. Often, specialized modules are combined in order to produce projections for different scenario assumptions. One of the presently established hybrid model systems was applied in the ADAM project (Adaptation and Mitigation Strategies), a European energy model [40]. The model system in ADAM combines a macroeconomic model (E3ME) [39], with a set of bottom-up models for the four final energy sectors (industry, services, transport, and the residential sector which was split up into buildings and electrical appliances).

Challenges facing hybrid energy modeling include, among others, the need to keep such combined model systems theoretically consistent and empirically valid without constructing huge models that are incomputable. Additionally, the endogenous consideration of structural change (inter-sectoral as well as intra-sectoral) and technological progress are important issues that require further attention and research. The development of hard-links between process-oriented energy models and macroeconomic models has been difficult. This is due to the disciplinary cultures in which each type of energy model has been developed. Linking the models is mostly limited to a manual transfer of a few major drivers (e.g., population, gross value added of the economic sectors, or energy prices on world markets) and to investment figures (often only from the energy conversion sector of bottom-up models being transferred to macroeconomic models). On top of this, the few existing links are sometimes not implemented in electronically based transformation modules ("hard-links"), but are manually transferred by the researchers and teams involved.

The need to link these two "worlds" is the challenge presently facing energy demand and supply modeling. Analysts have to simulate the projected futures in both types of models in a consistent way, which may induce the need for one or two iterative runs between the two types of models simulating those policy scenarios that deviate substantially from a reference scenario (e.g., a climate change scenario with substantial reductions of greenhouse gas emissions during the next few decades, assuming substantial increases in energy and material efficiency and intensive use of renewable energies). For example, the TRANSFORM module of the ADAM project translates monetary production data of basic goods industries into physical production. It also incorporates improvements in material efficiency and material substitution and saturation by the MATEFF model, a bottom-up model for simulating those effects on basic products like steel, cement, non-ferrous metals, paper, glass, etc. Finally, the IMPULSE module of this hybrid model system collects all data of the bottom-up models on investments, changing operating cost (including energy cost), and the program costs, stemming from the policies assumed in a particular scenario [40,42]. To relate the specific energy demand of basic materials to their physical production and not to monetary production data is very important in order to arrive at realistic and transparent results in energy demand projections of basic product industries.

Additional policy efforts in energy and material efficiency as well as in renewable energies imply a substitution of energy uses by increasing capital investments and related employment. These policies generally induce program costs for governments and industrial associations which can be relevant in real terms—or at least in the political debate. Hybrid models can analyze essential questions for detailed climate change policies reflected in bottom-up models on the one hand and the impacts of those mitigation scenarios on the economy by macroeconomic models on the other hand. The success of hybrid models is ultimately dependent upon the modules created for "hard-link" between microscopic (bottom-up) and macroscopic (top-down) models. The modular approach for these links is once again essential in the successful development of hybrid energy system model [2–43].

In summary, hybrid energy system models help understand the advantages and limitations of the existing bottom-up and top-down energy models and to improve the consultation process of the energy analysts for decision-makers in governments, international institutions (e.g., IEA, UNEP), and large energy supply companies as well as energy technology producers. **While the energy models on both levels (bottom-up and top-down) are further improved by more detailed structures, more empirically based equations and adding multi-agent aspects, the progress of the development of hard-links (in modular forms) of the two modeling levels is of crucial interest.**

In the near future, **transformation modules (i.e., hard-links)** should intensify the interaction between process-oriented models and macroeconomic models by implementing computer-based hard-links; example are as follows: (a) the development of living areas of the residential sector derived from relationships of demographic variables, income per capita, and other preferences of private households; (b) the mileage of cars, trucks, ship, or public transport depending on demographic variables, per capita income, foreign trade and industrial production, and inter-industrial structural change. In addition, the different impacts on material substitution or material efficiency in energy-intensive industries should be modeled in more detail based on numeric factors and relationships. In this context, export/import ratios and detailed recycling data of the different basic products should be taken into account.

In the more distant future, company size (e.g., small, medium, big) and the influences of barriers and supporting factors of energy efficiency measurements should also be implemented in bottom-up models in order to improve the transparency between potentials, obstacles, and impacts of sector- or technology-oriented policies. The progress expected will lead to a more transparent simulation of sector- and technology-oriented policies by governments and trade associations and to more reliable information of the impacts of those policies at the economic and societal level. Many attempts for the development of hybrid models for different energy sectors are reported in the literature.

10.8.1 INTEGRATED HYBRID MODELS

A subsequent step after hard-linking could be to integrate the models. The distinction between hard-linking and integration can be less clear. Integrated models are run in one common format with a single model formulation, instead of exchanging information between separate models. However, when an integrated model is solved using a decomposition approach, one may observe a data exchange between the partial problems, which is very much alike that in hard-linked models. Still, we should look at such a model as an integrated model and not as a hard-linked model. Integration of the models is not always a desired next step. There are substantial benefits from keeping the models stand-alone, such as development, maintenance, diverse foci, and relative strengths of the models, and allowing for different levels of spatial and temporal disaggregation. Integrated hybrid models include functionality both from top-down and from bottom-up approaches. The choice for which sectors are specified bottom-up is driven by the research questions.

As discussed above, the cost minimization problem of the energy system models can be incorporated in an extended version of a computable general equilibrium model. This leads to an integrated model. Böhringer [5] has described such a model, and Böhringer and Rutherford [12] have provided an approach for implementation. Kulmer [31] used a dynamic CGE to study the economic impacts of a carbon tax on Austrian transport sector. A bottom-up representation of passenger transport technologies is used, and endogenous and directed technical changes are included in the model as well.

10.9 LARGE-SCALE MODELING SYSTEMS COMBINING MULTIPLE MODULES

Large-scale modeling systems that are made up of several—soft and/or hard-linked—modules take a particular place in the spectrum of hybrid models. The National Energy Modeling System (NEMS) is developed by the Energy Information Administration of the U.S. Department of Energy [55,56]. The primary use of the NEMS is to generate the projections in the Annual Energy Outlook [55,56]. NEMS is a linked modeling system consisting of several demand, supply, and market modules (see Figure 10.1). Most of these modules are bottom-up models, but some are top-down. Predecessors to NEMS are the PIES (Project Independence Evaluation System) and the IFFS (Intermediate Future Forecasting System) systems.

PIES is a combination of a linear programming model and econometric demand equations used to determine valid prices and quantities of fuels. The model is solved for a supply–demand equilibrium in energy markets by iterating between the linear program and a reduced-form representation of end-use demand models. Shadow prices for fuels from the linear program were used as prices for end-user in each sector and region. The reduced-form demand representation was evaluated at these prices, and the new end-use demands were entered into the linear program, which was re-optimized. This iterative process continued until the end-use prices and demands were not changing between iterations, within a specified tolerance. The NEMS model is solved by iterating through different modules in order to reach an equilibrium.

Simulation, Modeling, and Design of Systems

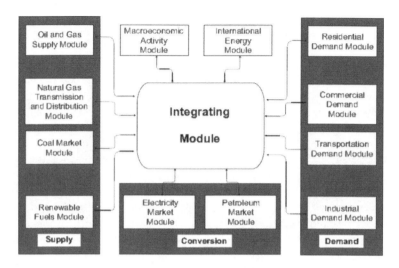

FIGURE 10.1 NEMS—National Energy Modeling System. (EIA DOE [55] The National Energy Modeling System.)

Gabriel et al. [26,27] argue that the problem could be modeled as a complementarity problem and thus implemented as an integrated hybrid model instead of a modular, linked hybrid model. However, the resulting mixed complementarity problem (MCP) would be too big to solve in acceptable time limits with the current state-of-the-art solution algorithms for MCP (i.e., PATH). As such, already in its current form, NEMS could benefit from more robust solution methods, as it regularly encounters convergence problems.

The Canadian Integrated Modeling System (CIMS) is another example of a modular linked model that is solved by an iterative process. CIMS projects energy consumption in order to forecast greenhouse gas emissions caused by the combustion of fossil fuel products. CIMS combines the strengths of the top-down and bottom-up approaches. It has the technological richness of a bottom-up model; however, it simulates technology choices by firms and households using empirically estimated behavioral parameters instead of portraying these agents as utility or financial cost optimizers. As such, it can be categorized as a behavioral simulation model.

CIMS' equilibrium solution is found by iterating first between energy supply and energy demand and subsequently between these components and the macroeconomic module. Changes in energy demand can result in changes in energy supply and consequently adjustments to energy prices, which in turn require updating energy demand for these new prices. Once energy supply and demand have reached an equilibrium, production cost changes may result in adjustments to demand for traded goods and services at the macroeconomic level, requiring additional iterations using these new demand levels. Given this simulation protocol, the more linked systems are integrated in a modeling system like CIMS, the more difficult it becomes to reach an overall equilibrium solution (to converge) [29]. This situation is quite similar to NEMS. Murphy, Rivers et al. [21,43] stated that alternative solving algorithms for CIMS will be explored in the future expansion of the model.

10.10 BRIEF REVIEW OF LITERATURE FOR MODULAR APPROACH FOR SIMULATION AND OPTIMIZATION

The effectiveness of modular approach for design and optimization has been demonstrated in the literature for a large number of energy systems. The purpose here is not to present an exhaustive assessment of the simulation, modeling, and optimization studies of various energy systems. Instead, here we focus on five topics to illustrate an effective use of modular approach in theoretical analysis of the energy systems. The topics considered are (a) typical energy system equipment and process, (b) simulation of various aspects of building construction and energy consumption, (c) simulation of power supply using various renewable sources, (d) simulation for energy consumption and efficiency processes, and (e) novel modular methods used to simulate and optimize energy systems. These five topics give a broad picture of the usefulness of modular approach in theoretical analysis of various types of energy systems. Since the main purpose of this very brief review is to demonstrate usefulness of modular system modeling for energy systems, no significant details on simulations are given. The readers are encouraged to get the details from the relevant references.

10.10.1 Modular Simulations of Energy System Equipment and Process

Fluidized bed is widely used for coal and biomass combustion and gasification. Numerous theoretical studies for modular simulation of this reactor are worth noting. Eslami et al. [57] developed a sequential modeling of coal volatile combustion in fluidized bed reactors. The study presented a sequential model to predict the combustion behavior of a coal's volatile matter during coal gasification. For this, an industrial fluidized bed combustor was divided into several sub-multiphase reactors based on the physical behavior of interacting phases. This work opens up a new way of modeling coal's volatile matter combustion, especially for its optimization and scale-up. Jafari et al. [58] presented a different basis for modular simulation of fluidized bed reactors. Simulation of chemical processes involving nonideal reactors is essential for process design, optimization, control, and scale-up. Various industrial process simulation programs are available for chemical process simulation. Most of these programs are being developed based on the sequential modular approach. They contain only standard ideal reactors but provide no module for non-ideal reactors, e.g., fluidized bed reactors. In the study by Jafari et al. [58], a new model was developed for the simulation of fluidized bed reactors by once again a sequential modular approach.

Liu et al. [59] evaluated simulation of pressurized ash agglomerating fluidized bed gasifier using Aspen Plus. A model of a pilot-scale pressurized ash agglomerating fluidized bed (AFB) gasifier was developed. The model was once again based on a sequential modular method with a recycle loop. Both hydrodynamics and reaction kinetics were considered simultaneously through FORTRAN codes if necessary. The physical properties of related gas and solid substance were calculated dynamically by the Peng–Robinson equation of state with Boston–Mathias alpha function (PR-BM) with in-line powerful physical property database in Aspen Plus, and then, the calculated results were passed immediately to the fluid-dynamic correlations for space division. Therefore, only a few parameters were needed in this model. Rafati et al. [60] also evaluated sequential modular simulation of hydrodynamics and reaction kinetics in a biomass bubbling fluidized-bed gasifier using Aspen Plus. A sequential modular simulation (SMS) approach was used to simulate hydrodynamics and detailed kinetics of a fluidized-bed biomass gasifier in Aspen Plus. The kinetics of tar cracking reactions were taken into account in the simulation. Finally, Huang and Wang [61] presented a modular design approach for coal-fired power plant control system which used a high-pressure fluidized bed. A high-performance control system is greatly important for a high-pressure and high-temperature coal-fired power plant unit. Simulation results showed that this approach can significantly reduce the workload and complexity of system design, and graph theory was used to validate these benefits.

Another equipment extensively examined in the literature is fuel cell [62]. The paper by Segura and Andújar [63] presented a supervision, control, data acquisition and simulation (SCADA and Simulator) system that allows for real-time training in the actual operation of a modular polyelectrolyte membrane (PEM) fuel cell system. This SCADA and simulator system consists of a free software tool that operates in real time and simulates real situations like failures and breakdowns in the system.

Carcascia et al. [64] evaluated a modular tool for the simulation of compressor trains for oil and gas applications. In order to keep both energy prices and environmental impact of industrial oil and gas activities low, extraction, processing, and transportation efficiency has become one of the most important objectives for companies. Used in drilling operations, in many production operations, and extensively used in surface transportation via pipelines, compression trains are widespread in the oil and gas sector. Compression plants for oil and gas applications are generally designed for near-peak hydrocarbon production of fields, and they become inevitably less performing, besides the normal process of components' aging, when production rates and operating conditions change. The study by Carcascia [64] provided a brief description of a new modular tool for compression trains simulation, named CMT (Compressors Modular Tool), based on a modular code thoroughly described on a previous work [64]. This tool is based on a mathematical solver implementing a trust-region Gauss–Newton method, called TRESNEI [64]. The whole code for simulation was implemented using the ANSI Fortran 90 standard. Each unit was developed as an independent subroutine ensuring maximum flexibility of maintenance and upgrade. The study showed that this modular approach for modeling worked well in predicting and optimizing compressor performance.

Because the drilling costs of horizontal wells are 1.4–4 times more than those of vertical wells, it is imperative to conduct a reservoir simulation and engineering study of the recovery economics of horizontal wells before drilling. A reservoir simulator with horizontal well capabilities can provide guidance into the design of well lengths, locations, optimal flow rates to prevent water coning or gas cusping and can predict the increase in recovery over that of conventional wells. There are several modeling efforts for horizontal wells, which have been published in the literature. Brekke et al. [65] evaluated a new modular approach for comprehensive simulation of horizontal wells. The model consisted of a detailed horizontal well flow simulator [66,67] coupled to a reservoir simulator. The well bore simulator was based on a network model which is capable of solving a broad range of possible flow configurations in the horizontal part of the well. The modular approach was facilitated by a coupling technique that reduces the requirement for iteration between the modules. The simulator was applied to high flow rate horizontal wells in thin oil zones with high permeability. A detailed flow simulator for horizontal well completions was also developed for modular integration with reservoir simulators [65].

A modular, comprehensive simulation of the reservoir and the well has shown to be feasible through a coupling project involving a detailed horizontal well bore simulator (HOSIM) and a 3D, two-phase reservoir simulator (FRONTSIM). Based on the experience from this coupling project, a modular iterative coupling approach was developed between HOSIM and a general black oil reservoir simulator, which is referred to as RESIM. The HOSIM is based on a general network solver for the calculation of steady state flow through well bore completions. The network solver which was originally developed for use with gas pipeline systems was modified to perform steady-state simulation of horizontal wells. As the network solver also was capable of determining the direction of flow, the simulator can be used for injection wells as well as for indicating cross-flow between formations [65].

10.10.2 MODULAR SIMULATION OF BUILDINGS CONSTRUCTION AND ENERGY CONSUMPTION

The study by Wetter [68–70] showed how modular approach can facilitate simulation, and design of building operation. At each step of the design or refurbishing of a building, decisions need taking, based on some goal and on the state of the project, i.e., the decisions taken in the past. Like any engineering process, building (re)design is in fact a cyclic activity, where any choice has to be reconsidered when its effects—no matter how later observed—are found to be unsatisfactory. Such an approach to simulation is very different from those adopted by typical engineering tools. Most are domain-specific (e.g., electrical, energy system or ES, computational fluid dynamics or CFD, and so forth), or have limited flexibility (e.g., there is a library of prebuilt "boiler" models, and creating a new one is very far from trivial), or do not allow for a structured management of the models and simulations within a project, or any combination thereof. Needless to say, adopting the Object-Oriented Modeling and Simulation (OOMS) paradigm, and in particular the Modelica language [71], is a very promising idea. This idea was examined in the study by Wetter [68–70] and Burhenne et al. [72]. While several Modelica libraries for building simulation already exist, the use of such libraries as a decision aid *along* the evolution of a project still experiences some difficulties. The study by Fritzson [73] proposes a possible *modus operandi* to solve the encountered problems.

Model predictive control (MPC) [74–81] is the only control methodology that can systematically take into account future predictions during the control design stage while satisfying the system operating constraints. The study by Ma [82] and others [74–81] focused on the design and implementation of MPC for building cooling and heating systems. The objective was to develop a control methodology that can (a) reduce building energy consumption while maintaining indoor thermal comfort by using predictive knowledge of occupancy loads and weather information,

(b) easily and systematically take into account the presence of storage devices, demand response signals from the grid, and occupants' feedback, (c) be implemented on existing inexpensive and distributed building control platform in real time, and (d) handle model uncertainties and prediction errors at both the design and implementation stage. With an inherent ability to handle constraints, MPC has led to the system reliability and improvement of user's comfort at the same time when compared with state-of-the-art adopted control methods. This is also expected to contribute to productivity within commercial buildings.

The study was focused on the integration of several energy systems for major mutual cost-saving opportunities, applied to building climate control with integrated microgrid. The following contributions are highlighted: (a) proposition of modular hierarchical MPC method for system coordination through predicted energy consumption and corresponding prices, and (b) detailed realistic simulations performed on the chosen test site of building with integrated microgrid in real (non-averaged) weather and market conditions and numerous scenarios. Results give a reliable estimate of building climate and microgrid operation costs with applied simple and more sophisticated controllers, including the proposed coordinated modular MPC. The proposed method was applied to a particular case of a building with integrated microgrid, where building zone heating/cooling is coordinated with renewable energy production, storage state of charge, and volatile electricity pricing from the electricity distribution grid. The modularity is also suitable for coordination extension to new vertical modules, e.g., central heating, ventilation and cooling (HVAC) unit level between the current levels or aggregator and district levels on the grid side above the microgrid level. Following the same established methodology, levels are coordinated by information on a predicted consumption in the up direction and issuing a price for the received predicted consumption in the down direction. In case of absent modules, the remaining ones are still coordinated via the matching interfaces for achieving the economical optimum in the current system configuration. The approach is readily applicable for buildings and grids operating in various conditions: different climates, building zone topologies, thermal actuator types, or additional microgrid components, including controllable or non-controllable loads.

Commercial buildings represent a significant portion of energy consumption and environmental emissions worldwide. To help mitigate the environmental impact of building operations, building energy management systems and behavior-based campaigns designed to reduce energy consumption are becoming increasingly popular. The study by Gulbinas et al. [37] described the development of a modular sociotechnical energy management system, *BizWatts*, which combines the two approaches by providing real-time, appliance-level power management and socially contextualized energy consumption feedback. The study describes in detail the physical and virtual architecture of the system, which simultaneously engages building occupants and facility managers, as well as the main principles behind the interface design and component functionalities. A discussion about how the data collection capabilities of the system enable insightful commercial building energy efficiency studies and quantitative network analysis is also included.

Chowdhary [83] showed that with the increased practice of modularization and prefabrication, the construction industry gained the benefits of quality management, improved completion time, reduced site disruption and vehicular traffic, and improved overall safety, security, and efficiency. Whereas industrialized construction methods, such as modular and manufactured buildings, have evolved over decades, core techniques used in prefabrication plants vary only slightly from those employed in traditional site-built construction. With a focus on energy and cost-efficient modular construction, this study presents the development of a simulation, measurement, and optimization system for energy consumption in the manufacturing process of modular construction. The system was based on Lean Six Sigma principles and loosely coupled system operation to identify the non-value adding tasks and possible causes of low energy efficiency. The proposed system in this study also included visualization functions for the demonstration of energy consumption in modular construction. The benefits of implementing this system include a reduction in the energy consumption in production cost, decrease of energy cost in the production of lean-modular construction, and increase in profit. In addition, the visualization functions provide a detailed information about energy efficiency and operation flexibility in modular construction.

10.10.3 Examples of Modular Approach for Renewable Power Supply

Over the past few years, the FAST wind turbine (WT) simulation tool has undergone a major restructuring. FAST is now, at its core, an algorithm and software framework for coupling time-dependent multi-physics modules relevant to computer-aided engineering (CAE) of WTs. Each module, which represents one or more turbine components or physics control volumes, is constituted by a mathematical model composed of time-dependent constraint and/or differential equations that are typically nonlinear. Under this new modular form, modules can interact through matching or non-matching spatial meshes and can be time advanced with different time steps and different time integrators. Sharing of data between modules is accomplished with a predictor-corrector approach, which allows for either implicit or explicit time integration within each module. This new modularity positions FAST as a backbone for coupling both high-fidelity and engineering-level WT physics models. The study of Sprague et al. [84] described new features of the FAST modular framework. In particular, the study described a new mixed-time-step algorithm, sparse-matrix storage, a direct solver for sparse linear systems, and interpolation of rotation fields in space for mesh mapping and in time for time advancement. Sprague et al. [84] showed several numerical examples that demonstrate the performance and flexibility of the FAST framework, and they use those results to provide modeling guidance to users.

The study by Dhakal et al. [85] discussed comparisons between centralized and modular topologies using dc/dc converters. The most common type of photovoltaic (PV) installation in residential applications is the centralized architecture [86,87]. This realization aggregates a number of solar panels into a single power converter for power processing. The performance of a centralized architecture is adversely affected when it is subject to partial shading effects due to clouds or surrounding obstacles, such as trees. An alternative modular approach can be implemented using several power converters with partial throughput power processing capability. This study presented a detailed study of these two architectures for the same throughput power level. The study compared the overall efficiency of these two different topologies, using a set of rapidly changing real solar irradiance data collected by the Solar Radiation Research Laboratory (SRRL) at the National Renewable Energy Laboratory (NREL). A detailed power loss analysis was also presented in the paper. Analytical results were validated through detailed computer simulations using the Matlab/Simulink mathematical software package. The study showed that modular architecture resulted in more overall efficiency than its counterpart CMPPT (centralized maximum power point tracking) under shading conditions such as clouds.

Tie [88] examined general hierarchical modular structure for hydropower station auxiliary system simulation. The study by Tie [88] proposes general hierarchical modular structure models to simulate hydropower station auxiliary system according to its process characteristics in real job. The simulation models in this modular structure are constructed with independent equipment and different technical constraints. The study introduces the method how to form modular structure and construct the models systematically. Based on modular modeling system (MMS), the study divides models into different levels and integrates these levels into an organic system according to certain rules. The system consisted of several interrelated subsystems in different levels. Equipment systems and technical constraints are established to adapt to different functions in the various subsystems. The study constructs auxiliary system models of Longtan hydropower station through "three-layer structure" by Fortran and achieves the whole process and full-range simulation. The models cover all auxiliary devices and reflect most status information of them concluding normal statements, abnormal statements, and fault statements from central computer system to field equipment. The "three-layer structure" can be used as general structure for auxiliary system simulation of hydropower station [89–91].

The advantage of hybrid power systems is the combination of the continuously available diesel power and locally available, pollution-free wind energy. Simulation system needs to facilitate an application-specific and low-cost study of the system dynamics for wind–diesel hybrid power systems. The simulation study can help in the development of control strategies to balance the system power flows under different generation/load conditions. Using the typical modules provided, it is easy to set up a particular system configuration. In the manual by NREL, Muljadi and Bialasiewicz [92] and Baring-Gould et al. [93] present the principal modules of the simulator. Using case studies of a hybrid system, the study also demonstrates some of the benefits that can be gained from understanding the effects of the designer's modifications to these complex dynamic systems. The study points out that the user must include the point of common coupling (PCC) module, as a node where all power sources and power sinks are connected, in every simulation diagram. The other principal modules are the diesel generator (DG), the AC WT with the induction generator and the wind speed time series as the input, the rotary converter (RC) with the battery bank (BB), the village load (VL), the dump load (DL), the inverter, and the PV array (PV). In addition, the system contains the transmission-line impedance and the power factor-correcting capacitors. In all electrical simulations, the study uses the d-q axis convention and synchronous reference. Using the typical modules provided, it is easy to set up a system of a desired configuration. More details of this modular approach to hybrid power control are described in the manual by NREL [92,93].

10.10.4 MODULAR SIMULATION OF ENERGY CONSUMPTION AND EFFICIENCY

Rees et al. [94] used a modular approach for integrated energy distribution analysis for CHP operation. The integration of electricity and heat generation with district heating (CHP) and electricity distribution networks can potentially provide significant carbon savings compared to the use of individual gas boilers with imported electricity. The study by Rees et al. [94] presented a modular approach to combined steady-state energy distribution analysis for use within detailed financial appraisal and project life emissions evaluation. A case study was used to demonstrate application of the model to the study of two potential supply options for a proposed new build community redevelopment in the UK. The model was used to compare the performance of two community generation options integrating biomass with natural gas CHP for a case study based on a proposed new building development in the UK.

In the last decades, industries around the world were faced by serious efforts to reduce energy consumption and increase energy efficiency [95,96] on national and international levels. Main improvement areas include waste heat recovery, the optimization of heat treatment, processes and cast pieces, the substitution of raw material, and residual gas recovery [97]. The focus on an extremely heterogeneous sector like the foundry industry advocates an integration problem of technical, economic, and ecologic methods and assessment proceedings to enable an integrated view on energy efficiency measures on a product level. In order to confirm the hypotheses and investigate the efficiency potential in the foundry industry, the modular-based, multilevel approach was developed by Coss et al. [98]. The approach enables the derivation of actual energy consumption of processes and corresponding manufactured products, and leads therefore to a better understanding of cost generation. Moreover, the methodology identifies energy efficiency potentials and merges them to a model-based approach for the planning, evaluation, and optimization of energy consumption in the foundry industry. The model was based on three scientific aspects: heterogeneity aspect, benchmarking aspect, and lifecycle aspect.

The model broke down processes into primary (like pattern making, melting and metal treatment, etc.) and secondary or support modules (like sand making and preparation processes, etc.).

The modules are part of the whole production process, where the use and connection of the modules completely describes the production site of a foundry. This level is set to be level 2, whereas the production site is set to be level 1. The different modules are no physical units and are completely depending on the production units which they contain (level 3). Furthermore, the modules contain a group of different production units (e.g., heat treatment for heat production). This model design offers the possibility to analyze all three given systems. Due to this modular representation, it is possible to define various key processes. Furthermore, it is possible to analyze key performance indicators, like energy efficiency, on a multilevel basis (offers the possibility to benchmark production units), key processes (modules), or even the whole production site. In addition, modular design allows the usage of the economic (top-down) and a thermodynamic approach (bottom-up) to analyze energy efficiency [95,96].

As the hierarchic composition of the approach is defined, the next step was to develop the model design. Here, the principle of using economic data in order to characterize energy utilization and energy efficiency analysis was integrated into the model design through the top-down approach. On the other hand, a bottom-up approach is applied in order to determine the actual energy utilization based on thermodynamic relationships and physical properties. In comparison with the top-down approach, which uses aggregated data on level 1 and level 2, the bottom-up analysis was applied on the production units of the site (level 3). The objective thereby is to determine (a) the actual energy consumption and (b) the theoretical energy consumption of the foundry production units based on a unit model for different foundry products in order to characterize the production site based on thermodynamic calculations. The calculated production units can then be used to determine the corresponding module key indicators through either an economic allocation or a thermodynamic calculation, energy operations, heat treatment, finishing operations, etc. (European Commission, 2009). Coss et al. [98] applied this modular approach to three different foundries in Austria. The application of the model approach in foundry companies enables to (a) calculate energy consumption for various important modules, (b) transfer economic and technical data to a process-oriented picture of the energy use, (c) calculate energy demand of various products, and (d) provide the first basis for energy efficiency potential analysis.

Mesap (Modular Energy System Analysis and Planning Environment) is an energy system analysis toolbox, and PlaNet (Planning Network) is a linear network module for Mesap that is designed to analyze and simulate energy demand, supply, costs, and environmental impacts for local, regional, and global energy systems. It was originally developed by the Institute for Energy Economics and the Rational Use of Energy (IER) at the University of Stuttgart in 1997, but it is now maintained by the German company Seven2one Informations systeme GmbH. In total, 15 versions of Mesap PlaNet have been released and it has approximately 20 users [99,100]. Mesap PlaNet calculates energy and emission balances for any kind of RESs. A detailed cost calculation determines the specific production cost of all commodities in the RES based on the annuity of investment cost and the fixed and variable O&M cost. The model uses a technology-oriented modeling approach where several competitive technologies that supply energy services are represented by parallel processes. All thermal generation, renewable, storage and conversion, and transport technologies are considered in the simulation. The simulation is carried out in a user-specified time step, which ranges from 1 min to multiple years, and the total time period is unlimited. Mesap PlaNet has previously been used to simulate global energy supply strategies and to compare energy-efficient strategies in Slovenia. The model has also simulated a 100% renewable energy system. Al-Mansour et al. [101,102] and Tomsic et al. [103] examined comparison of energy efficiency strategies in the industrial sector of Slovenia using Mesap. Two strategies for arc furnaces in steel industry, each with different intensities of implementation of energy-efficient technologies and measures, were compared (reference and intensive) for the period 1997–2020. Penetration of improved technologies was higher in the intensive strategy, based on the economic conditions established by the energy policy invoked. The study presented a comparison of the results of energy use and emissions and an evaluation of potential energy savings for both strategies.

10.10.5 NOVEL MODULAR APPROACHES FOR SIMULATION AND OPTIMIZATION

In the design process of technical systems, always optimizations are executed to improve the system behavior. Usually, these optimization steps are based on the experiences of the designer, who analyzes the system performance and modifies individual system parameters. The importance of MEMS optimization concerning performance, power consumption, and reliability has increased. In the MEMS design flow, a variety of specialized tools is available. For simulation on component level, finite element model (FEM) tools (e.g., ANSYS, CFD-ACE+) are widely used. Simulations on system level are carried out with simplified models using simulators like Saber, ELDO, or Spice. A few simulators offer tool-specific optimization capabilities, but there is a lack of simulator-independent support of MEMS optimization.

The main modules are simulation, error calculation, optimization, and model generation. The approach used by Schneider et al. [104] aims at a flexible combination of these modules. This new method is translated into a modular optimization system named Moscito implemented in Java. Interfaces to the simulators ANSYS, ELDO, Saber, SPICE, and METLAB are implemented. Thus, the optimization task can be solved on different levels of model abstraction (FEM, ordinary differential equations, generalized networks, block diagrams, etc.). Several optimization algorithms are available: methods without derivatives, methods using derivatives, and stochastic approaches. A graphical user interface (GUI) supports control and visualization of the optimization. The modules of the optimization system are able to communicate via the Internet to allow a distributed, web-based optimization. The study covers the partitioning of optimization cycle, the interaction between the modules of the optimization system, first experiences in web-based optimization, and the application of the approach to MEMS optimization.

Many frameworks have also been developed in the context of multidisciplinary design optimization (MDO) [105,106] using bottom-up approach. Padula and Gillian [107] review these and note that modularity, data handling, parallel processing, and user interface are the most important features of a framework that facilitates optimization with multiple disciplines. Prior to this, Salas and Townsend [108] had performed a similar survey and listed more detailed requirements for an MDO framework, categorized into the framework's architectural design, problem construction, problem execution, and data access. The existing MDO frameworks are susceptible to the same missing features as the commercial and high-performance computing (HPC)-focused frameworks. One notable exception is pyMDO [109], which automatically computes derivatives using analytic methods though in a less general form. Moreover, pyMDO focuses on facilitating the implementation of MDO architectures, so it lacks several other features such as built-in solvers for HPC and parallel data transfer. These should be added for workable modular analysis.

Hanson and Cunningham [110] examined modular approach to simulation with automatic sensitivity calculator. They used this for optimizing energy system design. The study indicated that when using simulation codes, one often has the task of minimizing a scalar objective function with respect to numerous parameters. This situation occurs when trying to fit (assimilate) data or trying to optimize an engineering design. For simulations in which the objective function to be minimized is reasonably well behaved, that is, is differentiable, and does not contain too many multiple minima, gradient-based optimization methods can reduce the number of function evaluations required to determine the minimizing parameters. However, gradient-based methods are only advantageous if one can efficiently evaluate the gradients of the objective function. Adjoint differentiation efficiently provides these sensitivities. One way to obtain code for calculating adjoint sensitivities is to use special compilers to process the simulation code. However, this approach is not always so "automatic." The study by Hanson and Cunningham [110] describes a modular approach to constructing simulation codes, which permits adjoint differentiation to be incorporated with relative ease.

For fuzzy systems, suitable notions of simulation, and logics which characterize them, become increasingly challenging. A framework that allows the automatic derivation of such definitions, together with proofs of expressiveness, would therefore prove valuable in the treatment of complex system types. The study by Cîrstea [111,112] develops such a framework, using *coalgebras* as a general setting. Coalgebras have, in recent years, been shown to provide suitable abstract models for a large class of state-based systems, which includes non-deterministic systems, probabilistic systems, and various kinds of automata [111,112]. The emphasis in such modeling is on the observations which can be performed on a system in one step. Cîrstea [111,112] proposed a modular approach to defining notions of simulation, and modal logics which characterize them. She used *coalgebras* to model state-based systems, *relators* to define notions of simulation for such systems, and inductive techniques to define the syntax and semantics of modal logics for coalgebras. She showed that the expressiveness of an inductively defined logic for coalgebras with reference to a notion of simulation follows from an expressivity condition involving one step in the definition of the logic, and the relator inducing that notion of simulation. Moreover, she showed that notions of simulation and associated characterizing logics for increasingly complex system types can be derived by lifting the operations used to combine system types, to a relational level as well as to a logical level. She used these results to obtain Baltag's logic for coalgebraic simulation, as well.

Modularization and hierarchical construction are key factors for stepwise refinement of complex system models. Modularity can favor incremental system development allowing the modeler to focus on separate parts with functional and temporal behaviors linked by interfaces exposed to the rest of the model. Letia and Kilyen [113] examined Fuzzy Logic Enhanced Time Petri Net models (FLETPN) for hybrid control systems (HCSs). A HCS model has to implement the reaction to discrete events that occur in a controlled plant and as well as to control the continuous time parts according to required behaviors. A certain type of models that includes discrete event and discrete time features is proposed in this study. More than this, the currently proposed models are capable of describing the synchronous and asynchronous reactions to discrete events and continuous modifications of plant outputs. The proposed models endow the Enhanced Time Petri Nets (ETPNs) with Fuzzy Logic (FL) rules leading to the FLETPN model. An example of application to the control of the longitudinal move of a vehicle is used to show the model utilization and its benefits.

Tordrup et al. [114] used a modular approach to inverse modeling of a district heating facility with seasonal thermal energy storage. Specifically, they used a modular approach to develop a TRNSYS model for a district heating facility by applying inverse modeling to 1 year of operational data for individual components. They assembled the components into a single TRNSYS model for the full system using the accumulation tanks as a central hub connecting all other components. The study compared predictions of the total heat delivered to the district heating network to observed values and found a model error of 7.1% for one simulation year.

Kim et al. [115] examined modular inter-cell power management (IPM) strategy for energy conservation in densely deployed networks. They considered a 5G smallcell model where the cells are deployed in a way to cover users in the neighbor cells in order to introduce cell active-sleep control, so-called IPM. Using a Markov decision process (MDP), one can obtain an optimal policy for IPM, but with limited scalability. This paper proposes a modular IPM strategy, which is a suboptimal policy to group only a few neighbor cells. The modular IPM strategy can mitigate the computational complexity of MDP, while it achieves a near-optimal solution to save the energy usage in small cells. It is verified to be a feasible solution through the memory usage estimation and simulation study. The modular IPM strategy with fine-step quantized states can save even more power than the non-modular strategy with coarse-step quantized states. Consequently, the proposed modular IPM strategy becomes much favored in the base station management for power saving in a large small-cell cluster.

REFERENCES

1. Shah, Y.T. (2019). *Modular Systems for Energy and Fuel Recovery and Conversion*. CRC Press, New York.
2. Wene, C.O. (1996). Energy-economy analysis: Linking the macroeconomic and systems engineering approaches. *Energy*, 21(9), 809–824.
3. Hourcade, J.C., Jaccard, M., Bataille, C., and Ghersi, F. (2006). Hybrid Modeling: New Answers to Old Challenges—Introduction to the Special Issue of The Energy Journal, 2(Special issue), 1–12.
4. Böhringer, C. and Rutherford, T.F. (2007). Combining Top-Down and Bottom-up in Energy Policy Analysis, A Decomposition Approach. Discussion Paper No. 06-007, Download this ZEW Discussion Paper from our ftp server: ftp://ftp.zew.de/pub/zew-docs/dp/dp06007.pdf. www.mpsge.org/qpdecomp.pdf.
5. Böhringer, C. (1998). The synthesis of bottom-up and top-down in energy policy modeling. *Energy Economics*, 20(3), 233–248.
6. Bohringer, C., Müller, A. and Wickart, M. (2003). Economic impacts of a premature nuclear phase-out in Switzerland. *Swiss Journal of Economics and Statistics*, 139(4), 461–505.
7. Bohringer, C. and Rutherford, T.F. (2005). Integrating Bottom-Up into Top-Down: A Mixed Complementarity Approach. Discussion Paper No. 05-28, ZEW, Mannheim.
8. Johansen, L. (1960). *A Multi-Sector Study of Economic Growth*. North-Holland Publishing Company, Amsterdam.
9. Johnsen, T.A. and Unander, F.F. (1996). Norwegian residential energy demand: Coordinated use of a system engineering and a macroeconomic model. *Modeling Identification and Control*, 17(3), 183–192.
10. Jorgenson, D.W. (1982). An econometric approach to general equilibrium analysis. In: Hazewinkel, M. and Rinnooy Kan, A.H.G. (eds) *Current Developments in the Interface: Economics, Econometrics, Mathematics*. Springer, Netherlands, pp. 125–155.

11. Bohringer, C. and Loschel, A. (2006). Computable general equilibrium models for sustainability impact assessment: Status quo and prospects. *Ecological Economics*, 60(1), 49–64.
12. Böhringer, C. and Rutherford, T.F. (2008). Combining bottom-up and top-down. *Energy Economics*, 30(2), 574–596.
13. Böhringer, C. and Rutherford, T.F. (2009). Integrated assessment of energy policies: Decomposing top-down and bottom-up. *Journal of Economic Dynamics & Control*, 33(9), 1648–1661.
14. Hoffman, K.C. and Jorgenson, D.W. (1977). Economic and technological models for evaluation of energy-policy. *Bell Journal of Economics*, 8(2), 444–466.
15. Berg, C., Krook-Riekkola, A., Ahlgren, E., and Söderholm, P. (2012). Mjuklänkning mellan EMEC och TIMES-Sweden - en metod för att förbättra energipolitiska underlag. Specialstudier. Konjunkturinstitutet/National Institute of Economic Research, Sweden, p. 73.
16. Manne, A. and Wene, C.-O. (1992). MARKAL-Macro: A Linked Model for Energy-Economy Analysis. Report BNL-47161, Brookhaven National Laboratory.
17. Bjertnæs, G., Tsygankova, M. and Martinsen, T. (2012). The Double Dividend in the Presence of Abatement Technologies and Local External Effects.
18. Martinsen, T. (2011). Introducing technology learning for energy technologies in a national CGE model through soft links to global and national energy models. *Energy Policy*, 39(6), 3327–3336.
19. Schafer, A. and Jacoby, H.D. (2005). Technology detail in a multisector CGE model: Transport under climate policy. *Energy Economics*, 27(1), 1–24.
20. Schafer, A. and Jacoby, H.D. (2006). Experiments with a hybrid CGE-MARKAL model. *Energy Journal*, 171–177.
21. Murphy, R., Rivers, N. and Jaccard, M. (2007). Hybrid modeling of industrial energy consumption and greenhouse gas emissions with an application to Canada. *Energy Economics*, 29(4), 826–846.
22. Abrell, J. and Rausch, S. (2016). Cross-country electricity trade, renewable energy and European transmission infrastructure policy. *Journal of Environmental Economics and Management*, 79, 87–113.
23. Chen, W.Y. (2005). The costs of mitigating carbon emissions in China: Findings from China MARKALMACRO modeling. *Energy Policy*, 33(7), 885–896.
24. Contaldi, M., Gracceva, F., and Tosato, G. (2007). Evaluation of green-certificates policies using the MARKAL-MACRO-Italy model. *Energy Policy*, 35(2), 797–808.
25. Facchinei, F. and Pang, J. (2003). *Finite-Dimensional Variational Inequalities and Complementarity Problems*. Springer, New York.
26. Gabriel, S.A., Conejo, A.J., Fuller, J.D., Hobbs, B.F., and Ruiz, C. (2012). *Complementarity Modeling in Energy Markets*, Springer, New York.
27. Gabriel, S.A., Kydes, A.S., and Whitman, P. (2001). The National Energy Modeling System: A large scale energy-economic equilibrium model. *Operations Research*, 49(1), 14–25.
28. Helgesen, P.I. (2013). Top-Down and Bottom-Up: Combining Energy System Models and Macroeconomic General Equilibrium Models. CenSES working paper 1/2013. www.ntnu.no/censes/working-papers.
29. Jaccard, M., Nyboer, J., Bataille, C., and Sadownik, B. (2003). Modeling the cost of climate policy: Distinguishing between alternative cost definitions and long-run cost dynamics. *The Energy Journal*, 49–73.
30. Klaassen, G. and Riahi, K. (2007). Internalizing externalities of electricity generation: An analysis with MESSAGE-MACRO. *Energy Policy*, 35(2), 815–827.
31. Kulmer, V. (2012). Directed Technological Change in a Bottom-Up/Top-Down CGE model: Analysis of Passenger Transport. Ecomod Conference Paper.

32. Messner, S. and Schrattenholzer, L. (2000). MESSAGE-MACRO: Linking an energy supply model with a macroeconomic module and solving it iteratively. *Energy*, 25(3), 267–282.
33. Strachan, N. and Kannan, R. (2008). Hybrid modelling of long-term carbon reduction scenarios for the UK. *Energy Economics*, 30(6), 2947–2963.
34. Strachan, N., Pye, S., and Kannan, R. (2009). The iterative contribution and relevance of modelling to UK energy policy. *Energy Policy*, 37(3), 850–860.
35. Holz, F., Ansari, D., Egging, R., and Helgesen, P.I. Issue Paper on Hybrid Modelling: Linking and Integrating Top-Down and Bottom-Up Models. Issue Paper prepared for the Modelling Workshop "Top-Down Bottom-Up Hybrid Modelling" 24–25 November 2016 at NTNU Trondheim (November 2016).
36. Bataille, C., Jaccard, M., Nyboer, J., and Rivers, N. (2006). Towards general equilibrium in a technology-rich model with empirically estimated behavioral parameters. Hybrid modeling of energy-environment policies: Reconciling bottom-up and top-down—Special Issue of the Energy Journal, 27, 93–112.
37. Gulbinas, R., Jain, R.K., and Taylor, J.E. (31 December 2014). *BizWatts*: A modular socio-technical energy management system for empowering commercial building occupants to conserve energy. *Applied Energy*, 136, 1076–1084. doi: 10.1016/j.apenergy.2014.07.034.
38. Catenazzi, G. (2009). Advances in Techno-Economic Energy Modeling: Costs, Dynamics and Hybrid Aspects. Dissertation, Swiss Federal Institute of Technology (ETH), Zürich.
39. E3Mlab. (2007). The Primes Model: Version used for the 2007 scenarios for the European Commission including newsub-models recently added, website: www.e3mlab.ntua.gr/DEFAULT.HTM.
40. Jochem, E., Barker, T., Scrieciu, S., Schade, W., Helfrich, N.,Edenhofer, O., Bauer, N., Marchand, S., Neuhaus, J., Mima, S., Criqui, P., Morel, J., Chateau, B., Kitous, A., Nabuurs, G.J., Schelhaas, M.J., Groen, T., Riffeser, L., Reitze, F., Jochem, E., Catenazzi, G., Jakob, M., Aebischer, B., Kartsoni, K., Eichhammer, W., Held, A., Ragwitz, M., Reiter, U., Kypreos, S., and Turton, H. (2007). EU-Project ADAM: Adaption and Mitigation Strategies: Supporting European Climate Policy—Deliverable M1.1: Report of the Base Case Scenario for Europe and full description of the modelsystem. Fraunhofer ISI, Karlsruhe (November 2007).
41. Jochem, E., Barker, T., Scrieciu, S., Schade, W., Helfrich, N.,Edenhofer, O., Bauer, N., Marchand, S., Neuhaus, J.,Mima, S., Criqui, P., Morel, J., Chateau, B., Kitous, A.,Nabuurs, G.J., Schelhaas, M.J., Groen, T., Riffeser, L., Reitze, F., Jochem, E., Catenazzi, G., Jakob, M., Aebischer, B., Kartsoni, K., Eichhammer, W., Held, A., Ragwitz, M., Reiter, U., Kypreos, S., and Turton, H. (2008). EU-Project ADAM: Adaption and Mitigation Strategies: Supporting European Climate Policy—Deliverable M1.2: Report of the Reference Case Scenario for Europe. Fraunhofer ISI, Karlsruhe (November 2008).
42. Schade, W., Jochem, E., Barker, T., Catenazzi, G., Eichhammer, W., Fleiter, T., Held, A., Helfrich, N., Jakob, M., Criqui, P., Mima, S., Quandt, L., Peters, A., Ragwitz, M., Reiter U., Reitze, F., Schelhaas, M., Scrieciu, S., and Turton, H. (2009). ADAM—2 Degree Scenario for Europe - Policies and Impacts. Deliverable M1.3 of ADAM (Adaptation and Mitigation Strategies: Supporting European Climate Policy). Project funded by the European Commission 6thRDT Programme, Karlsruhe, Germany.
43. Rivers, N. and Jaccard, M. (2005). Combining top-down and bottom-up approaches to energy-economy modeling using discrete choice methods. *The Energy Journal*, 26, 83–106. doi: 10.2307/41323052.
44. Brandsma, A., Ivanova, O., and Kancs, D. (2011). RHOMOLO—A Dynamic Spatial General Equilibrium Model. *19th International Input-Output Conference*, Alexandria, VA, p. 22.

45. Brocker, J. and Korzhenevych, A. (2013). Forward looking dynamics in spatial CGE modelling. *Economic Modelling*, 31, 389–400.
46. Connolly, D., Lund, H., Mathiesen, B.V., and Leahy, M. (2010). A review of computer tools for analysing the integration of renewable energy into various energy systems. *Applied Energy*, 87(4), 1059–1082.
47. Fleiter, T., Worrell, E., and Eichhammer, W. (2011). Barriers to energy efficiency in industrial bottom-up energy demand models—A review. *Renewable & Sustainable Energy Reviews*, 15(6), 3099–3111.
48. Herbst, A., Reitze, F., Toro, F.A., and Jochem, E. (2012). Bridging Macroeconomic and Bottom Up Energy Models—The Case of Efficiency in Industry. ECEEE 2012 Industrial Summer Study. Arnhem, The Netherlands, The European Council for an Energy Efficient Economy.
49. Hertel, T. (2013). Chapter 12—Global applied general equilibrium analysis using the global trade analysis project framework. In: Peter, B.D. and Dale, W.J. (eds) *Handbook of Computable General Equilibrium Modeling*. Elsevier, Waltham, MA, Volume 1, pp. 815–875.
50. Helgesen, I. Top-Down and Bottom-Up: Combining Energy System Models and Macroeconomic General Equilibrium Models, Project: Regional Effects of Energy Policy (RegPol) CenSES working paper, NTNU, Norway 1/2013 www.ntnu.no/.../2013...models.../4252b320-d68d-43df-81b8-e8c72ea1bfe1.
51. Bhattacharyya, S. and Timilsina, G.R. (2010). A review of energy system models. *The International Journal of Energy Sector Management*, 4(4), 494–518. www.researchgate.net/...review...energy_system_models/.../A-review-of-energy....
52. Bergmann, L. (1990). The development of computable general equilibrium models. In Bergman, L., Jorgenson, D. W., and Zalai, E. (eds) *General Equilibrium Modeling and Economic Policy Analysis*. Blackwell, Cambridge, pp. 3–30.
53. Frei, C.W., Hadi, P.-A., and Sarlos, G. (2003). Dynamic formulation of a top-down and bottom-up merging energy policy model. *Energy Policy*, 31(10), 1017–1031.
54. Östblom, G. and Berg, C. (November 2006). The EMEC Model: Version 2.0, Working Paper No. 96 Published by, The National Institute of Economic Research Stockholm 2006.
55. The National Energy Modeling System: An Overview. (2009). EIA. A website report www.eia.gov/outlooks/aeo/nems/overview/index.html.
56. Daniels, D. (31 August 2017). Overview of the National Energy Modeling System (NEMS). Presented at the *University of Bergamo/Georgia Tech Environment and Sustainability Workshop, Energy Information Administration*, 29 August 2017.
57. Eslami, A., Hashemisohi, A., Sheikhi, A., and Sotudeh-Gharebagh, R. (2012). Sequential modeling of coal volatile combustion in fluidized bed reactors. *Energy & Fuels*, 26, 5199–5209. doi: 10.1021/ef300710j.
58. Jafari, R., Sotudeh-Gharebagh, R., and Mostoufi, N. (2004). Modular simulation of fluidized bed reactors. *Chemical Engineering & Technology*, 27, 123–129. doi: 10.1002/ceat.200401908.
59. Liu, Z., Fang, Y., Deng, S., Huang, J., Zhao, J., and Cheng, Z. (2012). Simulation of pressurized ash agglomerating fluidized bed gasifier using ASPEN PLUS. *Energy & Fuels*, 26, 1237–1245. doi: 10.1021/ef201620t.
60. Rafati, N., Hashemisohi, A., Wang, L., and Shahbazi, G. (2015). Sequential modular simulation of hydrodynamics and reaction kinetics in a biomass bubbling fluidized-bed gasifier using aspen plus. *Energy & Fuels*, 29. doi: 10.1021/acs.energyfuels.5b02097.
61. Huang, S. and Wang, W. A Modular Design Approach for Coal-Fired Power Plant Control System. Publisher: IEEE Published in: *2013 10th IEEE International Conference on Control and Automation (ICCA)*, 12–14 June 2013, Hangzhou, China. INSPEC Accession Number: 13680106. doi: 10.1109/ICCA.2013.6565192.

62. Hans, R., Panik, F., and Reuss, H.C. (2014). Chapter 14: Modular modeling of a PEM fuel cell system for automotive applications. In: Bargende, M., Reuss, H.C., and Wiedemann, J. (eds) *Internationales Stuttgarter Symposium. Proceedings.* Springer Vieweg, Wiesbaden. First Online: 8 May 2014. doi: 10.1007/978-3-658-05130-3_13. Print ISBN 978-3-658-05129-7, Online ISBN: 978-3-658-05130-3.
63. Segura, F. and Andújar, J.M. (September 2015). Modular PEM fuel cell SCADA & simulator system. *Resources*, 4, 692–712. doi: 10.3390/resources4030692.
64. Carcasci, C., Marini, L., Morini, B., Porcelli, M., Micio, M., and Luigi Di Pillo, P. (December 2015). Modular tool for the simulation of compressor trains for oil and gas applications. *Energy Procedia*, 82, 546–553. doi: 10.1016/j.egypro.2015.11.868. ScienceDirect ATI 2015-70th Conference of the ATI Engineering Association.
65. Brekke, K., Johansen, T.E., and Olufsen, R. (1993). A New Modular Approach to Comprehensive Simulation of Horizontal Wells. doi: 10.2118/26518-MS. Document ID *SPE-26518-MS Publisher Society of Petroleum Engineers, SPE Annual Technical Conference and Exhibition*, 3–6 October, Houston, TX.
66. Collins, D., Nghiem, L., Sharma, R., Li, Y., and Jha, K. (January 1992). Field-scale simulation of horizontal wells. *Journal of Canadian Petroleum Technology*, 31(1). doi: 10.2118/92-01-01. Document ID PETSOC-92-01-01, Publisher Petroleum Society of Canada, Source.
67. de Souza, G., Pires, A.P., and de Abreu, E. (2014). Well-Reservoir Coupling on the Numerical Simulation of Horizontal Wells in Gas Reservoirs. Document ID: *SPE-169386-MS Publisher Society of Petroleum Engineers Source SPE Latin America and Caribbean Petroleum Engineering Conference*, 21–23 May, Maracaibo, Venezuela. doi: 10.2118/169386-MS.
68. Wetter, M. (2011). Co-simulation of building energy and control systems with the Building Controls Virtual Test Bed. *Journal of Building Performance Simulation*, 4, 185–203. doi: 10.1080/19401493.2010.518631.
69. Wetter, M. and Wetter, M. (2009). Modelica-based modeling and simulation to support research and development in building energy and control systems. *Journal of Building Performance Simulation*, 2(2), 143–161. doi: 10.1080/19401490902818259.
70. Wetter, M. (2009). A Modelica-Based Model Library for Building Energy and Control Systems.
71. Thiele, B. (January 2004). bernhard.thiele@liu.se Researcher at PELAB, Linköping Universit Principles of Object Oriented Modeling and Simulation with Modelica 2.1 Book. doi: 10.1109/9780470545669. Publisher: ISBN 0-471-471631. Publisher: Wiley IEEE Press.
72. Burhenne, S., Wystrcil, D., Elci, M., Narmsara, S., and Herkel, S. Building Performance Simulation Using Modelica: Analysis of the Current State and Application Areas. Institute for Solar Energy Systems ISE, Freiburg, Germany *Proceedings of BS2013: 13th Conference of International Building Performance Simulation Association*, Chambéry, France, August 26–28, pp. 3259–3266. www.ibpsa.org/proceedings/BS2013/p_1328.pdf.
73. Fritzson, P. (2 February 2016). Introduction to Object-Oriented Modeling, Simulation, Debugging and Dynamic Optimization with Modelica using Open Modelica Linköping University, peter.fritzson@liu.se Director of the Open Source Modelica Consortium Vice Chairman of Modelica Association.
74. Zhang, T., Wan, M.P., Ng, B.F., and Yang, S. (2018). Model Predictive Control for Building Energy Reduction and Temperature Regulation. *IEEE Green Technologies Conference (GreenTech)*, 4–6 April 2018, Austin, TX. Date Added to IEEE Xplore: 07 June 2018. Electronic ISSN: 2166-5478. INSPEC Accession Number: 17823910. doi: 10.1109/GreenTech.2018.00027.

75. Lamoudi, M.Y., Alamir, M., and Béguery, P. Model Predictive Control for Energy Management in Buildings. Part 1: Zone Model Predictive Control. *4th IFAC Nonlinear Model Predictive Control Conference International Federation of Automatic Control*, 23–27 August 2012, Noordwijkerhout, Netherlands.
76. Lamoudi, M.Y., Alamir, M., and Béguery, P. Model Predictive Control for Energy Management in Buildings. Part 2: Distributed Model Predictive Control. doi: 10.3182/20120823-5-NL-3013.00036.
77. Dobbs, J. Model Predictive Control of Building Energy Management Systems in a Smart Grid Environment, Dissertations, Master's Theses and Master's Reports, Michigan Technological University, Michigan jsdobbs@mtu.edu Copyright 2015.
78. Serale, G., Fiorentini, M., Capozzoli, A., Bernardini, D., and Bemporad, A. (2018). Model predictive control (MPC) for enhancing building and HVAC system energy efficiency: Problem formulation, applications and opportunities, review. *Energies*, 11(3), 631. doi: 10.3390/en11030631.
79. Martinčević, A. and Lesic, V. (January 2016). Model Predictive Control for Energy-saving and Comfortable Temperature Control in Buildings. Conference Paper.
80. Zakula, T. Model Predictive Control for Energy Efficient Cooling and Dehumidification. Submitted to the Department of Architecture in Partial Fulfillment of the Requirements for the Degree of Doctor of Philosophy in Building Technology at the Massachusetts Institute of Technology.
81. Zakula, T., Armstrong, P.R., and Norford, L. (2014). Modeling environment for model predictive control of buildings. *Energy and Buildings*, 85, 549–559. www.elsevier.com/locate/enbuild.
82. Ma, Y. (Fall 2012). Model Predictive Control for Energy Efficient Buildings, Ph.D. Thesis, University of California, Berkeley.
83. Chowdhury, M. (2016). Simulation of Value Stream Mapping and Discrete Optimization of Energy Consumption in Modular Construction, Theses and Dissertations, 610. http://ir.library.illinoisstate.edu/etd/610.
84. Sprague, M.A., Jonkman, J.M., and Jonkman, B.J. National Renewable Energy Laboratory, FAST Modular Framework for Wind Turbine Simulation: New Algorithms and Numerical Examples. *Presented at AIAA SciTech 2015: 33rd Wind Energy Symposium Kissimmee*, Florida, 5–9 January 2015.
85. Dhakal, B., Mancilla–David, F., and Muljadi, E. Centralized and Modular Architectures for Photovoltaic Panels with Improved Efficiency. Preprint *National Renewable Energy Laboratory, Golden Colorado, To be presented at the North American Power Symposium*, 9–11 September 2012, Urbana, IL. Conference Paper NREL/CP-5500-55894 July 2012 Contract No. DE-AC36-08GO28308. www.nrel.gov/docs/fy12o-sti/55894.pdf.
86. Dondi, D., Brunelli, D., Benini, L., Pavan, P., Bertacchini, A., and Larcher, L. (2007). Photovoltaic Cell Modeling for Solar Energy Powered Sensor Networks. 1-4244-1245-5/07/$25.00 ©2007 IEEEciteseerx.ist.psu.edu/viewdoc/download?doi=10.1.1.301.6889&rep=rep1....
87. Petreus, D., Farcas, C., and Ciocan, I. (2008). Modelling and simulation of photovoltaic cells, 49.
88. Tie, C. (2012). General hierarchical modular structure for hydropower station auxiliary system simulation. *Energy Procedia*, 14, 996–1001. doi: 10.1016/j.egypro.2011.12.1045.
89. Bialasiewicz, J.T., Muljadi, E., Drouilhet, S., and Nix, G. (27 April–1 May 1998). Nix Modular Simulation of a Hybrid Power System with Diesel and Wind Turbine Generation, National Wind Technology Center NREL/CP-500-24681. UC Category: 1213 Presented at Windpower '98 Bakersfield, CA.

90. Milivojevic, N., Divac, D., Vukosavic, D., Vukovic, D., and Milivojevic, V. (2009). Computer-aided optimization in operation planning of hydropower plants—Algorithms and examples. *Journal of the Serbian Society for Computational Mechanics*, 3(No. 1), 273–297. UDC: 621.311.21:519.863.
91. Hulea, M., Miron, R., and Letia, T. Modular Hybrid Control System with Petri-Nets Enhanced Components for Hydro Power Plant Control. *2016 IEEE International Conference on Automation, Quality and Testing, Robotics (AQTR)*, 19–21 May 2016. Date Added to IEEE Xplore: 30 June 2016, INSPEC Accession Number: 16107985, doi: 10.1109/AQTR.2016.7501335. Publisher: IEEE, Conference Location: Cluj-Napoca, Romania.
92. Muljadi, E. and Bialasiewicz, J.T. (November 2003). Hybrid Power System with a Controlled Energy Storage. NREL/CP-500-34692, Roanoke, VA, 2–6 November 2003.
93. Baring-Gould, E.I., Newcomb, C., Corbus, D., and Kalidas, R. Field Performance of Hybrid Power Systems. Presented at the *American Wind Energy Association's WINDPOWER 2001 Conference*, 4–7 June 2001, Washington, D.C., August 2001 NREL/CP-500-30566.
94. Rees, M., Zhong, J., Awad, B., Ekanayake, J., and Jenkins, N. A Modular Approach to Integrated Energy Distribution System Analysis. A paper Presented at: *17th Power Systems Computation Conference (PSCC '11)*, 22–26 August 2011, Stockholm, Sweden. (PSCC 2011 STOCKHOLM) Stockholm, Sweden, August 22–26 2011, Volume 1 of 2, 97–107 Official URL: www.pscc-central.org/uploads/tx_ethpublicat....
95. Bonvini, M. and Leva, A. (2011). Scalable-Detail Modular Models for Simulation Studies on Energy Efficiency. doi: 10.3384/ecp1106339.
96. Verl, A., Abele, E., Heisel, U., Dietmair, A., Eberspächer, P., Rahäuser, R., Schrems, S., and Braun, S. (2011). Modular modeling of energy consumption for monitoring and control. In: Hesselbach, J. and Herrmann, C. (eds) *Globalized Solutions for Sustainability in Manufacturing*. Springer, Berlin, Heidelberg. First Online: 7 March 2011. doi: 10.1007/978-3-642-19692-8_59. Springer, Berlin, Heidelberg, Print ISBN: 978-3-642-19691-1, Online ISBN978-3-642-19692-8.
97. Krause, M., Thiede, S., Hermann, C., and Butz, F. (2012). *A Material and Energy Flow Oriented Method for Enhancing Energy and Resource Efficiency in Aluminum Foundries*. Springer, Berlin.
98. Coss, S., Topić, M., Tschiggerl, K., and Raupenstrauch, H. (2015). Development and application of a modular- based, multi- level approach for increasing energy efficiency. *Journal of Thermal Engineering*, 1, 355–366. doi: 10.18186/jte.72061.
99. Schlenzig, C. (1999). Energy planning and environmental management with the information and decision support system MESAP. *International Journal of Global Energy Issues*, 12(1–6), 81–91.
100. Schlenzig, C. (2009). Seven2one modelling: Erstellung integrierter Ener-giekonzepte mit Mesap/PlaNet (Seven2one modelling: Building integrated energy concepts with Mesap/PlaNet). *Seven2one*. www.seven2one.de/de/technologie/mesap.html.
101. Al-Mansour, F., Tomsic, M., and Merse, S. Industrial Model for Energy Efficiency Strategies in Sloivenia. A website report, 495–506. https://aceee.org/files/proceedings/2001/data/papers/SS01_Panel1_Paper43.pdf (1999).
102. Al-Mansour, F., Merse, S., and Tomsic, M. (2003). Comparison of energy efficiency strategies in the industrial sector of Slovenia. *Energy*, 28(5), 421–440.
103. Tomsic, M., Urbanci, A., Al Mansour, F., and Merse, S. (1997). Energy Supply and Demand Planning Aspects in Slovenia. *Proceedings of the Sixth Forum: Energy day in Croatia*, RN:29064347, INIS-HR-98-004 https://inis.iaea.org/search/search.aspx?orig_q=RN:29064347.
104. Schneider, P., Huck, E., Reitz, S., Parodat, S., Schneider, A., and Schwarz, P. (28–30 November 2000). A Modular Approach for Simulation-Based Optimization of MEMS, Design, Modeling, and Simulation in Microelectronics, Singapore, 71–82 SPIE

Proceedings Series Volume 4228, Published 2002. doi: 10.1016/S0026-2692(01)00101-XProceedings Volume 4228, Design, Modeling, and Simulation in Microelectronics; (2000). doi: 10.1117/12.405441. Event: International Symposium on Microelectronics and Assembly, 2000, Singapore, Singapore.

105. Eynard, J., Grieu, S., and Polit, M. (2011). Modular approach for modeling a multi-energy district boiler. *Applied Mathematical Modelling*, 35(8), 3926–3957. doi: 10.1016/j.apm.2011.02.006.

106. Martins, J.R.R.A. and Lambe, A.B. Multidisciplinary Design Optimization: A Survey of Architectures, Multidisciplinary Design Optimization: A Survey of...—MDO Lab. mdolab.engin.umich.edu/sites/default/files/Martins-Lambe-AIAAJ-MDO-Survey.pdf.

107. Padula, S.L. and Gillian, R.E. Multidisciplinary environments: A History of Engineering Framework Development. *Proceedings of the 11th AIAA/ISSMO Multidisciplinary Analysis and Optimization Conference*, September 2006, Porthsmouth, VA, AIAA 2006-7083.

108. Salas, A.O. and Townsend, T. (1998). Framework Requirements for MDO Application Development. doi: 10.2514/6.1998-4740. www.semanticscholar.org/...MDO...Salas-Townsend/91ca27e260d4f39ceea6f0fa9.... Published Online: 22 August 2012. doi: 10.2514/6.1998-4740 AIAA-98-4740.

109. Martins, J.R.R.A., Marriage, C., and Tedford, N.P. (2009). pyMDO: An object-oriented framework for multidisciplinary design optimization. *ACM Transactions on Mathematical Software*, 36, 20:1–20:25. doi: 10.1145/1555386.1555389.

110. Hanson, K.M. and Cunningham, G.S. (2001). A Modular Approach to Simulation with Automatic Sensitivity Calculation. *Proceedings of Third International Symposium on Sensitivity Analysis of Model Output*, LA-UR-01-0802 http://lib-www.lanl.gov/la-pubs/00357125.pdf.

111. Cîrstea, C. and Pattinson, D. (2004). Modular Construction of Modal Logics. *International Conference on Concurrency Theory CONCUR 2004: CONCUR 2004-Concurrency Theory.* pp. 258–275.

112. Cîrstea, C. (2006). A modular approach to defining and characterizing notions of simulation. *Information and Computation*, 204, 469–502.

113. Letia, T. and Kilyen, A. (2016). Fuzzy Logic Enhanced Time Petri Net Models for Hybrid Control Systems, pp. 1–6. doi: 10.1109/AQTR.2016.7501322.

114. Tordrup, K.W., Poulsen, U.V., and Nielsen, C. (October 2017). A modular approach to inverse modeling of a district heating facility with seasonal thermal energy storage. *Energy Procedia*, 135, 263–271. doi: 10.1016/j.egypro.2017.09.518.

115. Kim, K., Song, N.-O., Kim, J., Rhee, J.-K.K., and Kong, P.-Y. Modular IPM Strategy for Energy Conservation in Densely Deployed Networks. *2015 IEEE International Conference on Communication Workshop (ICCW)*, 8–12 June 2015, London, UK. ISBN Information: Print ISSN: 2164-7038. INSPEC Accession Number: 15454412. doi: 10.1109/ICCW.2015.7247174.

Index

A

ABB, 231, 232, 233, 234, 235
ABE, Aspen basic engineering, 532
AC, 412, 413
AC/DC, 489
ACCE, 531
ADAM, 93
Adaptable, 134, 140
Additive Manufacturing Integrated Energy, 191
ADPN, 476
Adverse airflow, 41
Aerial solar vehicles, 205
AFDX, 52, 53
AgriPower module, 77
AHU, 101
AMPS project, 171, 172, 173
APWS-80, 362
ASHRAE, 129
ASME, 501
Aspen HYSYS, 531
Aspen platform, 528
Autonomy vehicles, 182
AWEC, 353

B

BAA, 54
Barge, 512
Batteries, 46, 181, 186
BESS, 394, 395
BiMOS, 32
Biomass, 74, 77, 80, 266, 269, 271, 273, 495
Biomass CHP, 269
Biomass gasification, 77
BioMax technology, 74, 76
BIST system, 148
BM3, 395
BMS, 397
BN-350, 358, 359
Bottom up approach, 525, 534
BPP, 180
Bright Build home, 146
BTES, 312
Building(s), 7, 129, 302
Building construction, 546
Building Security, 166
Bulk gravitational storage, 391
BWR reactor, 84

C

CAE, 548
CAES, 418
CAIFI, 478
Calvera modular hydrogen storage system, 445
Canary Islands, 442
CANDESAL, 363
CANDU, 87
CAPEX, 531
CAREM, 360
Catalyst, 51
CDEA, 260
CDN, 488
CE, 369
CES, 399
CFCT, 43
CFD, 546
CGE, 542
Challenges, 539
Chiller system, 154
Chiller thermal energy storage cell, 427
CHP projects, 95
CHP, 70, 267
CIBSE, 102
CIMS, 543
CMC, 105
CNG, 449, 450, 502, 503
CO_2nserve, 140
COD, 369
Cogeneration, 81, 113
Combined heat and power (CHP), 70, 77, 80
Commercial, 97, 98, 108, 140
Community energy storage, 398
COMPACT Modular, 244, 245
Complementarity framework, 538
Compressed air energy storage, 416
Compressed liquid and solid natural gas, 502
Computer, 217
Concentrating solar system, 139
Conductors and connectors, 35
Connection(s), 527
Connection types, 527
Containerized energy storage, 401
Control, 392
Converters, 31
Cooling properties for lighting system, 244
COP21, 356
Coselle technology, 503
Costs, 220
CPC, 76

CPIOMS, 52
CPU, 181
CRP, 359
CRPV technology, 505
Cryogenic liquid trucks, 512
CSC, 307, 478
CSP/MED, 342
CSP/RO, 342
CTES, 101
Customer success, 532

D

DAB, 396
Data center(s), 103, 217, 236
Data center from Delta, 227
DC, 413
Decarbonizing district heating, 279
Decomposition method, 538
DER, 479
Desalination by SMR, 355
Desalination, 72, 90, 325
Design and construction, 22, 500, 523
DG, 478
DGSs, 489
DH, 259
DHAPP, 286
DHN, 488
Distributed power, 108
District heating, 114, 259, 261
　by biomass, 266
　by industrial waste heat, 293
　by modular hybrid sources, 299
　by modular solar thermal energy, 279
　networks, 302
　by SMR, 285
　by wind energy, 284
DIW, 54
DM, 173
DMS, 489
DOI, 352
DOT, 501
Double layer capacitors, 412
DPA modules, 234
DPSE, 105
DRS modular fuel system, 499
DSM, 532

E

Energy star, 17
Energy central, 313
Energy harvest, 23, 248
Electric and hybrid vehicles, 29, 31, 38, 41
Electric motors, 32
Energy storage, 33, 54
Electrical drives, 34

Exhaust emission, 51
Energy industry, 118
Energy source, 132
Energy consumption, 186, 365, 546, 550
Electrical/electronic applications, 217
Energy optimum, 238
Electronic(s), 252
Electronic applications, 246
Electrical and electromagnetic energy storage
　　systems, 410
Envirogen modular wastewater
　　treatment, 367
Evoqua modular water technologies, 369
Evoqua modular desalination technology, 370
Energy and fuel storage, 385
Electrochemical storage, 386
Electrochemical capacitors, 388
Energy and fuel transport, 471
Expandable modular microgrids, 477
ETT project, 496, 497
Energy system models, 535
EPPA, 537
EMEC, 537
Equipment, 544
E&C, 531
ESS, 485
ELP, 510
EV, 398
EDLCs, 389
EPA, 372
ENERCON, 351
ED, 351
EA, 373
ESM, 9
ENCON program, 54
EBMUD, 71, 72
EVA, 93
ECO-WILL, 117
Evacuated tube solar collector, 137
Energy conservation, 3, 5, 28, 65, 97
Energy efficiency, 3, 5, 28, 29, 31, 32, 44, 54, 65,
　　97, 98, 102, 103, 115, 550
Energy saving modules, 9, 15

F

FAM, 478
FAST wind turbines, 548
FCV, 46
FDP, 36
Fire, 449
Flat plate solar collectors, 137
FNPS, 361
FO, 328
FOPID, 489
FSM, 478
FTN, 486

Index

Fuel cells, 34, 42, 43, 48, 186
Fuel storage, 392
FVB, 271, 272, 273

G

Gas compressor modules, 501
GCR, 93
GDN, 486
Geothermal district heating, 276
Geothermal heat pump, 159, 162, 301
Geothermal, 114
GHG, 6, 365
GHP, 299
GHTHR300, 328
GHX, 299
Global assessment, 288
Global landscape of modular desalination by SMR, 357
Global market for CHP, 117
Grid friendly water technology for desalination, 351
Ground heat exchanger, 300
GTG, 54
GTS, 508
GTT, 455
GVE, green village electricity, 481
GVW, 501

H

Hard links, 538, 541
Heat exchangers, 162
Heterogeneous biomass, 80
HEV, 181
HMEP, 479
HOSIM, 546
Hospital facilities, 113
HPC, 237
HPE, 237
HTGR, 87
HTTR, 94
HTTW, 65
HVAC, 133, 134, 151, 312
Hybrid buses, 414
Hybrid compressed air energy storage, 419
Hybrid global energy system models, 535
Hybrid models, 536
Hybrid renewable energy, 302
Hybrid solar and biomass system, 309
Hybrid solar heating, 144
Hybrid thermal membrane desalination, 327
Hybrid wall, 15
Hydrates storage and transport, 508
Hydrides, 444
HydroDI module, 371
Hydrogen produced on-site, 514

Hydrogen production, 92
Hydrogen storage modules, 444
Hydron modules, 153, 510, 513

I

IAEA, 359, 362, 485
ICE, 116, 117
IDE chemical free RO desalination plant, 331
IDE PROGREEN Plant, 332
IEA, 541
IGBT, 32
IMA, 52
INCHARGE POWER, 399
Industrial, 65, 83
Industrial waste heat, 291
INET, 360
INRADA GROUP, 399
Integrated energy storage, 412
Integrated hybrid models, 542
Integrated modular avionics, 51
Integrated power conversion, 252
Integrated tanks, 506
Intelligent cooling technology, 231
ISO containers, 508
ISO tank rack design, 500

J

JAEA, 95
Jenoptik modular energy system for military platform, 245

K

KANUPP, 360
KERS, 31, 39
Kinetic energy recovery, 39
KLT-40, 361
KSB modular desalination plant, 335
kWh, 396

L

Landfill gas, 81
Large scale modeling systems, 542
LED, 5, 22, 36
LEED, 4
LIANA+, 396
Lighting, 36
Liquid hydrogen, 513
Liquid hydrogen trailers, 513
LNG, 451, 455, 457, 506, 507, 508
LNG ISO containers, 508
LNG transport fleet, 507
LNG transport trailers, 508
LOR, 173

Index

Low pressure CAES, 418
LRU, 52
LTDH, 315
Lubrication oils, 45
LWR, 84

M

MACRO, 537, 538, 539
Manufacturing industries, 95
MARKAL, 536, 537, 538
MCP, mixed complementarity problem, 543
MCU computing, 52
MD, 328, 339
Mechanical storage, 390
MED, 360
MEG, 484
Melbourne case study, 333
Membrane biological reactor, 368
Membrane modular LNG tank, 455
MESSAGE, 537, 539
Methane hydrates, 508
MF, 327
MFC, 369
MHR, 93, 94
Micro and mini scale energy systems, 246
Microbial fuel cell, 369
Micro CHP, 115, 117
Micro combustor radiators, 251
Micro energy, Systems, 471
Micro fabricated hydrogen storage module, 446
Microgrids, 108
Micro power management, 251
Micropower systems, 446
Micro TPV, 251
MIT study on modular desalination plant, 333
MMC, 254, 395
MMGU, 478
MMSEV, 175
MNSs, 247
Mobile power system, 171, 179
Modeling, 523
Models, 533, 535, 538
Modular, 1
Modular automobiles, 181
Modular battery storage, 392
Modular battery technology, 178
Modular canisters, 500
Modular CHP, 102, 108
Modular CNG storage station, 449
Modular CNG tank, 450
Modular compressed natural gas storage, 449
Modular desalination by biomass and geothermal energy, 347
Modular desalination by wind energy, 345
Modular energy management, 183
Modular flow cell batteries, 405

Modular flywheel energy storage, 419
Modular fuel cell stack, 194
Modular fuel cell, 42, 44, 175, 179
Modular fuel system, 498
Modular geothermal heat storage system, 435
Modular geothermal measurement, 156
Modular global energy models, 533
Modular heat batteries, 428
Modular heat exchanger, 443
Modular heat pump, 23
Modular HISEER, 158
Modular hydrogen storage, 442
Modular hydropower storage system, 436
Modular IceBrick, 427
Modular integrated energy systems, 189
Modular LFG power generation, 81
Modular liquid fuel storage, 448
Modular LNG tank, 451
Modular metal hydride hydrogen storage system, 442
Modular micro grids, 471, 474
Modular microenergy grids, 484
Modular microenergy internet, 485
Modular multilevel converter, 254
Modular nanogrid, 488
Modular nuclear reactors, 83, 86
Modular open energy system, 491
Modular package, field erected wastewater treatment plant, 369
Modular passive solar heating, 142
Modular plant design, 240
Modular portable hydrogen storage, 448
Modular power generation, 174
Modular power management, 173, 183
Modular power plants, 239
Modular pumped hydropower system, 442
Modular PV based desalination, 343
Modular renewable heating and cooling grids, 264
Modular roofing solar technologies, 140
Modular roofing, 140
Modular shipboard energy storage, 54, 209
Modular simulation and modeling, 525, 544, 546, 550
Modular solar envelop house, 13
Modular solar thermal, 147
Modular solar vehicles, 197
Modular solid based thermal energy storage, 419
Modular thermochemical heat storage, 432
Modular timber transport project, 496
Modular transport of biomass, 495
Modular UPS system, 231
Modular verd2GO-EV connection, 189
Module types, 526
MODV, 181
Molten salt thermal energy storage, 429
MOS, 32

Index   567

MOSFET, 123
MPC, model predictive control, 546
MSF, 328, 360
MSG, 536
MTTR, 214
Multi level DC grid system, 493
Multiple modules, 542
Multiple vehicle applications, 187
Multi-vehicle space mission, 171
Municipal energy supply system, 304
MVC, 360
MWP, 352

N

Nano technology, 45, 50, 51, 118
NASA, 120
NASAs module based supercomputer, 217
Natural gas transport, 501
Naval vessel electrification, 210
NEMS, 543
Net zero cost, 21
Net zero energy buildings, 18, 25, 28
Newterra technology for wastewater
    treatment, 367
NEXED, 371
NexGenDESAL modular desalination plant, 334
NGH, 471, 508
NGL, 471
Novel modular approaches, 552
NPP, 362
NREL, 365
Nuclear process heat, 86
Nuclear reactors, 83, 86
NUScale reactor, 357
NZEB, 18

O

O&M, 349
Off-the-grid, 21
Oil and gas facilities, 484
Oil recovery, 87
Oil refinery, 89
OOMS, 546
OPEX, 234
Optimization, 544, 552
Organic loading, 369
OWS, 373

P

Package system from ClearBlu, 371
Passive heating and cooling, 7
Passivhaus construction, 144
PATH, 543
PCM materials, 8, 10, 12, 13

PCM modular system, 426
PCS package wastewater treatment plants,
    372, 373
PDN, 488
PDU, 218
Peaking boiler, 301
Peloton wheel turbine, 329
PEM fuel cell, 44, 179
Perspectives, 539
PG&E, 390
Phase change composite material, 424
Phase change materials, 10, 11
PHE, 101
PHWR, 360
PID, 489
Plug in hybrid and solar car, 199
Plug in hybrid vehicles, 185
PMAD, 173, 174
PNG technology, 505
Portable applications, 448
Portable energy systems, 252
Power converter, 412
Power density, 231
Power plants, 107
Power station, 401
Process, 544
Properties, 508
PUE, 221
Pulsed hydrogen supply, 44
PV, 249, 489
PV technologies, 249
PWM, 395
PWR, 84, 360

R

Rail, 202, 512
Rapid deployment, 433
Rapid transit vehicles, 202
RBMK, 287
Renewable energy desalination
    processes, 348
Renewable power supply, 548
Renewable resources, 478
Residential, 142
RESIM, 546
Reversible fuel cell, 43
RO, 371
ROI, 109
ROSA, 175
RTACP, 477

S

Sadara, 532
SAGD, 87
SAIFI, 478

Index

SCE, 390
School buildings, 144
Scientific modeling, 525
SDP, 36
Self powered MNSs, 249
Self supporting tanks, 507
SEP vehicle, 175
Sewer heat recovery, 301
Ships, 512
SIL2, 397
Simulation, 302, 523, 552
SMART reactor, 360
SMR (small modular reactors), 83, 86
SOC, 397
Soft Links, 536
SOH, 396
Solar array technology, 174
Solar assist/solar taxi, 199
Solar assisted modular desalination systems, 336, 337
Solar buses, 200
Solar cells, 38, 50
Solar electric aircraft, 204
Solar electric car, 197
Solar envelop house, 13, 14
Solar heating, 142
Solar impulse, 205
Solar photovoltaic canopy system, 202
SolarPod's Grid Tied solar array, 141
Solar pond, 136
Solar power tiles, 141
Solar powered boats, 207
Solar powered spacecraft, 206
Solar shingles, 141
Solar thermal applications, 337, 342
Solar thermal technology, 147
Solar vehicles, 197
Solid oxide fuel cell, 182
SoloPower flexible solar panel, 142
Space vehicles, 174
Spent nuclear fuel, 500
SRAM, 389
ST project, 497, 498
Standard connections, 527
Standardize modular design, 528
Sucrose wastewater treatment, 369
Suns computer operation, 228
Supercapacitor module SAM, 414
Supercapacitors, 33, 34, 47
Supercomputer, 237
Surface heating and cooling, 15
Sustainable, 140
Sweden, 496
SWRO, 356
Synthetic fuel, 186
System and technology, 55

T

Tankers for LNG, 506
Tankers with integrated tanks, 506
Tankers with self supporting tanks, 507
TCO, 234
TDS, 332, 349, 371
Techno-economic/social model, 533
TES, 312
Thermal capacity on demand, 433
Thermal energy storage in district heating, 311
Thermal energy storage in LTDH, 315
Thermal energy storage tanks, 428
Thermal energy storage, 310, 390
Thermal energy storage-solar energy-district heating partnership, 316
TIMBR system, 368
Tires, 37
TOHYCO Rider Bus, 414
Top down model, 533, 534
TPESs, 114
Trailers, 513
Transformation modules, 541
Transpired solar air collector, 136
Transportation, 28
Transport modules for hydrogen, 510
Transport of CNG, 502
Transport of LNG, 506
Transport of NGH and GTS technology, 508
Transport trailers, 508
TRL, 56
Tube trailers, 512
TVA, 436
TVC, 327
Two wheelers, 208

U

UAE, 329
UF, 327
Ultracapacitors, 410
Unglazed solar collector, 135
Unmanned aerial solar vehicles, 205
UPS, 54, 217, 230
USSR, 87

V

Vacuum insulated LNG modular storage system, 457
VAR, 390, 478
VC, 327, 351
VCD, 328
Vehicle system controller, 188
Vehicles, 38, 171
VHTGR, 85

Votrans technology, 504
VSC, 189
VSM, 478
VVER-1000, 287

W

Wastewater treatment, 325
Wastewater, 66, 69, 70
Water, 66, 70
Water desalination, 72, 90
Water treatment AdEdge technology, 366
Wave energy for desalination, 352

Wheel hub motors, 181
Wind energy, 284
Wind energy storage, 418
Wind powered pumped hydropower, 442
Wood based prefabricated façade, 15
WT, wind turbines, 548

Z

ZCP, 36
Zero carbon emission, 144
Zero energy buildings, 24
ZNE, 18

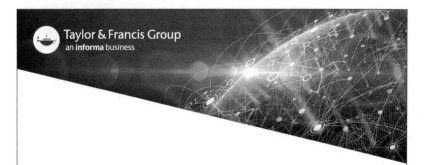